Fundamentals of Decision Analysis

SERIES IN QUANTITATIVE METHODS FOR DECISION MAKING

Robert L. Winkler, Consulting Editor

Brown, R. V., A. S. Kahr, and *C. Peterson* **DECISION ANALYSIS: AN OVERVIEW**

Brown, R. V., A. S. Kahr, and *C. Peterson* **DECISION ANALYSIS FOR THE MANAGER**

Chou, Ya-lun **PROBABILITY AND STATISTICS FOR DECISION MAKING**

Horowitz, I. **DECISION MAKING AND THE THEORY OF THE FIRM**

LaValle, I. H. **AN INTRODUCTION TO PROBABILITY, DECISION, AND INFERENCE**

LaValle, I. H. **FUNDAMENTALS OF DECISION ANALYSIS**

Lippman, S. A. **ELEMENTS OF PROBABILITY AND STATISTICS**

Miller, R. B., and *D. W. Wichern* **INTERMEDIATE BUSINESS STATISTICS: ANALYSIS OF VARIANCE, REGRESSION, AND TIME SERIES**

Miller, R. E. **MODERN MATHEMATICAL METHODS FOR ECONOMICS AND BUSINESS**

Press, S. J. **APPLIED MULTIVARIATE ANALYSIS**

Spivey, W. A., and *R. M. Thrall* **LINEAR OPTIMIZATION**

Trueman, R. E. **AN INTRODUCTION TO QUANTITATIVE METHODS FOR DECISION MAKING,** Second Edition

Winkler, R. L. **AN INTRODUCTION TO BAYESIAN INFERENCE AND DECISION**

Winkler, R. L., and W. L. Hays **STATISTICS: PROBABILITY, INFERENCE, AND DECISION,** Second Edition

Fundamentals of Decision Analysis

Irving H. LaValle

Tulane University

Holt, Rinehart and Winston

New York Chicago San Francisco Atlanta Dallas Montreal Toronto London Sydney

To

three of the many great professors
under whom it has been my privilege to study

George B. Cooper

Howard Raiffa

Lawrence W. Towle

Library of Congress Cataloging in Publication Data

LaValle, Irving H
Fundamentals of decision analysis.

Includes bibliographical references and index.
1. Decision-making. I. Title.
T57.95L38 511'.65 77-5838

ISBN 0-03-085408-3

Printed in the United States of America

8 9 0 1 038 9 8 7 6 5 4 3 2 1

PREFACE

General decision analysis has been maturing rapidly over the past few years. Although it has long been known that the various axiomatizations of consistent choice behavior apply to very general sorts of choice situations, until recently the main thrust of texts has been on statistical decision problems, involving finding optimal sample sizes under various assumptions about the economics of sampling and subsequent actions and about the sampling model. My earlier text [**78**] is a typical example.

But a number of excellent books emphasizing the *generality* of decision analysis have begun to appear. Among these are, of course, Raiffa [**113**] and Schlaifer [**128**], as well as Brown, Kahr, and Peterson [**14**] and the Stanford Research Institute readings [**65**] edited by Howard, Matheson, and Miller. Few of these are texts in the standard sense, and with good reason—the field has been changing so rapidly that a codifying text would soon have been dated.

At this point, however, it appears that the main thrusts of work in this field are discernible in sufficient clarity to warrant a standard text, provided that it is modest in its pretensions as to comprehensiveness. I hope that this book fills that role.

My objectives are to emphasize the generality of applicability of the basic decision-analytic model, to emphasize principles by showing that all the basic ideas arise in cases with finite numbers of states and acts, to show that many topics such as stochastic simulation and "classical" inference can be brought within the purview of our philosophy, and to do these things in a standard text format, with theorems, definitions, proofs, examples, and exercises set apart from the body of the text.

In addition, I have designed the book to be quite flexible as to coverage. For example, it is possible to read only the first two sections of Chapter 3 before proceeding to Chapter 4; and it is natural to omit the final three sections of Chapter 3 in courses assuming no calculus prerequisite, since these sections concern analytic probability models cited only rarely and optionally in the sequel. All material marked with an asterisk may similarly be omitted without jeopardizing the reader's understanding of any subsequent unasterisked material.

This book consists of four parts. Part I, comprising the first four chapters, concerns the fundamentals of decision-tree analysis. In Chapter 1 are given a number of examples of decision problems under uncertainty (abbreviated "dpuu"), some of which are definitely nonbusiness in nature, with consequences which differ largely in their nonmonetary attributes. Chapter 2 develops the principles of decision-tree construction in some detail, culminating in showing how any decision tree can be reformulated, without strategic misrepresentation, in a standard form admitting an n-stage interpretation and facilitating the subsequent proof of validity of the recursion analysis (Raiffa's averaging out and folding back). Chapter 3 introduces lotteries, the Pratt-Raiffa-Schlaifer axiomatization (in [**112**], for example), elementary methods for assessing utility and probability, proofs of the expected-utility theorems and the usual properties of probability, and, in optional sections, the common probability mass and density functions and Bayes' Theorem for some of the usual models. Despite its length, this chapter is flexible: natural termination points are the ends of Section 3.8 (omitting the usual probability models), Section 3.7 (omitting also proofs of the fundamental results), and Section 3.2 (being content with a survey of the next six sections). Chapter 4 extends the analysis from lotteries to decision trees, proving that the recursion analysis is justified by the axioms of consistent choice adopted in Chapter 3.

Part II, Chapters 5 and 6, reviews and extends the material in Chapter 3 on the quantification of preferences and judgments. Chapter 5, on utility, contains sections on the usual topics—assessing utility on a finite consequence set and on (an interval of) monetary returns—as well as three sections on utility functions with Cartesian-product domains (the multiattribute situation): first, additive separability and, more generally, the advantages of exploiting special structure in preference quantification; second, "extreme preferences" and lexicographic utility in a very restricted context; and third, the constant-risk-aversion utility functions on real n-space. Chapter 6, on probability, is similar, in that it contains sections on assessing probability on a finite event set and assessing probability laws of real- (and vector-) valued random variables. But among so-called "indirect methods" of assessment, I introduce the rudiments of sampling-theoretic inference, including Chebyshev's Inequality, confidence intervals, and (optionally) the Central Limit and Glivenko-Cantelli theorems. The final section is an introduction to stochastic simulation, viewed as a computational methodology aiding assessments; students seem to enjoy particularly the picturesque queueing example of Dr. Marcus Illby's clinic.

Part III is a mixed bag of further topics in individual decision making. Chapter 7 concerns normal-form analysis and includes sections on obtaining (economically) the normal form of a decision—no conditional expectations being taken yet, however; *cardinal* admissibility and Bayes strategies (with utility fully quantified, but probabilities not yet quantified at all); the same topics given that some probabilities have been specified (here we take the usual statistical decision-theoretic conditional expectations); *stochastic* and *ordinal* admissibilities and their relationships to Bayes strategies; and what happens if the decision maker is subjected to constraints on his choice of a randomized strategy. Exercises introduce the maximum criterion and justify its occasional use within our framework. Again, a rather comprehensive introductien and natural break points (after each section) provide flexibility in use. Chapter 8 concerns the monetary evaluation of opportunities—exchange values, reservation prices, bid ceiling, and the like—when utility for money is not necessarily linear. Two final sections introduce the (Bayesian) statistical decision model, involving choice of experiment. Except for some proofs, these sections are independent of the general material in the initial sections, which can therefore be omitted in shorter courses. Some of the elementary optimal-sample-size models are introduced in optional exercises and remarks. Chapter 9 concerns statistical inference, both sampling-theoretic and Bayesian, culminating in an exposition of the Conditionality, Sufficiency, and Likelihood Principles and their implications for inference. Chapter 10 introduces, typically without the difficult proofs, finite- and infinite-stage Markovian decision processes, with and without discounting.

Part IV concerns situations in which more than one "decision maker" must be explicitly considered. Chapter 11, on risk sharing and monetary group decisions, presents the basic notions of Pareto-optimal strategy-sharing rule pairs and frontiers in utility space, and then introduces an intuitively appealing, special mechanism called the allocation-function-solution approach by which the group can "agree to disagree" on a unique outcome. A final section merely mentions with references the general collective-choice problem and social welfare functions. Chapter 12 is a very brief introduction to game theory, passing rapidly over information structures and normalization, maximin and equilibrium in general noncooperative games, the two-person zerosum game as a well-behaved special case, and two final sections on cooperative games. The first, restricted to two-person games, covers principally the Nash and extended Nash solutions, with explicit directions on how to find the extended Nash solution point. The second, on n-person cooperative games, defines characteristic functions without assuming side payments, but then assumes them in all examples and allows for specification of just which coalitions are permitted to form, such specifications affording some perspective on the nature of coalition, core, stable sets, etc.

Five appendices cover the Greek alphabet, elementary set theory in some detail, certain useful topics in analysis dealing chiefly with convexity and separating hyperplanes, the halfway, or bisection, technique for finding a root of a monotone function, and References.

For several years I have used drafts of Chapters 1, 2, 3 through section 3.8, 4, 5, 6, 7 through section 7.4, 8, 9, and 12 in a fast-paced, one-semester course at the Tulane University Graduate School of Business Administration, sometimes in the second year but more often in the first or second semester of the student's experience. Because of its flexibility, this book is suitable for courses of varying lengths up to one year in business schools, as well as in economics and applied mathematics courses at the upper undergraduate and lower graduate levels.

A word on notation. There were two basic choices: Keep it simple and change it all the time, or develop a system which, while complex, could be maintained throughout. My experience has been that the latter is pedagogically preferable if the ingredients are relatively mnemonic. I think that the system in this book works reasonably well in that, for example, u, U, **U**, and $\mathcal{U}$ are all read "utility," and in P_S, P_T, $P_{2|1}$, and $P_C^\dagger$, the P's all stand for "probability."

Several years of students have contributed generously and, occasionally, not so generously, with suggestions and corrections. My former research assistants Roger D. Eck and John R. Page were of invaluable assistance, as have been Joan Steinberg and Dennis E. Seereiter. My colleagues, Professors Evan E. Anderson, Larry R. Arnold, Joseph L. Balintfy, Richard E. Beckwith, Charles B. Bell, Seymour S. Goodman, and John R. Moroney, have all contributed valuable advice and gratifying encouragement at various points in the process, as have also Professors Robert L. Winkler and Robert M. Thrall, who read portions of the manuscript. Because I did my own typing of the manuscript, I shall merely express thankfulness for the reliability of the instrument under the often frustrated pounding of its operator.

If I have not entirely avoided obscurity, I hope at least not to have entirely eschewed the humorous touch!

New Orleans, Louisiana — I. H. L.
December 1977

CONTENTS

Chapter Three FOUNDATIONS OF DECISION ANALYSIS 46

Chapter Four ANALYSIS OF DECISIONS IN EXTENSIVE FORM 158

II MORE ON PREFERENCE AND JUDGMENT QUANTIFICATIONS

Chapter Five QUANTIFICATION OF PREFERENCES 187

Chapter Nine APPROACHES TO STATISTICAL INFERENCE 414

Chapter Ten INTRODUCTION TO MARKOVIAN DECISION PROCESSES 454

IV SEVERAL-PERSON DECISIONS

Chapter Eleven MONETARY GROUP DECISION PROBLEMS 481

Chapter Twelve A GLIMPSE AT GAME THEORY 518

A: Introduction

B: Noncooperative Games

APPENDICES

I

Fundamentals of Decision Analysis in Extensive Form

"*Every* man has a grand chance."
Rudyard Kipling

1

INTRODUCTION AND PREVIEW

> Jesse: Are we lost, Owen?
> Owen: Aww—I wouldn't call it that.
> Jesse: Well, what *would* you call it?
> Owen: Undecided.
>
> *Bonanza* telecast, November 16, 1969

1.1 DECISIONS

A *decision*, or *choice*, or *act* is a commitment of resources. Resources need not be financial, natural, or even material. You, the reader, are making a decision right now, whether to continue reading this book. Time is a resource which each one of us commits at every instant of his existence. Hence the making of decisions is a ubiquitous activity.

1.2 PERCEPTION, FORMULATION, AND ANALYSIS OF DECISIONS

Most decisions, however, are made on a very casual basis. Many people do not deem it worthwhile to spend too much time thinking about the precise number of bites that should be applied to a tender morsel, taking into account total eating time, digestion, etc. There are so many decisions confronting a person that he must make most of them casually or unconsciously, so as to have time for very careful analyses of those few which appear really important to him.

We shall not attempt to set any standard as to what attributes should make a problem appear really important to a decision maker. The reflective reader will have noted that the allocation of analysis time to decision problems is itself a decision problem, one which generally is best left to informal analysis. At present, there are no safe and universally applicable rules for *perceiving* decision problems worthy of careful consideration.

This book, however, is concerned not with the *perception* of decision problems, but rather with their *formulation* and *analysis* once they have been selected for careful attention. Chapter 2 concerns a straightforward method of formulating any decision problem and previews a consistent method for its subsequent analysis. An alternative method of formulation is introduced in Chapter 7. All chapters are concerned with the analysis of decision problems (hence the title of this book).

1.3 HOW MANY DECISION MAKERS?

Through Chapter 10 we assume that a single individual is faced with the decision or sequence of decisions under consideration. The analytical methodologies in such situations are usually considerably simpler than methodologies regarded as appropriate when two or more individuals are playing decision-making roles.

Except in Chapters 11 and 12, we shall always take the viewpoint that there is a *single decision maker*, and our exposition will be devoted to helping him determine a desirable course of action in his decision problem. Even if the decision problem as broadly conceived involves other individuals in "starring" roles, our decision maker will be our client and all others will be treated as part of the general environment.

To emphasize our exclusive concern with a single decision maker and to help the reader become involved with the subject matter, we shall often phrase statements in the second person singular. *The reader should take "you" very personally*, imagining that he really confronts the situation in question and that I, the author, am his consultant.

We shall not use the second person extensively in Chapters 11 and 12, which concern group decision theory and game theory respectively.

Chapter 11 discusses decisions confronting a committee, group, or syndicate of individuals who must ultimately come up with some course of action despite their potential disagreements on matters of preference and/or of fact. Naturally, each individual in the group wishes the group's decision to go his way, and hence the group's decision procedure must take into account the conflict within the group.

Chapter 12 is a brief introduction to game theory, which is the formal study of conflict resolution in decision problems involving more than one decision maker. Since entire books have been devoted to this subject, Chapter 12 is necessarily only an overview and a guide to further reading.

1.4 DECISIONS UNDER UNCERTAINTY

For a decision problem to be of any genuine interest or concern, it is clearly necessary that the decision maker have more than one available course of action. Grammarians are justifiably adamant in noting that "alternatives" implies at least two available choices.

If the decision maker can predict *for certain* the outcome, or consequence, of *each* available course of action, then we say that he has a decision problem *under certainty*. Selecting a course of action in such situations depends solely upon the decision maker's relative preferences for the consequences: he should choose that course of action which will produce the most desirable consequence.

Many significant decision problems do not satisfy the criterion for decisions under certainty. The consequences of one (or more) of the available courses of action *cannot* be foreseen with certainty; they may be contingent upon forces of the environment such as market demand, actions of competitors, the weather, court rulings, and so forth, that are not subject to the absolute control of the decision maker. Even King Canute could not turn back the tide. Such decision problems are said to be decisions *under uncertainty*.

■ Some authors make the classical distinction among decisions under certainty, under risk, and under uncertainty. For them, a decision is under risk if the decision maker knows *the probabilities of* the possible consequences of each course of action, and under uncertainty if he does not know these probabilities. We combine the categories of risk and uncertainty under the latter term, because the distinction between them begs the question of what constitutes "knowing probabilities," our answer being that *all* probabilities are personal or subjective and that such probabilities can *always* be obtained when needed by the expenditure of more or less effort by the decision maker. ■

■ Much attention is paid in the scientific literature and in operations research curricula to decision problems under certainty, for three reasons. First, many problems in this area arise in such a way that the decision maker *does not know explicitly* what his available courses of action are; they are defined only *implicitly* as those potential courses of action which do not violate a number of constraints. Hence, even defining the available courses of action can be difficult. Second, when the consequences of each available course of action are "almost" certain, much work can be eliminated at little risk by regarding each consequence as certain. And third, many decision problems under uncertainty *give rise to* equivalent problems under certainty. Hence the emphasis on methodology for solving decisions under certainty is fully justified. ■

We have noted that the decision maker's relative *preference* for consequences is an essential ingredient in the analysis of decisions under certainty. Preference specification is even more essential in decisions under uncertainty, which also require the decision maker to specify his *judgments* regarding events on which the consequence of his course of action is contingent. The behavioral and logical principles guiding the specification and use of preferences and judgments are introduced in Chapter 3. Chapter 4 furnishes a general procedure for solving decision problems under uncertainty; Chapters 5 and 6 concern the quantification of preferences and judgments respectively.

Further discussion of generalities would be unproductive at this point. We turn now to the description of a number of prototypical examples.

1.5 EXAMPLES

1.5.1a Electronic Subsystem

You have contracted to deliver a special-purpose analog computer to the government at a price that will yield you a profit of $250,000—barring unforeseen failure of the computer to perform its function on a space vehicle. The only possible cause of failure would be defectiveness of a crucial electronic subsystem which you have subcontracted to another firm for political reasons despite the occasionally slipshod practices in the final-assembly operations of that company, which claims that the probability of defect in the subsystem is .001 (one tenth of 1 percent), a figure which you suspect was pulled from a hat and which you would revise to .10 (= 10 percent). If the computer malfunctions during its mission, your company will be subject to a $500,000 penalty and will also lose prestige in the industry. After careful consideration, you decide that your available courses of action are (1) to install the subsystem as is and take your chances with it and (2) to have your own people tear it down, inspect it, and carefully rebuild it, at a cost of $100,000, thus ensuring that the system will function properly. Which would you choose? Why?

1.5.1b

Now suppose you are informed that the subsystem can be pretested at a cost of $20,000 without affecting its subsequent reliability. If the subsystem is not defective, then this test will surely so indicate; but if the subsystem is defective, then the probability of the test's so indicating is .80 (= 80 percent; alternatively, the odds in favor of the test's so indicating are 4 to 1). This pretest can be performed only once. Now your available courses of action include not only (1) and (2) in 1.5.1a, but others involving pretesting. What would you do? Why?

1.5.1c

Now suppose there is also available a less reliable testing device which at a cost of $5,000 can be applied to the subsystem once without affecting its subsequent reliability, and, either before or after rebuilding and/or the pretesting described in 1.5.1b, without affecting the results of rebuilding and/or pretesting. If the subsystem is not defective, then the probability of this test's so indicating is .75; but if the subsystem is defective, then the probability of this test's so indicating is .65. Now you have many available courses of action. List them, indicate your choice, and give the reasoning by which you justify your choice.

1.5.2a Attorney's Problem

You are an attorney to whom a prospective client has brought a good case involving his attempt to recover $100,000 from a company on the grounds of fraud. He asks you to state your fee for taking the case. You have narrowed your consideration of fee proposals down to three: (1) $12,500 outright, (2) 25 percent of the $100,000 contingent on winning the case, and (3) $5,000 outright plus 15 percent of the $100,000 contingent on winning the case. Your time expenditure and out-of-pocket costs of $5,000 would be unaffected by the fee. You can present only one fee proposal, which the prospective client will either accept or reject and retain another attorney. You have assessed the odds at 2 to 1 against his accepting the fixed-fee plan (thus implying probability 1/3 of his accepting it), 2 to 1 in favor of his accepting the contingent-fee plan (thus implying probability 2/3 of his accepting it), and 4 to 3 in favor of his accepting the mixed-fee plan (thus implying probability 4/7 of his accepting it). If he accepts whatever proposal you make, then your odds in favor of winning the case are 2 to 1 (so that the probability of winning it is 2/3). Which fee would you propose? Why?

1.5.2b

Your bright junior partner now suggests an alternative development of the case if you should be retained. Instead of your original plan, costing $5,000 and successful with (your) probability 2/3, his plan would cost you $7,200; after careful thought, you assess the odds of winning the case with it at 9 to 1 (and hence your probability of winning with his plan is .90). Which plan should be used if the client accepts your proposal? Does the existence of your partner's alternative plan affect your choice of fee proposal; and if so, how and why? Conversely, if you obtain the case, does the agreed-upon fee affect your choice between your original plan and your partner's plan? Finally, which should come first, the fee proposal decision or the decision between the two case plans—Does it matter?

1.5.2c

Just when you have reasoned your way through the choice of fee proposal and case plan, you receive a telephone call from the prospective client, who tells you that the odds are 2 to 1 in favor of his having a similar case, involving recovery of $50,000, after the first case is resolved, and that he will retain you "under similar circumstances" provided you are retained for the first case and win it. He explains "under similar circumstances" to mean "at exactly half the amounts involved in the first case." You determine that the incremental cost of adapting your plan for the first case to fit the particular circumstances of the second would be only \$1,000 if the plan for the first case was successful. Reanalyze the choice of fee proposal and case plan for the current case as necessary.

1.5.3 Competitive Sealed Bidding

You are one of two contractors in a position to bid on a new government building. You and your competitor have received the specifications for the building, and each must now determine a bid (= price you would charge for constructing the building according to specifications) without collusion. Bids will be submitted in sealed envelopes, and the contract will be awarded to the contractor who submits the lower bid; in case of a tie the contract will be awarded by the flip of a coin. After some analysis, you arrive at the following conclusion: if $v(b, \alpha, \beta)$ denotes *your* net profit from submitting a bid of b, given that your competitor submits a bid of β and that the cost of construction by you is α, then

$$v(b, \alpha, \beta) = \begin{Bmatrix} 0 \\ b - \alpha \end{Bmatrix} \quad \text{if} \quad b \begin{Bmatrix} > \\ < \end{Bmatrix} \beta,$$

and $v(b, \alpha, \beta)$ is equally likely to be 0 or $b - \alpha$ if $b = \beta$. (Why?) What additional information would you wish to have in order to determine (1) whether to submit a bid, and (2) if you do, what bid to submit?

1.5.4 Portfolio Problem

Suppose that you have a sum, T, of money to invest for a fixed period of exactly one year, and that you have narrowed your choice to n securities—symbolically denoted by the integers $1, 2, \ldots, i, \ldots, n$—as candidates for inclusion in your portfolio. Let $b_1, b_2, \ldots, b_i, \ldots, b_n$ denote the amounts you will invest in securities $1, 2, \ldots, i, \ldots, n$ respectively. Your investment problem is to choose the n numbers $b_1, \ldots, b_n$ so as to obtain the best, or "optimal" portfolio. Assume that selling short is prohibited, and hence none of the b's can be negative (but some can be zero). Also assume that buying on margin is prohibited, so that $b_1, \ldots, b_n$ must be chosen to sum to not more than your capital, T. Since the optimality of a portfolio is (for most people anyway) related to its profitability, we define $\theta_1, \theta_2, \ldots, \theta_i, \ldots \theta_n$ to be the after-tax profit (capital gains plus dividends and interest) *per dollar* invested in security $1, 2, \ldots, i, \ldots, n$ respectively. Hence the after-tax return $v(b_1, \ldots, b_n, \theta_1, \ldots, \theta_n)$ from the portfolio is given by

$$v(b_1, \ldots, b_n, \theta_1, \ldots, \theta_n) = \sum_{i=1}^{n} b_i\theta_i,$$

where $\Sigma_{i=1}^{n} b_i\theta_i$ is shorthand for $b_1\theta_1 + b_2\theta_2 + \cdots + b_n\theta_n$. What additional judgments and/or preferences must you specify in order to make your portfolio decision under uncertainty?

1.5.5 Newsboy Problem

You are a newsstand owner who must decide on the number, b, of copies of next month's *Bon Vivant* magazine to order from your wholesaler. Each copy costs you \$0.50, and the retail price is \$0.75. All copies not sold during next month can be sold for \$0.15 each to a wholesaler of the centerfold. On the basis of past experience and the preview of next month's contents, you have assessed the following probabilities, $P(\theta)$, of the number θ of copies which your customers will demand:

θ	$P(\theta)$	θ	$P(\theta)$	θ	$P(\theta)$
<20	.000	27	.045	35	.090
20	.010	28	.030	36	.130
21	.040	29	.030	37	.090
22	.050	30	.025	38	.070
23	.050	31	.020	39	.020
24	.060	32	.020	40	.010
25	.065	33	.040	≥41	.000
26	.055	34	.050		

How many copies of *Bon Vivant* should you order?

1.5.6 Impulse Buyer

You wandered into an antique weapons store and were attracted to a purportedly ancient Arabian sword, for sale at \$40, in excellent condition. Not being an expert, you cannot tell if it is a genuine antique. To obtain information on the reasonableness of the price, you went to a nearby dealer, who responded to your request to see Arabian swords of that vintage by saying, "No Arabian swords of that period are on the market. All we see are cheap fakes. Now, let me show you a very fine Viking battleaxe. . . ." You wonder about the proximity of (and competition between) these dealers, and you also note that good hunting knives sell for over \$20. Moreover, genuine or not, the Arabian sword is a handsome wall decoration. You decide that you would be willing to pay \$30 for this sword even if you knew that it was not genuine. Would you buy it for \$40 *without* knowing? What additional judgments and/or preferences would you wish to render explicit in making your decision?

1.5.7 Marine Insurance

Assume that you are a nonseagoing, eighteenth-century English shipowner contemplating paying £2,000 to insure a round trip of your ship, *Pride of Old Sarum*, to the Colonies. The insurance is against the only significant risks—sinking (for any cause) or capture by French privateers—and covers the loss of the ship (£5,000) and profit (£7,000) on the cargoes. Ten out of 100 recent, similar voyages ended in disaster caused by the elements or the French. Would you buy the insurance?

1.5.8 Biologist's Problem

You are a biologist contemplating accepting one or the other of two positions. Position A is with a chemical-biological warfare group that is well funded and would offer you ample support of all kinds for your research on a particular strain of bacteria, a subject having great relevance to the mission of this group. Position B is with a small college which candidly admits that support for your research would be scanty, but that you would have complete control over the uses of (and publicity on) your results. Both appointments are for a period of one year; neither is renewable. You have carefully assessed the odds at 3 to 1 of substantially completing your project if you accept position A, and at only 2 to 3 in favor of completion in position B. The salaries offered are the same, and both positions promise equally arduous relocations and comparable living conditions. What additional factors, if any, do you wish to consider in making your decision?

1.5.9 Realpolitik

You are a highly placed official in the State Department; your recommendations have never been rejected. You have been asked to advise on the desirability of giving $10 million in military assistance to Alfredo Feodorovich, the repressive "president" of a reputedly staunchly allied country, with, however, a strong underground regarded as slightly inimical to our government. If Alfredo receives the foreign aid, the underground will be very inimical to our interests and you would put the odds in favor of its being crushed at 2 to 1. If Alfredo does not receive the foreign aid, the underground will be friendly, Alfredo will be inimical to us, and your odds in favor of the underground's being crushed would be only 1 to 2. Would you recommend giving the foreign aid to Alfredo?

1.5.10 Overbooking Problem

You are the organizer of a charter flight venture and have chartered a plane with 96 passenger seats, together with crew, for $7,500 which includes the salaries of the crew. If the number of requests for reservations at $100 each is 96 or fewer, then all should be granted. But if more than 96 requests come in, you may wish to overbook by making more than 96 reservations because of the possibility of no-shows, i.e., people who make reservations and then do not honor them. In your particular situation, however, overbooking may be dangerous, because you are required by law to provide transportation for all reservation holders. Thus, if more than 96 reservations are made and more than 96 people show up, you would then have to charter an additional plane at the last minute. The only such plane available is a 12-passenger model which can be chartered for $2,000. In addition to reservation holders, you might be able to pick up some standby travelers from the regular airlines who would pay $75 for the trip. Your problem is to determine the upper limit, $n°$, on the number of reservations to make. In *no* case can you make more than 108 reservations. After careful consideration you have assessed the following probabilities $P(r|n)$ of the possible numbers, r, who would show up out of n reservation holders, given any n exceeding 95.

r	$P(r\|n)$	r	$P(r\|n)$
n	.10	$n-6$	.08
$n-1$	.11	$n-7$	.05
$n-2$	.13	$n-8$	.03
$n-3$	.15	$n-9$	.02
$n-4$	.20	$n-10$	.01
$n-5$	.12	$\leq n-11$	.00

Also, you have assessed the following probabilities $P(x)$ of the number x of standbys who could be picked up.

x	$P(x)$	x	$P(x)$
0	.20	6	.06
1	.20	7	.05
2	.15	8	.05
3	.09	9	.03
4	.08	10	.02
5	.07	≥ 11	.00

You do not believe that any relationship exists between r and x. What limit $n°$ should you set? Why?

1.5.11a Political Dilemma

You are a community leader called upon to support publicly one of two candidates, A and B. Candidate A, the incumbent, has compiled a most mediocre and venal record and is known to be extremely vindictive. Your inclinations are toward B, a young challenger who has compiled a good record in a lesser office and is waging a clean and community-spirited campaign. If you declare for B and he wins, your prestige will be enhanced, but more importantly, you foresee much better government in your community. If you support B and he loses, your character will be defamed, your property taxes will be raised by $600 a year, and A will make running your business much more difficult by seeing to it that nuisance laws are obeyed to the letter. Your reputation for civicmindedness will take a pratfall if you support A regardless of whether he wins, but you will be safe from the costly harassment, because B is not vindictive. Your community standing is such that you assess odds of 2 to 1 in favor of A's winning with your endorsement and 6 to 5 in favor of A's losing if you endorse B. Whom would you endorse?

1.5.11b

Suppose that the candidates are equally mediocre and venal, and that your odds are still as given above. Thus the consequences of your having supported the winner are the same regardless of which is the winner, and similarly for the consequences of your having supported the loser. You think it more desirable to have supported the winner than to have supported the loser. Whom would you support?

***1.5.12a General Inventory Problem**

You are trying to decide on the number of widgets to carry in your inventory during each of N time periods $1, 2, \ldots, i, \ldots, N$. Your net pretax profit is p for every widget sold. Widgets are very valuable, and you must insure and guard these items in your inventory at a cost of h per widget per time period held. If you restock b_i widgets in period i, you have to pay transportation and clerical costs of $K + k \cdot b_i$, where K and k are positive numbers not depending upon the period number, i. Your orders are filled almost immediately by armored car from the nearby wholesaler. The total number, γ_i, of widgets demanded by your customers in period i is uncertain, and demands in different periods are completely unrelated. All your transactions with the wholesaler and your customers are on a cash basis, and you have a discount factor r (meaning that \$1 at the beginning of any period $i+1$ is worth $\$r < \$1$ at the beginning of period i). You make your reorder decision at the beginning of each period i, then receive the shipment and pay the cost, and then γ_i widgets are demanded, the demand occurring right after the b_i widgets are delivered. Now, if you should run out of widgets, you lose whatever the excess demand might be; that is, you cannot "back-order" for your customers. How would you go about determining $b_1, b_2, \ldots, b_N$?

***1.5.12b**

Suppose that the number of future periods is infinite rather than some positive integer (= positive whole number) N. Reanalyze as necessary.

1.6 ENCOURAGEMENT

If you experienced a sinking feeling when reading the questions in some of these examples, take heart! You are not expected to cope with them intelligently at this point. The intent of this book is to help you do so.

These examples are included for two reasons. First, they will be invoked throughout the book to exemplify our general approach; second, they illustrate a wide variety of applications and suggest the generality of our approach. Many beginners in decision analysis acquire the deplorable belief that our approach is pertinent only when all possible consequences of a decision are monetary and nothing else matters. Examples 1.5.6, 1.5.8, 1.5.9, and 1.5.11 should expunge that notion.

1.7 ROLE OF DECISION ANALYSIS

The intent of most of our efforts is to break up a *large*, complicated decision problem under uncertainty (such as Example 1.5.1c) into a number of much *smaller* problems, on which you can focus your judgments and preferences much more clearly. *Analysis* means just such a decomposition. Once you have solved all the smaller problems, you can *synthesize* these solutions into a solution of the large problem.

Decision analysts generally take on faith the superiority of the analysis-solution-synthesis approach over intuitive, "Gestalt" alternatives. We claim that a person's judgments and preferences are most easily focused, without intuitive logical slips, on the small subproblems, and that logical consistency then dictates the appropriateness of the subsequent synthesis.

But really, the proof of the pudding is in the eating. It will be up to *you* to determine the worth, to you, of decision analysis.

You are the decision maker. *You* are concerned with your own decisions. All we decision analysts can do is to present our approach to decision making under uncertainty for your consideration, in the belief that it will be of value and in the knowledge that it has been of value to others.

Our approach does *not* eliminate the need for you to make judgments and to express preferences; anyone who so claims is a charlatan. As you read the examples and will have occasion to really appreciate in subsequent chapters, your judgments and preferences enter the picture in crucial ways. Let there be no misunderstanding on this point.

2

PROBLEM FORMULATION: DECISIONS IN EXTENSIVE FORM

Look cautious round, your Genius nicely know,
And mark how far its utmost Stretch will go.
Bickham's Universal Penman, 1741

A: FUNDAMENTALS

2.1 PARLOR-GAME MODEL

A decision problem under uncertainty (= dpuu) may be regarded as a parlor game with two players, the decision maker (= $\mathscr{D}$) and his environment (= $\mathscr{E}$). This game consists of a sequence of choices by $\mathscr{D}$ and by $\mathscr{E}$ that ultimately determine the consequence to $\mathscr{D}$ of playing.

The game usually begins with $\mathscr{D}$ making an initial choice of an *act* (e.g., fee proposal, bid, portfolio, stock level of *Bon Vivants*, purchase of the sword or of insurance, which position to accept, a recommendation on foreign aid, his limit on reservations, or his political candidate).

After $\mathscr{D}$ has made his initial choice, the buck usually passes to $\mathscr{E}$, who then chooses some event to reveal to $\mathscr{D}$ (e.g., the prospective client's reaction to the fee proposal, the competitor's bid β and the cost α of fulfilling the contract, the year-end return θ_i per dollar invested in security i for $i = 1, 2, \ldots, n$, next month's demand for *Bon Vivants*, whether the sword is genuine or whether the *Pride of Old Sarum* will be lost, whether $\mathscr{D}$ will complete his research, whether Alfredo will be overthrown, the number of reservation holders who will appear, and which candidate wins).

More choices may ensue, depending upon the choices made so far by $\mathscr{D}$ and by $\mathscr{E}$.

It is occasionally desirable to allow a player to make two choices in a row. For instance, if the attorney $\mathscr{D}$ proposes a fee, and if the prospective client (a *part* of $\mathscr{E}$) accepts, then $\mathscr{E}$ determines, first, whether the case is won, and second, whether the client will have the similar case come up.

Ultimately, however, the game ends, and the decision maker experiences a *consequence* c which depends, in general, upon the entire sequence of his and $\mathcal{E}$'s choices. For example, the consequence of $\mathcal{D}$'s not insuring the *Pride of Old Sarum* and $\mathcal{E}$'s invocation of catastrophe is $\mathcal{D}$'s ending up with £0 value, whereas his buying the insurance guarantees him £10,000 value by the end of the voyage regardless of what $\mathcal{E}$ reveals.

2.2 $\mathcal{D}$-MOVES AND $\mathcal{E}$-MOVES

We now start to develop a system of symbolism appropriate for discussion of dpuu's in the abstract.[a]

Definition 2.2.1. A $\mathcal{D}$*-move* is the set $G = \{g^1, g^2, \ldots, g^m\}$ of *all* mutually exclusive *acts* available for choice by $\mathcal{D}$ at some point in the dpuu.

In Definition 2.1.1, m is some positive integer (whole number), and the numbers $1, 2, \ldots, m$ are not powers, but rather *superscripts*, used to index the acts. By "mutually exclusive acts" we mean that $\mathcal{D}$ can choose *only one* available act in that $\mathcal{D}$-move, and not two or more.

$\mathcal{D}$-moves are often represented by branching diagrams in which the individual acts g^j (for $j = 1, \ldots, m$) are denoted by straight lines all emanating from a square, □, which denotes the point at which $\mathcal{D}$ must choose one of the acts. See Figure 2.1.

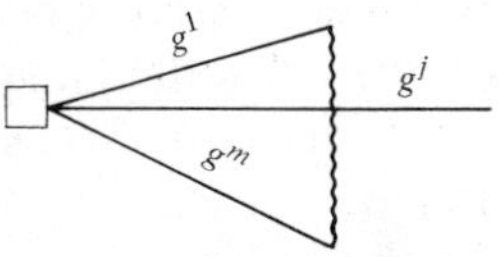

Figure 2.1 Branching diagram of a $\mathcal{D}$-move. The wavy line suggests the existence of some branches which are not drawn in the diagram explicitly.

Example 2.2.1

In Example 1.5.7 (the insurance example) there is only one $\mathcal{D}$-move. Let g^1 denote the act of not buying the insurance and let g^2 denote the act of buying the insurance. Then you have a two-branched $\mathcal{D}$-move.

Example 2.2.2

In Example 1.5.1a, there is only one $\mathcal{D}$-move; namely, {"install as is," "rebuild and then install"}. In Example 1.5.1b there are several $\mathcal{D}$-moves. The first in point of time is {"apply pretest," "do not test"}. Then if you choose "do not test," you have another $\mathcal{D}$-move: {"rebuild and then install," "install as is"}. If you choose "apply pretest" in the initial $\mathcal{D}$-move, then *for each of the possible outcomes of the pretest* you have

[a] Readers whose understanding of the notation and concepts of elementary set theory is shaky are referred to Appendix 2.

a $\mathscr{D}$-move {"rebuild and then install," "install as is"}. Many more $\mathscr{D}$-moves are present in Example 1.5.1c.

Exercise 2.2.1

Describe *each* $\mathscr{D}$-move (1) by a set and (2) by a branching diagram in:

(A)	Example 1.5.11a	(G)	Example 1.5.5
(B)	Example 1.5.11b	*(H)	Example 1.5.2c
(C)	Example 1.5.8	*(I)	Example 1.5.4
(D)	Example 1.5.6	*(J)	Example 1.5.3
(E)	Example 1.5.9	*(K)	Example 1.5.12a
(F)	Example 1.5.10	*(L)	Example 1.5.12b

The concepts we have introduced regarding $\mathscr{D}$'s (= "your") choices have perfect analogues for $\mathscr{E}$'s choices of event to reveal to $\mathscr{D}$.

Definition 2.2.2. An *$\mathscr{E}$-move* is the set $\Gamma = \{\gamma^1, \gamma^2, \ldots, \gamma^n\}$ of *all* mutually exclusive *elementary events* available for choice by $\mathscr{E}$ to reveal to $\mathscr{D}$ at some point in the dpuu.

Again, $1, 2, \ldots, n$ are superscripts; and $\mathscr{E}$ can choose *only one* elementary event in a given $\mathscr{E}$-move.

For reasons which will become apparent later on, we call a choice by $\mathscr{D}$ an "act" and a choice by $\mathscr{E}$ an "elementary event" (= ee). Moreover, whenever we use symbols we shall always denote ee's by *Greek* letters and always denote acts by *Latin* letters. Furthermore, every lower-case Greek letter denotes an ee, and every upper-case Greek letter denotes a *set* of ee's (often the $\mathscr{E}$-move itself). This convention will help you to keep things straight. The Greek alphabet is shown in Appendix 1.

$\mathscr{E}$-moves are also often represented by branching diagrams, in which the individual ee's, γ^ℓ (for $\ell = 1, \ldots, n$), are denoted by straight-line segments all emanating from a circle, O, which denotes the relative time point at which $\mathscr{E}$ intervenes by choosing one and only one ee to reveal to $\mathscr{D}$. See Figure 2.2.

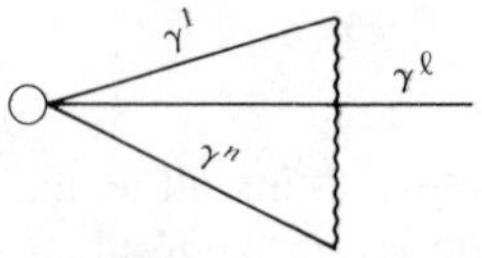

Figure 2.2 Branching diagram for an $\mathscr{E}$-move. Again the wavy line suggests the existence of some branches which are not drawn in the diagram explicitly.

Example 2.2.3

In the insurance example (Example 1.5.7), there is only one $\mathscr{E}$-move. Let γ^1 denote the ee that the *Pride of Old Sarum* is sunk or captured, and γ^2 the ee that the ship is neither sunk nor captured. Then $\mathscr{E}$ has a two-branched $\mathscr{E}$-move.

Example 2.2.4

In Example 1.5.1a, there is only one $\mathscr{E}$-move; namely, {"subsystem defective," "subsystem not defective"}. In Example 1.5.1b, there are several $\mathscr{E}$-moves of the above form and also {"pretest indicates defective," "pretest indicates nondefective"}, describing the possible outcomes of the pretest if $\mathscr{D}$ has chosen to pretest. There are many more $\mathscr{E}$-moves in Example 1.5.1c, some of which are of the form {"inaccurate test indicates defective," "inaccurate test indicates nondefective"}.

Exercise 2.2.2

Describe *each* $\mathscr{E}$-move (1) by a set and (2) by a branching diagram in the following.

(A)	Example 1.5.11a	(G)	Example 1.5.5
(B)	Example 1.5.11b	*(H)	Example 1.5.2c
(C)	Example 1.5.8	*(I)	Example 1.5.4
(D)	Example 1.5.6	*(J)	Example 1.5.3
(E)	Example 1.5.9	*(K)	Example 1.5.12a
(F)	Example 1.5.10	*(L)	Example 1.5.12b

2.3 CONCATENATING MOVES—DECISION TREES

Every dpuu starts with a $\mathscr{D}$-move.[b] Once $\mathscr{D}$ has chosen an act in that $\mathscr{D}$-move, one of three things will occur: either $\mathscr{D}$ experiences a consequence, or $\mathscr{E}$ has an $\mathscr{E}$-move, or $\mathscr{D}$ has another $\mathscr{D}$-move. Occasionally, which thing occurs depends upon the act $\mathscr{D}$ chose in his initial move.

Example 2.3.1

In the insurance example (Example 1.5.7), if $\mathscr{D}$ initially chooses "insure," he guarantees the consequence "end up with £10,000 value"; whereas if he chooses "do not insure," then $\mathscr{E}$ chooses one of the ee's, γ^1 or γ^2, and $\mathscr{D}$ experiences the consequence "end up with £0 value" if $\mathscr{E}$ chooses γ^1 or the consequence "end up with £12,000 value" if $\mathscr{E}$ chooses γ^2.

Example 2.3.2

In the preceding example we could let $\mathscr{E}$ choose between γ^1 and γ^2, given that $\mathscr{D}$ chose "insure," but $\mathscr{D}$'s consequence is still "end up with

[b] By an artifice introduced in Section 2.10, the truth of this statement can be guaranteed without doing violence to the underlying decision problem.

£10,000 value" regardless of whether γ^1 or γ^2 is revealed to $\mathscr{D}$. It often seems a needless complication to give $\mathscr{E}$ a move when the result of that move cannot affect $\mathscr{D}$'s consequence. More about this later.

We represent the contingent arrival of $\mathscr{D}$ and $\mathscr{E}$ at a given ($\mathscr{D}$ or $\mathscr{E}$) move M by concatenation of branching diagrams. If at a given point in the play of the game a choice must be made from a given move M, we place M (represented by its branching diagram) at the right-hand tip of the (unique) branch in the preceding move which leads up to M. So, concatenating all moves in the dpuu produces a figure called a *decision tree.*

Example 2.3.3

From the description of the insurance problem in Example 2.3.1, we obtain by concatenation the decision tree in Figure 2.3.

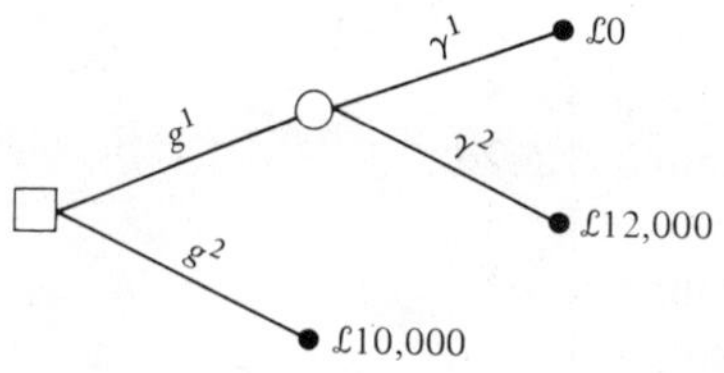

Figure 2.3 Decision tree for insurance problem.

Example 2.3.4

From the more elaborate description in Example 2.3.2 of the insurance problem, we obtain by concatenation the decision tree in Figure 2.4.

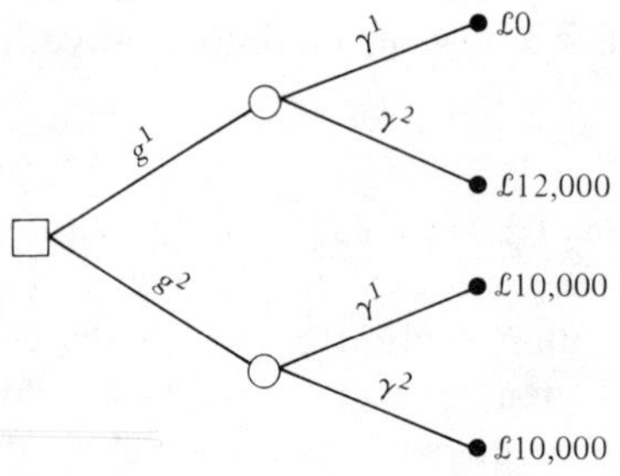

Figure 2.4 Alternative decision tree for insurance problem.

In our definition of a decision tree, we stipulated tangentially that every move M other than the initial move has a unique predecessor branch. This stipulation appears to do violence to problems in which (say) the choices available to $\mathscr{D}$ at a given point are unaffected by $\mathscr{E}$'s previous choice. But Example 2.3.4 and Figure 2.4 indicate the way to circumvent this difficulty: simply *duplicate* the unaffected move, adjoining one copy of it to each of its admissible predecessors. What is by definition *not allowed* in decision trees is the presence of diamondlike structures, called *loops*, as depicted in Figure 2.5.

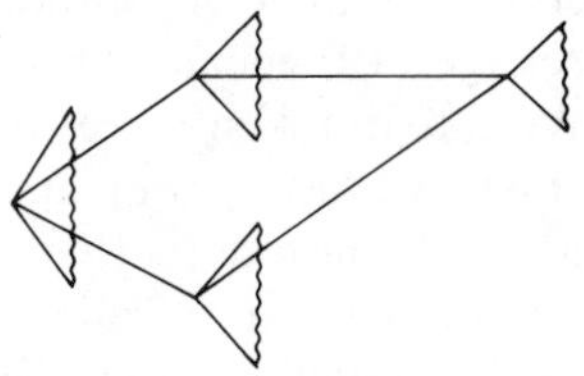

Figure 2.5 Forbidden structures for decision trees.

The following two definitions of features of decision trees will be important in our subsequent analysis.

Definition 2.3.1. An *endpoint* of a decision tree is a right-hand tip of a branch to which no successor move is attached.

Endpoints designate possible terminations of the decision situation. We denote them by heavy dots. (See Figures 2.3 and 2.4.)

Definition 2.3.2. A *full history* of the dpuu is a connected path (sequence of contiguous straight-line segments) through the tree, from the base of the initial move up to an endpoint.

We often omit the adjective "full" and speak of a history of the dpuu. There are, for example, three histories of the insurance dpuu as represented by Figure 2.3; namely, (g^1, γ^1), (g^1, γ^2), and g^2 alone.

In terms of histories, the requirement that loops be absent in decision trees can be imposed in the following form: *For every endpoint there is exactly one full history up to that endpoint.*

Each such history is a complete description of a sequence of choices by $\mathscr{D}$ and $\mathscr{E}$ in the dpuu. At the endpoint of each such history is attached a *consequence*. When symbols are necessary, we denote a consequence by the letter c, and use C to denote the set of *all* consequences conceivably realizable in the dpuu via appropriate choices by $\mathscr{D}$ and $\mathscr{E}$.

Example 2.3.5

From Examples 2.3.1 and 2.3.3 we have noted that there are three histories with three endpoints, and that C can be (succinctly) described as

$$C = \{£0,\ £12{,}000,\ £10{,}000\}.$$

The more elaborate description of the insurance problem in Examples 2.3.2 and 2.3.4 leads to four histories, but C does not require enlargement.

Example 2.3.6

Now let us tackle a more complicated dpuu; viz., Example 1.5.1b, a decision tree for which is depicted in Figure 2.6. The consequences attached to the endpoints are abbreviations for the ultimate net profit and the effect (if any) on your company's prestige; LP means "lose prestige"; MP means "maintain prestige." We have taken care to cumulate all profits and costs. For example, the history ("apply pretest," "pretest indicates defective," "rebuild and install," "not defective") produces \$250(000) profit minus \$20(000) cost of pretesting minus \$100(000) cost of rebuilding, for a net profit of \$+130(000); and the history ("do not test," "install as is," "defective") produces \$250(000) profit minus \$500(000) penalty for a net profit of \$−250(000), together with loss of prestige, which is suffered whenever the system is defective, whereas prestige is maintained otherwise. You should examine Figure 2.6 carefully, and satisfy yourself that it really does represent adequately the dpuu described in Examples 1.5.1a and 1.5.1b. Note also that the

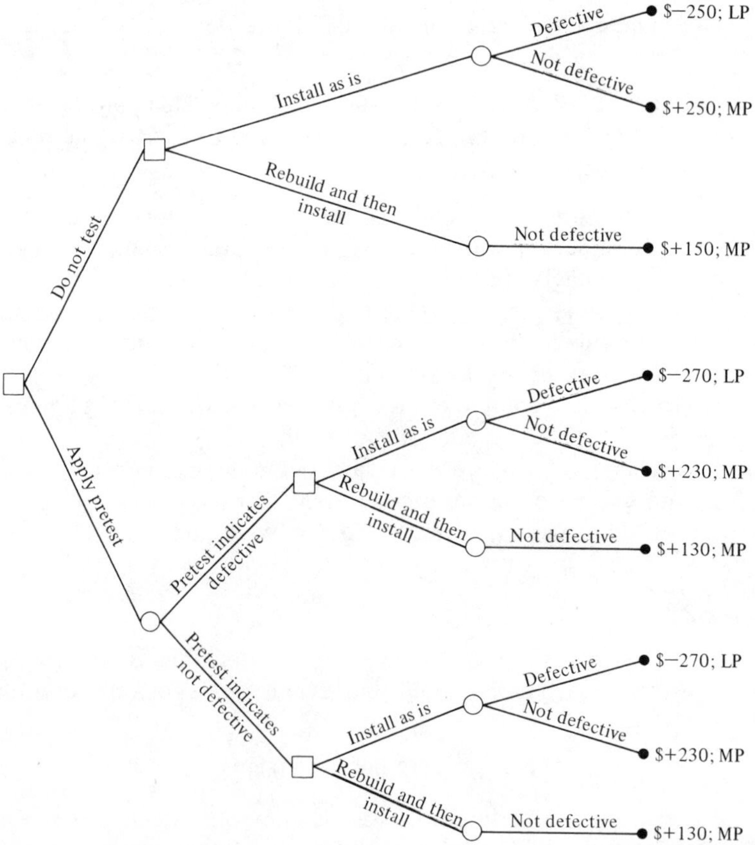

Figure 2.6 Decision tree for Example 1.5.1b. (Dollar amounts are in thousands.)

simpler dpuu of Example 1.5.1a is obtained from Figure 2.6 by merely pruning off the lower branch (and its successors) in the initial $\mathscr{D}$-move. Finally, note that Figure 2.6 contains three copies of each of the $\mathscr{E}$-moves {defective, not defective} and {not defective}, and three copies each of the $\mathscr{D}$-move {install as is, rebuild and then install}.

One final definition that is in general use.

Definition 2.3.3. The parlor-game model of a dpuu, as depicted by a decision tree, is called a *decision in extensive form.*

Exercise 2.3.1

Formulate the following dpuu's in extensive form (i.e., as decision trees).

(A)	Example 1.5.11a	(G)	Example 1.5.5
(B)	Example 1.5.11b	*(H)	Example 1.5.2c
(C)	Example 1.5.8	*(I)	Example 1.5.4
(D)	Example 1.5.6	*(J)	Example 1.5.3
(E)	Example 1.5.9	*(K)	Example 1.5.12a
(F)	Example 1.5.10	*(L)	Example 1.5.12b

2.4 GENERAL PRINCIPLES IN FORMULATING DECISION TREES

The general principles in formulating decision trees are simple to describe. As you probably already appreciate from working Exercise 2.3.1, formulating a decision tree involves three things:

(1) being careful to put moves in a correct order;
(2) being careful to describe the elements in each move (recall that a move is a set) accurately, so that:
  (a) in $\mathscr{D}$-moves, $\mathscr{D}$ will choose one and only one act, and
  (b) in $\mathscr{E}$-moves, $\mathscr{E}$ will choose one and only one ee;
(3) being careful to describe all relevant features of each potential consequence.

Cautionary note (1) involves two important points about sequencing. The first concerns the fact that the relative time, flowing from left to right in the decision tree, marks epochs at which *the decision maker* does something (chooses an act) or learns something (perceives an ee). Such time is not necessarily consistent with the *chronological* time at which various real "actors" in the environment perform. For example, the sword is *already* either a genuine antique or a fake, but $\mathscr{D}$ *will not know* which until after he purchases it, if he ever does find out; $\mathscr{D}$'s lack of knowledge dictates the appropriate sequencing of the $\mathscr{E}$-move {fake, genuine} as posterior to the $\mathscr{D}$-move {purchase, do not purchase}.

Appropriate sequencing of moves is facilitated by always regarding ee's as *possible revelations to* $\mathscr{D}$.

■ If the dpuu extends over a long chronological time period, then chronological time may be relevant to $\mathscr{D}$. If so, its relevant aspects should be reflected in the descriptions of the potential consequences. For example, if the voyage of the *Pride of Old Sarum* will take one year if completed and the insurance premium is payable on embarkation whereas a claim would be settled one year after the ship's scheduled return, then our previous descriptions fail to account adequately for the time value of money. If £1 one year hence is worth £.8 currently to $\mathscr{D}$ (not a very conservative assumption for Georgian shipping) and if we evaluate all transactions as of the scheduled termination of the voyage, then the potential consequences of not buying the insurance remain as given in Example 2.3.1; viz., "end up with £0 value one year hence" and "end up with £12,000 value one year hence." But £2,000 now is equivalent to £2,000/.8 ≐ £2,500 one year hence, and £12,000 two years hence is equivalent to £(.8)(12,000) = £9,600 one year hence, so that buying the insurance yields $\mathscr{D}$ the equivalent of £7,100 (= £9,600 − £2,500) if disaster strikes, and £9,500 (= £12,000 − £2,500) if disaster does not strike. Note how this introduction of the time value of money *requires* us to use the more complicated description of the dpuu in Examples 2.3.2 and 2.3.4. ■

The second point about the appropriate sequencing of moves concerns the relative placement of $\mathscr{D}$-moves together with associated, subsequent $\mathscr{E}$-moves when the relative placement is ambiguous in the verbal description of the problem. In such situations you should question whether the sequencing is *constrained by realities* of the decision situation. In realistic contexts, as distinguished from the already abstracted problems in cases and texts, you can always investigate this issue. If reality does constrain the sequencing, then the ambiguity vanishes. On the other hand, *if* $\mathscr{D}$ *has a choice of sequencing, then that choice should be included as a* $\mathscr{D}$*-move in the decision tree.* Be very careful about prejudging the relative desirabilities of the alternative sequencings.

Exercise 2.4.1

Check your decision tree for Example 1.5.2c carefully in light of cautionary notes (1)–(3).

Cautionary notes (2) and (3) are discussed further in Sections 2.5 and 2.6, respectively.

You now know all the *basic* principles of decision-tree formulation (arborculture?). The remainder of this chapter is devoted to refinements and niceties of formulation and reformulation.

B: ON FORMULATING DECISION TREES

2.5 ON DESCRIBING THE ELEMENTS OF MOVES

Suppose that you are playing a game of poker and it is now your turn to bid. Your acts *can* be described as "pass," "check," and "raise x" for x an admissible raise. But every poker player knows that there are many *ways* to do each of these; you can "check smugly," "check dubiously," "check impassively," etc.[c] Hence, should you include in your $\mathcal{D}$-move the *crude* description "check" or all the *finer* descriptions of "check" that you can think of? Similarly for your description of the elementary events in $\mathcal{E}$-moves.

There is no pat solution to the problem of how finely to describe acts and ee's. Observations that are relevant to your resolution of this problem in particular cases are the following.

(1) The more finely you describe acts and ee's, the more clear-cut will be the ultimate descriptions of the consequences.
(2) The more finely you describe acts and ee's, the easier it will be to express your preferences between any two consequences.
(3) The more finely you describe acts and ee's, the easier it will be to express your judgments regarding the relative odds of $\mathcal{E}$'s choice between any two ee's in subsequent $\mathcal{E}$-moves.

On the negative side, however:

(4) The more finely you describe acts and ee's, the *much* more numerous will become the preferences and judgments that you will have to express.
(5) The more finely you describe acts and ee's, the *much* more unwieldy will become the decision tree.

Raiffa [**113**] discusses (4) and (5) in a section entitled, "When Does a Tree Become a Bushy Mess?"

In many complicated dpuu's it is a good idea to *start* with a rather crude description of acts, ee's, and preferences, express crudely your preferences and judgments, and perform the analysis described in Chapters 3 and 4; then eliminate drastically poor courses of action and their successor moves, thus obtaining a smaller tree. This process is called "pruning" the tree—for obvious reasons. Now, in the *pruned* tree you can better afford the work involved in formulating finer descriptions and judgments more carefully. Then reanalyze. Perhaps after so doing you will want to reprune, redescribe, reexpress, and reanalyze.

Recall, however, that you can choose *only* one act in a given $\mathcal{D}$-move,

[c] In tournament bridge, such expressiveness is a violation of the rules.

and hence your act descriptions must not be so brief that they permit you to choose more than one. This can always be forced, by regarding as an act a combination of activities.

Example 2.5.1

Suppose at a given point that $\mathscr{D}$ can eat and/or drink or do neither. The appropriate formulation of his move is to make it consist of the acts "eat," "drink," "eat and drink," and "abstain." Alternatively, one could formulate a first $\mathscr{D}$-move consisting of "abstain" and "do not abstain," with the latter followed by a $\mathscr{D}$-move consisting of "eat," "drink," and "eat and drink." In this second $\mathscr{D}$-move, choosing one of the acts is equivalent to choosing either or both of the acts in the set (not a $\mathscr{D}$-move!) consisting of "eat" and "drink," since the act "eat and drink" in the $\mathscr{D}$-move is the same as choosing both "eat" and "drink"—obviously.

We have required that the acts in a $\mathscr{D}$-move must be *mutually exclusive*, meaning that the choice of any one of them precludes the choice of any other. They are also *collectively exhaustive*, in the sense that[d] $\mathscr{D}$ will not choose an act not belonging to that $\mathscr{D}$-move.

Much the same considerations pertain to the description of ee's and $\mathscr{E}$-moves. In a given $\mathscr{E}$-move, *one and only one* ee will be chosen for revelation to $\mathscr{D}$; that is, the ee's are *mutually exclusive* (the revelation of one of the ee's precludes the revelation of any of the others) and *collectively exhaustive* (no event *not* in the $\mathscr{E}$-move will be revealed to $\mathscr{D}$).

You can guarantee mutual exclusivity of ee's in an $\mathscr{E}$-move by precisely the same technique as suggested for $\mathscr{D}$-moves; *viz.*, regarding each ee as a combination of revelations.

Example 2.5.2

In Example 1.5.2c, one appropriate formulation contains the $\mathscr{E}$-move consisting of the ee's "the case is won and the client has a similar case," "the case is won and the client does not have a similar case," "the case is lost and the client has a similar case," and "the case is lost and the client does not have a similar case." The set consisting of "the case is won, "the case is lost," "the client has a similar case," and "the client does not have a similar case" is *not* an $\mathscr{E}$-move, because one of the first two listed elements *and* one of the last two listed elements will be revealed to $\mathscr{D}$.

Collective exhaustiveness of the ee's in an $\mathscr{E}$-move is more delicate than collective exhaustiveness of acts in a $\mathscr{D}$-move, because the decision maker has complete control over his choice of act and, by will power, can

---

[d] At the current stage of analysis, at any rate.

constrain that choice to the $\mathscr{D}$-move as he has defined it (once he has exercised some creativity in determining its constituent acts). But $\mathscr{D}$ has *no* control over the choice of ee, and hence his failure to achieve collective exhaustiveness in an $\mathscr{E}$-move can lead to what is often referred to as a "totally unforeseen contingency."

Thus you should be very careful in formulating each $\mathscr{E}$-move. Do not exclude any *logical possibility*—no matter how unlikely you may consider it. In this regard, two points require mention.

First, the inclusion of all logical possibilities in an $\mathscr{E}$-move can always be forced, if necessary, by including an ee entitled "something else," "none of the above," or some such phrase. You should use this device only as a last resort, however, since it renders the description of subsequent moves and consequences extremely difficult and also betrays a failure to think through the problem carefully.

Second, the inclusion of all logical possibilities in an $\mathscr{E}$-move is *not* inconsistent with your latitude in choosing the level of fineness with which to describe ee's. If all logical possibilities are included in a fine description, then any coarser description such that each finely described ee is included in exactly one coarsely described ee will also include all logical possibilities.

■ That is to say, the crude $\mathscr{E}$-move is a *partition* of the fine $\mathscr{E}$ move; see Appendix 2. If $\{\gamma: \gamma \in \Gamma\}$ is an $\mathscr{E}$-move, finely described, and Δ is a set of subsets δ of Γ with the properties that

$$(1) \quad \delta^1 \cap \delta^2 = \emptyset \quad \text{whenever} \quad \delta^1 \neq \delta^2$$

and

$$(2) \quad \cup\{\delta: \delta \in \Delta\} = \Gamma,$$

then by definition Δ is a partition of Γ and is an $\mathscr{E}$-move, since (1) guarantees mutual exclusivity and (2) guarantees collective exhaustiveness because Γ itself has these properties (since it was assumed to be an $\mathscr{E}$-move). ■

*■ Readers sophisticated in set theory will note the hierarchical difference between Γ as a set of elements and Δ as a set of sets of elements. This difference is purely relative, because Δ is a set of elements that happen to be sets from the standpoint of the finely described Γ. In fact, the entire issue evaporates if we regard *every* $\mathscr{E}$-move Γ as a partition of some incredibly finely described set Υ of all possible states υ of the environment. ■

Exercise 2.5.1: Discussion Question

Some people have asked whether $\mathscr{D}$'s decisions regarding fineness of descriptions can be handled by decision analysis itself and represented in decision tree form. Would doing so not *necessitate* using the finest descriptions under consideration? Or would it?

2.6 ON THE EXTENSION AND TRUNCATION OF DECISION TREES

We noted in Section 2.4 that the description of the potential consequences in a dpuu affects the appropriate formulation of the decision tree—which is only natural, because the aim of decision analysis is to aid $\mathscr{D}$ in attaining maximally desirable consequences in dpuu's. We also noted in Section 2.5 that some complexity in describing consequences can be eliminated by using reasonably fine descriptions of acts and ee's.

In general, consequences can be made more clear-cut by *extending* the decision tree to the right, so as to render explicit the reality beyond the horizon in your present formulation.

No important decision problem is completely isolated from other activities of the decision maker, nor do its ramifications ever terminate completely. Dpuu's that will affect your financial position also affect your ability to undertake future ventures, and similarly for dpuu's that may affect your prestige, standing, power, etc. Hence a completely thorough formulation of a dpuu is to imbed it in the gigantic tree of all present and future decisions.

Such a formulation, while thorough, is entirely impractical. In a largely intuitive manner, $\mathscr{D}$ should apply a threshold criterion of relevance to narrow down his problem to manageable proportions. Part of that narrowing down consists in describing acts and ee's less finely perhaps than is logically possible; this was considered in Section 2.5.

Another part, however, consists in choosing an appropriate point in relative time beyond which $\mathscr{D}$ will not elaborate any further on his and $\mathscr{E}$'s potential choices, that is, in choosing an appropriate *horizon*.[e]

A horizon that is too myopic and fails to elaborate at all on the ramifications of early moves will produce a decision tree having consequences that are complicated to describe.

Example 2.6.1

In Example 1.5.7, if $\mathscr{D}$ is myopic and does not carry the horizon out to the conclusion of the voyage for his decision tree, then he will have a very simple tree consisting solely of his initial move {"insure," "do not insure"}. But the consequence of "do not insure" is "end up with £12,000 value if disaster does not strike, and with £0 value if disaster strikes." Compare this consequence with the two simpler consequences "end up with £12,000 value" and "end up with £0 value" with respect to ease of comparing their desirabilities with that of "end up with £10,000 value."

The more remote the horizon, the more moves the decision tree will contain and hence the more endpoints and consequences. But the increase

[e] Only when the basic dpuu is sufficiently repetitious, as in Example 1.5.12b, can one hope to get somewhere with an infinite horizon.

in the number of consequences may *reduce* the task of expressing preferences among them, as suggested by Example 2.6.1, since extending the horizon generally leads to *simpler* consequences which do not contain statements of conditionality ("if . . . , then . . .") and potentiality ("we may be able . . .").

Hence there is a trade-off between complexity of consequence descriptions and complexity of the decision tree, nearby horizons necessitating the former and remote horizons producing the latter. Horizons are usually given at least implicitly in text problems and cases, which have *already* been subject to abstraction by an author; but in real life you must resolve such trade-offs for yourself.

In striking these balances, you will often want to modify your originally sketched tree, either (1) by replacing some remote moves by more complex consequences, which is the *truncation* process, or (2) by replacing some complex consequences with moves leading to less complex consequences, which is the *extension* process.

Example 2.6.2

Suppose that instead of the decision tree in Figure 2.6, you decided to use only the initial $\mathscr{D}$-move {"apply pretest," "do not test"}. Then the consequence attached to the right-hand tip of "do not test" is "have the option of installing as is or rebuilding and then installing, in the former case losing \$250(000) and prestige if the subsystem is defective, and making \$250(000) and maintaining prestige otherwise; and in the latter case making \$150(000) and maintaining prestige because the rebuilt system will not be defective." This consequence is devilish to ponder, from the standpoint of preference. You are invited to describe verbally the consequence attached to the right-hand tip of "apply pretest."

Example 2.6.2 makes clear that the complexity of consequences in problems with nearby horizons arises at least in part from the embodiment in these consequences of moves. Rendering these moves explicit in tree form and thus simplifying the consequence descriptions is the extension process.

Exercise 2.6.1: Discussion Question

Same question as in Exercise 2.5.1, but pertaining to choice of horizon.

2.7 CONSEQUENCE DESCRIPTIONS COMPATIBLE WITH THE DECISION TREE

In Sections 2.5 and 2.6 we related $\mathscr{D}$'s choice of fineness levels for describing acts and ee's, and of remoteness of horizon, to the clarity with which $\mathscr{D}$ defines the potential consequences. This discussion tacitly implied an innate desirability of clear consequence descriptions.

What is really necessary for subsequent developments, however, is that the descriptions of the consequences be *compatible* with the decision tree, in the sense that $\mathscr{D}$'s preferences between any two consequences are independent of the means of realizing them. Put another way, all *preference-relevant* aspects of the history h leading to a consequence c must be included in the description of c.

Example 2.7.1

In the insurance problem with the decision tree in Figure 2.4, consider defining the consequences as

c^1 = "have voyage uninsured," and
c^2 = "have voyage insured,"

with g^1 leading to c^1 and g^2 leading to c^2, regardless of which γ occurs. But clearly, you *strongly* prefer c^2 to c^1, given γ^1 (= "disaster"), whereas you *strongly* prefer c^1 to c^2 given γ^2 (= "no disaster"). Hence these crude consequence descriptions are *not* compatible with the decision tree in Figure 2.4. They *are* compatible with the shorter horizon tree consisting solely of the initial $\mathscr{D}$-move $G = \{g^1, g^2\}$, with g^1 leading to c^1 and g^2 to c^2.

Compatibility of consequence descriptions with the decision tree is essential in our subsequent developments, because we require that the decision maker *compare* various hypothetical gambles with the consequences of his dpuu as possible prizes. If his consequences are incompatible with his decision tree, then he will experience a great deal of mental anguish in making such comparisons, much more anguish than is inherent in the task.

Thus in putting the finishing touches on your decision trees with consequences attached to endpoints, you should ask yourself the following question: "Have I defined the potential consequences in sufficient detail for my preferences between any two of them to be independent of their preceding histories?" If so, then, in Chapter 3 and the following chapters, you will experience minimal difficulty in conceptualizing hypothetical gambles with the consequences as prizes.

Sections 2.5 and 2.6 held that clarity of consequence descriptions is *facilitated* by elaborateness of the decision tree; this section furnishes the obverse of the same coin by indicating that a minimal clarity of consequence descriptions is *necessitated* by elaborateness of the decision tree. The whole coin is therefore one of *compatibility*.

Exercise 2.7.1: Discussion Question

Decision analysis has been attacked on the philosophical ground that it is ends-oriented, or goal-oriented, rather than means-oriented, or process-oriented; and that therefore it is not applicable in ethics-fraught decisions by people of moral sensibility. I believe that this attack is unwarranted. Do you agree with me? You might wish to refer to your structuring of Example 1.5.8 in extensive form.

2.8 PREVIEW OF DECISION ANALYSIS

We have noted that truncation of decision trees, or adoption of less remote horizons, leads in general to less clear-cut, more complex consequence descriptions. This is true at the purely qualitative level of this stage of our exposition. It will *not* be true later on, once we have developed some analytical machinery in Chapter 3 for circumventing such difficulties.

In this section we give a capsule preview of that machinery. First, a technical definition.

Definition 2.8.1. A *terminal* $\mathscr{D}$*-move* (respectively: *terminal* $\mathscr{E}$*-move*) of a decision tree is a $\mathscr{D}$-move (respectively: $\mathscr{E}$-move) of which every branch [act (respectively: ee)] leads to an endpoint, with a consequence attached, rather than to a subsequent move.

In other words, a terminal move can lead only to a consequence. Terminal moves figure prominently in our analytical procedure, which will be a successive truncation of the original tree.

The main features of our analytical procedure are:

(1) the *quantification of relative preferences* for the ultimate consequences of the dpuu, in the form of what amounts to a subjective currency;
(2) the *quantification of judgments*, in the form of specifying probabilities, or relative odds, on $\mathscr{E}$'s choice of ee in each $\mathscr{E}$-move;

and, because of the powerful properties possessed by quantifications (1) and (2),

(3a) the quantification of relative preferences for terminal $\mathscr{E}$-moves by a simple calculation (a weighted average of quantitative preferences for the consequences of the ee's with their probabilities as weights gives the quantification for the terminal $\mathscr{E}$-move);
(3b) the quantification of relative preferences for terminal $\mathscr{D}$-moves by a simple comparison (the quantification for the terminal $\mathscr{D}$-move is the maximum quantitative preference for any of its consequences).

The gist of (3a) and (3b) is that *terminal moves may be replaced by numbers reflecting their relative desirabilities to $\mathscr{D}$.* Doing so yields a smaller, truncated tree in which formerly nonterminal moves become terminal. Quantifications (3a) and (3b) are then reapplied, yielding a still more truncated tree. This goes on until all that is left is the initial move.

If we keep track of that act in each $\mathscr{D}$-move (terminal at some stage in the procedure) which yields the highest valued preference, then we will have a complete prescription for action, a prescription which will be *optimal in light of $\mathscr{D}$'s judgments and preferences as expressed in* (1) *and* (2).

Now, just why this is so requires a detailed methodological development, which we start in Chapter 3 and conclude in Chapter 4. You should take it on faith for the present.

Example 2.8.1

Suppose, in Example 2.3.3, for simplicity that the quantification of preferences yields the preference numbers 0 for £0, 1 for £12,000, and .95 for £10,000. Also suppose that you use the data in Example 1.5.7 to quantify your judgments regarding disaster on the voyage by assessing probability .1 to γ^1 (disaster) and probability .9 to γ^2 (safe completion). By (3a) above, the relative preference of the terminal $\mathscr{E}$-move in Figure 2.3 is $(.1)(0)+(.9)(1)=.9$. Now truncate the tree in Figure 2.3, leaving only the initial $\mathscr{D}$-move with the numerical preferences .9 and .95 attached to the tips of its branches (Figure 2.7). Now apply (3b) to this terminal (and initial) $\mathscr{D}$-move, thus seeing that g^2 (= insure) is more desirable than g^1 (= do not insure).

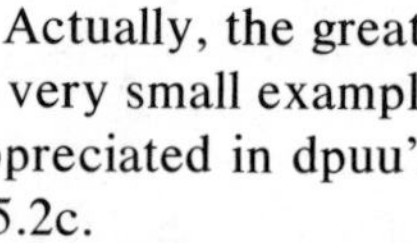

Figure 2.7 Truncation of Figure 2.3.

Actually, the great power of decision analysis is not reflected adequately in very small examples such as the insurance problem. Its full power is best appreciated in dpuu's such as those in Examples 1.5.1b, 1.5.1c, 1.5.2b, and 1.5.2c.

All that you need appreciate at this point is that any decision tree, no matter how large, can be systematically truncated by repeated application of (3a) and (3b), and that recording of the "best" act in each $\mathscr{D}$-move will ultimately produce that complete course of action which is optimal, given $\mathscr{D}$'s quantitatively expressed preferences and judgments.

2.9 CLARITY ACHIEVED WITH THE EXTENSIVE FORM

I venture to say that the greatest benefits one can reap from decision analysis accrue from simply *formulating* ill-structured and complex dpuu's in extensive form. Using the parlor-game model, defining moves, and properly sequencing them *forces* the decision maker to do a great deal of "analysis" in the dictionary sense of that word; he must think hard about his available options, the vicissitudes of the environment, how far into the future he should carry his explicit analysis, and so forth.

In purely discursive confrontations with dpuu's, it is altogether too easy to be nebulous about such matters as thresholds of relevance, horizons, and especially about alternatives to the recommended course of action. Such *unclear thinking is discouraged* by the definitional requirements of moves and decision trees.

Decision trees also offer distinct *communicational advantages.* They expose most succinctly the decision maker's formulational assumptions for the scrutiny of all, by rendering perfectly explicit assumptions which in discursive presentations are often not mentioned or even reconstructible by reading between the lines. A picture is worth a thousand words; and a large tree for a complex dpuu may be worth 25,000 words even though it may have to be drawn on a wall.

The general definition of "analysis" is the breaking down of a whole into its constituent parts. The smallest constituent part of a dpuu is the move. Moves and their interrelationships are accurately mirrored in the decision tree. Hence, *simply by formulating his dpuu in extensive form, the decision maker performs significant analysis* (and synthesis; the decision tree is a synthesis of moves).

In subsequent chapters we present a logically consistent and intuitively meaningful way by which the decision maker can make decisions about *very small subproblems,* involving comparisons of consequences, acts, or ee's only two at a time, and then *synthesize* these small (and, we think, relatively clear-cut) decisions into an entire course of action in the dpuu itself.

We now turn to a somewhat more abstract discussion, Part C, consisting of three sections concerning the equivalent reformulations of a dpuu and culminating with a *standard extensive form* of the dpuu. Readers shaky in elementary set theory might well turn to Appendix 2 before continuing.

C: EQUIVALENT REFORMULATION TECHNIQUES

2.10 INSERTING AND DELETING DUMMY MOVES

An $\mathscr{E}$-move (or a $\mathscr{D}$-move) consisting of only one ee (or act) is clearly of no *strategic* importance, because $\mathscr{E}$ (or $\mathscr{D}$) has no alternative. Such moves should not affect the analysis of a dpuu in any way, and they may be inserted or deleted at will from any decision tree. For this reason, they serve as important devices for making decision trees symmetrical (with all histories of the same length, for example). This importance warrants a title.

Definition 2.10.1. A *dummy move* is a move which consists of precisely one element (= act, or ee).

Figure 2.8 depicts a dummy $\mathscr{D}$-move and a dummy $\mathscr{E}$-move.

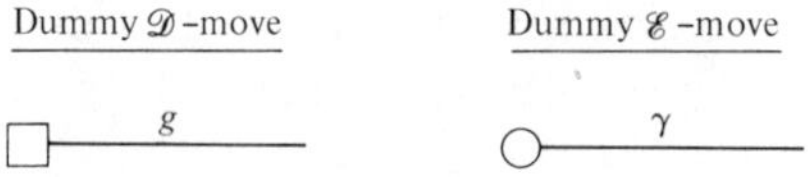

Figure 2.8

In footnote b, we noted that by using an artifice every decision tree can be made to start with a $\mathscr{D}$-move. The necessary artifice is a dummy $\mathscr{D}$-move; Figure 2.9 indicates how a tree beginning with an $\mathscr{E}$-move can be modified by preceding that $\mathscr{E}$-move with a dummy $\mathscr{D}$-move.

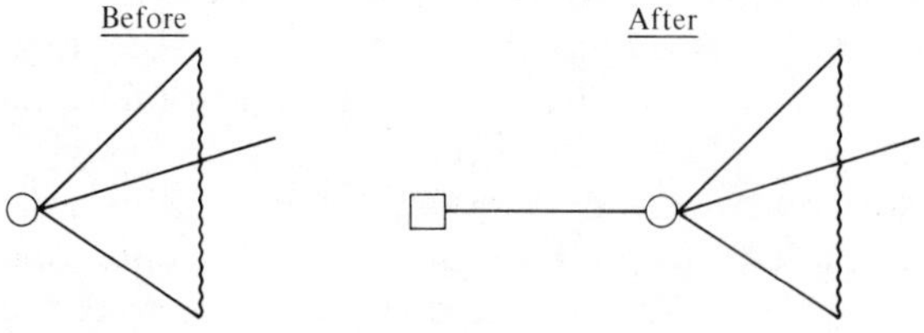

Figure 2.9

In a similar vein, every dpuu can be made to end with an $\mathscr{E}$-move by inserting a dummy $\mathscr{E}$-move between an act branch and the consequence at its tip (which after the insertion will no longer be an endpoint). See Figure 2.10.

Dummy moves can also be inserted at any point in a decision tree *without affecting judgments, preferences, or salient strategic features of the problem.*

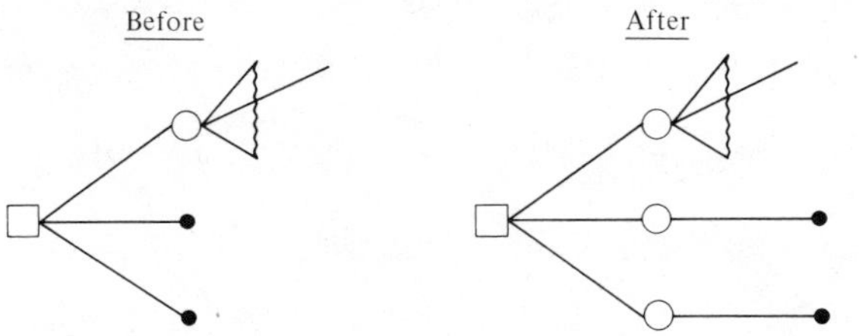

Figure 2.10

Such insertions are often used to separate two consecutive choices by the same player. Thus a dummy $\mathscr{E}$-move can be inserted to separate two consecutive act-choices by $\mathscr{D}$, and a dummy $\mathscr{D}$-move can be inserted to separate two consecutive ee-choices by $\mathscr{E}$. See Figure 2.11 for an illustration of separating consecutive act-choices; the analogous figure for separating consecutive ee-choices is obtained by replacing the boxes in Figure 2.11 with circles and vice versa.

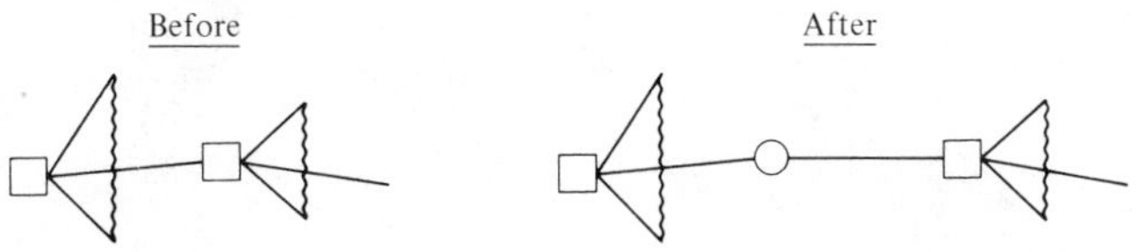

Figure 2.11

Example 2.10.1

In Figure 2.6 we may regard the initial "do not test" option as actually a degenerate test, nontest, or untest, which surely produces a "null outcome" that can be represented by a dummy $\mathscr{E}$-move. If inserted, this dummy $\mathscr{E}$-move serves to separate consecutive choices of act. Once such an $\mathscr{E}$-move has been inserted, the tree admits of an interesting interpretation in terms of relative time sequencing: *first*, $\mathscr{D}$ chooses a test (perhaps the nontest); *second*, $\mathscr{E}$ reveals the outcome of the chosen test; *third*, $\mathscr{D}$ chooses a final act based on the test outcome; and *fourth*, $\mathscr{E}$ reveals the true state of the subsystem in operation, whereupon $\mathscr{D}$ suffers (or enjoys) the consequence. Every history is of length four and consists of act, event, act, and event in that order. Finally, the dpuu can be regarded as a *two-stage* problem (consisting of a testing stage and an operating decision stage), in which in each stage $\mathscr{D}$'s choice precedes $\mathscr{E}$'s choice. See Figure 2.12.

Exercise 2.10.1

Insert dummy moves as necessary to make history begin with an act, end with an ee, and alternate (act, ee, act, ee, . . . , act, ee) in your tree for the following.

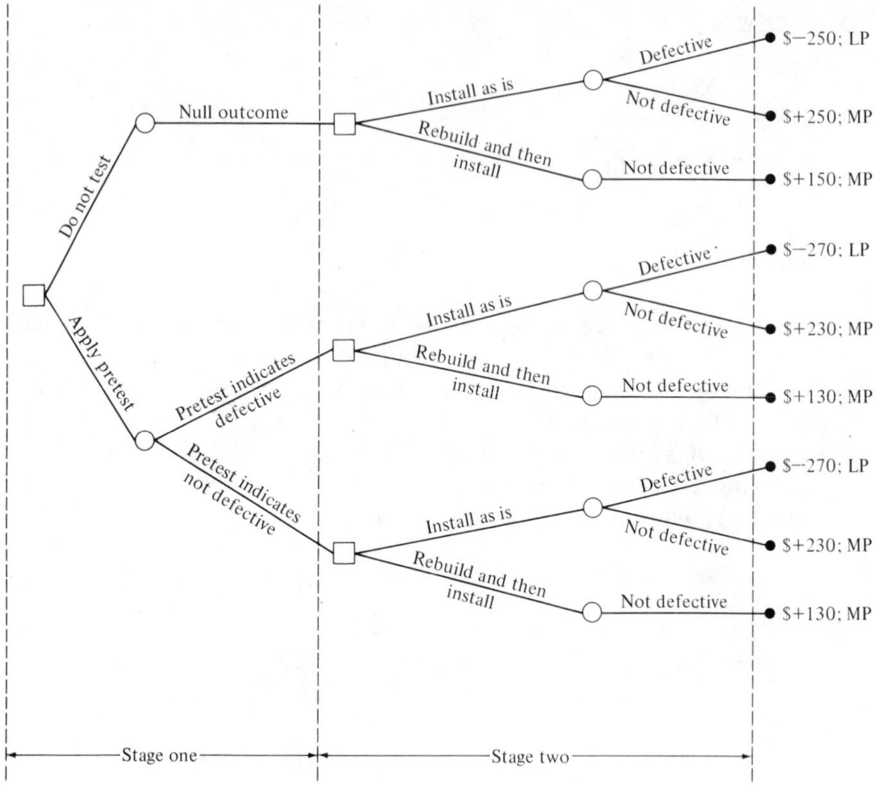

Figure 2.12 Revision of Figure 2.6.

(A)	Example 1.5.11a	(G)	Example 1.5.5
(B)	Example 1.5.11b	*(H)	Example 1.5.2c
(C)	Example 1.5.8	*(I)	Example 1.5.4
(D)	Example 1.5.6	*(J)	Example 1.5.3
(E)	Example 1.5.9	*(K)	Example 1.5.12a
(F)	Example 1.5.10	*(L)	Example 1.5.12b

Dummy moves can also be deleted. Simply erase the dummy move and join its successor move or consequence to its predecessor branch. So doing is just the reverse of the procedure for inserting dummy moves.

2.11 COMBINING SUCCESSIVE CHOICES BY THE SAME PLAYER

Another device used to ensure alternation between acts and ee's in every history is to combine successive choices by the same player. Such combining may be performed before or after inserting or deleting dummy moves.

Two successive choices of act, say g and then g', can be combined into one, (g, g'), meaning "g and g'." Similarly for two successive choices by $\mathscr{E}$ of ee's, say γ and γ', which can be combined as (γ, γ'), meaning "γ and γ'."

Whenever such combining is undertaken, it should be performed for *all* the choices in the successor *move*. Then that move can be eliminated altogether.

The decision tree in Figure 2.13 contains a number of successive choices by the same player. By combining such choices wherever possible, we obtain the decision tree in Figure 2.14.

It is important to note that *combining successive choices by the same player and inserting and/or deleting dummy moves do not change the total number of histories of the dpuu.* Once you have formulated a dpuu in extensive form to your satisfaction, as far as fineness of descriptions and remoteness of horizon are concerned, you can fiddle with dummy moves and combine successive choices by the same player all you like without affecting any salient feature of the dpuu or its ensuing analysis.

The devices discussed here and in the preceding section are purely technical tools for taking a dpuu in extensive form and reformulating it in *standard extensive form*, as defined in the following section.

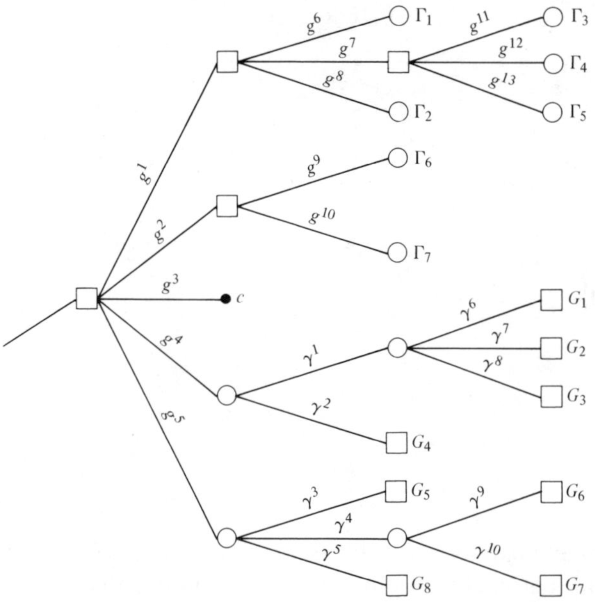

Figure 2.13 Subtree with successive choices by $\mathscr{D}$ and $\mathscr{E}$.

■ The meticulous reader will be concerned with whether combining all successive choices by the same player results in a genuine move, with acts or ee's being mutually exclusive and collectively exhaustive. The

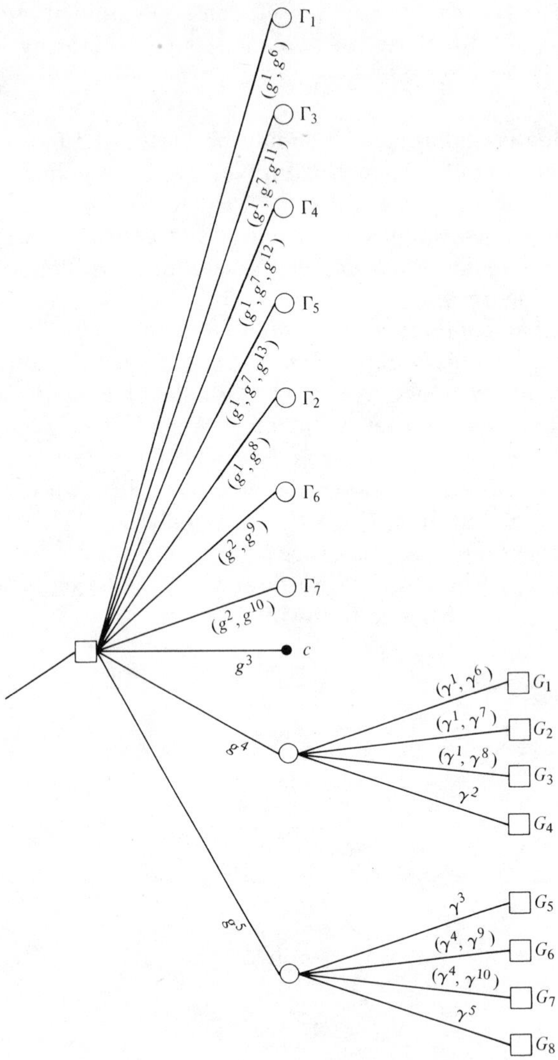

Figure 2.14 Figure 2.13 after combining.

answer is yes; mutual exclusivity follows from mutual exclusivity of the earlier choices and mutual exclusivity of the choices in each successor move. Similarly for collective exhaustiveness. ■

Now recall from Section 2.10 that dummy moves can be deleted as well as inserted, the former being the reverse of the latter conceptually. There is a procedure converse to the combining of successive choices by the same player; two examples of it are left to you as Exercises 2.11.5 and 2.11.6.

Exercise 2.11.1

To the fullest extent possible, use combination of successive choices by the same player in your decision trees for the preceding exercises.

Exercise 2.11.2

Combine as many successive choices by the same player as possible in the decision tree of Figure 2.15. [*Hint*: Start as far to the right as possible and work back successively. The symbols $(a, (b, c))$, $((a, b), c)$, and (a, b, c) need not be distinguished from each other.]

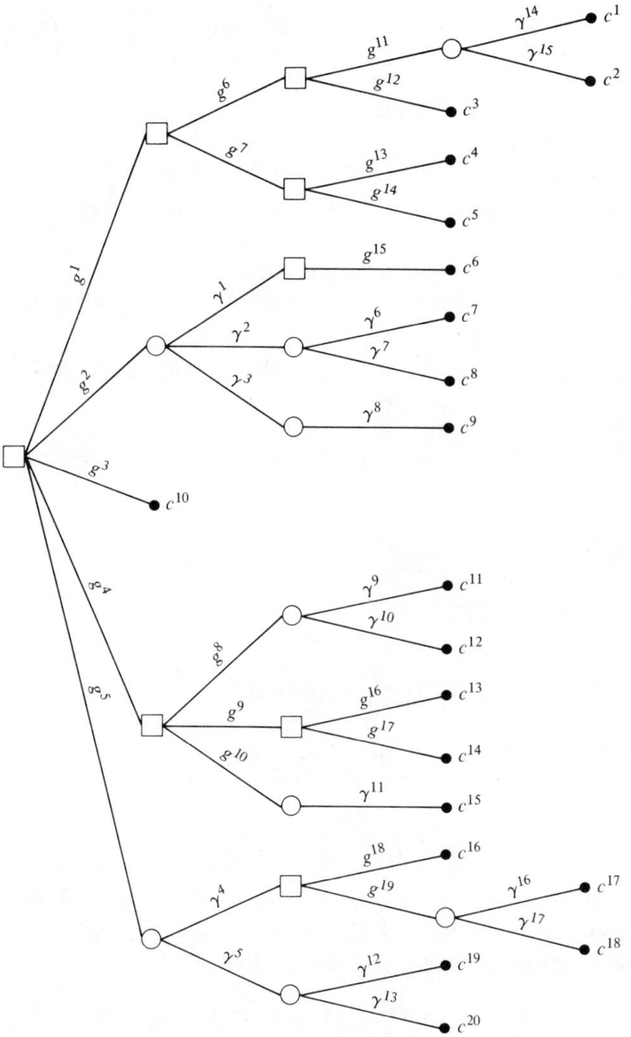

Figure 2.15 Decision tree for Exercise 2.11.2.

Exercise 2.11.3

Redo Exercise 2.11.2, this time *first* deleting all dummy moves and *then* combining all successive choices by the same player.

Exercise 2.11.4

Combine successive choices by the same player wherever possible in the subsystem decision tree of Figure 2.6. Compare the result with Figure 2.12.

Exercise 2.11.5

Let $G = \{g^1, g^2, g^3\}$. Show that G can be "decombined" in the form of $G_1 = \{g^1, \text{not } g^1\}$, with "not g^1" followed by $G_2 = \{g^2, g^3\}$.

Exercise 2.11.6

Let $G = \{(g_1{}^i, g_2{}^j)\colon i = 1, \ldots, m,\ j = 1, \ldots, n\}$; that is, G is the Cartesian product $G_1 \times G_2$ of $G_1 = \{g_1{}^i\colon i = 1, \ldots, m\}$ and $G_2 = \{g_2{}^j\colon j = 1, \ldots, n\}$.

(A) "Decombine" G by attaching one copy of G_2 to each branch $g_1{}^i$ of G_1.
(B) "Decombine" G by attaching one copy of G_1 to each branch $g_2{}^j$ of G_2.
(C) Verify that the successive choices you obtained in (A) can be combined to yield the original G; similarly for the successive choices you obtained in (B).

[*Note*: The technique introduced in this exercise is used frequently in Chapter 3.]

2.12 DECISIONS IN STANDARD EXTENSIVE FORM

The two preceding sections have furnished ample justification for the following dogmatic statement.

Theorem 2.12.1. *Any* dpuu can be represented in all its salient features by a decision tree in which each full history consists of exactly $2N$ choices for N some positive integer or $+\infty$ (depending upon the dpuu), and in which each full history is of the alternating form (act, ee, act, ee, . . . , act, ee), beginning with an act and ending with an ee.

Definition 2.12.1. A dpuu represented by a decision tree with the attributes specified in Theorem 2.12.1 is said to be in *standard extensive form*, with N *stages*.

Example 2.12.1

Figure 2.12 represents Example 1.5.1b in standard extensive form with two stages. Figure 2.6 is *not* a standard-extensive-form representation of Example 1.5.1b. Why? Figure 2.4 is a one-stage-standard-extensive-form representation of Example 1.5.7, but Figure 2.3 is not. Why? Insert a dummy $\mathscr{E}$-move in Figure 2.3 to obtain a one-stage-standard-extensive-form representation of this dpuu which is not identical with Figure 2.4.

As indicated in Example 2.12.1, there may be many, equally valid, standard-extensive-form representations of the same dpuu.

What do we gain by going through all this work, totaling over two sections, in order to reformulate dpuu's in standard extensive form? After all, the standard extensive form is equivalent in all salient, strategic features to your original decision tree, which represents the dpuu in (possibly nonstandard) extensive form.

There is one significant advantage with the standard extensive form. *It permits a standardized notation which can be applied to all dpuu's.* This is a very distinct advantage indeed, for the babel of notation in decision theory is probably the greatest stumbling block to neophytes.

The remainder of this section is devoted to developing ideas and notation for a general dpuu in standard extensive form.

The first idea is *stage.* By definition, a dpuu in standard extensive form has exactly N stages. In each stage i (for $i = 1, \dots. N$), $\mathscr{D}$ chooses first, and then $\mathscr{E}$ chooses. $\mathscr{D}$'s move in stage i depends upon the sequence of choices by the players in stages 1 through $i-1$.

Definition 2.12.2. Notation: For $i = 1, \dots, N$,

- Γ_i denotes the set of all conceivable choices by $\mathscr{E}$ in stage i.
- γ_i denotes a *typical,* or *generic,* element of Γ_i, and hence of some ith stage $\mathscr{E}$-move.
- G_i denotes the set of all conceivable choices of act by $\mathscr{D}$ in stage i.
- g_i denotes a *typical,* or *generic,* element of G_i, and hence of some ith stage $\mathscr{D}$-move.

Now you see why we use a superscript rather than a subscript to index acts and ee's; subscripts are reserved for denoting stages. More useful notations:

Definition 2.12.3. A *partial history* h_{i-1} (of the dpuu in standard extensive form) *through stage* $i-1$ is a connected path of the form $(g_1, \gamma_1, g_2, \gamma_2, \dots, g_{i-1}, \gamma_{i-1})$, in which γ_j belongs to the successor $\mathscr{E}$-move of g_j (for $j = 1, \dots, i-1$) and g_{j+1} belongs to the successor $\mathscr{D}$-move of γ_j (for $j = 1, \dots, i-2$). H_{i-1} denotes the set of *all* partial histories h_{i-1}.

For purely formal purposes, it is desirable to denote the *status quo ante* of the dpuu by h_0 and to define H_0 as the set $\{h_0\}$, consisting solely of h_0.

Then each of the stages $1, 2, \ldots, N$ commences from a partial history $h_0, h_1, \ldots, h_{N-1}$ of its predecessor stages.

Exercise 2.12.1

Why should h_N denote a full history of the dpuu, and H_N denote the set of *all* full histories of the dpuu?

Exercise 2.12.2

Prove that the following statement is true: For $i = 0, 1, \ldots, N-1$, every $h_i \in H_i$ leads to a unique $\mathscr{D}$-move, and every $\mathscr{D}$-move at the $(i+1)$st stage is preceded by exactly one $h_i \in H_i$.

Exercise 2.12.3

Show that any h_i can be written in the form

$$h_i = (h_{i-1}, g_i, \gamma_i),$$

where g_i belongs to the $\mathscr{D}$-move succeeding h_{i-1} and γ_i belongs to the $\mathscr{E}$-move succeeding h_{i-1} and g_i, for $i = 1, 2, \ldots, N$.

Definition 2.12.4. More notation: For $i = 1, 2, \ldots, N$, $G_i(h_{i-1})$ denotes the (unique!) $\mathscr{D}$-move preceded by h_{i-1}, and $\Gamma_i(h_{i-1}, g_i)$ denotes the (unique!) $\mathscr{E}$-move preceded by h_{i-1} and g_i.

From Definitions 2.12.2, 2.12.3, and 2.12.4, it is clear that we have:

Theorem 2.12.2. For $i = 1, 2, \ldots, N$,

$$G_i = \cup \{G_i(h_{i-1}) \colon h_{i-1} \in H_{i-1}\},$$

and

$$\Gamma_i = \cup \{\Gamma_i(h_{i-1}, g_i) \colon g_i \in G_i(h_{i-1}), h_{i-1} \in H_{i-1}\}.$$

■ Recall that the sets $\{a, a, b, a, c, b, d\}$ and $\{a, b, c, d\}$ are identical; repetitions of elements in a set do not change that set. ■

Theorem 2.12.2 simply states formally the obvious facts (1) that the set G_i of all conceivable choices by $\mathscr{D}$ at stage i can be gotten by lumping together all choices available, given the particular partial histories h_{i-1} of play through stage $i-1$, and (2) that the set Γ_i of all conceivable choices of ee by $\mathscr{E}$ at stage i can be gotten by lumping together all choices of ee available, given the particular partial histories h_{i-1} of play through stage $i-1$ and the particular choices g_i by $\mathscr{D}$ available, given h_{i-1}.

Exercise 2.12.4

Prove that the following statements are true:

(A) G_i is a $\mathcal{D}$-move if and only if there is some h_{i-1}, say h_{i-1}^*, such that

$$G_i(h_{i-1}) \subset G_i(h_{i-1}^*)$$

for every $h_{i-1} \in H_{i-1}$.

(B) Γ_i is an $\mathcal{E}$-move if and only if there is some h_{i-1}, say h_{i-1}^*, and some g_i, say g_i^*, such that

$$\Gamma_i(h_{i-1}, g_i) \subset \Gamma_i(h_{i-1}^*, g_i^*)$$

for every $g_i \in G_i(h_{i-1})$ and every $h_{i-1} \in H_{i-1}$.

Example 2.12.2

To illustrate the preceding great swatch of notation, we shall apply all of it to Figure 2.12. $H_0 = \{h_0\}$ always. Let g_1^1 = "do not test" and g_1^2 = "apply pretest." Then

$$G_1 = G_1(h_0) = \{g_1^1, g_1^2\}.$$

Let γ_1^1 = "null outcome," γ_1^2 = "pretest indicates defective," and γ_1^3 = "pretest indicates not defective." Then

$$\Gamma_1(h_0, g_1^1) = \{\gamma_1^1\},$$
$$\Gamma_1(h_0, g_1^2) = \{\gamma_1^2, \gamma_1^3\},$$

and

$$\Gamma_1 = \{\gamma_1^1, \gamma_1^2, \gamma_1^3\}.$$

Let $h_1^1 = (g_1^1, \gamma_1^1)$, $h_1^2 = (g_1^2, \gamma_1^2)$, and $h_1^3 = (g_1^2, \gamma_1^3)$. Then

$$H_1 = \{h_1^1, h_1^2, h_1^3\}.$$

Let g_2^1 = "install as is" and g_2^2 = "rebuild and then install." Then

$$G_2 = G_2(h_1^1) = G_2(h_1^2) = G_2(h_1^3) = \{g_2^1, g_2^2\}.$$

Let γ_2^1 = "defective (in operation)" and γ_2^2 = "not defective (in operation)." Then

$$\Gamma_2(h_1^j, g_2^1) = \{\gamma_2^1, \gamma_2^2\} = \Gamma_2,$$

while

$$\Gamma_2(h_1^j, g_2^2) = \{\gamma_2^2\} \qquad \text{for } j = 1, 2, 3.$$

There are nine full histories of the form h_2^j, because each full history corresponds to an endpoint of the tree and conversely.

Example 2.12.3

At last we should feel confident enough to tackle Example 1.5.1c. I shall define notation and exhibit a three-stage decision tree, Figure 2.16, and as an exercise you may verify that these abstractions faithfully

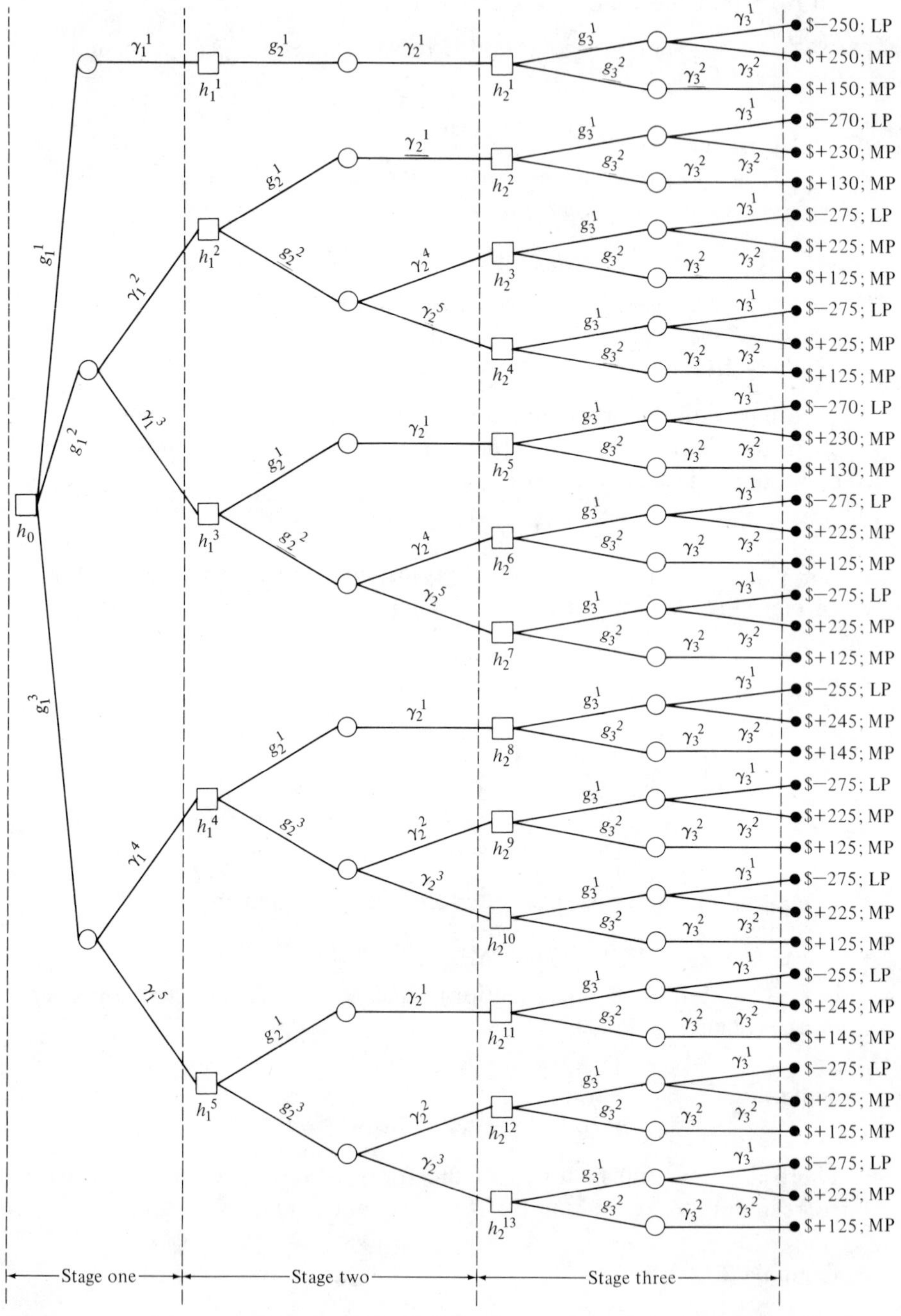

Figure 2.16 Standard extensive form of Example 1.5.1c.

capture all the salient features of the underlying dpuu. Let g_1^1 = "do not test," g_1^2 = "apply pretest," and g_1^3 = "apply less reliable test"; and let γ_1^1 = "null outcome," γ_1^2 = "pretest indicates defective," γ_1^3 = "pretest indicates not defective," γ_1^4 = "less reliable test indicates defective," and γ_1^5 = "less reliable test indicates not defective." Let

$$\begin{aligned} G_1 = G_1(h_0) &= \{g_1^1, g_1^2, g_1^3\}, \\ \Gamma_1(h_0, g_1^1) &= \{\gamma_1^1\}, \\ \Gamma_1(h_0, g_1^2) &= \{\gamma_1^2, \gamma_1^3\}, \\ \Gamma_1(h_0, g_1^3) &= \{\gamma_1^4, \gamma_1^5\}, \end{aligned}$$

and

$$\Gamma_1 = \{\gamma_1^1, \gamma_1^2, \gamma_1^3, \gamma_1^4, \gamma_1^5\}.$$

Let $h_1^1 = (g_1^1, \gamma_1^1)$, $h_1^2 = (g_1^2, \gamma_1^2)$, $h_1^3 = (g_1^2, \gamma_1^3)$, $h_1^4 = (g_1^3, \gamma_1^4)$, $h_1^5 = (g_1^3, \gamma_1^5)$, and

$$H_1 = \{h_1^1, h_1^2, h_1^3, h_1^4, h_1^5\}.$$

Let g_2^1 = "do not test (further)," g_2^2 = "apply less reliable test," and g_2^3 = "apply pretest"; and let $\gamma_2^j = \gamma_1^j$ for $j = 1, 2, 3, 4, 5$. Then

$$\begin{aligned} G_2(h_1^1) &= \{g_2^1\}, \\ G_2(h_1^2) = G_2(h_1^3) &= \{g_2^1, g_2^2\}, \end{aligned}$$

and

$$G_2(h_1^4) = G_2(h_1^5) = \{g_2^1, g_2^3\},$$

because: (1) no advantage can possibly be gained by postponing a test for one stage, and hence choice of g_1^1 should necessitate choice of g_2^1; and (2) you will recall that each of the tests can be applied only once. Also,

$$\begin{aligned} \Gamma_2(h_1^j, g_2^1) &= \{\gamma_2^1\} \qquad \text{for } j = 1, 2, 3, 4, 5, \\ \Gamma_2(h_1^j, g_2^2) &= \{\gamma_2^4, \gamma_2^5\} \qquad \text{for } j = 2, 3, \end{aligned}$$

and

$$\Gamma_2(h_1^j, g_2^3) = \{\gamma_2^2, \gamma_2^3\} \qquad \text{for } j = 4, 5.$$

Now, for any one of the thirteen (count them) partial histories $h_2 \in H_2$, the third stage is the same as that for any other partial history. Hence we can define g_3^1 = "install as is," g_3^2 = "rebuild and then install," γ_3^1 = "defective (in operation)," γ_3^2 = "not defective (in operation),"

$$\begin{aligned} G_3 = G_3(h_2) &= \{g_3^1, g_3^2\} \qquad \text{for all } h_2 \in H_2, \\ \Gamma_3(h_2, g_3^1) &= \{\gamma_3^1, \gamma_3^2\} \qquad \text{for all } h_2 \in H_2, \end{aligned}$$

and

$$\Gamma_3(h_2, g_3^2) = \{\gamma_3^2\} \qquad \text{for all } h_2 \in H_2.$$

Exercise 2.12.5

Verify the appropriateness of all constituents of Figure 2.16, including the consequences.

Exercise 2.12.6

Reformulate (as necessary) in standard extensive form, applying all notation defined in this section.

(A)	Example 1.5.11a	(G)	Example 1.5.5
(B)	Example 1.5.11b	*(H)	Example 1.5.2c
(C)	Example 1.5.8	*(I)	Example 1.5.4
(D)	Example 1.5.6	*(J)	Example 1.5.3
(E)	Example 1.5.9	*(K)	Example 1.5.12a
(F)	Example 1.5.10	*(L)	Example 1.5.12b

[*Hint*: (K) and (L) have a very special structure. Ask yourself what is the *least* you need to know about h_{i-1} in order to make an intelligent choice from your ith stage $\mathscr{D}$-move $G_i(h_{i-1})$. Assume that demands in different periods are entirely unrelated.]

D: REALISM AND INFINITUDE

2.13 INFINITE DECISION TREES

Definition 2.13.1. A dpuu in standard extensive form, or its associated decision tree, is said to be *finite* provided: (1) every move contains "only" a finite number of elements, *and* (2) the tree contains "only" a finite number of moves. If a dpuu (or tree) is not finite it is said to be *infinite.*

Most of our examples, and indeed our theoretical development, will be oriented primarily toward *finite* dpuu's, which as an immediate implication of Definition 2.13.1 will have "only" a finite number of potential consequences c in C.

The quotes around "only" are intended to emphasize the humor attending its use in a dpuu with $10^{100,000}$ moves, each containing $10^{100,000}$ elements. We shall omit the quotes in the future.

But implicit in the preceding paragraph is the point that, *practically speaking, every dpuu is finite.* Mechanical and human perceptions are necessarily both *discrete* and *bounded.* By discreteness we mean that *every* phenomenon is measurable only to a fixed number of decimal places (given extant technology), and by boundedness we mean that *every* phenomenon can yield a measurement only within a numerical interval of finite width (again given extant technology); measuring instruments, including ourselves, have saturation levels!

If realism dictates that all dpuu's are finite, why be concerned with infinite dpuu's (having trees with infinitely many stages and/or with moves containing infinitely many elements)? The answer is: for precisely the same reason that motivates engineers and physicists to study the mathematics of continuous variables; e.g., differential equations, the calculus of variations, etc. Remember, engineers and physicists are subject to the same discreteness and boundedness in their work as is any decision maker.

The reason is simply that by passing to the infinite case one can often drastically simplify the analysis of a dpuu (or of a physical phenomenon), thereby more than making up for the cost of slight unrealism by a drastic reduction in analytical effort.

In Examples 1.5.12a and 1.5.12b, it often pays to regard an inventory system with $N = 1{,}000$ stages (periods) to go *as if* $N = +\infty$, for when $N = +\infty$, powerful techniques can be brought to bear which are not applicable when $N < +\infty$.

Again, in Example 1.5.3, it may be much easier to suppose that α and β can vary continuously over some intervals of real numbers than to pick, say, 1,000 possible values of each and end up with *one million* possible pairs of values. Similarly, in Example 1.5.4, only twenty possible values of each θ_i produces 20^n possible n-tuples of values. Suppose $n = 100$; then 20^{100} is approximately 10^{130}. The magnitude of this number can be appreciated by noting that the age of the earth is said to be less than 10^{21} seconds. Enough said.

In future chapters we shall use notation that is sufficiently flexible to embrace infinite (including continuous) moves, but substantive consideration of the nuances of infinite cases will be largely relegated to starred comments, sections, and parts which the reader may omit without missing any ideas essential to a complete understanding of the principles of decision analysis.

3

FOUNDATIONS OF DECISION ANALYSIS

Good practice necessarily follows from good theory.
Professor Lawrence W. Towle, Lecture, 1960

A: PRELIMINARIES AND PREVIEW

3.1 PRELIMINARIES

In this section we shall establish mild generalizations of the concepts of "elementary event" and "act in a terminal $\mathscr{D}$-move." These generalizations provide the vocabulary necessary for the main task of this chapter, which is to establish the foundations for the systematic analysis of dpuu's which we promised in Section 2.9.

Since we included the adjective in defining "elementary event," it was only to be expected that a more general concept was forthcoming. As we noted in Section 2.5, each "elementary event" in a *coarse* description of an $\mathscr{E}$-move is a *set* of "elementary events" in a *finer* description of that $\mathscr{E}$-move. This motivates our defining an *event* to be a set of elementary events.

For our present purposes, we shall fix a fine level of description of each $\mathscr{E}$-move and occasionally consider coarser levels of description. The elementary events in the fixed, fine level will be denoted by a small Greek letter, such as ω; the $\mathscr{E}$-move itself will be denoted by the corresponding Greek capital, such as Ω; and events in (= subsets of) the finely described $\mathscr{E}$-move will be denoted by the corresponding Greek capital with superscripts, such as Ω'.

Moreover, it is convenient to have a notation for the expression "the as-yet unrevealed ee ω." We shall use $\tilde{\omega}$ as shorthand for this expression; it is read "omega tilde" and is also referred to (for primarily historical reasons) as "the *random variable* omega."

We shall occasionally use the expression "$\tilde{\omega} \in \Omega'$" instead of the less graphic "Ω'" to stress that we mean "the event that omega tilde will be found in Ω'," or "the event Ω'."

Example 3.1.1

Suppose that you are engaged in a wager in which you win \$10 if the outcome ω of the cast of an ordinary die[a] is 2, 4, or 6, and you lose \$10 if ω is 1, 3, or 5. Then the natural (but "fine") definition of the relevant $\mathscr{E}$-move is $\Omega = \{1, 2, 3, 4, 5, 6\}$. The "event that you win the wager" is $\Omega' = \{2, 4, 6\}$, or that $\tilde{\omega} \in \{2, 4, 6\}$; while the "event that you lose the wager" is $\Omega'' = \{1, 3, 5\}$, or that $\tilde{\omega} \in \{1, 3, 5\}$. Note also that Ω' corresponds to δ' = "win" and that Ω'' corresponds to δ'' = "lose," in the coarser description $\Delta = \{\delta', \delta''\}$ of the relevant $\mathscr{E}$-move.

In Section 2.2 we defined an $\mathscr{E}$-move to be the set of *all, mutually exclusive* choices available to $\mathscr{E}$ at a given point in a dpuu. That $\mathscr{E}$'s choices at a given point should be mutually exclusive, and that your descriptions of them should be collectively exhaustive, is independent of the fineness with which you have defined them. Hence in our new event terminology, an $\mathscr{E}$-move Ω in a decision tree should consist of mutually exclusive and collectively exhaustive events $\tilde{\omega} \in \Omega^i$, where "mutual exclusivity" of $\Omega^1, \ldots, \Omega^n$ means that *only* one of the Ω^i's can occur and "collective exhaustiveness" means that one of the Ω^i's *must* occur.

Example 3.1.2

Let Ω' and Ω'' be as defined in Example 3.1.1 for your wager. Clearly, Ω' and Ω'' are mutually exclusive, since the die cannot come up both odd and even; and Ω' and Ω'' are collectively exhaustive, since (if we exclude cavils about its coming up on edge or going into orbit) it *must* come up either odd (Ω'') or even (Ω'). Indeed, that Ω' and Ω'' are mutually exclusive and collectively exhaustive is clear from the fact that δ' and δ'' are mutually exclusive and collectively exhaustive in the coarser description $\Delta = \{\delta', \delta''\}$ of this $\mathscr{E}$-move, where $\delta' = \Omega'$ and $\delta'' = \Omega''$ as in Example 3.1.1.

If $\Omega^1, \ldots, \Omega^n$ are mutually exclusive and collectively exhaustive events in (or subsets of) an $\mathscr{E}$-move Ω, then we say that $\Omega^1, \ldots, \Omega^n$ constitute a *partition* of Ω. Other convenient definitions: (1) The empty subset $\emptyset$ of an $\mathscr{E}$-move Ω is called an *impossible event.* (2) The improper subset Ω of an $\mathscr{E}$-move Ω is called a *sure event.*

Example 3.1.3

In your die wager, Ω' and Ω'' constitute a partition of the sure event $\Omega = \{1, 2, 3, 4, 5, 6\}$. Because we have ruled out the die's standing on edge or perpetrating some such crime against reasonableness, the event that

---

[a] A *die* is one-half of a pair of dice. An *ordinary* die is a cube whose six faces are numbered 1, 2, 3, 4, 5, and 6.

$\tilde{\omega} \in \Omega$ is sure, or certain, and therefore the event that $\tilde{\omega} \in \emptyset$, or $\tilde{\omega} \notin \Omega$, is impossible.

We collect all this terminology in a formal definition, to facilitate review and subsequent reference.

Definition 3.1.1: Event Terminology

(A) An *event* is a subset Ω' of an $\mathscr{E}$-move Ω.
(B) A *sure event* is the improper subset Ω of an $\mathscr{E}$-move Ω.
(C) An *impossible event* is the empty subset $\emptyset$ of an $\mathscr{E}$-move Ω.
(D) Events $\Omega^1, \ldots, \Omega^n$ in Ω are *mutually exclusive* if $\Omega^i \cap \Omega^j = \emptyset$ whenever $i \neq j$.
(E) Events $\Omega^1, \ldots, \Omega^n$ in Ω are *collectively exhaustive* if $\cup_{i=1}^n \Omega^i = \Omega$.
(F) Events $\Omega^1, \ldots, \Omega^n$ in Ω constitute a *partition* of Ω if they are mutually exclusive and collectively exhaustive.
(G) $\tilde{\omega}$ denotes the as-yet unrevealed elementary event ω, or the *random variable* ω.

So much for concepts associated with generalizing the notion of elementary events. We now generalize the concept of "act in a terminal $\mathscr{D}$-move."

To begin with, we note that the fundamental *strategic* essence of an act in a terminal $\mathscr{D}$-move is that $\mathscr{D}$ cannot make any subsequent choices of act, because acts in terminal $\mathscr{D}$-moves lead directly to endpoints of the decision tree, with consequences attached.

Acts *other* than those in a terminal $\mathscr{D}$-move also possess this strategic essence. If an act leads to an $\mathscr{E}$-move and if all events in that $\mathscr{E}$-move lead to endpoints, then again $\mathscr{D}$'s choice of that act is a "final" step in exerting his influence over the consequence he will experience. Similarly, if an act leads to $\mathscr{E}$-moves and possibly also to *dummy* $\mathscr{D}$-moves (in which his act choices are trivial), then yet again is his choice of that act a final step in controlling his consequence.

The acts that are "ultimate," in that the "subtree" emanating from one of them contains no nondummy $\mathscr{D}$-move, are precisely those with which we shall be exclusively concerned in this chapter. They are called lotteries.

Definition 3.1.2. A *lottery* is an act which cannot lead to a subsequent nondummy $\mathscr{D}$-move.

Example 3.1.4

In Figure 2.15, the following acts are lotteries: g^{11}, g^{12}, g^{13}, g^{14}, g^3, g^{15}, g^8, g^{16}, g^{17}, g^{10}, g^{18}, and g^{19}. Moreover, g^2 is a lottery because it can lead only to $\mathscr{E}$-moves and, via γ^1, to the dummy $\mathscr{D}$-move $\{g^{15}\}$. On the other hand, g^5 is not a lottery; it can lead to the nondummy $\mathscr{D}$-move $\{g^{18}, g^{19}\}$. Note that the lotteries g^{11}, g^2, g^8, g^{10}, and g^{19} are *not* terminal $\mathscr{D}$-moves, but they are "last chances" for $\mathscr{D}$ to exert any control over his consequence.

Because of our complete flexibility regarding insertion and/or deletion of dummy moves, the existence of dummy $\mathscr{D}$-moves which can follow a lottery g does not alter the strategic essence of g as an act that, once chosen, leaves control over the consequence solely up to $\mathscr{E}$. For subsequent purposes, it will sometimes be convenient to insert and/or delete subsequent dummy $\mathscr{D}$-moves in drawing those subtrees which commence with a lottery. Such a subtree, commencing with an act branch (not necessarily a dummy $\mathscr{D}$-move), is called a *lottery tree.*

Four strategically equivalent lottery trees for g^2 in Figure 2.15 are depicted in Figure 3.1. All four of these lottery trees represent the same opportunity to $\mathscr{D}$, who should therefore not regard the desirability of g^2 vis-à-vis its alternatives g^1, g^3, g^4, and g^5 as dependent upon which tree in Figure 3.1 is used to represent g^2.

■ Note that (c) and (d) in the figure represent g^2 as a "simple," or one-stage, lottery with only one subsequent $\mathscr{E}$-move, whereas (a) and (b) represent g^2 as a "compound" lottery with two successive $\mathscr{E}$-moves. Moreover, (d) is obtained from (c), and (b) from (a), by insertion of dummy $\mathscr{D}$-moves. Each of d^5 through d^{12} is a lottery; and hence we see that *any consequence may be regarded as a lottery*, albeit of a degenerate sort, in which $\mathscr{D}$ will surely experience that consequence. ■

Now, a cursory examination of the representations of g^2 in Figure 3.1 suffices to establish that, regardless of the lottery representation, *each potential consequence of a lottery can be expressed uniquely as a function of the sequence of ee's resulting in that consequence.*

This observation leads to an alternative, somewhat more formal definition of a lottery than that in Definition 3.1.2.

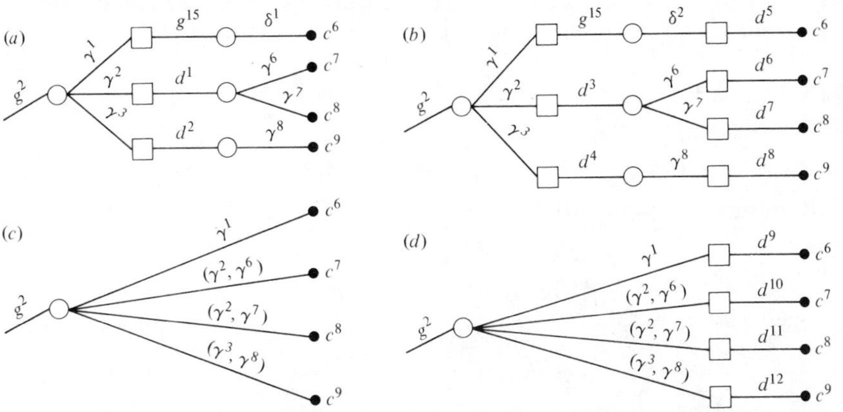

Figure 3.1 Lottery trees for g^2 in Figure 2.15. (*Key*: d denotes the act in a dummy $\mathscr{D}$-move not appearing in Figure 2.15; and δ denotes the ee in a similar dummy $\mathscr{E}$-move.)

Definition 3.1.3. Let Ω be an $\mathscr{E}$-move consisting of mutually exclusive and collectively exhaustive ee's ω, and let X be a nonempty set of elements x. A *lottery* ℓ *on* Ω *with outcomes in* X is a function $\ell: \Omega \to X$. The outcome x, given that $\tilde{\omega} = \omega$, is denoted by $\ell(\omega)$.

Example 3.1.5

Figure 3.1 shows that the lottery $\ell = g^2$ may be regarded as a function $\ell: \Omega \to X$ in four different ways. In representation (a), ℓ may be regarded as a lottery on $\Gamma = \{\gamma^1, \gamma^2, \gamma^3\}$ with outcomes in $X = \{g^{15}, d^1, d^2\}$, with $\ell(\gamma^1) = g^{15}$, $\ell(\gamma^2) = d^1$, and $\ell(\gamma^3) = d^2$. In representation (b), ℓ may be regarded as a lottery on the same $\mathscr{E}$-move Γ with outcomes in $X' = \{g^{15}, d^3, d^4\}$, with $\ell(\gamma^1) = g^{15}$, $\ell(\gamma^2) = d^3$, and $\ell(\gamma^3) = d^4$. In representation (c), ℓ may be regarded as a lottery on $\Omega = \{\gamma^1, (\gamma^2, \gamma^6), (\gamma^2, \gamma^7), (\gamma^3, \gamma^8)\}$ with outcomes in $X'' = \{c^6, c^7, c^8, c^9\}$, with $\ell(\gamma^1) = c^6$, $\ell(\gamma^2, \gamma^6) = c^7$, $\ell(\gamma^2, \gamma^7) = c^8$, and $\ell(\gamma^3, \gamma^8) = c^9$. And in representation (d), ℓ may be regarded as a lottery on the same $\mathscr{E}$-move Ω with outcomes in $X''' = \{d^9, d^{10}, d^{11}, d^{12}\}$. Note that d^9, d^{10}, d^{11}, and d^{12} differ only in a purely formal sense from c^6, c^7, c^8, and c^9 respectively.

If Example 3.1.5 leaves you with the impression that we have great flexibility in representing a lottery, it has served its purpose admirably. This flexibility comes in very handy in the work of this chapter. That is why we have taken the space to introduce the many faces of a lottery.

Exercise 3.1.1

Let $\Omega = [0, 1] = \{\omega: 0 \leq \omega \leq 1\}$, the "closed unit interval." Let $\Omega^1 = (.4, .7) = \{\omega: .4 < \omega < .7\}$; $\Omega^2 = [0, .3) \cup [.7, .85)$; and $\Omega^3 = [.3, .4] \cup [.85, 1]$. Are Ω^1, Ω^2, and Ω^3 mutually exclusive? Are they collectively exhaustive? If not, define an event Ω^4 which, together with Ω^1, Ω^2, and Ω^3, will produce collective exhaustiveness.

Exercise 3.1.2

Let $\Omega = (-\infty, +\infty)$, $\Omega^1 = (-100, 0)$, $\Omega^2 = [0, 25) \cup \{70\} \cup [100, +\infty)$, and $\Omega^3 = (-\infty, -100] \cup [25, 70) \cup (70, 100)$. Show that Ω^1, Ω^2, and Ω^3 constitute a partition of Ω.

Exercise 3.1.3

Show that in a dpuu in standard extensive form (Section 2.12), every act g_N in every Nth stage $\mathscr{D}$-move $G_N(h_{N-1})$ is a lottery.

Exercise 3.1.4 (Continuation of Exercise 3.1.3)

Show that an $(N-1)$th stage act, g^*_{N-1}, in $G_{N-1}(h^*_{N-2})$ is a lottery if and only if $G_N(h^*_{N-2}, g^*_{N-1}, \gamma_{N-1})$ is a dummy $\mathscr{D}$-move for every γ_{N-1} in $\Gamma_{N-1}(h^*_{N-2}, g^*_{N-1})$.

Exercise 3.1.5

State which acts are lotteries in your decision tree for the following.

(A)	Example 1.5.11a	(G)	Example 1.5.5
(B)	Example 1.5.11b	*(H)	Example 1.5.2c
(C)	Example 1.5.8	*(I)	Example 1.5.4
(D)	Example 1.5.6	*(J)	Example 1.5.3
(E)	Example 1.5.9	*(K)	Example 1.5.12a
(F)	Example 1.5.10	*(L)	Example 1.5.12b

Exercise 3.1.6

Suppose that Ω consists of precisely n elements, where $n \geq 1$. Show that there are precisely 2^n subsets of Ω, including $\emptyset$ and Ω itself. [*Hint*: Forming a subset Ω' of Ω is equivalent to deciding whether to include or exclude each element of Ω. Since there are two choices for ω^1 and two choices for ω^2, it follows that there are $2^2 = 4$ choices concerning both ω^1 and ω^2. For each of these, there are two choices for ω^3.]

Exercise 3.1.7 (*Generalization of Hint for Exercise 3.1.6*)

Suppose for $i = 1, \ldots, n$ that the set B_i has $\#(B_i)$ elements. Show that the Cartesian product $B_1 \times \cdots \times B_n = \{(b_1, \ldots, b_n): b_i \in B_i \text{ for every } i\}$ of $B_1, \ldots, B_n$ has a total of

$$\#(B_1 \times \cdots \times B_n) = \#(B_1) \cdot \cdots \cdot \#(B_n) = \prod_{i=1}^{n} \#(B_i)$$

elements.

3.2 PREVIEW OF THE BASIC RESULTS IN THIS CHAPTER

You were asked in Exercise 3.1.3 to verify that every Nth stage act g_N in any dpuu in standard extensive form is a lottery. A typical such lottery ℓ^t is depicted in Figure 3.2, where we utilize the event terminology of Definition 3.1.1 in assuming that $\ell^t(\omega_t) = c_t^i$ for every ω_t in Ω_t^i, with $\Omega_t^1, \Omega_t^2, \ldots, \Omega_t^{n_t}$ constituting a partition of the $\mathscr{E}$-move Ω_t on which the outcome of ℓ^t depends.

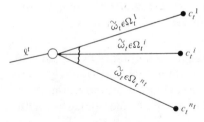

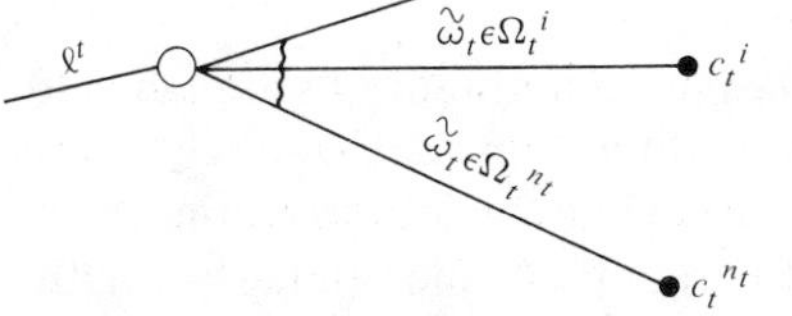

Figure 3.2 A typical lottery.

*■ In infinite idealizations of reality, it is assumed that ℓ^t may have infinitely many possible outcomes rather than the finite number n_t suggested in Figure 3.2. In such cases we shall drop the partition notation and think of ℓ^t as simply a lottery on an arbitrary $\mathscr{E}$-move Ω_t resulting in the typical consequence $c = \ell^t(\omega_t)$. ■

Now, this chapter shows *how $\mathscr{D}$ can make systematic choices among lotteries* of the form ℓ^t as depicted in Figure 3.2. Chapter 4 then continues by synthesizing these choices into a solution to the entire dpuu in question.

But this synthesis is preceded by analysis, as noted in Chapter 2. Our philosophy is to ask that $\mathscr{D}$ make *many small and clear-cut decisions* of such a nature that they can be processed to yield $\mathscr{D}$'s logically consistent choice among much less simple lotteries of the form ℓ^t. All these simple decisions are of the "oculist variety": would you prefer this—or that? No oculist presents the patient with all his test lenses and asks him to try one after the other in various positions until he finds the one which effects the greatest improvement in his vision. On the contrary, he has the patient compare lenses and positions two at a time, thus working toward the best by process of elimination. We shall ask $\mathscr{D}$ to behave likewise in comparing very simple lotteries.

These small and clear-cut decisions enable $\mathscr{D}$ to *quantify his relative preferences* for the consequences of the dpuu and to *quantify his judgments* about the events Ω' in each $\mathscr{E}$-move Ω in the dpuu. It should be clear that $\mathscr{D}$'s tastes and judgments are both germane to his preferences among lotteries with consequences as outcomes. But what is not yet clear is that the quantifications of $\mathscr{D}$'s preferences and judgments are all the results of his making *decisions* which reveal his preferences between ancillary, hypothetical lotteries of special, simple types.

Now, decisions involving preferences are so central to decision analysis that we use curvaceous versions of numerical inequalities to save much space in denoting desirability inequalities.

Definition 3.2.1. Let x and y be consequences or lotteries. Then:

(A) $x \succsim y$ means "$\mathscr{D}$ regards x as *at least as* desirable as y."
(B) $x \succ y$ means "$\mathscr{D}$ regards x as *more* desirable *than* y."
(C) $x \sim y$ means "$\mathscr{D}$ regards x and y as *equally* desirable."

Similarly, $x \prec y$ means that $\mathscr{D}$ regards x as less desirable than y; and $x \precsim y$ means that $\mathscr{D}$ regards x as no more desirable than y.

■ This notation for preferences or relative desirabilities is *intended* to suggest the corresponding numerical inequalities! In fact, you will do no violence to fundamental ideas by thinking of "$x \succsim y$" as meaning "x is greater than or equal to y in relative desirability," and similarly for the other parts of Definition 3.2.1. ■

■ By analogy with numerical inequalities or directly from Definition 3.2.1, it should be clear that $x \sim y$ if and only if both $x \succsim y$ and $y \succsim x$; that is, x and y are equally desirable if each is at least as desirable as the other. Also, $x \succ y$ if and only if $x \succsim y$ but not $y \succsim x$ (or not $x \sim y$); that is, x is more desirable than y if it is at least as desirable as y but not equally as desirable as y. ■

We now proceed to preview the primary results in this chapter. In Section 3.5, we show how $\mathscr{D}$ can make "oculist decisions" which result in his *quantifying his preferences* for all the potential consequences c of his dpuu, in terms of a numerical *utility function* $u: C \to (-\infty, +\infty)$. The utility numbers $u(c)$ play an important role in the ensuing analysis, as we shall soon see.

In Section 3.6, we show how $\mathscr{D}$ can make other "oculist decisions" which result in his *quantifying his judgments* about all events Ω_t^i—or $\tilde{\omega}_t \in \Omega_t^i$—in each $\mathscr{E}$-move Ω_t in his dpuu, in terms of numerical *probabilities* $P(\Omega_t^i)$—or $P(\tilde{\omega}_t \in \Omega_t^i)$. The result of the judgment quantifications is that each branch Ω_t^i in each $\mathscr{E}$-move Ω_t is assigned a probability number $P(\Omega_t^i)$ which represents $\mathscr{D}$'s judgments about how likely $\mathscr{E}$ is to choose that branch in that $\mathscr{E}$-move. The more likely $\mathscr{D}$ regards Ω_t^t, the higher will be his probability $P(\Omega_t^i)$.

Now, the probabilities $P(\Omega_t^1), \ldots, P(\Omega_t^{n_t})$ that $\mathscr{D}$ assigns to the respective branches $\Omega_t^1, \ldots, \Omega_t^{n_t}$ of $\mathscr{E}$-move Ω_t are numerical *relative weights*, in the sense that they conform to two stipulations.

$$\textit{stipulation } 1: \quad 0 \le P(\Omega_t^i) \le 1 \qquad \text{for every } i \text{ and } t;$$

and

$$\textit{stipulation } 2: \quad \sum_{i=1}^{n_t} P(\Omega_t^i) = 1 \qquad \text{for every } t.$$

■ Recall the capital-sigma notation for summations: by definition, $\sum_{i=1}^{k} a_i = a_1 + a_2 + \cdots + a_k$. ■

In Section 3.7, we discuss further aspects of the consistent quantification of judgments and state probability relationships that are essential in evaluating "compound" lotteries such as (a) and (b) of Figure 3.1.

*Section 3.8 is devoted to careful proofs of the basic results in this chapter. These proofs are not essential to one's understanding of how to *apply* the basic results, and hence this section is optional. But you cannot become *convinced* of the ultimate truth of our results without perusing their derivations, and hence you are urged not to skip over this section. The only mathematical prerequisite for understanding everything therein is ability to follow a logical argument.

Now, the gist of Sections 3.5 and 3.6 is that $\mathscr{D}$ can assign utility numbers $u(c_t^i)$ to all consequences c_t^i and probability numbers $P(\Omega_t^i)$ to all event branches Ω_t^i in any lottery ℓ^t such as depicted in Figure 3.2. The roles of these numbers are crucial: we shall show that the probability-weighted average

$$U(\ell^t) = \sum_{i=1}^{n_t} u(c_t^i)P(\Omega_t^i)$$

of $\mathscr{D}$'s utilities is an *index of the relative desirability to* $\mathscr{D}$ of ℓ^t vis-à-vis other lotteries ℓ^s, in the sense that

$$\ell^t \gtrsim \ell^s \qquad \text{if and only if} \qquad U(\ell^t) \geq U(\ell^s).$$

In other words, $\mathscr{D}$'s *desirability* inequalities between lotteries are fully reflected by the *numerical* inequalities between their associated desirability indices U!

■ This result, which we state below as Theorem 3.2.1, conveys the spirit of decision analysis perfectly. Although it may be extremely difficult for $\mathscr{D}$ to say offhand whether or not $\ell^t \gtrsim \ell^s$, it is generally much easier for him to make all the oculist decisions that determine all the $P(\Omega_t^i)$'s, $P(\Omega_s^j)$'s, $u(c_t^i)$'s, and $u(c_s^j)$'s. From there on it is an exercise in simple arithmetic to calculate $U(\ell^t) = \Sigma_{i=1}^{n_t} u(c_t^i)P(\Omega_t^i)$ and $U(\ell^s) = \Sigma_{j=1}^{n_s} u(c_s^j)P(\Omega_s^j)$, and then to compare $U(\ell^t)$ with $U(\ell^s)$ to see which (if either) is larger. The basic result implies, of course, that $\ell^t > \ell^s$ if $U(\ell^t) > U(\ell^s)$, that $\ell^t \sim \ell^s$ if $U(\ell^t) = U(\ell^s)$, and that $\ell^t < \ell^s$ if $U(\ell^t) < U(\ell^s)$. ■

We now give an example which illustrates the "outputs" of $\mathscr{D}$'s preference and judgment quantifications according to Sections 3.5 and 3.6. It shows how these quantifications may be recorded directly on the lottery trees and how the calculations of the desirability indices may also be recorded on the trees.

Example 3.2.1

Suppose that ℓ^t and ℓ^s are as depicted in Figure 3.3, in which $\Omega_t^1, \ldots, \Omega_t^4$ constitutes a partition of Ω_t and $\Omega_s^1, \Omega_s^2, \Omega_s^3$ constitutes a partition of Ω_s. Suppose that $\mathscr{D}$'s utilities of the seven consequences appearing in Figure 3.3 are as they are recorded in parentheses next to each consequence, and that his probabilities of the event branches are as they appear in brackets on those branches. Thus $u(c_s^2) = -1.2$ and $P(\Omega_t^3) = .47$, for example. Note that the sum $.23 + .18 + .47 + .12$ of the branch probabilities $P(\Omega_t^i)$ of ℓ^t is one, as is the sum $.20 + .32 + .48$ of the branch probabilities $P(\Omega_s^j)$ of ℓ^s. Now, finding $U(\ell^t)$ and $U(\ell^s)$ is easy, once the utilities and probabilities have been determined:

$$U(\ell^t) = (8.2)[.23] + (4.6)[.18] + (-3.1)[.47] + (2.5)[.12] = 1.557$$

and

$$U(\ell^s) = (4.0)[.20] + (-1.2)[.32] + (1.2)[.48] = .992.$$

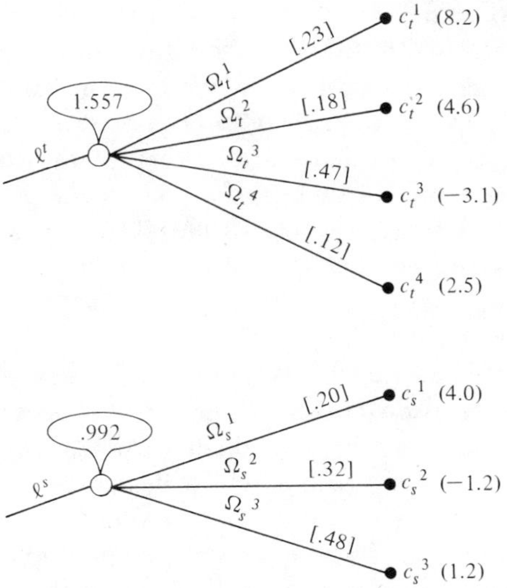

Figure 3.3 Two lotteries with quantifications made.

Since $U(\ell^t) = 1.557 > .992 = U(\ell^s)$, it follows that $\ell^t > \ell^s$ and hence that $\mathscr{D}$ should choose ℓ^t over ℓ^s if that opportunity should arise because ℓ^t and ℓ^s are in the same $\mathscr{D}$-move. Note how $U(\ell^t)$ and $U(\ell^s)$ can be recorded in cartoon "bubbles" over the bases of the $\mathscr{E}$-moves of these lotteries.

There is an alternative notation for the relative-desirability indices $U(\ell^t)$ of lotteries ℓ^t; namely,

$$U(\ell^t) = E\{u(\ell^t(\tilde{\omega}_t))\},$$

which is read as "the *expectation* of the function[b] $u(\ell^t(\cdot))$ of the random variable $\tilde{\omega}_t$." In this book the term "expectation" will *always* refer to such probability-weighted averagings.

The expectation notation for probability-weighted averagings is very common in probability and statistics in particular and in decision analysis in general. It will be very useful in our subsequent work. But in the sequel we shall wish to embellish the probability notation $P(\cdot)$ by adding subscripts and sundry affixes, such as $P\;\;(\cdot|\;\;)$. In such cases, we add the identical embellishments to $E\{\cdot\}$. Thus

$$E\;\;\{u(\ell^t(\tilde{\omega}_t))|\;\;\} = \sum_{i=1}^{n_t} u(c_t^i)P\;\;(\Omega_t^i|\;\;).$$

■ The use of the cats and the Dachshunds is *intended* to connote generality! ■

---

[b] Note from Definition 3.1.3 and Figure 3.2 that $\ell^t(\omega_t)$ is the consequence outcome of ℓ^t if $\tilde{\omega}_t = \omega_t$. In Figure 3.2,

$$\ell^t(\omega_t) = c_t^i \quad \text{if and only if} \quad \omega_t \in \Omega_t^i.$$

Let us pause for a moment to summarize up to this point. We have introduced utilities $u(c)$, or quantified preferences. We have introduced probabilities, or quantified judgments. And we have shown how the utilities and probabilities may be combined by a probability-weighted-average, relative-desirability index U in order to evaluate any lottery of the form depicted in Figure 3.2. The formal assertion that $U(\ell^t)$ is an index of the relative desirability of ℓ^t vis-à-vis other lotteries constitutes Theorem 3.2.1 below; but before stating that theorem we must point out that $\mathscr{D}$'s quantifications of preferences and judgments are *not sufficient*, in and of themselves, for the validity of our basic result.

Indeed, $\mathscr{D}$ must agree in principle that he wishes to abide by a few very plausible *principles of consistent decision-making behavior*—consistency principles, in short. These consistency principles, which we introduce in Section 3.3, serve as the glue for synthesizing the oculist decisions into a decision between any two lotteries and, in Chapter 4, into a solution of his entire dpuu. An additional assumption (not really a consistency principle) is introduced in Section 3.4, which introduces hypothetical devices useful in calibrating $\mathscr{D}$'s preferences and judgments in Sections 3.5 and 3.6 respectively.

Now for the formal statement of the basic result in this chapter.

Theorem 3.2.1. Let ℓ^r and ℓ^s be lotteries of the form ℓ^t as depicted in Figure 3.2. *If* $\mathscr{D}$:

(1) quantifies his relative *preferences* for all consequences c in C of his dpuu, in terms of a utility function $u: C \to (-\infty, +\infty)$ (as described in Section 3.5),

(2) quantifies his *judgments* regarding $\tilde{\omega}_r$ and $\tilde{\omega}_s$ in terms of *probabilities* $P(\Omega_r^i)$ and $P(\Omega_s^i)$ such that every $P(\Omega_t^i) \geq 0$ and $\sum_{i=1}^{n_t} P(\Omega_t^i) = 1$ for each of $t = r$ and $t = s$ (as described in Section 3.6), and

(3) accepts the Transitivity and Substitutability Principles (in Section 3.3), the Monotonicity Principle, and the Canonicity Assumption (both in Section 3.4),

then

$$\ell^r \begin{Bmatrix} > \\ \sim \\ < \end{Bmatrix} \ell^s \quad \text{if and only if} \quad U(\ell^r) \begin{Bmatrix} > \\ = \\ < \end{Bmatrix} U(\ell^s),$$

where

$$U(\ell^t) = E\{u(\ell^t(\tilde{\omega}_t))\} \qquad \text{for each of } t = r \text{ and } t = s.$$

■ Recall that $U(\ell^t) = E\{u(\ell^t(\tilde{\omega}_t))\} = \sum_{i=1}^{n_t} u(c_t^i)P(\Omega_t^i)$. ■

Examine Theorem 3.2.1 closely, as it summarizes almost everything we have introduced so far in this preview. Example 3.2.1 indicated how one operates with Theorem 3.2.1 in practice, once he has accomplished quantifications (1) and (2) and has accepted the "glue" cited by name in (3).

Theorem 3.2.1 has an important corollary. If $\mathscr{D}$ can evaluate *any* lottery ℓ^t by an index $U(\ell^t)$ of relative desirability, then he can cope with situations in which he is confronted with an entire $\mathscr{D}$-move $\{\ell^t: t \in T\}$ of lotteries. Clearly, he should choose any lottery ℓ^{t° which maximizes the index $U(\ell^t)$. That is, in the language of decision analysis, an *optimal* lottery ℓ^{t° satisfies the condition that

$$U(\ell^{t^\circ}) = \max_{t \in T} U(\ell^t),$$

or equivalently, the condition that

$$U(\ell^{t^\circ}) \geq U(\ell^t) \qquad \text{for every } t \text{ in } T.$$

■ Sometimes in infinite idealizations of $\mathscr{D}$-moves an optimal lottery does not exist! Suppose $T = (0, 1)$, the *open* unit interval; and suppose for every t in $(0, 1)$ that ℓ^t has desirability index $U(\ell^t) = t$. Then you are tempted to say that $\ell^{t^\circ} = \ell^1$, but ℓ^1 is *not* available for choice! You can choose ℓ^t for t close to 1, but for every such t there is a t' closer to 1 and $\ell^{t'}$ is more desirable than ℓ^t. Now, this pathology can arise *only* in infinite idealizations of real choice situations. In any *real* problem, T and hence also $\{\ell^t: t \in T\}$ are finite sets, and in such cases ℓ^{t° *always* exists. ■

We conclude this preview with a second (but somewhat less fundamental) result of great usefulness in the sequel. It concerns how $\mathscr{D}$ can evaluate a "compound" lottery without first reformulating it as a "simple" lottery and then applying Theorem 3.2.1. Essentially, $\mathscr{D}$ may compute the relative desirability index $U(\ell(\omega^i))$ of each contingent lottery $\ell(\omega^i)$—see Figure 3.4—and then treat the indices $U(\ell(\omega^1)), \ldots, U(\ell(\omega^n))$ just like the utilities $u(c_t^1), \ldots, u(c_t^{n_t})$ in Theorem 3.2.1 for purposes of evaluating the initial part of the lottery ℓ: multiply the utility $U(\ell(\omega^i))$ of the outcome $\ell(\omega^i)$ given $\tilde{\omega} = \omega^i$ by the probability $P_1(\omega^i)$ of ω^i, for every i, and then add up all the products. The result is the relative desirability index $U(\ell)$ of ℓ itself, as noted in the legend for Figure 3.4.

Theorem 3.2.2. Suppose that $\Omega = \{\omega^1, \ldots, \omega^n\}$ is an $\mathscr{E}$-move, L is a set of lotteries each with outcomes in C, and $\ell: \Omega \rightarrow L$ is therefore a "compound" lottery. Suppose that $\mathscr{D}$ has quantified all judgments and preferences as indicated in Figure 3.4, and that he satisfies the requirements of (3) of Theorem 3.2.1. Then

(1) $$U(\ell(\omega^i)) = E_{2|1}\{u(\ell(\tilde{\xi}, \omega^i))|\omega^i\} \qquad \text{for } i = 1, \ldots, n;$$

and

(2) $$U(\ell) = E_1\{U(\ell(\tilde{\omega}))\} = E_1\{E_{2|1}\{u(\ell(\tilde{\xi}, \tilde{\omega}))\}\}.$$

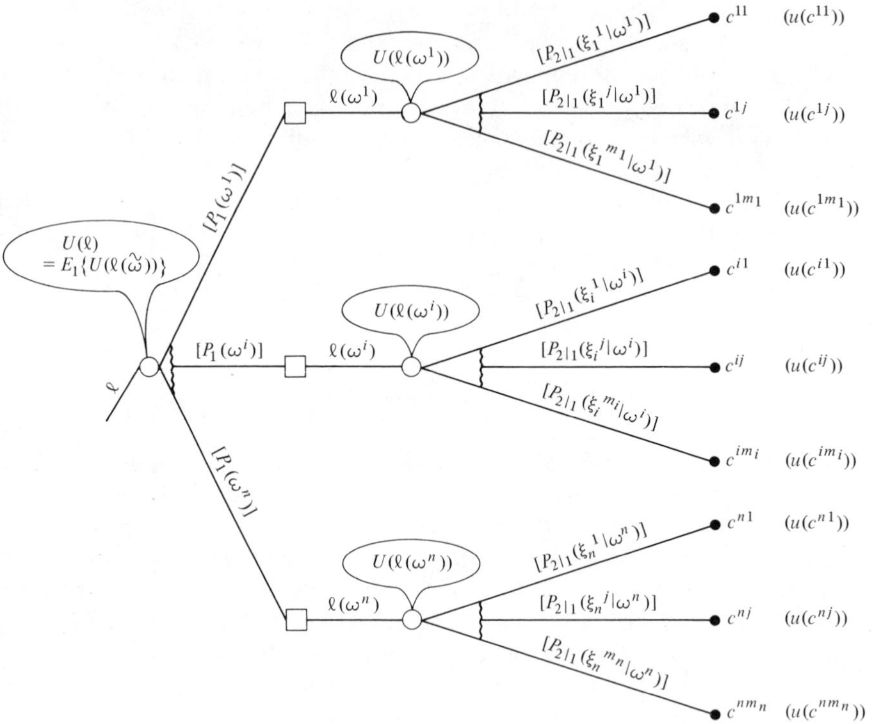

Figure 3.4 A compound lottery and Theorem 3.2.2. (*Key*: $\ell(\omega^i)$ results in c^{ij} if $\tilde{\xi} = \xi_i^j$, for $j = 1, \ldots, m_i$ and $i = 1, \ldots, n$.)

$$\sum_{j=1}^{m_i} P_{2|1}(\xi_i^j|\omega^i) = 1 \qquad \text{for } i = 1, \ldots, n.$$

$$U(\ell(\omega^i)) = \sum_{j=1}^{m_i} u(c^{ij})P_{2|1}(\xi_i^j|\omega^i) = E_{2|1}\{u(\ell(\tilde{\xi}, \omega^i))|\omega^i\}.$$

$$U(\ell) = \sum_{i=1}^{n} U(\ell(\omega^i))P_1(\omega^i) = E_1\{E_{2|1}\{u(\ell(\tilde{\xi}, \tilde{\omega}))|\tilde{\omega}\}\}$$

$$= \sum_{i=1}^{n}\sum_{j=1}^{m_i} u(c^{ij})P_{2|1}(\xi_i^j|\omega^i)P_1(\omega^i).$$

■ Recall that $E_1\{U(\ell(\tilde{\omega}))\} = \sum_{i=1}^{n} U(\ell(\omega^i))P_1(\omega^i)$, where $U(\ell(\omega^i)) = \sum_{j=1}^{m_i} u(c^{ij})P_{2|1}(\xi_i^j|\omega^i)$ for every i; and hence $E_1\{E_{2|1}\{u(\ell(\tilde{\xi}, \tilde{\omega}))|\tilde{\omega}\}\} = \sum_{i=1}^{n}\sum_{j=1}^{m_i} u(c^{ij})P_{2|1}(\xi_i^j|\omega^i)P_1(\omega^i)$. ■

The notation in Theorem 3.2.2 may seem slightly complicated, but the underlying idea is simple. Figure 3.5 gives a numerical example with most of the notation deleted; you may verify all "bubbled" calculations as Exercise 3.2.3.

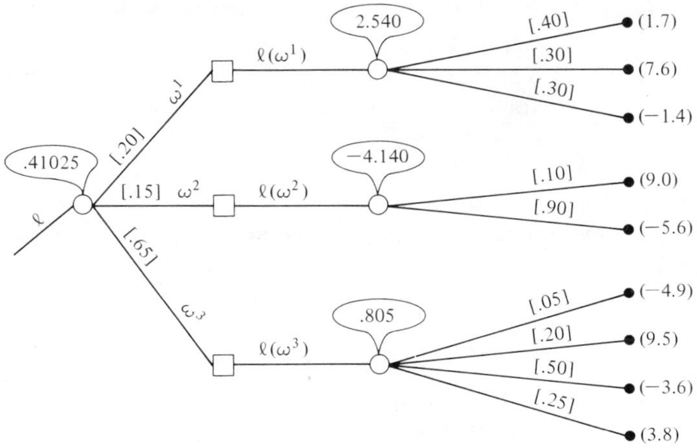

Figure 3.5 A numerical example of Theorem 3.2.2.

Since each of the three $\ell(\omega^i)$'s in Figure 3.5 is a "simple" lottery such as those depicted in Figure 3.3 and dealt with in Theorem 3.2.1, it follows that Theorems 3.2.2 and 3.2.1 simply *tell $\mathscr{D}$ to perform some multiplications and additions once he has quantified his preferences and judgments.* It is difficult to imagine anything methodologically respectable which is simpler than calculating these (branch-probability-) weighted averages of (branch-tip-attached) utilities.

■ Novices in decision analysis are usually perplexed and sometimes irritated by the exposition of such simple arithmetic operations in such complicated notation. But these operations appear in many different contexts throughout this book, and the diversity of these contexts precludes using a single, simple notation system. The only viable alternative to a complicated but flexible notation system is to develop *several* simpler systems. All things considered, this alternative places a greater burden on the learner. Moreover, our informal discussion of these operations should have dispelled by now whatever obscurity was occasioned by the notation. ■

We have now previewed all sections in Part B of this chapter and all the basic results. Part C concerns certain lotteries whose $\mathscr{E}$-moves are infinite idealizations, and none of the material in this part is essential to your understanding of the fundamental ideas of decision analysis. Only optional, starred material in subsequent chapters depends upon Part C. Anyone with a knowledge of elementary calculus and elementary linear algebra will have no trouble understanding all of Part C.

Exercise 3.2.1

Suppose $G = \{\ell^1, \ell^2, \ell^3, \ell^4\}$ is a $\mathscr{D}$-move consisting solely of lotteries, with $U(\ell^1) = 8.42$, $U(\ell^2) = 3.87$, $U(\ell^3) \doteq 7.76$, and $U(\ell^4) = 8.44$. Which lottery should $\mathscr{D}$ choose?

Exercise 3.2.2

Suppose $G = \{\ell^t : 0 \le t \le 2\}$ is a $\mathcal{D}$-move consisting solely of lotteries, with $U(\ell^t) = 10 - (t-1)^4$ for every t. Which lottery should $\mathcal{D}$ choose?

Exercise 3.2.3

Verify all the calculations which produced the numbers in the bubbles in Figure 3.5.

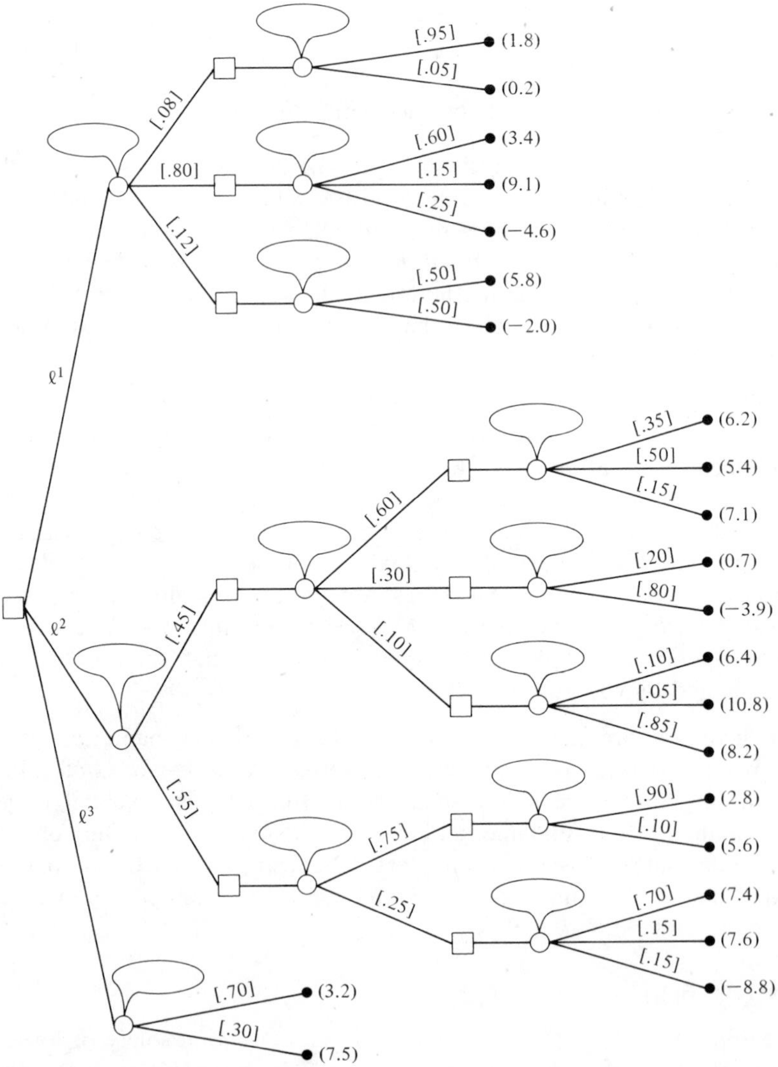

Figure 3.6 A $\mathcal{D}$-move of lotteries.

Exercise 3.2.4

Figure 3.6 depicts a $\mathscr{D}$-move $G = \{\ell^1, \ell^2, \ell^3\}$ of lotteries. Find $U(\ell^1)$, $U(\ell^2)$, and $U(\ell^3)$; use the remark following Theorem 3.2.1 to determine an *optimal* lottery, and verify that the sum of the probabilities attached to the branches of each $\mathscr{E}$-move is one.
[*Hint*: For ℓ^2, keep working backward! Fill in all bubbles.]

B: INGREDIENTS OF DECISION ANALYSIS

3.3 PRINCIPLES OF CONSISTENT CHOICE AMONG LOTTERIES

Now that you are duly impressed with the mechanical power of our brand of decision analysis in resolving complicated choices among lotteries, it is time to consider the inescapable and subjective ingredients which the decision maker must provide.

Our first requirement of $\mathscr{D}$ is that he accept the validity of the "glue" which enables him to synthesize his preference quantifications and judgment quantifications, via Theorems 3.2.1 and 3.2.2, into choices among lotteries and, in Chapter 4, into a solution of his entire dpuu. That glue is a set of consistency principles.

We contend that there are three principles which $\mathscr{D}$ *should want* to have characterize his choices among lotteries. These principles *do not* characterize his run-of-the-mill, casual choices, because we all make many choices in life in at most a semiattentive state. But we are concerned *here* with *important* choices, ones which $\mathscr{D}$ *wants* to analyze carefully. The unimportant ones fall outside the purview of this book.

The first consistency principle that we ask $\mathscr{D}$ to accept, in principle, is called *transitivity*. Suppose that $\mathscr{D}$ has considered three lotteries, ℓ', ℓ'', and ℓ''', and that he regards ℓ' as at least as desirable as ℓ'', and ℓ'' as at least as desirable as ℓ'''. Then transitivity requires that he *should* regard ℓ' as at least as desirable as ℓ'''. The following is a formal statement[c] of this principle.

Transitivity Principle. $\ell' \gtrsim \ell'''$ whenever $\ell' \gtrsim \ell''$ and $\ell'' \gtrsim \ell'''$.

Transitivity is a property of many relations, such as the relation "$\geq$" for real numbers. You know that if x, y, and z are real numbers, then $x \geq z$ whenever $x \geq y$ and $y \geq z$. We are simply asking that $\mathscr{D}$ want his relation "$\gtrsim$" to behave like "$\geq$".

But "$\gtrsim$" is not "$\geq$," and hence $\mathscr{D}$ might not want to accept transitivity of

[c] Recall that $\ell' \gtrsim \ell''$ means that $\mathscr{D}$ regards ℓ' as at least as desirable as ℓ'', etc.

"$\succsim$"—might not, that is, until he is confronted with the following potentially horrible consequence of being obstinately intransitive.

■ Suppose $\mathscr{D}$ regards ℓ' at least as desirable as ℓ'', and ℓ'' at least as desirable as ℓ''', but ℓ''' as *more* desirable than ℓ'. Also suppose that you have ℓ' and ℓ'' in a "bank" and have given ℓ''' to $\mathscr{D}$. Since ℓ'' is for $\mathscr{D}$ at least as desirable as ℓ''', he should be glad to exchange his ℓ''' for your ℓ'' at no charge. Now he holds ℓ'' and you hold ℓ' and ℓ'''. But since ℓ' is for $\mathscr{D}$ at least as desirable as ℓ'', he should be glad to make another exchange, of ℓ'' for ℓ'. Now he holds ℓ', and you hold ℓ'' and ℓ'''. Since $\mathscr{D}$ regards ℓ''' as more desirable than ℓ', he should be willing to *pay* a tiny amount $\$x$—say, a millionth of a mil—to exchange his ℓ' for your ℓ'''. Otherwise his preference statements are just chitchat. *But now he is back where he started and poorer to boot.* Continuing this cycling indefinitely will bankrupt him. He will *want* to revise his preference statements so as not to violate the Transitivity Principle! ■

Several stronger versions of the Transitivity Principle, involving strict preference "$>$" and equidesirability "$\sim$," are derivable from the preceding Transitivity Principle. See Exercise 3.3.1.

The second consistency principle that we ask $\mathscr{D}$ to accept, again *in principle*, is *dominance*. Assume throughout the discussion of this principle and its corollary, Substitutability, that $\mathscr{E}$ will make choices in complete ignorance of any choices that $\mathscr{D}$ might make. Dominance requires that if the outcome of a lottery ℓ is *certain* to be at least as desirable to $\mathscr{D}$ as the outcome of a lottery ℓ', then $\mathscr{D}$ *should* regard ℓ *itself* as at least as desirable as ℓ' *itself*. The following is a formal statement.

Dominance Principle. Suppose that $\ell: \Omega \to X$ and $\ell': \Omega \to X'$ are two lotteries on the same $\mathscr{E}$-move, from which $\mathscr{E}$ will choose in ignorance of $\mathscr{D}$'s behavior, with (possibly different) outcome sets X and X' respectively. If $\ell(\omega) \succsim \ell'(\omega)$ for every ω in Ω, then $\ell \succsim \ell'$.

It is hard to see how an allegedly levelheaded person could prefer ℓ' to ℓ and yet maintain that $\ell(\omega)$ is at least as desirable as $\ell'(\omega)$ for every ω, provided that $\mathscr{E}$'s choice from Ω will be made in ignorance of the choice between ℓ and ℓ'.

Actually, it is only in infinite idealizations that we need the full power of the Dominance Principle. Instead, we shall frequently use its easily derived corollary, the Substitutability Principle, which says that $\mathscr{D}$ should regard ℓ and ℓ' as *equally* desirable if the outcomes of ℓ and ℓ' are *certain* to be equally desirable.

Substitutability Principle (Corollary of Dominance). Let $\ell: \Omega \to X$ and $\ell': \Omega \to X'$ be two lotteries on the same $\mathscr{E}$-move. If $\ell(\omega) \sim \ell'(\omega)$ for every ω in Ω, then $\ell \sim \ell'$.

Deriving Substitutability from Dominance is an easy task left to you as Exercise 3.3.2. In fact, Transitivity and a *very* weak Substitutability Principle suffice; see Exercise 3.3.3.

A third consistency principle is needed to establish the validity of our anticipated procedure of starting the analysis at the last stage of a dpuu and working back to the first. Consider the simple dpuu in Figure 3.7; we shall indicate the need for an additional assumption by exemplifying obviously inconsistent behavior on $\mathscr{D}$'s part: Suppose that, before learning δ or $\overline{\{\delta\}}$, $\mathscr{D}$ expresses a preference for ℓ^1 over ℓ^2, then δ occurs, and then $\mathscr{D}$ announces a change of mind and a preference for ℓ^2 over ℓ^1. You would tend to accuse him of having been a casual conversationalist beforehand, capricious, addle pated, irresolute, or any combination of these.[d]

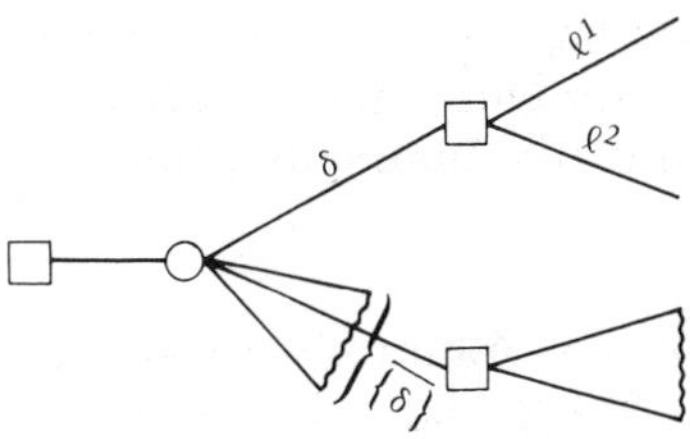

Figure 3.7 A simple dpuu.

■ In real life, an appreciable timespan between the pre-$\mathscr{E}$-move analysis and the contingent $\mathscr{D}$-move $\{\ell^1, \ell^2\}$ may contain revelations of information relevant to the choice between ℓ^1 and ℓ^2 which $\mathscr{D}$ did not *anticipate* in his decision tree. Such revelations may *validly* motivate $\mathscr{D}$ to change his prior preferences. But the key word here is "anticipate." The decision tree *presumably* includes all relevant features of the dpuu that $\mathscr{D}$ can anticipate. Hence $\mathscr{D}$ should want *in principle* to abide by his prior preferences. Otherwise he should improve his decision tree! ■

The necessary principle which rules out capricious behavior and/or unwillingness (in principle) to use one's decision tree will be called the invariant preference principle.

Invariant Preference Principle. Let δ be any ee, and let ℓ^1 and ℓ^2 be lotteries in the $\mathscr{D}$-move G immediately following δ. Then $\mathscr{D}$ *should* regard ℓ^1 as at least as desirable as ℓ^2 *after* the revelation of δ if and only if he would have regarded ℓ^1 as at least as desirable as ℓ^2 *before* this revelation but with the understanding that any choice from G is irrelevant unless δ is revealed.

The Transitivity, Dominance, and Invariant Preference principles, together with the Monotonicity Principle and the Canonicity Assumption

[d] Unless you were $\mathscr{D}$'s consultant, in which case you would seek another way of making this point.

introduced in Section 3.4, constitute sufficient glue for synthesizing preference and judgment quantifications in all but infinite idealizations of dpuu's. We conclude this section with an optional, starred introduction of the additional principle, called *continuity*, needed for dpuu's in which some $\mathscr{E}$-moves are infinite idealizations.

*■ The Continuity Principle essentially says that $\mathscr{D}$ can "zero in" on the evaluation $U(\ell)$ of a lottery $\ell: \Omega \to X$ with infinitely many distinct outcomes $\ell(\omega)$ in X by taking limits of appropriate sequences of evaluations $\{U(\ell_+{}^n): n = 1, 2, \ldots\}$ and $\{U(\ell_-{}^n): n = 1, 2, \ldots\}$ of lotteries $\ell_+{}^n: \Omega \to X$ and $\ell_-{}^n: \Omega \to X$ each of which has only a *finite* number of distinct outcomes in X. It says that if the sequences $\{\ell_-{}^n: n = 1, 2, \ldots\}$ and $\{\ell_+{}^n: n = 1, 2, \ldots\}$ of "finite" lotteries on Ω "trap ℓ in desirability" and have desirability-index sequences converging to a common, intermediate limit, then that limit should be the desirability index of ℓ. How else *could* $\mathscr{D}$ reasonably define $U(\ell)$? Formally, the Continuity Principle is as follows.

Continuity Principle. Suppose that $\ell: \Omega \to X$ has arbitrarily many mutually distinct outcomes in X and that $\ell_+{}^n: \Omega \to X$ and $\ell_-{}^n: \Omega \to X$ are lotteries with finitely many mutually distinct outcomes in X for $n = 1, 2, \ldots$ *ad inf* such that:

(1) $$\ell_-{}^n \precsim \ell_-^{n+1} \precsim \ell \precsim \ell_+^{n+1} \precsim \ell_+{}^n \qquad \text{for } n = 1, 2, \ldots \textit{ ad inf};$$

and

(2) $$\lim_{n\to\infty} U(\ell_-{}^n) = \lim_{n\to\infty} (\ell_+{}^n).$$

Then $\mathscr{D}$ should define $U(\ell)$ as the common limit in (2).

We repeat that the Continuity Principle is needed *only* to establish the validity of variants of Theorems 3.2.1 and 3.2.2 in infinite idealizations. These variants add nothing to one's understanding of the basic ideas of decision analysis and are therefore relegated to the optional Part C of this chapter. ■

Exercise 3.3.1: Variants of Transitivity

Assume that $\ell' \succsim \ell'''$ whenever $\ell' \succsim \ell''$ and $\ell'' \succsim \ell'''$. Show that, therefore,

(A) $\ell' \sim \ell'''$ whenever $\ell' \sim \ell''$ and $\ell'' \sim \ell'''$.
(B) $\ell' \succ \ell'''$ whenever $\ell' \succ \ell''$ and $\ell'' \sim \ell'''$.
(C) $\ell' \succ \ell'''$ whenever $\ell' \sim \ell''$ and $\ell'' \succ \ell'''$.
(D) $\ell' \succ \ell'''$ whenever $\ell' \succ \ell''$ and $\ell'' \succ \ell'''$.

[*Hint*: Recall that $\ell^a \sim \ell^b$ if and only if both $\ell^a \succsim \ell^b$ and $\ell^b \succsim \ell^a$; and that $\ell^a \succ \ell^b$ if and only if $\ell^a \succsim \ell^b$ but not $\ell^a \sim \ell^b$.]

Exercise 3.3.2: Dominance Implies Substitutability

Show that the Substitutability Principle is implied by the Dominance Principle, and hence that "Substitutability" imposes no restrictions on $\mathscr{D}$'s choice of behavior not already imposed by "Dominance."
[*Hint*: The hint for Exercise 3.3.1 about $\ell^a \sim \ell^b$ pertains here significantly.]

Exercise 3.3.3

In Chapter 2 of Pratt, Raiffa, and Schlaifer [**112**] (and elsewhere) the following, weak Substitutability Principle is used:

SP_1. If $\ell(\omega) = \ell'(\omega)$ for all $\omega \neq \omega^*$ and if $\ell(\omega^*) \sim \ell'(\omega^*)$, then $\ell \sim \ell'$.

Apply the Transitivity Principle of Exercise 3.3.1(A) n times to obtain the *finite* version of our Substitutability Principle:

SP_n. If $\Omega = \{\omega^1, \ldots, \omega^n\}$ for some (finite!) positive integer n and if $\ell(\omega^i) \sim \ell'(\omega^i)$ for $i = 1, 2, \ldots, n$, then $\ell \sim \ell'$.

[*Hint*: "Replace" the outcomes of ℓ one at a time, creating a chain $\ell^1, \ell^2, \ldots, \ell^{n+1}$ of lotteries such that

$$\ell^j(\omega^i) = \begin{cases} \ell(\omega^i) & \text{for all } i \geq j \\ \ell'(\omega^i) & \text{for all } i < j. \end{cases}$$

Use SP_1 to show that $\ell^j \sim \ell^{j+1}$ for $j = 1, 2, \ldots, n$, and note that $\ell = \ell^1$ and $\ell' = \ell^{n+1}$.]

Exercise 3.3.4: Ramifications of Substitutability

Suppose that (1) $\mathscr{D}$ regards ℓ' and ℓ'' as equally desirable, and (2) $\mathscr{D}$ regards the events in the $\mathscr{E}$-moves of ℓ' and of ℓ'' as completely unrelated to the events Ω^* and $\overline{\Omega^*}$ in an $\mathscr{E}$-move Ω. Examine the dpuu in Figure 3.8.

(A) Argue that $\mathscr{D}$ *should* regard acts g and ℓ''' as equally desirable.
(B) Argue that if $\mathscr{D}$ did *not* regard ℓ' and ℓ'' as equally desirable, then he *should* regard g as *at least as* desirable as ℓ'''.

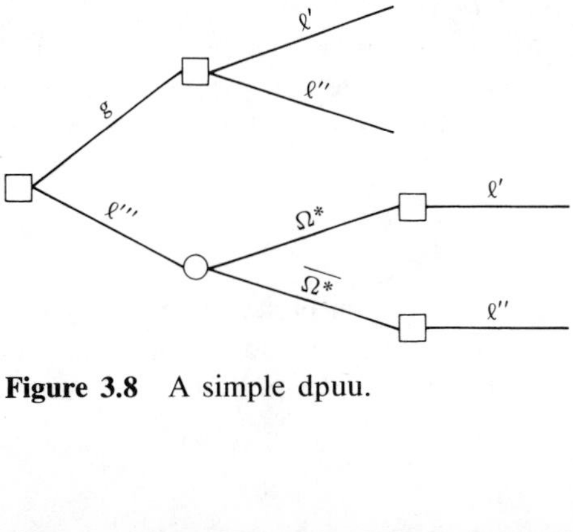

Figure 3.8 A simple dpuu.

*(C) In (B), under what circumstances would $\mathcal{D}$ regard g as *more desirable than* ℓ'''?

3.4 CANONICAL RANDOMIZATIONS

In Section 3.2 we alluded to the pending introduction of hypothetical, auxiliary chance mechanisms unrelated to the dpuu under analysis. Such mechanisms are generators of ee's which $\mathcal{D}$ regards as equally likely in a precise, decision-oriented sense.

Definition 3.4.1. A canonical randomization Λ is an $\mathscr{E}$-move Λ such that $\mathcal{D}$ regards the ee's λ in Λ as *equally likely*, in the sense that *if*:

(1) p is any given number in $[0, 1]$,
(2) Λ^1 and Λ^2 are any two events in Λ each containing proportion p of the elements in Λ,
(3) A and B are any two outcomes whatsoever, and
(4) for each $i \in \{1, 2\}$, ℓ^i is a lottery on Λ such that $\ell^i(\lambda) = A$ for all $\tilde{\lambda} \in \Lambda^i$ and $\ell^i(\lambda) = B$ for all $\tilde{\lambda} \notin \Lambda^i$;

then $\mathcal{D}$ regards ℓ^1 and ℓ^2 as *equally desirable.*

What determines the *canonical* (= authorized, accepted) nature of an $\mathscr{E}$-move is the decision maker's *judgment*, as operationally, and not just conversationally, manifested by his indifference in a choice between two simple lotteries which can be regarded as alternative wagers about $\tilde{\lambda}$.

Notice that Definition 3.4.1 makes no reference whatever to "objective probability" (whatever that might be), to the opinions of other people, or to any record of past repetitions of this $\mathscr{E}$-move.

Example 3.4.1

If otherwise identical balls are numbered from 1 through M and placed in an opaque urn; if this urn is stirred and shaken vigorously; and if a disinterested and blindfolded umpire then draws one of the balls from this urn; then I personally would regard the situation as canonical, in the sense that if Λ^1 and Λ^2 are *any* two subsets of $\{1, \ldots, M\}$ each consisting of k elements (for k any integer in $\{1, \ldots, M\}$), and if A and B are any outcomes, then I would as soon have the lottery which yields me A if $\tilde{\lambda} \in \Lambda^1$ and B otherwise as I would have the lottery which yields me A if $\tilde{\lambda} \in \Lambda^2$ and B otherwise.

Example 3.4.2

Just to underscore the subjectivity of canonicity, suppose that in Example 3.4.1 the ball has *already* been drawn and given to a silent, unblindfolded scorekeeper. For *me*, the randomization is *still* canonical, because the outcome has not been revealed to me. For the scorekeeper, it is now certainly not canonical!

Definition 3.4.2. Let Λ' be a subset of a canonical randomization which contains proportion p of the elements of Λ. We say that p is the *canonical probability* of the event Λ', and we write

$$P(\Lambda') = p.$$

Example 3.4.3

Consider a delicately balanced spinner over a circular dial. I would regard a spin of this spinner as a canonical randomization, in which every arc Λ' of length a proportion p of the dial circumference has canonical probability p of containing the point $\tilde{\lambda}$ at which the spinner will come to rest.

Example 3.4.4

There is a symmetric, icosahedral (= twenty-sided) die on the market. Each of the digits $0, 1, \ldots, 9$ appears on exactly two sides of this die. See Figure 3.9. With such a die, one can "generate" arbitrarily precise decimals—say, m-place decimals $.\lambda_1\lambda_2 \ldots \lambda_m$—by simply rolling the die m times and letting λ_i be the outcome of roll i. Now, I would call the ten digits possible on each roll equally likely regardless of the previous outcomes, and hence I would regard the 10^m possible decimals from $.00\ldots0$ through $.99\ldots9$ as equally likely. (Why 10^m? Think about it!) Hence I would assign canonical probability $k/10^m$ to every subset λ' consisting of exactly k of the 10^m m-place decimals.

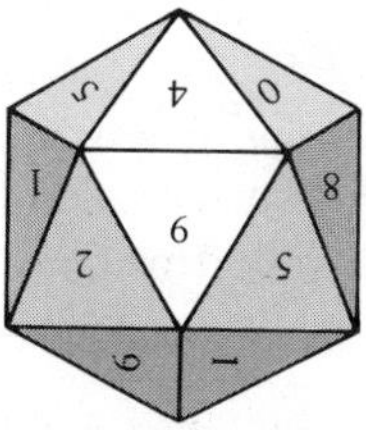

Figure 3.9 A random-digit-generating icosahedral die.

A finite number of casts of a canonical icosahedral die can be thought of as an approximation to the single spin of the spinner. In principle, the latter is an $\mathscr{E}$-move with events having canonical probability p for p any rational or irrational number in $[0, 1]$, whereas the former is an $\mathscr{E}$-move having only events with canonical probability p for p an m-place (rational) decimal. But reality is measured in rational numbers, as noted in Section 2.13. Hence, by making m sufficiently large, one can use the icosahedral die to create $\mathscr{E}$-moves Λ with events of canonical probability p for p as accurate a decimal as one could wish.

Example 3.4.5

Suppose that I wish to create an $\mathscr{E}$-move Λ with two mutually exclusive and collectively exhaustive events Λ^1 and Λ^2 having respective canonical probabilities .2987 and .7013. I can cast a canonical icosahedral die four times, letting Λ^1 consist of the 2,987 four-place decimals from .0000 through .2986 (inclusive), and Λ^2 consist of the 7,013 four-place decimals from .2987 through .9999 (inclusive).

By considering an *idealization* of reality such as a canonical spinner or infinitely many casts of a canonical icosahedral die, we obtain the following important concept, used extensively in the latter parts of this chapter.

Definition 3.4.3. A *basic* canonical randomization is a canonical randomization in which $\Lambda = [0, 1]$.

*■ Suppose that a canonical randomization is performed *twice* in such a way that the outcome $\tilde{\lambda}$ of the first randomization can have no bearing on the outcome, say $\tilde{\psi}$, of the second. Denoting the second canonical randomization by Ψ, it follows that all elements (λ, ψ) of $\Lambda \times \Psi$ are equally likely for $\mathscr{D}$. When a *basic* canonical randomization is performed twice with outcomes $\tilde{\lambda}$ of the first performance and $\tilde{\psi}$ of the second performance entirely unrelated, we call $\Lambda \times \Psi = [0, 1] \times [0, 1] =$ "the unit square" a *paired* basic canonical randomization. A fundamental fact about paired basic canonical randomizations is furnished in *Exercise 3.4.5 and used extensively in the proofs in *Section 3.8. ■

Next, we present an obvious definition for the sake of completeness and which hints at the role of canonical randomizations as aids to the quantification of preferences and judgments.

Definition 3.4.4. A *canonical lottery* is a lottery the outcome of which is determined solely by a canonical randomization (which, you will recall, is an $\mathscr{E}$-move); similarly for basic canonical lottery (*and for paired basic canonical lottery).

Finally, we noted in Theorem 3.2.1 and in Section 3.3 that we would make more assumptions about $\mathscr{D}$'s decision-making behavior in this section. The first of these is the canonicity assumption.

Canonicity Assumption. $\mathscr{D}$ can *imagine* the existence of a *basic* canonical randomization which can be performed repeatedly with the property that the outcomes of the individual performances are completely unrelated to each other and to the events in the dpuu under analysis.

Our previous and extended discussion of icosahedral dice and other randomization devices was intended to convince you of the reasonableness of the canonicity assumption. Even though it refers to a necessarily idealized basic canonical randomization, it is not difficult to imagine the number m of casts of an icosahedral die growing without bound, and thus creating an infinite sequence of digits that would represent *any* point in [0, 1].

Example 3.4.6

Repeated and unrelated, or *independent*, performances of a basic canonical randomization are not difficult to visualize. Suppose that we use an m-place-decimal approximation to the outcome of a basic canonical randomization. Such approximation can be realized by casting a canonical icosahedral die m times. Each such "block" of m casts constitutes *one* performance. Since individual casts are unrelated to each other,[e] the m-cast blocks are also unrelated to each other.[f] Moreover, it is difficult to see how the operation of a randomization device could affect or be affected by events in the dpuu itself, particularly in view of the *explicitly auxiliary* nature of the randomization!

Now we have been rather explicit in our discussion of icosahedral dice and other *physical* randomization devices, but this explicitness is intended *only* to dispel the fear that a canonical randomization is a chimera. No *physical* randomization device is actually required if $\mathscr{D}$ has acquired a sufficiently good "feel" for canonical probabilities. Otherwise we would have included an icosahedral die with this book!

The least[g] requirement of $\mathscr{D}$ is that he accept the *monotonicity principle,* which simply says that if each of two canonical lotteries will result in either c'' or c', then $\mathscr{D}$ prefers whichever of these lotteries has the *higher* canonical probability of resulting in the *more desirable* of the consequences.

Example 3.4.7

Suppose that ℓ results in $\mathscr{D}$'s winning \$10 with canonical probability .51 and losing \$5 with canonical probability .49, while ℓ' results in $\mathscr{D}$'s winning \$10 with probability .50 and losing \$5 with complementary canonical probability .50. The Monotonicity Principle requires that $\mathscr{D}$ prefer ℓ to ℓ', provided only that he is the ordinary sort who prefers winning money to losing it.

In quantifying preferences and judgments and in providing a formal statement of the Monotonicity Principle, it is highly desirable to have a

---

[e] See Example 3.4.4; all digits on each cast are equally likely irrespective of the outcomes of the other casts.

[f] The m-place decimals are equally likely in each performance irrespective of the outcomes of other performances.

[g] "Least" is a Freudian typographical error for "last," which we do not correct because it is quite appropriate!

succinct notation for simple canonical lotteries with only two possible outcomes:

$\ell^*(p; c'', c')$ = canonical lottery resulting in c'' with canonical probability p and resulting in c' otherwise (with canonical probability $1-p$).

See Figure 3.10 for a tree representation of $\ell^*(p; c'', c')$.

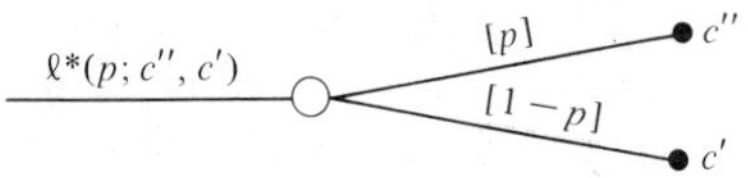

Figure 3.10 Tree representation of $\ell^*(p; c'', c')$.

Finally, a formal statement of the innocuous Monotonicity Principle.

Monotonicity Principle. Suppose that $c'' > c'$. Then

$$\ell^*(p^1; c'', c') \begin{Bmatrix} > \\ \sim \\ < \end{Bmatrix} \ell^*(p^2; c'', c') \qquad \text{if and only if} \qquad p^1 \begin{Bmatrix} > \\ = \\ < \end{Bmatrix} p^2.$$

In short, all the Monotonicity Principle requires is that $\mathcal{D}$ prefer that $\ell^*(p; c'', c')$ which offers him the higher probability of experiencing the more desirable outcome.

Exercise 3.4.1

Think of two randomization techniques, not discussed in this section, which *you* would regard as canonical.

Exercise 3.4.2

Consider casting a pair of ordinary, six-sided dice, one red and one white. Let Λ and Ψ be the sets of outcomes of the red and the white dice respectively. If each of Λ and Ψ is canonical, would you regard $\Lambda \times \Psi$ as canonical? If so, what is the canonical probability that the sum of the outcomes is seven?

Exercise 3.4.3: Elementary Events in a Basic Canonical Randomization

(A) Verify that the length of $[a, a]$ is *zero* for any a in $(-\infty, +\infty)$.
(B) Verify that the lengths of (a, b), $[a, b]$, $(a, b]$, and $[a, b)$ are all $b - a$ (for $b > a$).
(C) Using (B), verify that the events $\tilde{\lambda} \in [0, p)$ and $\tilde{\lambda} \in [p, 1]$ in a basic canonical randomization Λ have canonical probabilities p and $1-p$ respectively.
(D) Show that the event $\tilde{\lambda} = \lambda$ in a basic canonical randomization has p probability *zero* for every $\lambda \in [0, 1]$.

(E) Since (D) may seem contraintuitive to you, consider an urn containing N balls, numbered from 1 through N. Argue that if the ball selection is canonical, then the probability of drawing ball #1 is $1/N$. Now let e be any tiny positive number. N can be made so large that $1/N < e$. Hence the canonical probability of drawing ball #1 dwindles to zero as N grows to $+\infty$.

Exercise 3.4.4

Show that $\ell^*(p; c'', c')$ is strategically identical to (in fact, equals) $\ell^*(1-p; c', c'')$.
[*Hint*: Does it matter which branch-cum-attached-consequence is on top in Figure 3.10?]

***Exercise 3.4.5**

In a paired basic canonical randomization, let Δ' be the union of mutually exclusive subrectangles $\Delta^1, \Delta^2, \ldots, \Delta^r$ of the unit square, where (for $i = 1, \ldots, r$) $\Delta^i = \Lambda^i \times \Psi^i$, Λ^i is a subinterval of $\Lambda = [0, 1]$ of length x_i, and Ψ^i is a subinterval of $\Psi = [0, 1]$ of length y_i. Show that

(A) $$P(\Delta^i) = x_i y_i \qquad \text{for } i = 1, \ldots, r;$$

(B) $$P(\Delta') = \sum_{i=1}^{r} P(\Delta^i) = \sum_{i=1}^{r} x_i y_i.$$

[*Hint*: See Figure 3.11.]

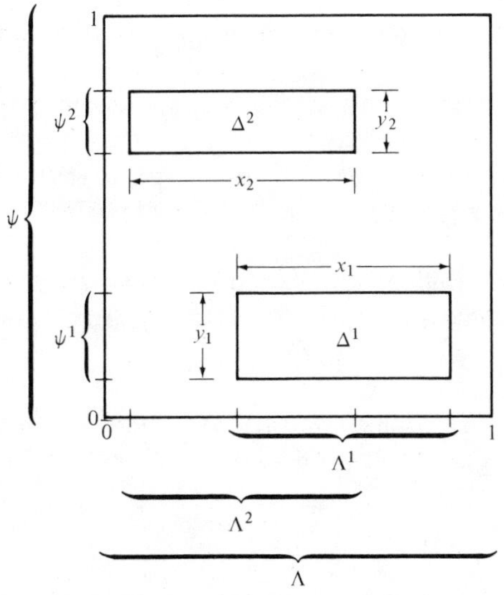

Figure 3.11 Venn diagram illustrating Exercise 3.4.5.

3.5 QUANTIFICATION OF PREFERENCES

At this point it remains for us "only" to show how $\mathscr{D}$ can quantify his preferences and judgments in order to complete our discussion of the prerequisites for invoking Theorems 3.2.1 and 3.2.2 in the analysis of dpuu's.

This section introduces the quantification of *preferences*; the quantification of judgments is introduced in Sections 3.6 and 3.7.

The *first* step is for $\mathscr{D}$ to examine the set C of all potential consequences in his dpuu, and to select two consequences, denoted by c^{B} and c^{W}, with the following properties: c^{B} is at least as desirable to $\mathscr{D}$ as is any consequence c in C, and c^{W} is at least as *un*desirable to $\mathscr{D}$ as is any c in C. Hence,

$$c^{\mathrm{B}} \gtrsim c \gtrsim c^{\mathrm{W}} \qquad \text{for every } c \text{ in } C.$$

The superscripts B and W are intended to connote "best" and "worst" respectively.

We shall assume hereafter that c^{B} is *more* desirable than c^{W}, since the entire dpuu is trivial if $c^{\mathrm{B}} \sim c^{\mathrm{W}}$. (Why? Exercise 3.5.1.)

■ If the consequences c are very complicated and radically different qualitatively, it may be hard for $\mathscr{D}$ to select a c^{B} and a c^{W} directly. Nevertheless, he would have to make such a selection even if the decision problem were under certainty. To aid him in this task, $\mathscr{D}$ might imagine that he is faced with the following simple decision problems under certainty.

(1) *Hypothetical problem for selecting* c^{B}. You, $\mathscr{D}$, may choose any c in C and you will receive (or experience) that c with certainty. Which will you choose?

(2) *Hypothetical problem for selecting* c^{W}. $\mathscr{E}$ will somehow select some c in C for you, $\mathscr{D}$, to receive (or experience), but you may first choose *one* of the elements of C *to be excluded* from $\mathscr{E}$'s selection. Which element would you bar $\mathscr{E}$ from choosing?

If choosing a c^{B} and a c^{W} is still difficult, $\mathscr{D}$ should consider the points made in Section 2.5 about complexity of his decision tree versus complexity of the consequence descriptions. ■

The *second* step is for $\mathscr{D}$ to choose utility numbers $u(c^{\mathrm{B}})$ and $u(c^{\mathrm{W}})$ for c^{B} and c^{W} respectively. $\mathscr{D}$'s choice is constrained *only* by the requirement that

$$u(c^{\mathrm{B}}) > u(c^{\mathrm{W}}),$$

which follows readily from the assumption that $c^{\mathrm{B}} > c^{\mathrm{W}}$ if we agree that utility numbers are supposed to reflect relative desirabilities.

■ Popular choices are: $u(c^B) = 100$ and $u(c^W) = 0$; $u(c^B) = 10$ and $u(c^W) = -10$; and $u(c^B) = 1$ and $u(c^W) = 0$. The $u(c^B) = 1$, $u(c^W) = 0$ choices have a particularly nice interpretation, which we shall develop presently. ■

In the *third* step, $\mathscr{D}$ quantifies his preference for *each* consequence c (other than c^B and c^W) by determining that value $w(c)$ of p for which

$$c \sim \ell^*(p; c^B, c^W).$$

In Figure 3.12 we present a hypothetical choice situation which $\mathscr{D}$ can exploit in determining $w(c)$. By varying p and continually asking himself (in oculist fashion) whether he would prefer c or $\ell^*(p; c^B, c^W)$, he will ultimately reach a value $w(c)$ of p for which c and $\ell^*(w(c); c^B, c^W)$ are equally desirable. Then he should record $w(c)$ and pass on to the next consequence.

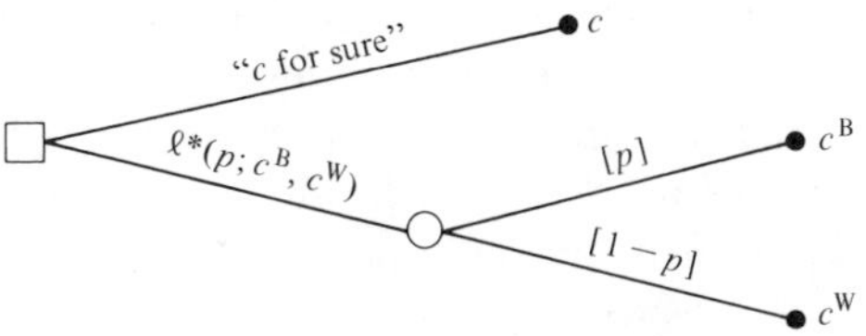

Figure 3.12 Hypothetical choice situation for preference quantification.

■ The comparisons of c with $\ell^*(p; c^B, c^W)$ and variations of p can be recorded in a convenient tabular fashion as scratch work. In a particular dpuu, $\mathscr{D}$'s determining that $c \sim \ell^*(.77; c^B, c^W)$, and hence that $w(c) = .77$ might have entailed the following comparisons.

p	My choice	Next p should be
0	c	larger
1	$\ell^*(1; c^B, c^W)$	smaller
.50	c	larger
.90	$\ell^*(.90; c^B, c^W)$	smaller
.80	$\ell^*(.80; c^B, c^W)$	smaller
.75	c	larger
.77	either	$w(c) = .77$. ■

■ Note that $\ell^*(1; c^B, c^W) = c^B$, and hence $w(c^B) = 1$ naturally. Hence also, $w(c) < 1$ if $c^B > c$. Similarly, $\ell^*(0; c^B, c^W) = c^W$, implying that $w(c^W) = 0$ and hence that $w(c) > 0$ if $c > c^W$. ■

■ $\mathscr{D}$ is occasionally tempted to be somewhat lazy in making these oculist choices and may wish to name *two* p values, say $w_L(c)$ and $w_U(c)$

exceeding $w_L(c)$, such that

(1) $c > \ell^*(p; c^B, c^W)$ whenever $p < w_L(c)$;
(2) $c \sim \ell^*(p; c^B, c^W)$ whenever $w_L(c) \leq p \leq w_U(c)$;
(3) $c < \ell^*(p; c^B, c^W)$ whenever $p > w_U(c)$.

This temptation should be resisted, however, because the Monotonicity and Transitivity Principles imply that there should be a *unique p* value $w(c)$ for which $\ell^*(p; c^B, c^W) \sim c$. (You may prove this as Exercise 3.5.2.) But this uniqueness does not mean that $\mathscr{D}$ should split hairs in trying to determine $w(c)$ to umpteen decimal places. Rather, he should regard determining $w(c)$ as a *very practical* exercise in decision making, since, in the last analysis, $w(c)$ represents $\mathscr{D}$'s *decision* that he would as soon experience c as "gamble" between c^B and c^W with odds $w(c)/[1 - w(c)]$ in favor of c^B. ■

The *final* step commences when $\mathscr{D}$ has determined $w(c)$ for every c in C, and now the time has come to characterize his utilities $u(c)$ of the consequences c. It is easy to show that $\mathscr{D}$'s choices of $u(c^B)$ and $u(c^W)$ in the second step and his choice of $w(c)$ for each c in the third step imply that we should *define* each utility value $u(c)$ by the equation

$$u(c) = w(c)u(c^B) + [1 - w(c)]u(c^W)$$

for every c in C; or equivalently,

$$u(c) = [u(c^B) - u(c^W)]w(c) + u(c^W)$$

for every c in C.

■ *Proof.* If Theorem 3.2.1 is to hold at all, it must hold for the comparison of the lotteries "c for sure" and $\ell^*(p; c^B, c^W)$ in Figure 3.12. Therefore we must have $U(\text{“}c\text{ for sure”}) = U(\ell^*(w(c); c^B, c^W))$ because $c \sim \ell^*(w(c); c^B, c^W)$. But clearly $U(\text{“}c\text{ for sure”}) = u(c)$ and $U(\ell^*(w(c); c^B, c^W)) = w(c)u(c^B) + [1 - w(c)]u(c^W)$, by Theorem 3.2.1. Hence setting $u(c)$ equal to $w(c)u(c^B) + [1 - w(c)]u(c^W)$ is *the only possible choice consistent with Theorem* 3.2.1. The second equation follows from the first by a very little bit of algebra. ■

Example 3.5.1

Suppose that ℓ^t and ℓ^s in Figure 3.3 are the only two lotteries in a dpuu; that $c^B = c_t^1$; that $c^W = c_t^3$; and that $\mathscr{D}$ has chosen $u(c^B) = 8.2$ and $u(c^W) = -3.1$. By the preceding equations relating $u(c)$ to $u(c^B)$ and $u(c^W)$, it follows that $c_s^1 \sim \ell^*(71/113; c^B, c^W)$, since only for $w(c_s^1) = 71/113$ do we have $u(c_s^1) = w(c_s^1)u(c^B) + [1 - w(c_s^1)]u(c^W)$ with $u(c_s^1) = 4.0$, $u(c^B) = 8.2$, and $u(c^W) = -3.1$. As Exercise 3.5.3, you may compute all $w(c_t^i)$'s and all other $w(c_s^j)$'s in Figure 3.3.

For future reference, we summarize the preceding argument in the form of a theorem.

Theorem 3.5.1. Let $c^{\mathrm{B}} \gtrsim c \gtrsim c^{\mathrm{W}}$ for every c in C, with $c^{\mathrm{B}} > c^{\mathrm{W}}$, and suppose that $\mathscr{D}$ has

(1) chosen $u(c^{\mathrm{W}})$ and $u(c^{\mathrm{B}}) > u(c^{\mathrm{W}})$; and
(2) for every c, determined $w(c)$ in $[0, 1]$ such that $c \sim \ell^*(w(c); c^{\mathrm{B}}, c^{\mathrm{W}})$.

Then $\mathscr{D}$'s utility function $u\colon C \to (-\infty, +\infty)$ should have the property that

$$u(c) = [u(c^{\mathrm{B}}) - u(c^{\mathrm{W}})]w(c) + u(c^{\mathrm{W}})$$

for every c in C.

Once $\mathscr{D}$ has chosen $u(c^{\mathrm{W}})$ and $u(c^{\mathrm{B}}) > u(c^{\mathrm{W}})$, his utilities $u(c)$ are uniquely determined in terms of $u(c^{\mathrm{W}})$, $u(c^{\mathrm{B}})$, and the desirability-equating probabilities $w(c)$. From the equation in Theorem 3.5.1, it is clear that different choices of $u(c^{\mathrm{W}})$ and $u(c^{\mathrm{B}})$ lead to different values of $u(c)$ for all c's.

Example 3.5.2

If $\mathscr{D}$ chooses $u(c^{\mathrm{W}}) = 0$ and $u(c^{\mathrm{B}}) = 1$, then Theorem 3.5.1 implies that $u(c) = w(c)$ for every c in C. That is why choosing $u(c^{\mathrm{W}}) = 0$ and $u(c^{\mathrm{B}}) = 1$ make the utility function particularly easy to interpret.

Since $\mathscr{D}$ may choose $u(c^{\mathrm{W}})$ and $u(c^{\mathrm{B}})$ arbitrarily except for the requirement that $u(c^{\mathrm{B}})$ exceed $u(c^{\mathrm{W}})$, it is clear that there are *many* numerical utility functions $u\colon C \to (-\infty, +\infty)$ which represents $\mathscr{D}$'s preferences—as many, in fact, as there are choices of $u(c^{\mathrm{W}})$ and $u(c^{\mathrm{B}})$. But each is related to the function $w\colon C \to (-\infty, +\infty)$, by the equation in Theorem 3.5.1, and therefore, all are related to one another. A precise statement of this relationship constitutes Theorem 3.5.2.

Theorem 3.5.2. If $u\colon C \to (-\infty, +\infty)$ is a utility function for $\mathscr{D}$ on C, then the function $u^*\colon C \to (-\infty, +\infty)$ is *also* a utility function for $\mathscr{D}$ on C if and only if there exist constants $a^* > 0$ and $b^* \in (-\infty, +\infty)$ such that

$$u^*(c) = a^*u(c) + b^* \qquad \text{for every } c \text{ in } C.$$

■ *Proof.* Let $a = u(c^{\mathrm{B}}) - u(c^{\mathrm{W}})$, $b = u(c^{\mathrm{W}})$, $a' = u^*(c^{\mathrm{B}}) - u^*(c^{\mathrm{W}})$, and $b' = u^*(c^{\mathrm{W}})$. By Theorem 3.5.1, $u(c) = aw(c) + b$ for every c in C. Hence,

(i) $$w(c) = [u(c) - b]/a$$

for every c in C. But Theorem 3.5.1 also implies that $u^*(\cdot)$ is also a utility function for $\mathscr{D}$ on C if and only if

(ii) $$u^*(c) = a'w(c) + b'$$

for every c in C. Now replace $w(c)$ in (ii) with the right-hand side of (i), define $a^* = a'/a$, and define $b^* = (ab' - a'b)/a$, in order to obtain the desired result. ■

The relationship between utility functions described in Theorem 3.5.2 is often expressed by the statement "any two utility functions are related via change of scale and/or origin," since multiplying all values of $u(\cdot)$ by a positive number a^* $(\neq 1)$ is called changing the scale of $u(\cdot)$, whereas adding any constant b^* $(\neq 0)$ to all values of $u(\cdot)$ is called changing the *origin* of $u(\cdot)$.

Example 3.5.3

By multiplying all utilities in Figure 3.3 by 10 and then adding 31, we obtain the new (u^*-type) utilities given in the following table.

i	$u^*(c_t^i)$	j	$u^*(c_s^j)$
1	113	1	71
2	77	2	19
3	0	3	43
4	56		

Now it is easy to verify that the new, probability-weighted average, relative-desirability indices of ℓ^t and ℓ^s are $U^*(\ell^t) = 46.57$ and $U^*(\ell^s) = 40.92$, respectively. Indeed, $U^*(\ell^r) = 10U(\ell^r) + 31$ for each of $r = t$ and $r = s$; that is, precisely the same scale and origin changes are produced on $U(\cdot)$ as are applied to $u(\cdot)$. This fact is a consequence of properties of the expectation operation $E\{\cdot\}$, which you may derive as Exercises 3.7.5 through 3.7.8.

Exercise 3.5.1

Why is the entire dpuu trivial if $c^B \sim c^W$?

Exercise 3.5.2

Show that if $c \sim \ell^*(p^1; c^B, c^W)$ and $c \sim \ell^*(p^2; c^B, c^W)$, then the Monotonicity and Transitivity Principles imply that $p^1 = p^2$.

Exercise 3.5.3

Compute $w(c_t^i)$ for $i = 1, 2, 3, 4$ and $w(c_s^j)$ for $j = 2, 3$ from the utilities in Figure 3.3.

Exercise 3.5.4

Show that if $u: C \to (-\infty, +\infty)$ is a utility function for $\mathscr{D}$ on C, then

$$c'' \begin{Bmatrix} > \\ \sim \\ < \end{Bmatrix} c' \quad \text{if and only if} \quad u(c'') \begin{Bmatrix} > \\ = \\ < \end{Bmatrix} u(c').$$

[*Hint*: $u(c) = U(\text{“}c \text{ for sure”})$.]

***Exercise 3.5.5: Ordinal Utility**

For decision making under *certainty*, the only requirement of a utility function is that it satisfy the conclusion of Exercise 3.5.4. Show that if $u(\cdot)$ satisfies Exercise 3.5.4, then so does $u^*(\cdot)$ if and only if

$$u^*(c) = T[u(c)]$$

for every c in C, where $T: \{u(c): c \in C\} \to (-\infty, +\infty)$ is some *strictly increasing* function (that is, $T[x''] > T[x']$ whenever $x'' > x'$).

■ Note that if $u(\cdot)$ and $u^*(\cdot)$ are both required to satisfy Theorem 3.2.1, which is much more stringent than Exercise 3.5.4, then Theorem 3.5.2 shows that $T[\cdot]$ must be of the more restricted form $T[x] = a^*x + b^*$ for a^* positive, a strictly increasing *linear* function. Utility functions that mirror only $\mathscr{D}$'s preferences between consequences, in the sense of Exercise 3.5.4, but do not necessarily reveal anything about $\mathscr{D}$'s preferences between nontrivial lotteries with these consequences as possible outcomes, are called *ordinal* utility functions for $\mathscr{D}$. A utility function that satisfies Theorem 3.5.1 and hence possesses the stronger properties required for Theorem 3.2.1 is often called a *cardinal* utility function, or a *von Neumann-Morgenstern* utility function; but we shall omit such qualifiers, since Theorem 3.2.1 and decision making under uncertainty are our objectives. ■

Exercise 3.5.6

Let $C = [\$0, \$1{,}000]$, with c in C denoting a potential amount of money you may receive, with no strings attached and no ethical compromises. It is reasonable to assume that you would put $c^{\mathrm{B}} = \$1{,}000$ and $c^{\mathrm{W}} = \$0$.

(A) Use the lottery-comparison method depicted in Figure 3.12 to determine $w(c)$ for:

(i) $c = \$100$
(ii) $c = \$250$
(iii) $c = \$500$
(iv) $c = \$750$
(v) $c = \$900$

(B) Name that amount of money c^p such that *you* would regard "c^p for sure" and $\ell^*(p; \$1{,}000, \$0)$ as equally desirable for:

(i) $p = 1/2$ (v) $p = 3/8$
(ii) $p = 1/4$ (vi) $p = 5/8$
(iii) $p = 3/4$ (vii) $p = 7/8$
(iv) $p = 1/8$

(C) Show that $w(c^p) = p$ for every p in $[0, 1]$, where c^p is as defined in (B).

(D) For the c^p's you determined in (B), graph the points $(\$c^p, p)$ in the plane and draw a smooth curve $\{(c, w(c)) : 0 \le c \le 1{,}000\}$ through them, through the origin $(0, 0)$, and through the point $(1{,}000, 1)$, with the additional property[h] that $w(c'') > w(c')$ whenever $c'' > c'$.

■ This exercise hints at how $\mathscr{D}$ can use some common-sense assumptions about continuity and other *qualitative* properties of his preferences to circumvent the impossibility of assessing $w(c)$ directly for every c when C is an infinite set. We introduce a number of such properties and methods in Chapter 5. ■

Exercise 3.5.7

In the dpuu in Figure 3.6, change the scale and origin of $\mathscr{D}$'s utility function so that the resulting utility function $u^*(\cdot)$ has $u^*(c^{\mathrm{B}}) = 1.0$ instead of 10.8 and $u^*(c^{\mathrm{W}}) = 0$ instead of -8.8. Recalculate all "bubbled" figures in Figure 3.6, and show that using the new utility function yields the exact same optimal lottery as did using the original utility function in your solution to Exercise 3.2.4.

3.6 QUANTIFICATION OF JUDGMENTS

The "oculist choices" which $\mathscr{D}$ makes to quantify his judgments are very similar to those he makes to quantify his preferences.

Suppose that Ω' is any event[i] in an $\mathscr{E}$-move Ω and that $\mathscr{D}$ wishes to quantify his judgments regarding the event Ω'—or, equivalently, the event that $\tilde{\omega} \in \Omega'$. As in the third step of preference quantification, $\mathscr{D}$ may make desirability comparisons between lotteries: (1) a variable canonical lottery $\ell^*(p; c^{\mathrm{B}}, c^{\mathrm{W}})$ defined exactly as before and (2) in place of "c for sure," the lottery $\ell^{\wedge}(\Omega'; c^{\mathrm{B}}, c^{\mathrm{W}})$ which by definition results in c^{B} if $\tilde{\omega} \in \Omega'$ and in c^{W} if $\tilde{\omega} \notin \Omega'$. This lottery choice is depicted in Figure 3.13.

Now, $\mathscr{D}$ may vary p in $\ell^*(p; c^{\mathrm{B}}, c^{\mathrm{W}})$ to ascertain that value $P(\Omega')$ of p for which $\ell^{\wedge}(\Omega'; c^{\mathrm{B}}, c^{\mathrm{W}})$ and $\ell^*(p; c^{\mathrm{B}}, c^{\mathrm{W}})$ are equally desirable. Exactly

---

[h] This property reflects the (reasonable) assumption that you always prefer receiving more to receiving less money.

[i] Recall that an *event in* an $\mathscr{E}$-move Ω is a *subset of* Ω.

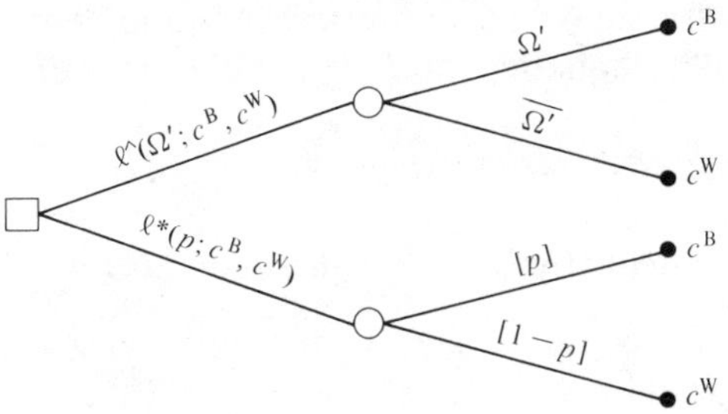

Figure 3.13 Hypothetical choice situation for judgment quantification.

the same tabular format as introduced for preference quantifications may be used.

We *define* $\mathscr{D}$'s probability $P(\Omega')$ of the event Ω' to be that p value such that

$$\ell\hat{}(\Omega'; c^{\mathrm{B}}, c^{\mathrm{W}}) \sim \ell^*(p; c^{\mathrm{B}}, c^{\mathrm{W}}).$$

■ As in the discussion of our defining $u(c)$ so that $u(c) = w(c)u(c^{\mathrm{B}}) + [1 - w(c)]u(c^{\mathrm{W}})$, we shall appeal to the goal of Theorem 3.2.1 to justify our definition of $P(\Omega')$. *If* Theorem 3.2.1 is to obtain for the simple dpuu in Figure 3.13 and *if* $\mathscr{D}$ is to assign probabilities q and $1-q$ to the respective branches Ω' and $\overline{\Omega'}$ of the $\mathscr{E}$-move of $\ell\hat{}(\Omega'; c^{\mathrm{B}}, c^{\mathrm{W}})$, then $\ell\hat{}(\Omega'; c^{\mathrm{B}}, c^{\mathrm{W}}) \sim \ell^*(P(\Omega'); c^{\mathrm{B}}, c^{\mathrm{W}})$ implies that

(i) $$U(\ell\hat{}(\Omega'; c^{\mathrm{B}}, c^{\mathrm{W}})) = U(\ell^*(P(\Omega'); c^{\mathrm{B}}, c^{\mathrm{W}})),$$

if Theorem 3.2.1 is to hold. But

(ii) $$U(\ell\hat{}(\Omega'; c^{\mathrm{B}}, c^{\mathrm{W}})) = qu(c^{\mathrm{B}}) + [1-q]u(c^{\mathrm{W}})$$

and

(iii) $$U(\ell^*(P(\Omega'); c^{\mathrm{B}}, c^{\mathrm{W}})) = P(\Omega')u(c^{\mathrm{B}}) + [1 - P(\Omega')]u(c^{\mathrm{W}}),$$

by definition of $U(\cdot)$. But (i) implies that the right-hand sides of (ii) and (iii) must be equal, and this can be true *only if*[j] $q = P(\Omega')$. Thus Theorem 3.2.1 requires that $\mathscr{D}$ should assign branch Ω' relative weight q equal to his desirability-equating canonical probability $p = P(\Omega')$. ■

But the preceding argument actually establishes *more* than we have asserted. We have argued, in fact, that $\mathscr{D}$'s probability $P(\Omega')$ should *not* depend upon the choice of reference consequences c^{B} and c^{W}, and hence that $\mathscr{D}$ may quantify his judgments about *any* event Ω' as follows.

(1) Choose two unequally desirable reference consequences, say $x =$ "receive \$1,000 with no strings attached" and $y =$ "receive \$0 with no strings attached."

---

[j] This is true because $u(c^{\mathrm{B}}) \neq u(c^{\mathrm{W}})$, by assumption.

(2) Compare $\ell^{\wedge}(\Omega'; x, y)$ with canonical lotteries of the form $\ell^*(p; x, y)$ until the desirability-equating value $P(\Omega')$ of p is found, for which

$$\ell^{\wedge}(\Omega'; x, y) \sim \ell^*(P(\Omega'); x, y).$$

For future reference, we state this result formally as Theorem 3.6.1.

Theorem 3.6.1. Let x and y be any two outcomes such that either $x > y$ or $y > x$. Let Ω' be any event. Then

$$\ell^{\wedge}(\Omega'; x, y) \sim \ell^*(p; x, y)$$

if and only if

$$\ell^{\wedge}(\Omega'; c^B, c^W) \sim \ell^*(p; c^B, c^W).$$

As a matter of convenience, $\mathscr{D}$ should select reference outcomes x and y which are easily conceptualizable and sufficiently distinct to encourage him to think *seriously* about the hypothetical choices between $\ell^{\wedge}(\Omega'; x, y)$ and $\ell^*(p; x, y)$.

■ As in his quantification of preferences, $\mathscr{D}$ may be tempted to be lazy and hence may wish to settle for two numbers, $P_L(\Omega')$ and $P_U(\Omega') > P_L(\Omega')$, such that:

(1) $\ell^{\wedge}(\Omega'; x, y) > \ell^*(p; x, y)$ whenever $p < P_L(\Omega')$;
(2) $\ell^{\wedge}(\Omega'; x, y) \sim \ell^*(p; x, y)$ whenever $P_L(\Omega') \le p \le P_U(\Omega')$; and
(3) $\ell^{\wedge}(\Omega'; x, y) < \ell^*(p; x, y)$ whenever $p > P_U(\Omega')$.

Again this temptation should be resisted, since the Monotonicity and Transitivity Principles imply that the desirability-equating p value $P(\Omega')$ should be unique. (See Exercise 3.6.2.) But here too $\mathscr{D}$ should not split hairs. ■

Example 3.6.1

Let Ω' denote the event that I will have at least one flat tire within the next year. I choose reference outcomes $x = \$1{,}000$ and $y = \$0$ and think about the state of my tires, the amount of driving I plan on, etc. Then I start comparing $\ell^{\wedge}(\Omega'; x, y)$ and $\ell^*(p; x, y)$, with the following results.

p	My choice	Next p should be
1	$\ell^*(1; x, y)$	smaller
0	$\ell^{\wedge}(\Omega'; x, y)$	larger
.80	$\ell^*(.80; x, y)$	smaller
.30	$\ell^{\wedge}(\Omega'; x, y)$	larger
.50	$\ell^*(.50; x, y)$	smaller
.40	$\ell^{\wedge}(\Omega'; x, y)$	larger
.45	$\ell^{\wedge}(\Omega'; x, y)$	larger
.48	either	$P(\Omega') = .48$

In Section 3.2 we remarked that the probabilities which $\mathcal{D}$ assigns to the branches of any $\mathcal{E}$-move should obey two stipulations. We shall now state as Theorem 3.6.2 three fundamental properties, or *laws*, which $\mathcal{D}$'s quantified judgments should obey and which *imply* the Section 3.2 stipulations. These laws of probability are usually taken as *axioms* in the formal, mathematical theory of probability, but they are implied by our definition of probability and by the consistency principles.

Theorem 3.6.2: Laws of Probability. Let Ω be an $\mathcal{E}$-move, and let Ω', Ω^1, and Ω^2 be events in Ω. Then

(1) $P(\Omega') \geq 0$ for every event Ω'.
(2) $P(\Omega^1 \cup \Omega^2) = P(\Omega^1) + P(\Omega^2)$ if Ω^1 and Ω^2 are mutually exclusive.
(3) $P(\Omega) = 1$.

Law (1) is obvious from the definition of $P(\Omega')$ and the fact that canonical probabilities are in [0, 1]. Law (3) is also obvious: Ω is the sure event, and hence, $\ell\hat{\ }(\Omega; x, y) = x = \ell^*(1; x, y)$, implying that $P(\Omega) = 1$. But Law (2) is more difficult to prove rigorously. Its proof is deferred to the beginning of Section *3.8.

■ But the essentiality of Law (2) to Theorem 3.2.1 can be readily appreciated by examining the lotteries ℓ' and ℓ'' in Figure 3.14. Both represent exactly the same strategic opportunity, and hence $U(\ell')$ should equal $U(\ell'')$. But Theorem 3.2.1 implies that

$$U(\ell') = u(x)[P(\Omega^1) + P(\Omega^2)] + u(y)P(\Omega^3),$$

whereas

$$U(\ell'') = u(x)P(\Omega^1 \cup \Omega^2) + u(y)P(\Omega^3).$$

These three equations are simultaneously valid for all x and y [and hence for all $u(x)$ and $u(y)$] if and only if $P(\Omega^1) + P(\Omega^2) = P(\Omega^1 \cup \Omega^2)$. ■

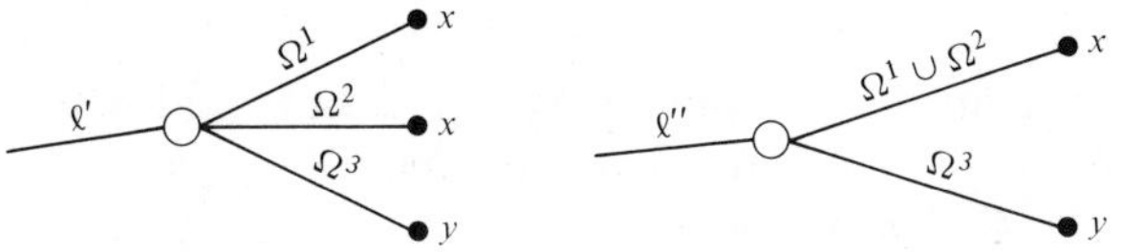

Figure 3.14 Two strategically equivalent lotteries.

Example 3.6.2

Let Ω^1 denote the event that I will have *exactly one* flat tire within the next year, and let Ω^2 denote the event that I will have *more than* one flat tire within the next year. Then Ω' in Example 3.6.1 is $\Omega^1 \cup \Omega^2$, and Law (2) requires that my probabilities of Ω^1 and Ω^2 be such that $P(\Omega^1) + P(\Omega^2) = .48$.

Stipulation 1 in Section 3.2 follows directly from the facts that canonical probabilities are in the interval [0, 1] and that every $P(\Omega_t^i)$ is a canonical probability. *Stipulation* 2 follows from the n-fold generalization of (2), which we state as property (8) in Theorem 3.6.3—that $P(\cup_{i=1}^k \Omega_t^i) = \Sigma_{i=1}^k P(\Omega_t^i)$ because $\Omega_t^1, \ldots, \Omega_t^{n_t}$ are mutually exclusive—and the fact that $P(\cup_{i=1}^k \Omega_t^i) = P(\Omega) = 1$ by property (3) and collective exhaustiveness of $\Omega_t^1, \ldots, \Omega_t^{n_t}$.

Properties (1)–(3) in Theorem 3.6.2 imply numerous other properties of probability, some of which we state as Theorem 3.6.3, the proof of which is left to you as *Exercise 3.6.3.

Theorem 3.6.3: More Properties of Probability. Let Ω be an $\mathscr{E}$-move, and let $\Omega', \Omega'', \Omega^1, \ldots, \Omega^k$ be events in Ω. Then[k]

(4) $P(\emptyset) = 0$;
(5) $P(\Omega') \le P(\Omega'')$ whenever $\Omega' \subset \Omega''$;
(6) $0 \le P(\Omega') \le 1$ for every Ω' in Ω;
(7) $P(\overline{\Omega'}) = 1 - P(\Omega')$;
(8) $P(\cup_{i=1}^k \Omega^i) = \Sigma_{i=1}^k P(\Omega^i)$ if $\Omega^1, \ldots, \Omega^k$ are mutually exclusive;
(9) $P(\cup_{i=1}^k \Omega^i) \le \Sigma_{i=1}^k P(\Omega^i)$; and
(10) $P(\cap_{i=1}^k \Omega^i) \ge (1-k) + \Sigma_{i=1}^k P(\Omega^i)$.

■ In applying the Figure 3.13 comparisons n_t times to quantify your judgments regarding each branch Ω_t^i in an $\mathscr{E}$-move Ω_t, you would most likely come up with n_t probabilities $P(\Omega_t^1), \ldots, P(\Omega_t^{n_t})$ which do *not* add up to one, as required by Stipulation 2. There is a way of assessing probabilities[l] which skirts this inconsistency: You may assess $P(\Omega_t^1)$ directly, then assess $P(\Omega_t^1 \cup \Omega_t^2)$ directly, then assess $P(\Omega_t^1 \cup \Omega_t^2 \cup \Omega_t^3)$ directly, and so forth. Then set $P(\Omega_t^2) = P(\Omega_t^1 \cup \Omega_t^2) - P(\Omega_t^1)$, $P(\Omega_t^3) = P(\Omega_t^1 \cup \Omega_t^2 \cup \Omega_t^3) - P(\Omega_t^1 \cup \Omega_t^2)$, and so forth. The details of this approach are included in Chapter 6. ■

Example 3.6.3

Let $\Omega' = \Omega^1 \cup \Omega^2$ as defined in Examples 3.6.1 and 3.6.2. If I had also assessed $P(\Omega^1) = .39$, then I could solve for $P(\Omega^2)$ by noting that $P(\Omega^1 \cup \Omega^2) = .48 = .39 + P(\Omega^2)$, whence $P(\Omega^2) = .09$.

$\mathscr{D}$'s probabilities, or quantified judgments, are the result of his making hypothetical decisions. If $\mathscr{D}$ is to behave in accordance with our consistency principles, his probabilities must satisfy all parts of Theorems 3.6.2 and 3.6.3. In the following section we introduce other properties that his probabilities should possess if he is to satisfy the consistency principles.

---

[k] Recall that $\overline{\Omega'}$ denotes the *complement* of Ω'; that is, $\omega \in \overline{\Omega'}$ if and only if $\omega \notin \Omega'$.

[l] "Quantifying judgments" is often referred to as "assessing probabilities." Similarly, "assessing utilities" means "quantifying preferences."

Exercise 3.6.1

Define Ω', Ω^1, and Ω^2 as in Examples 3.6.1 and 3.6.2 but for *your* car. Assess *your* probabilities of Ω^1 and of Ω^2. [If you do not have a car, your motorcycle or bicycle will do. If you have none of these, replace "flat tire" with "loose horseshoe."]

Exercise 3.6.2

Show that if $\ell^{\wedge}(\Omega'; c^B, c^W) \sim \ell^*(p^1; c^B, c^W)$ and $\ell^{\wedge}(\Omega'; c^B, c^W) \sim \ell^*(p^2; c^B, c^W)$, then the Monotonicity and Transitivity Principles imply that $p^1 = p^2$.

***Exercise 3.6.3**

Derive Theorem 3.6.3 from only Theorem 3.6.2 and the rules of ordinary logic.

Exercise 3.6.4

Let Ω^x denote the event that, one year from today, the closing value of the Dow-Jones Industrial Average will be no greater than x percent of its closing value yesterday.

(A) Assess your $P(\Omega^x)$ for the following values of x.

(i) $x = 70\%$	(iv) $x = 100\%$	(vii) $x = 130\%$
(ii) $x = 80\%$	(v) $x = 110\%$	(viii) $x = 140\%$
(iii) $x = 90\%$	(vi) $x = 120\%$	

(B) Plot the pairs $(x, P(\Omega^x))$ on a graph, and sketch a smooth (= continuous) function $\{(x, F(x)): 0 \leq x \leq 140\}$ through these points, with $F(x)$ having the two properties

(I) $F(0) = 0$, and
(II) $F(x') \leq F(x'')$ whenever $x' < x''$.

(C) Does $F(125)$ adequately approximate your judgments regarding the event Ω^{125}?

(D) Does $F(125) - F(100)$ adequately approximate your judgments regarding the event $\Omega^{125} \cap \overline{\Omega^{100}}$ (i.e., the event that the DJIA will not fall, but rather will rise by not more than 25 percent)?

■ This exercise is analogous to Exercise 3.5.6 in introducing a method for dealing in a reasonable fashion with infinite idealizations. See also Part C of this chapter and Chapter 6. ■

Exercise 3.6.5: Extending a Probability Consistently to a Superset

Suppose that P is a probability on subsets of Ω and that Ω is a subset of $\Omega^{\uparrow}$—equivalently, $\Omega^{\uparrow}$ is a superset of Ω. Define a function $P^{\uparrow}$ on subsets of $\Omega^{\uparrow}$ by

$$P^{\uparrow}(\Omega') = P(\Omega' \cap \Omega) \qquad \text{for every } \Omega' \subset \Omega^{\uparrow}.$$

(A) Show that $P^{\uparrow}(\Omega') = P(\Omega')$ whenever $\Omega' \subset \Omega$.
(B) Show that $P^{\uparrow}$, regarded as a function on all events in $\Omega^{\uparrow}$, is a probability [that is, $P^{\uparrow}$ possesses properties (1)–(3) in Theorem 3.6.2].

3.7 COMBINING SUCCESSIVE $\mathscr{E}$-MOVES AND CONDITIONAL PROBABILITY

In Section 3.1 we noted that there are several strategically equivalent ways of representing the same lottery. In particular, any "compound," two-stage lottery such as ℓ in Figure 3.4 can be represented as a one-stage lottery by combining its successive $\mathscr{E}$-moves. (See Section 2.11.)

Example 3.7.1

Combining the successive $\mathscr{E}$-moves in the lottery ℓ of Figure 3.4 produces a lottery with $\mathscr{E}$-move $\{(\omega^i, \xi_i^j): j = 1, \ldots, m_i; i = 1, \ldots, n\}$, resulting in consequence c^{ij} if $(\tilde{\omega}, \tilde{\xi}) = (\omega^i, \xi_i^j)$.

But if we combine successive $\mathscr{E}$-moves in a lottery, what relationship should the probabilities on the combined branches have to the probabilities on the original, constituent branches in the "compound" representation? The answer, Theorem 3.7.1, is that the probability on a combined branch should be the *product* of the probabilities of its constituent branches.

Example 3.7.2

In the context of Example 3.7.1 and Figure 3.4, $\mathscr{D}$'s probability $P_{12}(\omega^i, \xi_i^j)$ of combined branch (ω^i, ξ_i^j) should equal $P_{2|1}(\xi_i^j|\omega^i)P_1(\omega^i)$, for every i and j.

Theorem 3.7.1. Let $\Omega_1{}^i$ be an event branch in an $\mathscr{E}$-move Ω_1; let Ω_2 be an $\mathscr{E}$-move attached to branch $\Omega_1{}^i$; and let $\Omega_2{}^j$ be an event in Ω_2. Then $\mathscr{D}$'s probabilities $P_{2|1}(\Omega_2{}^j|\Omega_1{}^i)$, $P_1(\Omega_1{}^i)$, and $P_{12}(\Omega_1{}^i, \Omega_2{}^j)$ of Ω_2^j, $\Omega_1{}^i$, and the combined event $(\Omega_1{}^i, \Omega_2{}^j)$ respectively should satisfy

$$P_{12}(\Omega_1{}^i, \Omega_2{}^j) = P_{2|1}(\Omega_2{}^j|\Omega_1{}^i)P_1(\Omega_1{}^i).$$

Figure 3.15 is a graphic portrayal of Theorem 3.7.1.

■ It is easy to see that Theorem 3.7.1 is a *necessary* condition for Theorem 3.2.2. Lotteries ℓ' and ℓ'' in Figure 3.16 obviously represent exactly the same opportunity to $\mathscr{D}$, and hence should have the same relative-desirability index U. Hence $U(\ell') = U(\ell'')$. Now suppose (without loss of generality) that $u(c^{\mathrm{B}}) = 1$ and $u(c^{\mathrm{W}}) = 0$. Then $U(\ell'') = P_{12}(\Omega_1{}^i, \Omega_2{}^j)$, whereas, by Theorem 3.2.2, $U(\ell') = U(\ell\dagger)P_1(\Omega_1{}^i) =$

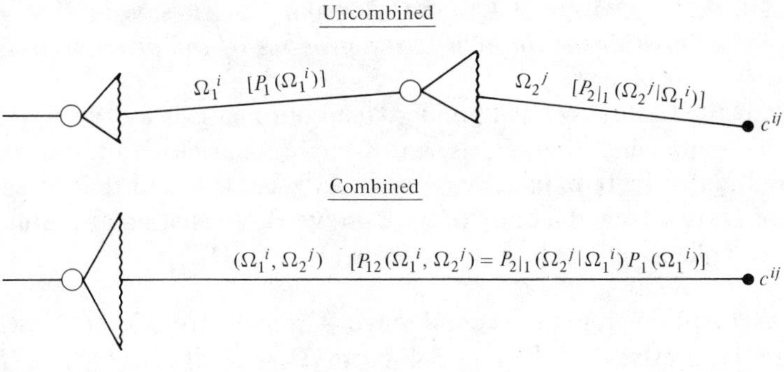

Figure 3.15 Visualization of Theorem 3.7.1.

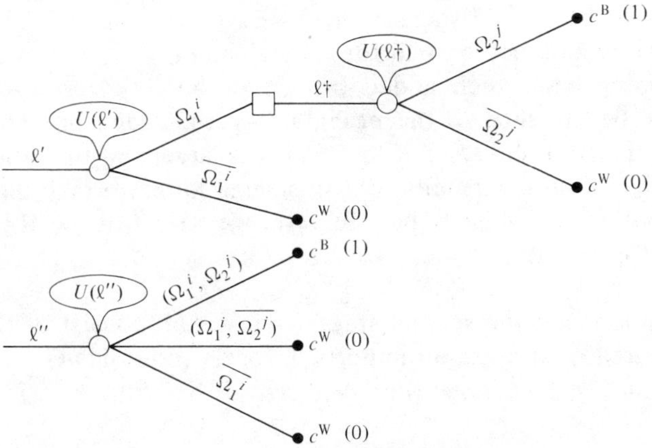

Figure 3.16 Lotteries motivating Theorem 3.7.1.

$P_{2|1}(\Omega_2^j|\Omega_1^i)P_1(\Omega_1^i)$. Hence $P_{12}(\Omega_1^i, \Omega_2^j) = P_{2|1}(\Omega_2^j|\Omega_1^i)P_1(\Omega_1^i)$. But this argument assumes Theorem 3.2.2 to be true, whereas in fact we need Theorem 3.7.1 to prove Theorem 3.2.2. An argument that avoids this circularity will be given in *Section 3.8. ■

A useful corollary of Theorem 3.7.1, called the "chain rule," is obtained by induction and extends Theorem 3.7.1 to more than two successive $\mathscr{E}$-moves. (See *Exercise 3.7.8.)

Corollary 3.7.1: Chain Rule. Let $\Omega_1^{i_1}$ be an event branch with probability $P_1(\Omega_1^{i_1})$, and suppose that for $j = 2, \ldots, n$ an event branch $\Omega_j^{i_j}$ is attached to branch $\Omega_{j-1}^{i_{j-1}}$ and given probability $P_{j|1\cdots j-1}(\Omega_j^{i_j}|\Omega_1^{i_1}, \ldots, \Omega_{j-1}^{i_{j-1}})$. Then $\mathscr{D}$'s probability $P_{1\cdots n}(\Omega_1^{i_1}, \ldots, \Omega_n^{i_n})$ should satisfy

$$P_{1\cdots n}(\Omega_1^{i_1}, \ldots, \Omega_n^{i_n}) = P_1(\Omega_1^{i_1})P_{2|1}(\Omega_2^{i_2}|\Omega_1^{i_1}) \cdot \cdots \cdot P_{n|1\cdots n-1}(\Omega_n^{i_n}|\Omega_1^{i_1}, \ldots, \Omega_{n-1}^{i_{n-1}})$$

$$= {}^{\mathrm{m}}P_1(\Omega_1^{i_1}) \prod_{j=2}^{n} P_{j|1\cdots j-1}(\Omega_j^{i_j}|\Omega_1^{i_1}, \ldots, \Omega_{j-1}^{i_{j-1}}).$$

[m] Recall the capital pi, product notation: by definition, $\Pi_{i=1}^{n} a_i = a_1 a_2 \cdots a_n$.

In fact, all that Theorem 3.7.1 and Corollary 3.7.1 say is that *the probability of a "combined" branch is the product of the probabilities of its "constituent" branches.*

So much for the basics. We shall now extend our analysis a little further. We begin by supposing that Ω_1 is an $\mathscr{E}$-move consisting of mutually exclusive and collectively exhaustive events $\Omega_1{}^1, \ldots, \Omega_1{}^n$, and that to each branch $\Omega_1{}^i$ of Ω_1 is attached a copy of an $\mathscr{E}$-move Ω_2 consisting of mutually exclusive and collectively exhaustive events $\Omega_2{}^1, \ldots, \Omega_2{}^m$.

*■ The assumption that the second-stage $\mathscr{E}$-moves are all the same is not really restrictive. In Figure 3.4 (with $\Omega_1 = \Omega$ and $\Omega_1{}^i = \omega^i$), for example, let $\Omega_2 = \bigcup_{i=1}^n \Xi_i = \{\xi_i^j : j = 1, \ldots, m_i;\ i = 1, \ldots, n\}$. Then Ω_2 may consist of fewer than $\Sigma_{i=1}^n m_i$ elements because the same event may appear in several of the second-stage $\mathscr{E}$-moves Ξ_i. Now "enlarge" each Ξ_i to all of Ω_2: (1) by adding a branch for each element of Ω_2 *not* in Ξ_i, and (2) by assigning each such added branch probability zero. *Any* consequence may be attached to the probability-zero, added branches without affecting the desirability of ℓ to $\mathscr{D}$. This enlargement operation on the second-stage $\mathscr{E}$-moves results in a two-stage lottery which does satisfy the assumption introduced above. (Here, we would define $\Omega_2{}^j = \{\omega_2{}^j\}$ for every $\omega_2{}^j$ in Ω_2.) ■

Given our assumption that the second-stage $\mathscr{E}$-moves all consist of the same events, we are ready to make an important observation; namely, *the ordering of the first and second stages may be reversed*! See Figure 3.17; in

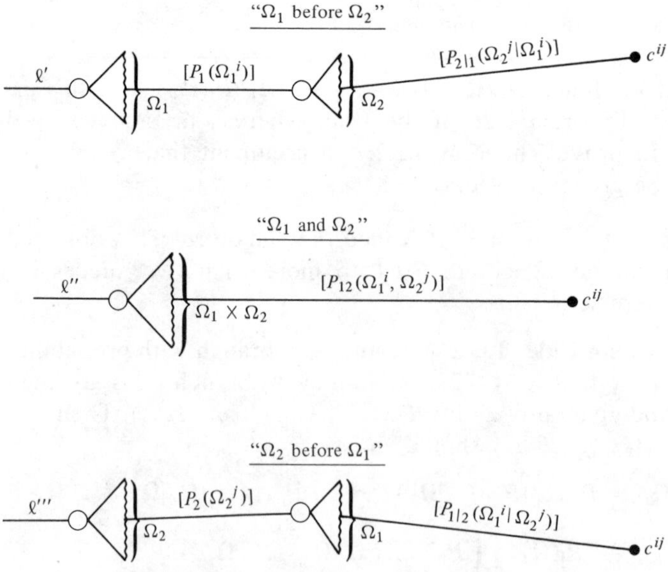

Figure 3.17 Reversal of order in a compound lottery.

all three representations ℓ', ℓ'', and ℓ''' of a lottery, 𝒟's consequence is c^{ij} if both $\Omega_1{}^i$ and $\Omega_2{}^j$ obtain, for $i = 1, \ldots, n$ and $j = 1, \ldots, m$—regardless of which comes first. That is, ℓ' and ℓ''' are equally good alternatives to the one-stage-lottery representation ℓ''.

■ The probability notation for the "Ω_2-before-Ω_1" lottery, ℓ''', is completely analogous to that previously introduced for the "Ω_1-before-Ω_2" lottery, ℓ'. ■

There are several analytic relations among the probabilities appearing in Figure 3.17. They will be introduced and derived following our definition of some important probability terminology.

Definition 3.7.1: Joint, Conditional, and Marginal Probabilities

Let $\Omega_1 \times \Omega_2$ be an ℰ-move consisting of mutually exclusive and collectively exhaustive events $(\Omega_1{}^i, \Omega_2{}^j)$, where $\Omega_1{}^1, \ldots, \Omega_1{}^n$ are mutually exclusive and collectively exhaustive in Ω_1 and $\Omega_2{}^1, \ldots, \Omega_2{}^m$ are mutually exclusive and collectively exhaustive in Ω_2. Refer to Figure 3.17.

(A) $P_{12}(\Omega_1{}^i, \Omega_2{}^j)$ is called the *joint probability of* $\Omega_1{}^i$ *and* $\Omega_2{}^j$;
(B) $P_{2|1}(\Omega_2{}^j|\Omega_1{}^i)$ and $P_{1|2}(\Omega_1{}^i|\Omega_2{}^j)$ are called respectively the *conditional probability of* $\Omega_2{}^j$ *given* $\Omega_1{}^i$ and the *conditional probability of* $\Omega_1{}^i$ *given* $\Omega_2{}^j$; and
(C) $P_1(\Omega_1{}^i)$ and $P_2(\Omega_2{}^j)$ are called respectively the *marginal* (or *unconditional*) *probability of* $\Omega_1{}^i$ and the *marginal* (or *unconditional*) *probability of* $\Omega_2{}^j$.

■ The meaning of the adjectives "conditional" and "joint" in Definition 3.7.1 is evident from Figure 3.17. The meaning of "marginal" will be clarified shortly. ■

We now state as Theorem 3.7.2 some important relations among the probabilities in Figure 3.17. Part (B) is the famous "Bayes' Theorem," which plays a significant role in that special class of dpuu's involving the acquisition of information. After stating, proving, and exemplifying the numerical application of Theorem 3.7.2, we shall briefly discuss information-acquisition decisions and the role of Theorem 3.7.2 in them. Such decisions are introduced in much greater depth in Chapter 8.

Theorem 3.7.2: Some Probability Relationships

Let $\Omega_1 \times \Omega_2$ be as in Definition 3.7.1. Then

(A) $$P_2(\Omega_2{}^j) = \sum_{i=1}^{n} P_{12}(\Omega_1{}^i, \Omega_2{}^j) \qquad \text{for } j = 1, \ldots, m;$$

and

(B) $$P_{1|2}(\Omega_1{}^i|\Omega_2{}^j) = \frac{P_{12}(\Omega_1{}^i, \Omega_2{}^j)}{P_2(\Omega_2{}^j)} = \frac{P_{2|1}(\Omega_2{}^j|\Omega_1{}^i)P_1(\Omega_1{}^i)}{\sum_{q=1}^{n} P_{2|1}(\Omega_2{}^j|\Omega_1{}^q)P_1(\Omega_1{}^q)}$$

for $i = 1, \ldots, n$ and for every j such that the denominator is positive.

■ *Proof.* To prove (A), note that the event that $\Omega_2{}^j$ obtains is the same as the event that $\Omega_1{}^1$ and $\Omega_2{}^j$, *or* $\Omega_1{}^2$ and $\Omega_2{}^j$, *or* $\cdots$ *or* $\Omega_1{}^n$ and $\Omega_2{}^j$ obtains. But this event is just $\bigcup_{i=1}^n (\Omega_1{}^i, \Omega_2{}^j)$. Hence $P_2(\Omega_2{}^j) = P_{12}(\bigcup_{i=1}^n (\Omega_1{}^i, \Omega_2{}^j))$. But since $(\Omega_1{}^1, \Omega_2{}^j), \ldots, (\Omega_1{}^n, \Omega_2{}^j)$ are mutually exclusive, property (8) of probability in Theorem 3.6.3 implies that $P_{12}(\bigcup_{i=1}^n(\Omega_1{}^i, \Omega_2{}^j)) = \Sigma_{i=1}^n P_{12}(\Omega_1{}^i, \Omega_2{}^j)$. Hence we obtain (A). To obtain (B), we note that Theorem 3.7.1 implies that (*) $P_{12}(\Omega_1{}^i, \Omega_2{}^j) = P_{1|2}(\Omega_1{}^i|\Omega_2{}^j)P_2(\Omega_2{}^j)$ when applied to the "Ω_2-before-Ω_1" lottery. Dividing both sides of (*) by $P_2(\Omega_2{}^j)$ is permissible if $P_2(\Omega_2{}^j)$ is positive and results in the first equality of (B). But applying Theorem 3.7.1 to the "Ω_1-before-Ω_2" lottery yields (**) $P_{12}(\Omega_1{}^i, \Omega_2{}^j) = P_{2|1}(\Omega_2{}^j|\Omega_1{}^i)P_1(\Omega_1{}^i)$, the numerator of the second equality in (B). Now change i to q and substitute the right-hand side of (**) into (A) to obtain the denominator of the second equality in (B) from the denominator of the first. Thus both numerators are equal, and so are both denominators. ■

■ By reversing the roles of 1 and 2, of $\Omega_1{}^i$ and $\Omega_2{}^j$, etc., we obtain

(A*) $$P_1(\Omega_1{}^i) = \sum_{j=1}^{m} P_{12}(\Omega_1{}^i, \Omega_2{}^j) \qquad \text{for } i = 1, \ldots, n,$$

and

(B*) $$P_{2|1}(\Omega_2{}^j|\Omega_1{}^i) = \frac{P_{12}(\Omega_1{}^i, \Omega_2{}^j)}{P_1(\Omega_1{}^i)} = \frac{P_{1|2}(\Omega_1{}^i|\Omega_2{}^j)P_2(\Omega_2{}^j)}{\sum_{k=1}^{m} P_{1|2}(\Omega_1{}^i|\Omega_2{}^k)P_2(\Omega_2{}^k)}$$

for $j = 1, \ldots, m$ and for every i such that the denominator is positive. ■

Example 3.7.3

The joint probabilities $P_{12}(\Omega_1{}^i, \Omega_2{}^j)$ are often arranged in tabular fashion, with rows $\Omega_1{}^i$, columns $\Omega_2{}^j$, and table entries $P_{12}(\Omega_1{}^i, \Omega_2{}^j)$. For example, with $n = 2$ and $m = 3$ we might have the following joint probabilities.

$P_{12}(\Omega_1{}^i, \Omega_2{}^j)$

	$j = 1$	2	3	$\Sigma = P_1(\Omega_1{}^i)$
$i = 1$	.12	.48	.00	.60
2	.32	.04	.04	.40
$\Sigma = P_2(\Omega_2{}^j)$	.44	.52	.04	= 1.00

Here we have applied (A*) in summing each row i to obtain $P_1(\Omega_1{}^i)$ and (A) in summing each column j to obtain $P_2(\Omega_2{}^j)$. Since these probabilities appear in the right-hand and lower *margins* of the joint-probability table respectively, we see the justification of the adjective "marginal".

Example 3.7.4

Both sets of conditional probabilities are readily obtainable from the table of joint probabilities. They too can be recorded in tabular fashion. From the joint and marginal probabilities tabulated in Example 3.7.3, we readily obtain

$P_{1|2}(\Omega_1{}^i|\Omega_2{}^j)$

	$j=1$	2	3
$i=1$	12/44	48/52	0/4
2	32/44	4/52	4/4
$=1$	1	1	1

from (B) and

$P_{2|1}(\Omega_2{}^j|\Omega_1{}^i)$

	$j=1$	2	3	$=1$
$i=1$	12/60	48/60	0/60	1
2	32/40	4/40	4/40	1

from (B*).

As we mentioned above, Theorem 3.7.2 is extremely useful in the special types of dpuu's involving 𝒟's contemplated acquisition of information, say Ω_2, to improve a decision he must make which hinges upon, say, Ω_1. The events $\Omega_2{}^j$ represent the possible results of an experiment, readings on an imperfectly reliable instrument, etc. In such a situation, 𝒟's judgments concerning the decision-relevant ℰ-move Ω_1 *prior* to performing the experiment are represented by the $P_1(\Omega_1{}^i)$'s, whereas his judgments *posterior* to performing the experiment and observing its outcome $\Omega_2{}^j$ are represented by the $P_{1|2}(\Omega_1{}^i|\Omega_2{}^j)$'s.

Now *all* the relevant probability relations in such a case are implied (via Theorem 3.7.2) by 𝒟's *joint* probabilities $P_{12}(\Omega_1{}^i, \Omega_2{}^j)$, and hence one of his initial tasks is to obtain these joint probabilities. Frequently he will find it most *natural* to assess two sets of probabilities: all marginal probabilities $P_1(\Omega_1{}^i)$ of the Ω_1-events; and all conditional probabilities $P_{2|1}(\Omega_2{}^j|\Omega_1{}^i)$ of the Ω_2 events given each of the possible Ω_1-events. After having made these assessments, it is a simple matter for 𝒟 to apply Theorem 3.7.1 and thus calculate all the needed joint probabilities by straightforward multiplication. Then he can apply (A) to obtain his marginal probabilities of the Ω_2-events and (B) to obtain his conditional probabilities of the decision-relevant Ω_1-events, given each of the informational Ω_2-events. Applying (A) and (B) may appear too mechanical for prudence, but *these formulae indicate what $\mathscr{D}$'s judgments must be if $\mathscr{D}$ is to abide by the consistency principles*; they are logically necessary deductions from Theorem 3.7.1, which itself follows from the consistency principles.

Example 3.7.5

Suppose that $n = 3$, $m = 2$, and $\mathscr{D}$ has assessed probabilities $P_1(\Omega_1{}^1) = .3$, $P_1(\Omega_1{}^2) = .6$, and $P_1(\Omega_1{}^3) = .1$ of the Ω_1-events, as well as the following, tabulated conditional probabilities of the Ω_2-events given each of the Ω_1-events.

	$P_{2\|1}(\Omega_2{}^j\|\Omega_1{}^i)$			[reference:
	$j = 1$	2	$\Sigma = 1$	$P_1(\Omega_1{}^i)$]
$i = 1$	.80	.20	1.00	[.3]
2	.50	.50	1.00	[.6]
3	.40	.60	1.00	[.1]

By multiplying each row i of this table by $P_1(\Omega_1{}^i)$, we readily obtain the following table of joint (and marginal) probabilities.

	$P_{12}(\Omega_1{}^i, \Omega_2{}^j)$		
	$j = 1$	2	$\Sigma = P_1(\Omega_1{}^i)$
$i = 1$	.24	.06	.30
2	.30	.30	.60
3	.04	.06	.10
$\Sigma = P_2(\Omega_2{}^j)$	.58	.42	1.00

By dividing each column j of *this* table by its column total $P_2(\Omega_2{}^j)$, we readily obtain the following table of conditional probabilities of Ω_1-events given each of the possible, informational Ω_2-events.

	$P_{1\|2}(\Omega_1{}^i\|\Omega_2{}^j)$	
	$j = 1$	2
$i = 1$	24/58	6/42
2	30/58	30/42
3	4/58	6/42
$\Sigma = 1$	1	1

Finally, we compare how $\mathscr{D}$'s judgments about each of the Ω_1^i's are affected by each of the Ω_2^j's, by expressing all $P_{1|2}$'s and P_1's in decimals.

$\mathscr{D}$'s Judgments about Ω_1^i

	(a) Given Ω_2^1	(b) Prior	(c) Given Ω_2^2
$i = 1$	.414	.300	.143
2	.517	.600	.714
3	.069	.100	.143

It is clear that observing Ω_2^1 should increase $\mathscr{D}$'s probability of Ω_1^1 at the expense of Ω_1^2 and Ω_1^3, whereas the reverse is true of observing Ω_2^2.

Two final examples introduce opposite ends of a spectrum of "informativeness" of Ω_2 about Ω_1.

Example 3.7.6: Independence, or Useless Information

Suppose that $\mathscr{D}$'s conditional probabilities of the Ω_2-events, given each of the Ω_1-events, do not actually depend upon the Ω_1-event, in the sense that

$$P_{2|1}(\Omega_2^j|\Omega_1^i) = q^j \qquad \text{for } i = 1, \ldots, n \text{ and } j = 1, \ldots, m,$$

where $q^1, \ldots, q^m$ are m nonnegative numbers. Under this assumption, you may show as Exercise 3.7.9 that

$$(*) \qquad q^j = P_2(\Omega_2^j) \qquad \text{for } j = 1, \ldots, m;$$

and

$$(**) \qquad P_{1|2}(\Omega_1^i|\Omega_2^j) = P_1(\Omega_1^i) \qquad \text{for } i = 1, \ldots, n \text{ and } j = 1, \ldots, m.$$

That is, observing Ω_2^j cannot change any of $\mathscr{D}$'s "prior" judgments $P_1(\Omega_1^i)$. Therefore, Ω_2-information is useless in a decision hinging upon the Ω_1-event.

Example 3.7.7: Perfect Information

Suppose that $m = n$ and that the Ω_2-events have been so labeled that $\mathscr{D}$'s conditional probabilities satisfy

$$P_{2|1}(\Omega_2^j|\Omega_1^i) = \begin{cases} 1, & j = i \\ 0, & j \neq i. \end{cases}$$

Under this assumption, you may show as Exercise 3.7.10 that

$$(\dagger) \qquad P_2(\Omega_2^j) = P_1(\Omega_1^j) \qquad \text{for } j = 1, \ldots, m$$

and

$$(\dagger\dagger) \qquad P_{1|2}(\Omega_1{}^i|\Omega_2{}^j) = \begin{cases} 1, & i = j \\ 0, & i \neq j \end{cases}$$

for $i = 1, \ldots, m$ and $j = 1, \ldots, m$. Therefore, observing the Ω_2-event is sure to eliminate all of $\mathscr{D}$'s uncertainty about which Ω_1-event will obtain, and hence Ω_2-information is perfectly informative about Ω_1.

Exercise 3.7.1

In Figure 3.6, reexpress each of ℓ^1 and ℓ^2 as a one-stage lottery, by combining successive $\mathscr{E}$-moves to the fullest extent possible and attaching the appropriate probability to each "combined" branch.

Exercise 3.7.2

Suppose that $n = 4$, that $m = 3$, and that $\mathscr{D}$ has quantified his judgments in the form of $P_1(\Omega_1{}^1) = .1$, $P_1(\Omega_1{}^2) = .3$, $P_1(\Omega_1{}^3) = .1$, and $P_1(\Omega_1{}^4) = .5$, and $\{P_{2|1}(\Omega_2{}^j|\Omega_1{}^i)$: all $i, j\}$ as in the following table.

$P_{2|1}(\Omega_2{}^j|\Omega_1{}^i)$

	$j = 1$	2	3	$\Sigma = 1$
$i = 1$	.75	.15	.10	1.00
2	.20	.60	.20	1.00
3	.10	.80	.10	1.00
4	.15	.10	.75	1.00

(A) Tabulate all joint probabilities $P_{12}(\Omega_1{}^i, \Omega_2{}^j)$.
(B) Calculate all marginal probabilities $P_2(\Omega_2{}^j)$.
(C) Calculate and tabulate all conditional probabilities $P_{1|2}(\Omega_1{}^i|\Omega_2{}^j)$.
(D) Repeat (A)–(C) under the assumption that $P_1(\Omega_1{}^i) = 1/4$ for $i = 1, 2, 3, 4$.
(E) Repeat (A)–(C) under the assumption that $P_1(\Omega_1{}^3) = 1$ and $P_1(\Omega_1{}^i) = 0$ for $i \neq 3$.
(F) What effect do the $P_1(\Omega_1{}^i)$'s appear to have on the $P_{1|2}(\Omega_1{}^i|\Omega_2{}^j)$'s?

Exercise 3.7.3: Variants of Bayes' Theorem

Suppose for convenience that $\Omega_1 = \{\omega_1{}^1, \ldots, \omega_1{}^n\}$, $\Omega_2 = \{\omega_2{}^1, \ldots, \omega_2{}^m\}$, $\Omega_1{}^i = \omega_1{}^i$ for every i, and $\Omega_2{}^j = \omega_2{}^j$ for every j. Consider each of $P_1(\cdot)$, $P_{2|1}(\omega_2{}^j|\cdot)$ for each j, and $P_{1|2}(\cdot|\omega_2{}^j)$ for each j as a real-valued function on Ω_1.

(A) Show that, given $\omega_2{}^j$,

$$(*) \qquad P_{1|2}(\cdot|\omega_2{}^j) \propto P_{2|1}(\omega_2{}^j|\cdot)P_1(\cdot)$$

provided that there is some $\omega_1{}^i$ such that $P_{2|1}(\omega_2{}^j|\omega_1{}^i)P_1(\omega_1{}^i)>0$. [That is, show that there is some (positive) real number $K(j)$—not dependent upon i—such that

$$P_{1|2}(\omega_1{}^i|\omega_2{}^j) = K(j)P_{2|1}(\omega_2{}^j|\omega_1{}^i)P_1(\omega_1{}^i)$$

for every i.]

(B) Show that

(**) $$O_{1|2}(\omega_1{}^i : \omega_1{}^k|\omega_2{}^j) = O^*(\omega_1{}^i : \omega_1{}^k|\omega_2{}^j)O_1(\omega_1{}^i : \omega_1{}^k),$$

provided that O^* is defined (not of the form 0/0) and $0 \cdot \infty = 0$ by agreement, where

$$O_1(\omega_1{}^i : \omega_1{}^k) = P_1(\omega_1{}^i)/P_1(\omega_1{}^k)$$

= *unconditional* relative odds of $\omega_1{}^i$ to $\omega_1{}^k$;

$$O_{1|2}(\omega_1{}^i : \omega_1{}^k|\omega_2{}^j) = P_{1|2}(\omega_1{}^i|\omega_2{}^j)/P_{1|2}(\omega_1{}^k|\omega_2{}^j)$$

= *conditional* relative odds of $\omega_1{}^i$ to $\omega_1{}^k$ given $\omega_2{}^j$; and

$$O^*(\omega_1{}^i : \omega_1{}^k|\omega_2{}^j) = P_{2|1}(\omega_2{}^j|\omega_1{}^i)/P_{2|1}(\omega_2{}^j|\omega_1{}^k)$$

= *likelihood* relative odds of $\omega_1{}^i$ to $\omega_1{}^k$, given $\omega_2{}^j$.

[Note that $O^*(\omega_1{}^i : \omega_1{}^k|\omega_2{}^j)$ denotes how many times as likely $\omega_1{}^i$ makes $\omega_2{}^j$ as does $\omega_1{}^k$. It is undefined if $P_{2|1}(\omega_2{}^j|\omega_1{}^i) = P_{2|1}(\omega_2{}^j|\omega_1{}^k) = 0$.]

Exercise 3.7.4: Properties of Expectation (I)

The probability-weighted averaging operation on numerical functions of random variables that we introduced in Section 3.2 and called *expectation* has a number of important properties. Let $\Omega = \{\omega^1, \ldots, \omega^n\}$ be an $\mathscr{E}$-move, let $P(\cdot)$ be $\mathscr{D}$'s probability function on events in Ω, and let $x: \Omega \to (-\infty, +\infty)$ be a real-valued function on Ω. By definition,

$$E\{x(\tilde{\omega})\} = \sum_{j=1}^{n} x(\omega^j)P(\omega^j).$$

Let $y: \Omega \to (-\infty, +\infty)$ be another real-valued function on Ω, and let k be any real number. Prove the following facts about $E\{\cdot\}$.

(A) $E\{x(\tilde{\omega})\} = k$ if $x(\omega) = k$ for every ω in Ω.
(B) $E\{x(\tilde{\omega})\} \geq 0$ if $x(\omega) \geq 0$ for every ω in Ω.
(C) $E\{kx(\tilde{\omega})\} = kE\{x(\tilde{\omega})\}$.
(D) $E\{x(\tilde{\omega}) + y(\tilde{\omega})\} = E\{x(\tilde{\omega})\} + E\{y(\tilde{\omega})\}$.
(E) $E\{x(\tilde{\omega})\} \geq E\{y(\tilde{\omega})\}$ if $x(\omega) \geq y(\omega)$ for every ω in Ω.

Exercise 3.7.5: Properties of Expectation (II)

Suppose that $k_1, \ldots, k_r$ are constants and that $x_1(\cdot), \ldots, x_r(\cdot)$ are real-valued functions on Ω. Define $[\min_i x_i](\cdot)$ and $[\max_i x_i](\cdot)$ on Ω by

$$[\min_i x_i](\omega) = \min [x_\ell(\omega), \ldots, x_r(\omega)]$$

and

$$[\max_i x_i](\omega) = \max [x_1(\omega), \ldots, x_r(\omega)]$$

for every ω in Ω. Show that

(A) $$E\left\{\sum_{i=1}^{r} k_i x_i(\tilde{\omega})\right\} = \sum_{i=1}^{r} k_i E\{x_i(\tilde{\omega})\}.$$

(B) $$E\{[\min_i x_i](\tilde{\omega})\} \leq \min_i [E\{x_i(\tilde{\omega})\}].$$

(C) $$E\{[\max_i x_i](\tilde{\omega})\} \geq \max_i [E\{x_i(\tilde{\omega})\}].$$

Exercise 3.7.6: Properties of Expectation (III)

Suppose that $\Omega_1 = \{\omega_1{}^1, \ldots, \omega_1{}^n\}$ and $\Omega_2 = \{\omega_2{}^1, \ldots, \omega_2{}^m\}$ are $\mathscr{E}$-moves and that $x(\cdot, \cdot)$ is a real-valued function on $\Omega_1 \times \Omega_2$. Now there are three types of expectations which turn out to be closely related; namely,

(A) $$E_{2|1}\{x(\omega_1{}^i, \tilde{\omega}_2)|\omega_1{}^i\} = \sum_{j=1}^{m} x(\omega_1{}^i, \omega_2{}^j) P_{2|1}(\omega_2{}^j|\omega_1{}^i)$$

for each $\omega_1{}^i$ in Ω_1;

(B) $$E_{12}\{x(\tilde{\omega}_1, \tilde{\omega}_2)\} = \sum \{x(\omega_1{}^i, \omega_2{}^j) P_{12}(\omega_1{}^i, \omega_2{}^j)\text{: all } i \text{ and } j\};$$

and

(C) $$E_1\{E_{2|1}\{x(\tilde{\omega}_1, \tilde{\omega}_2)|\tilde{\omega}_1\}\} = \sum_{i=1}^{n} E_{2|1}\{x(\omega_1{}^i, \tilde{\omega}_2)|\omega_1{}^i\} P_1(\omega_1{}^i).$$

Show that (B) = (C); that is, prove

(D) $$E_{12}\{x(\tilde{\omega}_1, \tilde{\omega}_2)\} = E_1\{E_{2|1}\{x(\tilde{\omega}_1, \tilde{\omega}_2)|\tilde{\omega}_1\}\}.$$

[Note in (c) that the outside expectation is of the *function*

$$E_{2|1}\{x(\cdot, \tilde{\omega}_2)|\}\colon \Omega_1 \to (-\infty, +\infty),$$

and hence (c) makes sense according to the definition of $E\{\cdot\}$ in Exercise 3.7.4.]

Exercise 3.7.7: Properties of Expectation (IV)

There is an important alternative to (d) in Exercise 3.7.6 for the special case in which $x(\cdot, \cdot)$ does not depend upon ω_1, that is, in which there exists a function $y\colon \Omega_2 \to (-\infty, +\infty)$ such that $x(\omega_1, \omega_2) = y(\omega_2)$ for every (ω_1, ω_2) in $\Omega_1 \times \Omega_2$. Show that

(E) $$E_2\{y(\tilde{\omega}_2)\} = E_1\{E_{2|1}\{y(\tilde{\omega}_2)|\tilde{\omega}_1\}\}.$$

***Exercise 3.7.8**

Prove Corollary 3.7.1, assuming the truth of Theorem 3.7.1. [*Hint*: Use finite induction.]

Exercise 3.7.9

Prove (*) and (**) in Example 3.7.6.

Exercise 3.7.10

Prove (†) and (††) in Example 3.7.7.

Exercise 3.7.11

The *variance* of a function $x(\cdot)$ of a random variable $\tilde{\omega}$ is defined by the equation

$$V\{x(\tilde{\omega})\} = E\{[x(\tilde{\omega}) - E\{x(\tilde{\omega})\}]^2\}.$$

(A) Show that

$$V\{x(\tilde{\omega})\} = E\{[x(\tilde{\omega})]^2\} - [E\{x(\tilde{\omega})\}]^2.$$

(B) Show that if k_1 and k_2 are any real numbers, then

$$V\{k_1 + k_2 x(\tilde{\omega})\} = [k_2]^2 V\{x(\tilde{\omega})\}.$$

*(C) Show that if $x_1(\cdot), \ldots, x_r(\cdot)$ are mutually *independent*, in the sense that

$$P(\cap_{i=1}^{r} \{\omega: x_i(\omega) \in X_i\}) = \prod_{i=1}^{r} P(\{\omega: x_i(\omega) \in X_i\})$$

for every r subsets $X_1, \ldots, X_r$ of $(-\infty, +\infty)$, then

$$V\left\{\sum_{i=1}^{r} x_i(\tilde{\omega})\right\} = \sum_{i=1}^{r} V\{x_i(\tilde{\omega})\}.$$

(D) Suppose that $\Omega = \{\omega^1, \omega^2, \omega^3, \omega^4\}$ with $P(\omega^j) = 1/4$ for every j. Calculate $V\{x_i(\tilde{\omega})\}$ for each of the following functions $x_i: \Omega \rightarrow (-\infty, +\infty)$:

$j =$	1	2	3	4
$x_1(\omega^j) =$	0	8	7	5
$x_2(\omega^j) =$	0	−8	−7	−5
$x_3(\omega^j) =$	1	−1	−1	1
$x_4(\omega^j) =$	−1	1	1	−1

Exercise 3.7.12

The *covariance* $V\{x_i(\tilde{\omega}), x_t(\tilde{\omega})\}$ of two functions $x_i(\cdot)$ and $x_t(\cdot)$ of a random variable is defined by the equation

$$V\{x_i(\tilde{\omega}), x_t(\tilde{\omega})\} = E\{[x_i(\tilde{\omega}) - E\{x_i(\tilde{\omega})\}][x_t(\tilde{\omega}) - E\{x_t(\tilde{\omega})\}]\}.$$

(A) Show that

$$V\{x_i(\tilde{\omega}), x_t(\tilde{\omega})\} = E\{x_i(\tilde{\omega}) \cdot x_t(\tilde{\omega})\} - E\{x_i(\tilde{\omega})\} \cdot E\{x_t(\tilde{\omega})\}.$$

(B) Show that if k_1, k_2, k_3, and k_4 are any real numbers, then

$$V\{k_1 + k_2 x_i(\tilde{\omega}), k_3 + k_4 x_t(\tilde{\omega})\} = k_2 k_4 V\{x_i(\tilde{\omega}), x_t(\tilde{\omega})\}.$$

(C) Show that

$$V\{x(\tilde{\omega}), x(\tilde{\omega})\} = V\{x(\tilde{\omega})\};$$

that is, the covariance of $x(\tilde{\omega})$ with itself is its variance.

(D) Show that

$$\begin{aligned} V\left\{\sum_{i=1}^{r} x_i(\tilde{\omega})\right\} &= \sum_{i=1}^{r}\sum_{t=1}^{r} V\{x_i(\tilde{\omega}), x_t(\tilde{\omega})\} \\ &= \sum_{i=1}^{r} V\{x_i(\tilde{\omega})\} + \sum_{i=1}^{r}\sum_{t\neq i} V\{x_i(\tilde{\omega}), x_t(\tilde{\omega})\} \\ &= \sum_{i=1}^{r} V\{x_i(\tilde{\omega})\} + 2\sum_{i=1}^{r-1}\sum_{t=i+1}^{r} V\{x_i(\tilde{\omega}), x_t(\tilde{\omega})\}. \end{aligned}$$

(E) Calculate the covariance of $x_i(\tilde{\omega})$ and $x_t(\tilde{\omega})$ for every $\{i, t\} \subset \{1, 2, 3, 4\}$, with the functions defined in part (D) of Exercise 3.7.11.

*■ Parts (B)–(D) of Exercise 3.7.12 and (A) of Exercise 3.7.5 can be summarized nicely in matrix-vector notation. If we have r functions $x_i: \Omega \to (-\infty, +\infty)$, let **V** denote the r-by-r matrix whose (i, t)th element is $V\{x_i(\tilde{\omega}), x_t(\tilde{\omega})\}$, and let **m** denote the r-component column vector whose ith component is $E\{x_i(\tilde{\omega})\}$. Any set $y_1(\cdot), \ldots, y_k(\cdot)$ of *linear* functions $y_q(\tilde{\omega}) = b_q + \Sigma_{i=1}^{r} a_{qi} x_i(\tilde{\omega})$ of the $x_i(\cdot)$'s has expectations, variances, and covariances which can be readily determined from **m** and **V**. Let **b** denote the k-component column vector whose qth component is b_q, and let **A** denote the k-by-r matrix whose (q, i)th element is a_{qi}. Then $E\{y_q(\tilde{\omega})\}$ is the qth component of **b** + **Am**; and $V\{y_q(\tilde{\omega}), y_s(\tilde{\omega})\}$ is the (q, s)th element of **AVA**′, where **A**′ is the transpose of **A**. ■

Exercise 3.7.13: Conditional Probability on Subsets

Suppose that Ξ is an $\mathscr{E}$-move and that Ξ' and Ξ^* are events in Ξ. $\mathscr{D}$'s *conditional probability* $P(\Xi'|\Xi^*)$ *of* Ξ', *given* Ξ^*, is defined to be his quantified judgments regarding the event that $\tilde{\xi} \in \Xi'$, given that he has ascertained that $\tilde{\xi} \in \Xi^*$.

(A) Show that $\mathscr{D}$ should assess conditional and unconditional probabilities in such a way that

$$P(\Xi' \cap \Xi^*) = P(\Xi'|\Xi^*)P(\Xi^*)$$

always obtains.
[*Hint*: Define $\Omega_1{}^1 =$ "Ξ^* obtains," $\Omega_1{}^2 = \overline{\Omega_1{}^1}$, $\Omega_2{}^1 =$ "Ξ' obtains," and $\Omega_2{}^2 = \bar{\Omega}_2{}^1$. Identify $\Xi' \cap \Xi^*$ with $(\Omega_1{}^1, \Omega_2{}^1)$.]

(B) Show that if $\Xi^1, \ldots, \Xi^n$ are any events in Ξ, then $\mathscr{D}$ should assess conditional and unconditional probabilities in such a way that

$$P(\cap_{i=1}^n \Xi^i) = P(\Xi^1) \prod_{t=2}^{n} P(\Xi^t | \cap_{i=1}^{t-1} \Xi^i).$$

***Exercise 3.7.14: Chebyshev's Inequality**

Let $x(\cdot)$ be a real-valued function of a random variable $\tilde{\omega}$, and let $E\{x(\tilde{\omega})\} = m$ and $V\{x(\tilde{\omega})\} = v$. Show that

$$P(-b \leq x(\tilde{\omega}) - m \leq +b) \geq 1 - v/b^2$$

for every $b > 0$.

[*Hints*: (1) Show that $v \geq \Sigma \{(x(\omega) - m)^2 P(\omega): x(\omega) \notin [m-b, m+b]\}$.
(2) Show that $(x(\omega) - m)^2 \geq b^2$ if $x(\omega) \notin [m-b, m+b]$.
(3) Show that $v \geq b^2[1 - P(-b \leq x(\tilde{\omega}) - m \leq +b)]$.]

*3.8 PROOFS OF PREVIOUSLY ASSERTED FACTS

Evidently a generous application of glue will be needed in synthesizing all the small decisions determining the $u(c_t^i)$'s and $P(\Omega_t^i)$'s into Theorems 3.2.1 and 3.2.2.

In this section, we shall prove all hitherto merely asserted basic facts, in the following order:

(1) property, or "Law" (2) of probability, in Theorem 3.6.2;
(2) a Fundamental Lemma[n] from which Theorem 3.2.1 easily follows;
(3) Theorem 3.2.1;
(4) Theorem 3.7.1; and finally,
(5) Theorem 3.2.2.

Since our arguments frequently rely on manipulations with canonical lotteries, you would do well to review Section 3.4 (including *Exercise 3.4.5) before proceeding.

■ *1. Proof of Theorem 3.6.2(2).* Assume that Ω^1 and Ω^2 are mutually exclusive. The assertion that

$$P(\Omega^1 \cup \Omega^2) = P(\Omega^1) + P(\Omega^2)$$

is true if and only if

$$(1/2)P(\Omega^1 \cup \Omega^2) = (1/2)[P(\Omega^1) + P(\Omega^2)],$$

which is true if and only if

$$(\#) \quad \ell^*((1/2)P(\Omega^1 \cup \Omega^2); c^B, c^W) \sim \ell^*((1/2)[P(\Omega^1) + P(\Omega^2)]; c^B, c^W),$$

---

[n] A *lemma* is a preliminary mathematical deduction that paves the way for an important deduction, but which is considered to be of little interest per se.

by virtue of the Monotonicity Principle. We shall prove (#) by lottery comparisons, Substitutability, and Transitivity of "∼" [Exercise 3.3.1(A)]. First, define $\Omega^3 = \overline{\Omega^1 \cup \Omega^2}$, so that Ω^1, Ω^2, and Ω^3 constitute a partition of Ω. Now refer to Figure 3.18 and verify the following facts.

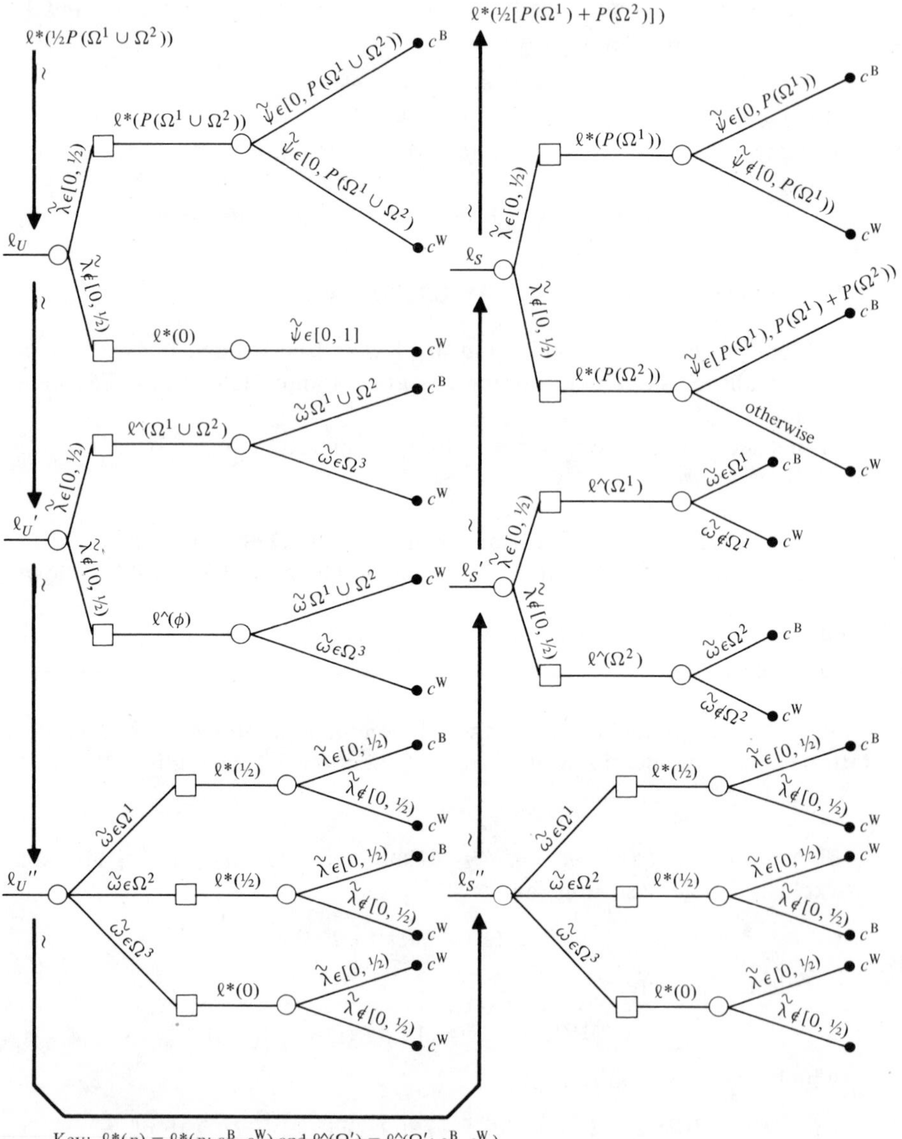

Figure 3.18 Lotteries for proof of Theorem 3.6.2(2). [*Key*: $\ell^*(p) = \ell^*(p; c^B, c^W)$ and $\ell^\wedge(\Omega') = \ell^\wedge(\Omega'; c^B, c^W)$.]

(1) ℓ_U is a paired basic canonical lottery which yields c^B if and only if $(\tilde{\lambda}, \tilde{\psi})$ belongs to a subset I of the unit square having total area $(1/2)P(\Omega^1 \cup \Omega^2)$. Hence $\ell^*((1/2)P(\Omega^1 \cup \Omega^2); c^B, c^W) \sim \ell_U$, because these lotteries are essentially identical.

(2) By substituting $\ell\hat{}(\Omega^1 \cup \Omega^2; c^B, c^W)$ for $\ell^*(P(\Omega^1 \cup \Omega^2); c^B, c^W)$ and $\ell\hat{}(\emptyset; c^B, c^W)$ for $\ell^*(0; c^B, c^W)$ in ℓ_U, we have substituted equidesirables to create ℓ'_U. Thus the Substitutability Principle implies that $\ell_U \sim \ell'_U$.

(3) ℓ''_U is simply ℓ'_U with the order of the $\tilde{\lambda}$- and the $\tilde{\omega}$-$\mathscr{E}$-moves reversed. (See Figure 3.17.) Since the desirability of a lottery is obviously unaffected by (permissibly) arbitrary choices in its tree representation, it follows that $\ell'_U \sim \ell''_U$.

(4) ℓ''_S is created from ℓ''_U by substituting one basic canonical lottery $\ell^*(1/2; c^B, c^W)$ for another, equally desirable basic canonical lottery $\ell^*(1/2; c^B, c^W)$. Hence the Substitutability Principle implies that $\ell''_U \sim \ell''_S$.

(5) ℓ'_S is obtained from ℓ''_S by reversing the order of the $\tilde{\omega}$- and the $\tilde{\lambda}$-$\mathscr{E}$-moves. Hence ℓ''_S and ℓ'_S are essentially identical, so that $\ell''_S \sim \ell'_S$.

(6) By substituting $\ell^*(P(\Omega^1); c^B, c^W)$ and $\ell^*(P(\Omega^2); c^B, c^W)$ for the equally desirable lotteries $\ell\hat{}(\Omega^1; c^B, c^W)$ and $\ell\hat{}(\Omega^2; c^B, c^W)$ respectively, we obtain ℓ_S from ℓ'_S. [Pause to convince yourself that the lottery yielding c^B if the outcome $\tilde{\psi}$ of a basic canonical randomization falls in the interval $[P(\Omega^1), P(\Omega^1) + P(\Omega^2))$, and yields c^W otherwise, is indeed an $\ell^*(P(\Omega^2); c^B, c^W)$.] Hence the Substitutability Principle implies that $\ell'_S \sim \ell_S$.

(7) Observe that ℓ_S is a paired basic canonical lottery which yields c^B if and only if $(\tilde{\lambda}, \tilde{\psi})$ belongs to a subset I^* of the unit square having total area $(1/2)[P(\Omega^1) + P(\Omega^2)]$. Hence ℓ_S is an $\ell^*((1/2) \times P(\Omega^1) + P(\Omega^2)]; c^B, c^W)$, and so these lotteries must be equally desirable.

Now use transitivity of "$\sim$" (six times) to obtain the fundamental equidesirability (#). ■

We remark that property (8) of probability in Theorem 3.6.3 follows directly from property (2) by finite induction.

Next, we prove the Fundamental Lemma.

Fundamental Lemma. Let ℓ be a lottery which yields consequence c^i if and only if $\tilde{\omega} \in \Omega^i$, for $i = 1, \ldots, n$, where $\Omega^1, \ldots, \Omega^n$ constitute a partition of Ω. For every i, let $w(c^i)$ be as defined in Section 3.5. Then

$$\ell \sim \ell^* \left(\sum_{i=1}^{n} w(c^i) P(\Omega^i); c^B, c^W \right).$$

■ 2. *Proof of Fundamental Lemma.* The basic device in our proof is to use the Substitutability Principle several times, creating a sequence of four lotteries ℓ', ℓ'', ℓ''', *and* ℓ'''' such that $\ell \sim \ell'$, $\ell' \sim \ell''$, . . . , and $\ell'''' \sim \ell^*(\Sigma_{i=1}^{n} w(c^i)P(\Omega^i); c^{\mathrm{B}}, c^{\mathrm{W}})$. As in the proof of Theorem 3.6.2(2), transitivity of "$\sim$" will then yield the Fundamental Lemma readily. As a preliminary step, we may assume that the consequences c^i and events Ω^i have been so labeled that $w(c^1) \le w(c^2) \le \cdots \le w(c^n)$. This assumption entails no loss of generality, since if it were not satisfied we could always *re*label consequences and events, with obviously no effect on the desirability of ℓ to $\mathscr{D}$.

First step.[o] By the Canonicity Assumption, $\mathscr{D}$ can imagine the existence of a basic canonical randomization $\Lambda = [0, 1]$ such that $c^i \sim \ell^*(w(c^i); c^{\mathrm{B}}, c^{\mathrm{W}})$, where $\ell^*(w(c^i); c^{\mathrm{B}}, c^{\mathrm{W}})$ results in $c^{\mathrm{B}} \gtrsim c^{\mathrm{n}}$ if $\tilde{\lambda} \in [0, w(c^i))$ and in $c^{\mathrm{W}} \lesssim c^1$ if $\tilde{\lambda} \in [w(c^i), 1]$. Using the Substitutability Principle, we conclude that $\ell \sim \ell'$, where ℓ' is the lottery depicted on the right-hand side in Figure 3.19.

Second step. Partition $\Lambda = [0, 1]$ by defining $\Lambda^1 = [0, w(c^1))$, $\Lambda^i = [w(c^{i-1}), w(c^i))$ for $i = 2, \ldots, n$, and $\Lambda^{n+1} = [w(c^n), 1]$. Then $\Lambda^1, \ldots, \Lambda^{n+1}$ constitute a partition of Λ; and clearly

$$\tilde{\lambda} \in [0, w(c^i)) \qquad \text{if and only if} \qquad \tilde{\lambda} \in \cup_{j=1}^{i} \Lambda^j$$

for $i = 1, 2, \ldots, n$. Thus $\ell^*(w(c^i); c^{\mathrm{B}}, c^{\mathrm{W}})$ may be regarded as the lottery which results in c^{B} if Λ^1, in c^{B} if $\Lambda^2, \ldots,$ in c^{B} if Λ^i, in c^{W} if Λ^{i+1}, in c^{W} if $\Lambda^{i+2}, \ldots,$ and in c^{W} if Λ^{n+1}. Making this representation of the

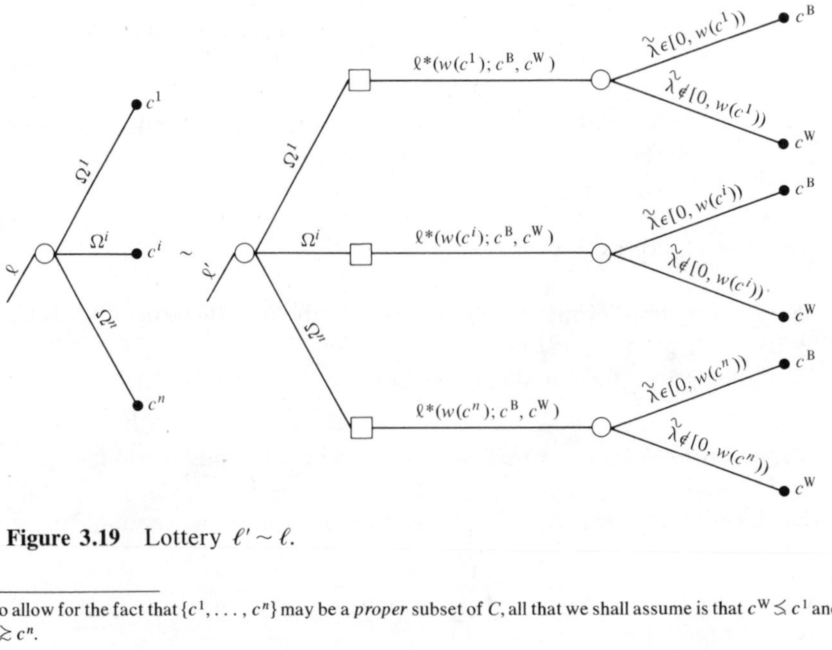

Figure 3.19 Lottery $\ell' \sim \ell$.

---

[o] To allow for the fact that $\{c^1, \ldots, c^n\}$ may be a *proper* subset of C, all that we shall assume is that $c^{\mathrm{W}} \lesssim c^1$ and $c^{\mathrm{B}} \gtrsim c^n$.

$\ell^*(w(c^i); c^B, c^W)$'s explicit yields the lottery ℓ'' depicted in Figure 3.20. Since ℓ'' is just another representation of ℓ', it is clear that $\ell' \sim \ell''$.

Third step. Just as we interchanged the orders of the $\mathscr{E}$-moves in ℓ'_U and ℓ''_S in the proof of Theorem 3.6.2(2) to obtain ℓ''_U and ℓ'_S respectively, we may interchange the orders of the $\mathscr{E}$-moves Ω and Λ here, in ℓ'', to obtain ℓ''' as depicted in Figure 3.21. Note that in ℓ'' the consequence is c^B if $\tilde{\omega} \in \Omega^i$ for some i *and* $\tilde{\lambda} \in \Lambda^j$ for some $j \leq i$, whereas in ℓ''' the consequence is c^B if $\tilde{\lambda} \in \Lambda^j$ for some j *and* $\tilde{\omega} \in \Omega^i$ for some $i \geq j$. But these describe the same event; namely, c^B if $(\tilde{\omega}, \tilde{\lambda}) \in \Omega^i \times \Lambda^j$ with $i \geq j$. Hence $\ell''' \sim \ell''$.

Fourth step. Let $\Psi = [0, 1]$ be a basic canonical randomization completely unrelated to either Ω or Λ. Now $\mathscr{D}$ quantifies his judgments regarding $\tilde{\omega}$-events of the form $\bigcup_{j=i}^{n} \Omega^j$ for $i = 1, \ldots, n$ by noting that $\ell^{(i)}$ in Figure 3.21 is an abbreviation for $\ell^{\wedge}(\bigcup_{j=i}^{n} \Omega^j; c^B, c^W)$, and by finding—according to Section 3.6—a number $P(\bigcup_{j=i}^{n} \Omega^j)$ such that

$$\ell^{(i)} \sim \ell^*(P(\textstyle\bigcup_{j=i}^{n} \Omega^j); c^B, c^W).$$

Since

$$\Omega^n \subset \textstyle\bigcup_{j=n-1}^{n} \Omega^j \subset \cdots \subset \bigcup_{j=2}^{n} \Omega^j \subset \bigcup_{j=1}^{n} \Omega^j = \Omega,$$

it is clear that

$$P(\Omega^n) \leq P(\textstyle\bigcup_{j=n-1}^{n} \Omega^j) \leq \cdots \leq P(\bigcup_{j=2}^{n} \Omega^j) \leq P(\bigcup_{j=1}^{n} \Omega^j) = P(\Omega) = 1.$$

Hence there exists a partition $\Psi^1, \ldots, \Psi^n$ of $\Psi = [0, 1]$ into intervals,

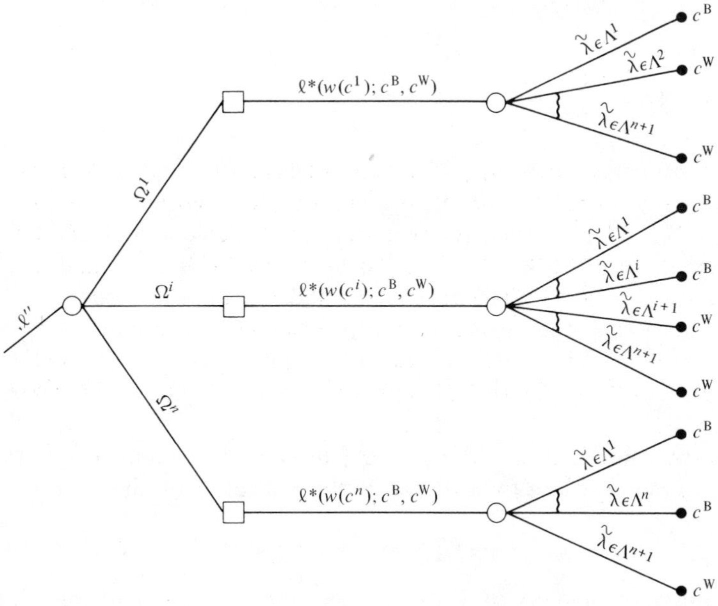

Figure 3.20 Lottery $\ell'' \sim \ell'$.

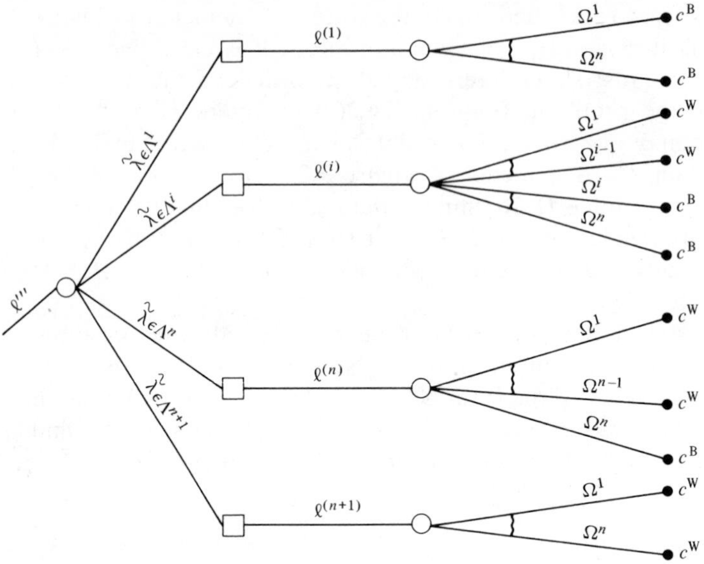

Figure 3.21 Lottery $\ell''' \sim \ell''$.

denoted by

$$\Psi^1 = [0, P(\Omega^1)),$$

$$\Psi^i = \left[\sum_{j=1}^{i-1} P(\Omega^j), \sum_{j=1}^{i} P(\Omega^j)\right) \qquad \text{for } i = 2, \ldots, n-1,$$

and

$$\Psi^n = \left[\sum_{j=1}^{n-1} P(\Omega^j), 1\right],$$

with the property that $\cup_{j=i}^n \Psi^j$ is an interval of length $P(\cup_{j=i}^n \Omega^j) = \Sigma_{j=i}^n P(\Omega^j)$—by property (8) of probability in Theorem 3.6.3. Hence $\ell^*(P(\cup_{j=i}^n \Omega^j); c^B, c^W)$ may be regarded as a basic canonical lottery which results in c^W if $\Psi^1, \ldots,$ in c^W if Ψ^{i-1}, in c^B if $\Psi^i, \ldots,$ and in c^B if Ψ^n, for $i = 2, \ldots, n-1$. (The cases $i = 1$ and $i = n$ are obvious.)

Fifth step. Use the Substitutability Principle to replace $\ell^{(1)}$ with $\ell^*(1; c^B, c^W)$, $\ell^{(i)}$ with $\ell^*(P(\cup_{j=i}^n \Omega^j); c^B, c^W)$ for $i = 2, \ldots, n$, and $\ell^{(n+1)}$ with $\ell^*(0; c^B, c^W)$ in Figure 3.21, thus obtaining the equally desirable ℓ'''' in Figure 3.22.

Sixth step. Recognize that ℓ'''' is a paired basic canonical lottery in which $\mathscr{D}$ obtains c^B if $(\tilde{\lambda}, \tilde{\psi})$ falls in the subset I of the unit square defined by

$$I = \cup_{i=1}^n [(\cup_{j=1}^i \Lambda^j) \times \Psi^i],$$

where mutual exclusivity of $\Psi^1, \Psi^2, \ldots, \Psi^n$ implies mutual exclusivity of the rectangles $(\cup_{j=1}^i \Lambda^j) \times \Psi^i$ for different values of i. Hence ℓ'''' is a

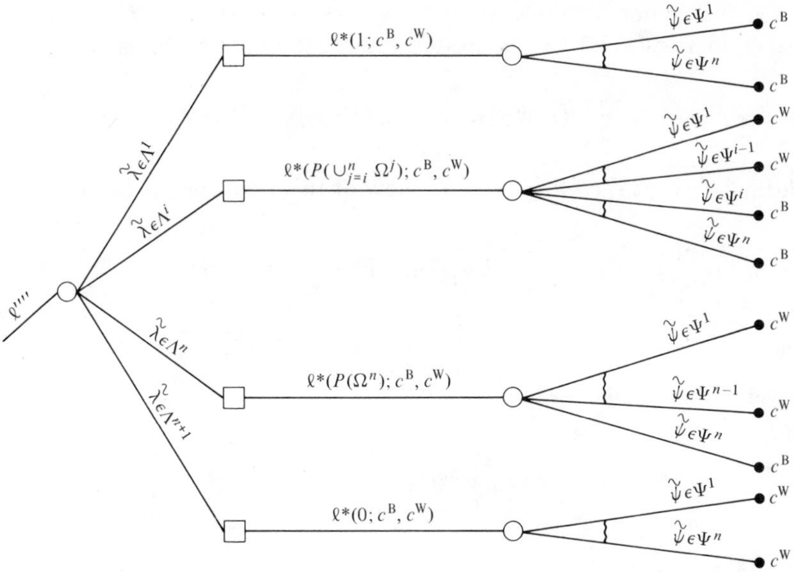

Figure 3.22 Lottery $\ell'''' \sim \ell'''$.

canonical lottery in which the probability of $\mathscr{D}$'s obtaining c^B is the area of I, which is the sum of the areas of the rectangles $(\cup_{j=1}^{i} \Lambda^j) \times \Psi^i$. The length of the "base" $\cup_{j=1}^{i} \Lambda^j$ is $w(c^i)$, by definition of the Λ^j's; and the length of the "side" Ψ^i is $P(\Omega^i) = \Sigma_{j=1}^{i} P(\Omega^j) - \Sigma_{j=1}^{i-1} P(\Omega^j)$; and hence

$$\text{area } ((\cup_{j=1}^{i} \Lambda^j) \times \Psi^i) = w(c^i)P(\Omega^i).$$

Therefore,

$$\text{area } (I) = \sum_{i=1}^{n} w(c^i)P(\Omega^i).$$

Thus ℓ'''' is the same as (or, another representation of) the canonical lottery $\ell^*(\Sigma_{i=1}^{n} w(c^i)P(\Omega^i);\ c^B, c^W)$. Hence surely

$$\ell'''' \sim \ell^* \left(\sum_{i=1}^{n} w(c^i)P(\Omega^i);\ c^B, c^W \right).$$

Seventh step. As in the preceding proof, we apply the Transitivity Principle, this time four times.

$\ell \sim \ell''$ because $\ell \sim \ell'$ and $\ell' \sim \ell''$;
$\ell \sim \ell'''$ because $\ell \sim \ell''$ and $\ell'' \sim \ell'''$;
$\ell \sim \ell''''$ because $\ell \sim \ell'''$ and $\ell''' \sim \ell''''$; and finally,
$\ell \sim \ell^*(\Sigma_{i=1}^{n} w(c^i)P(\Omega^i);\ c^B, c^W)$
because $\ell \sim \ell''''$ and $\ell'''' \sim \ell^*(\Sigma_{i=1}^{n} w(c^i)P(\Omega^i);\ c^B, c^W)$.

This conclude the proof. ■

Now, the Fundamental Lemma says that to any lottery ℓ^t of the form depicted in Figure 3.2 can be associated a canonical probability

$$\sum_{i=1}^{n_t} w(c_t^{\,i})P(\Omega_t^{\,i}) = E\{w(\ell^t(\tilde{\omega}_t))\} = W(\ell^t)$$

such that $\ell^t \sim \ell^*(W(\ell^t); c^{\mathrm{B}}, c^{\mathrm{W}})$. In view of this fact, the proof of Theorem 3.2.1 is easy.

■ *3. Proof of Theorem 3.2.1.* The chain of relations

$$\ell^r$$

$$\sim \quad \ell^*(W(\ell^r); c^{\mathrm{B}}, c^{\mathrm{W}})$$

[by Fundamental Lemma]

$$\begin{Bmatrix} > \\ \sim \\ < \end{Bmatrix} \ell^*(W(\ell^s); c^{\mathrm{B}}, c^{\mathrm{W}})$$

if and only if

$$W(\ell^r) \begin{Bmatrix} > \\ = \\ < \end{Bmatrix} W(\ell^s) \qquad \text{[Monotonicity Principle]}$$

$$\sim \quad \ell^s$$

[by Fundamental Lemma]

implies, by the Transitivity Principle, that

$$\ell^r \begin{Bmatrix} > \\ \sim \\ < \end{Bmatrix} \ell^s \qquad \text{if and only if} \qquad W(\ell^r) \begin{Bmatrix} > \\ = \\ < \end{Bmatrix} W(\ell^s).$$

It remains for us only to show that

$$W(\ell^r) \begin{Bmatrix} > \\ = \\ < \end{Bmatrix} W(\ell^s) \qquad \text{if and only if} \qquad U(\ell^r) \begin{Bmatrix} > \\ = \\ < \end{Bmatrix} U(\ell^s).$$

But this is clear from Theorem 3.5.1 or from the facts that (1) $w: C \to [0, 1]$ is the particular utility function with c^{B} and c^{W} assigned the values 1 and 0 respectively, and (2) any utility function is as good a representative of $\mathscr{D}$'s preferences as is any other. ■

Our next task is to prove Theorem 3.7.1, which relates conditional and unconditional probabilities via the formula

$$P_{12}(\Omega_1^{\,i}, \Omega_2^{\,j}) = P_{2|1}(\Omega_2^{\,j}|\Omega_1^{\,i})P_1(\Omega_1^{\,i}).$$

■ *4. Proof of Theorem 3.7.1.* Examine Figure 3.23. We see that $\ell^*(P_{12}(\Omega_1{}^i, \Omega_2{}^j); c^B, c^W) \sim \ell\hat{}((\Omega_1{}^i, \Omega_2{}^j); c^B, c^W)$, by definition of $P_{12}(\Omega_1{}^i, \Omega_2{}^j)$. Since ℓ' is merely another representation of $\ell\hat{}((\Omega_1{}^i, \Omega_2{}^j); c^B, c^W)$, these lotteries must be equally desirable. Now, $\ell\hat{}(\Omega_2{}^j|\Omega_1{}^i; c^B, c^W)$ in ℓ' and $\ell^*(P_{2|1}(\Omega_2{}^j|\Omega_1{}^i); c^B, c^W)$ are equally desirable by definition of $P_{2|1}(\Omega_2{}^j|\Omega_1{}^i)$, and naturally $\ell\hat{}(\emptyset|\Omega_1{}^i; c^B, c^W) \sim$

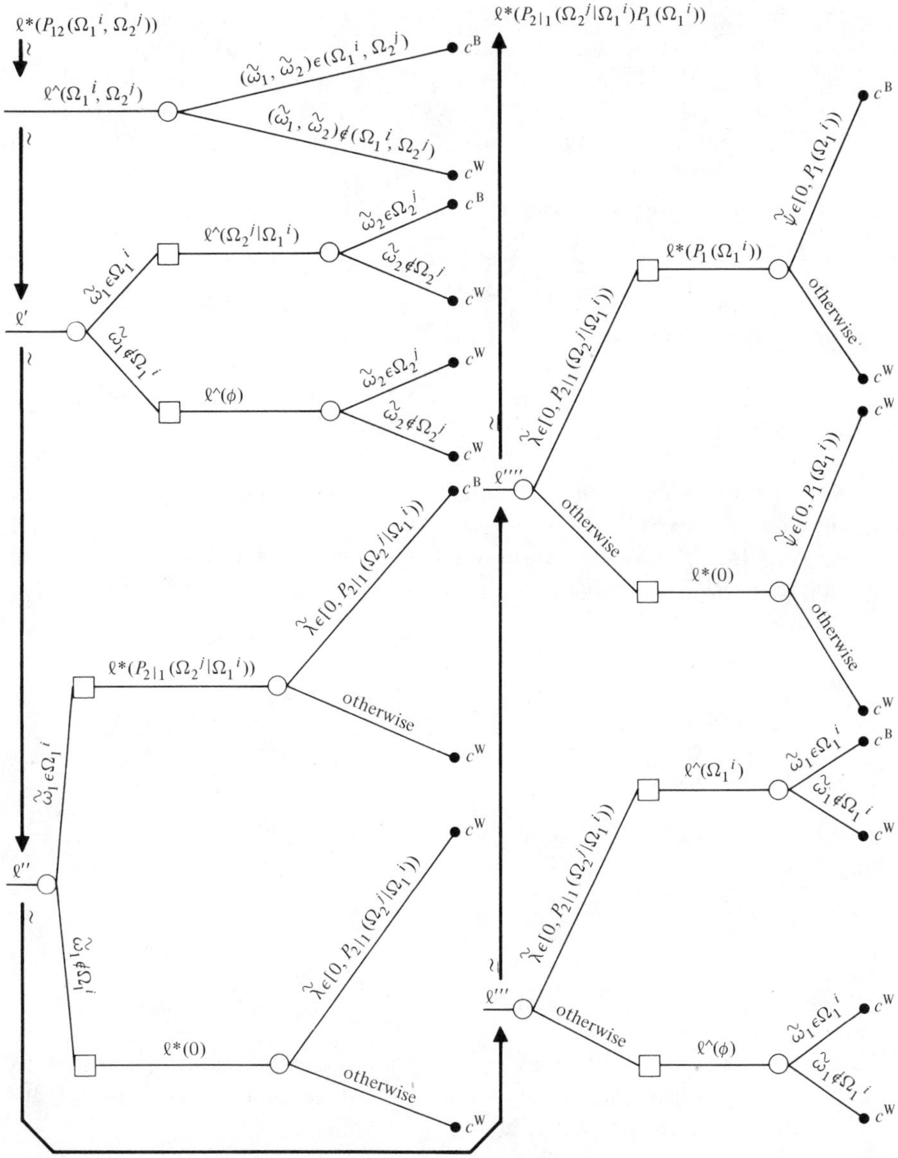

Figure 3.23 Lotteries for proof of Theorem 3.7.1. [*Key*: $\ell^*(p) = \ell^*(p; c^B, c^W)$; $\ell\hat{}(\Omega') = \ell\hat{}(\Omega'; c^B, c^W)$; and $\ell\hat{}(\Omega_2{}^j|\Omega_1{}^i) = \ell\hat{}(\Omega_2{}^j|\Omega_1{}^i; c^B, c^W)$.]

$\ell^*(0; c^B, c^W)$, because both result in c^W for sure. Hence the Substitutability Principle implies that $\ell' \sim \ell''$. Next, ℓ''' is simply ℓ'' with the order of the $\tilde{\lambda}$- and $\tilde{\omega}_1$-$\mathscr{E}$-moves reversed. Since ℓ''' is merely another representation of ℓ'', it is clear that $\ell''' \sim \ell''$. Next, $\ell^*(P_1(\Omega_1{}^i); c^B, c^W) \sim \ell\hat{}(\Omega_1{}^i; c^B, c^W)$ and $\ell^*(0; c^B, c^W) \sim \ell\hat{}(\emptyset; c^B, c^W)$, and hence the Substitutability Principle implies that $\ell'''' \sim \ell'''$. Finally, it is clear that ℓ'''' is a paired basic canonical lottery which results in c^B if and only if $(\tilde{\lambda}, \tilde{\psi})$ falls in the rectangle $[0, P_{2|1}(\Omega_2{}^j|\Omega_1{}^i)) \times [0, P_1(\Omega_1{}^i))$, the area of which is $P_{2|1}(\Omega_2{}^j|\Omega_1{}^i)P_1(\Omega_1{}^i)$. Hence certainly $\ell'''' \sim \ell^*(P_{2|1}(\Omega_2{}^j|\Omega_1{}^i)P_1(\Omega_1{}^i); c^B, c^W)$. By applying the Transitivity Principle five times, we conclude that

$$\ell^*(P_{12}(\Omega_1{}^i, \Omega_2{}^j); c^B, c^W) \sim \ell^*(P_{2|1}(\Omega_2{}^j|\Omega_1{}^i)P_1(\Omega_1{}^i); c^B, c^W),$$

which, by the Monotonicity Principle, is true if and only if

$$P_{12}(\Omega_1{}^i, \Omega_2{}^j) = P_{2|1}(\Omega_2{}^j|\Omega_1{}^i)P_1(\Omega_1{}^i),$$

the desired result. ■

Our final task is to prove Theorem 3.2.2, and it is an easy one in view of Theorem 3.7.1.

■ *5. Proof of Theorem 3.2.2.* Glance at Figure 3.4 again. By deleting all dummy $\mathscr{D}$-moves $\{\ell(\omega^i)\}$ and then combining successive $\mathscr{E}$-moves, it follows that ℓ may be represented as a one-stage lottery in which the outcome is c^{ij} if $(\tilde{\omega}, \tilde{\xi}) = (\omega^i, \xi_i^j)$, for $j = 1, \ldots, m_i$ and $i = 1, \ldots, n$. Hence Theorem 3.2.1 implies that

$$U(\ell) = \sum_{i=1}^{n} \sum_{j=1}^{m_i} u(c^{ij})P(\omega^i, \xi_i^j).$$

But by Theorem 3.7.1,

$$P(\omega^i, \xi_i^j) = P_{2|1}(\xi_i^j|\omega^i)P_1(\omega^i)$$

for all i and j, and therefore

$$U(\ell) = \sum_{i=1}^{n} \sum_{j=1}^{m_i} u(c^{ij})P_{2|1}(\xi_i^j|\omega^i)P_1(\omega^i).$$

Assertion (2) now follows by rearrangement and substituting notation, as indicated in the bottom of Figure 3.4. Assertion (1) is a direct application of Theorem 3.2.1. ■

This concludes our derivation of the basic results in Part B. All others there were derived simply and briefly from those we have proved here.

C: ANALYSIS OF $\mathscr{E}$-MOVES WITH INFINITELY MANY BRANCHES

In the three sections included in this part we show how $\mathscr{E}$-moves with infinitely many branches are typically and practicably handled. All such $\mathscr{E}$-moves are, of course, idealizations of finite reality, but they are *useful* idealizations in that a suitably chosen infinite $\mathscr{E}$-move may be not only a good approximation to the (large-) finite real $\mathscr{E}$-move of interest, but also very easy to handle mathematically.

*Section 3.9 concerns $\mathscr{E}$-moves containing a countable[p] number of k-tuples $(\omega_1, \ldots, \omega_k) = \boldsymbol{\omega}$ of real numbers. *Section 3.10 concerns $\mathscr{E}$-moves containing a continuum of k-tuples of real numbers, for k any positive integer. In each of these sections we consider (1) how to characterize probabilities of events in Ω, and (2) how to find expectations $E\{x(\tilde{\boldsymbol{\omega}})\}$ of real-valued functions $x(\cdot)$ of $\tilde{\boldsymbol{\omega}}$. Both of these problems are satisfactorily solved by introducing appropriate functions which *represent* the probability function on events in Ω, in the sense that by using the representative function one can calculate $E\{x(\tilde{\boldsymbol{\omega}})\}$ and also $P(\Omega')$ for $x(\cdot)$ any function on Ω and Ω' any event in Ω.

Whenever the $\mathscr{E}$-move Ω is a set of k-tuples $\boldsymbol{\omega} = (\omega_1, \ldots, \omega_k)$ with k greater than 1, we can regard Ω as a "combined" $\mathscr{E}$-move in the sense of, say, Figure 3.17, with Ω_1 consisting of all possible $(\omega_1, \ldots, \omega_j)$ for $j < k$ and Ω_2 consisting of all possible $(\omega_{j+1}, \ldots, \omega_k)$. Then the question of how to characterize marginal and conditional probabilities arises. This question is also dealt with in *Sections 3.9 and *3.10.

But frequently we do not *start* with Ω and then worry about probabilities on Ω_1 and Ω_2; rather, we start with Ω_1 and Ω_2 and then worry about Ω, in the spirit of Bayes' Theorem [= Theorem 3.7.2(B)]. In *Section 3.11 we show how to "combine" and "reverse" $\mathscr{E}$-moves Ω_1 and Ω_2 when each is potentially infinite. Examples in *Section 3.11 include some of those which have proved most useful in (statistical) practice.

Throughout Part C, we shall assume that each branch of Ω is an *elementary* event. Actually, this assumption entails no loss of generality, in view of the considerations in Section 2.5, but it makes the essential facts easier to state and interpret.

The only prerequisite for understanding all of the theory in Part C is an acquaintance (but not necessarily an intimacy) with elementary integral calculus. A bit of elementary linear algebra is introduced in examples, and the change-of-variable theorem for multiple integrals[q] is required for Exercise 3.10.10. Anyone frightened by this paragraph should approach Part C in the spirit of the quotation heading of Chapter 2!

---

[p] "Countable" means "either finite or denumerably infinite," where "denumerably infinite" means "being in one-to-one correspondence with the set of integers."

[q] Consult any good intermediate calculus text.

We shall not consume the space necessary to prove rigorously that the Continuity Principle, in conjunction with the other consistency principles, implies the results of the ensuing sections. Since our viewpoint is that infinite $\mathscr{E}$-moves are always approximations, we content ourselves with seeing that the corresponding analytical results are approximations to their finitistic counterparts.

Although we shall develop several new concepts and introduce many significant examples, the most important fact to remember from this part is that *Theorems 3.2.1 and 3.2.2 remain true when their $\mathscr{E}$-moves are infinite and expectations are calculated according to Definition 3.9.2 or Definition 3.10.2.*

*3.9 DISCRETE RANDOM REAL k-TUPLES

In the portfolio problem (Example 1.5.4), in the competitive sealed bidding problem (Example 1.5.3), and in many other realistic problems there is an $\mathscr{E}$-move Ω which can be regarded as a subset of the set R^k of *all* k-tuples $(\omega_1, \ldots, \omega_k)$ of real numbers, for some positive integer k.

In this section we shall assume that the number of ee's in Ω is *countable.* Given this assumption, our first task is to give $\tilde{\boldsymbol{\omega}}$ a name and to define a function which characterizes the probabilities $P(\Omega')$ of events Ω' in Ω.

Definition 3.9.1. $\tilde{\boldsymbol{\omega}}$ is called a *discrete random real* k*-tuple,* or a *discrete random vector,* (or, when $k = 1$, a *discrete random variable*) if

(A) Ω is a subset of R^k.

(B) There is a function f: $R^k \to R^1$, called the *mass function* of $\tilde{\boldsymbol{\omega}}$, such that

  (B1) $f(\boldsymbol{\omega}) \geq 0$ for every $\boldsymbol{\omega}$ in R^k;

  (B2) $\Sigma\{f(\boldsymbol{\omega})\colon \boldsymbol{\omega} \in R^k\} = 1$; and

  (B3) $P(\tilde{\boldsymbol{\omega}} \in \Omega') = \Sigma\{f(\boldsymbol{\omega})\colon \omega \in \Omega'\}$ for every event Ω' *in* R^k.

A function f: $R^k \to R^1$ is the mass function of *some* discrete random real k-tuple if it satisfies (B1) and (B2).

■ Notice that in (B3) we are defining events to be subsets of R^k rather than subsets of the subset Ω of R^k. This is quite permissible, in view of the conclusion of Exercise 3.6.5. It means that we shall hereafter regard the $\mathscr{E}$-move as all of R^k, and this will be convenient subsequently. ■

■ Also notice that the summations in (B2) and (B3) *could not* be well defined *unless* Ω were countable and $f(\boldsymbol{\omega}) = 0$ for all $\boldsymbol{\omega} \notin \Omega$. *All* summations of the form in (B2) and (B3) are to be *understood* as summations only over those $\boldsymbol{\omega}$'s in the indicated set for which $f(\boldsymbol{\omega}) > 0$. Thus, for example,

$$\sum \{f(\boldsymbol{\omega})\colon \boldsymbol{\omega} \in \Omega'\} = \sum \{f(\boldsymbol{\omega})\colon \boldsymbol{\omega} \in \Omega', f(\boldsymbol{\omega}) > 0\}. \blacksquare$$

It is clear from the definition of the mass function $f(\cdot)$ of $\tilde{\boldsymbol{\omega}}$ that

$$f(\boldsymbol{\omega}) = P(\tilde{\boldsymbol{\omega}} = \boldsymbol{\omega})$$

for every $\boldsymbol{\omega}$ in R^k. Thus the mass function of a discrete random real k-tuple is simply the probability function on its conceivable values $\boldsymbol{\omega}$ (but not on its *subsets* of conceivable values). For those $\boldsymbol{\omega}$ in R^k which are *not* branches (= ee's here) of Ω, it is obvious that $P(\tilde{\boldsymbol{\omega}} = \boldsymbol{\omega})$ is zero; and hence, so is $f(\boldsymbol{\omega})$.

Example 3.9.1

Suppose that $k = 1$, that $\Omega = \{0, 1, 2, 3, 4\}$, and that $f(\omega)$ is defined as follows.

ω	$f(\omega)$	ω	$f(\omega)$
0	1/16	3	4/16
1	4/16	4	1/16
2	6/16	$\omega \notin \{0, 1, \ldots, 4\}$	0

Verifying (B2) is easy: $\Sigma\{f(\omega)\colon \omega \in R^1\} = \Sigma\{f(\omega)\colon \omega \in \{0, 1, 2, 3, 4\}\} = 16/16 = 1$. To illustrate the use of (B3), it is easy to verify that $P(\tilde{\omega} \in [2.5, 10])$, the probability of $\tilde{\omega}$ being found in the *closed interval* [2.5, 10] *of real numbers*, is just the probability that $\tilde{\omega}$ is one or the other of the two possible values 3 and 4, in that interval; namely, $P(\tilde{\omega} \in \{3, 4\}) = P(\tilde{\omega} = 3) + P(\tilde{\omega} = 4) = 5/16$. This is $\Sigma\{f(\omega)\colon \omega \in [2.5, 10]\}$. Figure 3.24 is a graphical depiction of $f(\cdot)$.

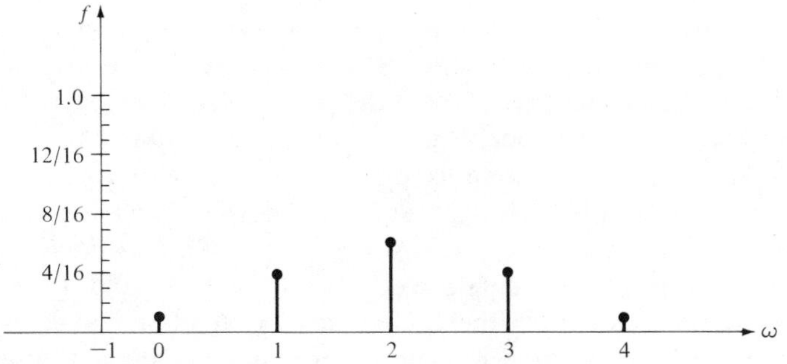

Figure 3.24 The mass function $f(\cdot)$ in Example 3.9.1.

Example 3.9.2

Suppose: (1) that in each of n *instances* we can observe either the *occurrence* or the *nonoccurrence* of some *phenomenon*, and (2) that the occurrences or nonoccurrences in the different instances are entirely unrelated to each other. [In Example 1.5.10, an "instance" would be a reservation holder, and an "occurrence" would be his showing up; in

Example 1.5.5, an "instance" would be a potential customer of the newsboy's, and an "occurrence" would be his buying next month's *Bon Vivant*.] Next, suppose that there is a probability p of the phenomenon's occurring in any given instance. Then you may show as Exercise 3.9.1 that the *total* number $\tilde{\omega}$ of occurrences in the n instances is a discrete random real 1-tuple—or discrete real random variable—with mass function the so-called *binomial* mass function, defined by

$$f_b(\omega|n, p) = \begin{cases} C_\omega^n p^\omega (1-p)^{n-\omega}, & \omega \in \{0, 1, \ldots, n\} \\ 0, & \omega \notin \{0, 1, \ldots, n\}, \end{cases}$$

where C_ω^n is the total number of ω-element subsets of an n-element set; it is numerically defined by

$$C_\omega^n = \frac{n!}{\omega!(n-\omega)!},$$

where

$$j! = \begin{cases} 1, & j = 0 \\ \prod_{i=1}^{j} i, & j \in \{1, 2, 3, \ldots\}. \end{cases}$$

The mass function in Example 3.9.1 and Figure 3.24 is $f_b(\cdot|n, p)$ for $n = 4$ and $p = 1/2$. You may argue that the assumptions which imply that $\tilde{\omega}$ has the binomial mass function are indeed appropriate for the overbooking problem, but perhaps less so for the newsboy problem because the newsboy probably has only vague ideas about his total number of customers.

Example 3.9.3

We shall elaborate a bit on Example 3.9.2 by supposing that in each "instance" the phenomenon is in one of $k+1$ different "states," state $k+1$ being "nonoccurrence (say)" and the first k states being various modes or degrees of occurrence. Again suppose that there are n "instances," in each of which there is probability $p_i > 0$ of finding the phenomenon in state i, for $i = 1, \ldots, k+1$. Mutual exclusivity and collective exhaustiveness of the states will be assumed, and they imply that $\Sigma_{i=1}^{k+1} p_i = 1$. Let $\mathbf{p}$ denote the vector $(p_1, \ldots, p_k)$; and let $\boldsymbol{\omega} = (\omega_1, \ldots, \omega_k)$, where ω_i denotes the total number out of n instances in which the phenomenon is found in state i. Then one can show[r] that $\tilde{\boldsymbol{\omega}}$ is a discrete random real k-tuple with the *multinomial* mass function $f_{mu}^{(k)}(\cdot|n, \mathbf{p})$, defined by

$$f_{mu}^{(k)}(\boldsymbol{\omega}|n, \mathbf{p}) = \begin{cases} C_{\omega_1 \ldots \omega_k}^n \prod_{i=1}^{k+1} p_i^{\omega_i}, & \boldsymbol{\omega} \in \Omega \\ 0, & \boldsymbol{\omega} \notin \Omega, \end{cases}$$

[r] See LaValle [78], Chapter 11, for example.

where

$$\Omega = \{(\omega_1, \ldots, \omega_k) : \omega_i \in \{0, \ldots, n\} \quad \text{for every } i, \ \sum_{i=1}^{k} \omega_i \leq n\},$$

$$\omega_{k+1} = n - \sum_{i=1}^{k} \omega_i$$

$$p_{k+1} = 1 - \sum_{i=1}^{k} p_i,$$

and

$$C^n_{\omega_1 \ldots \omega_k} = n! \Big/ \prod_{i=1}^{k+1} (\omega_i!).$$

Note that for $k = 1$ the multinomial mass function *is* the binomial. It has important applications in sampling theory, in which p_i denotes the proportion of the sampled population with attribute i (out of $k+1$ attributes), and ω_i denotes the total number of respondents out of n who are found to possess attribute i.

Example 3.9.4

This last assertion about sampling is true *only* if each member of the sampled population is equally likely to be picked in *any* of the n instances as is any other member.[s] Many samples (military conscription, for instance) are conducted in such a way that once a member is picked, he can never be picked again. Under *this* assumption, population proportions p_i no longer suffice. Let N denote the total number of members in the *population*; assume that R_i of them possess attribute i for $i = 1, \ldots, k+1$; define $\mathbf{R} = (R_1, \ldots, R_k)$; and let $\tilde{\boldsymbol{\omega}}$ and n be as in Example 3.9.3. Then it can be shown that $\tilde{\boldsymbol{\omega}}$ is a discrete random real k-tuple with the *hypergeometric* mass function

$$f_h^{(k)}(\boldsymbol{\omega} | n, \mathbf{R}, N) = \begin{cases} \dfrac{\prod_{i=1}^{k+1} C^{R_i}_{\omega_i}}{C^N_n}, & \boldsymbol{\omega} \in \Omega \\ 0, & \boldsymbol{\omega} \notin \Omega, \end{cases}$$

where

$$\omega_{k+1} = n - \sum_{i=1}^{k} \omega_i,$$

$$R_{k+1} = N - \sum_{i=1}^{k} R_i,$$

and where Ω is the set of all k-tuples $(\omega_1, \ldots, \omega_k)$ of *integers* such that $0 \leq \omega_i \leq \min[n, R_i]$ for every i and $\Sigma_{i=1}^{k} \omega_i \geq n - R_{k+1}$. (This last condition guarantees that ω_{k+1} is a nonnegative integer.)

[s] In statistical language, this is called "sampling with replacement." We always replace the sampled person in the pool for the next instance.

Example 3.9.5

To illustrate the relationship of the multinomial and hypergeometric mass functions (corresponding to sampling with and without replacement respectively), suppose a population consists of $N = 8$ members, $R_1 = 2$ of which favor candidate 1 and $R_2 = 4$ of which favor candidate 2. Then $p_1 = R_1/N = 1/4$ and $p_2 = R_2/N = 1/2$. Suppose a sample of $n = 4$ is to be taken. *If* the sampling is *with* replacement, $\tilde{\boldsymbol{\omega}} = (\tilde{\omega}_1, \tilde{\omega}_2)$ has the mass function $f_{mu}^{(2)}(\boldsymbol{\omega}|4, (1/4, 1/2)) = f(\omega_1, \omega_2)$, which we tabulate.

$\omega_1 \backslash \omega_2$	0	1	2	3	4	*row sum*
0	1/256	8/256	24/256	32/256	16/256	81/256
1	4/256	24/256	48/256	32/256	0	108/256
2	6/256	24/256	24/256	0	0	54/256
3	4/256	8/256	0	0	0	12/256
4	1/256	0	0	0	0	1/256
column sum	1/16	4/16	6/16	4/16	1/16	

But if the sampling is *without* replacement, then $\tilde{\boldsymbol{\omega}} = (\tilde{\omega}_1, \tilde{\omega}_2)$ has the hypergeometric mass function $f_h^{(2)}(\boldsymbol{\omega}|4, (2, 4), 8) = f(\omega_1, \omega_2)$ described by the following table.

$\omega_1 \backslash \omega_2$	0	1	2	3	4	*row sum*
0	0	0	6/70	8/70	1/70	15/70
1	0	8/70	24/70	8/70	0	40/70
2	1/70	8/70	6/70	0	0	15/70
3	0	0	0	0	0	0
4	0	0	0	0	0	0
column sum	1/70	16/70	36/70	16/70	1/70	

It is clear that there is a substantial difference between these two tables—caused entirely by the replacement versus nonreplacement assumption.

Example 3.9.6

If the total time $\tilde{\tau}$ which one must wait for the next occurrence of a given phenomenon does not depend upon when the last occurrence was, and if $P(\tilde{\tau} \leq \tau'') > P(\tilde{\tau} \leq \tau') > P(\tilde{\tau} \leq 0) = 0$ whenever $\tau'' > \tau' > 0$, then it can be shown that the total number $\tilde{\omega}$ of occurrences of the phenomenon observed during a fixed time period of length t is a discrete real random variable with the *Poisson* mass function $f_{Po}(\cdot|st)$, defined by

$$f_{Po}(\omega|st) = \begin{cases} (st)^{\omega}e^{-st}/\omega!, & \omega \in \{0, 1, 2, \ldots\} \\ 0, & \omega \notin \{0, 1, 2, \ldots\}, \end{cases}$$

where s is some *positive* real number. It is called the (rate of) *intensity* of the occurrences (in number per unit time). Note that $f_{Po}(\omega|st)$ is positive if ω is *any* nonnegative integer. Here, Ω is a countably infinite set.

We have now introduced a number of important examples of discrete random real k-tuples. The next item on our agenda is to define expectations of functions of $\tilde{\boldsymbol{\omega}}$.

Definition 3.9.2. If $\tilde{\boldsymbol{\omega}}$ is a discrete random real k-tuple with mass function f: $R^k \to R^1$, and if x: $R^k \to R^1$, then

$$E\{x(\tilde{\boldsymbol{\omega}})\} = \sum \{x(\boldsymbol{\omega})f(\boldsymbol{\omega}): \boldsymbol{\omega} \in R^k\},$$

provided that this sum converges absolutely. Otherwise, $E\{x(\tilde{\boldsymbol{\omega}})\}$ is undefined, and this infinite idealization is inadequate for any reasoning involving $E\{x(\tilde{\boldsymbol{\omega}})\}$.

■ Again, this summation is understood as being taken only over those $\boldsymbol{\omega}$ for which $f(\boldsymbol{\omega}) > 0$. ■

■ By "absolute convergence" of this sum, we mean that $0 \le \Sigma\, \{|x(\boldsymbol{\omega})| \cdot f(\boldsymbol{\omega}): f(\boldsymbol{\omega}) > 0\} < \infty$, where "$|\cdot|$" denotes absolute value (= distance from zero). When $f(\boldsymbol{\omega}) > 0$ for infinitely many $\boldsymbol{\omega}$'s, the summation may indeed fail to converge absolutely. Suppose that $k = 1$, that

$$f(\omega) = \begin{cases} (1/2)^{\omega}, & \omega \in \{1, 2, \ldots\} \\ 0, & \omega \notin \{1, 2, \ldots\}, \end{cases}$$

and that $x(\omega) = 2^{\omega}$ for every ω. Then $\Sigma\, \{x(\omega)f(\omega): \omega \in R^1\} = \Sigma_{\omega=1}^{\infty} 2^{\omega}(1/2)^{\omega} = \Sigma_{\omega=1}^{\infty} 1 = +\infty$. ■

Example 3.9.7

Suppose that $\tilde{\omega}$ has the Poisson mass function $f_{Po}(\cdot|st)$ defined in Example 3.9.6, and that $x(\omega) = \omega$ for every ω. Then[t]

[t] We apply the same embellishments to the symbol $E\{\cdot\}$ as we do to $f(\cdot)$.

$$
\begin{aligned}
E\{x(\tilde{\omega})\} = E_{\mathrm{Po}}\{\tilde{\omega}|st\} &= \sum_{\omega=0}^{\infty} \omega \cdot \frac{(st)^{\omega} e^{-st}}{\omega!} \\
&=^1 \sum_{\omega=1}^{\infty} \omega \cdot \frac{(st)^{\omega} e^{-st}}{\omega!} \\
&=^2 (st) \sum_{\omega=1}^{\infty} \frac{(st)^{\omega-1} e^{-st}}{(\omega-1)!} \\
&=^3 (st) \sum_{\alpha=0}^{\infty} \frac{(st)^{\alpha} e^{-st}}{\alpha!} \\
&=^4 (st) \sum_{\alpha=0}^{\infty} f_{\mathrm{Po}}(\alpha|st) \\
&=^5 st,
\end{aligned}
$$

where equality (1) is because the first ($\omega = 0$) summand is zero; equality (2) is factoring out an st and noting that $\omega/\omega! = 1/(\omega-1)!$; equality (3) is changing summation variable from $\omega - 1$ to α; equality (4) is by definition of f_{Po}; and equality (5) is (B2) of Definition 3.9.1.

Deriving a number of other expectations is left to you as Exercise 3.9.3.

The final topic on our agenda is to characterize the marginal and conditional probability functions for subvectors of $\tilde{\boldsymbol{\omega}}$. Suppose that $\tilde{\boldsymbol{\omega}}_1 = (\tilde{\omega}_1, \ldots, \tilde{\omega}_j)$ and that $\tilde{\boldsymbol{\omega}}_2 = (\tilde{\omega}_{j+1}, \ldots, \tilde{\omega}_k)$ for some integer j in $\{1, \ldots, k-1\}$. Then $\tilde{\boldsymbol{\omega}}$ can be written as the vector of vectors

$$\tilde{\boldsymbol{\omega}} = (\tilde{\boldsymbol{\omega}}_1, \tilde{\boldsymbol{\omega}}_2),$$

and its mass function $f(\cdot)$ can be written as

$$f_{12}(\boldsymbol{\omega}_1, \boldsymbol{\omega}_2) = f(\boldsymbol{\omega}_1, \boldsymbol{\omega}_2) = f(\boldsymbol{\omega})$$

for every $\boldsymbol{\omega} = (\boldsymbol{\omega}_1, \boldsymbol{\omega}_2)$ in R^k. The subscripting in f_{12} is intended to conform with that in Section 3.7.

Now note that if we think of the $\tilde{\boldsymbol{\omega}}$-$\mathscr{E}$-move in the context of Figure 3.17, with $\tilde{\boldsymbol{\omega}}_1$ occurring before $\tilde{\boldsymbol{\omega}}_2$, then we should be able to derive the *marginal* mass function $f_1(\cdot)$ of $\tilde{\boldsymbol{\omega}}_1$ by "summing $\boldsymbol{\omega}_2$ out" of $f_{12}(\cdot, \cdot)$, so that

$$f_1(\boldsymbol{\omega}_1) = \sum \{f_{12}(\boldsymbol{\omega}_1, \boldsymbol{\omega}_2) \colon \boldsymbol{\omega}_2 \in R^{k-j}\};$$

and then we should be able to derive the *conditional* mass function $f_{2|1}(\cdot|\boldsymbol{\omega}_1^*)$ of $\tilde{\boldsymbol{\omega}}_2$ given any $\boldsymbol{\omega}_1^*$ such that $f_1(\boldsymbol{\omega}_1^*) > 0$ by requiring that the multiplication rule (Theorem 3.7.1)

$$f_{12}(\boldsymbol{\omega}_1^*, \boldsymbol{\omega}_2) = f_{2|1}(\boldsymbol{\omega}_2|\boldsymbol{\omega}_1^*) f_1(\boldsymbol{\omega}_1^*)$$

hold for every $\boldsymbol{\omega}_2$ in R^{k-j}. That is,

$$f_{2|1}(\boldsymbol{\omega}_2|\boldsymbol{\omega}_1^*) = f_{12}(\boldsymbol{\omega}_1^*, \boldsymbol{\omega}_2)/f_1(\boldsymbol{\omega}_1^*)$$

for every $\boldsymbol{\omega}_2$ and every $\boldsymbol{\omega}_1^*$ such that $f_1(\boldsymbol{\omega}_1^*) > 0$.

■ If our infinite idealizations are to hold as decent approximations, these relationships certainly must hold for Theorem 3.2.2 to hold. ■

But the roles of $\tilde{\omega}_1$ and $\tilde{\omega}_2$ could be reversed, by thinking of $\tilde{\omega}_2$ as coming before $\tilde{\omega}_1$. Thus the marginal mass function $f_2(\cdot)$ of $\tilde{\omega}_2$ must satisfy

$$f_2(\boldsymbol{\omega}_2) = \sum \{f_{12}(\boldsymbol{\omega}_1, \boldsymbol{\omega}_2): \boldsymbol{\omega}_1 \in R^j\};$$

and the conditional mass function $f_{1|2}(\cdot|\boldsymbol{\omega}_2^*)$ of $\tilde{\omega}_1$, given any $\boldsymbol{\omega}_2^*$ such that $f_2(\boldsymbol{\omega}_2^*) > 0$ must satisfy

$$f_{12}(\boldsymbol{\omega}_1, \boldsymbol{\omega}_2^*) = f_{1|2}(\boldsymbol{\omega}_1|\boldsymbol{\omega}_2^*)f_2(\boldsymbol{\omega}_2^*),$$

implying that

$$f_{1|2}(\boldsymbol{\omega}_1|\boldsymbol{\omega}_2^*) = f_{12}(\boldsymbol{\omega}_1, \boldsymbol{\omega}_2^*)/f_2(\boldsymbol{\omega}_2^*)$$

for every $\boldsymbol{\omega}_1$ and every $\boldsymbol{\omega}_2^*$ such that $f_2(\boldsymbol{\omega}_2^*) > 0$.

Example 3.9.8

Suppose that $\tilde{\omega}_1$ denotes the first j components of $\tilde{\omega}$, where $\tilde{\omega}$ has the multinomial mass function $f_{mu}^{(k)}(\cdot|n, \mathbf{p})$; and let $\mathbf{p}_1$ denote the first j components of $\mathbf{p}$. By some fancy analysis it can be shown that the mass function $f_1(\cdot)$ of $\tilde{\omega}_1$ is also multinomial:

$$f_1(\boldsymbol{\omega}_1) = f_{mu}^{(j)}(\boldsymbol{\omega}_1|n, \mathbf{p}_1)$$

for every $\boldsymbol{\omega}_1$ in R^j; but we can obtain the same obvious result by regarding the states $j+1, j+2, \ldots, k+1$ of the phenomenon as being one catchall class—"state other than one of the first j." Then there are $j+1$ states; and the probabilities of events exclusively concerning the first j states should not depend on how we have renamed the remainders!

Example 3.9.9

We continue the preceding example by finding the conditional mass function $f_{2|1}(\cdot|\boldsymbol{\omega}_1^*)$ of $\tilde{\omega}_2$. It is

$$\begin{aligned} f_{2|1}(\boldsymbol{\omega}_2|\boldsymbol{\omega}_1^*) &= f_{12}(\boldsymbol{\omega}_1^*, \boldsymbol{\omega}_2)/f_1(\boldsymbol{\omega}_1^*) \\ &= f_{mu}^{(k)}(\boldsymbol{\omega}_1^*, \boldsymbol{\omega}_2|n, (\mathbf{p}_1, \mathbf{p}_2))/f_{mu}^{(j)}(\boldsymbol{\omega}_1^*|n, \mathbf{p}_1) \\ &= f_{mu}^{(k-j)}(\boldsymbol{\omega}_2|n^*(\boldsymbol{\omega}_1^*), \mathbf{p}_2^*), \end{aligned}$$

where $n^*(\boldsymbol{\omega}_1^*) = n - \Sigma_{i=1}^j \omega_i^*$ and $p_i^* = p_i/(1 - \Sigma_{h=1}^j p_h)$ for $i = j+1, \ldots, k$. This result can be obtained by dividing the two multinomial mass functions. It has a nice intuitive justification, though: $n^*(\boldsymbol{\omega}_1^*)$ is the number of remaining instances; and in any instance the *conditional* probability of the state's being i, given that it is not in $\{1, \ldots, j\}$ is p_i^*, for $i = j+1, \ldots, k$, and $k+1$.

Exercise 3.9.1

Work out the derivation of the binomial mass function, in Example 3.9.2.
[*Hint*: There are C_ω^n sequences of n "occurrences" and "nonoccurrences" in which precisely ω are "occurrences."]

Exercise 3.9.2

Under the assumptions of Example 3.9.2, suppose that you are to keep observing instances until the phenomenon has occurred in precisely r of them. Show that the number $\tilde{\omega}$ of instances which will be observed is a discrete real random variable with the *Pascal* mass function $f_{\text{Pa}}(\cdot|r, p)$, defined by

$$f_{\text{Pa}}(\omega|r, p) = \begin{cases} C_{r-1}^{\omega-1} p^r (1-p)^{\omega-r}, & \omega \in \{r, r+1, \ldots \textit{ad inf}\} \\ 0, & \omega \notin \{r, r+1, \ldots \textit{ad inf}\}, \end{cases}$$

where r is a positive integer.
[*Hint*: The last instance observed will be an occurrence! The number of such sequences of length ω, containing r occurrences, in which the last is an occurrence, should equal the number of sequences of length $\omega - 1$ containing $r - 1$ occurrences, should it not?]

Exercise 3.9.3

Define $x(\tilde{\boldsymbol{\omega}}; \mathbf{q})$ by

$$x(\boldsymbol{\omega}; \mathbf{q}) = \prod_{i=1}^{k+1} M_i(\omega_i; q_i),$$

where every q_i is a nonnegative integer and $M_i(\cdot, \cdot)$ is defined by

$$M_i(\omega_i; q_i) = \begin{cases} 0, & \omega_i < q_i \\ \omega_i!/(\omega_i - q_i)!, & \omega_i \geq q_i. \end{cases}$$

Expectations of the functions $x(\cdot; \mathbf{q})$ of $\tilde{\boldsymbol{\omega}}$ are very useful in obtaining expectations of other functions of $\tilde{\boldsymbol{\omega}}$ when $\tilde{\boldsymbol{\omega}}$ has a Poisson, multinomial (including binomial), or hypergeometric mass function. For example, $x(\tilde{\boldsymbol{\omega}}; \mathbf{q}) = \tilde{\omega}_i$ for $q_i = 1$ and all other $q_j = 0$; and $x(\tilde{\boldsymbol{\omega}}; \mathbf{q}) = \tilde{\omega}_i\tilde{\omega}_m$ for $q_i = q_m = 1$ and all other $q_j = 0$. It should be rather fun for you to show that

(A) $$E_{mu}^{(k)}\{x(\tilde{\boldsymbol{\omega}}; \mathbf{q})|n, \mathbf{p}\} = \begin{cases} \dfrac{n!}{(n-Q)!} \prod_{i=1}^{k+1} p_i^{q_i}, & Q \leq n \\ 0, & Q > n, \end{cases}$$

where

$$Q = \sum_{i=1}^{k+1} q_i;$$

(B) $$E_h^{(k)}\{x(\tilde{\boldsymbol{\omega}};\mathbf{q})|n,\mathbf{R},N\}=\begin{cases}\dfrac{n!(N-Q)!\prod_{i=1}^{k+1}R_i!}{N!(n-Q)!\prod_{i=1}^{k+1}(R_i-q_i)!}, & \begin{cases}Q\le n\\ \text{and}\\ q_i\le R_i\\ \text{for every } i,\end{cases}\\ 0, & \text{otherwise,}\end{cases}$$

where again

$$Q=\sum_{i=1}^{k+1}q_i;$$

(C) $$E_{\mathrm{Po}}\{x(\tilde{\omega};q_1)|st\}=(st)^{q_1};$$

(D) $$E_{\mathrm{Pa}}\left\{\frac{(\tilde{\omega}+q-1)!}{(\tilde{\omega}-1)!}\,\middle|\,r,p\right\}=\frac{(r+q-1)!}{(r-1)!}p^{-q};$$

(E) $$E_b\{\tilde{\omega}|n,p\}=np;$$

(F) $$E_{\mathrm{Pa}}\{\tilde{\omega}|r,p\}=r/p;$$

(G) $$E_h^{(1)}\{\tilde{\omega}_1|n,R_1,n\}=n(R_1/N).$$

[*Hints*: (1) Refer to Exercise 3.9.2 for $f_{\mathrm{Pa}}(\cdot|r,p)$; (2) (E)–(G) follow from (A), (B), and (D); and (3) (A)–(D) are proved by using straightforward adaptations of the following derivation.

$$\begin{aligned}E_b\{\tilde{\omega}(\tilde{\omega}-1)|n,p\}&=\sum_{\omega=0}^{n}\omega(\omega-1)\frac{n!}{\omega!(n-\omega)!}p^{\omega}(1-p)^{n-\omega}\\&=\sum_{\omega=2}^{n}\omega(\omega-1)\frac{n!}{\omega!(n-\omega)!}p^{\omega}(1-p)^{n-\omega}\\&=\sum_{\omega=2}^{n}\frac{n!}{(\omega-2)!(n-\omega)!}p^{\omega}(1-p)^{n-\omega}\\&=\sum_{\zeta=0}^{n-2}\frac{n!}{(n-2)!}\cdot\frac{(n-2)!}{\zeta!(n-2-\zeta)!}p^2p^{\zeta}(1-p)^{n-2-\zeta}\\&=\frac{n!}{(n-2)!}p^2\sum_{\zeta=0}^{n-2}f_b(\zeta|n-2,p)\\&=\frac{n!}{(n-2)!}p^2,\end{aligned}$$

provided that $n\ge 2$, necessary for $f_b(\cdot|n,p)$ to make sense.]

Exercise 3.9.4

The *variance* of a real-valued function of a random variable was defined in Exercise 3.7.11. When $\tilde{\omega}$ is itself real-valued, $V\{\tilde{\omega}\}$ is well defined by setting $x(\omega)=\omega$ for every ω. Show that[u]

(A) $$V_b\{\tilde{\omega}|n,p\}=np(1-p)$$

(B) $$V_{\mathrm{Pa}}\{\tilde{\omega}|r,p\}=r(1-p)/p^2$$

---

[u] We apply the same embellishments to the symbol $V\{\cdot\}$ as we do to $f(\cdot)$.

(C) $$V_{\mathrm{Po}}\{\tilde{\omega}|st\} = st$$

[*Hint*: From Exercise 3.7.5(A) and 3.7.11(A),

$$V\{\tilde{\omega}\} = E\{\tilde{\omega}(\tilde{\omega}-1)\} + E\{\tilde{\omega}\} - [E\{\tilde{\omega}\}]^2 = E\{\tilde{\omega}(\omega+1)\} - E\{\omega\} - [E\{\omega\}]^2.]$$

Exercise 3.9.5

The *covariance* $V\{x'(\tilde{\omega}), x''(\tilde{\omega})\}$ of two real-valued functions of a random variable was defined in Exercise 3.7.12. By setting $x'(\omega_1, \ldots, \omega_k) = \omega_i$ and $x''(\omega_1, \ldots, \omega_k) = \omega_j$ for all $\boldsymbol{\omega} = (\omega_1, \ldots, \omega_k)$, it follows that $V\{\tilde{\omega}_i, \tilde{\omega}_j\}$ is well defined whenever $\tilde{\boldsymbol{\omega}}$ is a random real k-tuple. Show that

(A) $$V^{(k)}_{mu}\{\tilde{\omega}_i, \tilde{\omega}_j|n, \mathbf{p}\} = \begin{cases} np_i(1-p_i), & j = i \\ -np_ip_j, & j \neq i \end{cases}$$

for every $\{i, j\} \subset \{1, \ldots, k+1\}$, provided that $n \geq 2$; and

(B) $$V^{(k)}_h\{\tilde{\omega}_i, \tilde{\omega}_j|n, \mathbf{R}, N\} = \begin{cases} \dfrac{N-n}{N-1}\left[n\left(\dfrac{R_i}{N}\right)\left(1-\dfrac{R_i}{N}\right)\right], & j = i \\ \dfrac{N-n}{N-1}\left[-n\left(\dfrac{R_i}{N}\right)\left(\dfrac{R_j}{N}\right)\right], & j \neq i \end{cases}$$

for every $\{i, j\} \subset \{1, \ldots, k+1\}$, provided that $N \geq n \geq 2$ and every $R_i \geq 1$. [*Hint*: From Exercise 3.7.12(A),

$$V\{\tilde{\omega}_i, \tilde{\omega}_j\} = E\{\tilde{\omega}_i \cdot \tilde{\omega}_j\} - E\{\tilde{\omega}_i\} \cdot E\{\tilde{\omega}_j\}.]$$

■ Compare (A) and (B) of Exercise 3.9.5, defining $p_m = R_m/N$ for every m. What does "sampling without replacement" do to variances? ■

Exercise 3.9.6: Analogue of Examples 3.9.8 and 3.9.9 for Hypergeometric Mass Functions

Argue that if $\tilde{\boldsymbol{\omega}} = (\tilde{\boldsymbol{\omega}}_1, \tilde{\boldsymbol{\omega}}_2)$ has the hypergeometric mass function $f_h^{(k)}(\cdot, \cdot|n, (\mathbf{R}_1, \mathbf{R}_2), N)$, then

(A) The marginal mass function of $\tilde{\boldsymbol{\omega}}_1$ *is* $f_h^{(j)}(\cdot|n, \mathbf{R}_1, N)$.
(B) The conditional mass function of $\tilde{\boldsymbol{\omega}}_2$, given $\boldsymbol{\omega}_1^*$ such that $f_h^{(j)}(\boldsymbol{\omega}_1^*|n, \mathbf{R}_1, N) > 0$ is $f_h^{(k-j)}(\cdot|n^*(\boldsymbol{\omega}_1^*), \mathbf{R}_2, N^*)$, where $n^*(\boldsymbol{\omega}_1^*) = n - \Sigma_{i=1}^j \omega_i$ and $N^* = N - \Sigma_{i=1}^j R_i$.

Exercise 3.9.7

Graph the mass function of the number $\tilde{\theta}$ of copies of *Bon Vivant* that will be demanded from the newsboy in Example 1.5.5.

Exercise 3.9.8

Graph the following mass function $f(\cdot)$ of a discrete real random variable $\tilde{\omega}$.

ω	$f(\omega)$	ω	$f(\omega)$	ω	$f(\omega)$	ω	$f(\omega)$
1	.007	14	.018	27	.036	39	.019
2	.007	15	.021	28	.034	40	.018
3	.008	16	.022	29	.033	41	.017
4	.008	17	.023	30	.030	42	.016
5	.009	18	.025	31	.027	43	.015
6	.010	19	.028	32	.025	44	.014
7	.010	20	.030	33	.025	45	.013
8	.011	21	.031	34	.024	46	.012
9	.012	22	.035	35	.023	47	.012
10	.012	23	.037	36	.022	48	.009
11	.013	24	.038	37	.021	49	.005
12	.017	25	.039	38	.020	50	.004
13	.017	26	.038				

Exercise 3.9.9: Cumulative Distribution Functions

Let $k = 1$, and suppose that $\tilde{\omega}$ has mass function $f(\cdot)$. The *cumulative distribution function* $F(\cdot)$ of $\tilde{\omega}$ is defined by

$$F(\omega) = P(\tilde{\omega} \leq \omega) = P(\tilde{\omega} \in (-\infty, \omega])$$

for every ω in $(-\infty, +\infty)$. The cumulative distribution function $F_b(\cdot|n, p)$ corresponding to the binomial mass function graphed in Figure 3.24 is depicted in Figure 3.25.

(A) Show that $F(\omega) = \Sigma\{f(\omega'): \omega' \leq \omega\}$ for every ω in $(-\infty, +\infty)$.

(B) Show that if $F(\cdot)$ is the cumulative distribution function of *some* discrete real random variable, then
  (i) $F(\cdot)$ is a step function (that is, constant on the open interval between adjacent discontinuity points ω);
  (ii) $F(\cdot)$ is nondecreasing, in that $F(\omega') \leq F(\omega'')$ whenever $\omega' < \omega''$; and
  (iii) $\lim_{\omega \to +\infty} F(\omega) = 1$ and $\lim_{\omega \to -\infty} F(\omega) = 0$.

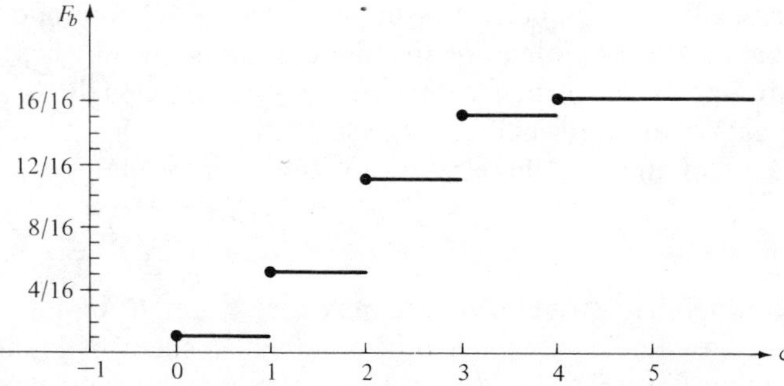

Figure 3.25 A binomial cumulative distribution function.

Exercise 3.9.10

Graph the cumulative distribution function of the discrete real random variable $\tilde{\omega}$ whose mass function is tabulated in Exercise 3.9.8.

■ Exercises 3.9.8 and 3.9.10 are *intended* to make you impatient with the clumsiness inherent in being realistic about large-finite $\mathscr{E}$-moves in general! Help is on the way, in the following section. ■

■ Cumulative distribution functions $F(\cdot)$ are defined when $k > 1$ by

$$F(\omega_1, \ldots, \omega_k) = P(\tilde{\boldsymbol{\omega}} \in \{\boldsymbol{\omega}': \omega_i' \leq \omega_i \text{ for every } i\}),$$

but they are clumsy to use and we shall refrain from so doing. ■

*3.10 CONTINUOUS RANDOM REAL k-TUPLES

Many important $\mathscr{E}$-moves in practice are "almost naturally continuously infinite," in the sense that:

(1) There are vastly many possible values $\boldsymbol{\omega}$ of $\tilde{\boldsymbol{\omega}}$.
(2) The possible values of $\tilde{\boldsymbol{\omega}}$ are very close together.
(3) Each possible value of $\tilde{\boldsymbol{\omega}}$ has very small probability.

Your graph for Exercise 3.9.8 probably prompts you to say that 50 possible values are vastly many, that a spacing of one unit is very close, and that all probabilities not exceeding .039 are very small.

But we might consider a potentially much more drastic example; namely, the $\mathscr{E}$-move Ω consisting of all 10^m possible m-place-decimal outcomes ω of rolling a canonical icosahedral die m times. (See Section 3.4.) If m is large enough, 10^m is a vast number of possible values ω of $\tilde{\omega}$, the possible values are very close together because they are only 10^{-m} apart, and their individual probabilities of 10^{-m} are very small. Just set $m = 500$, for example!

When an $\mathscr{E}$-move is almost naturally continuously infinite, in the preceding sense, it pays analytically to suppose that it *is* continuously infinite. The advantages in so doing are the same as those in physics and engineering, where one is concerned with vast numbers of closely spaced potential measurements of physical quantities. In many such situations, analytical techniques of the calculus are an invaluable boon. Here too.

Example 3.10.1

We begin with a fairly straightforward physical example of the sort frequently found in calculus texts. A thin rectangular sheet of metal is an imperfect alloy, in that the constituent metals were not mixed homogeneously in the casting process. These metals have different masses, and hence the total mass of the sheet is distributed irregularly across the sheet. In particular, its center of gravity (= balance point) need not be the center of the rectangle. It is then assumed that we are given a (decent) function $m(\cdot, \cdot)$ of length and width which expresses the *density* of the mass at the point (ℓ, w) of the rectangle—say, in grams per square centimeter at (ℓ, w). With this function one can then say that the *approximate* mass of the tiny subrectangle $(\ell, \ell + d\ell] \times (w, w + dw]$ is $m(\ell, w)d\ell dw$; that is, the mass density at (ℓ, w) times the area $d\ell dw$ of the subrectangle. The expression $m(\ell, w)d\ell dw$ is sometimes called the *mass element* at (ℓ, w). Then, to find the total mass of any (reasonably decent) region A of the sheet, we may "add up" all the mass elements $m(\ell, w)d\ell dw$ in A. In this context, of course, "adding up" is integrating, and we have

$$(*) \qquad \text{MASS}(A) = \iint_A m(\ell, w)d\ell dw.$$

(See Figure 3.26.) But the density-of-mass function is itself an idealization of reality, since the sheet could *only* be chopped up and weighed in pieces of finite rather than infinitesimal area. Brutal realism would require that we partition the length and width of the sheet into smallest achievable intervals of the forms $(\ell, \ell + D\ell]$ and $(w, w + Dw]$ respectively, thus partitioning the sheet into small subrectangles $(\ell, \ell + D\ell] \times (w, w + Dw]$ each of area $D\ell Dw$. Then we would let M_i denote the total mass of subrectangle S_i, and we would define the *average* mass m_i of S_i by the obvious relation $m_i D\ell Dw = M_i$, or $m_i = M_i/(D\ell Dw)$. Now we can

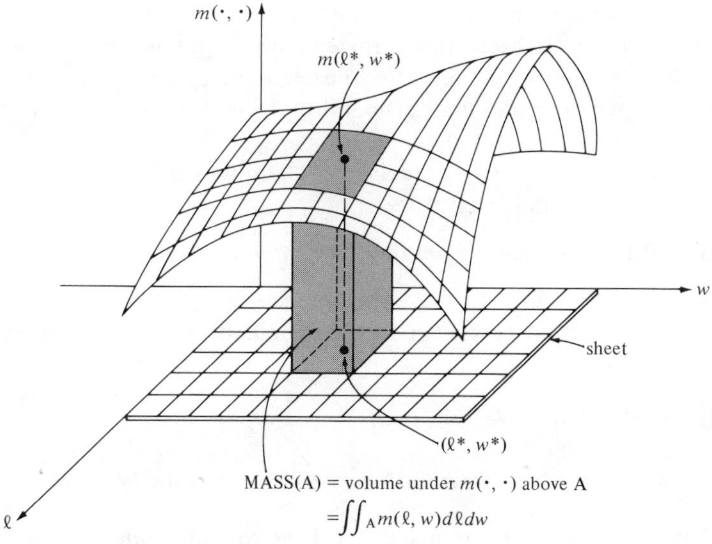

Figure 3.26 Density-of-mass function in Example 3.10.1.

calculate the total mass of an area A by

(**) $$\text{MASS}(A) = \sum \{m_i D\ell Dw\colon S_i \subset A\}.$$

Summations of the form (**) are usually regarded as approximations to integrals of the form (*), but because of the idealization of the measurement problems inherent in defining the density-of-mass function $m(\cdot, \cdot)$, it is equally valid to think of (*) as an approximation to (**). To obtain such an approximation, one "smoothes" the discrete numbers m_i to create the integrable function $m(\cdot, \cdot)$.

Example 3.10.2

Precisely the same reasoning pertains when the metal is not assumed to be thin, and hence vertical nonhomogeneity may be troublesome. If it is in the shape of a brick, let h denote the height variable. Assume given a density-of-mass function $m(\cdot, \cdot, \cdot)$ of the points (ℓ, w, h) of the block. Mass elements are of the form $m(\ell, w, h)d\ell dwdh$ and are "added up" to produce masses MASS(V) of subvolumes V of the block

$$\text{MASS}(V) = \int\int\int_V m(\ell, w, h)d\ell dwdh.$$

The discrete analogue is

$$\text{MASS}(V) = \sum \{m_i D\ell DwDh\colon S_i \subset V\},$$

where $m_i = M_i/(D\ell DwDh)$ for M_i, the mass of subvolume S_i of V.

The point of these long examples is that we intend to treat *probability* just as these examples treated *mass*. Instead of assuming, as we did in Section 3.9, that probability adheres to the elementary events $\boldsymbol{\omega} \in R^k$ in discrete packets, we assume that probability is distributed over R^k in greater or lesser density.

This idea is made much more precise in Definition 3.10.1, which you should compare with Definition 3.9.1.

Definition 3.10.1. $\tilde{\boldsymbol{\omega}}$ is called a *continuous random real k-tuple* if

(A) Ω is a subset of R^k.

(B) There is a function $f\colon R^k \to R^1$, called a *density function* of $\tilde{\boldsymbol{\omega}}$, such that

  (B1) $f(\boldsymbol{\omega}) \geq 0$ for every $\boldsymbol{\omega}$ in R^k;

  (B2) $\int_{R^k} f(\boldsymbol{\omega})d\boldsymbol{\omega} = 1$; and

  (B3) $P(\tilde{\boldsymbol{\omega}} \in \Omega') = \int_{\Omega'} f(\boldsymbol{\omega})d\boldsymbol{\omega}$ for every event Ω' in R^k.

A function $f\colon R^k \to R^1$ is a density function of *some* continuous random real k-tuple if it satisfies (B1) and (B2).

■ For $k > 1$, the symbol $\int_{\Omega'} f(\boldsymbol{\omega})d\boldsymbol{\omega}$ is convenient shorthand for $\int \cdots \int_{\Omega'} f(\omega_1, \ldots, \omega_k)d\omega_1 \cdots d\omega_k$. ■

Example 3.10.3

When $k = 1$, we call $\tilde{\omega}$ a *continuous real random variable.* The probability $P(\Omega')$ of $\tilde{\omega}$ falling in Ω' is the area under the graph of the density function $f(\cdot)$ of $\tilde{\omega}$ above Ω'. See Figure 3.27. When $k = 2$, probability is volume under a density-function surface, analogous to Figure 3.26 for mass. For $k > 2$, probability is hypervolume under a density-function hypersurface.

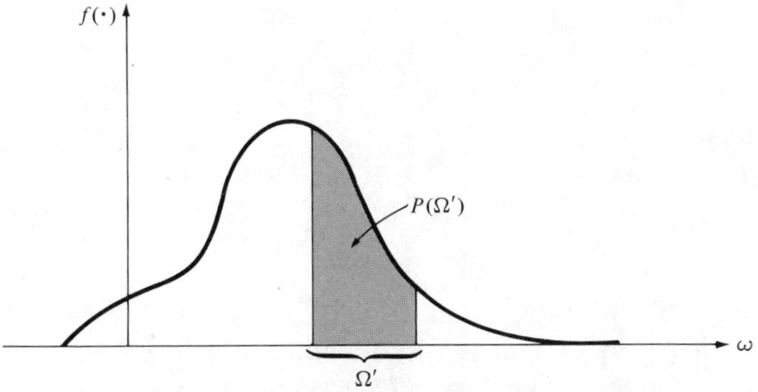

Figure 3.27 Finding probabilities from a density function.

Example 3.10.4

In Section 3.4 we introduced the basic canonical randomization as an idealization of equally likely m-place decimals to infinitely many decimal places. It is easy to verify that a density function for a basic canonical randomization is the *uniform* density function $f_u(\cdot|0, 1)$, defined by

$$f_u(\omega|0, 1) = \begin{cases} 1, & \omega \in [0, 1] \\ 0, & \omega \notin [0, 1]. \end{cases}$$

According to (B3),

$$P(\tilde{\omega} \in (y, z]) = \int_y^z 1 d\omega = z - y$$

whenever $0 \leq y < z \leq 1$, so that $f_u(\cdot|0, 1)$ is indeed a density function for the basic canonical randomization. Somewhat more generally, the *uniform density function over* $[a, b]$ is defined by

$$f_u(\omega|a, b) = \begin{cases} 1/(b - a), & \omega \in [a, b] \\ 0, & \omega \notin [a, b]. \end{cases}$$

Naturally, the parameters must be such that $-\infty < a < b < +\infty$.

Example 3.10.5

If $\tilde{\omega}$ must necessarily be a positive real number, such as a timespan, then a good approximation to one's judgments about $\tilde{\omega}$ might be obtained by assuming that $\tilde{\omega}$ is a continuous real random variable with the so-called *gamma* density function $f_\gamma(\cdot|r, t)$, defined by

$$f_\gamma(\omega|r, t) = \begin{cases} 0, & \omega \leq 0 \\ \dfrac{te^{-t\omega}(t\omega)^{r-1}}{(r-1)!}, & \omega > 0, \end{cases}$$

where r and t are *positive* real-valued parameters. [*Caution*: Unless x is a nonnegative *integer*, $x!$ is not as defined in Example 3.9.2; rather, it is defined by

$$x! = \int_0^\infty e^{-\xi}\xi^x d\xi,$$

which equals

$$\int_0^\infty te^{-t\omega}(t\omega)^x d\omega,$$

by changing variable of integration from ξ to $\omega = \xi/t$.] Figure 3.28 depicts several shapes of gamma density functions obtainable by various choices of the parameter r. The parameter t is called a *scale* parameter: It follows from Exercise 3.10.9 that $\tilde{\omega}$ has a density function $f_\gamma(\cdot|r, t)$ if and only if $\tilde{\zeta} = t\tilde{\omega}$ has a density function $f_\gamma(\cdot|r, 1)$.

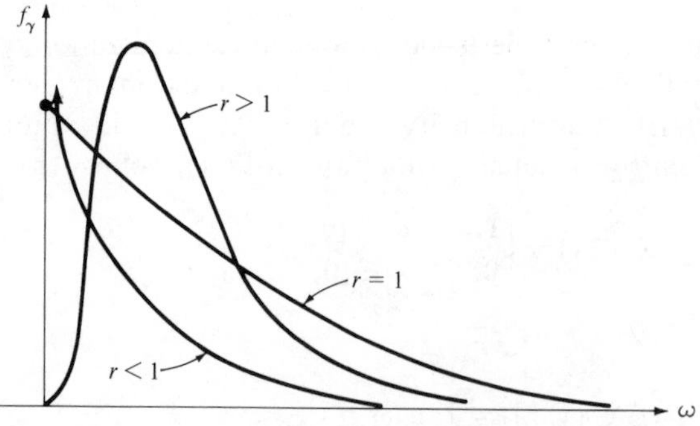

Figure 3.28 Possible shapes of gamma density functions.

Example 3.10.6

If $\tilde{\omega}$ must necessarily be in [0, 1], because $\tilde{\omega}$ denotes a proportion or probability of a phenomenon's occurrence, then one's judgments about $\tilde{\omega}$ might be well approximated by assuming that $\tilde{\omega}$ is a continuous real random variable with the *beta* density function $f_\beta(\cdot|r, n)$, defined by

$$f_\beta(\omega|r, n) = \begin{cases} \dfrac{(n-1)!}{(r-1)!(n-r-1)!}\,\omega^{r-1}(1-\omega)^{n-r-1}, & \omega \in (0, 1) \\ 0, & \omega \notin (0, 1), \end{cases}$$

where n and r are real-valued parameters such that $n > r > 0$. The beta density function can assume many shapes depending upon the relationships of n and r to each other, to 1, and to 2. The shapes range from U-shaped (for $0 < r < 1$ and $0 < n - r < 1$), which places highest probability on ω-intervals of fixed width just above zero and just below one, to uniform on (0, 1) (for $r = n - r = 1$), which places equal probability on all subintervals of (0, 1) having equal widths, to upside-down parabolic shaped, to J-shaped and backward-J-shaped, to bell-shaped. See Figure 3.29.

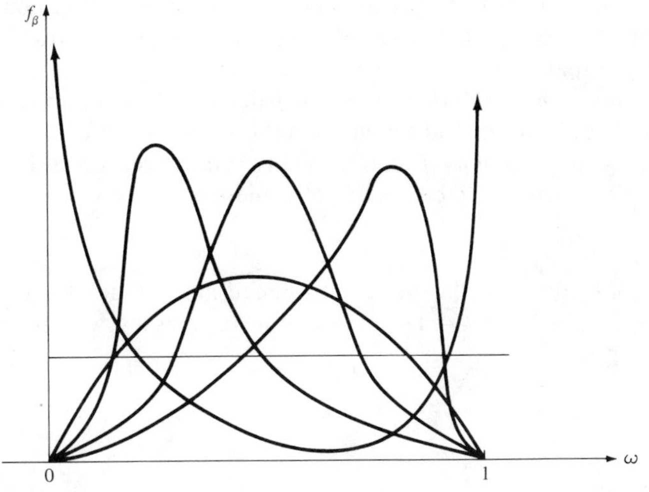

Figure 3.29 Some general shapes of beta density functions.

Example 3.10.7

The *most important* density function of a continuous real random variable, in theory and in application, is the so-called (*univariate*) *Normal* density function $f_N(\cdot|m, V)$, defined by

$$f_N(\omega|m, V) = (2\pi V)^{-1/2} e^{-(\omega-m)^2/(2V)}$$

for every ω in R^1, where M and V are parameters, with m any real number but V constrained to be a *positive* real number. Normal density functions are very frequently used in expressing judgments about numerical quantities and measurements. The parameter m denotes the "center" of the probability distribution in three respects: (1) m is the "most likely" value of $\tilde{\omega}$, or *mode* of $\tilde{\omega}$, in the sense that the probability of an interval of fixed width is maximized by centering that interval on m; (2) it splits the probability in half, with $\tilde{\omega}$ as likely to exceed m as to fall below m—that is,[v] $P_N(\tilde{\omega} \geq m|m, V) = P_N(\tilde{\omega} \leq m|m, V) = 1/2$; and (3), $E_N\{\tilde{\omega}|m, V\} = m$. [See Exercise 3.10.5(D).] The parameter V is an indicator of the *diffuseness* with which probability is distributed over the $\mathscr{E}$-move $\Omega = R^1$, in the sense[w] that the larger the V, the smaller $\mathscr{D}$'s probability of $\tilde{\omega}$ falling in any interval centered at m, and hence the larger $\mathscr{D}$'s probability of $\tilde{\omega}$ departing substantially from m. Mathematically, if $a > 0$ and $V'' > V' > 0$, then $P_N(m - a \leq \tilde{\omega} \leq m + a|m, V'') < P_N(m - a \leq \tilde{\omega} \leq m + a|m, V')$. Two Normal density functions, with the same m but different V's are depicted in Figure 3.30. When people speak of a "bell curve," they are almost invariably referring to a univariate Normal density function. It is frequently used to represent judgments about numerical measurements, as noted above, when those measurements satisfy the three criteria for almost naturally continuously infinite $\mathscr{E}$-moves. The reasons for the widespread use of Normal density functions in this context are twofold. First, a set of theorems, called "central limit theorems," imply that under various conditions some types of measurements *should* have *approximately* a Normal density function. Second, many analytical procedures and calculations are very easy to conduct when $\tilde{\omega}$ has a Normal density function. (This second reason is hardly despicable, in view of our objectives of convenience sought with infinite cases!)

Examples 3.10.3 through 3.10.7 have introduced important classes of density functions for the case $k = 1$. Three important classes for the case $k > 1$ will now be presented.

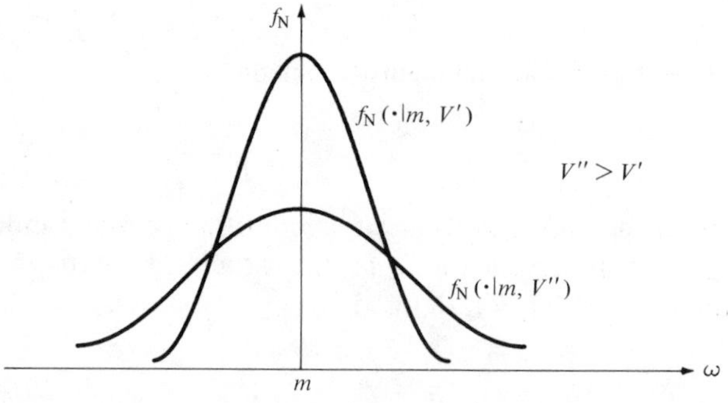

Figure 3.30 Two Normal density functions.

[v] We apply the same embellishments to the symbol $P(\cdot)$ as we do to $f(\cdot)$.

[w] V is often simply called a *measure of $\mathscr{D}$'s uncertainty* about $\tilde{\omega}$. But there is a serious flaw in this line of thinking. See the remark following Exercise 3.10.9.

Example 3.10.8

Let $\mathbf{a}$ and $\mathbf{b}$ in R^k be such that $a_i < b_i$ for every i, and denote by $[\mathbf{a}, \mathbf{b}]$ the hyper-rectangle defined by

$$[\mathbf{a}, \mathbf{b}] = \{\boldsymbol{\omega}: a_i \leq \omega_i \leq b_i \text{ for } i = 1, \ldots, k\}.$$

The (kth order) *uniform density function on* $[\mathbf{a}, \mathbf{b}]$ is defined by

$$f_u^{(k)}(\boldsymbol{\omega}|\mathbf{a}, \mathbf{b}) = \begin{cases} \left[\prod_{i=1}^{k} (b_i - a_i)\right]^{-1}, & \boldsymbol{\omega} \in [\mathbf{a}, \mathbf{b}] \\ 0, & \boldsymbol{\omega} \notin [\mathbf{a}, \mathbf{b}]. \end{cases}$$

Figure 3.31 depicts the uniform density function on $[\mathbf{0}, \mathbf{1}]$ for $k = 2$; this is the density function of a paired basic canonical randomization, since the probability that $(\tilde{\lambda}, \tilde{\psi}) = \tilde{\boldsymbol{\omega}}$ falls in a region of area A in the unit square $[\mathbf{0}, \mathbf{1}]$ is just A. For $k \geq 3$, we call $[\mathbf{0}, \mathbf{1}]$ the (closed) *unit hyper-cube*; and in this case $f_u^{(k)}(\cdot|\mathbf{0}, \mathbf{1})$ is the density function of a "k-aired" basic canonical randomization (= one performed k times with the outcomes of all performances completely unrelated to each other).

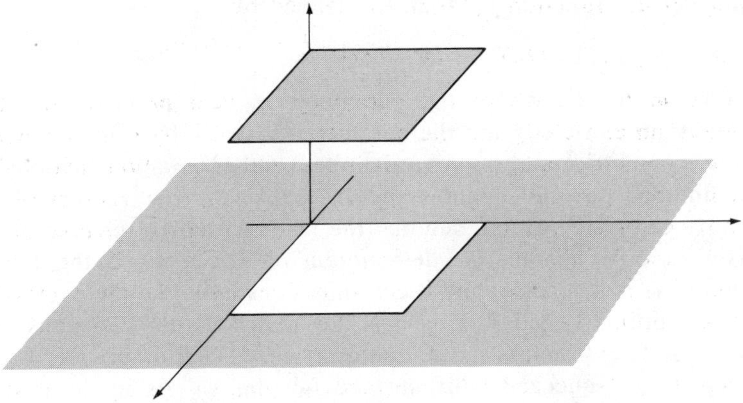

Figure 3.31 The uniform density function on the unit square.

Example 3.10.9

There is a generalization to $k > 1$ of the beta density function in Example 3.10.6. This generalization is useful in expressing judgments about the k-tuple $\tilde{\boldsymbol{\omega}} = (\tilde{\omega}_1, \ldots, \tilde{\omega}_k)$ of proportions of a population, where $\tilde{\omega}_i$ denotes the proportion who possess attribute i and there are $k+1$ mutually exclusive and collectively exhaustive attributes.[x,y] Here, $\Omega = \{(\omega_1, \ldots, \omega_k): 0 < \omega_i < 1 \text{ for every } i, \Sigma_{i=1}^{k} \omega_i < 1\}$; and the appropriate

[x] If there were only k attributes, then necessarily $\omega_k = 1 - \Sigma_{i=1}^{k-1} \omega_i$ and $\omega_{k+1} = 0$, so that ω_k could be dropped from $\boldsymbol{\omega}$ without jeopardizing the completeness of our description of the characteristics of the sampled population.

[y] In our Example 3.9.3 discussion of the multinomial mass function, we denoted $\boldsymbol{\omega}$ by $\mathbf{p}$.

generalization of Example 3.10.6 is to suppose that $\tilde{\boldsymbol{\omega}}$ has the k-*variate beta* density function, or *Dirichlet* density function $f_\beta^{(k)}(\cdot|\mathbf{r}, n)$, defined by

$$f_\beta^{(k)}(\boldsymbol{\omega}|\mathbf{r}, n) = \begin{cases} \dfrac{(n-1)!}{\prod_{i=1}^{k+1}(r_i-1)!} \prod_{i=1}^{k+1} \omega_i^{r_i-1}, & \boldsymbol{\omega} \in \Omega \\ 0, & \boldsymbol{\omega} \notin \Omega, \end{cases}$$

where $r_{k+1} = n - \sum_{i=1}^k r_i$, $\omega_{k+1} = 1 - \sum_{i=1}^k \omega_i$, $\mathbf{r} = (r_1, \ldots, r_k)$, and the parameters n and $\mathbf{r}$ must satisfy the constraints $n > 0$ and $r_i > 0$ for every i (including $i = k+1$). Figure 12.8 in LaValle [**78**] furnishes qualitative information about some of the possible shapes of $f_\beta^{(2)}(\cdot|\mathbf{r}, n)$, where $\mathbf{r}$ and n are denoted there by $\boldsymbol{\rho}$ and ν.

Example 3.10.10

The *most important* density function for expressing judgments about continuous random real k-tuples, such as vectors of measurements or returns on potential securities in a portfolio, is the (nonsingular) k-*variate Normal* density function $f_N^{(k)}(\cdot|\mathbf{m}, \mathbf{V})$, defined by

$$f_N^{(k)}(\boldsymbol{\omega}|\mathbf{m}, \mathbf{V}) = (2\pi)^{-k/2}|\mathbf{V}|^{-1/2} e^{-(1/2)(\boldsymbol{\omega}-\mathbf{m})'\mathbf{V}^{-1}(\boldsymbol{\omega}-\mathbf{m})}$$

for every $\boldsymbol{\omega}$ in R^k, where the parameter $\mathbf{m}$ is a point in R^k (and completely unrestricted) and the parameter $\mathbf{V}$ is a k-by-k matrix which must be symmetric[z] and positive definite.[aa] Other notation involved in this definition: $(\boldsymbol{\omega} - \mathbf{m})'$ denotes the *transpose* (a *row* vector) of the (column) vector $\boldsymbol{\omega} - \mathbf{m}$; $\mathbf{V}^{-1}$ denotes the (k-by-k matrix) *inverse* of the matrix $\mathbf{V}$; and $|\mathbf{V}|$ denotes the *determinant* of $\mathbf{V}$. For $k = 2$, the density function of $\tilde{\boldsymbol{\omega}}$ is a surface; and it certainly does look like the exterior of a bell—an ordinary bell if $\mathbf{V}$ is a scalar matrix (with V_{ii} = const. for every i and $V_{ij} = 0$ unless $i = j$), and a squeezed bell otherwise. Figure 3.32 depicts a "squeezed bell" surface $f_N^{(2)}(\cdot|\mathbf{m}, \mathbf{V})$. As in the case of $k = 1$ discussed in Example 3.10.7, $\mathbf{m}$ is the most likely value of $\tilde{\boldsymbol{\omega}}$; and $\mathbf{V}$ is an indicator of the diffusion of probability over R^k in the sense that $f_N^{(k)}(\cdot|\mathbf{m}, t\mathbf{V})$ is lower and squatter than $f_N^{(k)}(\cdot|\mathbf{m}, \mathbf{V})$ when $t > 1$. In Chapter 6 we shall discuss how $\mathscr{D}$ can quantify his judgments about $\tilde{\boldsymbol{\omega}}$ in the form of a k-variate Normal density function.

We have now presented some of the very important density functions arising in practice. Others can be derived from these by techniques such as those introduced in Exercise 3.10.9 and Exercise 3.10.10, as well as in others.

The next topic on our agenda is, as in Section 3.9, to define the

---

[z] A k-by-k matrix $\mathbf{M} = [m_{ij}]$ is *symmetric* if $m_{ij} = m_{ji}$ for every i and every j.

[aa] A k-by-k matrix $\mathbf{M}$ is *positive definite* if $\mathbf{x}'\mathbf{M}\mathbf{x} > 0$ whenever $\mathbf{x} \neq \mathbf{0}$.

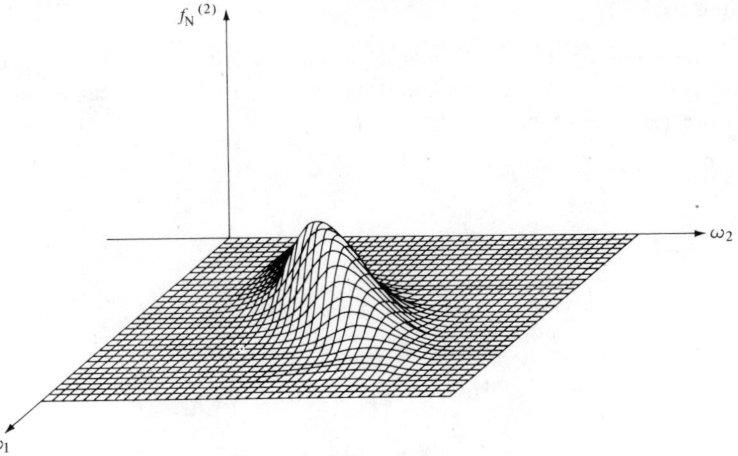

Figure 3.32 A bivariate (i.e., $k = 2$) Normal density function.

expectation $E\{x(\tilde{\omega})\}$ of a function $x(\cdot)$ of a continuous random real k-tuple. As you might suspect from our intuitive characterization of integration as addition in Example 3.10.1, the sum in Definition 3.9.2 is to be replaced here with an integral.

Definition 3.10.2. If $\tilde{\boldsymbol{\omega}}$ is a continuous random real k-tuple with density function $f: R^k \to R^1$, and if $x: R^k \to R^1$, then

$$E\{x(\tilde{\boldsymbol{\omega}})\} = \int_{R^k} x(\boldsymbol{\omega})f(\boldsymbol{\omega})d\boldsymbol{\omega},$$

provided that this integral converges absolutely. Otherwise, $E\{x(\tilde{\boldsymbol{\omega}})\}$ is undefined and this infinite idealization of Ω is inadequate for any reasoning involving $E\{x(\tilde{\boldsymbol{\omega}})\}$.

■ Recall the shorthand introduced after Definition 3.10.1 for multiple integrals! ■

■ By "absolute convergence" of this integral, we mean that $0 \le \int_{R^k}|x(\boldsymbol{\omega})|f(\boldsymbol{\omega})d\boldsymbol{\omega} < +\infty$, where "$|\cdot|$" denotes absolute value (or distance from zero). It is very easy to find cases where the integral does not converge absolutely. For example, let $k = 1$, $x(\omega) = e^{\omega}$ for every ω, and

$$f(\omega) = \begin{cases} e^{-\omega}, & \omega > 0 \\ 0, & \omega \le 0. \end{cases}$$

Then

$$\int_{-\infty}^{\infty} e^{\omega}f(\omega)d\omega = \int_{-\infty}^{0} e^{\omega} \cdot 0 d\omega + \int_{0}^{\infty} e^{\omega} \cdot e^{-\omega}d\omega = \int_{0}^{\infty} 1 \cdot d\omega = +\infty. ■$$

■ Our comments in Example 3.10.1 about regarding integrals as approximations to summations rather than vice versa are equally valid in the context of Definition 3.10.2. ■

Example 3.10.11

Suppose that $k = 1$, that $\tilde{\omega}$ has a gamma density function $f_\gamma(\cdot|r, t)$ as defined in Example 3.10.5, and that $x(\omega) = e^{z\omega}$ for every real number ω and some real number z. Then

$$
\begin{aligned}
E\{x(\tilde{\omega})\} &= E_\gamma\{e^{z\tilde{\omega}}|r, t\} \\
&= \int_{-\infty}^{\infty} e^{z\omega} f(\omega)d\omega \\
&= \int_{-\infty}^{0} e^{z\omega} \cdot 0 d\omega + \int_{0}^{\infty} e^{z\omega} f_\gamma(\omega|r, t)d\omega \\
&= 0 + \int_{0}^{\infty} e^{z\omega}[te^{-t\omega}(t\omega)^{r-1}/(r-1)!]d\omega \\
&=^1 [t/(t-z)]^r \int_{0}^{\infty} [(t-z)e^{-(t-z)\omega}[(t-z)\omega]^{r-1}/(r-1)!]d\omega \\
&=^2 [t/(t-z)]^r \int_{0}^{\infty} f_\gamma(\omega|r, t-z)d\omega \\
&= [t/(t-z)]^r,
\end{aligned}
$$

where equality (1) follows from straightforward algebra, together with moving constant factors outside of the integral, and equality (2) is by definition of $f_\gamma(\cdot|\cdot, \cdot)$ in Example 3.10.5. But this expectation exists *only* if $t - z > 0$, since the second parameter of a gamma density function must be positive. Thus $E_\gamma\{e^{z\tilde{\omega}}|r, t\}$ exists for all z less than t and equals $[t/(t-z)]^r$.

Example 3.10.12

Suppose that $k = 1$, that $\tilde{\omega}$ has the Normal density function $f_N(\cdot|m, V)$, and that $x(\omega) = e^{z\omega}$ for every ω and some real number z. Then

$$
\begin{aligned}
E\{x(\tilde{\omega})\} &= E_N\{e^{z\tilde{\omega}}|m, V\} \\
&= \int_{-\infty}^{\infty} e^{z\omega} f_N(\omega|m, V)d\omega \\
&= \int_{-\infty}^{\infty} e^{z\omega}(2\pi V)^{-1/2} e^{-(1/2)(\omega-m)^2/V} d\omega \\
&=^1 e^{zm+(1/2)z^2V} \int_{-\infty}^{\infty} (2\pi V)^{-1/2} e^{-(1/2)(\omega-[m+zV])^2} d\omega \\
&=^2 e^{zm+(1/2)z^2V} \int_{-\infty}^{\infty} f_N(\omega|m+zV, V)d\omega \\
&= e^{zm+(1/2)z^2V},
\end{aligned}
$$

where equality (1) arises from combining (adding) exponents of e in the integrand, then completing the square in the exponent, and then factoring outside of the integral the exponential not depending upon ω; and equality (2) follows from the definition of $f_N(\cdot|\cdot,\cdot)$ in Example 3.10.7. Note that $m+zV$ is an admissible value of the first parameter of a Normal density function for *any* real number z. Hence $E_N\{e^{z\tilde{\omega}}|m, V\}$ exists for every z in $(-\infty, +\infty)$.

Example 3.10.13

A result similar to and generalizing that in Example 3.10.12 obtains when k is *any* nonnegative integer, $\tilde{\boldsymbol{\omega}}$ has the k-variate Normal density function $f_N^{(k)}(\cdot|\mathbf{m}, \mathbf{V})$, and $x(\boldsymbol{\omega}) = e^{\mathbf{z}'\boldsymbol{\omega}}$ for $\mathbf{z}$ any point in R^k and $\mathbf{z}'$ denoting the transpose of $\mathbf{z}$ (so that $\mathbf{z}'\boldsymbol{\omega} = \Sigma_{i=1}^k z_i\omega_i$). We have

$$\begin{aligned} E\{x(\tilde{\boldsymbol{\omega}})\} &= \int_{R^k} e^{\mathbf{z}'\boldsymbol{\omega}} f_N^{(k)}(\boldsymbol{\omega}|\mathbf{m}, \mathbf{V})d\boldsymbol{\omega} \\ &= \int_{R^k} e^{\mathbf{z}'\boldsymbol{\omega}}(2\pi)^{-n/2}|\mathbf{V}|^{-1/2} e^{-(1/2)(\boldsymbol{\omega}-\mathbf{m})'\mathbf{V}^{-1}(\boldsymbol{\omega}-\mathbf{m})} d\boldsymbol{\omega} \\ &\overset{1}{=} e^{\mathbf{z}'\mathbf{m}+(1/2)\mathbf{z}'\mathbf{V}\mathbf{z}} \int_{R^k} f_N^{(k)}(\boldsymbol{\omega}|\mathbf{m}+\mathbf{V}\mathbf{z}, \mathbf{V})d\boldsymbol{\omega} \\ &= e^{\mathbf{z}'\mathbf{m}+(1/2)\mathbf{z}'\mathbf{V}\mathbf{z}}, \end{aligned}$$

with equality (1) obtaining by completing the (multivariate) square in the exponent of e: By some tedious multiplications, cancellations, and rearrangements of terms, those of you who are interested in linear algebra may verify that

$$\begin{aligned} \mathbf{z}'\boldsymbol{\omega} - (1/2)(\boldsymbol{\omega}-\mathbf{m})'\mathbf{V}^{-1}(\boldsymbol{\omega}-\mathbf{m}) &= \mathbf{z}'\mathbf{m} + (1/2)\mathbf{z}'\mathbf{V}\mathbf{z} \\ &\quad - (1/2)(\boldsymbol{\omega} - [\mathbf{m}+\mathbf{V}\mathbf{z}])'\mathbf{V}^{-1}(\boldsymbol{\omega} - [\mathbf{m}+\mathbf{V}\mathbf{z}]). \end{aligned}$$

But the upshot of all this is that $E_N^{(k)}\{e^{\mathbf{z}'\tilde{\boldsymbol{\omega}}}|\mathbf{m}, \mathbf{V}\}$ exists for *every* z in R^k and equals $e^{\mathbf{z}'\mathbf{m}+(1/2)\mathbf{z}'\mathbf{V}\mathbf{z}}$.

You may derive expectations of some other functions of continuous random real k-tuples in Exercise 3.10.5.

We now proceed to the last item on our agenda for continuous random real k-tuples $\tilde{\boldsymbol{\omega}}$; namely, characterizing marginal and conditional density functions of subvectors of $\tilde{\boldsymbol{\omega}}$. As in Section 3.9, we suppose that $\tilde{\boldsymbol{\omega}} = (\tilde{\boldsymbol{\omega}}_1, \tilde{\boldsymbol{\omega}}_2)$ with $\tilde{\boldsymbol{\omega}}_1$ a j-tuple for $j \in \{1, \ldots, k-1\}$ and $\tilde{\boldsymbol{\omega}}_2$ a $(k-j)$-tuple.

To make a potentially long story short, *all of the discussion on this matter in Section 3.9 remains valid once we substitute "density function" for "mass function," and "integral" for "sum."* That is, if $(\tilde{\boldsymbol{\omega}}_1, \tilde{\boldsymbol{\omega}}_2)$ has the density function $f_{12}(\cdot,\cdot)$, then

(1) $\tilde{\boldsymbol{\omega}}_1$ is a continuous random real j-tuple with *marginal* density function $f_1(\cdot)$ given by

$$f_1(\boldsymbol{\omega}_1) = \int_{R^{k-j}} f_{12}(\boldsymbol{\omega}_1, \boldsymbol{\omega}_2)d\boldsymbol{\omega}_2$$

for every $\boldsymbol{\omega}_1$ in R^j;

(2) $\tilde{\boldsymbol{\omega}}_2$ is a continuous random real $(k-j)$-tuple with *marginal* density function $f_2(\cdot)$ given by

$$f_2(\boldsymbol{\omega}_2) = \int_{R^j} f_{12}(\boldsymbol{\omega}_1, \boldsymbol{\omega}_2)d\boldsymbol{\omega}_1$$

for every $\boldsymbol{\omega}_2$ in R^{k-j};

(3) conditional upon $\tilde{\boldsymbol{\omega}}_1 = \boldsymbol{\omega}_1^*$ with $f_1(\boldsymbol{\omega}_1^*) > 0$, $\tilde{\boldsymbol{\omega}}_2$ is a continuous random real $(k-j)$-tuple with density function $f_{2|1}(\cdot|\boldsymbol{\omega}_1^*)$ given by

$$f_{2|1}(\boldsymbol{\omega}_2|\boldsymbol{\omega}_1^*) = \frac{f_{12}(\boldsymbol{\omega}_1^*, \boldsymbol{\omega}_2)}{f_1(\boldsymbol{\omega}_1^*)}$$

for every ω_2 in R^{k-j};

(4) conditional upon $\tilde{\boldsymbol{\omega}}_2 = \boldsymbol{\omega}_2^*$ with $f_2(\boldsymbol{\omega}_2^*) > 0$, $\tilde{\boldsymbol{\omega}}_1$ is a continuous random real j-tuple with density function $f_{1|2}(\cdot|\boldsymbol{\omega}_2^*)$ given by

$$f_{1|2}(\boldsymbol{\omega}_1|\boldsymbol{\omega}_2^*) = \frac{f_{12}(\boldsymbol{\omega}_1, \boldsymbol{\omega}_2^*)}{f_2(\boldsymbol{\omega}_2^*)}$$

for every $\boldsymbol{\omega}_1$ in R^j; and hence

(5) (analogue of Theorem 3.7.1) the Multiplication Rule obtains:

$$f_{1|2}(\boldsymbol{\omega}_1|\boldsymbol{\omega}_2)f_2(\boldsymbol{\omega}_2) = f_{12}(\boldsymbol{\omega}_1, \boldsymbol{\omega}_2) = f_{2|1}(\boldsymbol{\omega}_2|\boldsymbol{\omega}_1)f_1(\boldsymbol{\omega}_1)$$

whenever $f_{12}(\boldsymbol{\omega}_1, \boldsymbol{\omega}_2) > 0$.

■ Formal proofs of these facts tend to be long winded. Anyone acquainted with the calculus should be able to convince himself that the Multiplication Rule is "almost necessary" for Theorem 3.2.2, in that for

$$E_{12}\{x(\tilde{\boldsymbol{\omega}}_1, \tilde{\boldsymbol{\omega}}_2)\} = E_1\{E_{2|1}\{x(\tilde{\boldsymbol{\omega}}_1, \tilde{\boldsymbol{\omega}}_2)|\tilde{\boldsymbol{\omega}}_1\}\}$$

to obtain when expectations are found according to Definition 3.10.2, it follows that

$$f_{12}(\boldsymbol{\omega}_1, \boldsymbol{\omega}_2) = f_{2|1}(\boldsymbol{\omega}_2|\boldsymbol{\omega}_1)f_1(\boldsymbol{\omega}_1)$$

must be true at least for all $(\boldsymbol{\omega}_1, \boldsymbol{\omega}_2)$ in an event Ω^* of probability one.[bb] ■

Example 3.10.14

Suppose that $\tilde{\boldsymbol{\omega}}$ has the k-variate beta density function $f_\beta^{(k)}(\cdot|\mathbf{r}, n)$ for $k > 1$, that $j \in \{1, \ldots, k-1\}$, and that we want the marginal density function of $\tilde{\boldsymbol{\omega}}_1$. By a tedious integration which involves some clever reexpression of $f_{12}(\boldsymbol{\omega}_1, \boldsymbol{\omega}_2) = f_\beta^{(k)}(\boldsymbol{\omega}_1, \boldsymbol{\omega}_2|(\mathbf{r}_1, \mathbf{r}_2), n)$, it follows that

$$f_1(\boldsymbol{\omega}_1) = f_\beta^{(j)}(\boldsymbol{\omega}_1|\mathbf{r}_1, n) \qquad \text{for every } \boldsymbol{\omega}_1 \text{ in } R^j.$$

This fact is often expressed by saying that "marginals of a k-variate beta

[bb] When $\tilde{\boldsymbol{\omega}}$ is a *continuous* random real k-tuple, probability one does *not* imply certainty. See Exercise 3.4.3, which made this point in connection with basic canonical randomizations.

are j-variate beta." Now, dividing $f_\beta^{(k)}(\boldsymbol{\omega}_1^*, \boldsymbol{\omega}_2|(\mathbf{r}_1, \mathbf{r}_2), n)$ by $f_\beta^{(j)}(\boldsymbol{\omega}_1^*|\mathbf{r}_1, n)$ gives us $f_{2|1}(\boldsymbol{\omega}_2|\boldsymbol{\omega}_1^*)$, according to assertion (3) above. This density function is not one that we have introduced. But as Exercise 3.10.11 you may show that the continuous random real $(k-j)$-tuple $\tilde{\boldsymbol{\omega}}_2^\uparrow = [1-\Sigma_{i=1}^j \omega_i^*]^{-1}\tilde{\boldsymbol{\omega}}_2$—a scaled-up version of $\tilde{\boldsymbol{\omega}}_2$—has the conditional density function

$$f_{2\uparrow|1}(\boldsymbol{\omega}_2^\uparrow|\boldsymbol{\omega}_1^*) = f_\beta^{(k-j)}(\boldsymbol{\omega}_2^\uparrow|\mathbf{r}_2, n-\Sigma_{i=1}^j r_i)$$

given that $\tilde{\boldsymbol{\omega}}_1 = \boldsymbol{\omega}_1^*$ with $f_\beta^{(j)}(\boldsymbol{\omega}_1^*|\mathbf{r}_1, n) > 0$. Upon a little thought, this result is quite reasonable.

Example 3.10.15

For our most important, k-variate Normal case, quite a bit of linear algebra is needed to establish that if $\tilde{\boldsymbol{\omega}}$ has the k-variate Normal density function $f_N^{(k)}(\cdot|\mathbf{m}, \mathbf{V})$, then

(1_N) the marginal density function of $\tilde{\boldsymbol{\omega}}_1$ is

$$f_N^{(j)}(\cdot|\mathbf{m}_1, \mathbf{V}_{11}),$$

where

$$\mathbf{m} = (\mathbf{m}_1, \mathbf{m}_2)$$

and

$$\mathbf{V} = \begin{bmatrix} \mathbf{V}_{11} & \mathbf{V}_{12} \\ \mathbf{V}_{21} & \mathbf{V}_{22} \end{bmatrix}$$

have been partitioned in conformity with $\tilde{\boldsymbol{\omega}} = (\tilde{\boldsymbol{\omega}}_1, \tilde{\boldsymbol{\omega}}_2)$—that is, $\mathbf{m}_1 = (m_1, \ldots, m_j)$ and $\mathbf{V}_{11}$ is the j-by-j upper left corner of $\mathbf{V}$; and

(2_N) conditional upon $\tilde{\boldsymbol{\omega}}_1 = \boldsymbol{\omega}_1^*$ for any $\boldsymbol{\omega}_1^*$ in R^j, the density function of $\tilde{\boldsymbol{\omega}}_2$, given $\boldsymbol{\omega}_1^*$, is

$$f_N^{(k-j)}(\cdot|\mathbf{m}_2 + \mathbf{V}_{21}\mathbf{V}_{11}^{-1}(\boldsymbol{\omega}_1^* - \mathbf{m}_1), \mathbf{V}_{22} - \mathbf{V}_{21}\mathbf{V}_{11}^{-1}\mathbf{V}_{12}).$$

A derivation of these results may be found in Section 12.7 of LaValle [**78**], as well as in almost any reasonably rigorous text on probability and statistics. For doubting Thomases and masochists, a direct verification may be obtained by multiplying $f_N^{(k-j)}$ in (2_N) by $f_N^{(j)}$ in (1_N) and then invoking a great deal of linear algebra to show that the resulting product is identically equal to $f_N^{(k)}(\boldsymbol{\omega}_1^*, \boldsymbol{\omega}_2|\mathbf{m}, \mathbf{V})$.

We have now completed our formal discussion of the continuous random real k-tuple idealizations. As a summary comment, we offer the suggestion that *when in doubt about interpreting and working with a density function* $f(\boldsymbol{\omega}) = f(\omega_1, \ldots, \omega_k)$, *do two things: first, think instead about the corresponding probability elements* $f(\omega)d\omega = f(\omega_1, \ldots, \omega_k)d\omega_1, \ldots, d\omega_k$; *and second, remember that adding up probability elements is integrating density functions.*

Exercise 3.10.1

For $k = 1$, suppose that a function $f_{\mathrm{Lo}}(\cdot|m, S)\colon R^1 \to R^1$ is defined by

$$f_{\mathrm{Lo}}(\omega|m, S) = \frac{e^{-(\omega-m)/S}}{S[1+e^{-(\omega-m)/S}]^2}$$

for every ω in R^1, where m is a parameter that can be any real number, and S is a parameter that must be a *positive* real number.

(A) Show that $f_{\mathrm{Lo}}(\cdot|m, S)$ is the density function of some continuous real random variable $\tilde{\omega}$. That is, show that (B1) and (B2) in Definition 3.10.1 obtain.
[*Hint*: You will want to change variable from ω to $\xi = (\omega - m)/S$.]

(B) Find $P_{\mathrm{Lo}}(\tilde{\omega} \in \Omega'|m, S)$ for $m = 0$, $S = 1$, where Ω' is:

(i) $[-1, +1]$	(iv) $(0, 1] \cup (3, 8]$
(ii) $(0, +\infty)$	(v) $[-2, +2]$
(iii) $[-.6745, +.6745)$	(vi) $[-3, +3]$

[*Hint*: You will want a table of natural logarithms for this task.]

■ $f_{\mathrm{Lo}}(\cdot|m, S)$ is called the *Logistic* density function. It is frequently used as an approximation to the Normal density function. Like the Normal density functions, the Logistics are symmetric about the parameter m. Just sketch the graph of $f_{\mathrm{Lo}}(\cdot|0, 1)$. Does it not look like $f_{\mathrm{N}}(\cdot|0, V)$ for some V? ■

Exercise 3.10.2

For $k = 1$, suppose that a function $f_{\Delta}(\cdot|a, t, b)$ is defined by

$$f_{\Delta}(\omega|a, t, b) = \begin{cases} 0, & \omega \le a \\ \dfrac{2}{b-a}\cdot\dfrac{\omega-a}{t-a}, & a \le \omega \le t \\ \dfrac{2}{b-a}\cdot\dfrac{b-\omega}{b-t}, & t \le \omega \le b \\ 0, & \omega \ge b, \end{cases}$$

where a, t, and b are real-valued parameters which are so constrained as to satisfy $-\infty < a < t < b < +\infty$.

(A) Show that $f_{\Delta}(\cdot|a, t, b)$ satisfies (B1) and (B2) in Definition 3.10.1 and is therefore the density function of some continuous real random variable $\tilde{\omega}$.

(B) Find $P_{\Delta}(\tilde{\omega} \in \Omega'|0, 1, 4)$, where Ω' is:

(i) $(-5, 0]$	(iv) $(0, 2)$
(ii) $(0, 1]$	(v) $(1/2, 3/2)$
(iii) $[3, 8]$	(vi) $(1/4, 1/2) \cup (5/4, 7/4)$

■ Although we have beta and gamma density functions, $f_\Delta(\cdot|a, t, b)$ is not called the delta density function, but rather the *triangular* density function with peak at t and endpoints at $a < t$ and $b > t$. ■

Exercise 3.10.3

For $k = 2$, suppose that a function f_{12}: $R^2 \to R^1$ is defined by

$$f_{12}(\omega_1, \omega_2) = \begin{cases} (1/5)\omega_1^{-3/5}\omega_2^{-4/5}, & 0 < \omega_1^3 < \omega_2 < \omega_1^{1/2} < 1 \\ 0, & \text{elsewhere in } r^2. \end{cases}$$

Here, of course, all the superscripts denote powers.

(A) Show that $f_{12}(\cdot, \cdot)$ satisfies (B1) and (B2) in Definition 3.10.1 and is therefore the density function of some continuous random real 2-tuple $(\tilde{\omega}_1, \tilde{\omega}_2)$.

(B) Find $P_{12}(\Omega')$, where Ω' is:
  (i) $\{(\omega_1, \omega_2): \omega_1 \geq 1/2, \omega_2 \geq 1/2\}$;
  (ii) $\{\omega_1, \omega_2): \omega_1 \geq \omega_2\}$;
  *(iii) $\{(\omega_1, \omega_2): \omega_1 \leq x_1, \omega_2 \leq x_2\}$ for (x_1, x_2) an arbitrary point in R^2.

(C) Find $f_1(\cdot)$, $f_2(\cdot)$, $f_{2|1}(\cdot|\omega_1^*)$ for every ω_1^* in $(0, 1)$, and $f_{1|2}(\cdot|\omega_2^*)$ for every ω_2^* in $(0, 1)$.
[*Hint*: For (B) and (C), you will want to sketch the set Ω on which $f_{12}(\omega_1, \omega_2) > 0$.]

Exercise 3.10.4

For $k = 2$, suppose that a function f_{12}: $R^2 \to R^1$ is defined by

$$f_{12}(\omega_1, \omega_2) = \begin{cases} \dfrac{1}{2\omega_2}, & 0 < \omega_1 < \omega_2 < 1/\omega_1 \textit{ and } \omega_1 < 1 \\ 0, & \text{elsewhere.} \end{cases}$$

(A) Show that $f_{12}(\cdot, \cdot)$ satisfies (B1) and (B2) of Definition 3.10.1 and is therefore a density function.

(B) Find $P_{12}(\Omega')$, where Ω' is:
  (i) $\{(\omega_1, \omega_2): \omega_1 \geq 1/2, \omega_2 \geq 1/2\}$;
  (ii) $\{(\omega_1, \omega_2): \omega_1 \geq \omega_2\}$;
  *(iii) $\{(\omega_1, \omega_2): \omega_1 \leq x_1, \omega_2 \leq x_2\}$ for (x_1, x_2) an arbitrary point in R^2.

(C) Find $f_1(\cdot)$, $f_2(\cdot)$, $f_{2|1}(\cdot|\omega_1^*)$ for every ω_1^* in $(0, 1)$, and $f_{1|2}(\cdot|\omega_2^*)$ for every ω_2^* in $(0, +\infty)$.
[*Hint*: For (B) and (C) you will want to sketch the set Ω on which $f_{12}(\omega_1, \omega_2) > 0$.]

Exercise 3.10.5

Show that:

(A) $$E_u\{(\tilde{\omega})^z|a, b\} = \frac{(b^{z+1} - a^{z+1})}{[(z + 1)(b - a)]}$$

for every positive integer z;

(B) $$E_\gamma\{(\tilde{\omega})^z|r, t\} = \frac{(r+z-1)!}{(r-1)!}\, t^{-z}$$

for every real $z > -r$, and

$$E_\gamma\{(\tilde{\omega})^z|r, t\} = \frac{t^{-z}}{r+z} \prod_{i=0}^{z} (r+z-i)$$

whenever z (though not necessarily r) is a nonnegative integer;

(C) $$E_\beta\{(\tilde{\omega})^{z_1}(1-\tilde{\omega})^{z_2}|r, n\} = \frac{(n-1)!(r+z_1-1)!(n-r+z_2-1)!}{(n+z_1+z_2-1)!(r-1)!(n-r-1)!}$$

for every real $z_1 > -r$ and $z_2 > -(n-r)$, and

$$E_\beta\{(\tilde{\omega})^{z_1}(1-\tilde{\omega})^{z_2}|r, n\} = K \frac{\left[\prod_{j=0}^{z_1} (r+z_1-j)\right]\left[\prod_{i=0}^{z_2} (n-r+z_2-i)\right]}{\prod_{k=0}^{z_1+z_2} (n+z_1+z_2-k)}$$

for $K = (n+z_1+z_2)/[(r+z_1)(n-r+z_2)]$ whenever z_1 and z_2 are both nonnegative integers;

(D) $$E_{\mathrm{N}}\{\tilde{\omega}|m, V\} = m;$$

(E) $$E_{\mathrm{Lo}}\{\tilde{\omega}|m, S\} = m \qquad \text{(see Exercise 3.10.1);}$$

and

(F) $$E_\Delta\{(\tilde{\omega})^z|a, t, b\} = \frac{2[(b-t)a^{z+2} + (t-a)b^{z+2} + (b-a)t^{z+2}]}{(b-t)(t-a)(b-a)(z+1)(z+2)}$$

for every nonnegative integer z. (See Exercise 3.10.2.)
[*Hints*: $f_\gamma(\omega|r+z, t)$ appears in (B); $f_\beta(\omega|r+z_1, n+z_1+z_2)$ appears in (C); and, for (D) and (E), try evaluating

$$\int_{-\infty}^{\infty} (\omega - m) f(\omega)\, d\omega,$$

recalling that if $g: R^1 \to R^1$ is symmetric about zero, then

$$\int_{-\infty}^{\infty} x g(x)\, dx = 0.]$$

Exercise 3.10.6

The *variance* $V\{x(\tilde{\omega})\}$ of a real-valued function of a random variable was defined in Exercise 3.7.11. When $\tilde{\omega}$ is itself real valued, then $V\{\tilde{\omega}\}$ is well defined by setting $x(\omega) = \omega$ for every ω. Show that

(A) $$V_u\{\tilde{\omega}|a, b\} = \frac{(b-a)^2}{12}$$

(B) $$V_\gamma\{\tilde{\omega}|r, t\} = \frac{r}{t^2}$$

(C) $$V_\beta\{\tilde{\omega}|r, n\} = \frac{r(n-r)}{n^2(n+1)}$$

(D) $$V_N\{\tilde{\omega}|m, V\} = V$$

Exercise 3.10.7

The *covariance* $V\{x'(\tilde{\omega}), x''(\tilde{\omega})\}$ of two real-valued functions of a random variable was defined in Exercise 3.7.12. By setting $x'(\omega_1, \ldots, \omega_k) = \omega_i$ and $x''(\omega_1, \ldots, \omega_k) = \omega_j$ for all $\boldsymbol{\omega} = (\omega_1, \ldots, \omega_k)$, it follows that $V\{\tilde{\omega}_i, \tilde{\omega}_j\}$ is defined whenever $\tilde{\boldsymbol{\omega}}$ is a random real k-tuple. Show that

(A) $$V_\beta^{(k)}\{\omega_i, \tilde{\omega}_j|\mathbf{r}, n\} = \begin{cases} \dfrac{r_i(n-r_i)}{n^2(n+1)}, & i = j \\ -\dfrac{r_i r_j}{n^2(n+1)}, & i \neq j \end{cases}$$

for every $\{i, j\} \subset \{1, \ldots, k+1\}$; and

(B) $$V_N^{(k)}\{\tilde{\omega}_i, \tilde{\omega}_j|\mathbf{m}, \mathbf{V}\} = V_{ij}[= (i, j)\text{th element of } \mathbf{V}]$$

for every $\{i, j\} \subset \{1, \ldots, k\}$.

Exercise 3.10.8

The *cumulative distribution function* $F\colon R^1 \to R^1$ of a continuous real random variable ($k = 1$) is defined by

$$F(\omega) = P(\tilde{\omega} \leq \omega) = P(\tilde{\omega} \in (-\infty, \omega])$$

for every ω in R^1, just as $F(\cdot)$ was defined in Exercise 3.9.9 for a discrete real random variable.

(A) Show that

$$F(\omega) = \int_{-\infty}^{\omega} f(\omega)d\omega$$

for every ω in R^1.

(B) Show that if $F(\cdot)$ is the cumulative distribution function of a continuous real random variable, then

(i) $F(\cdot)$ is a continuous function;
(ii) $F(\cdot)$ is nondecreasing, in that $F(\omega') \leq F(\omega'')$ whenever $\omega' < \omega''$; and
(iii) $\lim_{\omega \to +\infty} F(\omega) = 1$ and $\lim_{\omega \to -\infty} F(\omega) = 0$.

(C) Derive the cumulative distribution function corresponding to:

(i) $f_u(\cdot|a, b)$;
(ii) $f_\Delta(\cdot|a, t, b)$ [in Exercise 3.10.2]; and
(iii) $f_{Lo}(\cdot|m, S)$ [in Exercise 3.10.1].

■ As in the discrete case, cumulative distribution functions are defined when $k > 1$ by

$$F(\omega_1, \ldots, \omega_k) = P(\tilde{\boldsymbol{\omega}} \in \{\boldsymbol{\omega}': \omega_i' \le \omega_i \text{ for every } i\}).$$

We shall not have occasion to use them. ■

Exercise 3.10.9: Changes of Variable: $k = 1$

Suppose that $\tilde{\xi}$ is a continuous real random variable with density function $f_\Xi(\cdot)$ and cumulative distribution function $F_\Xi(\cdot)$. Let $\omega: R^1 \to R^1$ be a differentiable function and either strictly increasing or strictly decreasing; and define $\tilde{\omega} = \omega(\tilde{\xi})$. Let $f_\Omega(\cdot)$ and $F_\Omega(\cdot)$ be the density and cumulative distribution functions of $\tilde{\omega}$ respectively.

(A) Show that

$$F_\Omega(\omega) = \begin{cases} F_\Xi(\xi(\omega)), & \omega: R^1 \to R^1 \text{ is increasing} \\ 1 - F_\Xi(\xi(\omega)), & \omega: R^1 \to R^1 \text{ is decreasing} \end{cases}$$

for every ω in R^1. [Note that $\xi(\cdot)$ is the inverse function of $\omega(\cdot)$.]

(B) Show that

$$f_\Omega(\omega) = f_\Xi(\xi(\omega)) \left| \frac{d\xi}{d\omega} \right|$$

for every ω in R^1.

(C) Show that $\tilde{\omega}$ has the Normal density function $f_N(\cdot|m, V)$ if and only if $\tilde{\xi} = (\tilde{\omega} - m)/\sqrt{V}$ has the "standardized" Normal density function $f_N(\cdot|0, 1)$.

(D) Show that $f_u(\cdot|a, b)$ is the density function of $\tilde{\omega}$ if and only if $f_u(\cdot|0, 1)$ is the density function of $\tilde{\xi} = (\tilde{\omega} - a)/(b - a)$.

(E) Show that the density function of $\tilde{\omega}$ is $f_u(\cdot|0, 1)$ if and only if the density function of $\tilde{\xi} = (\tilde{\omega})^{1/10}$ is given by

$$f_\Xi(\xi) = \begin{cases} 10\xi^9, & \xi \in [0, 1] \\ 0, & \xi \notin [0, 1]. \end{cases}$$

(F) If $\tilde{\omega}$ has the density function $f_u(\cdot|a, b)$ for $0 < a < b$, and if $\tilde{\xi} = 1/\tilde{\omega}$, show that

(i)

$$f_\Xi(\xi) = \begin{cases} \dfrac{1}{(b-a)\xi^2}, & \xi \in [1/b, 1/a] \\ 0, & \xi \notin [1/b, 1/a]; \end{cases}$$

and

(ii) the variance $V_\Xi\{\tilde{\xi}\}$ of $\tilde{\xi}$ is given by

$$V_\Xi\{\tilde{\xi}\} = \frac{1}{ab} - \frac{[\log(b/a)]^2}{(b-a)^2}.$$

■ It has been suggested that $\mathscr{D}$'s complete ignorance about $\tilde{\omega}$ should be represented by a uniform density function. But if $\mathscr{D}$ is completely ignorant about $\tilde{\omega}$, then he must also be completely ignorant about the tenth root of $\tilde{\omega}$! But according to (E), the tenth root of $\tilde{\omega}$ has a decidedly nonuniform density function. Therefore it is incorrect to state that a uniform density function represents "complete ignorance." No better is the suggestion that large variance is a manifestation of great ignorance, since letting a tend to zero in (F) yields $\lim_{a\to 0} V_\Xi\{\tilde{\xi}\} = +\infty$, whereas $\lim_{a\to 0} V_u\{\tilde{\omega}|a, b\} = b^2/12$ for $\tilde{\omega} = 1/\tilde{\xi}$ [from Exercise 3.10.6(A)]. Even worse, now let $b\to 0$ also. Then $\lim_{b\to 0}[\lim_{a\to 0} V_\Xi\{\tilde{\xi}\}] = +\infty$, while $\lim_{b\to 0}[\lim_{a\to 0} V_u\{\tilde{\omega}|a, b\}] = 0$. Clearly, in (F) one's ignorance about $\tilde{\omega}$ should equal his ignorance about $\tilde{\xi} = 1/\tilde{\omega}$; and hence, as a measure of ignorance, variance does not measure up! ■

Exercise 3.10.10: Changes of Variable: Arbitrary k

Suppose that $\tilde{\boldsymbol{\xi}}$ is a continuous random real k-tuple with density function $f_\Xi: R^k \to R^1$. Let $\boldsymbol{\omega}: \Xi \to R^k$ be a continuously differentiable, one-to-one function which possesses the additional property that the absolute value

$$\left|\frac{\partial(\omega_1, \ldots, \omega_k)}{\partial(\xi_1, \ldots, \xi_k)}\right| = \left|\frac{\partial \boldsymbol{\omega}}{\partial \boldsymbol{\xi}}\right|$$

of the Jacobian determinant $\det[\partial\omega_i/\partial\xi_j]$ is *positive* for every $(\xi_1, \ldots, \xi_k)$ in Ξ. Finally, let $\tilde{\boldsymbol{\omega}} = \boldsymbol{\omega}(\tilde{\boldsymbol{\xi}})$, and denote the density function of $\tilde{\boldsymbol{\omega}}$ by $f_\Omega(\cdot)$.

(A) Show that

$$f_\Omega(\boldsymbol{\omega}) = f_\Xi(\boldsymbol{\xi}(\boldsymbol{\omega}))\left|\frac{\partial \boldsymbol{\xi}}{\partial \boldsymbol{\omega}}\right|$$

for every $\boldsymbol{\omega}$ in R^k. [Note that $\boldsymbol{\xi}(\cdot)$ is the inverse function of $\boldsymbol{\omega}(\cdot)$. Note also that $|\partial\boldsymbol{\xi}/\partial\boldsymbol{\omega}|$ is the absolute value of the Jacobian determinant $\det[\partial\xi_i/\partial\omega_j]$; it is naturally a function of $\boldsymbol{\omega}$.]

(B) Show that $\tilde{\boldsymbol{\omega}}$ has the k-variate Normal density function $f_N^{(k)}(\cdot|\mathbf{m}, \mathbf{V})$ if and only if $\tilde{\boldsymbol{\xi}} = \mathbf{Q}(\tilde{\boldsymbol{\omega}} - \mathbf{m})$ has the k-variate Normal density function $f_N^{(k)}(\cdot|\mathbf{0}, \mathbf{I})$, where $\mathbf{0} = (0, 0, \ldots, 0)$ in R^k, $\mathbf{I}$ is the k-by-k identity matrix, and $\mathbf{Q}$ is any k-by-k matrix such that $\mathbf{Q}'\mathbf{V}^{-1}\mathbf{Q} = \mathbf{I}$. (This is a *very* important result in "simulation" applications.)

(C) Show that $\tilde{\boldsymbol{\xi}}$ has the density function $f_u^{(2)}(\cdot, \cdot|(0, 0), (1, 1))$ if and only if $\tilde{\boldsymbol{\omega}}$ has the density function $f_{12}(\cdot, \cdot)$ defined in Exercise 3.10.3, where $\tilde{\boldsymbol{\omega}}$ is defined by $\tilde{\omega}_1 = (\tilde{\xi}_1)^2\tilde{\xi}_2$ and $\tilde{\omega}_2 = \tilde{\xi}_1(\tilde{\xi}_2)^3$.

(D) Show that $\tilde{\boldsymbol{\xi}}$ has the density function $f_u^{(2)}(\cdot, \cdot|(0, 0), (1, 1))$ if and only if $\tilde{\boldsymbol{\omega}}$ has the density function $f_{12}(\cdot, \cdot)$ defined in Exercise 3.10.4, where $\tilde{\boldsymbol{\omega}}$ is defined by $\tilde{\omega}_1 = \tilde{\xi}_1 \cdot \tilde{\xi}_2$ and $\tilde{\omega}_2 = \tilde{\xi}_1/\tilde{\xi}_2$.

■ A minor difficulty arises here if $f_u^{(2)}(\cdot, \cdot|(0, 0), (1, 1))$ is regarded as positive on the *closed* unit square, since then $f_\Xi(\xi_1, 0) = 1$ rather than 0. But this difficulty is easily handled because $P(\tilde{\xi}_2 = 0) = 0$ and events of probability zero do not matter. Just redefine $f_u^{(2)}$ to be zero outside the *open* unit interval. So doing changes the probability of *no* event. In fact, it is permissible to redefine a density function on *any* event of probability zero. Hence, unlike mass functions in the discrete case, a continuous random real k-tuple does *not* have a *unique* density function. In fact, $f'(\cdot)$ and $f''(\cdot)$ are both density functions for $\tilde{\boldsymbol{\omega}}$ if and only if they both satisfy (B3) of Definition 3.10.1. Incidentally, compare the definite and indefinite articles "the" and "a" in Definitions 3.9.1 and 3.10.1! ■

■ Exercises 3.10.9 and 3.10.10 pursue calculus-amenable special cases of the following, nearly obvious fact. If $P_\Xi(\cdot)$ is a probability function on events Ξ' in an $\mathscr{E}$-move Ξ and if $\omega(\cdot)$ is a function with domain Ξ and codomain Ω (or, with domain Ξ and values in Ω), then Ω may be regarded as an $\mathscr{E}$-move in which events Ω' have probabilities given by the probability function $P_\Omega(\cdot)$ which satisfies

$$P_\Omega(\Omega') = P_\Xi(\{\xi\colon \omega(\xi) \in \Omega'\})$$

for every event Ω' in Ω. ■

Exercise 3.10.11

Let all notation be as in Example 3.10.14. Show that

$$f_{2\uparrow|1}(\boldsymbol{\omega}_2^{\updownarrow}|\boldsymbol{\omega}_1^*) = f_\beta^{(k-j)}\left(\boldsymbol{\omega}_2^{\updownarrow}|\mathbf{r}_2, n - \sum_{i=1}^{j} r_i\right).$$

[*Hint*: First find the density function $f_{2|1}(\cdot|\boldsymbol{\omega}_1^*)$ of $\tilde{\boldsymbol{\omega}}_2$, and then apply Exercise 3.10.10(A).]

Exercise 3.10.12: Linear Functions of Normals Are Normal

The following facts are of great importance in statistics and indicate the exemplary behavior of Normal random real k-tuples.

(A) Prove that if $\tilde{\boldsymbol{\omega}} = \mathbf{B}\tilde{\boldsymbol{\xi}} + \mathbf{b}$ for $\mathbf{B}$ an r-by-k matrix of rank r and $\mathbf{b} \in R^r$, and if $\tilde{\boldsymbol{\xi}}$ has the k-variate Normal density function $f_N^{(k)}(\cdot|\mathbf{m}, \mathbf{V})$, then $\tilde{\boldsymbol{\omega}}$ has the r-variate Normal density function $f_N^{(r)}(\cdot|\mathbf{Bm} + \mathbf{b}, \mathbf{BVB}')$.

(B) Prove that if $\tilde{\boldsymbol{\xi}}$ has the n-variate Normal density function $f_N^{(n)}(\cdot|\mathbf{m}, \mathbf{V})$ for $\mathbf{m} = (\mu, \mu, \ldots, \mu)$ and $\mathbf{V} = \upsilon\mathbf{I}$, where $\mu \in R^1$, $\upsilon > 0$, and $\mathbf{I}$ is the nth order identity matrix, then the density function of $\tilde{\omega} = \bar{\tilde{\xi}} = (1/n)\Sigma_{j=1}^n \tilde{\xi}_j$ is $f_N(\cdot|\mu, \upsilon/n)$.

[*Hints*: (B) follows directly from (A). To prove (A), proceed as follows.

(1) First prove (A) for $r = k$, using
    (a) Exercise 3.10.10(A),
    (b) $\boldsymbol{\xi} - \mathbf{m} = \mathbf{B}^{-1}[\boldsymbol{\omega} - (\mathbf{Bm} + \mathbf{b})]$,
    (c) $|\partial\boldsymbol{\xi}/\partial\boldsymbol{\omega}| =$ constant, independent of $\boldsymbol{\omega}$, and
    (d) $(\mathbf{B}^{-1})'\mathbf{V}^{-1}\mathbf{B}^{-1} = (\mathbf{BVB}')^{-1}$.

(2) $r > k$ is impossible, given the assumption that $\mathbf{B}$ is of rank r.

(3) If $r < k$, let $\tilde{\zeta}^* = (\tilde{\omega}', \tilde{\zeta}')'$, $\mathbf{A} = [\mathbf{B}', \mathbf{C}']'$, and $\mathbf{a} = (\mathbf{b}', \mathbf{c}')'$ for any $\mathbf{c} \in R^{k-r}$ and any $(k-r)$-by-k matrix $\mathbf{C}$ such that $\mathbf{A}$ is kth order of rank k. Then $\tilde{\zeta}^* = \mathbf{A}\tilde{\xi} + \mathbf{a}$; apply your proof for $r = k$ to obtain $f_N^{(k)}(\cdot|\mathbf{Am} + \mathbf{a}, \mathbf{AVA}')$ as the density function of $\tilde{\zeta}^*$. Then apply (1_N) in Example 3.10.15 to obtain the desired result.]

■ In Exercise 3.10.12(B), the queer, bar superfix in the synonym for $\bar{\omega}$ is introduced in conformity with standard statistical usage. It reappears in Chapters 6 and 9. ■

*3.11 COMBINING SUCCESSIVE $\mathscr{E}$-MOVES AND BAYES' THEOREM

Our closing comment in *Section 3.10 about thinking in terms of probability elements with continuous random real k-tuples is the key to appreciating what we shall discuss here about reversing the order of two successive $\mathscr{E}$-moves, from "Ω_1 before Ω_2" to "Ω_2 before Ω_1."

Suppose that $\tilde{\boldsymbol{\omega}}_1$ is a random real k-tuple, either discrete with mass function $f_1(\cdot)$ or continuous with density function $f_1(\cdot)$. Also suppose that a copy of the (extended) $\mathscr{E}$-move R^m of a random real m-tuple $\tilde{\boldsymbol{\omega}}_2$ is attached to each of the branches $\boldsymbol{\omega}_1$. This is the "Ω_1-before-Ω_2" viewpoint. Conditional probabilities of $\tilde{\boldsymbol{\omega}}_2$-events should be derivable from a mass function or a density function, but we must be a bit more specific. To rule out rather confusing situations which arise only occasionally in practical applications, exemplified by Exercise 3.11.6, we shall impose a convenient assumption that is almost always satisfied in practice.

Convenient Assumption. Either:

(C) Given every $\boldsymbol{\omega}_1$ such that $f_1(\boldsymbol{\omega}_1)$ is positive, $\tilde{\boldsymbol{\omega}}_2$ is a *continuous* random real m-tuple with conditional density function $f_{2|1}(\cdot|\boldsymbol{\omega}_1)$; *or*

(D) Given every $\boldsymbol{\omega}_1$ such that $f_1(\boldsymbol{\omega}_1)$ is positive, $\tilde{\boldsymbol{\omega}}_2$ is a *discrete* random real m-tuple with conditional mass function $f_{2|1}(\cdot|\boldsymbol{\omega}_1)$, and moreover, there is a *countable* subset $\Omega_2^\dagger$ of R^m such that $f_{2|1}(\boldsymbol{\omega}_2|\boldsymbol{\omega}_1) = 0$ for every $\boldsymbol{\omega}_1$ such that $f_1(\boldsymbol{\omega}_1) > 0$ *unless* $\boldsymbol{\omega}_2$ belongs to $\Omega_2^\dagger$.

From *Section 3.9, it is clear that if $\tilde{\boldsymbol{\omega}}_1$ is discrete with mass function $f_1(\cdot)$ and $\tilde{\boldsymbol{\omega}}_2$ is "fully discrete" in that it satisfies (D) of the Convenient Assumption, then the "multiplication rule" implies that $(\tilde{\boldsymbol{\omega}}_1, \tilde{\boldsymbol{\omega}}_2)$ is a *discrete* random real $(k+m)$-tuple with mass function $f_{12}(\cdot, \cdot)$ readily obtained from the multiplication rule:

$$(\#) \qquad f_{12}(\boldsymbol{\omega}_1, \boldsymbol{\omega}_2) = \begin{cases} f_{2|1}(\boldsymbol{\omega}_2|\boldsymbol{\omega}_1) f_1(\boldsymbol{\omega}_1), & f_1(\boldsymbol{\omega}_1) > 0 \\ 0, & f_1(\boldsymbol{\omega}_1) = 0. \end{cases}$$

■ The multiplication rule only related f_{12}, $f_{2|1}$, and f_1 for those $\boldsymbol{\omega}_1$ such that $f_1(\boldsymbol{\omega}_1) > 0$—the only $\boldsymbol{\omega}_1$ for which we have assumed in (D) that

$f_{2|1}(\cdot|\boldsymbol{\omega}_1)$ is defined. But if $f_1(\boldsymbol{\omega}_1^*) = 0$, then every $f_{12}(\boldsymbol{\omega}_1^*, \boldsymbol{\omega}_2)$ must be zero, since otherwise, summing $\boldsymbol{\omega}_2$ out would yield $f_1(\boldsymbol{\omega}_1^*) > 0$, thus producing a contradiction. ■

But an almost identical line of reasoning establishes (#) when $\tilde{\boldsymbol{\omega}}_1$ is *continuous* with *density* function $f_1(\cdot)$ and $\tilde{\boldsymbol{\omega}}_2$ is "fully continuous" in that it satisfies (C) of the Convenient Assumption.

Now we shall argue that (#) should also hold when $f_1(\cdot)$ is a density function and the $f_{2|1}(\cdot|\boldsymbol{\omega}_1)$'s are mass functions. But in this case, $f_{12}(\cdot, \cdot)$ is neither a density function nor a mass function. It is a *hybrid* which must be integrated with respect to $\boldsymbol{\omega}_1$ and summed with respect to $\boldsymbol{\omega}_2$ in finding probabilities of $(\tilde{\boldsymbol{\omega}}_1, \tilde{\boldsymbol{\omega}}_2)$-events and expectations of functions x: $R^{k+m} \to R^1$. That is,

$$E_{12}\{x(\tilde{\boldsymbol{\omega}}_1, \tilde{\boldsymbol{\omega}}_2)\} = \int_{R^k} \sum_{R^m} x(\boldsymbol{\omega}_1, \boldsymbol{\omega}_2) f_{12}(\boldsymbol{\omega}_1, \boldsymbol{\omega}_2)\, d\boldsymbol{\omega}_1$$

for expectations and

$$P_{12}(\Omega') = \int \sum_{\Omega'} f_{12}(\boldsymbol{\omega}_1, \boldsymbol{\omega}_2)\, d\boldsymbol{\omega}_1$$

for probabilities, the "integral-sum-over-Ω'" symbol being interpreted as the integral over $\{\boldsymbol{\omega}_1: (\boldsymbol{\omega}_1, \boldsymbol{\omega}_2) \in \Omega'$ for some $\boldsymbol{\omega}_2 \in R^m\}$ of the sum over $\{\boldsymbol{\omega}_2: (\boldsymbol{\omega}_1, \boldsymbol{\omega}_2) \in \Omega'\}$ of the positive values of $f_{12}(\boldsymbol{\omega}_1, \cdot)$.

■ A little reflection in conjunction with Figure 3.33 should convince you

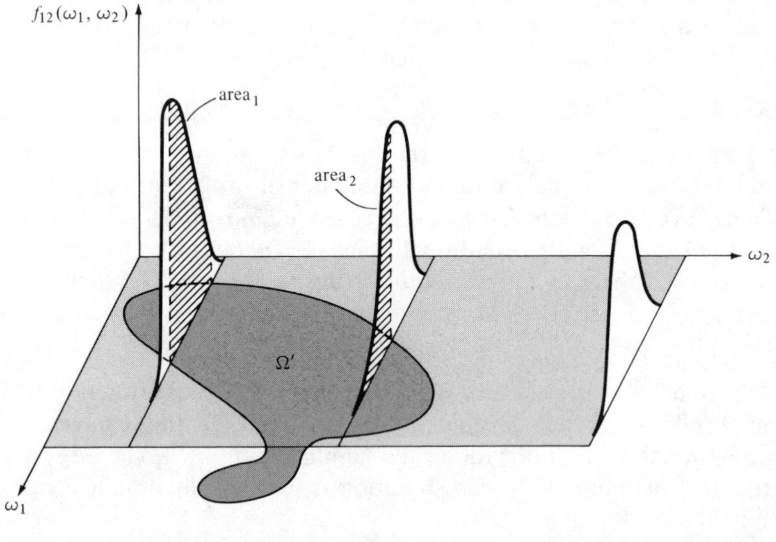

Figure 3.33 Probabilities of $(\tilde{\omega}_1, \tilde{\omega}_2)$-events with $\tilde{\omega}_1$ continuous and $\tilde{\omega}_2$ conditionally always discrete. [*Key*: Lines parallel to the ω_1 axis denote ω_2's in $\Omega_2^\dagger$. Lightly shaded strip denotes $\{\omega_1: f_1(\omega_1) > 0\}$. $f_{12}(\omega_1, \omega_2) = 0$ unless (ω_1, ω_2) belongs to a line parallel to the ω_1 axis, in which case it equals the ordinate of the curve on that line. Then $P_{12}(\Omega') = \text{area}_1 + \text{area}_2$.]

that $f_{12}(\cdot, \cdot)$ as defined by (#) and interpreted in these formulae for expectations and probabilities is the *only* way to make Theorem 3.2.2 work in this case ($\tilde{\boldsymbol{\omega}}_1$ continuous and $\tilde{\boldsymbol{\omega}}_2$ fully discrete). ■

A similar sort of hybrid function is defined by (#) when $f_1(\cdot)$ is a mass function and the $f_{2|1}(\cdot|\boldsymbol{\omega}_1)$'s are density functions, and again the definition (#) must be operated on by a combination of summation and integration. To be consistent with Theorem 3.2.2, expectations of functions $x: R^{k+m} \to R^1$ are given by

$$E_{12}\{x(\tilde{\boldsymbol{\omega}}_1, \tilde{\boldsymbol{\omega}}_2)\} = \sum_{R^k} \int_{R^m} x(\boldsymbol{\omega}_1, \boldsymbol{\omega}_2) f_{12}(\boldsymbol{\omega}_1, \boldsymbol{\omega}_2) d\boldsymbol{\omega}_2,$$

and probabilities of events Ω' in R^{k+m} are given by

$$P_{12}(\Omega') = \sum \int_{\Omega'} f_{12}(\boldsymbol{\omega}_1, \boldsymbol{\omega}_2) d\boldsymbol{\omega}_2,$$

the "sum-integral-over-Ω'" symbol being interpreted as the sum over $\{\boldsymbol{\omega}_1: (\boldsymbol{\omega}_1, \boldsymbol{\omega}_2) \in \Omega' \text{ for some } \boldsymbol{\omega}_2 \in R^m\}$ of the positive values of the integral over $\{\boldsymbol{\omega}_2: (\boldsymbol{\omega}_1, \boldsymbol{\omega}_2) \in \Omega'\}$ of $f_{12}(\boldsymbol{\omega}_1, \cdot)$.

■ Figure 3.34 is the analogue of Figure 3.33 for the case in which $\tilde{\boldsymbol{\omega}}_1$ is discrete and $\tilde{\boldsymbol{\omega}}_2$ is fully continuous. ■

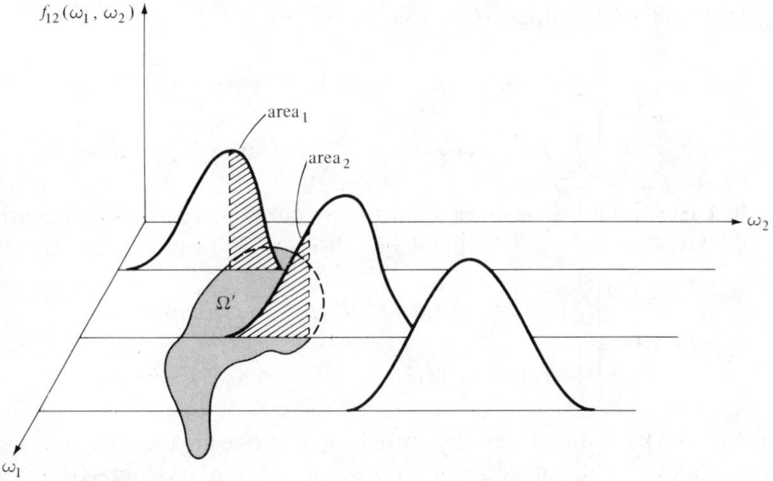

Figure 3.34 Probabilities of $(\tilde{\omega}_1, \tilde{\omega}_2)$-events with $\tilde{\omega}_1$ discrete and $\tilde{\omega}_2$ conditionally always continuous. [*Key*: $f_{12}(\omega_1, \omega_2) = 0$ unless (ω_1, ω_2) belongs to a horizontal line segment (corresponding to one of the countably many possible values of $\tilde{\omega}_1$), in which case it equals the ordinate of the curve on that horizontal line. Then $P_{12}(\Omega') = \text{area}_1 + \text{area}_2$.]

From here on, everything is easy. We have argued that (#) is true in each of the four cases of interest; namely:

Case 1: $\tilde{\boldsymbol{\omega}}_1$ discrete, $\tilde{\boldsymbol{\omega}}_2$ given $\boldsymbol{\omega}_1$ discrete in the sense of Convenient Assumption (D)

Case 2: $\tilde{\boldsymbol{\omega}}_1$ discrete, $\tilde{\boldsymbol{\omega}}_2$ given $\boldsymbol{\omega}_1$ continuous in the sense of Convenient Assumption (C)

Case *3*: $\tilde{\boldsymbol{\omega}}_1$ continuous, $\tilde{\boldsymbol{\omega}}_2$ given $\boldsymbol{\omega}_1$ discrete in the sense of Convenient Assumption (D)

Case 4: $\tilde{\boldsymbol{\omega}}_1$ continuous, $\tilde{\boldsymbol{\omega}}_2$ given $\boldsymbol{\omega}_1$ continuous in the sense of Convenient Assumption (C)

We shall now reverse the order, from "Ω_1 before Ω_2" to "Ω_2 before Ω_1." The first task in so doing is to characterize the marginal density function (in cases 2 and 4) or marginal mass function (in cases 1 and 3) of $\tilde{\boldsymbol{\omega}}_2$. Theorem 3.11.1(A) is the counterpart of Theorem 3.7.2(A).

Theorem 3.11.1(A). Suppose that $\tilde{\boldsymbol{\omega}}_1$ and $\tilde{\boldsymbol{\omega}}_2$ in the "Ω_1-before-Ω_2" representation of $\Omega_1 \times \Omega_2$ satisfy the assumptions of any of cases 1 through 4. Suppose that f_{12}: $R^{k+m} \to R^1$ is defined for every $(\boldsymbol{\omega}_1, \boldsymbol{\omega}_2)$ by

$$f_{12}(\boldsymbol{\omega}_1, \boldsymbol{\omega}_2) = \begin{cases} f_{2|1}(\boldsymbol{\omega}_2|\boldsymbol{\omega}_1)f_1(\boldsymbol{\omega}_1), & f_1(\boldsymbol{\omega}_1) > 0 \\ 0, & f_1(\boldsymbol{\omega}_1) = 0. \end{cases}$$

Then, *unconditionally as regards* $\tilde{\boldsymbol{\omega}}_1$, *either*:

(AD) In cases 1 and 3, $\tilde{\boldsymbol{\omega}}_2$ is a *discrete* random real m-tuple with mass function $f_2(\cdot)$ defined for all $\boldsymbol{\omega}_2$ in R^m by

$$f_2(\boldsymbol{\omega}_2) = \begin{cases} \sum \{f_{12}(\boldsymbol{\omega}_1, \boldsymbol{\omega}_2): \boldsymbol{\omega}_1 \in R^k\}, & \text{case 1 pertains} \\ \int_{R^k} f_{12}(\boldsymbol{\omega}_1, \boldsymbol{\omega}_2)\, d\boldsymbol{\omega}_1, & \text{case 2 pertains; } \textit{or} \end{cases}$$

(AC) in cases 2 and 4, $\tilde{\boldsymbol{\omega}}_2$ is a *continuous* random real m-tuple with density function $f_2(\cdot)$ defined for all $\boldsymbol{\omega}_2$ in R^m by

$$f_2(\boldsymbol{\omega}_2) = \begin{cases} \sum \{f_{12}(\boldsymbol{\omega}_1, \boldsymbol{\omega}_2): \boldsymbol{\omega}_1 \in R^k\}, & \text{case 2 pertains} \\ \int_{R^k} f_{12}(\boldsymbol{\omega}_1, \boldsymbol{\omega}_2)\, d\boldsymbol{\omega}_1, & \text{case 4 pertains.} \end{cases}$$

Theorem 3.11.1(A) characterizes the initial $\tilde{\boldsymbol{\omega}}_2$-$\mathscr{E}$-move in the "$\Omega_2$-before-$\Omega_1$" representation. The subsequent copies of Ω_1 are characterized by Theorem 3.11.1(B), the counterpart of Theorem 3.7.2(B).

Theorem 3.11.1(B). Together with assumptions and definitions as in Theorem 3.11.1(A), also define $\Omega_2^* = \{\boldsymbol{\omega}_2: f_2(\boldsymbol{\omega}_2) > 0\}$. Then, *conditional upon each* $\boldsymbol{\omega}_2^*$ *in* Ω_2^*, *either*:

(BD) In cases 1 and 2, $\tilde{\boldsymbol{\omega}}_1$ is a *discrete* random real k-tuple with mass function $f_{1|2}(\cdot|\boldsymbol{\omega}_2^*)$ defined for all $\boldsymbol{\omega}_1$ in R^k by

$$f_{1|2}(\boldsymbol{\omega}_1|\boldsymbol{\omega}_2^*) = \frac{f_{12}(\boldsymbol{\omega}_1, \boldsymbol{\omega}_2^*)}{f_2(\boldsymbol{\omega}_2^*)},$$

or

(BC) In cases 3 and 4, $\tilde{\boldsymbol{\omega}}_1$ is a *continuous* random real k-tuple with density function $f_{1|2}(\cdot|\boldsymbol{\omega}_2^*)$ defined for all $\boldsymbol{\omega}_1$ in R^k by

$$f_{1|2}(\boldsymbol{\omega}_1|\boldsymbol{\omega}_2^*) = \frac{f_{12}(\boldsymbol{\omega}_1, \boldsymbol{\omega}_2^*)}{f_2(\boldsymbol{\omega}_2^*)}.$$

■* *Proof. For Theorem 3.11.1(A)*, we note that it is evident, with a little thought, that

$$(\dagger) \qquad P_2(\Omega_2') = E_1\{P_{2|1}(\Omega_2'|\boldsymbol{\omega}_1)\},$$

which in cases 1 and 3 becomes

$$(\dagger D) \qquad P_2(\Omega_2') = E_1\left\{\sum_{\Omega_2'} f_{2|1}(\boldsymbol{\omega}_2|\tilde{\boldsymbol{\omega}}_1)\right\}$$

and, in cases 2 and 4,

$$(\dagger C) \qquad P_2(\Omega_2') = E_1\left\{\int_{\Omega_2'} f_{2|1}(\boldsymbol{\omega}_2|\tilde{\boldsymbol{\omega}}_1)\,d\boldsymbol{\omega}_2\right\}.$$

Now,

$$E_1\left\{\sum_{\Omega_2'} f_{2|1}(\boldsymbol{\omega}_2|\tilde{\boldsymbol{\omega}}_1)\right\} =^1 \begin{cases} \sum_{R^k}\sum_{\Omega_2'} f_{2|1}(\boldsymbol{\omega}_2|\boldsymbol{\omega}_1) f_1(\boldsymbol{\omega}_1), & \text{case 1} \\ \int_{R^k}\sum_{\Omega_2'} f_{2|1}(\boldsymbol{\omega}_2|\boldsymbol{\omega}_1) f_1(\boldsymbol{\omega}_1)\,d\boldsymbol{\omega}_1, & \text{case 3} \end{cases}$$

$$=^2 \begin{cases} \sum_{\Omega_2'}\sum_{R^k} f_{2|1}(\boldsymbol{\omega}_2|\boldsymbol{\omega}_1) f_1(\boldsymbol{\omega}_1), & \text{case 1} \\ \sum_{\Omega_2'}\int_{R^k} f_{2|1}(\boldsymbol{\omega}_2|\boldsymbol{\omega}_1) f_1(\boldsymbol{\omega}_1)\,d\boldsymbol{\omega}_1, & \text{case 3} \end{cases}$$

$$=^3 \begin{cases} \sum_{\Omega_2'}\sum_{R^k} f_{12}(\boldsymbol{\omega}_1, \boldsymbol{\omega}_2), & \text{case 1} \\ \sum_{\Omega_2'}\int_{R^k} f_{12}(\boldsymbol{\omega}_1, \boldsymbol{\omega}_2)\,d\boldsymbol{\omega}_1, & \text{case 3} \end{cases}$$

$$=^4 \begin{cases} \sum_{\Omega_2'} f_2(\boldsymbol{\omega}_2), & \text{case 1} \\ \sum_{\Omega_2'} f_2(\boldsymbol{\omega}_2), & \text{case 3.} \end{cases}$$

(Equality 1 is by Definitions 3.9.2 and 3.10.2 of E_1-expectation; 2 is interchanging orders of summation or of summation and integration; 3 is by definition of f_{12}; and 4 is by definition of f_2.) Hence in cases 1 and 3, the probability $P_2(\Omega_2')$ of any $\tilde{\boldsymbol{\omega}}_2$-event Ω_2' is the sum of the $f_2(\boldsymbol{\omega}_2)$'s over all $\boldsymbol{\omega}_2$ in Ω_2'. Therefore, $\tilde{\boldsymbol{\omega}}_2$ is *discrete* with *mass* function $f_2(\cdot)$. This proves (AD) from (†D). But (AC) follows from (†C) by replacing "$\Sigma_{\Omega_2'}$" with "$\int_{\Omega_2'} \cdots d\boldsymbol{\omega}_2$," "case 1" with "case 2," "case 3" with "case 4," and "summation and summation" by "summation and integration." The conclusion is that $P_2(\Omega_2') = \int_{\Omega_2'} f_2(\boldsymbol{\omega}_2)d\boldsymbol{\omega}_2$ for all $\tilde{\boldsymbol{\omega}}_2$-events Ω_2', implying that $f_2(\cdot)$ is a density function and that $\tilde{\boldsymbol{\omega}}_2$ is unconditionally a continuous random real m-tuple. *For Theorem 3.11.1(B)*, note that $f_{1|2}(\boldsymbol{\omega}_1|\boldsymbol{\omega}_2^*)$ is not defined if $\boldsymbol{\omega}_2^* \notin \Omega_2^*$, but it does not *need* to be defined in order for $f_{12}(\boldsymbol{\omega}_1, \boldsymbol{\omega}_2^*)$ to be expressible unambiguously as

$$(\#\,\#) \qquad f_{12}(\boldsymbol{\omega}_1, \boldsymbol{\omega}_2^*) = \begin{cases} f_{1|2}(\boldsymbol{\omega}_1|\boldsymbol{\omega}_2^*)f_2(\boldsymbol{\omega}_2^*), & f_2(\boldsymbol{\omega}_2^*) > 0 \\ 0, & f_2(\boldsymbol{\omega}_2^*) = 0, \end{cases}$$

this equation being (essentially) the only one consistent with Theorem 3.2.2 applied to the "Ω_2-before-Ω_1" representation of $\Omega_1 \times \Omega_2$. From (# #) both (BD) and (BC) follow readily. ■

We now give a few examples of practical significance.

Example 3.11.1

Suppose that $\Omega_1 = \{\omega_1: 0 \le \omega_1 \le 1\}$, where ω_1 denotes the probability that a phenomenon will occur on any given instance. Suppose that n instances are to be observed. Then, conditional on n and ω_1, the number $\tilde{\omega}_2$ of occurrences has the binomial mass function $f_b(\cdot|n, \omega_1)$, defined in Example 3.9.2. This is $f_{2|1}(\cdot|\omega_1)$. Moreover, for every ω_1 in [0, 1] it is clear that $P_b(\{0, 1, \ldots, n\}|n, \omega_1) = 1$, so that Convenient Assumption (D) obtains. Now suppose that $\mathscr{D}$'s judgments about $\tilde{\omega}_1$ *prior* to observing instances were represented by a beta density function $f_\beta(\cdot|r', n')$ as defined in Example 3.10.6, with parameters $r' > 0$ and $n' > r'$. ("Priming" the parameters here is useful, as you will soon see.) This beta density function is $f_1(\cdot)$. Now, *case 3* pertains, and $(\tilde{\omega}_1, \tilde{\omega}_2)$ has the joint density-mass function $f_{12}(\cdot, \cdot)$, defined in Theorem 3.11.1(A) for this problem as $f_b(\omega_2|n, \omega_1)f_\beta(\omega_1|r', n')$, where $f_\beta(\omega_1|r', n') > 0$ and zero elsewhere. That is,

$$f_{12}(\omega_1, \omega_2) = \frac{(n'-1)!n!}{(r'-1)!(n'-r'-1)!\omega_2!(n-\omega_2)!}\,\omega_1^{r'+\omega_2-1}(1-\omega_1)^{[n'+n]-[r'+\omega_2]-1}$$

provided that $\omega_1 \in [0, 1]$ and $\omega_2 \in \{0, 1, \ldots, n\}$. A difficult integration with respect to ω_1 shows that $\tilde{\omega}_2$ is discrete with mass function

$$f_2(\omega_2) = \frac{(n'-1)!n!(r'+\omega_2-1)!(n'+n-r'-\omega_2-1)!}{(r'-1)!(n'-r'-1)!\omega_2!(n-\omega_2)!(n'+n-1)!}$$

for every ω_2 in $\{0, 1, \ldots, n\}$, and $f_2(\omega_2) = 0$ for all other ω_2. Then, by (BC) in Theorem 3.11.1(B),

$$f_{1|2}(\omega_1|\omega_2) = \frac{(n''-1)!}{(r''-1)!(n''-r''-1)!}\,\omega_1^{r''-1}(1-\omega_1)^{n''-r''-1} = f_\beta(\omega_1|r'', n'')$$

for all $\omega_1 \in [0, 1]$, equaling zero elsewhere, where $r'' = r' + \omega_2$ and $n'' = n' + n$. Hence $\mathscr{D}$'s judgments about $\tilde{\omega}_1$ *posterior* to observing $\omega_2 \in \{0, 1, \ldots, n\}$ occurrences of the phenomenon in the n instances are expressible by the beta density function $f_\beta(\cdot|r' + \omega_2, n' + n)$ if his *prior* judgments were expressed by the beta density function $f_\beta(\cdot|r', n')$. Note how easy it is to calculate the parameters r'' and n'' of the "posterior" density function once the observations of instances have been made. This implementationally important ease is a special property of the beta density function in conjunction with observations having the binomial mass function, a property called *natural conjugacy* in Raiffa and Schlaifer [**115**] and LaValle [**78**], for example.

■ There is a nice trick which enables one to avoid difficult integrations in certain cases, especially those in which natural conjugacy pertains. The trick is called the "recognition method" (see Roberts [**119**]), and it proceeds as follows. (1) Look at $f_{12}(\cdot, \cdot)$, and isolate *all* those factors in its formula which depend *only* upon $\boldsymbol{\omega}_1$ [e.g., $\omega_1^{r'+\omega_2-1}(1-\omega_1)^{[n'+n]-[r'+\omega_2]-1}$ in Example 3.11.1]. (2) Then see if you *recognize* those factors as the ones in some family of density functions (or mass functions) for $\tilde{\boldsymbol{\omega}}_1$ (e.g., the beta family in Example 3.11.1). (3) If so, then it follows that $f_{1|2}(\cdot|\boldsymbol{\omega}_2)$ *is* that member of the family having the parameters which give the isolated factors in (1) [e.g., $f_\beta(\cdot|r' + \omega_2, n' + n)$ in Example 3.11.1]. The reason for the conclusion (3) is that

$$f_{12}(\boldsymbol{\omega}_1, \boldsymbol{\omega}_2) = f_{1|2}(\boldsymbol{\omega}_1|\boldsymbol{\omega}_2)f_2(\boldsymbol{\omega}_2)$$

where positive, implying that $f_{12}(\cdot, \boldsymbol{\omega}_2)$ is *proportional to* $f_{1|2}(\cdot|\boldsymbol{\omega}_2)$ for each fixed $\boldsymbol{\omega}_2$, since the proportionality constant $f_2(\boldsymbol{\omega}_2)$ obviously cannot depend upon $\boldsymbol{\omega}_1$ [because $\boldsymbol{\omega}_1$ was summed or integrated out of $f_{12}(\cdot, \cdot)$ according to Theorem 3.11.1(A)]. Next, having determined $f_{1|2}(\cdot|\boldsymbol{\omega}_2)$, one can determine $f_2(\boldsymbol{\omega}_2)$ by the equation[cc]

$$f_2(\boldsymbol{\omega}_2) = \frac{f_{12}(\boldsymbol{\omega}_1, \boldsymbol{\omega}_2)}{f_{1|2}(\boldsymbol{\omega}_1|\boldsymbol{\omega}_2)},$$

obvious from (# #) in the proof of Theorem 3.11.1. This recognition trick works beautifully in all our examples here. ■

Example 3.11.2

In Example 3.9.6 we defined the Poisson mass function to characterize the number $\tilde{\omega}_2$ of occurrences of a given phenomenon during time span t under given assumptions about the "interoccurrence time." Suppose that the intensity rate $\tilde{\omega}_1$ is unknown to $\mathscr{D}$, who expresses his judgments about it in the form of a gamma density function $f_\gamma(\cdot|r', t')$, defined in Example 3.10.5. Hence $f_1(\cdot) = f_\gamma(\cdot|r', t')$, and $f_{2|1}(\cdot|\omega_1) = f_{Po}(\cdot|\omega_1 t)$ for

[cc] Determining where $f_2(\boldsymbol{\omega}_2) > 0$ is easy in view of the fact that $f_{1|2}(\cdot|\boldsymbol{\omega}_2)$ is defined *only* at such points.

every $\omega_1 > 0$. Therefore, $f_{12}(\omega_1, \omega_2) = 0$ unless $\omega_1 > 0$ and ω_2 is a nonnegative integer, in which case

$$f_{12}(\omega_1, \omega_2) = \frac{t' e^{-t'\omega_1}(t'\omega_1)^{r'-1}}{(r'-1)!} \cdot \frac{e^{-\omega_1 t}(\omega_1 t)^{\omega_2}}{\omega_2!}$$
$$= \frac{t'^{r'} t^{\omega_2}}{(r'-1)!\omega_2!} \cdot [e^{-(t'+t)\omega_1} \omega_1^{(r'+\omega_2)-1}],$$

the factor in $[\cdots]$ constituting the term of interest in determining $f_{1|2}(\cdot|\omega_2)$. That factor is proportional to $f_\gamma(\cdot|r'+\omega_2, t'+t)$, which *must therefore be* $f_{1|2}(\cdot|\omega_2)$, for every nonnegative integer ω_2. We can thus obtain

$$f_2(\omega_2) = \frac{f_{Po}(\omega_2|\omega_1 t) f_\gamma(\omega_1|r', t')}{f_\gamma(\omega_1|r'+\omega_2, t'+t)}$$
$$= \frac{\dfrac{t'^{r'} t^{\omega_2} e^{-(t'+t)\omega_1}(\omega_1)^{r'+\omega_2-1}}{(r'-1)!\omega_2!}}{\dfrac{(t'+t)e^{-(t'+t)\omega_1}[(t'+t)\omega_1]^{r'+\omega_2-1}}{(r'+\omega_2-1)!}}$$
$$= \frac{(r'+\omega_2-1)!\, t'^{r'} t^{\omega_2}}{(r'-1)!\omega_2!(t'+t)^{r'+\omega_2}}$$

for every nonnegative integer ω_2. [$f_2(\cdot)$ is called the *negative binomial* mass function $f_{nb}(\cdot|r', t'/[t'+t])$ with parameters r' and $t'/[t'+t]$. If $\tilde{\omega}_2$ has the mass function $f_{nb}(\cdot|r', t'/[t'+t])$ and if r' is a positive integer, then $\tilde{\omega} = \tilde{\omega}_2 + r'$ has the *Pascal* mass function $f_{Pa}(\cdot|r', t'/[t'+t])$, defined in Exercise 3.9.2.]

Our final example with $k = m = 1$ is the Normal case.

Example 3.11.3

Suppose that $\tilde{\omega}_1$ is some real-valued quantity of interest and that $\mathscr{D}$'s judgments about $\tilde{\omega}_1$ are expressed in the form of a Normal density function $f_N(\cdot|m', V')$, defined in Example 3.10.7. Suppose also that $\mathscr{D}$ can obtain information about $\tilde{\omega}_1$ in the form of an observation ω_2 whose conditional density function is Normal and centered at ω_1; that is, $f_{2|1}(\cdot|\omega_1) = f_N(\cdot|\omega_1, V)$. Then case 4 obtains, and

$$f_{12}(\omega_1, \omega_2) = f_N(\omega_2|\omega_1, V) f_N(\omega_1|m', V')$$
$$= (2\pi)^{-1}(V'V)^{-1/2} e^{-(1/2)[(\omega_2-\omega_1)^2/V + (\omega_1-m')^2/V']}$$

for every (ω_1, ω_2) in R^2. Now,

$$e^{-(1/2)[(\omega_2-\omega_1)^2/V+(\omega_1-m')^2/V']} = e^{-(1/2)[(\omega_2-m')^2/V^* + (\omega_1-m'')^2/V'']}$$

for $V'' = [(V)^{-1} + (V')^{-1}]^{-1}$, $m'' = V''[(V')^{-1}m' + (V)^{-1}\omega_2]$, and $V^* = V' + V$, by elementary algebra. Hence

$$f_{12}(\omega_1, \omega_2) = (2\pi)^{-1}(V'V)^{-1/2}e^{-(1/2)(\omega_2-m')^2/V^*}[e^{-(1/2)(\omega_1-m'')^2/V''}],$$

the factor in $[\cdots]$ again constituting the term of interest. It is proportional at every ω_1 to

$$f_N(\cdot|m'', V'') = (2\pi V'')^{-1/2}e^{-(1/2)(\omega_1-m'')^2/V''},$$

which *must therefore be* $f_{1|2}(\cdot|\omega_2)$. Hence $\mathscr{D}$'s judgments about $\tilde{\omega}_1$ *posterior* to observing ω_2 are expressed by the Normal density function $f_N(\cdot|m'', V'')$ if his judgments *prior* to observing ω_2 were expressed by $f_N(\cdot|m', V')$. Now,

$$\begin{aligned} f_2(\omega_2) &= \frac{f_{12}(\omega_1, \omega_2)}{f_{1|2}(\omega_1|\omega_2)} \\ &= \frac{(2\pi)^{-1}(V'V)^{-1/2}e^{-(1/2)(\omega_2-m')^2/V^*}}{[2\pi V'']^{-1/2}} \\ &= (2\pi V^*)^{-1/2}e^{-(1/2)(\omega_2-m')^2/V^*} \\ &= f_N(\omega_2|m', V^*), \end{aligned}$$

since $(V'V/V'')^{-1/2} = (V'V/[(V')^{-1} + (V)^{-1}]^{-1})^{-1/2} = (V' + V)^{-1/2} = (V^*)^{-1/2}$. Hence $\mathscr{D}$'s judgments about $\tilde{\omega}_2$, unconditionally as regards $\tilde{\omega}_1$, are expressed by the Normal density function $f_N(\cdot|m', V^*)$.

There are important generalizations to more arbitrary k and m of Examples 3.11.1 and 3.11.3.

Example 3.11.4

Suppose that $\tilde{\boldsymbol{\omega}}_1$ denotes the unknown k-tuple of probabilities of finding the phenomenon in states $1, \ldots, k$ out of $k + 1$ states in any given instance. Suppose that $\mathscr{D}$ has quantified his judgments about $\tilde{\boldsymbol{\omega}}_1$ in the form of a k-variate beta density function $f_1(\cdot) = f_\beta^{(k)}(\cdot|\mathbf{r}', n')$, defined in Example 3.10.9. If n instances are now to be observed, then the k-tuple $\tilde{\boldsymbol{\omega}}_2$ (in which the ith component denotes the number of instances out of n in which the phenomenon is to be found in state i) has the multinomial mass function $f_{mu}^{(k)}(\cdot|n, \boldsymbol{\omega}_1)$, *conditional upon* $\boldsymbol{\omega}_1$, as defined in Example 3.9.3. An adaptation of the reasoning in Example 3.11.1, abbreviated by using natural conjugacy, establishes that

$$f_{1|2}(\boldsymbol{\omega}_1|\boldsymbol{\omega}_2) = f_\beta^{(k)}(\boldsymbol{\omega}_1|\mathbf{r}' + \boldsymbol{\omega}_2, n' + n)$$

for every $\boldsymbol{\omega}_2$ with nonnegative-integer components which sum to at most

n; and that $\tilde{\omega}_2$ is unconditionally discrete with mass function

$$f_2(\omega_2) = \frac{(n'-1)!N!}{(n'+n-1)!} \prod_{i=1}^{k+1} \left[\frac{(r_i'+\omega_{2i}-1)!}{(r_i'-1)!(\omega_{2i})!}\right]$$
$$= \frac{\prod_{i=1}^{k+1} C_{\omega_{2i}}^{r_i'+\omega_{2i}-1}}{C_n^{n'+n}}$$

for all $\boldsymbol{\omega}_2 = (\omega_{21}, \ldots, \omega_{2k})$ with nonnegative-integer components summing to no more than n, with $\omega_{2(k+1)} = n - \Sigma_{i=1}^k \omega_{2i}$ and $r_{k+1}' = n' - \Sigma_{i=1}^k r_i'$. This mass function is called the *hypermultinomial* and is denoted by $f_{hmu}^{(k)}(\cdot|\mathbf{r}', n', n)$. You may derive these facts as Exercise 3.11.1. It is easy to verify that everything here reduces to Example 3.11.1 when $k = 1$.

Example 3.11.5

Our generalization of Example 3.11.3 consists in letting k exceed 1; in assuming that $\mathscr{D}$'s prior judgments about $\tilde{\omega}_1$ are expressed by the k-variate Normal density function $f_N^{(k)}(\cdot|\mathbf{m}', \mathbf{V}')$, defined in Example 3.10.10; and in supposing that $\mathscr{D}$ can obtain information about $\tilde{\omega}_1$ in the form of a t-tuple observation $\tilde{\omega}_2$ which has the conditional density function $f_N^{(t)}(\cdot|\mathbf{A}\boldsymbol{\omega}_1 + \mathbf{b}, \mathbf{V})$ given $\boldsymbol{\omega}_1$, where $\mathbf{A}$ is a fixed, t-by-k matrix parameter, $\mathbf{b}$ is a t-tuple parameter, and $\mathbf{V}$ is a t-by-t, positive definite, and symmetric matrix parameter. t may be equal to, greater than, or less than k. Then $\mathscr{D}$'s judgments about $\tilde{\omega}_1$ *posterior* to observing $\boldsymbol{\omega}_2$ are expressed by

$$f_{1|2}(\cdot|\boldsymbol{\omega}_2) = f_N^{(k)}(\cdot|\mathbf{m}'', \mathbf{V}''),$$

where

$$\mathbf{V}'' = [(\mathbf{V}')^{-1} + \mathbf{A}'(\mathbf{V})^{-1}\mathbf{A}]^{-1}$$

and

$$\mathbf{m}'' = \mathbf{V}''[(\mathbf{V}')^{-1}\mathbf{m}' + \mathbf{A}'(\mathbf{V})^{-1}(\boldsymbol{\omega}_2 - \mathbf{b})].$$

[*Caution*: $\mathbf{A}'$ denotes the *transpose* of $\mathbf{A}$. Elsewhere here the prime is just a convenient distinguishing affix.] Furthermore, $\mathscr{D}$'s judgments about $\tilde{\omega}_2$, unconditionally as regards $\tilde{\omega}_1$, are expressed by

$$f_2(\boldsymbol{\omega}_2) = f_N^{(t)}(\boldsymbol{\omega}_2|\mathbf{A}\mathbf{m}' + \mathbf{b}, \mathbf{A}\mathbf{V}'\mathbf{A}' + \mathbf{V}).$$

You may derive these facts as *Exercise 3.11.2.

Example 3.11.5 is only one of many generalizations of Example 3.11.3. Others may be found in Chapters 11–13 of Raiffa and Schlaifer [**115**] and in Chapter 14 of LaValle [**78**]; still others are contained in the papers cited in the preface to [**78**].

Exercise 3.11.1

Derive all asserted facts in Example 3.11.4.

***Exercise 3.11.2**

Derive all asserted facts in Example 3.11.5.

Exercise 3.11.3

Suppose that $\tilde{\omega}_1$ has the uniform density function $f_1(\cdot) = f_u(\cdot|a, b)$ defined in Example 3.10.4; and that, conditional upon ω_1, $\tilde{\omega}_2$ satisfies the Convenient Assumption, with a density or mass function $f_{2|1}(\cdot|\omega_1)$ for every ω_1 in $[a, b]$.

(A) Show that

$$f_{1|2}(\omega_1|\omega_2) = \begin{cases} \dfrac{f_{2|1}(\omega_2|\omega_1)}{\displaystyle\int_a^b f_{2|1}(\omega_2|\omega_1)d\omega_1}, & \omega_1 \in [a, b] \\ 0, & \omega_1 \notin [a, b], \end{cases}$$

for every ω_2 such that $\int_a^b f_{2|1}(\omega_2|\omega_1)d\omega_1 > 0$.

(B) Show that

$$f_2(\omega_2) = \frac{1}{(b-a)} \int_a^b f_{2|1}(\omega_2|\omega_1)d\omega_1$$

for every ω_2 in R^1.

(C) Verify that the density functions in parts (A) and (B) satisfy (B1) and (B2) of Definition 3.10.1.

(D) The assertions in (A) and (B) are valid for *any* a and b such that $a < b$. What does $f_{1|2}(\cdot|\omega_2)$ become as $a \to -\infty$ and $b \to +\infty$?

(E) What does $f_2(\cdot)$ become as $a \to -\infty$ and $b \to +\infty$? Is the result a density or mass function? Is $f_u(\cdot|-\infty, +\infty) = \lim \{f_u(\cdot|a, b): a \to -\infty, b \to +\infty\}$ a density function?

(F) Can you find a reasonable interpretation of the limit functions in (D) and (E)?

[*Hint*: In (D)–(F) you may wish to assume, provisionally at least, that $\int_{-\infty}^{\infty} f_{2|1}(\omega_2|\omega_1)d\omega_1$ is finite.]

Exercise 3.11.4: Chain Rule and Bayes' Theorem Extended

Suppose, for $i = 1, \ldots, n$, that $\tilde{\boldsymbol{\omega}}_i$ is a random real k_i-tuple, and suppose that $\tilde{\boldsymbol{\omega}}_1$ is either discrete with mass function $f_1(\cdot)$ or continuous with density function $f_1(\cdot)$. Further suppose, inductively, that, given each $(\boldsymbol{\omega}_1, \ldots, \boldsymbol{\omega}_i)$ such that

$$f_{1\cdots i}(\boldsymbol{\omega}_1, \ldots, \boldsymbol{\omega}_i) = f_1(\boldsymbol{\omega}_1) \cdot f_{2|1}(\boldsymbol{\omega}_2|\boldsymbol{\omega}_1) \cdot \cdots \cdot f_{i|1\cdots(i-1)}(\boldsymbol{\omega}_i|\boldsymbol{\omega}_1, \ldots, \boldsymbol{\omega}_{i-1})$$
$$= f_1(\boldsymbol{\omega}_1) \prod_{j=2}^{i} f_{j|1\cdots(j-1)}(\boldsymbol{\omega}_j|\boldsymbol{\omega}_1, \ldots, \boldsymbol{\omega}_{j-1})$$

is positive, $\tilde{\boldsymbol{\omega}}_{i+1}$ is either discrete with mass function $f_{i+1|1\cdots i}(\cdot|\boldsymbol{\omega}_1, \ldots, \boldsymbol{\omega}_i)$ or[dd] continuous with density function $f_{i+1|1\cdots i}(\cdot|\boldsymbol{\omega}_1, \ldots, \boldsymbol{\omega}_i)$.

(A) Let $k = \Sigma_{i=1}^n k_i$. Argue that probabilities of $(\tilde{\boldsymbol{\omega}}_1, \ldots, \tilde{\boldsymbol{\omega}}_n)$ events should be found by summing (over discrete $\boldsymbol{\omega}_i$) and integrating (over continuous $\boldsymbol{\omega}_i$) the function $f_{1\cdots n}$: $R^k \to R^1$ defined by

$$f_{1\cdots n}(\boldsymbol{\omega}_1, \ldots, \boldsymbol{\omega}_n) = f_1(\boldsymbol{\omega}_1) \prod_{j=2}^{n} f_{j|1\cdots(j-1)}(\boldsymbol{\omega}_j|\boldsymbol{\omega}_1, \ldots, \boldsymbol{\omega}_{j-1})$$

wherever this product is well defined, and zero elsewhere.

(B) Let x: $R^k \to R^1$. Argue that the expectation of $x(\tilde{\boldsymbol{\omega}}_1, \ldots, \tilde{\boldsymbol{\omega}}_n)$ should be found by summing (over discrete $\boldsymbol{\omega}_i$) and integrating (over continuous $\boldsymbol{\omega}_i$) the function $x(\cdot, \cdot, \ldots, \cdot)f_{1\cdots n}(\cdot, \cdot, \ldots, \cdot)$.

(C) Argue that $(\tilde{\boldsymbol{\omega}}_{j+1}, \ldots, \tilde{\boldsymbol{\omega}}_n)$ has the mass, density, or hybrid function $f_{(j+1)\cdots n}$ defined by

$$f_{(j+1)\cdots n}(\boldsymbol{\omega}_{j+1}, \ldots, \boldsymbol{\omega}_n) = \int_{R^{k_j}} \cdots \int_{R^{k_1}} f_{1\cdots n}(\boldsymbol{\omega}_1, \ldots, \boldsymbol{\omega}_n)\, d\boldsymbol{\omega}_1 \cdots d\boldsymbol{\omega}_j,$$

the integrals being interpreted as sums for those $\boldsymbol{\omega}_i$ (with $i \le j$) which are discrete.

(D) Argue that, conditionally on any $(\boldsymbol{\omega}_{j+1}^*, \ldots, \boldsymbol{\omega}_n^*)$ such that $f_{(j+1)\cdots n}(\boldsymbol{\omega}_{j+1}^*, \ldots, \boldsymbol{\omega}_n^*)$ is positive, $(\tilde{\boldsymbol{\omega}}_1, \ldots, \tilde{\boldsymbol{\omega}}_j)$ has the mass, density, or hybrid function $f_{1\cdots j|(j+1)\cdots n}(\cdot, \cdot, \ldots, \cdot|\boldsymbol{\omega}_{j+1}^*, \ldots, \boldsymbol{\omega}_n^*)$ defined by

$$f_{1\cdots j|(j+1)\cdots n}(\boldsymbol{\omega}_1, \ldots, \boldsymbol{\omega}_j|\boldsymbol{\omega}_{j+1}^*, \ldots, \boldsymbol{\omega}_n^*) = \frac{f_{1\cdots n}(\boldsymbol{\omega}_1, \ldots, \boldsymbol{\omega}_j, \boldsymbol{\omega}_{j+1}^*, \ldots, \boldsymbol{\omega}_n^*)}{f_{(j+1)\cdots n}(\boldsymbol{\omega}_{j+1}^*, \ldots, \boldsymbol{\omega}_n^*)}.$$

(E) Let the "is proportional to" symbol $\propto$ be interpreted in

$$y(\cdot) \propto w(\cdot)z(\cdot)$$

as meaning "there exists a K *which does not depend upon* ζ such that

$$y(\zeta) = Kw(\zeta)z(\zeta)$$

for every possible ζ." Then argue that

$$f_{1|2\cdots n}(\cdot|\boldsymbol{\omega}_2^*, \ldots, \boldsymbol{\omega}_n^*) \propto f_1(\cdot) \prod_{j=2}^{n} f_{j|1\cdots(j-1)}(\boldsymbol{\omega}_j^*|\cdot, \boldsymbol{\omega}_2^*, \ldots, \boldsymbol{\omega}_{j-1}^*).$$

[This is analogous to Exercise 3.7.3(A).]

Exercise 3.11.5: Change of Variable and Bayes' Theorem

[Refer to Exercise 3.11.4(E) for definition of "$\propto$" and to Exercise 3.10.10(A) for the change-of-variable theorem.] Let $\tilde{\boldsymbol{\xi}}_{\text{I}}$ be a discrete or

[dd] By analogy with the last part of Convenient Assumption (D), assume the existence of a *fixed countable* set Ω_{i+1}' in $R^{k_{i+1}}$ such that $f_{i+1|1\cdots i}(\boldsymbol{\omega}_{i+1}|\boldsymbol{\omega}_1, \ldots, \boldsymbol{\omega}_i)$ always equals zero unless $\boldsymbol{\omega}_{i+1}$ belongs to Ω_{i+1}'.

continuous random real k-tuple with mass or density function $f_{\mathrm{I}}(\cdot)$; and let $\tilde{\boldsymbol{\omega}}_1 = \boldsymbol{\omega}_1(\tilde{\boldsymbol{\xi}}_{\mathrm{I}})$ for $\boldsymbol{\omega}_1(\cdot)$ a one-to-one function on $\{\boldsymbol{\xi}_{\mathrm{I}}: f_{\mathrm{I}}(\boldsymbol{\xi}_{\mathrm{I}}) > 0\}$ which, if $\tilde{\boldsymbol{\xi}}_{\mathrm{I}}$ is continuous, is also continuously differentiable with nonvanishing Jacobian determinant. Similarly, let $\tilde{\boldsymbol{\xi}}_{\mathrm{II}}$ be *conditionally on* $\tilde{\boldsymbol{\xi}}_{\mathrm{I}}$ either continuous or discrete in the full sense of the Convenient Assumption, with conditional density or mass functions $f_{\mathrm{II}|\mathrm{I}}(\cdot|\boldsymbol{\xi}_{\mathrm{I}}^*)$. Also let $\tilde{\boldsymbol{\omega}}_2 = \boldsymbol{\omega}_2(\tilde{\boldsymbol{\xi}}_{\mathrm{II}})$ for $\boldsymbol{\omega}_2(\cdot)$ a one-to-one function on

$$\cup\{\{\boldsymbol{\xi}_{\mathrm{II}}: f_{\mathrm{II}|\mathrm{I}}(\boldsymbol{\xi}_{\mathrm{II}}|\boldsymbol{\xi}_{\mathrm{I}}) > 0\}: f_{\mathrm{I}}(\boldsymbol{\xi}_{\mathrm{I}}) > 0\},$$

which, if $\tilde{\boldsymbol{\xi}}_{\mathrm{II}}$ is continuous, is also continuously differentiable with nonvanishing Jacobian determinant. Define

$$M_{\mathrm{I}}(\boldsymbol{\omega}_1) = \begin{cases} 1, & \tilde{\boldsymbol{\xi}}_{\mathrm{I}} \text{ is discrete} \\ \left\| \dfrac{\partial \boldsymbol{\xi}_{\mathrm{I}}}{\partial \boldsymbol{\omega}_1} \right\|, & \tilde{\boldsymbol{\xi}}_{\mathrm{I}} \text{ is continuous} \end{cases}$$

and

$$M_{\mathrm{II}}(\boldsymbol{\omega}_2) = \begin{cases} 1, & \tilde{\boldsymbol{\xi}}_{\mathrm{II}} \text{ is (always) discrete} \\ \left\| \dfrac{\partial \boldsymbol{\xi}_{\mathrm{II}}}{\partial \boldsymbol{\omega}_2} \right\|, & \tilde{\boldsymbol{\xi}}_{\mathrm{II}} \text{ is (always) continuous.} \end{cases}$$

(A) Show that the mass or density functions f_1, $f_{2|1}$, f_2, and $f_{1|2}$ of $\tilde{\boldsymbol{\omega}}_1$ and $\tilde{\boldsymbol{\omega}}_2$ are related to their counterparts f_{I}, $f_{\mathrm{II}|\mathrm{I}}$, f_{II}, and $f_{\mathrm{I}|\mathrm{II}}$ for $\tilde{\boldsymbol{\xi}}_{\mathrm{I}}$ and $\tilde{\boldsymbol{\xi}}_{\mathrm{II}}$ via

$$\begin{aligned} f_1(\cdot) &= f_{\mathrm{I}}(\boldsymbol{\xi}_{\mathrm{I}}(\cdot))M_{\mathrm{I}}(\cdot); \\ f_{2|1}(\cdot|\boldsymbol{\omega}_1) &= f_{\mathrm{II}|\mathrm{I}}(\boldsymbol{\xi}_{\mathrm{II}}(\cdot)|\boldsymbol{\xi}_{\mathrm{I}}(\boldsymbol{\omega}_1))M_{\mathrm{II}}(\cdot); \\ f_2(\cdot) &= f_{\mathrm{II}}(\boldsymbol{\xi}_{\mathrm{II}}(\cdot))M_{\mathrm{II}}(\cdot); \text{ and} \\ f_{1|2}(\cdot|\boldsymbol{\omega}_2) &= f_{\mathrm{I}|\mathrm{II}}(\boldsymbol{\xi}_{\mathrm{I}}(\cdot)|\boldsymbol{\xi}_{\mathrm{II}}(\boldsymbol{\omega}_2))M_{\mathrm{I}}(\cdot) \\ &\propto f_{\mathrm{II}|\mathrm{I}}(\boldsymbol{\xi}_{\mathrm{II}}(\boldsymbol{\omega}_2)|\boldsymbol{\xi}_{\mathrm{I}}(\cdot))f_{\mathrm{I}}(\boldsymbol{\xi}_{\mathrm{I}}(\cdot))M_{\mathrm{I}}(\cdot), \end{aligned}$$

where $\boldsymbol{\xi}_{\mathrm{I}}(\cdot)$ and $\boldsymbol{\xi}_{\mathrm{II}}(\cdot)$ are the functions inverse to $\boldsymbol{\omega}_1(\cdot)$ and $\boldsymbol{\omega}_2(\cdot)$ respectively.

(B) Let $k = m = 1$ and suppose that ξ_{I} and ξ_{II} are the ω_1 and ω_2 respectively in Example 3.11.1. A popular alternative to the probability ξ_{I} of the phenomenon's occurring in any given instance is the *log odds* $\omega_1 = \log(\xi_{\mathrm{I}}/[1-\xi_{\mathrm{I}}])$ in favor of occurrence. Assume that $f_{\mathrm{I}}(\xi_{\mathrm{I}}) = f_\beta(\xi_{\mathrm{I}}|r', n')$, and $f_{\mathrm{II}|\mathrm{I}}(\xi_{\mathrm{II}}|\xi_{\mathrm{I}}) = f_b(\xi_{\mathrm{II}}|n, \xi_{\mathrm{I}})$, as in Example 3.11.1. Let $\tilde{\omega}_2 = \tilde{\xi}_{\mathrm{II}}/n$. Derive $f_1(\cdot)$, $f_{2|1}(\cdot|\omega_1)$, $f_2(\cdot)$, and $f_{1|2}(\cdot|\omega_2)$ under these assumptions.

Exercise 3.11.6

Suppose that $\tilde{\omega}_1$ is a continuous real random variable with density function $f_1(\cdot)$.

(A) Also suppose that, conditional upon ω_1, $\tilde{\omega}_2$ is a *discrete* real random variable with *mass* function $f_{2|1}$ defined by

$$f_{2|1}(\omega_2|\omega_1) = \begin{cases} 1, & \omega_2 = \omega_1 \\ 0, & \omega_2 \neq \omega_1. \end{cases}$$

[Convenient Assumption (D) is *not* satisfied.]

(i) Show that, *un*conditionally, $\tilde{\omega}_2$ is a *continuous* real random variable with *density* function $f_2(\cdot)$ satisfying

$$f_2(\omega_2) = f_1(\omega_2)$$

for every ω_2 in R^1.

(ii) Show that, conditional on ω_2, $\tilde{\omega}_1$ is a *discrete* real random variable with *mass* function $f_{1|2}(\cdot|\cdot)$ given by

$$f_{1|2}(\omega_1|\omega_2) = f_{2|1}(\omega_2|\omega_1).$$

(B) Suppose that, conditional on ω_1, $\tilde{\omega}_2$ is a discrete real random variable with mass function $f_{2|1}$ given by

$$f_{2|1}(\omega_2|\omega_1) = \begin{cases} p, & \omega_2 = \omega_1 + a \\ 1-p, & \omega_2 = \omega_1 + b \\ 0, & \omega_2 \notin \{\omega_1 + a, \omega_1 + b\}, \end{cases}$$

where $p \in [0, 1]$ and a and b are distinct real numbers.

(i) Show that $\tilde{\omega}_2$ is *un*conditionally a continuous real random variable with density function f_2 given by

$$f_2(\omega_2) = pf_1(\omega_2 - a) + (1-p)f_1(\omega_2 - b)$$

for every ω_2 in R^1.

(ii) Show that, conditional on ω_2, $\tilde{\omega}_1$ is a discrete real random variable with mass function $f_{1|2}(\cdot|\cdot)$ given by

$$f_{1|2}(\omega_1|\omega_2) = \begin{cases} p, & \omega_1 = \omega_2 - a \\ 1-p, & \omega_1 = \omega_2 - b \\ 0, & \omega_1 \notin \{\omega_2 - a, \omega_2 - b\}. \end{cases}$$

■ In the remainder of these exercises we *will* assume that Convenient Assumption (D) pertains, so as to rule out the queer phenomena manifested in Exercise 3.11.6. ■

Exercise 3.11.7: Independence

Let $\tilde{\boldsymbol{\omega}}_i$ be a real random k_i-tuple for $i = 1, \ldots, n$. Then $\tilde{\boldsymbol{\omega}}_1, \ldots, \tilde{\boldsymbol{\omega}}_n$ are said to be (mutually) *independent* if

$$P_{1\cdots n}((\tilde{\boldsymbol{\omega}}_1, \ldots, \tilde{\boldsymbol{\omega}}_n) \in \Omega_1' \times \cdots \times \Omega_n') = \prod_{i=1}^{n} P_i(\tilde{\boldsymbol{\omega}}_i \in \Omega_i')$$

for every n events Ω_1' in R^{k_1}, Ω_2' in $R^{k_2}, \ldots$, and Ω_n' in R^{k_n}, where $P_{1\cdots n}$ corresponds to $f_{1\cdots n}$ as defined in Exercise 3.11.4.

(A) Denote by $f_i(\cdot)$ the unconditional mass or density function of $\tilde{\boldsymbol{\omega}}_i$. Show that $\tilde{\boldsymbol{\omega}}_1, \ldots, \tilde{\boldsymbol{\omega}}_n$ are independent if

$$f_{1\cdots n}(\boldsymbol{\omega}_1, \ldots, \boldsymbol{\omega}_n) = \prod_{i=1}^{n} f_i(\boldsymbol{\omega}_i)$$

for every $(\boldsymbol{\omega}_1, \ldots, \boldsymbol{\omega}_n)$ in R^k, where $k = \Sigma_{i=1}^{n} k_i$.

*(B) Prove that the converse is also essentially true, in the sense that

$$P_{1\cdots n}\left(\left\{(\boldsymbol{\omega}_1, \ldots, \boldsymbol{\omega}_n) : f_{1\cdots n}(\boldsymbol{\omega}_1, \ldots, \boldsymbol{\omega}_n) \neq \prod_{i=1}^{n} f_i(\boldsymbol{\omega}_i)\right\}\right) = 0$$

if $\tilde{\boldsymbol{\omega}}_1, \ldots, \tilde{\boldsymbol{\omega}}_n$ are independent.
(This is hard. The only possible departures from equality involve density values being *somewhat* arbitrarily specifiable. Recall that the probability of *any specific* value of a continuous real random k-tuple is zero.)

Exercise 3.11.8: Probabilities of Extremal Events

Suppose that $\tilde{\omega}_1, \ldots, \tilde{\omega}_n$ are real random variables (all $k_i = 1$ in Exercise 3.11.7) and have (unconditional) mass or density functions $f_i(\cdot)$. Let $F_i(\cdot)$ denote the cumulative distribution function $P_i(\tilde{\omega}_i \leq \cdot)$ of $\tilde{\omega}_i$ (as studied in Exercises 3.9.9 and 3.10.8). Let $\tilde{\epsilon}_L = \min[\tilde{\omega}_1, \ldots, \tilde{\omega}_n]$ and $\tilde{\epsilon}_U = \max[\tilde{\omega}_1, \ldots, \tilde{\omega}_n]$. Assume that $\tilde{\omega}_1, \ldots, \tilde{\omega}_n$ are independent.

(A) Show that the cumulative distribution function $F_U(\cdot)$ of the maximum $\tilde{\epsilon}_U$ of $\tilde{\omega}_1, \ldots, \tilde{\omega}_n$ must satisfy

$$F_U(\epsilon_U) = \prod_{i=1}^{n} F_i(\epsilon_U)$$

for every ϵ_U in R^1.
[*Hint*: For every x in R^1, it is clear that $\tilde{\epsilon}_U \leq x$ if and only if $\tilde{\omega}_i \leq x$ for every i, or equivalently, if and only if $(\tilde{\omega}_1, \ldots, \tilde{\omega}_n) \in (-\infty, x] \times \cdots \times (-\infty, x]$.]

(B) Show that the cumulative distribution function $F_L(\cdot)$ of the minimum $\tilde{\epsilon}_L$ of $\tilde{\omega}_1, \ldots, \tilde{\omega}_n$ must satisfy

$$F_L(\epsilon_L) = 1 - \prod_{i=1}^{n} [1 - F_i(\epsilon_L)]$$

for every ϵ_L in R^1.
[*Hints*: (1) For any x in R^1, $\tilde{\epsilon}_L > x$ if and only if $\tilde{\omega}_i > x$ for every i; and (2) $1 - F_L(\epsilon_L) = P_L(\tilde{\epsilon}_L > \epsilon_L)$.]

(C) Suppose that $n = 2$, $\tilde{\omega}_1$ is a continuous real random variable with density function $f_u(\cdot|0, 1)$, and $\tilde{\omega}_2$ is a discrete real random variable with mass function $f_2(\omega_2) = 1/2$ for $\omega_2 = .30$ or $.70$ and $f_2(\omega_2) = 0$ for all other ω_2. Derive and graph $F_U(\cdot)$ and $F_L(\cdot)$.

(D) Verify that $\tilde{\epsilon}_U$ is neither a discrete nor a continuous real random variable; and similarly, neither is $\tilde{\epsilon}_L$.

Exercise 3.11.9: Mixed Real Random Variables

The phenomenon appearing in (C) and (D) of Exercise 3.11.8 is rather interesting. A real random variable $\tilde{\omega}$ is called a *mixed* real random variable if its cumulative distribution function $F(\cdot)$ is neither a step function [Exercise 3.9.9(B)(i) for discrete real random variables] nor a continuous function [Exercise 3.10.8(B)(i) for continuous real random variables].

(A) Show that if $\tilde{\omega}$ is a mixed real random variable, then there is a number p in (0, 1), a cumulative distribution function $F_c(\cdot)$ of a *continuous* real random variable, and a cumulative distribution function $F_d(\cdot)$ of a *discrete* real random variable, such that the cumulative distribution function $F(\cdot)$ of $\tilde{\omega}$ can be written as

$$F(\omega) = pF_d(\omega) + (1-p)F_c(\omega)$$

for every ω in R^1.
[*Hint*: Let $j(\omega) = F(\omega) - \lim_{x\uparrow\omega} F(x) =$ distance of the jump in $F(\cdot)$ at ω, for every ω in R^1. Let $J(\omega) = \Sigma\{j(\omega'): \omega' \leq \omega\}$, let $p = \lim_{\omega\to+\infty} J(\omega)$, and let $F_d(\cdot) = (1/p)J(\cdot)$.]

(B) Find p, $F_d(\cdot)$, and $F_c(\cdot)$ for

(i) $F(\cdot) = F_L(\cdot)$ in Exercise 3.11.8; and
(ii) $F(\cdot) = F_U(\cdot)$ in Exercise 3.11.8.

(C) Suppose that $\tilde{\omega}_1$ is a discrete real random variable with mass function

$$f_1(\omega_1) = \begin{cases} p, & \omega_1 = 0 \\ 1-p, & \omega_1 = 1 \\ 0, & \omega_1 \notin \{0, 1\}, \end{cases}$$

and that, *conditional on* $\tilde{\omega}_1 = 0$, $\tilde{\omega}_2$ is a discrete real random variable with cumulative distribution function $F_d(\cdot)$, but that, *conditional on* $\tilde{\omega}_1 = 1$, $\tilde{\omega}_2$ is a continuous real random variable with cumulative distribution function $F_c(\cdot)$. Show that the *unconditional* cumulative distribution function $F(\cdot)$ of $\tilde{\omega}_2$ is given by the formula in part (A) of this exercise. (*Every* mixed real random variable may be regarded as arising in this two-stage fashion.)

*(D) Given the "Ω_1-before-Ω_2" model of a mixed real random variable, introduced in part C of this exercise, in which judgments about $\tilde{\omega}_2$ are conditionally on ω_1 either discrete or continuous and unconditionally mixed, show that "posterior" judgments about $\tilde{\omega}_1$ conditional on ω_2 are characterized as follows. Let Ω_2^{**} denote the set of discontinuities (= points of positive probability) of $F_d(\cdot) = F_{2|1}(\cdot|\omega_1 = 0)$. Then

$$P_{1|2}(\tilde{\omega}_1 = 0|\omega_2) = \begin{cases} 1, & \omega_2 \in \Omega_2^{**} \\ 0, & \omega_2 \notin \Omega_2^{**}. \end{cases}$$

[*Hints*: (1) Require that

$$E_1\{E_{2|1}\{x(\tilde{\omega}_1, \tilde{\omega}_2)|\tilde{\omega}_1\}\} = E_2\{E_{1|2}\{x(\tilde{\omega}_1, \tilde{\omega}_2)|\tilde{\omega}_2\}\}$$

for all (decent) functions $x(\cdot, \cdot)$.

(2) $\int_{\overline{\Omega_2^{**}}} f_c(\omega_2)d\omega_2 = 0$.

(3) Let $x(1, \omega_2) = 0$ for all ω_2 and $x(0, \omega_2) = 0$ for all $\omega_2 \notin \Omega_2^{**}$ to verify $P_{1|2}$ for $\omega_2 \in \Omega_2^{**}$.

(4) Let $x(0, \omega_2) = 0$ for all ω_2 and $x(1, \omega_2) = 0$ for all $\omega_2 \in \Omega_2^{**}$ to verify $P_{1|2}$ for $\omega_2 \notin \Omega_2^{**}$.]

■ Part (D) essentially states that an observation ω_2 of zero probability "washes out the discrete part," whereas an observation of positive probability "washes out the continuous part." ■

■ Let $\tilde{\boldsymbol{\omega}}$ be a random real k-tuple for $k > 1$. Then $\tilde{\boldsymbol{\omega}}$ is said to be *mixed* if it can be regarded as having (1) a mass function $f_d(\cdot)$ *given* an event δ of some probability $p \in (0, 1)$, and (2) a density function $f_c(\cdot)$ *given* the complementary event $\bar{\delta}$ of probability $1 - p$. Then Assertion (D) continues to obtain. Moreover, it is clear from Theorem 3.2.2 that the expectation $E\{x(\tilde{\boldsymbol{\omega}})\}$ of a function $x(\cdot)$ of a mixed random real k-tuple $\tilde{\boldsymbol{\omega}}$ must be given by

$$E\{x(\tilde{\boldsymbol{\omega}})\} = p \sum_{R^k} x(\boldsymbol{\omega})f_d(\boldsymbol{\omega}) + (1-p)\int_{R^k} x(\boldsymbol{\omega})f_c(\boldsymbol{\omega})d\boldsymbol{\omega}. \blacksquare$$

4

ANALYSIS OF DECISIONS IN EXTENSIVE FORM

Knowledge may give weight but accomplishments give lustre, and many more people see than weigh.

Philip Dormer Stanhope, Fourth Earl of Chesterfield, Letters, May 8, 1750

4.1 INTRODUCTION

The best way for you to capture the spirit of this chapter at a glance is to review Section 2.8 briefly. What we shall do here is apply the analytical methodology developed (painfully?) in Chapter 3 to your dpuu.

Through Section 4.4 we shall assume that your dpuu has been expressed in *standard* extensive form. Section 4.2 consists of a summary of preliminary facts, mostly based on Chapter 3, needed in succeeding sections. In Section 4.3 we apply these facts to characterize our recursive (i.e., backward-forward) method of determining your optimal strategy in a dpuu in standard extensive form. This method is very much in the spirit of Section 2.8.

In Section 4.4 we apply this method to Example 1.5.1c, depicted in Figure 2.16 in standard extensive form, thus indicating the ability of decision analysis to cope successfully with complex dpuu's. To add to the complexity of this example, we are not given the necessary probabilities in the decision tree and must compute them from information in Examples 1.5.1a–c. So doing constitutes the majority of the work in this case.

In most actual applications of decision analysis, there is no particular need for a uniform notation system, and hence such dpuu's need not be reformulated in standard extensive form. Decision analysis is no less applicable in such situations, however; the comments necessary for establishing this fact are made in Section 4.5.

Section 4.6 consists of concluding remarks concerning internally consistent decision making under uncertainty.

Section 4.7 is comprised of exercises affording you an opportunity to cope with most of the prototypical examples in Section 1.5.

4.2 PRELIMINARIES

We cite here six preliminaries which follow almost immediately from the preceding chapters, chiefly Chapter 3. We do so for two reasons: (1) to isolate those facts which bear directly on the ensuing analysis; (2) to free you from frequent page shuffling by making this chapter more self-contained.

First Preliminary

From the definition of a lottery, Definition 3.1.2, every Nth stage $\mathscr{D}$-move $G_N(h_{N-1})$ of a dpuu in standard extensive form consists solely of lotteries.

Observation 4.2.1. Every act g_N in every nth stage $\mathscr{D}$-move $G_N(h_{N-1})$ is a lottery.

Second Preliminary

Theorem 3.2.1 showed that if $u(\cdot)$ is a utility function for $\mathscr{D}$ on C, then every lottery ℓ with outcomes in C has an *index* $U(\ell)$ *of relative desirability*, in the sense that $U(\ell'') \geq U(\ell')$ if and only if $\ell'' \gtrsim \ell'$. The relative-desirability index $U(\ell)$ is a probability-weighted average, or *expectation*, in the sense that

$$U(\ell) = E\{u(\ell(\tilde{\omega}))\}$$

if ℓ is a lottery on Ω with outcomes in C.

■ If ℓ is a lottery on $\Omega = \{\omega^1, \ldots, \omega^m\}$ with outcomes in C and if $P(\omega^i)$ denotes $\mathscr{D}$'s probability that $\tilde{\omega} = \omega^i$ for $i = 1, \ldots, m$, then

$$E\{u(\ell(\tilde{\omega}))\} = \sum_{i=1}^{m} u(\ell(\omega^i))P(\omega^i).$$

In the infinite idealizations considered in Sections *3.9–*3.11, $E\{x(\tilde{\omega})\}$ is a sum and/or integral: see Definitions 3.9.2 and 3.10.2 and the remark following Exercise 3.11.8. ■

Now suppose that L is an entire $\mathscr{D}$-move of lotteries and that ℓ° in L is such that $\ell^\circ \gtrsim \ell$ for every ℓ in L. Then obviously $\mathscr{D}$ can do no better than to choose ℓ°. In fact, the superfix "°" is intended to connote "optimal"! But Theorem 3.2.1 implies that

$$\ell^\circ \gtrsim \ell \quad \text{for every } \ell \text{ in } L$$

if and only if

$$U(\ell^\circ) \geq U(\ell) \qquad \text{for every } \ell \text{ in } L,$$

or equivalently, if and only if

$$U(\ell^\circ) = \max_{\ell \in L} U(\ell).$$

Thus an optimal lottery ℓ° in a $\mathscr{D}$-move L of lotteries is a maximizer (over L) of $U(\ell)$, and conversely. Moreover, the desirability to $\mathscr{D}$ of having to make a choice from L is not lessened if branches *other than* ℓ° are "pruned," since *actually* choosing ℓ° amounts to, in effect, pruning all other branches anyway. Thus, once $\mathscr{D}$ has determined an optimal lottery ℓ° in L, he may prune *all* other branches in L. Mathematically speaking, the subset $\{\ell^\circ\}$ of L which consists solely of ℓ° is equally as desirable a $\mathscr{D}$-move as is L itself.

Summarizing:

Observation 4.2.2. Let $u(\cdot)$ be a utility function for $\mathscr{D}$ on C, and let $U(\cdot)$ be the corresponding relative-desirability index on lotteries with outcomes in C. Let L be a $\mathscr{D}$-move of lotteries. Then ℓ° is an *optimal* lottery in L if and only if

$$U(\ell^\circ) = \max_{\ell \in L} U(\ell).$$

The decision opportunity represented by L is no more or no less desirable to $\mathscr{D}$ than that represented by $\{\ell^\circ\}$; hence, in tree form, all branches of L other than ℓ° may be "pruned."

Third Preliminary

By Theorem 3.2.2, any "compound" lottery may be evaluated by iterating the expectation, or weighted-averaging operation. See Figure 3.4. In slightly different notation, Theorem 3.2.2 amounts to:

Observation 4.2.3. Let $u(\cdot)$ be a utility function for $\mathscr{D}$ on C, and let $U(\cdot)$ be the corresponding relative-desirability index on lotteries with outcomes in C. Let ℓ be a lottery on an $\mathscr{E}$-move Ω with $\ell(\omega)$ a lottery on an $\mathscr{E}$-move Ξ_ω with outcomes $\ell(\xi, \omega)$ in C, for every ω in Ω. Let the relative-desirability index of the lottery $\ell(\omega)$ be denoted by

$$U(\ell(\omega)) = E\{u(\ell(\tilde{\xi}, \omega))|\omega\}.$$

Then

$$U(\ell) = E\{U(\ell(\tilde{\omega}))\} = E\{E\{u(\ell(\tilde{\xi}, \tilde{\omega}))|\tilde{\omega}\}\}.$$

Fourth Preliminary

By the Invariant Preference Principle in Section 3.3, preferences and relative-desirability indices are as valid when ℓ, or L, is realizable only contingently as when $\mathscr{D}$ actually confronts ℓ, or a choice from L. In particular, Observation 4.2.2 is valid for L denoting any Nth stage $\mathscr{D}$-move $G_N(h_{N-1})$ even when analyzed before a choice is made from $G_1(h_0)$.

Fifth Preliminary

By Theorem 3.6.1, $\mathscr{D}$'s probabilities of events in the $\mathscr{E}$-move of a lottery should not depend upon the reference outcomes used to assess them. Hence $\ell^{\wedge}(\Omega'; x, y) \sim \ell^*(P(\Omega'); x, y)$ *for any outcomes* x *and* y—where, you will recall, $\ell^{\wedge}(\Omega'; x, y)$ is the lottery resulting in x if $\tilde{\omega} \in \Omega'$ and in y otherwise; and $\ell^*(P(\Omega'); x, y)$ is a canonical lottery resulting in x with canonical probability $P(\Omega')$ and in y with remaining canonical probability $1 - P(\Omega')$. See Figure 3.13.

Therefore, $\mathscr{D}$ can quantify his judgments regarding *every* $\mathscr{E}$-move $\Gamma_i(h_{i-1}, g_i)$ in the decision tree by using convenient reference outcomes, say $x = \$1{,}000$ and $y = \$0$, *before* commencing the analysis. We now introduce appropriate notation for the resulting probabilities and their associated expectations (weighted averagings).

Definition 4.2.1. For any i in $\{1, \ldots, N\}$, any h_{i-1} in H_{i-1}, and any g_i in $G_i(h_{i-1})$, let

$$P_i(\cdot|h_{i-1}, g_i)$$

denote $\mathscr{D}$'s probability function on events in the ith stage $\mathscr{E}$-move $\Gamma_i(h_{i-1}, g_i)$. If $x(\cdot)$ is a real-valued function with domain $\Gamma_i(h_{i-1}, g_i)$, let

$$E_i\{x(\tilde{\gamma}_i)|h_{i-1}, g_i\}$$

denote expectation with respect to $P_i(\cdot|h_{i-1}, g_i)$.

Sixth Preliminary

Recall from Exercise 2.12.3 that for any i in $\{1, \ldots, N\}$ and any h_i in H_i, we may write h_i as

$$h_i = (h_{i-1}, g_i, \gamma_i)$$

for unique elements h_{i-1} of H_{i-1}, g_i of $G_i(h_{i-1})$, and γ_i of $\Gamma_i(h_{i-1}, g_i)$. Furthermore, recall that each full history h_N specifies the consequence of the dpuu uniquely; and hence we may write the typical (or generic) consequence c as a function $c(h_N)$ of the typical full history h_N.

We shall now define some subsequently important functions $U_i\colon H_i \to (-\infty, +\infty)$ *recursively*; that is, we first define $U_N(\cdot)$, and then define $U_{i-1}(\cdot)$, in terms of $U_i(\cdot)$, so that $U_{N-1}(\cdot)$ can be calculated from $U_N(\cdot)$, then $U_{N-2}(\cdot)$ from $U_{N-1}(\cdot)$, and so forth.

Definition 4.2.2. Let $u(\cdot)$ be a utility function for $\mathscr{D}$ on C.

(A) We define a function $U_N\colon H_N \to (-\infty, +\infty)$ by

$$U_N(h_N) = u(c(h_N))$$

for every h_N in H_N.

(B) We define, for every $i = N, N-1, \ldots, 2, 1$, a function $U_{i-1}\colon H_{i-1} \to$

$(-\infty, +\infty)$ by

$$U_{i-1}(h_{i-1}) = \max_{g_i \in G_i(h_{i-1})} [E_i\{U_i(h_{i-1}, g_i, \tilde{\gamma}_i)|h_{i-1}, g_i\}]$$

for every h_{i-1} in H_{i-1}.

By now you can probably guess that the functions U_i are relative-desirability indices, and that their recursive definition implies that we will analyze a dpuu in standard extensive form by working backward; that is, by dealing first with stage N, in which all acts g_N are lotteries, then dealing with stage $N-1$, and so on, until we have dealt with stage 1.

4.3 ANALYSIS OF A dpuu IN STANDARD EXTENSIVE FORM

We suppose that you have formulated your dpuu in standard extensive form, quantified your preferences among its possible consequences $c(h_N)$ in the form of a utility function $u(\cdot)$ on $C = \{c(h_N): h_N \in H_N\}$, quantified your judgments regarding events in every $\mathscr{E}$-move $\Gamma_i(h_{i-1}, g_i)$ in the form of a probability function $P_i(\cdot|h_{i-1}, g_i)$, are willing to abide by the Transitivity, Dominance (implying Substitutability), Continuity,[a] and Invariant Preference Principles, and accept the Canonicity Assumption and the Monotonicity Principle. Then all preliminaries in Section 4.2 are pertinent.

We begin the analysis with the final stage of the dpuu, stage N. By Observation 4.2.1, every act g_N in every Nth stage $\mathscr{D}$-move $G_N(h_{N-1})$ is a lottery, one in which $\mathscr{D}$ obtains consequence $c(h_N) = c(h_{N-1}, g_N, \gamma_N)$, having utility $U_N(h_N) = U_N(h_{N-1}, g_N, \gamma_N)$, if $\tilde{\gamma}_N = \gamma_N$. Hence g_N has relative-desirability index $U(g_N) = E_N\{U_N(h_{N-1}, g_N, \gamma_N)|h_{N-1}, g_N\}$.

Now this is true for every g_N in $G_N(h_{N-1})$, and hence we see by Observation 4.2.2 with $L = G_N(h_{N-1})$ that an *optimal* act $g_N^\circ(h_{N-1})$ in $G_N(h_{N-1})$ is any act such that

$$U(g_N^\circ(h_{N-1})) = \max_{g_N \in G_N(h_{N-1})} [U(g_N)];$$

or equivalently, such that

$$\begin{aligned} &E_N\{U_N(h_{N-1}, g_N^\circ(h_{N-1}), \tilde{\gamma}_N)|h_{N-1}, g_N^\circ(h_{N-1})\} \\ &= \max_{g_N \in G_N(h_{N-1})} [E_N\{U_N(h_{N-1}, g_N, \tilde{\gamma}_N)|h_{N-1}, g_N\}] \\ &= U_{N-1}(h_{N-1}), \end{aligned}$$

the second equality being by Definition 4.2.2 for $i = N$.

Once you have found a $g_N^\circ(h_{N-1})$, you may "prune" all *other* branches of $G_N(h_{N-1})$, by virtue of Observation 4.2.2.

The above calculations—of $U(g_N)$ for all g_N in $G_N(h_{N-1})$—, comparisons—of the $U(g_N)$'s for all g_N in $G_N(h_{N-1})$—, and prunings—of all $g_N \neq g_N^\circ(h_{N-1})$—must be performed for *every* h_{N-1} in H_{N-1}. Once this has

[a] The Continuity Principle is needed only when the dpuu contains an $\mathscr{E}$-move with infinitely many elements.

been done, the analysis of stage N is complete. We summarize in the form of a lemma.

Lemma 4.3.1: Analysis of Stage N. For every h_{N-1} in H_{N-1}, do the following.

(1) For every g_N in $G_N(h_{N-1})$, calculate

$$E_N\{U_N(h_{N-1}, g_N, \tilde{\gamma}_N)|h_{N-1}, g_N\}.$$

(2) Find an act $g_N^\circ(h_{N-1})$ in $G_N(h_{N-1})$ which maximizes the index calculated in (1).

(3) "Prune" all acts g_N in $G_N(h_{N-1})$ other than $g_N^\circ(h_{N-1})$, together with their ensuing $\mathscr{E}$-moves.

(4) Set $U_{N-1}(h_{N-1}) = \max_{g_N \in G_N(h_{N-1})} [E_N\{U_N(h_{N-1}, g_N, \tilde{\gamma}_N)|h_{N-1}, g_N\}]$.

Now we attack stage $N-1$. It is crucial to note that *once step (3) of Lemma 4.3.1 has been performed for every* h_{N-1} *in* H_{N-1}, *every act* g_{N-1} *in every* $(N-1)$*th stage* $\mathscr{D}$*-move* $G_{N-1}(h_{N-2})$ *becomes a lottery*, since the result of the pruning is that every $G_N(h_{N-1})$ is reduced to the dummy $\mathscr{D}$-move $\{g_N^\circ(h_{N-1})\}$.

For any h_{N-2} in H_{N-2} and any g_{N-1} in $G_{N-1}(h_{N-2})$, it follows that g_{N-1} is a lottery which results in a lottery with desirability index $U_{N-1}(h_{N-2}, g_{N-1}, \gamma_{N-1}) = U_{N-1}(h_{N-1})$ if $\tilde{\gamma}_{N-1} = \gamma_{N-1}$, for every γ_{N-1} in $\Gamma_{N-1}(h_{N-2}, g_{N-1})$. Hence by Observation 4.2.3, the desirability index of g_{N-1} is $U(g_{N-1})$, where

$$U(g_{N-1}) = E_{N-1}\{U_{N-1}(h_{N-2}, g_{N-1}, \tilde{\gamma}_{N-1})|h_{N-2}, g_{N-1}\}.$$

As in stage N, an *optimal* act $g_{N-1}^\circ(h_{N-2})$ in $G_{N-1}(h_{N-2})$ is any act such that

$$U(g_{N-1}^\circ(h_{N-2})) = \max_{g_{N-1} \in G_{N-1}(h_{N-2})} [U(g_{N-1})];$$

or equivalently, such that

$$\begin{aligned} &E_{N-1}\{U_{N-1}(h_{N-2}, g_{N-1}^\circ(h_{N-2}), \tilde{\gamma}_{N-1})|h_{N-2}, g_{N-1}^\circ(h_{N-2})\} \\ &= \max_{g_{N-1} \in G_{N-1}(h_{N-2})} [E_{N-1}\{U_{N-1}(h_{N-2}, g_{N-1}, \tilde{\gamma}_{N-1})|h_{N-2}, g_{N-1}\}] \\ &= U_{N-2}(h_{N-2}), \end{aligned}$$

the second equality being by Definition 4.2.2 for $i = N-1$.

Once you have found a $g_{N-1}^\circ(h_{N-2})$, you may proceed as in Stage N by pruning all other branches of $G_{N-1}(h_{N-2})$, together with the subtrees emanating from all these branches. Summarizing:

Lemma 4.3.2. Analysis of Stage N − 1. For every h_{N-2} in H_{N-2}, do the following.

(1) For every g_{N-1} in $G_{N-1}(h_{N-2})$, calculate

$$E_{N-1}\{U_{N-1}(h_{N-2}, g_{N-1}, \tilde{\gamma}_{N-1})|h_{N-2}, g_{N-1}\}.$$

(2) Find an act $g_{N-1}^\circ(h_{N-2})$ in $G_{N-1}(h_{N-2})$ which maximizes the index calculated in (1).

(3) Prune all acts g_{N-1} in $G_{N-1}(h_{N-2})$ other than $g^{\circ}_{N-1}(h_{N-2})$, together with the moves which follow those acts.

(4) Set

$$U_{N-2}(h_{N-2}) = \max_{g_{N-1} \in G_{N-1}(h_{N-2})}[E_{N-1}\{U_{N-1}(h_{N-2}, g_{N-1}, \tilde{\gamma}_{N-1})|h_{N-2}, g_{N-1}\}].$$

Once step (3) of Lemma 4.3.2 has been performed for every h_{N-2} in H_{N-2}, every act g_{N-2} in every $(N-2)$th stage $\mathscr{D}$-move becomes a lottery, and the preceding reasoning for stage $N-1$ remains valid for stage $N-2$. In particular, Lemma 4.3.2 remains valid if $N-1$ and $N-2$ are replaced by $N-2$ and $N-3$ respectively wherever they appear.

The same holds for the analysis of stage $N-3$ once stage $N-2$ has been analyzed. In fact, the general approach should now be clear. We start at the end of the dpuu, evaluating and pruning acts in stage N, then evaluate and prune acts in stage $N-1$, then . . . , then evaluate and prune acts in stage 1. We summarize in the form of a theorem.

Theorem 4.3.1. Recursion Analysis of dpuu's **in Standard Extensive Form.** $\mathscr{D}$'s optimal strategy in a dpuu in standard extensive form may be determined as follows. Recursively for $i = N, N-1, \ldots, 1$, suppose that

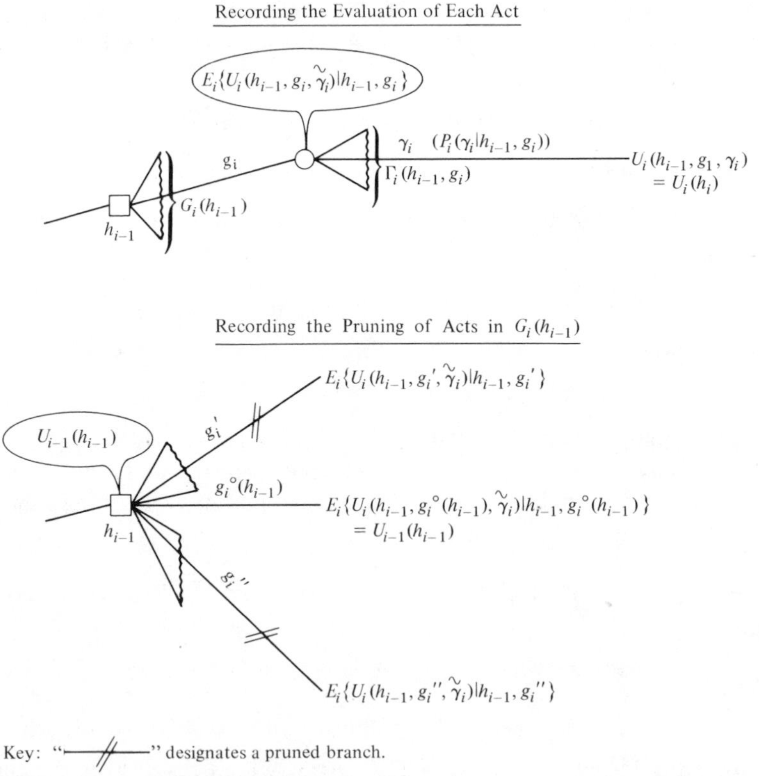

Figure 4.1 Recording calculations on the tree.

the desirability indices $U_i(h_i) = U_i(h_{i-1}, g_i, \gamma_i)$ have been determined for every h_{i-1} in H_{i-1}, g_i in $G_i(h_{i-1})$, and γ_i in $\Gamma_i(h_{i-1}, g_i)$. Then $\mathcal{D}$ can perform the following steps, for every h_{i-1} in H_{i-1}.

(1) For every g_i in $G_i(h_{i-1})$, calculate $E_i\{U_i(h_{i-1}, g_i, \tilde{\gamma}_i)|h_{i-1}, g_i\}$.
(2) Find an act $g_i^\circ(h_{i-1})$ in $G_i(h_{i-1})$ which maximizes the index calculated in (1).
(3) Prune all acts g_i in $G_i(h_{i-1})$ other than $g_i^\circ(h_{i-1})$, together with all moves which follow those acts.
(4) Set $U_{i-1}(h_{i-1}) = \max_{g_i \in G_i(h_{i-1})} [E_i\{U_i(h_{i-1}, g_i, \tilde{\gamma}_i)|h_{i-1}, g_i\}]$.

In Figure 4.1 we show that *all* the evaluations and prunings dictated by Theorem 4.3.1 can be recorded directly on the original decision tree.

Example 4.3.1

Figure 4.2 depicts a two-stage dpuu in standard extensive form with preferences u and all calculations recorded on the tree; symbols for all the consequences have been omitted for notational brevity. Parenthesized numbers on ee-branches are the respective probabilities of those branches. For example,

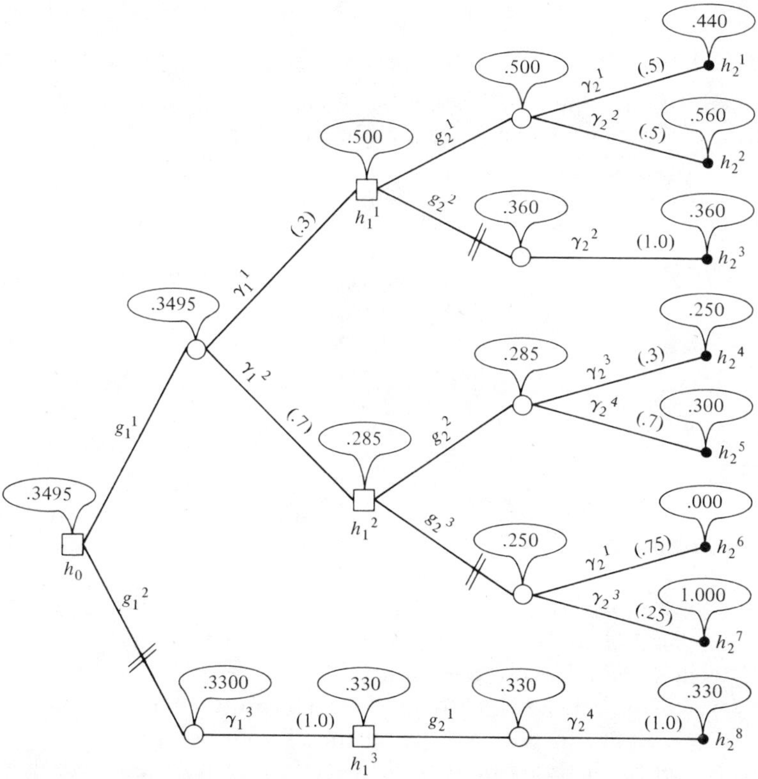

Figure 4.2 Analysis of a dpuu in standard extensive form.

$$U_2(h_2^{\,1}) = .440 = u(c(h_2^{\,1}));$$
$$P_2(\gamma_2^{\,1}|h_1^{\,1}, g_2^{\,1}) = P_2(\gamma_2^{\,2}|h_1^{\,1}, g_2^{\,1}) = .5,$$

and hence

$$E_2\{U_2(h_1^{\,1}, g_2^{\,1}, \tilde{\gamma}_2)|h_1^{\,1}, g_2^{\,1}\} = (.5)(.440) + (.5)(.560) = .500$$

and

$$U_1(h_1^{\,1}) = \max\,[.500, .360] = .500, \text{ implying that } g_2^0(h_1^{\,1}) = g_2^{\,1}.$$

You may verify the other "bubbled" evaluations and the prunings as Exercise 4.3.1. $\mathscr{D}$'s optimal strategy in the dpuu is to select $g_1^{\,1}$ in the initial $\mathscr{D}$-move, and then $g_2^{\,1}$ if $\tilde{\gamma}_1 = \gamma_1^{\,1}$ and $g_2^{\,2}$ if $\tilde{\gamma}_1 = \gamma_1^{\,2}$.

Exercise 4.3.1

Verify all calculations in Figure 4.2.

Exercise 4.3.2

Consider a dpuu which coincides with that in Figure 4.2 except that $P_1(\gamma_1^{\,1}|h_0, g_1^{\,1}) = .1$ instead of .3. What is $\mathscr{D}$'s optimal strategy in this dpuu?

Exercise 4.3.3

Reanalyze the dpuu in Figure 4.2 under the assumption that $\mathscr{D}$'s utilities of the consequences are as tabulated here.

j	$U_2(h_2^{\,j})$
1	1.000
2	0.000
3	2.860
4	−3.470
5	6.326
6	1.153
7	0.884
8	2.167

4.4 EXAMPLE 1.5.1c

The power of our methodical approach is best brought out by attacking a very complicated dpuu, such as Example 1.5.1c, which we depicted in Figure 2.16 in standard extensive form.

Your (= $\mathscr{D}$'s) first task is to quantify your preferences for the consequences. We shall suppose that the results of your doing so are as given in Table A below, in terms of the particular utility function u such that $u(c^{\mathrm{B}}) = 1$ and $u(c^{\mathrm{W}}) = 0$.

Table A Your Quantified Preferences

c	$u(c)$	c	$u(c)$
$+250; MP	1.000	$+130; MP	.960
$+245; MP	.999	$+125; MP	.950
$+230; MP	.993	$−250; LP	.140
$+225; MP	.990	$−255; LP	.110
$+150; MP	.970	$−270; LP	.030
$+145; MP	.965	$−275; LP	.000

The next step in the procedure is to attach appropriate probabilities to the ee's in Figure 2.16. But the probabilities of most of the ee's in Figure 2.16 are not stated in Examples 1.5.1a–c. They may be *derived*, however, from probabilities that *are* stated there, by applying the Chain Rule [Corollary 3.7.1] and concepts of marginal probability, particularly as expressed in Exercise 3.7.1.

Example 1.5.1a indicates that your probability of the subsystem's *arriving* defective is .10, implying probability .90 of its *arriving* nondefective. The outcome of each test depends upon the subsystem's arrival state. We shall begin by calculating the joint probabilities of both tests' outcomes and of the subsystem's arrival state. Since so many things are "going on" in this example, we shall develop a special, temporary notation to help us keep track of things by means of mnemonics.

Let

AD = "subsystem *a*rrived *d*efective,"
AN = "subsystem *a*rrived *n*ondefective,"
PD = "*p*retest indicates subsystem *d*efective,"
PN = "*p*retest indicates subsystem *n*ondefective,"
LD = "*l*ess reliable test indicates subsystem *d*efective,"

and

LN = "*l*ess reliable test indicates subsystem *n*ondefective."

From Examples 1.5.1a–c,

$$P(\text{AD}) = .10 \text{ and } P(\text{AN}) = .90\text{—Example 1.5.1a},$$

$$P(\text{PD}|\text{AD}) = .80 \text{ and } P(\text{PN}|\text{AD}) = .20,$$
$$P(\text{PD}|\text{AN}) = .00 \text{ and } P(\text{PN}|\text{AN}) = 1.00\text{—Example 1.5.1b};$$

and

$$P(\text{LD}|\text{AD}) = .65 \text{ and } P(\text{LN}|\text{AD}) = .35,$$
$$P(\text{LD}|\text{AN}) = .25 \text{ and } P(\text{LN}|\text{AN}) = .75\text{—Example 1.5.1c}.$$

Now the Chain Rule [Corollary 3.7.1] implies that

$$P(\text{PD, LD, AD}) = P(\text{PD}|\text{LD, AD})P(\text{LD}|\text{AD})P(\text{AD}),$$
$$P(\text{PD, LN, AD}) = P(\text{PD}|\text{LN, AD})P(\text{LN}|\text{AD})P(\text{AD}),$$

and so on, for the other six (why?) ee's involving the outcomes of both tests and the arrival state of the subsystem. But Example 1.5.1c indicates that the outcome of each test affects neither the outcome of the other test nor the subsystem's reliability (= arrival state). Hence,

$$P(\text{PD}|\text{LD, AD}) = P(\text{PD}|\text{LN, AD}) = P(\text{PD}|\text{AD}),$$
$$P(\text{PN}|\text{LD, AD}) = P(\text{PN}|\text{LN, AD}) = P(\text{PN}|\text{AD}),$$
$$P(\text{PD}|\text{LD, AN}) = P(\text{PD}|\text{LN, AN}) = P(\text{PD}|\text{AN}),$$

and

$$P(\text{PN}|\text{LD, AN}) = P(\text{PN}|\text{LN, AN}) = P(\text{PN}|\text{AN}).$$

Hence we conclude that

$$P(\pi, \lambda, \alpha) = P(\pi|\alpha)P(\lambda|\alpha)P(\alpha)$$

for $\pi \in \{\text{PD, PN}\}$, $\lambda \in \{\text{LD, LN}\}$, and $\alpha \in \{\text{AD, AN}\}$.

We are therefore able to obtain the desired table of joint probabilities.

Table B Joint Probabilities $P(\pi, \lambda, \alpha)$

π, λ \ α	AD	AN	$\Sigma = P(\pi, \lambda)$
PD, LD	.052	.000	.052
PD, LN	.028	.000	.028
PN, LD	.013	.225	.238
PN, LN	.007	.675	.682
$\Sigma = P(\alpha)$	.100	.900	1.000

Hence the table of conditional probabilities $P(\alpha|\pi, \lambda)$ is easily found.

Table C Conditional Probabilities $P(\alpha|\pi, \lambda)$

π, λ \ α	AD	AN	$\Sigma = 1$
PD, LD	1.00	.00	1
PD, LN	1.00	.00	1
PN, LD	13/238	225/238	1
PN, LN	7/682	675/682	1

We shall require analogues of Table C which express your judgments about $\tilde{\alpha}$, given only the outcomes $\tilde{\pi}$ of the pretest or only the outcomes $\tilde{\lambda}$ of the less reliable test. By addition in Table B we obtain:

Table D Joint Probabilities $P(\pi, \alpha)$

π \ α	AD	AN	$\Sigma = P(\pi)$
PD	.08	.00	.08
PN	.02	.90	.92
$\Sigma = P(\alpha)$	.10	.90	1.00

implying

Table E Conditional Probabilities $P(\alpha|\pi)$

π \ α	AD	AN	$\Sigma = 1$
PD	1.00	.00	1
PN	2/92	90/92	1

and also by addition in Table B we obtain:

Table F Joint Probabilities $P(\lambda, \alpha)$

λ \ α	AD	AN	$\Sigma = P(\lambda)$
LD	.065	.225	.290
LN	.035	.675	.710
$\Sigma = P(\alpha)$	.100	.900	1.000

implying

Table G Conditional Probabilities $P(\alpha|\lambda)$

λ \ α	AD	AN	$\Sigma = 1$
LD	65/290	225/290	1
LN	35/710	675/710	1

Finally, we shall need the conditional probabilities $P(\pi/\lambda)$ and $P(\lambda|\pi)$ of the outcomes of one test, given the outcomes of the other. From Table B we obtain immediately

Table H Joint Probabilities $P(\pi, \lambda)$

λ \ π	PD	PN	$\Sigma = P(\lambda)$
LD	.052	.238	.290
LN	.028	.682	.710
$\Sigma = P(\pi)$	.080	.920	1.000

from which we obtain:

Table I Conditional Probabilities $P(\pi|\lambda)$

λ \ π	PD	PN	$\Sigma = 1$
LD	52/290	238/290	1
LN	28/710	682/710	1

and

Table J Conditional Probabilities $P(\lambda|\pi)$

$\lambda \backslash \pi$	PD	PN
LD	52/80	238/920
LN	28/80	682/920
$\Sigma = 1$	1	1

We are now ready to assign all probabilities in the tree of Figure 2.16.

First, every dummy $\mathscr{E}$-move naturally has probability 1 attached to its sole branch.

Second, $P_1(\gamma_1^2|h_0, g_1^2) = P(\text{PD}) = .080$ and $P_1(\gamma_1^3|h_0, g_1^2) = P(\text{PN}) = .920$, from Table H; and similarly, $P_1(\gamma_1^4|h_0, g_1^3) = P(\text{LD}) = .290$ and $P_1(\gamma_1^4|h_0, g_1^3) = P(\text{LN}) = .710$.

Third, from Tables I and J we obtain the probabilities of ee's in nondummy, second-stage $\mathscr{E}$-moves, given the outcome of the first-stage test. We have:

$$P_2(\gamma_2^4|h_1^2, g_2^2) = P(\text{LD}|\text{PD}) = 52/80,\ P_2(\gamma_2^5|h_1^2, g_2^2) = P(\text{LN}|\text{PD}) = 28/80,$$
$$P_2(\gamma_2^4|h_1^3, g_2^2) = P(\text{LD}|\text{PN}) = 238/920, \text{ and } P_2(\gamma_2^5|h_1^3, g_2^2) = P(\text{LN}|\text{PN}) = 682/920,$$

from Table J; and

$$P_2(\gamma_2^2|h_1^4, g_2^3) = P(\text{PD}|\text{LD}) = 52/290,\quad P_2(\gamma_2^3|h_1^4, g_2^3) = P(\text{PN}|\text{LD}) = 238/290,$$
$$P_2(\gamma_2^2|h_1^5, g_2^3) = P(\text{PD}|\text{LN}) = 28/710, \text{ and } P_2(\gamma_2^3|h_1^5, g_2^3) = P(\text{PN}|\text{LN}) = 682/710,$$

from Table I.

Fourth, note that in each (nondummy) $\mathscr{E}$-move $\Gamma_3(h_2^j, g_3^1)$ (for $j = 1, \ldots, 13$) we have "defective in operation," $= \gamma_3^1$, if and only if $\tilde{\alpha} = \text{AD}$, = "arrived defective"; and "not defective in operation," $= \gamma_3^2$, if and only if $\tilde{\alpha} = \text{AN}$, = "arrived nondefective," because testing does not affect the reliability of the subsystem. Hence

$$P_3(\gamma_3^1|h_2^1, g_3^1) = P(\text{AD}) = .1 \text{ and } P_3(\gamma_3^2|h_2^1, g_3^1) = P(\text{AN}) = .9,$$

from Table B;

$$P_3(\gamma_3^1|h_2^2, g_3^1) = P(\text{AD}|\text{PD}) = 1 \text{ and } P_3(\gamma_3^2|h_2^2, g_3^1) = P(\text{AN}|\text{PD}) = 0,$$

from Table E;

$$P_3(\gamma_3^1|h_2^3, g_3^1) = P(\text{AD}|\text{PD, LD}) = 1,\ P_3(\gamma_3^2|h_2^3, g_3^1) = P(\text{AN}|\text{PD, LD}) = 0,$$
$$P_3(\gamma_3^1|h_2^4, g_3^1) = P(\text{AD}|\text{PD, LN}) = 1, \text{ and } P_3(\gamma_3^2|h_2^4, g_3^1) = P(\text{AN}|\text{PD, LN}) = 0,$$

from Table C;

$P_3(\gamma_3^1|h_2^5, g_3^1) = P(\text{AD}|\text{PN}) = 2/92$ and $P_3(\gamma_3^2(\gamma_3^2|h_2^5, g_3^1) = P(\text{AN}|\text{PN})$
$= 90/92$,

from Table E;

$P_3(\gamma_3^1|h_2^6, g_3^1) = P(\text{AD}|\text{PN, LD}) = 13/238$, $P_3(\gamma_3^2|h_2^6, g_3^1) = P(\text{AN}|\text{PN, LD})$
$= 225/238$,

$P_3(\gamma_3^1|h_2^7, g_3^1) = P(\text{AD}|\text{PN, LN}) = 7/682$, and $P_3(\gamma_3^2|h_2^7, g_3^1)$
$= P(\text{AN}|\text{PN, LN}) = 675/682$,

from Table C;

$P_3(\gamma_3^1|h_2^8, g_3^1) = P(\text{AD}|\text{LD}) = 65/290$ and $P_3(\gamma_3^2|h_2^8, g_3^1) = P(\text{AN}|\text{LD})$
$= 225/290$,

from Table G;

$P_3(\gamma_3^1|h_2^9, g_3^1) = P(\text{AD}|\text{LD, PD}) = 1$, $P_3(\gamma_3^2|h_2^9, g_3^1) = P(\text{AN}|\text{LD, PD}) = 0$,
$P_3(\gamma_3^1|h_2^{10}, g_3^1) = P(\text{AD}|\text{LD, PN}) = 13/238$, and $P_3(\gamma_3^2|h_2^{10}, g_3^1)$
$= P(\text{AN}|\text{LD, PN}) = 225/238$,

from Table C;

$P_3(\gamma_3^1|h_2^{11}, g_3^1) = P(\text{AD}|\text{LN}) = 35/710$ and $P_3^2|h_2^{11}, g_3^1) = P(\text{AN}|\text{LN})$
$= 675/710$,

from Table G; and

$P_3(\gamma_3^1|h_2^{12}, g_3^1) = P(\text{AD}|\text{LN, PD}) = 1$, $P_3(\gamma_3^2|h_2^{12}, g_3^1) = P(\text{AN}|\text{LN, PD}) = 0$,
$P_3(\gamma_3^1|h_2^{13}, g_3^1) = P(\text{AD}|\text{LN, PN}) = 7/682$, and $P_3(\gamma_3^2|h_2^{13}, g_3^1)$
$= P(\text{AN}|\text{LN, PN}) = 675/682$.

These calculations may seem to represent a great deal of work, but they all follow from probability relationships in Section 3.7. It should be remarkable that such a complicated situation is as tractable as it is!

We record the above probabilities and preferences in Figure 4.3, which summarizes all subsequent expectations and prunings of the tree according to Section 4.3. Figure 4.3 indicates that the optimal strategy for $\mathcal{D}$ is first to apply the less reliable test, then to rebuild the subsystem if the outcome indicates that it arrived defective, but to apply the pretest otherwise and then rebuild the subsystem only if the pretest indicates it arrived defective.

If you are tempted to protest that this strategy is simply common sense and that we have thus labored mightily to bring forth a mouse, you will find it instructive to work Exercises 4.4.2 through 4.4.4, which require reworking this example with different probabilities and preferences.

You will note that by far the greatest effort has been devoted to deriving from the probabilities "you" assessed in Examples 1.5.1a–c the pro-

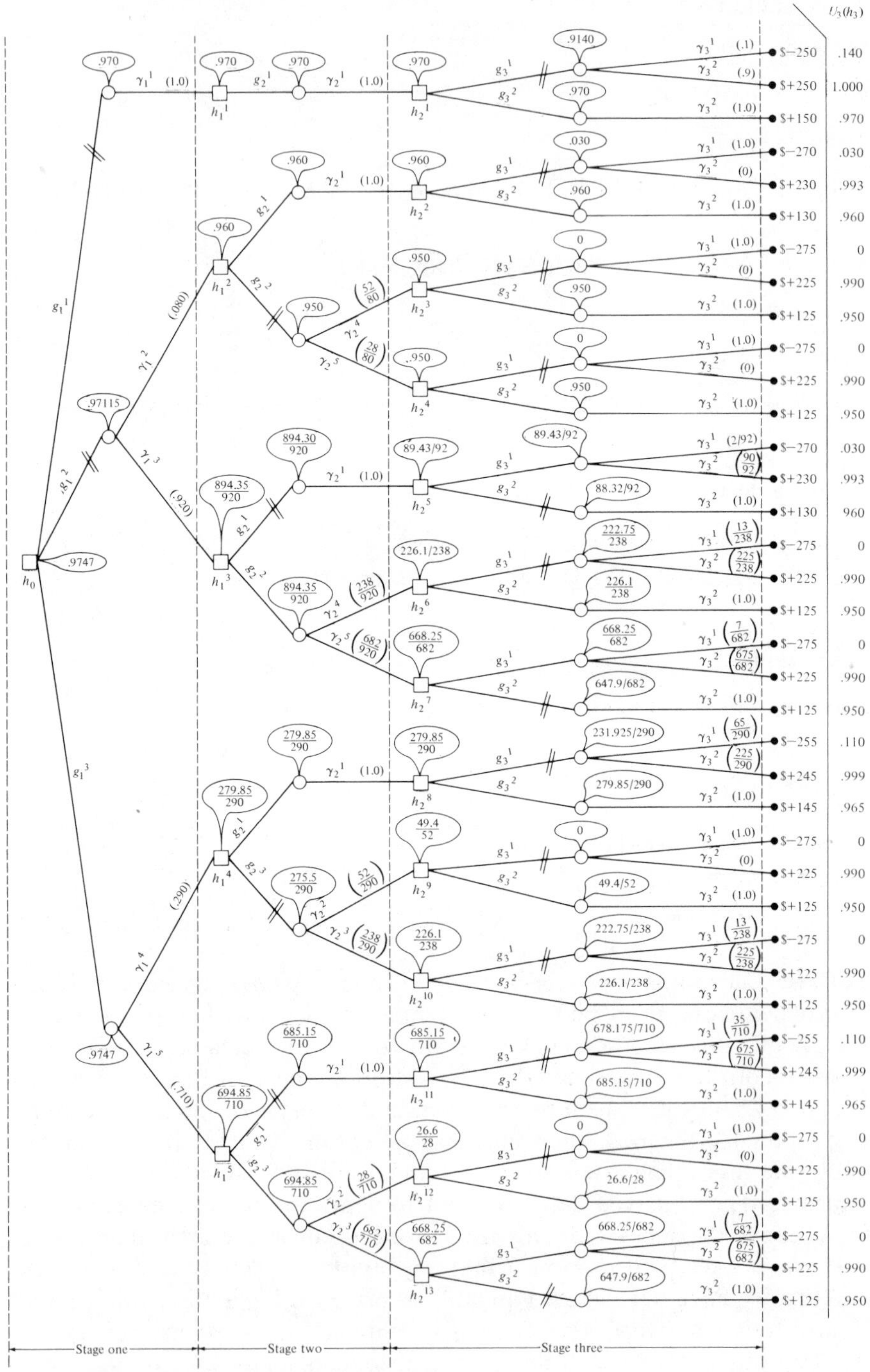

Figure 4.3 Analysis of Example 1.5.1c.

babilities required in the decision tree. These derivations required manifold application of the chain rule and the definition of conditional probability.

All these manipulations could be avoided in practice by forcing you ($=\mathscr{D}$) to assess the necessary probabilities *directly*. The trouble with so doing is that, in many cases such as the one at hand, *you can bring your judgments most easily to bear* on related events, the probabilities of which can be used to *calculate* the probabilities needed in the decision tree by invoking the probability relationships in Section 3.7, all of which follow via ordinary logic from our basic consistency principles. The spirit of our lengthy manipulations is thus to *put analysis to work for the decision maker.*

Exercise 4.4.1

Verify all calculations in Figure 4.3.

Exercise 4.4.2

Rework Example 1.5.1c, assuming instead of the preferences in Table A the preferences in Table A′ below.

Table A′

c	$u(c)$	c	$u(c)$
$+250; MP	1.000	$+130; MP	.518
$+245; MP	.969	$+125; MP	.508
$+230; MP	.877	$−250; LP	.013
$+225; MP	.857	$−255; LP	.010
$+150; MP	.579	$−270; LP	.003
$+145; MP	.564	$−275; LP	.000

Exercise 4.4.3

You now have second thoughts about revising your supplier's probability of defectiveness from .001 to .10; its recent and woeful record leads you to carefully assess $P(\text{AD}) = .30$ instead of $P(\text{AD}) = .10$.

(A) Show that Table B should be replaced by the following table.

Table B′ Joint Probabilities $P(\pi, \lambda, \alpha)$

π, λ \ α	AD	AN	$\Sigma = P(\pi, \lambda)$
PD, LD	.156	.000	.156
PD, LN	.084	.000	.084
PN, LD	.039	.175	.214
PN, LN	.021	.525	.546
$\Sigma = P(\alpha)$	.300	.700	1.000

(B) Use Table B′ to compute the analogues $C' - J'$ of Tables C–J.

(C) Use the probabilities in Tables B′–J′ in conjunction with Table A to reanalyze the dpuu.

Exercise 4.4.4

Reanalyze the dpuu in Example 1.5.1c with assumptions as manifested by Tables A′–J′.

4.5 ANALYSIS OF DECISIONS IN NONSTANDARD EXTENSIVE FORM

As noted in Section 4.1, actual applications do not usually require symbolic notation, and hence you have no particular motivation for reformulating them in standard extensive form.

But the methodology developed in Sections 4.2 and 4.3 is clearly applicable. You will have noted that the recursion analysis in Section 4.3 involved essentially only two operations.

(1) An *averaging operation* (expectation) over the branches emanating from lotteries, in which we evaluate each lottery ℓ on (say) Ω which has prizes with evaluations $U(\ell(\omega))$ by the number $U(\ell) = E\{U(\ell(\tilde{\omega}))\}$.

(2) A *pruning operation* on $\mathscr{D}$-moves L of lotteries, in which a lottery ℓ° with $U(\ell^\circ) = \max_{\ell \in L}[U(\ell)]$ is kept and all others are pruned, and the entire $\mathscr{D}$-move L is then evaluated by $U(\ell^\circ)$.

These operations are depicted in Figures 4.2(A) and 4.2(B) respectively in the notation for the standard extensive form.

Recall from Definition 2.8.1 that a *terminal* move is a move with each branch attached to an endpoint of the tree. In the nonstandard extensive form, start the analysis by applying the *pruning* operation to all terminal $\mathscr{D}$-moves, evaluating them by the maximum attainable utility therein. If the decision is under uncertainty, there will be preceding $\mathscr{D}$-moves, some of which now consist solely of lotteries. To each such lottery apply the *averaging* operation, and then apply the *pruning* operation to these $\mathscr{D}$-moves themselves. Now perhaps other $\mathscr{D}$-moves will consist solely of lotteries. Repeat the averaging and pruning until the initial $\mathscr{D}$-move itself has been evaluated.[b]

The basic ideas involved in the preceding paragraph coincide with those expressed formally in Section 4.3, but our eschewal of notation makes for a slight obfuscation of ideas. Figure 4.4 depicts a numerical example, the calculations in which you may verify as Exercise 4.5.1.

[b] If the initial move is an $\mathscr{E}$-move, precede it by a dummy $\mathscr{D}$-move and ultimately evaluate the (sole) lottery therein.

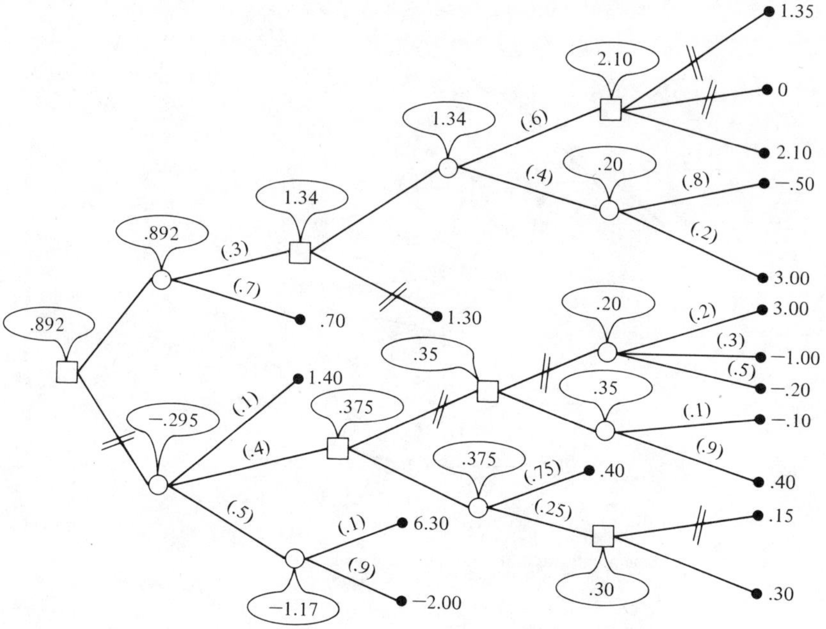

Figure 4.4 An analyzed decision in nonstandard extensive form.[1]

[1] Numbers attached to endpoints are utilities of consequences.

Exercise 4.5.1

Verify all calculations in Figure 4.4.

Exercise 4.5.2

Reformulate the decision in Figure 4.4 in standard extensive form, and verify that the solution to the reformulated decision agrees, except for dummy $\mathscr{D}$-moves, with the solution indicated in Figure 4.4.

Exercise 4.5.3

Solve the dpuu depicted in Figure 4.5.

Exercise 4.5.4

In Section 2.10 we argued on an intuitive basis that insertion and deletion of dummy moves should not affect the salient strategic features of the dpuu. Argue that $U = U'$ and $U'' = U'''$ in Figure 4.6, thus justifying that statement within the context of the analytical framework we have developed.

■ We have already justified (in Observation 4.2.3) the assertion in Section 2.11 that successive choices *by* $\mathscr{E}$ can be combined without

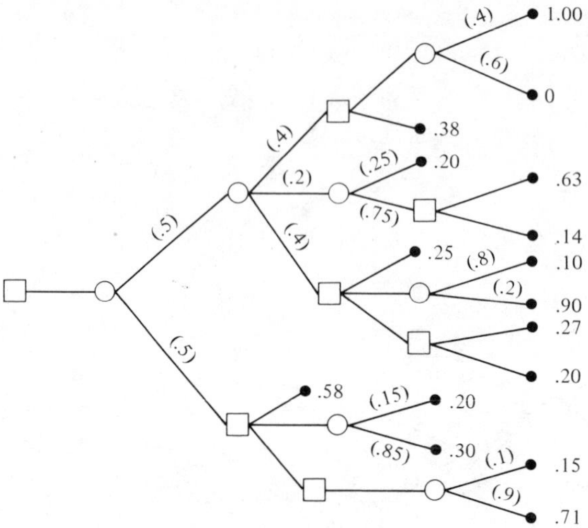

Figure 4.5 A dpuu in extensive form.[1]

[1] Numbers attached to endpoints are utilities of consequences.

affecting any salient strategic features of the dpuu, since combining successive choices by $\mathscr{E}$ does not affect evaluations of lotteries which precede those choices. Successive choices *by* $\mathscr{D}$ can also be combined without strategic import, as the next exercise indicates. ■

Exercise 4.5.5

Suppose G is a $\mathscr{D}$-move; G' is a subset of G such that every g in G' is a lottery with evaluation $U(g)$; $G'' = \overline{G'}$, and every g in G'' leads to a

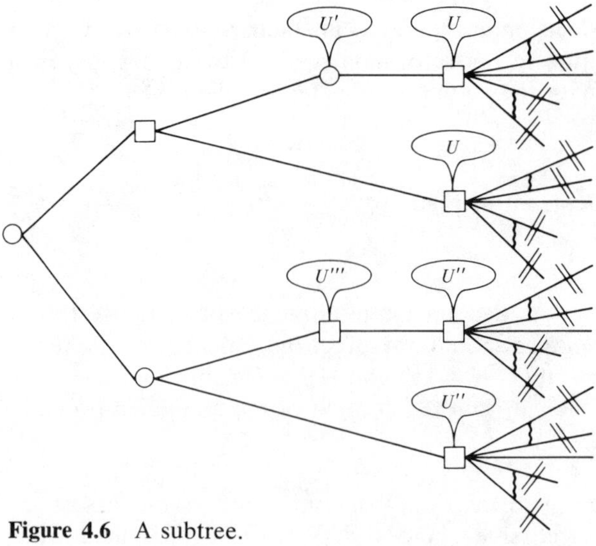

Figure 4.6 A subtree.

$\mathscr{D}$-move $L(g)$ of lotteries ℓ with evaluations $U(\ell)$. Define

$$G^* = G' \cup \{(g, \ell): \ell \in L(g), g \in G''\};$$

that is, G^* is the $\mathscr{D}$-move which results by combining all successive choices in G and its successor $\mathscr{D}$-moves $L(g)$. Assume that all maximizers required below exist.

(A) Let g^* denote the typical (or generic) element of G^*. Show that

$$U(g^*) = \begin{cases} U(g), & g^* = g \in G' \\ U(\ell), & g^* = (g, \ell) \quad \text{for } g \in G'' \text{ and } \ell \in L(g). \end{cases}$$

(B) Show that if $g \in G''$, then

$$U(g) = \max_{\ell \in L(g)} [U(\ell)].$$

(C) Show that

$$\max_{g \in G} [U(g)] = \max_{g^* \in G^*} [U(g^*)],$$

the left-hand side representing the ultimate evaluation of the "earlier" $\mathscr{D}$-move G, and the right-hand side representing the evaluation of the "combined" $\mathscr{D}$-move G^*.

4.6 CONSISTENT DECISION MAKING UNDER UNCERTAINTY

In Section 1.7 we claimed that our approach is intended to *help* decision makers make decisions *consistent* with their basic judgments and preferences, but *not* to obviate such judgments and preferences. By now you should appreciate this remark!

The decision maker is called upon to express *many* preferences, in the form of $u(c)$, and *many* judgments, in the form of probabilities. He is also asked to accept certain more or less philosophical principles of consistent choice in simple, hypothetical decision situations. Once he has agreed that he *would like to* abide by the consistency principles and once he has made the many simple, clear-cut choices in hypothetical decisions, described in Sections 3.5 and 3.6, then *and only then* can he delegate his task to a calculator or clerk with instructions to perform the requisite averaging and pruning operations that ultimately determine his optimal strategy in the dpuu.

Now, the preferences and judgments which $\mathscr{D}$ expresses affect the optimal strategy crucially. To appreciate this, just compare $\mathscr{D}$'s optimal strategy for Example 1.5.1c as derived in Section 4.4 with the three different, and *mutually* different, optimal strategies corresponding to different judgments and/or preferences which you (should have) derived in Exercises 4.4.2–4.4.4.

If decision makers always behave consistently with their basic preferences and judgments, there would be no need for decision analysis in general and hence for this book in particular—but they do not.

Exercise 4.6.1

Let $\mathscr{D}$ be someone you know very well. Imagine that $\mathscr{D}$ has been given Examples 1.5.1a–1.5.1c and Table A. Let Ω' denote the event that $\mathscr{D}$ correctly guesses (without calculation) the optimal strategy as derived in Section 4.4. Assess your $P(\Omega')$ according to the procedure you used in Exercise 3.6.1.

Exercise 4.6.2: Allais "Paradox"

(A) *Without performing any calculations whatsoever and without reading Part (B) of this exercise*, choose *intuitively* (a) between canonical lotteries ℓ^1 and ℓ^2 and (b) between canonical lotteries ℓ^3 and ℓ^4 in Figure 4.7, in which "\$$x$" is shorthand for "receive \$$x$ with no strings attached and no ethical compromises," and in which "(p)" means "this event occurs with canonical probability p."

(B) Let $u(\$25{,}000{,}000) = 1$ and $u(\$0) = 0$, and suppose $u(\$5{,}000{,}000) \in (0, 1)$. Show that $\ell^1 \succsim \ell^2$ if and only if $\ell^3 \succsim \ell^4$. [*Now* you may compute $U(\ell^i)$ for $i = 1, 2, 3, 4$!]

(C) Was *your* choice between ℓ^3 and ℓ^4 consistent with your choice between ℓ^1 and ℓ^2?

■ We have put the word "paradox" in quotes, because your probable behavior is really not paradoxical except to those who attempt to use the

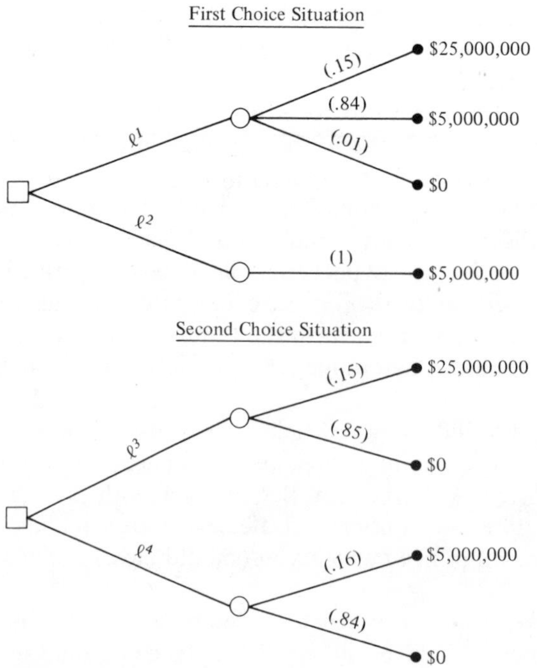

Figure 4.7 Allais "paradox."

conclusions of decision analysis to describe or forecast actual behavior. We are making no claims for the validity of decision analysis as a *de*scriptive theory, however; quite the contrary. We have taken some pains to establish decision analysis as a *pre*scriptive theory, an aid in the making of carefully reasoned and important decisions. Inconsistency is the hobgoblin of intuitive behavior! ■

4.7 EXERCISES

Believe it or not, you are now acquainted with the fundamentals of decision analysis in extensive form. This chapter concludes Part I of the book; Chapters 5 and 6, consisting of Part II, delve more deeply into the quantification of preferences (utility theory) and judgments (probability theory) respectively. But the basic concepts are now out in the open.

This section consists of exercises based on the now familiar examples in Section 1.5.

Exercise 4.7.1: Marine Insurance Problem

Reconsider Example 1.5.7, and reread in particular the indented comment in Section 2.4 about the time value of money and discounting. Suppose that your utility function for £x one year hence can be expressed as

$$u(x) = [x/12{,}000]^z$$

for all $x \in [£0, £+12{,}000]$, where z is a positive constant, or *parameter*, of the utility function.

(A) Sketch the utility function for the following.

(i) $z = .1$	(iv) $z = .8$
(ii) $z = .3$	(v) $z = 1.0$
(iii) $z = .5$	(vi) $z = 2.0$

(B) Assume that the premium of £2,000 must be paid *now*, before the *Pride of Old Sarum* embarks; that £x at any date is equally as desirable to you as £(.8)x one year prior to that date; and that the insurance company would settle your claim (if made) two years from now. Recall that the *POS* is scheduled to return one year from now, and suppose that if its voyage does not encounter disaster it will return very close to one year from now. Should you insure, if

(i) $z = .1$?	(iv) $z = .8$?
(ii) $z = .3$?	(v) $z = 1.0$?
(iii) $z = .5$?	(vi) $z = 2.0$?

(C) Now suppose that the insurance company would settle your claim *three* years from now instead of two. Other assumptions are as in (B). Same question as in (B).
(D) Same problem as (C) except that "*three*" is replaced by "*four.*"
(E) Redo (C) and (D), using "discount factor" .95 instead of .8; that is, assume that £x is equally as desirable to you as £(.95)x one year sooner.

Exercise 4.7.2: Attorney's Problem

Find your optimal strategy for the dpuu in Example 1.5.2c, assuming that the total time span involved is sufficiently short for you to safely ignore the time value of money.

(A) Use your own utility function for money.
(B) Use the utility function $u(\$x) = (x + k)^z$ for $k = \$10{,}000$ and

(i)	$z = .1$	(iv)	$z = .8$
(ii)	$z = .3$	(v)	$z = 1.0$
(iii)	$z = .5$	(vi)	$z = 2.0$

Exercise 4.7.3: Impulse Buyer

How much would you be willing to pay for the opinion of an infallible expert on the authenticity of the Arabian sword? Assume that your present information leads you to assess probability .5 that it is authentic.

(A) Use your own utility function.
(B) Same as Part (B) of Exercise 4.7.2, but with $k = \$100$.

Exercise 4.7.4: Newsboy Problem

Answer the question in Example 1.5.5, using

(A) your own utility function for money
(B) the utility functions in part (B) of Exercise 4.7.2, but with $k = \$15$.

Exercise 4.7.5: Realpolitik

Quantify your own preferences for the consequences of Example 1.5.9 and answer the question therein.

Exercise 4.7.6: Overbooking

Answer the question in Example 1.5.10, using

(A) your own utility function for money, and
(B) the utility functions in part (B) of Exercise 4.7.2, but with $k =$ \$10,000.

*Exercise 4.7.7: Competitive Sealed Bidding

(To be omitted by those who have not read *Section 3.10 and/or have not had a course in elementary integral calculus.) Suppose in the competitive sealed bidding problem: (1) that you quantify your judgments about the cost $\tilde{\alpha}$ of fulfilling the contract in the form of the density function

$$f_\alpha(\alpha) = \begin{cases} 0, & \alpha < 0 \\ te^{-t\alpha}, & \alpha \geq 0, \end{cases}$$

where t is a positive constant; (2) that you quantify your judgments about the bid $\tilde{\beta}$ of your competitor in the form of the density function

$$f_\beta(\beta) = \begin{cases} s[\beta/M]^{s-1}/M, & 0 \leq \beta \leq M \\ 0, & \beta \notin [0, M], \end{cases}$$

where s and M are positive constants; and (3) you regard $\tilde{\alpha}$ and $\tilde{\beta}$ as *independent*, in the sense that no information about $\tilde{\alpha}$ would cause you to modify your judgments about $\tilde{\beta}$, and vice versa.

(A) Assuming that $u(\$x) = x$ for all x, show that

$$U(b) = \begin{cases} b - t^{-1}, & b \leq 0 \\ (b - t^{-1})(1 - [b/M]^s), & 0 \leq b \leq M \\ 0, & b > M. \end{cases}$$

(B) Derive your optimal bid $b°$ under the above assumptions.

*■ **Portfolio Problem.** [This comment should be skipped by readers who (1) have not read *Section 3.10; (2) have not become acquainted with elementary multivariable calculus; and (3) have not learned the basic concepts of linear algebra.] In the following, let $b = (b_1, \ldots, b_n)'$ and $\boldsymbol{\theta} = (\theta_1, \ldots, \theta_n)'$; we convene that all vectors are of the column form unless transposed. Transposes are denoted by the superfix "prime." Assume that your utility function $u(x)$ for $\$x$ is expressible as

$$u(x) = 1 - e^{-zx}$$

for all real x, where z is a positive constant. [Recall that

$$e = \lim_{h\to\infty} (1 + h^{-1})^h \doteq 2.718281828459045.]$$

Since $v(\mathbf{b}, \boldsymbol{\theta}) = \Sigma_{i=1}^n b_i\theta_i = \mathbf{b}'\boldsymbol{\theta}$, it follows that

$$U(\mathbf{b}) = E\{u(\mathbf{b}'\tilde{\boldsymbol{\theta}})\} = 1 - E\{e^{-z\mathbf{b}'\tilde{\boldsymbol{\theta}}}\}.$$

Suppose that your judgments about the vector $\tilde{\boldsymbol{\theta}}$ of returns per dollar invested in the various securities are expressible by the joint "n-variate Normal" density function $f_N^{(n)}(\cdot|\mathbf{m}, \mathbf{V})$, defined by

$$f_N^{(n)}(\boldsymbol{\theta}|\mathbf{m}, \mathbf{V}) = (2\pi)^{-n/2}|\mathbf{V}|^{-1/2}e^{-(\boldsymbol{\theta}-\mathbf{m})'\mathbf{V}^{-1}(\boldsymbol{\theta}-\mathbf{m})/2}$$

for every n-tuple $\boldsymbol{\theta}$ of real numbers, where $\mathbf{m}$ is an n-tuple of constants,

$\mathbf{V}^{-1}$ is a positive definite, symmetric, nth-order matrix [and hence, $\mathbf{V} = (\mathbf{V}^{-1})^{-1}$ is also positive definite, symmetric, and nth order], and $|\mathbf{V}^{-1}|$ denotes the determinant of $\mathbf{V}^{-1}$. Under this assumption, it can be shown that

$$\begin{aligned} U(\mathbf{b}) &= 1 - E\{e^{-z\mathbf{b}'\tilde{\boldsymbol{\theta}}}\} \\ &= 1 - E_N^{(n)}\{e^{-z\mathbf{b}'\tilde{\boldsymbol{\theta}}}|\mathbf{m}, \mathbf{V}\} \\ &= 1 - e^{-(z/2)[2\mathbf{b}'\mathbf{m} - z\mathbf{b}'\mathbf{V}\mathbf{b}]}. \end{aligned}$$

Now, since $z > 0$, it follows that

$$U(\mathbf{b}_{(1)}) \geq U(\mathbf{b}_{(2)})$$

if and only if

$$2\mathbf{b}'_{(1)}\mathbf{m} - z\mathbf{b}'_{(1)}\mathbf{V}\mathbf{b}_{(1)} \geq 2\mathbf{b}'_{(2)}\mathbf{m} - z\mathbf{b}'_{(2)}\mathbf{V}\mathbf{b}_{(2)}.$$

Hence, an *optimal portfolio* $\mathbf{b}^\circ$ is any maximizer of

(†) $$2\mathbf{b}'\mathbf{m} - z\mathbf{b}'\mathbf{V}\mathbf{b}$$

subject to the constraints

(*) $$\mathbf{b} \geq \mathbf{0} \qquad \text{(no short selling)}$$

[where $\mathbf{0} = (0, 0, \ldots, 0)'$], and

(**) $$\mathbf{1}'\mathbf{b} \leq T \qquad \text{(total capital} = T\text{)}$$

[where $\mathbf{1} = (1, 1, \ldots, 1)'$]. This is a rather simple "quadratic programming" problem, and it is amenable to solution by rapid, computerizable, iterative procedures (algorithms)—provided n is not *too* large. ■

***Exercise 4.7.8: Portfolio Problem (concluded)**

(To be omitted by those who encountered difficulty with the preceding note.)

(A) Show that

$$2\mathbf{b}'\mathbf{m} - z\mathbf{b}'\mathbf{V}\mathbf{b} = z^{-1}\mathbf{m}'\mathbf{V}^{-1}\mathbf{m} - z(\mathbf{b} - z^{-1}\mathbf{V}^{-1}\mathbf{m})'\mathbf{V}(\mathbf{b} - z^{-1}\mathbf{V}^{-1}\mathbf{m}),$$

and hence $\mathbf{b}^\circ$ maximizes (†) if and only if $\mathbf{b}^\circ$ minimizes

(††) $$(\mathbf{b} - z^{-1}\mathbf{V}^{-1}\mathbf{m})'\mathbf{V}(\mathbf{b} - z^{-1}\mathbf{V}^{-1}\mathbf{m}),$$

in both cases subject to the constraints (*) and (**). [*Hint*: Recall that $z > 0$ and that $\mathbf{V}$ is symmetric if and only if $\mathbf{V} = \mathbf{V}'$; this equality is the defining condition of matrix symmetry.]

(B) The necessity for solving the quadratic programming problem "minimize (††) [or maximize (†)] subject to the constraints (*) and (**)" by iterative numerical methods arises because (*) and (**) are *in*equalities. For the remainder of this exercise suppose we drop the nonnegativity constraint (*); i.e., we shall permit short selling and full reinvestment of the proceeds. *Temporarily*, replace the total-capital constraint (**) by an equality constraint

$$(***) \qquad \mathbf{1}'\mathbf{b} = T_e,$$

where T_e is as yet unspecified. Show that the (unique) minimizer $\mathbf{b}°(T_e)$ of (††) subject to (***) is given by

$$\mathbf{b}°(T_e) = \mathbf{m}^* + K(T_e, \mathbf{m}^*)\mathbf{V}^{-1}\mathbf{1},$$

where

$$\mathbf{m}^* = z^{-1}\mathbf{V}^{-1}\mathbf{m}$$

and

$$K(T_e, \mathbf{m}^*) = [T_e - \mathbf{1}'\mathbf{m}^*]/[\mathbf{1}'(\mathbf{V}^{-1})\mathbf{1}].$$

[*Hint*: Introduce a Lagrange multiplier y for the equality constraint (***), and find the *un*constrained minimizer $(\mathbf{b}°(T_e), y°(T_e))$ of the Lagrangean function

$$L(\mathbf{b}, y) = (\mathbf{b} - \mathbf{m}^*)'\mathbf{V}(\mathbf{b} - \mathbf{m}^*) - y(T_e - \mathbf{1}'\mathbf{b}).]$$

(C) Show that the minimum of (††) subject to the constraint (***) is given by

$$L(\mathbf{b}°(T_e), y°(T_e)) = (T_e - \mathbf{1}'\mathbf{m}^*)^2/[\mathbf{1}'(\mathbf{V}^{-1})\mathbf{1}],$$

with $\mathbf{1}'(\mathbf{V}^{-1})\mathbf{1} > 0$ because of the positive definiteness of $\mathbf{V}^{-1}$.

(D) Show that the minimizer $\mathbf{b}°$ of (††) subject to the *in*equality constraint (**) is given by

$$\mathbf{b}° = \mathbf{b}°(T_e°),$$

where $T_e°$ satisfies

$$(T_e° - \mathbf{1}'\mathbf{m}^*)^2 = \min_{T_e \in [0, T]}[(T_e - \mathbf{1}'\mathbf{m}^*)^2].$$

(E) Verify that $\mathbf{b}°$ as characterized in (D) is your optimal portfolio, given your total capital T, your utility function $1 - e^{-zx}$, and your judgments $f_N^{(n)}(\boldsymbol{\theta}|\mathbf{m}, \mathbf{V})$.

■ Exercise 4.7.8 and the preceding comment on the portfolio problem constitute an excellent example of how a decision problem under uncertainty can give rise to a constrained optimization problem, usually called a "mathematical programming" problem, the solution to which coincides with the solution to the dpuu of interest. Mathematical programming problems are all expressible as decisions under certainty, in which $\mathscr{D}$ is to maximize (or minimize) his utility (or disutility) via choice of strategy from an explicitly or implicitly defined set. See the indented comment about decisions under certainty in Section 1.4. You will later appreciate that *all approaches to solving dpuu's ultimately give rise to constrained optimization problems.* ■

II

MORE ON PREFERENCE AND JUDGMENT QUANTIFICATIONS

"He who has best calculated his behavior, will win the advantage over those who act in less consequence than he."

Frederick the Great

5

QUANTIFICATION OF PREFERENCES

Retire unto thyself. The rational principle which rules has this nature, that it is content with itself when it does what is just, and so secures tranquility.

Marcus Aurelius Antoninus

5.1 INTRODUCTION

This chapter is concerned with the quantification of $\mathscr{D}$'s preferences in the form of a utility function. The basic definitions for and approach to this subject were given in Section 3.5, the essential contents of which are reviewed and extended in Section 5.2.

Section 5.3 concerns the quantification of preferences on a (small-) finite set of consequences; Section 5.4 is devoted to utility functions on an interval of real numbers representing possible net monetary returns. These two sections discuss actual procedures for quantifying preferences in some detail.

Sections 5.5 through 5.7 concern special topics in the quantification of preferences on a set C of consequences which has a Cartesian product structure. These three sections are of more peripheral than central interest and may therefore be omitted on a first reading. The results therein will be referred to only occasionally in subsequent chapters.

5.2 PRELIMINARIES

First Preliminary

From Section 3.5, a *utility function* for $\mathscr{D}$ on C is a function $u: C \to (-\infty, +\infty)$ such that every lottery $\ell: \Omega \to C$ with $\mathscr{E}$-move Ω and (a finite number of possible) outcomes in C has relative desirability index $U(\ell)$, expressed as a probability-weighted average (or "expectation")

$$U(\ell) = E\{u(\ell(\tilde{\omega}))\}, \tag{5.2.1}$$

vis-à-vis other[a] lotteries $\ell': \Omega \to C$, in the sense that the more desirable of ℓ and ℓ' to $\mathcal{D}$ is the lottery with the larger index U. Symbolically,[b]

$$\ell \succsim \ell' \quad \text{if and only if} \quad U(\ell) \geq U(\ell'). \tag{5.2.2}$$

Second Preliminary

We shall show that the $\mathcal{E}$-move domain Ω of a lottery $\ell: \Omega \to C$ is relevant to the relative desirability of ℓ *only through* the probabilities with which ℓ results in the various possible consequences. To see this, we note that ℓ results in a consequence c in the subset $C\dagger$ of C if and only if $\tilde{\omega} \in \{\omega: \ell(\omega) \in C\dagger\}$. Hence the probability $P(\tilde{c} \in C\dagger|\ell)$ that the as-yet unrevealed consequence[c] $\tilde{c}$ of ℓ falls in the subset $C\dagger$ of C is simply the probability $P(\tilde{\omega} \in \{\omega: \ell(\omega) \in C\dagger\})$ that $\tilde{\omega}$ falls in the set $\{\omega: \ell(\omega) \in C\dagger\}$ of elementary events ω yielding an outcome $\ell(\omega)$ in $C\dagger$.

Definition 5.2.1. Let ℓ be a lottery on Ω with outcomes in C; that is, $\ell: \Omega \to C$. The probability $P(\cdot|\ell)$ on subsets of C *induced by* ℓ is defined by

$$P(\tilde{c} \in C\dagger|\ell) = P(\{\omega: \ell(\omega) \in C\dagger\}) \tag{5.2.3}$$

for every subset $C\dagger$ of C.

The probability $P(\cdot|\ell)$ induced by ℓ embodies all relevant aspects of Ω, which can therefore be suppressed for desirability-index purposes. This assertion is an immediate consequence of the following observation.

Observation 5.2.1. Let $P(\cdot|\ell)$ be the probability on subsets of C induced by the lottery ℓ on an $\mathcal{E}$-move Ω. Then

$$U(\ell) = E\{u(\tilde{c})|\ell\}. \tag{5.2.4}$$

■ *Proof.* Suppose that ℓ is a finite lottery, resulting in c^i if and only if $\tilde{\omega} \in \Omega^i$ for $i = 1, \ldots, n$, where $\Omega^1, \ldots, \Omega^n$ constitutes a partition of Ω. Then Equation (5.2.1) and Theorem 3.2.1 imply that $U(\ell) = E\{u(\ell(\tilde{\omega}))\} = \Sigma_{i=1}^n u(c^i)P(\Omega^i)$. But for every $i \in \{1, \ldots, n\}$ the definition of Ω^i implies that $\Omega^i = \{\omega: \ell(\omega) = c^i\}$. Hence Definition 5.2.1 implies that

$$P(c^i|\ell) = P(\{\omega: \ell(\omega) = c^i\}) = P(\Omega^i)$$

for every i. Hence

$$U(\ell) = \sum_{i=1}^n u(c^i)P(c^i|\ell) = E\{u(\tilde{c})|\ell\},$$

the second equality being by definition of expectation. [*If ℓ is an

[a] In this section we start numbering mathematical expressions; $(x \,.\, y \,.\, z)$ denotes expression z of section $x \,.\, y$.

[b] Recall that "$\succsim$," "$\succ$," and "$\sim$" are desirability relations similar to the numerical relations "$\geq$," "$>$," and "$=$" respectively.

[c] We shall now regard subsets of C as events and hence violate our convention that all events are denoted by Greek letters.

idealization of reality, with infinitely many possible consequences, Observation 5.2.1 still holds, but the expectation is evaluated differently. Part C in Chapter 3 shows how expectations are to be evaluated in such cases; and Observation 5.2.1 follows by general results of the "change-of-variable-of integration" sort.] ■

For the (realistic) finite case, Observation 5.2.1 implies that the desirability index $U(\ell)$ of ℓ can be obtained as follows. (1) Determine the probabilities $P(c^i|\ell)$ of the possible consequences $c^1, \ldots, c^n$ of ℓ. (2) Determine the utilities $u(c^1), \ldots, u(c^n)$ of $c^1, \ldots, c^n$. (3) Calculate the probability-weighted average $\Sigma_{i=1}^n u(c^i)P(c^i|\ell) = U(\ell)$ of the consequences' utilities.

Now that we have shown that the relative desirability of a lottery ℓ with outcomes in C is completely determined by the probability $P(\cdot|\ell)$ induced by ℓ on C, it is possible to restate Theorem 3.2.1 in such a way that no reference is made to lottery domains.

Observation 5.2.2. Let ℓ^r and ℓ^s be lotteries with outcomes in C, and let $u\colon C \to (-\infty, +\infty)$ be a utility function for $\mathscr{D}$ on C. Then

$$\ell^r \begin{Bmatrix} > \\ \sim \\ < \end{Bmatrix} \ell^s$$

if and only if

$$U(\ell^r) \begin{Bmatrix} > \\ = \\ < \end{Bmatrix} U(\ell^s),$$

where

$$U(\ell^t) = E\{u(\tilde{c})|\ell^t\} \qquad \text{for } t = r, s.$$

It will occasionally be convenient in the following discussion to use $\tilde{c}$, $\tilde{c}'$, etc., to denote lotteries with outcomes in C. In such cases, the desirability indices are written as $U(\tilde{c}) = E\{u(\tilde{c})\}$, $U(\tilde{c}') = E\{u(\tilde{c}')\}$, etc.

Third Preliminary

Recall from Chapter 3 that $\ell^*(p; c'', c')$ denotes a canonical lottery which results in c'' with canonical probability p and in c' with complementary canonical probability $1-p$.

For purposes of actually assessing $\mathscr{D}$'s utility function u, it will become pleasantly apparent that $\mathscr{D}$ can make do with very simple desirability comparisons, of the form c'' versus c', and of the form c versus $\ell^*(p; c'', c')$. As an immediate corollary of Observation 5.2.2, we have the following observation.

Observation 5.2.3. If $c'' > c$ and $c > c'$, then $c \sim \ell^*(p; c'', c')$ for some (unique) p *strictly* between zero and one.

■ *Proof.* $U(c) = u(c)$ and $U(\ell^*(p; c'', c')) = pu(c'') + (1-p)u(c')$ by virtue of Theorem 3.2.1. Hence by Observation 5.2.2,

$$c \sim \ell^*(p; c'', c') \quad \text{if and only if} \quad u(c) = pu(c'') + (1-p)u(c').$$

That such a p exists in (0, 1) and is unique is now clear from the fact that $u(c'') > u(c) > u(c')$, these inequalities being obvious because $u(\cdot)$ reflects relative desirabilities of consequences to $\mathscr{D}$. ■

Fourth Preliminary

In Section 3.5 we showed that $\mathscr{D}$ does not have a *unique* utility function on a set C of consequences, but that all utility functions, necessarily satisfying Observation 5.2.2, are closely related by changes of scale and/or origin.

Observation 5.2.4. U and u^* are utility functions for $\mathscr{D}$ on C if and only if there are real numbers a^* and b^*, with a^* *positive*, such that

$$u^*(c) = a^*u(c) + b^*$$

for every c in C.

■ *Proof.* This is Theorem 3.5.2. ■

Fifth Preliminary

In the second preliminary we argued that references to the *domains* of lotteries can be suppressed. Here, we shall argue that references to the *ranges* C of lotteries can also be suppressed to a certain extent.

As a step in this direction, suppose that ℓ is a lottery with outcomes in a set C^* of consequences, and that C^* is a subset of a set C. Then ℓ is a lottery with outcomes in C. Furthermore, if $u: C \to (-\infty, +\infty)$ is a utility function for $\mathscr{D}$ on C, then the relative desirability of ℓ vis-à-vis any other lotteries with outcomes *in* C is given by $U(\ell) = E\{u(\tilde{c})|\ell\}$. But *every* lottery with outcomes in C^* is a lottery with outcomes in C, and therefore u must also be a utility function for $\mathscr{D}$ on the subset C^* of C. Hence a utility function on a given set is also a utility function on each of its subsets.

But the converse of this conclusion is what is really important: Any utility function u^* on a set C^* of consequences can be *extended* to a utility function u on a superset[d] C of C^*, in such a way that $u^*(c) = u(c)$ for every c in C^*. More formally:

Observation 5.2.5. If u^* is a utility function for $\mathscr{D}$ on C^* and if $C^* \subset C$, then there is a utility function u for $\mathscr{D}$ on C such that $u^*(c) = u(c)$ for every c in C^*.

[d] C is a superset of C^* if and only if C^* is a subset of C.

■ *Proof.* Let u^{**} be any utility function for $\mathscr{D}$ on C. Then u^{**} is also a utility function for $\mathscr{D}$ on C^*, and hence [by applying Observation 5.2.4 to C^* with u^{**} in place of u], there are numbers a^* and b^* with a^* positive, such that $u^*(c) = a^*u^{**}(c) + b^*$ for every c in C^*. Now, the function u on C defined by $u(c) = a^*u^{**}(c) + b^*$ for every c in C is a utility function for $\mathscr{D}$ on C, by Observation 5.2.4; and clearly $u^*(c) = u(c)$ for every c in C^*. ■

Observation 5.2.5 implies that, once $\mathscr{D}$'s utilities of two unequally desirable consequences have been specified, $\mathscr{D}$ has only one utility function, on the gargantuan set of all possible consequences of anything. This may seem like a very "puristic" result, but it has two important implications in the assessment of utility. First, a utility function on an infinite set of consequences is always determined by extending a utility function which $\mathscr{D}$ assesses *directly* on some small-finite subset. You used this approach in Exercise 3.5.6 to determine your utility of monetary returns in the interval [\$0, \$1,000], containing an infinite number of monetary returns.

The second important implication of the extensibility of a utility function is that it is permissible to replace some or all of the consequences c of a dpuu with equally desirable consequences $t(c)$ which may be much easier to think about and discuss. This topic constitutes the next preliminary.

Sixth Preliminary

Let C^* be the set of possible consequences of a dpuu. For every c in C^*, let $t(c)$ be a consequence, not necessarily in C^*, such that c and $t(c)$ are equally desirable. Now define C by

$$C = C^* \cup \{t(c): c \in C^*\},$$

and suppose that u is a utility function for $\mathscr{D}$ on C. Then equidesirability of c and $t(c)$ for every c in C^* implies that $u(c) = u(t(c))$ for every c in C^*. Furthermore, if ℓ is a lottery with outcomes in C^* and $\ell_{(t)}$ is the lottery which yields $t(c)$ if and only if ℓ yields c, then $U(\ell) = U(\ell_{(t)})$, implying that $\ell \sim \ell_{(t)}$, and hence that replacing the c's by their respective equally desirable $t(c)$'s results in an equally desirable lottery. But this is just the Substitutability Principle! It is convenient, however, to use just one utility function u for c's and for $t(c)$'s as well.

Definition 5.2.2. Let $C_{(t)}$ be a set of consequences, and suppose that $t: C^* \to C_{(t)}$ is such that $t(c) \sim c$ for every c in C^*. Then t is called a *substitute function* on C^*.

The following example is an important application of the concept of substitute functions in dpuu's whose consequences vary primarily, though not entirely, in their monetary attributes.

Example 5.2.1

Suppose that the consequences c in C^* of a given dpuu can be written as (c_1, c_2), where c_1 denotes net monetary return to be received at a fixed and given date, and c_2 denotes all other attributes of the consequences. Then C^* is a subset of the Cartesian product $C_1 \times C_2$, where C_i denotes the set of all possible c_i. Since $\mathscr{D}$ can probably focus his attention more easily on consequences differing solely in their net monetary returns as of the fixed reference point in time, he might like, first, to choose a *nonmonetary reference attribute* c_2^* in C_2; second, to determine for each c in C^* by introspection a *monetary adjustment* $N(c|c_2^*) = N(c_1, c_2|c_2^*)$ such that

$$(c_1, c_2) \sim (c_1 + N(c_1, c_2|c_2^*), c_2^*);$$

and third, to determine his utility function on the set

$$\{(c_1, c_2^*): -\infty < c_1 < +\infty\}$$

of consequences which differ *solely* in their net monetary returns. The substitute function corresponding to this monetary adjustment procedure is clearly defined by $t(c_1, c_2) = (c_1 + N(c_1, c_2|c_2^*), c_2^*)$ for every (c_1, c_2) in C^*. Since $c \sim t(c)$ for every c in C^*, it is clear that $u(c_1, c_2) = u(c_1 + N(c_1, c_2|c_2^*), c_2^*)$ for every (c_1, c_2) in C^*. This procedure was used in determining a utility function on the set C^* of consequences of Example 1.5.1c: the author took $c_2^* =$ MP (= "maintain prestige"); assumed that $(\$c_1; \text{LP}) \sim (\$c_1 - \$50(000); \text{MP})$ for every c_1; and then graphed a utility function over the interval $[\$-325(000), \$+250(000)]$ with "; MP" understood. Then he set $u(\$-325; \text{MP}) = u(\$-275; \text{LP})$, for example.

Example 5.2.2 extends Example 5.2.1 to the cases in which monetary returns may be received at different points in time.

Example 5.2.2

Suppose that the consequences c in C^* of a dpuu can be written as (c_1, c_2), with c_2 denoting nonmonetary attributes as before, but with c_1 denoting an entire cash flow, that is, a list of (monetary amount – date-of-occurrence pairs). Determining a utility function over such complicated consequences may be difficult if done directly. Hence, it may be much easier for $\mathscr{D}$ to proceed as follows. First, choose a nonmonetary reference attribute c_2^*; second, for every cash flow c_1 *cum* nonmonetary attribute c_2, determine that amount $N_0(c_1, c_2|c_2^*)$ of money to be received *right now* which makes (c_1, c_2) equally as desirable as $(N_0(c_1, c_2|c_2^*), c_2^*)$; and third, determine the utility of all ("present value," c_2^*) pairs which differ only in their present values N_0. The term "present value" is common in finance and usually signifies a discounting operation, but what is solely relevant here is that $(N_0(c_1, c_2|c_2^*), c_2^*)$ be *equally as* *desirable to* $\mathcal{D}$ as (c_1, c_2), for then and only then do we have $u(N_0(c_1, c_2|c_2^*), c_2^*) = u(c_1, c_2)$ for every (c_1, c_2) in C^*.

Seventh Preliminary

Suppose that ℓ' is a lottery with outcomes in some set C and that a consequence c in C satisfies the equidesirability $c \sim \ell'$. Convenient terminology: In this case we say that c is a *certainty equivalent* of ℓ', and we write

$$c = Q\{\ell'\}.$$

The concept of certainty equivalents is particularly useful in determining a utility function on the set of all possible monetary returns as of a given time, each return being accompanied by the same nonmonetary attribute.

Exercise 5.2.1

Let $\Omega = \{\omega^1, \omega^2, \omega^3\}$; $\Gamma = \{\gamma^1, \gamma^2, \gamma^3, \gamma^4, \gamma^5\}$; $P(\omega^1) = .2$, $P(\omega^2) = .3$, and $p(\omega^3) = .5$; and $P(\gamma^1) = .1$, $P(\gamma^2) = .4$, $P(\gamma^3) = .1$, $P(\gamma^4) = .1$, and $P(\gamma^5) = .3$. Let ℓ and ℓ' be lotteries with outcomes in $\{c^1, c^2, c^3\}$ defined on Ω and Γ respectively as follows: $\ell(\omega^i) = c^i$ for $i = 1, 2, 3$; and $\ell'(\gamma^1) = \ell'(\gamma^4) = c^1$, $\ell'(\gamma^2) = \ell'(\gamma^3) = c^3$, and $\ell'(\gamma^5) = c^2$.

(A) Show that $P(C\dagger|\ell) = P(C\dagger|\ell')$ for every subset $C\dagger$ of $\{c^1, c^2, c^3\}$.
(B) Calculate $E\{u(\tilde{c})|\ell\} = E\{u(\tilde{c})|\ell'\}$ if u is given by $u(c^1) = 1$, $u(c^2) = 3$, and $u(c^3) = -2$.
(C) Let $\Delta = \{\delta^1, \delta^2\}$ with $P(\delta^1) = .45$ and therefore $P(\delta^2) = .55$. Suppose $\ell'': \Delta \to \{c^1, c^2, c^3\}$ is defined by $\ell''(\delta^1) = c^2$ and $\ell''(\delta^2) = c^3$; and let u be as given in (B). Show that $E\{u(\tilde{c})|\ell''\} > E\{u(\tilde{c})|\ell'\}$, and hence that $\ell'' > \ell'$.

Exercise 5.2.2

Let u be a utility function on C. Show that if $c \sim \ell^*(p; c'', c')$ with $c'' \gtrsim c \gtrsim c'$, then the following equations obtain.

$$u(c) = pu(c'') + (1-p)u(c'); \tag{5.2.5}$$

$$p = \frac{u(c) - u(c')}{u(c'') - u(c')}; \tag{5.2.6}$$

$$u(c') = u(c) - \frac{p}{1-p}[u(c'') - u(c')] \text{ if } p < 1; \tag{5.2.7}$$

and

$$u(c'') = u(c') + \frac{1}{p}[u(c) - u(c')] \text{ if } p > 0. \tag{5.2.8}$$

Exercise 5.2.3

Let a utility function u^* be defined on $C^* = \{c^1, c^2, c^3\}$ by $u^*(c^1) = 4$, $u^*(c^2) = 7$, and $u^*(c^3) = 0$. Let $C = \{c^1, \ldots, c^6\}$, and let u^{**} be defined on

C by $u^{**}(c^1) = 14/5$, $u^{**}(c^2) = 17/5$, $u^{**}(c^3) = 2$, $u^{**}(c^4) = 11/5$, $u^{**}(c^5) = 19/5$, and $u^{**}(c^6) = 8/5$. Find another utility function u on C which coincides with u^* on C^*.

Exercise 5.2.4 (Continuation of Example 5.2.1)

Let assumptions and notation be as in Example 5.2.1. Let $N(c_1, c_2 | c_2^{**})$ be the monetary adjustment but with c_2^{**} as the non-monetary reference attribute. Assume the following "positive marginal utility of return" condition: For every c_2, *if* $c_1'' > c_1'$, then $(c_1'', c_2) > (c_1', c_2)$. Then show that

$$N(c_1, c_2 | c_2^{**}) = N(c_1, c_2 | c_2^*) + N(c_1 + N(c_1, c_2 | c_2^*), c_2^* | c_2^{**}).$$

[*Hint*: The left-hand side can be regarded as going *directly* from (c_1, c_2) to (c_1'', c_2^{**}); the right-hand side involves getting there via (c_1''', c_2^*).]

5.3 ASSESSING A UTILITY FUNCTION ON A (SMALL-) FINITE SET

This section presents two methods by which $\mathcal{D}$ can determine, or "assess," a utility function for himself on a set C of consequences which is finite. In fact, each method is cumbersome unless C contains only a few unequally desirable consequences.

Method F1 is a generalization of the approach we presented in Section 3.5, in that we do not give any prominence to "best" and "worst" consequences c^{B} and c^{W}, since if $C = (-\infty, +\infty)$ and represents all conceivable net monetary returns at a given point in time, then no such best and worst consequences exist. Admittedly, such cases are infinite idealizations, but we shall see that the utility-assessment machinery for them comes cheaply.

Method F1 consists of two steps, the second being repeated several times, with the exact number of repetitions depending upon the number of consequences in C. We shall suppose that "you" are the decision maker.

STEP F1-1: You select two *reference consequences*, say c^- and c^+, with $c^+ > c^-$. You assign to c^- and c^+ the utility numbers $u(c^-)$ and $u(c^+)$ arbitrarily, except for the requirement that $u(c^+) > u(c^-)$. *Any* two numbers satisfying this inequality may be used.

STEP F1-2 involves your assessing the utility $u(c)$ of a consequence $c \notin \{c^+, c^-\}$. Thus if C contains n elements, this step will be performed $n - 2$ times. Your assessment of $u(c)$ proceeds differently according to whether $c^+ \succsim c \succsim c^-$, $c > c^+$, or $c^- > c$.

CASE F1-2.1: $c^+ \succsim c \succsim c^-$. You compare c with canonical lotteries $\ell^*(p; c^+, c^-)$, so as to find that canonical probability $w_1(c)$ in [0, 1] such that c and $\ell^*(w_1(c); c^+, c^-)$ are equally desirable, that is,[e] such that $c \sim \ell^*(w_1(c); c^+, c^-)$. From Equation (5.2.5) it follows readily that

$$u(c) = [u(c^+) - u(c^-)]w_1(c) + u(c^-). \tag{5.3.1}$$

Naturally, $c \sim c^+$ implies $w_1(c) = 1$ and $u(c) = u(c^+)$, whereas $c \sim c^-$ implies $w_1(c) = 0$ and $u(c) = u(c^-)$.

CASE F1-2.2: $c > c^+$. In this case you compare c^+ with canonical lotteries of the form $\ell^*(p; c, c^-)$—note where c appears!—so as to find that canonical probability $w_2(c)$ for which $c^+ \sim \ell^*(w_2(c); c, c^-)$. From Equation (5.2.8) it follows readily that

$$u(c) = [u(c^+) - u(c^-)]\left(\frac{1}{w_2(c)}\right) + u(c^-). \tag{5.3.2}$$

[To apply Equation (5.2.8) it is necessary to note that c, c^+, and c^- here play the roles of c'', c, and c' in Exercise 5.2.2. Since $c > c^+ > c^-$, it necessarily follows from Observation 5.2.3 that $0 < p = w_2(c) < 1$ and hence that $u(c) > u(c^+)$.]

CASE F1-2.3: $c^- > c$. In this case you compare c^- with canonical lotteries of the form $\ell^*(p; c^+, c)$—note where c appears!—so as to find that canonical probability $w_3(c)$ for which $c^- \sim \ell^*(w_3(c); c^+, c)$. From Equation (5.2.7) it follows readily that

$$u(c) = -[u(c^+) - u(c^-)]\left(\frac{w_3(c)}{1 - w_3(c)}\right) + u(c^-). \tag{5.3.3}$$

[To apply Equation (5.2.7) it is necessary to note that c, c^+, and c^- here play the roles of c', c'', and c in Exercise 5.2.2. Since $c^+ > c^- > c$, it necessarily follows from Observation 5.2.3 that $0 < p = w_3(c) < 1$ and hence that $u(c) < u(c^-)$.]

Example 5.3.1

Suppose that $C = \{c^1, \ldots, c^8\}$, that you determine that $c^2 > c^7$, and that you choose $c^+ = c^2$ and $c^- = c^7$. Suppose also that you set $u(c^+) = 10$ and $u(c^-) = -10$. This completes step F1-1. Now you may take each c^i except c^2 and c^7 in turn and execute step F1-2. *For* c^1: Suppose that you determine that $c^+ > c^1 > c^-$; hence Case F1-2.1 obtains. Then suppose that you determine that $c^1 \sim \ell^*(.8; c^+, c^-)$. Then (5.3.1) implies that $u(c^1) = [20](0.8) + (-10) = 6$. *For* c^3: Suppose that $c^- > c^3$ and hence Case F1-2.3 obtains, and you find that $c \sim \ell^*(.4; c^+, c^3)$. Then (5.3.3) implies that $u(c^3) = -[20](.4/.6) + (-10) = -70/3$. *For* c^4: Suppose that $c^4 > c^+$ and hence Case F1-2.2 obtains, and you find that $c^+ \sim \ell^*(.2; c^4, c^-)$. Then $u(c^4) = [20](1/.2) + (-10) = 90$, by (5.3.2). Each of c^5, c^6, and c^8 is handled similarly.

---

[e] Note that if $c^+ = c^B$ and $c^- = c^W$, then $w_1(c) = w(c)$ and Case F1-2.1 always obtains. Moreover, your assessment task here coincides with that in Section 3.5.

Are you finished once you have determined $u(c)$ for every c in C by the preceding method? Most practically oriented decision analysts would think not. You should check the implications of your utility function for conformity with your directly expressible preferences. For example, you should check whether you really find c' and c'' equally desirable if you have assessed $u(c') = u(c'')$, since otherwise, $u(c')$ and $u(c'')$ should *not* be equal. Furthermore, you should select a few sets of three consequences such that $c'' > c > c'$ and check to see whether you really find c and $\ell^*(p; c'', c')$ equally desirable for p as given by Equation (5.2.6).

Example 5.3.2

The utilities determined in Example 5.3.1 imply that $c^1 \sim \ell^*(.16; c^4, c^-)$, $c^+ \sim \ell^*(1/6; c^4, c^1)$, and $c^+ \sim \ell^*(13/34; c^4, c^3)$ (among others). You may derive these facts as Exercise 5.3.1. Now, if any of the equidesirabilities above were in *serious* discord with your intuitive preferences, you would have made an "introspectional mistake" in assessing your utility function, because your utility function must be consistent with—and must faithfully represent—your true preferences.

One of the most common inconsistencies that crops up with beginners in our formal brand of decision analysis is to find that they have assessed $u(c'')$ greater than $u(c')$ but regard c' as more desirable than c''. In such cases, either their basic preference for c' over c'' is not very strong (in some qualitative sense), or else $u(c'')$ does not exceed $u(c')$ by very much, or both. But such a finding is a serious inconsistency. Our second assessment method guards against it from the outset.

Method F2 essentially involves *ranking* the elements of C from least to most desirable, then quantifying your preferences for each consequence in terms of its immediate neighbors in rank, and finally, determining your utility function by some computations.

STEP F2-1 requires that you rank all consequences from least to most desirable, with the relative placement of equally desirable consequences being arbitrary. Suppose that the consequences have been labeled so that $C = \{c^1, c^2, \ldots, c^n\}$ with $c^{i+1} \gtrsim c^i$ for $i = 1, \ldots, n-1$. Then, within each set of equally desirable consequences, eliminate all but one. Those eliminated will have exactly the same utility as the one remaining, because $c^i \sim c^{i+1}$ implies that $u(c^i) = u(c^{i+1})$. Call the remaining consequence set $C_* = \{c_*^1, \ldots, c_*^m\}$, where $m \leq n$. ($m = n$ only if $c^{i+1} > c^i$ for all i.)

*■ The mathematically sophisticated reader will recognize that C_* is a system of representatives of the equivalence relation "$\sim$". ■

Example 5.3.3

Suppose $C = \{c^1, \ldots, c^8\}$ with $c^W = c^1$ and $c^B = c^8$ (naturally) and $c^8 > c^7 \sim c^6 \sim c^5 > c^4 \sim c^3 > c^2 \sim c^1$. Then you may eliminate, say, c^1, c^3,

c^7, and c^5, leaving $C_* = \{c_*^1, c_*^2, c_*^3, c_*^4\}$ with $c_*^1 = c^2$, $c_*^2 = c^4$, $c_*^3 = c^6$, and $c_*^4 = c^8$.

The rest of Method F2 guides you through assessing your utility function on C_*, after which the eliminated consequences may be reintroduced. Naturally, $c \sim c_*^i$ implies $u(c) = u(c_*^i)$.

Now note that $c_*^{i+1} > c_*^i$ for $i = 1, \ldots, m-1$. This fact is important in what follows.

STEP F2-2 requires you to compare each c_*^i with canonical lotteries of the form $\ell^*(p; c_*^{i+1}, c_*^{i-1})$ for $i = 2, \ldots, m-1$, so as to find that number p_i such that $c_*^i \sim \ell^*(p_i; c_*^{i+1}, c_*^{i-1})$. Since $c_*^{i+1} > c_*^i > c_*^{i-1}$, it is clear that $0 < p_i < 1$. Note that each comparison is of a consequence with a canonical lottery offering only the immediate neighbors of that consequence as possible outcomes.

STEP F2-3 requires that you specify $u(c_*^1)$ and $u(c_*^m)$ in any way you like, provided only that $u(c_*^m) > u(c_*^1)$.

Now, the p_i's you assessed in step F2-2 and the numbers $u(c_*^m)$ and $u(c_*^1)$ suffice to determine your utility function on C_*, but this determination involves some slightly complicated calculations. The essentials are stated as theorem 5.3.1.

Theorem 5.3.1. Given $u(c_*^1)$, $u(c_*^m) > u(c_*^1)$, and the numbers $p_2, \ldots, p_{m-1}$ as defined in step F2-2. Define $r_1 = 0$, $p_m = 1$, and (successively)

$$r_i = \frac{p_i}{1 - (1 - p_i) r_{i-1}}$$

for $i = 2, 3, \ldots, m$. Then for every i, we have

$$u(c_*^i) = [u(c_*^m) - u(c_*^1)] \cdot \prod_{j=i}^{m} r_j + u(c_*^1).$$

■ *Proof.* In view of Observation 5.2.4 it suffices to prove the assertion

(a) $$w(c_*^i) = \prod_{j=i}^{m} r_j$$

for every i; $w(\cdot)$ is simply the utility function on C_* such that $w(c_*^1) = 0$ and $w(c_*^m) = 1$. Now, (a) is obvious for $i = 1$ because $r_1 = 0$ and hence $\Pi_{j=1}^m r_j = 0$. Moreover, (a) is true for $i = m$ because $p_m = 1$ (by definition) and therefore $r_m = 1$, implying $\Pi_{j=m}^m r_j = r_m = 1 = w(c_*^m)$. The cases $i \in \{2, \ldots, m-1\}$ are a little less obvious. Assume for the moment that each $w(c_*^i)$ satisfies

(b) $$w(c_*^i) = r_i w(c_*^{i+1}), \qquad i = 2, \ldots, m-1.$$

Then (a) follows from (b) by an easy argument, relegated to you as Exercise 5.3.2. An inductive argument establishes (b), which is obviously true for $i = 1$ because $r_1 = 0$. Assume (b) is true for $i = k - 1$. By definition of p_k and of $w(\cdot)$, $c_*^k \sim \ell^*(p_k; c_*^{k+1}, c_*^{k-1})$ and hence

$$\begin{aligned} w(c_*^k) &= p_k w(c_*^{k+1}) - (1-p_k) w(c_*^{k-1}) \\ &= p_k w(c_*^{k+1}) + (1-p_k) r_{k-1} w(c_*^k), \qquad \text{[by inductive assumption]} \end{aligned}$$

implying that

$$[1-(1-p_k)]w(c_*^k) = p_k w(c_*^{k+1}),$$

and hence

$$\begin{aligned} w(c_*^k) &= \frac{p_k}{1-(1-p_k)r_{k-1}} w(c_*^{k+1}) \\ &= r_k w(c_*^k). \qquad \text{[by definition of } r_k\text{]} \end{aligned}$$

This proves (b) and hence (a) for $i = 2, \ldots, m-1$. ■

Example 5.3.4 (Continuation of Example 5.3.3)

Suppose that you have assessed $p_2 = .2$ and $p_3 = .7$. Then $r_1 = 0$, $r_2 = .2/[1-(.8)(0)] = .2$, $r_3 = .7/[1-(.3)(.2)] = 35/47$, and $r_4 = 1$. Suppose that you chose $u(c_*^4) = 100$ and $u(c_*^1) = 0$. Then $u(c_*^3) = 100r_3r_4 = 100r_3 = 3500/47$ and $u(c_*^2) = 100r_2r_3r_4 = (.2)(3500/47) = 700/47$.

Example 5.3.5 (Continuation of Example 5.3.4)

Suppose that you want to define your utility function on C. Since $c^1 \sim c^2 = c_*^1$, we have $u(c^1) = u(c^2) = 0$ $[= u(c_*^1)]$. Since $c^3 \sim c^4 = c_*^2$, we have $u(c^3) = u(c^4) = 700/47$ $[= u(c_*^2)]$. Since $c^5 \sim c^6 = c_*^3 \sim c^7$, we have $u(c^5) = u(c^6) = u(c^7) = 3500/47$ $[= u(c_*^3)]$. And $u(c^8) = 100$ because $c^8 = c_*^4$.

Exactly the same cautionary remarks about testing the implications of your utility function for consistency with your basic preferences pertain to the result of using Method F2, except that "$u(c'') \geq u(c')$ if and only if $c'' \succsim c'$" is ensured by Method F2's initial ranking of the consequences.

The preceding methods are useful but certainly do not exhaust all approaches to assessing a utility function on a finite set. In Exercise 5.3.3 you may derive the utility function corresponding to a number of equidesirabilities that would not have arisen from either Method F1 or Method F2 as initial assessments.

Exercise 5.3.1

Prove the three equidesirabilities asserted in Example 5.3.2.

Exercise 5.3.2

Derive (a) from (b) in the proof of Theorem 5.3.1.

Exercise 5.3.3

Suppose that $C=\{c^1,\ldots,c^6\}$ and that you have determined that $c^B=c^6$, $c^W=c^1$, $c^3\sim c^4$, and also that $c^2\sim \ell^*(.4;c^6,c^1)$, $c^3\sim \ell^*(.25;c^5,c^2)$, and $c^5\sim \ell^*(.6;c^6,c^4)$. Let $u(c^6)=1$ and $u(c^1)=0$. Find $u(c^i)$ for $i\in\{2,\ldots,5\}$.

5.4 ASSESSING UTILITY ON AN INTERVAL OF MONETARY RETURNS

This section concerns the assessment of your utility function on an interval of real numbers, each real number c representing a net monetary return from your dpuu at a given, fixed point in time. We shall assume that all nonmonetary attributes of consequences and varying dates at which cash flows occur have been adjusted for by the use of a substitute function as described in the sixth preliminary, or, more particularly, in Examples 5.2.1 and 5.2.2. In what follows, we suppress notation for the fixed nonmonetary reference attribute and the subscript 1 for monetary returns.

Utility functions for monetary return (at a given date, etc.) appear prominently in decision-analytic literature for (at least) three reasons. (1) Our approach to decision analysis was first made practical in business decisions, a characteristic of which is the almost exclusive importance of monetary return in many cases. (2) The economic role of money as a standard of value and very common benchmark in most individuals' value-judgment processes makes monetary returns a *convenient* range for a substitute function in many cases. (3) Utility functions for monetary returns should possess (as we shall argue) several qualitative properties that make them a convenient introduction to the subject of assessing a utility function on an infinite (but well-structured) set of consequences.

Before proceeding to describe two methods of assessing a utility function for monetary returns, we shall discuss *three qualitative properties* which (we think) you should want your utility function to possess. These properties are:

(A) $u(\cdot)$ should be a *strictly increasing* function of returns c; that is, if $c''>c'$, then $u(c'')$ should exceed $u(c')$.

(B) $u(\cdot)$ should be a *continuous* function of returns c; that is,[f] one should be able to trace the graph $(\cdot, u(\cdot))$ of u without lifting his pencil from the paper.

(C) $u(\cdot)$ should be a *concave* function of returns c; that is, $u(.5c'+.5c'')\geq .5u(c')+.5u(c'')$ whenever c' and c'' both belong to the interval of returns in question.

Figure 5.1 depicts a strictly increasing, continuous, and concave utility

[f] A formal definition of continuity is: For every $e>0$ and every c in the domain of $u(\cdot)$, there is a $d(e,c)>0$ such that $u(c)-d(e,c)<u(c')<u(c)+d(e,c)$ whenever $c-e<c'<c+e$.

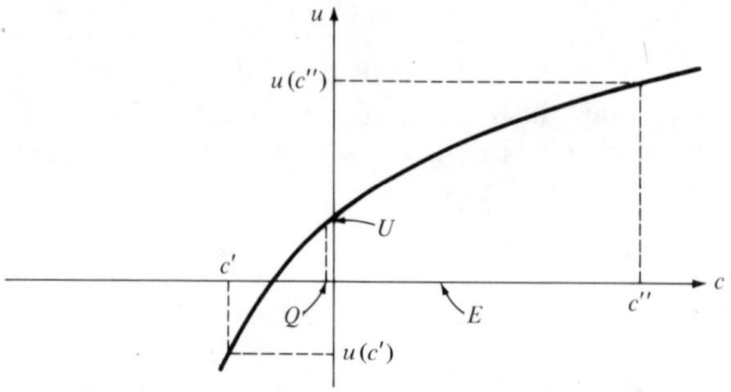

Figure 5.1 *u* Strictly increasing, continuous, and concave.

function. The function in Figure 5.2 is strictly increasing and continuous, but not concave. Even worse for utility purposes are the functions in Figures 5.3 and 5.4, which are respectively not strictly increasing and discontinuous. In each of these figures we have indicated how to find a certainty equivalent[g]

$$Q = Q\{\ell^*(.5; c'', c')\}$$

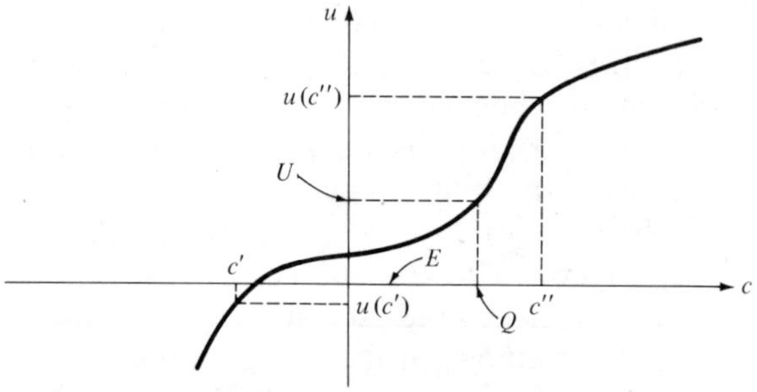

Figure 5.2 *u* Strictly increasing and continuous but not concave.

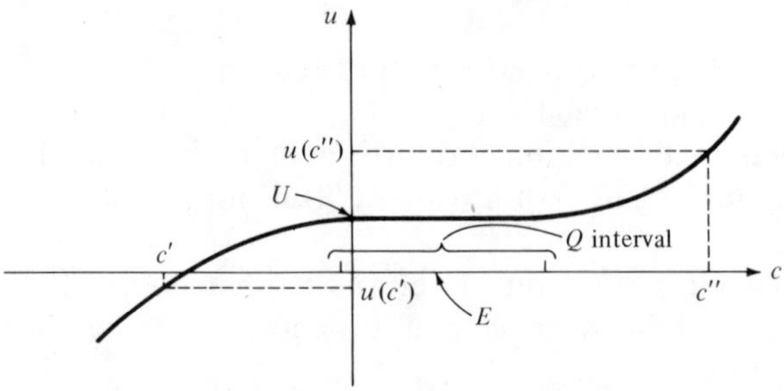

Figure 5.3 *u* Continuous but not strictly increasing.

[g] Recall from the seventh preliminary in Section 5.2 that $Q\{\ell\}$ is a *certainty equivalent* of ℓ if and only if "$Q\{\ell\}$ for sure" is equally as desirable as ℓ.

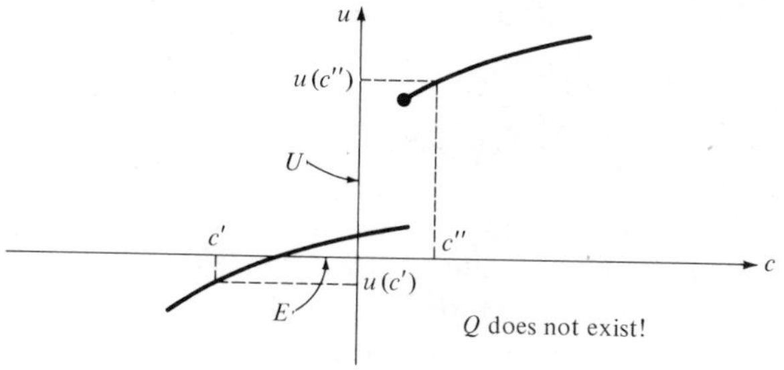

Figure 5.4 *u* Strictly increasing but not continuous.

of the simple canonical lottery which yields c'' and c' with equal probabilities. These figures also depict graphically the expectation

$$E = E\{\tilde{c}|\ell^*(.5; c'', c')\} = (c' + c'')/2$$

of the *return* and the expectation

$$U = E\{u(\tilde{c})|\ell^*(.5; c'', c')\} = [u(c') + u(c'')]/2 = u(Q)$$

of the *utility* (= the desirability index) of $\ell^*(.5; c'', c')$.

Figures 5.3 and 5.4 show that unless $u(\cdot)$ is strictly increasing, some lottery will have *many* certainty equivalents; and that unless $u(\cdot)$ is continuous, some lottery will have *no* certainty equivalent.

■ In the language of economics, strict increasingness of u is tantamount to positive marginal utility of return. If u is constant on some interval, say the interval indicated in Figure 5.3, then you would willingly *burn* a portion of any return falling in the middle of that interval! ■

Because you probably always prefer more net return to less and because you probably could state a certainty equivalent $Q = Q\{\ell^*(.5; c'', c')\}$ of any simple lottery of the form $\ell^*(.5; c'', c')$, it may be assumed that *you would want* your utility function for return to be strictly increasing and continuous; that is, you would want your utility function to possess properties (A) and (B).

Qualitative property (C) is more difficult to justify. We shall try[h] to do so by asking you to answer a simple decision question, and then by discussing the "nonconcave" answer.

Question: Are there any returns c' and $c'' > c'$ such that *you* would *prefer*:

(1) experiencing the lottery $\ell^*(.5; c'', c')$ in which, say, a balanced coin is flipped to determine whether you receive c'' or c'; *to*

[h] Whether we succeed is up to you; but we remark here that little in the remainder of this book depends upon u possessing property (C).

(2) receiving with certainty the average $.5(c' + c'') = E\{\tilde{c}|\ell^*(.5; c'', c')\}$ of c' and c''?

If your answer is "no," then your utility function is concave, since option (2) is simply "$.5(c' + c'')$ for sure" and has utility $u(.5(c' + c''))$, whereas option (1) has utility $.5u(c') + .5u(c'')$; and a "no" answer implies that $.5(c' + c'') \succsim \ell^*(.5; c'', c')$, which, by Observation 5.2.2, implies that $u(.5(c' + c'')) \geq .5u(c') + .5u(c'')$ for all c' and c''—the definition of concavity for u.

People who answer "yes" for certain return levels c' and c'' generally have two types of explanations.

The *first explanation* is that a person derives pleasure from gambling for small sums c' and c'' and hence prefers $\ell^*(.5; c'', c')$ when c' and c'' are small to the average $.5(c' + c'')$ for certain. But "joy of gambling" can be regarded as a *nonmonetary attribute* of consequences which differs at small c's from its level at large c's. A person who derives "joy of gambling" from $\ell^*(.5; c'', c')$ may (and indeed should) adjust c' and c'' to their "joyless" equivalents, say $c' + N'$ and $c'' + N''$, by the technique introduced in Example 5.2.1. Doing so may (and often will) result in the joyless equivalent lottery $\ell^*(.5; c'' + N'', c' + N')$ satisfying the concavity condition

$$\ell^*(.5; c'' + N'', c' + N') \precsim \text{“}.5(c' + N' + c'' + N'') \text{ for sure”}$$

even though

$$\ell^*(.5; c'' + N'', c' + N') \succ \text{“}.5(c' + c'') \text{ for sure.”}$$

Put briefly, differing joys of gambling violate our assumption at the beginning of this section that all returns are accompanied by the same nonmonetary attribute.

The *second explanation* for a "yes" answer is that the person is willing to "go for broke" in order to have a chance at a much more sybaritic life-style which receiving c'' would enable him to afford—a life-style for which $.5(c' + c'')$ would be insufficient. Raiffa [**113**, pages 94–97] notes that this explanation is cogent, but that such a person should still not *use* the resulting, nonconcave utility function for the analysis of his dpuu, because it fails to reflect the fact that, once he has received a return in the dpuu which falls in the nonconcave portion of the graph of u, he *should go out and gamble* with that return so as to increase his utility from $u(c)$ to some *higher* $U(\ell)$. This being the case, he should regard his utility of c for decision-making purposes not as $u(c)$ but rather as his utility $U(\ell)$ of getting c *and then gambling* it. In the framework of Chapter 2, this argument states that the person has not adopted a sufficiently distant horizon for his dpuu, by not considering the availability of "almost-fair" gambles in Las Vegas or Monte Carlo and of "subjectively fair or better than fair" gambles in the securities and commodities markets.

We shall not pursue the matter further here, but refer the interested reader to [**113**] for further details of an argument which implies that a

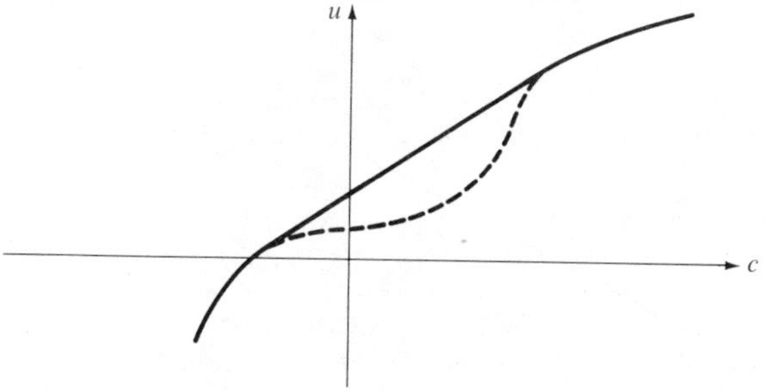

Figure 5.5 Concavification of u as given in Figure 5.2.

person whose assessed u appears as in Figure 5.2 should use the "concavification" of u as depicted in Figure 5.5 for the analysis of his dpuu.

One further qualitative property of some utility functions will be introduced at the end of this section. We shall now discuss two methods of assessing your utility function for returns.

All methods of assessing a utility function on an interval consisting of infinitely many net monetary returns involve making only a few actual equidesirability determinations of the now familiar form

$$c \sim \ell^*(p; c'', c') \tag{5.4.1}$$

for $c'' > c > c'$ (and therefore $c'' > c > c'$). Method \$1, reminiscent of Exercise 3.5.6, involves graphing the $(c, u(c))$ pairs corresponding to the equidesirabilities Equation (5.4.1) actually determined and then drawing a continuous and strictly increasing (and, it is hoped, also concave) curve through these points. Method \$2 involves using equidesirabilities (5.4.1) to determine the parameter (or parameters) of a function of predetermined mathematical form.

In both methods, the continuously variable nature of monetary returns—given the natural continuous idealization of discrete monetary units—enables us to determine an equidesirability of the form (5.4.1) by fixing *any three* of c, p, c'', and c' and then varying the remaining one. In the finite-C case discussed in Section 5.3, only p was continuously variable and hence every comparison of c and $\ell^*(p; c'', c')$ that you made ultimately required you to vary p until equidesirability, (5.4.1), obtained. Since most people find it easiest to think about equally likely events, *all of our comparisons here will be such that* $p = .5$; that is, we shall always vary c, c'', or c'.

Method \$1 is most easily discussed in terms of the notation in which $c^{.5(x+y)}$ denotes the certainty equivalent of $\ell^*(.5; c^y, c^x)$, and in which $c^y > c^x$ whenever $y > x$. Hence

$$c^{.5(x+y)} \sim \ell^*(.5; c^y, c^x),$$

which implies by Observation 5.2.2 and a modicum of algebra that

$$u(c^{.5(x+y)}) = u(c^x) + (\tfrac{1}{2})[u(c^y) - u(c^x)]. \tag{5.4.2}$$

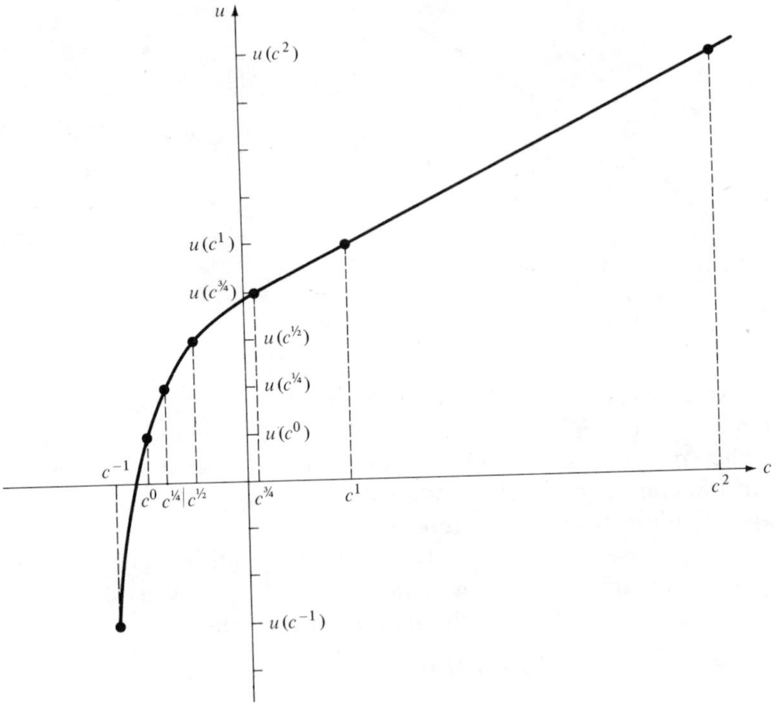

Figure 5.6 A utility function determined by Method \$1.

Now, you will wish to refer to Figure 5.6 frequently while following through the steps of Method \$1.

As we have noted, Method \$1 basically amounts to drawing a curve through a finite number of $(c^x, u(c^x))$ points in the plane.

STEP \$1-1 requires you to choose two initial amounts c^0 and c^1 of returns, with $c^1 > c^0$, and to choose their utilities $u(c^0)$ and $u(c^1) > u(c^0)$. The next two steps involve your determining $u(\cdot)$ on the interval $[c^0, c^1]$ of returns.

■ As with c^- and c^+ in Method F1, c^0 and c^1 need not be the worst and best returns c^W and c^B in your dpuu. We shall show later how to extend u to the "northeast" of $(c^1, u(c^1))$ and to the "southwest" of $(c^0, u(c^0))$. ■

After choosing c^0, $u(c^0)$, $c^1 > c^0$, and $u(c^1) > u(c^0)$, you should graph the two points $(c^0, u(c^0))$ and $(c^1, u(c^1))$ in the c-$u(c)$ plane.

STEP \$1-2[1/2] requires you to determine (judgmentally) *your* certainty equivalent $c^{1/2}$ of $\ell^*(.5; c^1, c^0)$. By Equation (5.4.2) with $x = 0$ and $y = 1$, it is clear that

$$u(c^{1/2}) = u(c^0) + (\tfrac{1}{2})[u(c^1) - u(c^0)], \qquad (5.4.3[1/2])$$

which enables you to calculate $u(c^{1/2})$ readily. Now you may add the point $(c^{1/2}, u(c^{1/2}))$ to your graph.

■ If $u(\cdot)$ is to be concave, it is necessary that $c^{1/2}$ not exceed $(c^0 + c^1)/2$. ■

STEP $1-2[i/4] requires that you determine (judgmentally) *your* certainty equivalents $c^{1/4}$ of $\ell^*(.5; c^{1/2}, c^0)$ and $c^{3/4}$ of $\ell^*(.5; c^1, c^{1/2})$. Since $c^{1/4} \sim \ell^*(.5; c^{1/2}, c^0)$ and $c^{3/4} \sim \ell^*(.5; c^1, c^{1/2})$, Equation (5.4.2) and a little additional algebra imply that

$$u(c^{i/4}) = u(c^0) + (i/4)[u(c^1) - u(c^0)] \qquad (5.4.3[i/4])$$

for $i = 0, 1, 2, 3, 4$. Since you already have $(c^{i/4}, u(c^{i/4}))$ for $i = 0, 2$, and 4, it only remains to add the points $(c^{1/4}, u(c^{1/4}))$ and $(c^{3/4}, u(c^{3/4}))$ to your graph.

■ If $u(\cdot)$ is to be concave, it is necessary that $c^{1/4} \leq (c^{1/2} + c^0)/2$, that $c^{3/4} \leq (c^1 + c^{1/2})/2$, and also that $c^{1/2} \leq (c^{3/4} + c^{1/4})/2$. ■

At this point the five graphed points may be sufficient for you to draw a strictly increasing, continuous, and perhaps also concave utility curve over $[c^0, c^1]$ with confidence. In Figure 5.6, for example, it is clear that *any* strictly increasing, continuous, and concave function which passes through all five points $(c^{i/4}, u(c^{i/4}))$ must pass very close to the function we have drawn. But if more points are needed, the foregoing process may be continued. The general step is as follows.

STEP $1-2[$i/2^n$] for $n = 3, 4, \ldots$ begins by assuming that you have determined all points $(c^{i/2^n}, u(c^{i/2^n}))$ for the *even-numbered* i's; namely, $i = 0, 2, 4, \ldots, 2^n$. This is the result of having gone through step $1-2[$i/2^{n-1}$]. Now you must determine (judgmentally) your certainty equivalent $c^{i/2^n}$ of $\ell^*(.5; c^{(i+1)/2^n}, c^{(i-1)/2^n})$ for each odd-numbered i. As before, it follows (by induction) that

$$u(c^{i/2^n}) = u(c^0) + (i/2^n)[u(c^1) - u(c^0)] \qquad (5.4.3[i/2^n])$$

for every $i \in \{0, 1, 2, \ldots, 2^n\}$. Now the new points $(c^{i/2^n}, u(c^{i/2^n}))$ may be added to your graph.

■ In order for $u(\cdot)$ to be concave, it is necessary that $c^{py+(1-p)x} \leq pc^y + (1-p)c^x$ whenever this inequality can be tested on the basis of your assessments of certainty equivalents. To see this, note first that $c^{py+(1-p)x} \sim \ell^*(p; c^y, c^x)$ if and only if $u(c^{py+(1-p)x}) = pu(c^y) + (1-p)u(c^x)$. But strict increasingness of u implies that $c^{py+(1-p)x} \leq pc^y + (1-p)c^x$ if and only if $u(c^{py+(1-p)x}) \leq u(pc^y + (1-p)c^x)$. Hence $c^{py+(1-p)x} \leq pc^y + (1-p)c^x$ if and only if $u(pc^y + (1-p)c^x) \geq pu(c^y) + (1-p)u(c^x)$, the definition of concavity of u if required for all c^x, c^y, and $p \in [0, 1]$ rather than just those corresponding to your assessed certainty equivalents. ■

STEP $1-3 requires you to graph your utility curve once you are satisfied that you have determined a sufficient number of points $(c^x, u(c^x))$ according to the various parts of step $1-2.

Since we did not assume that c^1 was the largest possible level of returns in which you might be interested, you may wish to extend your utility curve to the "northeast" of $(c^1, u(c^1))$.

STEP \$1-4[2] consists in your determining a level c^2 of returns such that $c^1 \sim \ell^*(0.5; c^2, c^0)$. Note that here c^1 and c^0 are fixed, whereas the more desirable outcome of the canonical lottery is varied until equidesirability obtains. This equidesirability implies, by Equation (5.4.2) and a little algebra, that

$$u(c^2) = u(c^0) + (2)[u(c^1) - u(c^0)]. \qquad (5.4.4[2])$$

STEP \$1-4[n] must be undertaken if you have determined $(c^{n-1}, u(c^{n-1}))$ and need $u(c)$ for some c exceeding c^{n-1}. In this step, you determine a return c^n such that $c^{n-1} \sim \ell^*(.5; c^n, c^{n-2})$. This equidesirability implies that

$$u(c^n) = u(c^0) + (n)[u(c^1) - u(c^0)]. \qquad (5.4.4[n])$$

Within each interval $[c^{n-1}, c^n]$, you may use step \$1-2, with c^{n-1} and c^n in place of c^0 and c^1 respectively, to obtain intermediate points on your graph.

Finally, we did not assume that c^0 was the smallest possible level of returns in which you might be interested, and hence you may wish to extend your utility curve to the "southwest" of $(c^0, u(c^0))$.

STEP \$1-5[−1] requires that you determine a level c^{-1} of returns such that $c^0 \sim \ell^*(.5; c^1, c^{-1})$—the less desirable outcome of the canonical lottery being varied until equidesirability obtains—and this equidesirability readily implies

$$u(c^{-1}) = u(c^0) + (-1)[u(c^1) - u(c^0)]. \qquad (5.4.5[-1])$$

STEP \$1-5[$-n$] must be undertaken if you have determined $(c^{-n+1}, u(c^{-n+1}))$ and need $u(c)$ for some c less than c^{-n+1}. In this step you determine a return c^{-n} such that $c^{-n+1} \sim \ell^*(.5; c^{-n+2}, c^{-n})$. This equidesirability implies that

$$u(c^{-n}) = u(c^0) + (-n)[u(c^1) - u(c^0)]. \qquad (5.4.5[-n])$$

Within each interval $[c^{-n}, c^{-n+1}]$, you may use step \$1-2, with c^{-n} and c^{-n+1} in place of c^0 and c^1 respectively, to obtain intermediate points on your graph.

Figure 5.6 involved steps \$1-1, \$1-2[1/2], \$1-2[i/4], \$1-4[2], \$1-5[−1], and graphing. Points $(c^x, u(c^x))$ arising from direct assessments appear as the heavy dots through which the curve was fitted.

Method \$2 requires using certainty-equivalent assessments to determine the parameter or parameters of a given mathematical function—defined by a formula involving c together with parameters—so that this function, with parameters specified numerically, represents adequately your preferences for lotteries with returns as outcomes.

This method is rather easy to describe on a somewhat cryptic level. We suggest your skimming over its two steps and then proceeding to a perusal of the extremely important special cases, Examples 5.4.1–5.4.5.

STEP $2-1 requires that you choose a function form $u(c|z_1, \ldots, z_n)$ with as-yet unspecified *essential* parameters[i] $z_1, \ldots, z_n$. Qualitative properties of the function form, discussed in the special cases which follow, guide you in this choice.

STEP $2-2 consists in your determining certainty equivalents $Q\{\ell^i\}$ of lotteries ℓ^i in order to obtain equations

$$u(Q\{\ell^i\}|z_1, \ldots, z_n) = E\{u(\tilde{c}|z_1, \ldots, z_n)|\ell^i\} \tag{5.4.6}$$

in the essential parameters $z_1, \ldots, z_n$. Each assessed certainty equivalent $Q\{\ell^i\}$ results in one equation. Hence it generally suffices to assess the certainty equivalents of n different lotteries, thus obtaining a system of n independent equations in the n essential parameters $z_1, \ldots, z_n$. If a solution $(z_1^*, \ldots, z_n^*)$ to this system of equations exists and is unique, this solution constitutes the desired specification of the parameters, and your utility function is $u(\cdot|z_1^*, \ldots, z_n^*)$.

The following examples should be examined carefully, because they introduce very important classes of utility functions which will serve as concrete examples in the latter part of this book, and which are widely used in practice.

Example 5.4.1

The *linear utility function* $u(\cdot)$ is defined for all returns $c \in (-\infty, +\infty)$ by

$$u(c) = c. \tag{5.4.7}$$

Its graph is the 45°, diagonal line through the origin. It is strictly increasing and continuous, and also concave: Since

$$u(.5(c' + c'')) = .5u(c') + .5u(c'')$$

when Equation (5.4.7) obtains, then also

$$u(.5(c' + c'')) \geq .5u(c') + .5u(c'').$$

It contains *no* essential parameters to be determined by step $2-2 and, partially because of this, it is the most popular assumption in practice. To determine whether it approximates your preferences, however, you must check that *your* directly assessed certainty equivalents $Q\{\ell\}$ of lotteries ℓ, with outcomes belonging to the interval $[c^W, c^B]$ of possible returns in your dpuu, are approximately equal to the corresponding expectations $E\{\tilde{c}|\ell\}$ of return, because if $u(\cdot)$ satisfies (5.4.7), then the defining equality

$$u(Q\{\ell\}) = E\{u(\tilde{c})|\ell\}$$

[i] By virtue of Observation 5.2.4, an *essential* parameter is one which is neither an additive constant nor a positive multiplicative constant. That is, if $u^*(c|z_1, \ldots, z_n, a^*, b^*) = a^*u(c|z_1, \ldots, z_n) + b^*$ for all c, then a^* and b^* are *unessential* parameters of u^*; they specify nothing about your preferences.

for certainty equivalents $Q\{\ell\}$ implies that

$$Q\{\ell\} = E\{\tilde{c}|\ell\}.$$

If you wish your utility function to be concave, it suffices to confine attention to lotteries ℓ of the form $\ell^*(p; c^B, c^W)$, and thus see (for a representative set of p's) whether your directly assessed certainty equivalents $Q\{\ell^*(p; c^B, c^W)\}$ are close to their corresponding expectations $pc^B + (1-p)c^W$. How close "close" should be is one of the few subjective matters that we shall not quantify.

Example 5.4.2a: Concave Exponential Utility (I)

The concave exponential utility function $u(\cdot)$ is defined for all return levels $c \in (-\infty, +\infty)$ by

$$u(c) = -e^{-zc}, \tag{5.4.8}$$

where z is an essential parameter which must be *positive*, and e is the base of the natural logarithms (approximately 2.71828). This is a strictly increasing, continuous, and concave utility function. It also possesses the following property of *constant risk aversion*: If $\ell + c^*$ denotes the lottery ℓ' which results in $c + c^*$ if and only if ℓ results in c, then

$$Q\{\ell + c^*\} = Q\{\ell\} + c^*.$$

All this says is that if all your potential returns in ℓ are increased by c^*, then your certainty equivalent of ℓ is also increased by c^*. Many people find the qualitative property of constant risk aversion very appealing; and a celebrated result of Pfanzagl [**109**] and of Pratt [**111**] implies that Equation (5.4.7) and Equation (5.4.8) are the *only* strictly increasing, continuous, and concave functions[j] which possess this property. Hence if you accept constant risk aversion as a desideratum of your utility function, and if your assessed $Q\{\ell\}$ is significantly less than $E\{\tilde{c}|\ell\}$ for some lotteries ℓ with outcomes all in the relevant range $[c^W, c^B]$, then your utility function *must* be of the form (5.4.8) for some positive real number z.

Example 5.4.2b: Concave Exponential Utility (II)

How can you determine the value z^* of z for which your preferences are expressible by $-e^{-z^*c}$? By step $2-2, one directly assessed certainty equivalent suffices. If you take ℓ as $\ell^*(.5; c'', c')$, then (5.4.6) and (5.4.8) imply that

$$e^{-zQ\{\ell^*(.5;\, c'',\, c')\}} = .5e^{-zc'} + .5e^{-zc''}, \tag{5.4.9}$$

a nonlinear equation in z which has a unique solution $z^* > 0$ provided

[j] Except for functions that differ from Equations (5.4.7) and (5.4.8) by inessential parameters, such as $u^*(c) = a^*c + b^*$ and $u^*(c) = b^* - a^*e^{-zc}$, with a^* positive.

that $Q\{\ell^*(.5; c'', c')\} < .5(c' + c'') = E\{\tilde{c}|\ell^*(.5; c'', c')\}$. This solution can be determined by numerical approximation techniques of the following sort.

(1) Guess initially that $z^* = z^0$ for $z^0 = $ (say) $E\{\tilde{c}|\ell^*(.5; c'', c')\}$.
(2) If the left-hand side LHS(z^0) of Equation (5.4.9) for $z = z^0$ is greater than the right-hand side RHS(z^0), guess $z^1 < z^0$, but guess $z^1 > z^0$ if LHS(z^0) < RHS(z^0).
(3) If LHS(z^1) > RHS(z^1), guess $z^2 < z^1$; guess $z^2 > z^1$ if LHS(z^1) < RHS(z^1), . . . , and so on, ultimately setting $z^* = z^n$ when LHS(z^n) and RHS(z^n) are "approximately" equal.

Numerical-analytic techniques, such as "regula falsi" or the Newton-Raphson method (see Hildebrand [**58**], Chapter 10), specify how the various guesses may be made in various intelligent ways. We present one simple method of searching for z^* in Appendix 4.

***Example 5.4.2c: Concave Exponential Utility (III)**

It is easier to assess your certainty equivalent $Q\{\ell\}$ of a lottery the as-yet undetermined outcome of which has the Normal distribution with density function $f_N(\cdot|m, V)$, defined in *Section 3.10. By Example 3.10.12 and a little algebra, (5.4.6) implies that

$$-e^{-zQ\{\ell\}} = -e^{-zm+(1/2)z^2V},$$

which (by taking negatives and then natural logs and then dividing by $-z$) implies that

$$z^* = 2[m - Q\{\ell\}]/V. \qquad (5.4.10)$$

But assessing your certainty equivalent of a Normally distributed return lottery requires some intuitive feel for Normal distributions—and, naturally, at least an understanding of *Section 3.10.

Example 5.4.3: Three-Parameter Concave Exponential Utility

The three- (essential-) parameter concave exponential utility function $u(\cdot)$ is defined for all returns $c \in (-\infty, +\infty)$ by

$$u(c) = -e^{-z_1c} - z_2e^{-z_3c}, \qquad (5.4.11)$$

where z_1, z_2, and z_3 are all *positive*, essential parameters and $z_1 \neq z_3$. It is a strictly increasing, continuous, and (strictly) concave function which also possesses the appealing (to some people) property of *decreasing risk aversion*: If $\ell + c^*$ is defined as in Example 5.4.2a, then

$$E\{\tilde{c}|\ell + c^*\} - Q\{\ell + c^*\} \qquad (5.4.12)$$

is a *decreasing*[k] function of c^* (and greater than zero if $u(\cdot)$ is concave). A person who would be *less* inclined to buy insurance against any given risk if he were wealthier has decreasing risk aversion. Many classes of functions possess the property of decreasing risk aversion, but that given by Equation (5.4.11) is convenient and also furnishes an adequate representation of many people's preferences. Certainty equivalents of three different lotteries must be directly assessed in order to specify the parameters of (5.4.11) via a system of three nonlinear equations. Finding the solution (z_1^*, z_2^*, z_3^*) such that every $z_i^* > 0$ is generally very difficult if done by hand, but a computer program [**127**] exists for this purpose.[l]

Example 5.4.4: (2*n* − 1)-Parameter Concave Exponential Utility

A generalization of (5.4.11) consists in assuming that your utility of any return $c \in (-\infty, +\infty)$ is expressible by

$$u(c) = -\sum_{i=1}^{n} z_{2i-1} e^{-z_{2i}c}, \tag{5.4.13}$$

where $z_1 = 1$ and $z_2, z_3, \ldots, z_{2n}$ are all *positive*, essential parameters, and where $z_2, z_4, z_6, \ldots, z_{2n}$ are all different. Very great flexibility can be obtained with utility functions of this form, which also possess the property of decreasing risk aversion for every $n > 1$. To determine the $2n - 1$ parameter values, you would have to assess directly the certainty equivalents of $2n - 1$ different lotteries. So far, no computer program for solving the resulting system of nonlinear equations when $n > 2$ has been publicized—no doubt because the cases $n = 1$ and $n = 2$ in Examples 5.4.2 and 5.4.3 respectively have sufficed for most persons.

Example 5.4.5: The Quadratic Utility Function

$u(\cdot)$ is defined for all returns $c \leq z$ (note this restriction) by

$$u(c) = -(z - c)^2, \tag{5.4.14}$$

where z is an essential parameter which must necessarily satisfy $z > c^B$ if you are to be able to use this function in analyzing your dpuu.[m] The parameter z is easily specified by one certainty equivalent. For example, your assessment of $Q = Q\{\ell^*(.5; c^B, c^W)\} < .5(c^B + c^W)$ implies, by

---

[k] An alternative definition of *constant* risk aversion is to require that $E\{\tilde{c}|\ell + c^*\} - Q\{\ell + c^*\}$ be *constant* in c^*, since such constancy implies that $E\{\tilde{c}|\ell\} - Q\{\ell\} = E\{\tilde{c}|\ell + c^*\} - Q\{\ell + c^*\}$; and this, together with $E\{\tilde{c}|\ell + c^*\} = E\{\tilde{c}|\ell\} + c^*$ [from Exercise 3.7.4 (D) and (A)], implies that $Q\{\ell + c^*\} = Q\{\ell\} + c^*$.

[l] *No* solution with every $z_i^* > 0$ exists unless the following condition is satisfied. Suppose that $c^1 < c^2 < c^3 < c^4$ with $c^4 - c^3 = c^3 - c^2 = c^2 - c^1 = L$, and that you determine $Q\{\ell^i\}$ for $\ell^i = \ell^*(.5; c^{i+1}, c^i)$ and $i = 1, 2, 3$. The condition is that $.5(c^{i-1} + c^i) > Q\{\ell^i\} - L > Q\{\ell^{i-1}\}$ for $i = 2, 3, 4$. In other words, as i increases and the return in ℓ^i is shifted up by L, $Q\{\ell^i\}$ gets *closer* to the average of c^i and c^{i+1} (= the expectation of the return from ℓ^i).

[m] The function $-(z - c)^2$ is decreasing in c if $c > z$ and hence does not represent a nonmasochist's preferences there.

Equations (5.4.14), (5.4.6), and a little algebra, that

$$z^* = (1/2)\frac{(c^B)^2 + (c^W)^2 - 2(Q)^2}{c^B + c^W - 2Q}, \tag{5.4.15}$$

from which it follows, again by a little algebra, that $z^* \geq c^B$ if and only if $c^W < Q < c^B$ (which is natural) *and*

$$Q \geq (1 - 1/\sqrt{2})c^B + (1/\sqrt{2})c^W. \tag{5.4.16}$$

Therefore, $Q\{\ell^*(.5; c^B, c^W)\}$ must fall at least .293 (and at most .500) of the way from c^W to c^B in order for u to be (concave and) strictly increasing on the interval $[c^W, c^B]$ of interest. The quadratic utility function (Equation 5.4.14) has the property of *increasing risk aversion*—(5.4.12) is an increasing function of c^*—and thus implies that one would be *more* inclined to buy insurance against risks if he were wealthier. Since most people do not feel this way, their use of a quadratic utility function is justifiable *only* by its convenience and adequacy in approximating their preferences.

The properties of constant, decreasing, and increasing risk aversion introduced in the preceding examples have been studied by Pfanzagl [**109**], Pratt [**111**], and Arrow [**2**], among others. Also see *Exercise 5.4.6 for an introduction to the "local risk aversion function."

Exercise 5.4.1

(A) Derive Equation (5.4.3[$i/4$]).
(B) Derive Equation (5.4.3[$i/2^n$]). [*Hint*: Use induction.]
(C) Derive Equation (5.4.4[2]).
(D) Derive Equation (5.4.4[n]). [*Hint*: Use induction.]
(E) Derive Equation (5.4.5[-1]).
(F) Derive Equation (5.4.5($-n$]). [*Hint*: Use induction.]

Exercise 5.4.2

A decision maker has furnished the following certainty equivalents, in the notation of Method $1: $c^0 = \$0$; $c^1 = \$50{,}000$; $c^{1/2} = \$15{,}000$; $c^{1/4} = \$6{,}000$; and $c^{3/4} = \$22{,}000$. He further states that he wishes his utility function to be concave over the interval [$0, $50,000]. Show that his statements are mutually inconsistent.

Exercise 5.4.3

Use Method $1 to assess *your* utility function for *immediate* returns c in the interval [$0, $1,000], with nonmonetary concomitant held fixed at the status quo. Use $u(\$1{,}000) = 1$ and $u(\$0) = 0$. [Compare the resulting curve with your previous response to Exercise 3.6.3(D). Are they very different?]

***Exercise 5.4.4**

If your utility function on [\$0, \$1,000] as determined in Exercise 5.4.3 was strictly concave and had constant risk aversion, then it would be of the form

$$u(c) = -ae^{-zc} + b \tag{5.4.17}$$

for all $c \in [\$0, \$1{,}000]$, where $a = b = (1 - e^{-1000z})^{-1}$. [These values of the inessential parameters make $u(\$0) = 0$ and $u(\$1{,}000) = 1$.]

(A) Approximate z correct to three decimal places, using your $c^{1/2}$ as assessed in Exercise 5.4.3.
(B) Graph the function specified by your answer to (A).
(C) Superimpose your graph from Exercise 5.4.3, and compare these functions.

Exercise 5.4.5

If your $c^{1/2}$ in Exercise 5.4.3 was between \$293 and \$500, your preferences for returns in [\$0, \$1,000] can be (at worst) caricatured by a quadratic utility function of the form $au(\cdot) + b$ for $u(\cdot)$ given by Equation (5.4.14).

(A) Solve for z by (5.4.15), putting $Q = c^{1/2}$, of course.
(B) Having solved for z, determine a and b such that $-a(z-0)^2 + b = 0$ and $-a(z-1000)^2 + b = 1$.
(C) Graph $u(c) = -a(z-c)^2 + b$ for z, a, and b as determined in (A) and (B), for all c in [\$0, \$1,000]. How does the resulting function compare with your graphs in Exercises 5.4.3 and 5.4.4?

***Exercise 5.4.6: Local Risk Aversion**

Let $u(\cdot)$ be a twice continuously differentiable utility function for net monetary return. In **[111]**, Pratt shows that the risk-aversion characteristics of $u(\cdot)$ are completely represented by the function

$$r(c) = -u''(c)/u'(c), \tag{5.4.18}$$

where $u'(\cdot)$ and $u''(\cdot)$ are the first and second derivatives of $u(\cdot)$. The function $r(\cdot)$ is called a local risk aversion function because, for "small" gambles, $Q\{\ell\}$ is well approximated by

$$E\{\tilde{c}|\ell\} - (1/2)V\{\tilde{c}|\ell\} \cdot r(E\{\tilde{c}|\ell\}),$$

where "small" refers to $V\{\tilde{c}|\ell\}$. Clearly, $u(\cdot)$ is strictly increasing if and only if $r(\cdot)$ is a nonnegative-valued function, but Pratt also proves that $u(\cdot)$ has the property of decreasing (or constant, or increasing) risk aversion if and only if $r(\cdot)$ is a decreasing (or constant, or increasing, respectively) function. These results indicate that one can readily deduce the risk-aversion characteristics of a given mathematical function by calculating the associated $r(\cdot)$ and studying its behavior.

(A) Show that the quadratic utility function (5.4.14) has the property of increasing risk aversion.

(B) Show that the three-parameter concave exponential utility function has the property of decreasing risk aversion.

(C) Show that, if $r(\cdot)$ is defined for all c in some interval (x, y), where $-\infty \leq x < y \leq +\infty$, then, except for inessential parameters,

$$u(c) = \int_x^c \exp\left[-\int_x^s r(t)dt\right]ds, \tag{5.4.19}$$

for all c in (x, y), where $\exp[z] = e^z$ for all real z.

(D) Show that if $r(\cdot)$ is constant, then $u(\cdot)$ is either linear (Example 5.4.1), concave exponential (Example 5.4.2), or *convex* exponential: $u(c) = +e^{+zc}$ for all c and some $z > 0$.

*■ In [**111**] Pratt also studies *proportional* risk-aversion behavior concerning gambles expressed as proportions of the decision maker's wealth prior to the gamble. Recently, Hammond [**51**] has shown how $\mathcal{D}$'s specifying some constraints on $r(\cdot)$ can suffice in certain situations to determine his preference among lotteries, without his having to assess specifically his utility function $u(\cdot)$. ■

5.5 UTILITY ON A CARTESIAN PRODUCT (I): GENERALITIES AND SEPARABLE UTILITY

This section is the first of three in which you have the opportunity of becoming exposed to the only partially arcane delights associated with the properties and assessments of utility functions on a set C of consequences which can be expressed as a Cartesian product

$$C = C_1 \times \cdots \times C_n$$

of n nonempty sets $C_1, \ldots, C_n$ called *factors* of the product, or *attribute sets*. Thus each consequence c in C can be written out as an n-tuple $(c_1, \ldots, c_n)$, in which each "component" c_i describes how c measures up according to the ith attribute[n] by which you have described the consequences of your dpuu.

This explanation of the components c_i of $c = (c_1, \ldots, c_n)$ justifies our calling the factors C_i attribute sets, and also suggests how Cartesian product consequence sets often arise in practice.

When we compare dissimilar consequences of a dpuu, it is natural in either formal or informal analysis to begin by making a list of those—and only those—attributes which cause the consequences to be dissimilar. On one hand, this means that we would not list an attribute with respect to

---

[n] Popular synonyms for "attribute" in this context are "aspect," "criterion," "factor," "feature," and "dimension," the last now appearing to be losing its prominence as a cliche to "interface" and "communication" (especially failure thereof).

which all consequences were identical. On the other hand, we would list every noticeably important attribute with respect to which any two or more consequences differed.

Example 5.5.1

In Example 1.5.1 (the subsystem problem), we would *not* list as an attribute what the company president had for breakfast yesterday, because it would be the same for all consequences of the dpuu. We would, on the other hand, (and did) list company prestige, because failure of the subsystem would damage prestige, whereas nonfailure would leave prestige intact.

Each attribute i thus gives rise to a nonempty° set C_i of attribute descriptions c_i.

Now, use of a substitute function as defined in the sixth preliminary (Definition 5.2.2) may be regarded as a means of reducing the number of attributes. The monetary adjustments $N(c_1, c_2|c_2^*)$ in Example 5.2.1 which compensate for c_2 in place of c_2^* allow us to ignore attribute-2 variability and focus instead on monetary returns, i.e., on consequences of the form (c_1, c_2^*).

Sometimes it would be very inconvenient or unnatural to eliminate all attributes of the consequences but one, especially if these attributes are drastically dissimilar, difficult to compare, or of extremely different orders of importance to the decision maker. When the differences in orders of importance are *too* extreme, utility may not even be expressible by a single real number, as we shall see in Section 5.6.

From Example 5.5.1 it is clear that not every C_i need be a set of real numbers. But when every C_i is an interval of real numbers, then $C_1 \times \cdots \times C_n$ is a subset of real n-space

$$R^n = \{(c_1, \ldots, c_n): -\infty < c_i < +\infty \text{ for every } i\}.$$

This is true in numerous cases; for example, the case in which each c_i is a cash inflow or outflow occurring on date i and in which the most remote flow occurs on date n. Section 5.7 presents a few facts about utility functions on R^n.

When each attribute set consists of many elements and the total number n of attributes exceeds one, your task in assessing a utility function on $C_1 \times \cdots \times C_n$ can be formidable. A primary purpose of Sections 5.6 and 5.7 is to ameliorate such difficulties by exploiting certain properties of utility functions on Cartesian products which follow from qualitative properties of your preferences.

The remainder of this section shows that if you accept in principle *one* such qualitative property, then your utility function $u: C_1 \times \cdots \times C_n \to$

° In fact, we have argued that C_i should consist of at least two elements.

$(-\infty, +\infty)$ is *separable*, in the sense that it can be written as the sum of n real-valued functions on the individual attribute sets.

Definition 5.5.1

A utility function $u: C_1 \times \cdots \times C_n \to (-\infty, +\infty)$ is called *separable* if there are n functions $u_i: C_i \to (-\infty, +\infty)$ such that

$$u(c_1, \ldots, c_n) = \sum_{i=1}^{n} u_i(c_i) \tag{5.5.1}$$

for every $(c_1, \ldots, c_n)$ in $C_1 \times \cdots \times C_n$.

If—and this is a big "if"—your utility is separable, then it is relatively easy to assess. The assessment method S, described at the end of this section, involves (essentially) separately assessing each $u_i(\cdot)$ on C_i, and then rescaling these functions so that they can be added and adequately reflect your preferences on $C_1 \times \cdots \times C_n$.

The "iffiness" of separability is clear from the fact that, when $n = 2$, the definition of separability clearly implies that

$$(c_1^1, c_2^1) \succsim (c_1^2, c_2^1) \qquad \text{if and only if} \qquad (c_1^1, c_2^2) \succsim (c_1^2, c_2^2) \tag{5.5.2}$$

for all c_1^1 and c_1^2 in C_1 and all c_2^1 and c_2^2 in C_2. That is, the relative desirabilities of the c_1's is unaffected by their c_2 concomitant. The same is true if the roles of c_1 and c_2 are reversed.

Example 5.5.2

Suppose that c_1^1 = "power fails," c_1^2 = "power does not fail," c_2^1 = "control fails," and c_2^2 = "control does not fail," and that *any* failure is equally undesirable. Then we should have

$$(c_1^2, c_2^2) \succ (c_1^1, c_2^1) \sim (c_1^1, c_2^2) \sim (c_1^2, c_2^1).$$

Hence u cannot satisfy Equation (5.5.1), since c_1^1 and c_1^2 are, so to speak, equally desirable when accompanied by c_2^1, whereas c_1^2 is more desirable than c_1^1 when accompanied by c_2^2. In economic language, c_1^2 and c_2^2 are *complementary* goods: The relative desirability of each is increased by the presence of the other.

Example 5.5.3

Suppose that c_1^1 = "eat a piece of pie now," c_1^2 = "do not eat a piece of pie now," c_2^1 = "eat(?) a double-thick milk shake now," and c_2^2 = "do not eat(?) a double-thick milk shake now." If (c_1^1, c_2^1) would cause indigestion, you might have

$$(c_1^2, c_2^1) \succ (c_1^1, c_2^2) \succ (c_1^2, c_2^2) \succ (c_1^1, c_2^1).$$

Here, c_1^2 is preferable to c_1^1 when combined with c_2^1, but c_1^1 is preferable to c_1^2 when combined with c_2^2. Similarly for the *sub*scripts interchanged. In economic language, c_1^1 and c_2^1 are *substitutable* goods: The relative desirability of each is lessened by the presence of the other. Again in this case, u cannot be represented in the form (5.5.1).

We shall introduce a general independence assumption which, if acceptable to you in a given context, implies that your utility on $C_1 \times \cdots \times C_n$ satisfies (5.5.1).

Definition 5.5.2. Attribute-Independence Condition. We say that your preferences are *attribute-independent* if

$$\ell^*(1/2; c^1, c^2) \sim \ell^*(1/2; c^3, c^4)$$

whenever $\{c_i^1, c_i^2\} = \{c_i^3, c_i^4\}$ for $i = 1, \ldots, n$, where $c^j = (c_1^j, \ldots, c_n^j)$ for $j = 1, 2, 3, 4$.

■ $\{c_i^1, c_i^2\} = \{c_i^3, c_i^4\}$ if and only if either (1) $c_i^1 = c_i^2$ and hence $c_i^3 = c_i^4 = c_i^1 = c_i^2$, or (2) $c_i^1 \neq c_i^2$, $c_i^3 \neq c_i^4$, and either (a) $c_i^3 = c_i^1$ and $c_i^4 = c_i^2$, or (b) $c_i^3 = c_i^2$ and $c_i^4 = c_i^1$. In any case, $\{c_i^1, c_i^2\} = \{c_i^3, c_i^4\}$ if and only if the "*marginal* lotteries" $\ell_i = \ell^*(1/2; c_i^1, c_i^2)$ and $\ell_i' = \ell^*(1/2; c_i^3, c_i^4)$ on each attribute set C_i yield precisely the same elements with equal probabilities, and hence coincide. ■

Now we prove that attribute-independence implies separability of utility.

Theorem 5.5.1. Your utility function u on $C_1 \times \cdots \times C_n$ is separable if and only if your preferences satisfy the attribute-independence condition.

■ *Proof.* That a separable utility function, of the form (5.5.1), satisfies the attribute-independence condition is an easy calculation and is left to you as Exercise 5.5.1. To prove that attribute-independence implies separability of utility, we shall show first that u can be written as the sum of two functions each defined on a subset of the attributes. For notational simplicity, define $X_i = C_1 \times \cdots \times C_i$, $Y_i = C_{i+1} \times \cdots \times C_n$, $x_i = (c_1, \ldots, c_i)$, and $y_i = (c_{i+1}, \ldots, c_n)$ for any $(c_1, \ldots, c_n)$ and any $i \leq n-1$. Every $(c_1, \ldots, c_n)$ can then be written in the form (x_i, y_i). Next, fix a consequence $c^- = (c_1^-, \ldots, c_n^-) = (x_i^-, y_i^-)$ and set $u(c^-) = u(x_i^-, y_i^-) = 0$ arbitrarily. (By virtue of Observation 5.2.4, this can always be done.) Now note that the attribute-independence condition implies that

$$\ell^*(1/2; (x_i, y_i), (x_i^-, y_i^-)) \sim \ell^*(1/2; (x_i, y_i^-), (x_i^-, y_i))$$

for every $c = (x_i, y_i)$ and every $i \in \{1, \ldots, n-1\}$. The U-indices of these lotteries must be equal, and hence

$$.5u(x_i, y_i) + .5u(x_i^-, y_i^-) = .5u(x_i, y_i^-) + .5u(x_i^-, y_i),$$

which can be rearranged to yield

$$\text{(a)} \qquad u(x_i, y_i) = u(x_i, y_i^-) + u(x_i^-, y_i),$$

since $u(x_i^-, y_i^-) = 0$. Now, (a) obtains for every $c = (x_i, y_i)$ and every i, which shows that, once i has been chosen, u can be written as the sum of the function $u(\cdot, y_i^-)$ on $X_i = C_1 \times \cdots \times C_i$ and the function $u(x_i^-, \cdot)$ on $Y_i = C_{i+1} \times \cdots \times C_n$. This concludes the proof if $n = 2$. If $n > 2$, we obtain the following equalities from (a).

(b$_1$) $u(c_1, \ldots, c_n) = u(c_1, c_2^-, \ldots, c_n^-) + u(c_1^-, c_2, \ldots, c_n)$;

(b$_2$) $u(c_1^-, c_2, \ldots, c_n) = u(c_1^-, c_2, c_3^-, \ldots, c_n^-) + u(c_1^-, c_2^-, c_3, \ldots, c_n)$;

[by (a) with $i = 1$ and by (a) with $i = 2$ respectively]; and, in general,

(b$_j$)
$$u(c_1^-, \ldots, c_{j-1}^-, c_j, \ldots, c_n) = u(c_1^-, \ldots, c_{j-1}^-, c_j, c_{j+1}^-, \ldots, c_n^-) + u(c_1^-, \ldots, c_j^-, c_{j+1}, \ldots, c_n)$$

for $j = 2, 3, \ldots, n-1$ [by (a) with $i = j$]. Now, substituting (b_{n-1}) into (b_{n-2}), then substituting the result into (b_{n-3}), and so on yields

$$u(c_1, \ldots, c_n) = \sum_{i=1}^{n} u(c_1^-, \ldots, c_{i-1}^-, c_i, c_{i+1}^-, \ldots, c_n^-),$$

which is the desired result once we *define*

$$u_1(\cdot),\ u_i(\cdot) \text{ for } i = 2, \ldots, n-1, \quad \text{and} \quad u_n(\cdot)$$

by $u_1(c_1) = u(c_1, c_2^-, \ldots, c_n^-)$,

$$u_i(c_i) = u(c_1^-, \ldots, c_{i-1}^-, c_i, c_{i+1}^-, \ldots, c_n^-),$$

and

$$u_n(c_n) = u(c_1^-, \ldots, c_{n-1}^-, c_n),$$

respectively. ■

Observation 5.2.4 specifies the set of all utility functions on $C_1 \times \cdots \times C_n$ which differ from each other only by unessential scale and origin parameters $a^* > 0$ and b^* respectively. This observation implies that the "utility" functions u_i on the attribute sets C_i may be incremented by any constants b_i^* independently of each other, but that changing the scale of u_i by multiplying all its values by $a^* > 0$ must *only* be done simultaneously with all the other u_j's.

Theorem 5.5.2. If $u(c_1, \ldots, c_n) = \sum_{i=1}^{n} u_i(c_i)$ is a utility function for you on $C_1 \times \cdots \times C_n$, then so is $u^*(c_1, \ldots, c_n)$ if and only if

$$u^*(c_1, \ldots, c_n) = \sum_{i=1}^{n} u_i^*(c_i),$$

where there are real numbers $b_1^*, \ldots, b_n^*$ and a positive real number a^* such that

$$u_i^*(c_i) = a^* u_i(c_i) + b_i^*$$

for every c_i in C_i and every $i \in \{1, \ldots, n\}$.

■ *Proof*. Exercise 5.5.2. ■

The proof of Theorem .5.5.1 provides a hint about how to assess a separable utility function on $C_1 \times \cdots \times C_n$ once you have ascertained that your preferences satisfy the attribute-independence condition. We present only one method; Fishburn discusses others in [**41**] and [**42**], which contain many additional references on utility independence (a synonym for separability).

STEP S-1 requires that you choose two consequences $(c_1^-, \ldots, c_n^-) = c^-$ and $(c_1^+, \ldots, c_n^+) = c^+$ such that the consequence c^i, defined by

$$c^i = (c_1^-, \ldots, c_{i-1}^-, c_i^+, c_{i+1}^-, \ldots, c_n^-),$$

satisfies the condition that

$$c^i > c^-$$

for every i. For notational convenience, then set $u(c^+) = 1$ and $u(c^-) = 0$.

■ c^+ and c^- can be chosen such that $c^i > c^-$ for every i *if* your preferences satisfy the attribute-independence condition and *if* every C_i consists of at least two unequally desirable elements c_i. If the latter condition fails, then you may either omit C_i from the Cartesian product and from further consideration or else set $u_i(c_i) = \text{constant}$ arbitrarily. ■

STEP S-2 requires that you assess a "pseudoutility" function u_i' on C_i for every i such that $u_i'(c_i^-) = 0$ and $u_i'(c_i^+) = 1$. [The adjective "pseudo" arises because of Theorem 5.5.2, which implies that the appropriate differences $u_i(c_i^+) - u_i(c_i^-)$ may be unequal. In step S-3 we rescale the pseudoutilities appropriately.] Any method discussed in Sections 5.3 and 5.4 may be used, but we shall describe your present task in terms of Method F1. To apply that method, we might note that c^- here coincides with c^- there, but that the function of c^+ in Method F1 is performed here by $c^i = (c_1^-, \ldots, c_{i-1}^-, c_i^+, c_{i+1}^-, \ldots, c_n^-)$. Also, in place of variable consequences c there, we use the variable consequence c_i^* defined by

$$c_i^* = (c_1^-, \ldots, c_{i-1}^-, c_i, c_{i+1}^-, \ldots, c_n^-).$$

Clearly, c_i^* is just c_i together with c_j^- for all $j \neq i$, and hence c_i^* and c_i may be identified with each other. There are three cases, corresponding to the three cases of Method F1.

CASE S-2.1. If $c^- \lesssim c_i^* \lesssim c^i$ and if $c_i^* \sim \ell^*(p; c^i, c^-)$, then $u_i'(c_i) = p$ $[= w_1(c_i^*)]$.

CASE S-2.2. If $c_i^* > c^i$ and if $c^i \sim \ell^*(p; c_i^*, c^-)$, then $u_i'(c_i) = 1/p$ $[= 1/w_2(c_i^*)]$.

CASE S-2.3. If $c^- > c_i^*$ and if $c^- \sim \ell^*(p; c^i, c_i^*)$, then $u_i'(c_i^-) = -p/(1-p)$ $[= -w_3(c_i^*)/(1 - w_3(c_i^*))]$.

It is easy to show (as Exercise 5.5.3) that

$$u_i'(c_i^+) = 1 \text{ and } u_i'(c_i^-) = 0 \text{ for every } i. \tag{5.5.3}$$

STEP S-3 involves your rescaling the pseudoutility functions $u_i'(\cdot)$ so as to constitute the appropriate summands $u_i(\cdot)$ in Equation (5.5.1). In the following, it will be convenient to specify that $u_i(c_i^-) = 0 = u_i'(c_i^-)$ for every i.

■ Theorem 5.5.2 implies that this can always be done without violating your preferences, by appropriate choices of the numbers b_i^*. ■

It then remains to determine scale factors $p_i > 0$ such that

$$u_i(c_i) = p_i u_i'(c_i)$$

for every c_i in C_i. This is accomplished by determining $n-1$ other numbers q_i in $(0, 1)$ such that

$$(c_1^+, \ldots, c_i^+, c_{i+1}^-, \ldots, c_n^-) \sim \ell^*(q_i; (c_1^+, \ldots, c_n^+), (c_1^-, \ldots, c_n^-)). \quad (5.5.4)$$

Since

$$(c_1^+, \ldots, c_{i+1}^+, c_{i+2}^-, \ldots, c_n^-) \succ (c_1^+, \ldots, c_i^+, c_{i+1}^-, \ldots, c_n^-)$$

for $i = 1, \ldots, n-1$, it follows that q_i should be less than q_{i+1} for every i. Then we set

$$p_i = q_i - q_{i-1} \quad (5.5.5)$$

for $i = 1, \ldots, n$, where we *define* $q_n = 1$ and $q_0 = 0$.

■ Since $u(c_1^+, \ldots, c_n^+) = 1$ and $u(c_1^-, \ldots, c_n^-) = 0$, the utility of the right-hand side of Equation (5.5.4) is q_i. By separability and $u_j(c_j^-) = 0$, the utility of the left-hand side of (5.5.4) is $\Sigma_{j=1}^{i} u_j(c_j^+)$, which therefore equals q_i by equidesirability and Observation 5.2.2. But

$$p_i = p_i u_i'(c_i^+) = u_i(c_i^+) = \sum_{j=1}^{i} u_j(c_j^+) - \sum_{j=1}^{i-1} u_j(c_j^+) = q_i - q_{i-1}. \ ■$$

Once you have determined by method S that separable utility function for which $u(c^+) = 1$ and $u(c^-) = 0$, you may apply Observation 5.2.2 to obtain any other utility function that also represents your preferences. Every such utility function can be written separably, by virtue of Theorem 5.5.2.

Example 5.5.4

Suppose that the consequences of a given dpuu have two relevant attributes: what you will have for supper tomorrow; and what book you will read tomorrow evening. Let $C_1 = \{c_1^1, c_1^2, c_1^3\}$, with $c_1^1 =$ "pork chop," $c_1^2 =$ "chicken," and $c_1^3 =$ "steak"; and let $C_2 = \{c_2^1, c_2^2, c_2^3\}$, with $c_2^1 =$ *Well Clad Came the Acquaintance*, $c_2^2 =$ *Uncountably Many Ways to Get Rich*, and $c_2^3 =$ *Fundamentals of Decision Analysis*. Suppose that you choose $c_1^- = c_1^1$, $c_2^- = c_2^2$, $c_1^+ = c_1^2$, and $c_2^+ = c_2^3$. You check to see that $c^1 = (c_1^+, c_2^-) = (c_1^2, c_2^2) \succ (c_1^1, c_2^2) = (c_1^-, c_2^-) = c^-$ and that $c^2 = (c_1^-, c_2^+) =$

$(c_1^1, c_2^3) > (c_1^1, c_2^2) = (c_1^-, c_2^-) = c^-$, as step S-1 requires. Then set $u(c^+) = u(c_1^2, c_2^3) = 1$ and $u(c^-) = u(c_1^1, c_2^2) = 0$. This completes step S-1. In step S-2, we shall determine $u_1'(\cdot)$ first. Since $u_1'(c_1^-) = u_1'(c_1^1) = 0$ and $u_1'(c_1^+) = u_1'(c_1^2) = 1$, it remains to determine $u_1'(c_1^3)$. If you prefer steak to chicken for supper tomorrow, then $(c_1^3, c_2^2) = c_1^* > c^1 = (c_1^2, c_2^2)$ and Case S-2.2 obtains. Suppose that your preferences are such that $c^1 = (c_1^2, c_2^2) \sim \ell^*(.4; (c_1^3, c_2^2), (c_1^1, c_2^2)) = \ell^*(.4; c^1, c^-)$. Then $u_1'(c_1^3) = 1/.4 = 2.5$. Next, we must determine $u_2'(\cdot)$. Since we know that $u_2'(c_2^2) = 0$ and $u_2'(c_2^3) = 1$, all that remains is $u_2'(c_2^1)$. We suppose that you possess the good taste to prefer FDA $= c_2^3$ to WCCTA $= c_2^1$, and that you prefer WCCTA to UMWTGR $= c_2^2$. Then Case S-2.1 obtains. Suppose also that $c_2^* = (c_1^1, c_2^1) \sim \ell^*(.2; (c_1^1, c_2^3), \ (c_1^1, c_2^2)) = \ell^*(.2; c^2, c^-)$. Then $u_2'(c_2^1) = .2$. Now for step S-3. You compare $c^1 = (c_1^2, c_2^2)$ with canonical lotteries of the form $\ell^*(q; c^+, c^-) = \ell^*(q; (c_1^2, c_2^3), (c_1^1, c_2^2))$, and you find that $c^1 \sim \ell^*(.15; c^+, c^-)$. Thus $p_1 = q_1 - 0 = q_1 = .15$, and hence $p_2 = 1 - p_1 = .85$. Therefore $u_1(c_1) = .15\, u_1'(c_1)$ and $u_2(c_2) = .85 u_2'(c_2)$ for all c_1 and c_2, and we have the following table.

j	$u_1(c_1^j)$	$u_2(c_2^j)$
1	.000	.170
2	.150	.000
3	.375	.850

By Equation (5.5.1), $u(c_1^j, c_2^k) = u_1(c_1^j) + u_2(c_2^k)$ for every j and k.

There are $3 \times 3 = 9$ elements of $C_1 \times C_2$ in Example 5.5.4. Hence *direct* application of Method F1 would have required $9 - 2 = 7$ comparisons[p] of c's with $\ell^*(p; c^+, c^-)$'s if you had not exploited attribute-independence and separability. But in Example 5.5.4 you made $3 - 2 = 1$ comparison of this form for u_1', also one comparison of this form for u_2', and $2 - 1 = 1$ comparison of this form for p_1, for a total of only 3 rather than 7 comparisons.

Such an economy is true in general. Suppose that C_i consists of $m_i \geq 2$ elements, for $i = 1, \ldots, n$. Then $C_1 \times \cdots \times C_n$ consists of $\Pi_{i=1}^n m_i = (m_1)(m_2) \cdots (m_n)$ elements, and hence Method F1 would require

$$\prod_{i=1}^{n} m_i - 2$$

lottery comparisons. On the other hand, method S requires $m_i - 2$ lottery comparisons to determine u_i' for each i, together with $n - 1$ lottery com-

[p] The arbitrary specification of $u(c^+)$ and $u(c^-)$ for two elements of C imply that Method F1 requires $m - 2$ comparisons of c with $\ell^*(p; c^+, c^-)$—or similar comparisons required by the other two cases, if C has m elements.

parisons to determine the q_i's (and hence the p_i's). Thus method S requires

$$\sum_{i=1}^{n}(m_i-2)+(n-1)=\sum_{i=1}^{n} m_i-n-1$$

lottery comparisons, generally a *very* much smaller number than $\Pi_{i=1}^{n} m_i - 2$.

Example 5.5.5

If $n=3$, $m_1=5$, $m_2=4$, and $m_3=10$, then 198 comparisons are required by Method F1, whereas only 15 are required by method S.

Example 5.5.6

Consequences of governmental (and other) expenditure programs are often described crudely in terms of which goals i out of n are achieved. In such cases, $C=C_1\times\cdots\times C_n$, with each $C_i=\{$"goal i is achieved," "goal i is not achieved"$\}$. Thus, $m_i=2$ for each i, and Method F1 would require 2^n-2 lottery comparisons to determine u on $C_1\times\cdots\times C_n$, whereas method S would require only the $n-1$ comparisons which specify the q_i's, since $u_i'(c_i)\in\{0, 1\}$ for both elements of C_i. It is often *merely assumed* that the program administrator's (or the public's) preferences satisfy the attribute-independence condition, and often this assumption is indeed justifiable. But it is probably more often *not* justifiable. Certainly, the attribute-independence condition and its implications should be reflected on before blithely assuming that it holds, for the saving in assessment effort with method S—and the ensuing analytical conveniences of separable utility functions—can be bought at too high a price in terms of violence done to the decision maker's basic preferences. Garbage in, garbage out. *Exercise 5.5.5 introduces a milder assumption than separability and asks you to show that the number of lottery comparisons required to determine the corresponding utility function is between the large number with Method F1 and the small number with method S. So there is a halfway house.

Exercise 5.5.1

Show that if $u(\cdot)$ is separable, then your preferences are attribute-independent.

Exercise 5.5.2

Prove Theorem 5.5.2.

Exercise 5.5.3

Let all terms be as defined in method S. Show that $u_i'(c_i^+)=1$ and $u_i'(c_i^-)=0$.

Exercise 5.5.4

Determine $a^* > 0$, b_1^*, and b_2^* so that your preferences in Example 5.5.4 are expressed by a separable utility function $u^*(\cdot, \cdot)$ with $u^*(c_1^1, c_2^2) = 10$, $u^*(c_1^3, c_2^3) = 25$, and $u_2^*(c_2^3) = 100$.

***Exercise 5.5.5: Quasi-Separable Utility**

This exercise introduces the more general concept of a quasi-separable utility function, after Keeney [**72**]. Suppose attributes have been so labeled that[q] $C = X \times Y$, and that your preferences for lotteries $(\tilde{x}, y)$ with y held fixed do not depend upon the fixed value of y. More formally: We say that your preferences are x-*from*-y-*quasi-separable* if

$$(x, y') \begin{Bmatrix} > \\ \sim \\ < \end{Bmatrix} \ell^*(p; (x'', y'), (x', y'))$$

for *some* y' in Y implies that

$$(x, y) \begin{Bmatrix} > \\ \sim \\ < \end{Bmatrix} \ell^*(p; (x'', y), (x', y))$$

for *every* y in Y.

(A) Show that if your preferences are x-from-y-quasi-separable, then your $u: X \times Y \to (-\infty, +\infty)$ can be written in the form

$$u(x, y) = u_{21}(y) + u_{22}(y)u_1(x), \tag{5.5.6}$$

for $u_1(\cdot)$ a function on X and $u_{21}(\cdot)$ and $u_{22}(\cdot)$ functions on Y, such that $u_{22}(y) > 0$ for every y.

[*Hints*: (1) Choose y^- in Y, and x^+ and x^- in X, such that $(x^+, y^-) > (x^-, y^-)$.

(2) Set $u(x^+, y^-) = 1$ and $u(x^-, y^-) = 0$.

(3) By comparing (x, y), (x^+, y), (x^-, y), and lotteries involving them, in the spirit of step F1-2 in Section 5.3, show that x-from-y-quasi-separability implies the existence of a function u^* of x *alone* such that

$$u(x, y) = [u(x^+, y) - u(x^-, y)]u^*(x) + u(x^-, y).$$

(4) Define $u_{21}(y) = u(x^-, y)$ and $u_{22}(y) = u(x^+, y) - u(x^-, y)$. Quasi-separability implies that $(x^+, y) > (x^-, y)$ for every y by virtue of $(x^+, y^-) > (x^-, y^-)$, and hence $u(x^+, y) > u(x^-, y)$, implying $u_{22}(y) > 0$.

(5) Note, for subsequent assessment purposes, that $u^*(x) = u(x, y^-) = u_1(x)$ for all x, $u_1(x^-) = 0$, and $u_1(x^+) = 1$.]

---

[q] Each of X and Y may itself be a Cartesian product.

(B) Hence, justify the following Method QS for assessing an x-from-y-quasi-separable utility function.

STEP QS-1. Choose (x^+, y^-) and (x^-, y^-) such that $(x^+, y^-) > (x^-, y^-)$, and set $u(x^+, y^-) = 1$ and $u(x^-, y^-) = 0$.

STEP QS-2: Choose a y^* in Y such that (x^+, y^*) and (x^+, y^-) are not equidesirable; and assess your utility of (x^+, y^*) and (x^-, y^*) by step F1-2 of Method F1, recalling from step QS-1 that $u(x^+, y^-) = 1$ and $u(x^-, y^-) = 0$.

STEP QS-3: Assess your utility function $u(\cdot, y^-): X \to (-\infty, +\infty)$, again recalling that $u(x^+, y^-) = 1$ and $u(x^-, y^-) = 0$. This is $u_1(\cdot)$ [because $u_{21}(y^-) = 0$ and $u_{22}(y^-) = 1$].

STEP QS-4: Assess your utility function $u(x^-, \cdot): Y \to (-\infty, +\infty)$ such that $u(x^-, y^-) = 0$ and $u(x^-, y^*)$ agrees with your assessment of it in step QS-2. Then $u_{21}(y) = u(x^-, y)$ for every y [because $u_1(x^-) = 0$].

STEP QS-5: Assess your utility function $u(x^+, \cdot): Y \to (-\infty, +\infty)$ such that $u(x^+, y^-) = 1$ and $u(x^+, y^*)$ agrees with your assessment of it in step QS-2. Then $u_{22}(y) = u(x^+, y) - u_{21}(y)$ for every y [because $u_1(x^+) = 1$].

(C) Show that if X and Y contain $m_1 \geq 2$ and $m_2 \geq 2$ elements, respectively, then the total number of lottery comparisons required by Method QS is at most $m_1 + 2m_2 - 4$.

(This should be compared with the smaller $m_1 + m_2 - 3$ for method S and separable utility; and with the larger $m_1 m_2 - 2$ for Method F1 and arbitrary utility.)

■ Utility functions having special structure, such as additive separability or quasi-separability, have been studied intensively in recent years, because exploiting such structure simplifies the quantification of preferences. See, for example, Fishburn **[38]** and Keeney **[69]**, where additional references may be found. Such structures have recently been put into a significantly more general framework by Farquhar **[31]**, **[32]**. ■

5.6 UTILITY ON A CARTESIAN PRODUCT (II): EXTREME PREFERENCES AND LEXICOGRAPHIC UTILITY FUNCTIONS

In some usually very grim dpuu's your preferences might violate Observation 5.2.3, in that the set C of possible consequences contains three consequences c'', c, and c' such that $c'' > c$ and $c > c'$, but either

$$c > \ell^*(p; c'', c') \qquad \text{for } every\ p < 1$$

or

$$c < \ell^*(p; c'', c') \qquad \text{for } every\ p > 0.$$

Example 5.6.1

Let $c'' =$ "receive \$2 and live," $c =$ "receive \$1 and live," and $c' =$ "receive \$0 and be killed." One's initial reaction is to say, "yes, $c'' \succ c$ and $c \succ c'$, but I wouldn't accept *any positive* probability $1-p$ at being killed in return for the remaining probability p of the extra dollar." A dpuu containing such consequences is grim but not unheard of.

But Observation 5.2.3 *must* be true if one has preferences for lotteries with outcomes in C which can be represented by a real-valued utility function. In other words, *either*

(a) your preferences *satisfy* Observation 5.2.3 and *can* be represented by a real-valued utility function $u: C \to (-\infty, +\infty)$; *or*
(b) your preferences *violate* Observation 5.2.3 and *cannot* be represented by a real-valued utility function $u: C \to (-\infty, +\infty)$.

Logic forbids violating Observation 5.2.3 and using a real-valued utility function to represent your preferences!

Something must give. If you find your initial reactions violating Observation 5.2.3, first *try hard* to convince yourself that there is *some tiny* probability $1-p$ of the horrible consequence c' which you will be willing to accept in conjunction with the virtual certainty p of the good consequence c'', in preference to the complete certainty of the pretty good consequence c.

Example 5.6.2

When life or death, or other fundamental values are at stake, it may be impossible to convince oneself that 10^{-9999} is so small as to be negligible, and that therefore $\ell^*(1-10^{-9999};\, c'', c') \gtrsim c$. This is true even though one may run red lights, thereby accepting a probability $1-p$ greater than 10^{-9999} of a consequence comparable to c' in return for probability p of a consequence c'' only marginally more desirable than c.

Example 5.6.3

There is a small probability $1-p$, surely exceeding 10^{-9999}, that I would be served a poisonous toadstool at my favorite gourmet restaurant here in New Orleans, and subsequently die prematurely $(= c')$. But the (admittedly) slight incremental pleasure of having mushroom sauces $(= c'')$, together with the substantial probability p of survival, make $\ell^*(p; c'', c')$ more desirable to me than $c =$ "no mushroom sauce."

Fortunately, all is not lost if you cannot convince yourself to accept a tiny probability of obtaining an exceptionally bad consequence. The "kicker" in statements (a) and (b) above is "real-valued"; and we shall argue that all of decision analysis developed in Chapter 4 continues to hold if your preferences are represented by a certain n-*tuple-valued* utility function $u: C \to R^n$, where

$$R^n = \{(u_1, \ldots, u_n): -\infty < u_i < +\infty \text{ for every } i\}. \tag{5.6.1}$$

To motivate the particulars of such a utility function, we begin by noting that extreme preferences that violate Observation 5.2.3 arise because of particular *attributes* of the consequences. What makes c' so extremely repugnant that $c > \ell^*(p; c'', c')$ for all $p < 1$? Or what makes c'' so extremely attractive that $c < \ell^*(p; c'', c')$ for all $p > 0$? Such questions can be handled more readily if you describe your consequences in terms of n-tuples $(c_1, \ldots, c_n)$ of attribute descriptions, with the property that each attribute i is *overridingly more important than* all subsequently listed attributes $j > i$. In Example 5.6.1, living versus dying is overridingly more important than small monetary returns. Definition 5.6.1 makes more precise our notion of overridingly greater importance.

Definition 5.6.1. Each attribute i is *overridingly more important than* all attributes $j > i$ if for every i and every pair of lotteries ℓ_i^1 and ℓ_i^2 with outcomes in $C_1 \times \cdots \times C_i$, it follows that $(\ell_i^1, c'_{i+1}, \ldots, c'_n) > (\ell_i^2, c''_{i+1}, \ldots, c''_n)$ whenever $(\ell_i^1, c^\dagger_{i+1}, \ldots, c^\dagger_n) > (\ell_i^2, c^\dagger_{i+1}, \ldots, c^\dagger_n)$ for some $(c^\dagger_{i+1}, \ldots, c^\dagger_n)$ in $C_{i+1} \times \cdots \times C_n$.

■ The notation in Definition 5.6.1 should not cause difficulty; for any ℓ with outcomes in $C_1 \times \cdots \times C_i$ and any $(c^*_{i+1}, \ldots, c^*_n)$, the symbol $(\ell, c^*_{i+1}, \ldots, c^*_n)$ denotes the lottery with outcomes in $C_1 \times \cdots \times C_i \times \{(c^*_{i+1}, \ldots, c^*_n)\}$ which results in $(c_1, \ldots, c_i, c^*_{i+1}, \ldots, c^*_n)$ if and only if ℓ results in $(c_1, \ldots, c_i)$. ■

According to Definition 5.6.1, the first i attributes are overridingly more important than the last $n - i$ if, whenever ℓ_i^1 is more desirable than ℓ_i^2 when each is accompanied by some $(c^\dagger_{i+1}, \ldots, c^\dagger_n)$, then there is no possibility of reversing or even nullifying this preferability by accompanying ℓ_i^1 with a "bad" $(c'_{i+1}, \ldots, c'_n)$ and accompanying ℓ_i^2 with a "good" $(c''_{i+1}, \ldots, c''_n)$.

If you adamantly refuse to accept Observation 5.2.3, then your first task in "n-tupally" quantifying your preferences is to describe the consequences in terms of attributes so arranged that Definition 5.6.1 is satisfied.

Before discussing your assessment task any further, we must define and discuss just what it is that you must assess. But for that definition, of lexicographic utility, a preliminary definition of the lexicographic order "$\geq_L$" is required. Roughly, we say that two n-tuples $\mathbf{x}$ and $\mathbf{y}$ are such that $\mathbf{x} \geq_L \mathbf{y}$ if either $\mathbf{x} = \mathbf{y}$ or if $x_i > y_i$ for the earliest unequal pair of corresponding components of $\mathbf{x}$ and $\mathbf{y}$.

Definition 5.6.2. Let $\mathbf{x} = (x_1, \ldots, x_n)$ and $\mathbf{y} = (y_1, \ldots, y_n)$ be points in R^n. We write $\mathbf{x} \geq_L \mathbf{y}$ and say "$\mathbf{x}$ is *lexicographically greater than or equal to* $\mathbf{y}$" if either

(1) $x_i = y_i$ for all i [and hence $\mathbf{x} = \mathbf{y}$] or

(2) for some $i < n$, $x_i > y_i$ and $x_k = y_k$ for all $k < i$.

Example 5.6.4

In R^3, $(5, 8, 3) \geq_L (4, 0, 10^{+9999})$ because $x_1 = 5 > 4 = y_1$, and hence the second and third components can be ignored. Also, $(5, 8, 3) \geq_L (5, 7, 10^{+9999})$ since $x_1 = y_1 = 5$ but $x_2 = 8 > 7 = y_2$ and hence the third components can be ignored. Similarly, $(5, 8, 3) \geq_L (5, 8, 2)$.

■ You may show as Exercise 5.6.1 that "$\geq_L$" in R^n possesses all the nice properties of "$\geq$" in R^1; namely:

$$\textit{reflexivity:} \quad \mathbf{x} \geq_L \mathbf{x} \quad \text{for every } x \text{ in } R^n; \tag{5.6.2}$$

$$\textit{antisymmetry:} \quad \mathbf{x} = \mathbf{y} \quad \text{if } \mathbf{x} \geq_L \mathbf{y} \text{ and } \mathbf{y} \geq_L \mathbf{x}; \tag{5.6.3}$$

$$\textit{transitivity:} \quad \mathbf{x} \geq_L \mathbf{z} \quad \text{if } \mathbf{x} \geq_L \mathbf{y} \text{ and } \mathbf{y} \geq_L \mathbf{z}; \tag{5.6.4}$$

$$\textit{completeness:} \text{ for all } \mathbf{x}, \mathbf{y}, \text{ either } \quad \mathbf{x} \geq_L \mathbf{y} \quad \text{and/or} \quad \mathbf{y} \geq_L \mathbf{x}. \tag{5.6.5}$$

We write "$\mathbf{x} >_L \mathbf{y}$" and say "$\mathbf{x}$ is lexicographically greater than $\mathbf{y}$" if $\mathbf{x} \geq_L \mathbf{y}$ but $\mathbf{x} \neq \mathbf{y}$. It is easily verified that "$>_L$" in R^n possesses all the nice properties of "$>$" in $(-\infty, +\infty)$. We also define $\mathbf{x} \leq_L \mathbf{y}$ to mean $\mathbf{y} \geq_L \mathbf{x}$ and $\mathbf{x} \leq_L \mathbf{y}$ to mean $\mathbf{y} >_L \mathbf{x}$. ■

The term "lexicographic" means "in dictionary fashion," which describes adequately the analogue between (1) testing if $\mathbf{x} \geq_L \mathbf{y}$ by checking down the list of components, and (2) looking up a word by locating its successive letters.

We are now ready to define a lexicographic utility function to represent your preferences for lotteries when you violate Observation 5.2.3.

Definition 5.6.3. $u: C \to R^n$ is a *lexicographic utility function* representing your preferences for lotteries with outcomes in C if

$$\ell^r \begin{Bmatrix} > \\ \sim \\ < \end{Bmatrix} \ell^s$$

if and only if

$$\mathbf{U}(\ell^r) \begin{Bmatrix} >_L \\ = \\ <_L \end{Bmatrix} \mathbf{U}(\ell^s),$$

where

$$\mathbf{U}(\ell^t) = (U_1(\ell^t), U_2(\ell^t), \ldots, U_n(\ell^t)) \text{ and } U_i(\ell^t) = E\{u_i(\tilde{c}) | \ell^t\}$$

for $i = 1, \ldots, n$ and $t = r, s$.

■ Since $u_i(c) = U_i(c)$ for every c, it follows that

$$c'' \begin{Bmatrix} < \\ \sim \\ > \end{Bmatrix} c'$$

if and only if

$$\mathbf{u}(c'')\begin{Bmatrix} >_L \\ = \\ <_L \end{Bmatrix}\mathbf{u}(c'),$$

provided that $\mathbf{u}(\cdot)$ is a lexicographic utility function for you on C. ■

Definition 5.6.3 says that *if* you have a lexicographic utility function on C, then you can ascertain the relative desirabilities of lotteries ℓ^r and ℓ^s by the following procedure. *First*, regard each "component function" $u_i: C \to (-\infty, +\infty)$ as a real-valued utility function, of the ordinary sort, and compute $U_i(\ell^t) = E\{u_i(\tilde{c})|\ell^t\}$ for $t = r, s$. Do this for each i. *Second*, then form the n-tuples $\mathbf{U}(\ell^t) = (U_1(\ell^t), \ldots, U_n(\ell^t))$ for $t = r, s$. *Third* and last, see whether $\mathbf{U}(\ell^r) >_L \mathbf{U}(\ell^s)$ [implying that $\ell^r > \ell^s$], $\mathbf{U}(\ell^r) = \mathbf{U}(\ell^s)$ [implying that $\ell^r \sim \ell^s$], or $\mathbf{U}(\ell^r) <_L \mathbf{U}(\ell^s)$ [implying that $\ell^r < \ell^s$].

Example 5.6.5

Suppose that c', c, and c'' are as in Example 5.6.1 and that $C = \{c', c, c''\}$. Also suppose that $\mathbf{u}: C \to R^2$ is defined by $\mathbf{u}(c') = (0, 0)$, $\mathbf{u}(c) = (1, q)$ with $0 < q < 1$, and $\mathbf{u}(c'') = (1, 1)$. Since $(1, 1) >_L (1, q) >_L (0, 0)$, Definition 5.6.3 implies (gratifyingly) that $c'' > c > c'$. Moreover, $\mathbf{U}(\ell^*(p; c'', c')) = (p, p)$. (Why?) Since $(1, q) >_L (p, p)$ for all $p < 1$, we see that $c > \ell^*(p; c'', c')$ for all $p < 1$, in line with our comments in Example 5.6.1. Roughly speaking, $u_1(\cdot)$ is your utility of the life-or-death attribute, and $u_2(\cdot)$ is your utility of the monetary-return attribute.

Since lexicographic utility enables choices among lotteries to be made on the basis of the lexicographic relation of their (n-tuple) utilities $\mathbf{U}(\ell)$, decision analysis proceeds just as before, except that instead of carrying single numbers back through the decision tree we must carry n-tuples of numbers back. Figure 5.7 furnishes a graphic example.

So much for the application of lexicographic utility (1) once you have determined your need for it (by refusing to accept Observation 5.2.3 as valid for all triples of consequences such that $c'' > c > c'$), and (2) once you have assessed your lexicographic utility function.

We have commented on determining your need for a utility function $\mathbf{u}: C \to R^n$ with $n > 1$; the need is present if C can be described as a subset of a Cartesian product $C_1 \times \cdots \times C_n$ of attribute sets with every attribute i overridingly more important than the attributes $j > i$. It can be shown that if none of these attributes can be subdivided into two attributes with one overridingly more important than the other,[r] then your utility function $\mathbf{u}$ should have exactly the same number n of components as there are attributes; and furthermore, each component $u_i(\cdot)$ of the utility should depend only upon the first i components of $(c_1, \ldots, c_n)$. That is, $\mathbf{u}(\cdot)$ can be

[r] This condition is almost always satisfied automatically by your initial definition of the attributes.

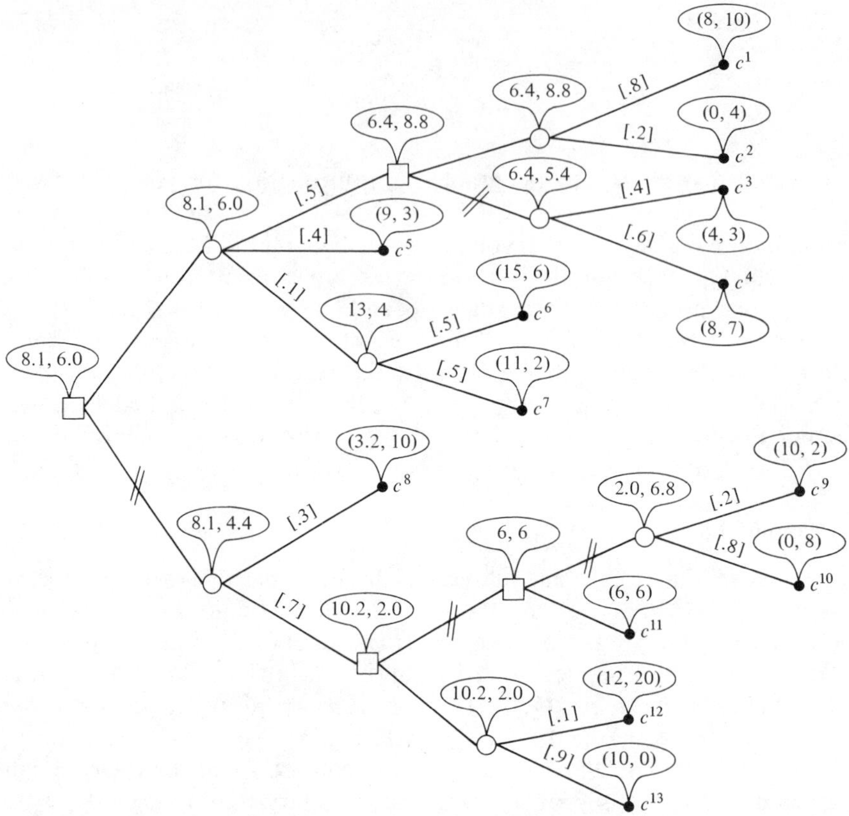

Figure 5.7 A dpuu analyzed with lexicographic utility.

written in the form

$$\mathbf{u}(c_1, \ldots, c_n) = (u_1(c_1), u_2(c_1, c_2), u_3(c_1, c_2, c_3), \ldots, u_n(c_1, \ldots, c_n)) \tag{5.6.6}$$

for every $(c_1, \ldots, c_n)$ in $C_1 \times \cdots \times C_n$.

*■ This assertion is not hard to prove. Note that if $u_1(\cdot)$ depended upon c_2, then a pair of lotteries ℓ_1^1 and ℓ_1^2 on C_1 could be constructed, and a $(c_2^\dagger, \ldots, c_n^\dagger)$, a c_2', and a c_2'' could be found such that $E\{u_1(\tilde{c}_1, c_2^\dagger)|\ell_1^1\} > E\{u_1(\tilde{c}_1, c_2^\dagger)|\ell_1^2\}$ but $E\{u_1(\tilde{c}_1, c_2')|\ell_1^1\} < E\{u_1(\tilde{c}_1, c_2'')|\ell_1^2\}$, these inequalities implying that $(\ell_1^1, c_2^\dagger, c_3^\dagger, \ldots, c_n^\dagger) > (\ell_1^2, c_2^\dagger, c_3^\dagger, \ldots, c_n^\dagger)$ but $(\ell_1^1, c_2', c_3^\dagger, \ldots, c_n^\dagger) < (\ell_1^2, c_2'', c_3^\dagger, \ldots, c_n^\dagger)$, which violates the assumption that attribute 1 must be overridingly more important than attribute 2. Similarly, for $i > 1$. ■

But assessing a utility function of the form of Equation (5.6.6) is still extremely difficult. How to assess utility functions of this form, without additional simplifying assumptions, appears to be an open topic for applied research at the time of this writing.[s]

The assessment task becomes very easy if we impose an additional simplifying assumption, namely, the *attribute-independence condition* (Definition 5.5.2). It is not hard to show that if your preferences also satisfy the attribute-independence condition, then each u_i can be written as a function of c_i alone, rather than in (5.6.6) as a function of all of c_1 through c_i. That is, (5.6.6) simplifies to

$$\mathbf{u}(c_1, \ldots, c_n) = (u_1(c_1), u_2(c_2), u_3(c_3), \ldots, u_n(c_n)) \tag{5.6.7}$$

for every $(c_1, \ldots, c_n)$.

Lexicographic utility functions of the form (5.6.7) are easy to assess—easier, in fact, than separable real-valued utility functions, because the u_i's here do not have to be scaled in mutual conformity.

Method LI, for assessing a lexicographic utility function of the form (5.6.7), is closely akin to method S, for assessing a separable real-valued utility function $u(c_1, \ldots, c_n) = \Sigma_{i=1}^n u_i(c_i)$.

STEP LI-1 requires that you express C as a subset of a Cartesian product $C_1 \times \cdots \times C_n$ of attribute sets C_i, in which (1) each attribute i is overridingly more important than its successor attributes $j > i$; and (2) no attribute i can be subdivided into two attributes i_1 and i_2 with $C_i \subset C_{i_1} \times C_{i_2}$ and i_1 overridingly more important than i_2. Once having performed this task, your utility function can be expressed in the form (5.6.7) *provided that* your preferences satisfy the attribute-independence condition.

STEP LI-2 requires that you assess each component utility function $u_i\colon C_i \to (-\infty, +\infty)$. First, pick a reference consequence $(c_1^*, \ldots, c_n^*)$, and then pick two attribute-i descriptions c_i^+ and c_i^- such that

$$(c_1^*, \ldots, c_{i-1}^*, c_i^+, c_{i+1}^*, \ldots, c_n^*) > (c_1^*, \ldots, c_{i-1}^*, c_i^-, c_{i+1}^*, \ldots, c_n^*).$$

Set $u_i(c_i^+) = 1$ and $u_i(c_i^-) = 0$. Now let $\{c_i\}^*$ be defined by

$$\{c_i\}^* = (c_1^*, \ldots, c_{i-1}^*, c_i, c_{i+1}^*, \ldots, c_n^*)$$

for every c_i in C_i. You may now use any method in Sections 5.3 and 5.4 to assess your utility function $u_i(\cdot)$ on the rest of C_i, so that it satisfies

$$u_i(c_i) = p u_i(c_i'') + (1-p) u_i(c_i')$$

if and only if

$$\{c_i\}^* \sim \ell^*(p; \{c_i''\}^*, \{c_i'\}^*).$$

Do this for every i so as to obtain all n component functions of $\mathbf{u}$.

Your assessment task is completed with step LI-2. No rescaling is necessary in view of the fact that if $\mathbf{u}\colon C \to R^n$ is a lexicographic utility

[s] July 1970.

function representing your preferences for lotteries with outcomes in C, then so is $\mathbf{u}^*: C \to R^n$ if

$$u_i^*(c) = a_i^* u_i(c) + b_i^*$$

for every c and i, where $b_1^*, \ldots, b_n^*$ are real numbers and $a_1^*, \ldots, a_n^*$ are *positive* real numbers. [This is true without any special assumptions about C such as those which imply (5.6.6) and (5.6.7).]

■ *Proof.* In order that $\mathbf{u}^*$ satisfy Definition 5.6.3 for all lotteries ℓ, it is obviously sufficient that $U_i^*(\ell^r) \geq U_i^*(\ell^s)$ if and only if $U_i(\ell^r) \geq U_i(\ell^s)$ for every i; that is, $\mathbf{u}^*$ must lead to exactly the same componentwise relationships as does $\mathbf{u}$. But "$U_i^*(\ell^r) \geq U_i^*(\ell^s)$ if and only if $U_i(\ell^r) \geq U_i(\ell^s)$" obtains precisely when $u_i^*(c) = a_i^* u_i(c) + b_i^*$ for $a_i^* > 0$, by Observation 5.2.4. ■

Now that we have described a decent method for assessing a lexicographic utility function of the form (5.6.7), we hasten to caution you regarding the general implausibility of the attribute-independence condition in this context. In Example 5.6.1, for instance, most people's preferences for monetary returns would depend rather *substantially* upon whether they would survive the dpuu. But such dependence of $u_2(\cdot)$ upon c_1 is ruled out by the attribute-independence condition. Hence you should be very careful about assuming that your lexicographic utility function takes the simple form of (5.6.7).

On the other hand, we are convinced that in any practical problem the consequences can be formulated in such a way that $\mathbf{u}(\cdot)$ can be written in the form of (5.6.6). To date, nothing appears to have been done to help one assess u_2 and higher numbered component functions of $\mathbf{u}(\cdot)$, but we may take comfort in the thought that step LI-2 can be used to assess $u_1: C_1 \to (-\infty, +\infty)$, and that the analysis of a dpuu using $u_1(\cdot)$ alone *may* lead to a unique optimal course of action without having to invoke $u_2(\cdot)$, etc.

■ In the dpuu depicted in Figure 5.7, u_2 is needed twice for breaking ties in the choice of an optimal act. But the numbers in that figure were contrived precisely for that purpose. In other, noncontrived dpuu's one can expect much less frequent use of utility components other than the first. ■

The foregoing is only a brief introduction to the subject of lexicographic utility. Readers interested in pursuing this subject further are referred first to Fishburn's treatment **[39]** of lexicographic utility on Cartesian products, of the form (5.6.7), and then to Hausner's **[55]** and Thrall's **[133]** original papers. (Example 5.6.1 is based on an example in Thrall's paper.) See also Fishburn's more recent survey paper **[37]**.

We now part company with the subject of lexicographic utility, with two closing remarks First, much of the work on this subject has had a primarily theoretical orientation rather than an applied one, and hence there is much room for attention to the practicalities of assessing a $\mathbf{u}(\cdot)$ of the form (5.6.6). The reason for this dearth of assessment methodology is that lexicographic utility is rarely if ever needed in the main areas (to date) of applications of decision analysis; namely, business and engineering dpuu's. Therefore, decision analysts with applied interests have assigned the assessment of (5.6.6) a low priority (lexicographically?). Budding applications in philosophy and in military and social planning, however—where ethical, survival, and social imperatives are adduced—may stimulate the development of assessment methodology for lexicographic utility.

Second, the fact that lexicographic utility is ignored in business and engineering contexts gives us the opportunity to caution you once again to *try* to convince yourself that 10^{-9999} is virtually zero!

Exercise 5.6.1

Prove the following equations,

(A) (5.6.2),
(B) (5.6.3),
(C) (5.6.4),
(D) (5.6.5).

Exercise 5.6.2

Re-solve the dpuu in Figure 5.7, replacing the $\mathbf{u}(c)$'s given there with those in the following table.

c	$\mathbf{u}(c)$	c	$\mathbf{u}(c)$	c	$\mathbf{u}(c)$
c^1	$(0, 0)$	c^6	$(7, 0)$	c^{10}	$(-3, 6)$
c^2	$(-2, 6)$	c^7	$(-8, 2)$	c^{11}	$(4, 6)$
c^3	$(8, -1)$	c^8	$(0, 10)$	c^{12}	$(0, 12)$
c^4	$(3, 3)$	c^9	$(-10, 5)$	c^{13}	$(10, 10)$
c^5	$(6, -4)$				

Exercise 5.6.3

Reanalyze Example 1.5.1c, now assuming that the prestige attribute is overridingly more important than the net-monetary-return attribute. Assume also that $u_1(\text{MP}) = 1$, $u_1(\text{LP}) = 0$, and $u_2(\$c_2) = c_2$ for every c_2 in question. Also assume the attribute-independence condition. [*Hint*: First guess $\mathscr{D}$'s optimal strategy.]

Exercise 5.6.4

In "Realpolitik" (Example 1.5.9), let $C_1 = \{c_1^1, c_1^2\}$, $C_2 = \{c_2^1, c_2^2\}$, and $C_3 = \{c_3^1, c_3^2\}$, with c_1^1 = "rightist government," c_1^2 = "leftist government," c_2^1 = "friendly government," c_2^2 = "unfriendly government,"

c_3^1 = "\$10 million spent," and c_3^2 = "nothing spent." Suppose that i is overridingly more important than all $j > i$ for $i = 1, 2$; also suppose that the attribute-independence condition is satisfied. Hence $n = 3$ and $\mathbf{u}$: $C_1 \times C_2 \times C_3 \to R^3$ satisfies Equation (5.6.7). Since each of the factors has only two elements, a_i^* and b_i^* can be so chosen that $u_i(c_i) \in \{0, 1\}$ for each $c_i \in C_i$ and $i \in \{1, 2, 3\}$. Suppose $\mathcal{D}$'s preferences are such that $u_1(c_1^1) = 1$, $u_1(c_1^2) = 0$, $u_2(c_2^1) = 1$, $u_2(c_2^2) = 0$, $u_3(c_3^1) = 0$, and $u_3(c_3^2) = 1$. Analyze "Realpolitik" under these assumptions. By the way, what do you think of the appropriateness of the assumptions about our diplomat's preferences?

Exercise 5.6.5

In the context of Example 5.6.5, with $\mathbf{u}$ as defined therein, suppose that ℓ^r is a lottery resulting in c'' with probability p_1^r and in c with probability p_2^r (and hence in c' with probability $1 - p_1^r - p_2^r$) for $r = 1, 2$.

(A) Show that $\ell^1 \succsim \ell^2$ if and only if $(p_1^1 + p_2^1,\ p_1^1 + qp_2^1) \geq_L (p_1^2 + p_2^2, p_1^2 + qp_2^2)$.

(B) Suppose $p_1^2 = .3$, $p_2^2 = .4$, and $q = .6$. Show that $\ell^1 \succsim \ell^2$ if and only if (p_1^1, p_2^1) falls in the shaded region of Figure 5.8.

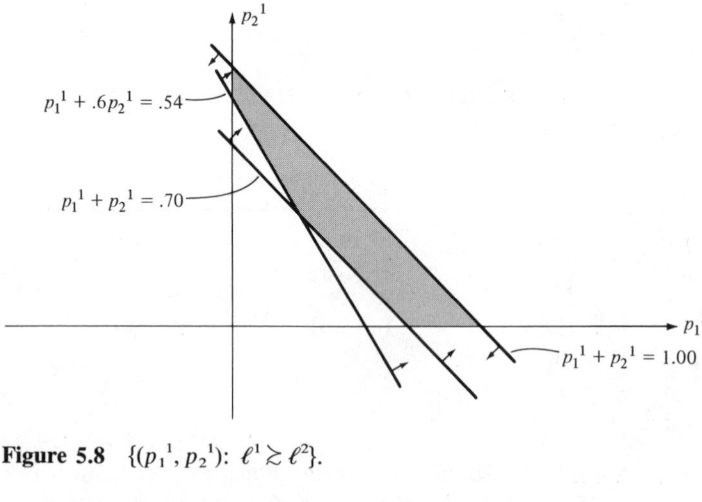

Figure 5.8 $\{(p_1^1, p_2^1)$: $\ell^1 \succsim \ell^2\}$.

5.7 UTILITY ON A CARTESIAN PRODUCT (III): UTILITY ON R^n AND MULTIVARIATE CONSTANT RISK AVERSION

In many dpuu's each possible consequence is expressible as an n-tuple of real numbers, and hence the set C of all possible consequences is a subset of the set R^n of *all* n-tuples of real numbers. Given the usual continuous idealizations of reality, C contains infinitely many elements, and hence assessing a real-valued utility function on C is a hopeless task unless we can simplify it by using qualitative properties similar to those introduced in Section 5.4.

Example 5.7.1

Suppose that the consequences are $(c_1, \ldots, c_n)$, where c_i represents a cash inflow or outflow on date i. If all conceivable sums could flow in or out on date i, then $C_i = (-\infty, +\infty)$ and $C_1 \times \cdots \times C_n = R^n$.

Example 5.7.2

Suppose that c_i denotes the net monetary return of individual i under consideration, and that the consequences of your dpuu are of the form $(c_1, \ldots, c_n)$; that is, your acts and the environment determine the monetary returns of n people. Again, $C = C_1 \times \cdots \times C_n = R^n$ if each C_i consists of all conceivable returns to i.

Example 5.7.3

Economic theory is concerned with situations in which c_i denotes the amount of *commodity* i one purchases for consumption. If there are n commodities, then the set of all conceivable "shopping lists" $(c_1, \ldots, c_n)$ is R^n and is called the "commodity space," under the assumption that all commodities are infinitely divisible.

In what follows, we shall speak of u as being defined on all of R^n, even though only a proper subset of R^n may be of interest. Doing so will simplify various statements.

One popular qualitative property of $u: R^n \to R^1$ is continuity. By "continuity" we mean that for every $e > 0$ and every $\mathbf{c} = (c_1, \ldots, c_n)$, there is a $d(e, \mathbf{c}) > 0$ such that $u(\mathbf{c}) - e < u(\mathbf{c}') < u(\mathbf{c}) + e$ whenever $c_i - d(e, \mathbf{c}) < c_i' < c_i + d(e, \mathbf{c})$ for every $i \in \{1, \ldots, n\}$, where $\mathbf{c}' = (c_1', \ldots, c_n')$.

Another popular qualitative property of $u: R^n \to R^1$ is strict increasingness. By "strict increasingness" we mean that $u(\mathbf{c}'') > u(\mathbf{c}')$ whenever $c_i'' \geq c_i'$ for every i and $c_i'' > c_i'$ for some i.

Example 5.7.4

If c_i denotes your net cash inflow (possibly negative) on date i, then $u: R^n \to R^1$ should be strictly increasing and continuous for reasons closely akin to those adduced in Section 5.4 for return on a single date. But if c_i denotes net return of individual i, then u should probably still be continuous, although it should be increasing in c_i *only* if you like individual i! If c_i denotes your consumption of commodity i, then u should not be increasing in c_i if i denotes garbage or some other undesirable.

A third popular assumption is risk aversion, or concavity. It is natural to say that you are *risk averse* if you always find "$\mathbf{E}\{\tilde{\mathbf{c}}|\ell'\}$ for sure" at least as desirable as the lottery ℓ' with outcomes in R^n, where

$$\mathbf{E}\{\tilde{\mathbf{c}}|\ell'\} = (E\{\tilde{c}_1|\ell'\}, E\{\tilde{c}_2|\ell'\}, \ldots, E\{\tilde{c}_n|\ell'\}).$$

(That is, the expectation of an uncertain vector of numbers is the corresponding vector of the components' expectations.) You are risk averse for all such lotteries with outcomes in R^n if and only if your utility function u on R^n is a *concave* function.

■ Your utility function does not have to be strictly increasing to be concave. For example, the quadratic function $u(c_1) = -(10 - c_1)^2$ is concave on *all* of $R^1 = (-\infty, +\infty)$, but it is decreasing when $c_1 > 10$. If u is concave, however, and if u is defined on *all* of R^n, then u is necessarily continuous. ■

A fourth and recently introduced qualitative property is *constant risk aversion.* To define it, we must first note that any lottery ℓ with outcomes in R^n typically has an infinite number of certainty equivalents $\mathbf{Q}\{\ell\}$, each satisfying

$$u(\mathbf{Q}\{\ell\}) = E\{u(\tilde{\mathbf{c}})|\ell\} \tag{5.7.1}$$

by definition $\mathbf{Q}\{\ell\} \sim \ell$ of certainty equivalent. Now let $\ell + \mathbf{c}^*$ denote the lottery which is sure to result in $\mathbf{c} + \mathbf{c}^*$ if ℓ results in $\mathbf{c}$. That is, $\ell + \mathbf{c}^*$ yields what ℓ does, plus a bonus of $\mathbf{c}^*$.

Definition 5.7.1. Your utility function $u: R^n \to R^1$ is said to possess *constant risk aversion* if $\mathbf{Q}\{\ell\} + \mathbf{c}^*$ is a certainty equivalent of $\ell + \mathbf{c}^*$ whenever $\mathbf{Q}\{\ell\}$ is a certainty equivalent of ℓ.

■ Definition 5.7.1 is a straightforward generalization of the definition of constant risk aversion in Example 5.4.2a. ■

Intuitively, the property of constant risk aversion implies that if the outcome $\tilde{\mathbf{c}}$ of a lottery ℓ is to be incremented by some constant $\mathbf{c}^*$, then incrementing any certainty equivalent $\mathbf{Q}\{\ell\}$ of ℓ by the same constant $\mathbf{c}^*$ should result in a certainty equivalent of the incremented lottery.

Constant risk aversion is a very powerful property, as it specifies the functional form of your utility function quite precisely, as shown by Theorem 5.7.1.

Theorem 5.7.1. If your utility function $u: R^n \to R^1$ is continuous and possesses constant risk aversion, then there is a (column) vector $\mathbf{z}$ in R^n such that $\mathbf{u}(\cdot)$ is expressible in one of the three forms

(A) $u(\mathbf{c}) = -e^{-\mathbf{z}'\mathbf{c}}$ for every $\mathbf{c}$ in R^n, and u is concave; (5.7.2)

(B) $u(\mathbf{c}) = \mathbf{z}'\mathbf{c}$ for every $\mathbf{c}$ in R^n, and u is linear (hence also concave); (5.7.3)

and

(C) $u(\mathbf{c}) = e^{\mathbf{z}'\mathbf{c}}$ for every $\mathbf{c}$ in R^n, and u is convex, (5.7.4)

where $\mathbf{z}'\mathbf{c} = \sum_{i=1}^{n} z_i c_i$.

■ Before proceeding to the proof itself, we must define indifference sets. Any subset of R^n of the form

$$N_u(t) = \{\mathbf{c}: u(\mathbf{c}) = t\}$$

is called an *indifference set* of u. Much of the following proof depends upon properties of the indifference sets of u. ■

*■ *Proof.* The first proof of a slightly restricted version of this theorem, in LaValle [**80**], used a generalization of Pfanzagl's proof [**109**] for $n = 1$. The following proof, due to Keeney [**73**], is easier to comprehend. First, note that if u is constant, then there is only one nonempty indifference set; namely, R^n itself. Put $\mathbf{z} = \mathbf{0}$ in any of Equations (5.7.2) through (5.7.4) to obtain such a utility function and hence to see that the theorem is true for this case. If u is not constant, then the proof is (unfortunately) longer. We shall show (as step 1) that indifference sets are hyperplanes, then (as step 2) that these hyperplanes are parallel, and finally (as step 3) that u is given by one of the forms (5.7.2)–(5.7.4). As a preliminary, however, you may show as *Exercise 5.7.1 that u has constant risk aversion if and only if

$$E\{u(\mathbf{t}+\tilde{\mathbf{c}})\}\begin{Bmatrix}\geq\\=\\\leq\end{Bmatrix}u(\mathbf{t}+\mathbf{Q}) \quad \text{implies} \quad E\{u(\mathbf{t}'+\tilde{\mathbf{c}})\}\begin{Bmatrix}\geq\\=\\\leq\end{Bmatrix}u(\mathbf{t}'+\mathbf{Q}) \tag{5.7.5}$$

for every $\{\mathbf{t}, \mathbf{t}', \mathbf{Q}\} \subset R^n$ and lottery $\tilde{\mathbf{c}}$ with outcomes in R^n.

STEP 1. We show that each indifference set is a hyperplane by showing that it contains the entire line $\{r\mathbf{c}_1 + [1-r]\mathbf{c}_2: -\infty < r < \infty\}$ through any two of its points $\mathbf{c}_1$ and $\mathbf{c}_2$. (Since u is real-valued, continuous, and nonconstant, this implies that the indifference set is a hyperplane rather than some linear set of topological dimension less than $n-1$.) Assume that $\mathbf{c}_1 \sim \mathbf{c}_2$ but that $\mathbf{c}_1 \neq \mathbf{c}_2$. Taking the lottery $\tilde{\mathbf{c}}$ in (5.7.5) as the sure lottery equaling $\mathbf{0}$ in R^n for certain, and taking $\mathbf{t} = \mathbf{c}_1$, $\mathbf{Q} = [1-r](\mathbf{c}_2 - \mathbf{c}_1)$, and $\mathbf{t}' = \mathbf{c}_1 - r(\mathbf{c}_1 - \mathbf{c}_2)$, we see that $\mathbf{t} + \tilde{\mathbf{c}} = \mathbf{c}_1$ for certain, $\mathbf{t} + \mathbf{Q} = r\mathbf{c}_1 + [1-r]\mathbf{c}_2$, $\mathbf{t}' + \tilde{\mathbf{c}} = [1-r]\mathbf{c}_1 + r\mathbf{c}_2$ for certain, and $\mathbf{t}' + \mathbf{Q} = \mathbf{c}_2$. Hence from (5.7.5) we obtain

(a) $u(\mathbf{c}_1) \geq u(r\mathbf{c}_1 + [1-r]\mathbf{c}_2)$ implies $u([1-r]\mathbf{c}_1 + r\mathbf{c}_2) \geq u(\mathbf{c}_2)$

for every real number r. But $u(\mathbf{c}_1) = u(\mathbf{c}_2)$ because $\mathbf{c}_1 \sim \mathbf{c}_2$ by assumption. Hence (a) implies that

(b) $u([1-r]\mathbf{c}_1 + r\mathbf{c}_2) \geq u(\mathbf{c}_2) = u(\mathbf{c}_1) \geq u(r\mathbf{c}_1 + [1-r]\mathbf{c}_2)$

for every r. But reversing the roles of r and $1-r$ in the preceding yields the conclusion that

(c) $$u(r\mathbf{c}_1+[1-r]\mathbf{c}_2)\geq u(\mathbf{c}_2)=u(\mathbf{c}_1)\geq u([1-r]\mathbf{c}_1+r\mathbf{c}_2)$$

for every r. Now, (b) and (c) together imply that

(d) $$u(r\mathbf{c}_1+[1-r]\mathbf{c}_2)=u(\mathbf{c}_1)=u(\mathbf{c}_2)=u([1-r]\mathbf{c}_1+r\mathbf{c}_2)$$

for every r, and hence that $r\mathbf{c}_1+[1-r]\mathbf{c}_2$ belongs to the same indifference set as do $\mathbf{c}_1$ and $\mathbf{c}_2$. Thus indifference sets are hyperplanes.

STEP 2. The indifference hyperplanes are parallel, since nonparallel hyperplanes intersect and different indifference sets cannot intersect. (See Exercise 5.7.2.) Hence each indifference set is the set of solutions $\mathbf{c}$ to the equation $\mathbf{y}'\mathbf{c}=k$ for some k, where $\mathbf{y}$ is an n-tuple of coefficients which do not depend upon the indifference set in question (because of parallelism).

STEP 3. Choose any i such that $y_i\neq 0$—say, $i=1$ for notational convenience. Then note that the function $\mathbf{t}$: $R^n\to R^n$ defined by

$$\mathbf{t}(\mathbf{c})=([1/y_1]\mathbf{y}'\mathbf{c},0,0,\ldots,0)$$

for every $\mathbf{c}$ in R^n is a substitute function in the sense of Definition 5.2.2, with $\mathbf{y}'\mathbf{t}(\mathbf{c})=\mathbf{y}'\mathbf{c}$ for every $\mathbf{c}$. [In effect, the substitute function maps n-tuple consequences $\mathbf{c}$ into equally desirable real-valued (first-component) consequences $[1/y_1]\mathbf{y}'\mathbf{c}$.] If u is to have constant risk aversion on R^n, it must have constant risk aversion on the one-dimensional subspace consisting of all n-tuples of the form $(c_1,0,0,\ldots,0)$. We may now ignore the now-irrelevant second through nth components and apply the theorem of Pfanzagl [**109**] and Pratt [**111**] for the case $n=1$ to see that u must be expressible as $u(\mathbf{t}(\mathbf{c}))=-e^{-z^*x}$, $u(\mathbf{t}(\mathbf{c}))=z^*x$, or $u(\mathbf{t}(\mathbf{c}))=e^{z^*x}$, where $x=\mathbf{y}'\mathbf{t}(\mathbf{c})$ and z^* is a real number. But $u(\mathbf{t}(\mathbf{c}))=u(\mathbf{c})$. This concludes the main part of the proof, once we define $\mathbf{z}=z^*\mathbf{y}$, so that $z^*x=\mathbf{z}'\mathbf{c}$. Proofs of the assertions about concavity, linearity, and convexity are immediate from an easy theorem in analysis to the effect that if $g: R^n\to R^1$ is concave and $f: R^1\to R^1$ is both concave and nondecreasing, then the composite function $f[g(\cdot)]$ is concave. Similarly for "convex" replacing "concave" in all three places. Here, $g(\mathbf{c})=\mathbf{z}'\mathbf{c}$ is both concave and convex for every $\mathbf{z}$ in R^n, whereas $f(x)=-e^{-x}$ is concave and nondecreasing and $f(x)=e^x$ is convex and nondecreasing. ∎

The preceding proof may have been complicated, but applying Theorem 5.7.1 is not. In fact, assessing a utility function of one of the three forms in Theorem 5.7.1 is not much harder than assessing a utility function for monetary returns. Method M exploits the fact that all indifference sets are of the form $\mathbf{z}'\mathbf{c}=k$.

Step M-1 requires that you determine $\mathbf{z}$ up to some positive multiplicative factor. Let $\mathbf{z}=z^*\mathbf{y}$ for some z^* to be determined later. Then $\mathbf{z}'\mathbf{c}=\mathbf{z}'\mathbf{c}^\dagger$ if and only if $\mathbf{y}'\mathbf{c}=\mathbf{y}'\mathbf{c}^\dagger$. (Here, a prime denotes transpose, and all untransposed vectors are *column* vectors.) Choose an i such that u is not constant[t] in c_i. If possible, choose i such that u is *increasing* in c_i. For every $j \neq i$, determine that (possibly negative) amount dc_j which would have to be added to c_j in order to precisely compensate you for a reduction of *one* unit of c_i. Then set y_j equal to y_i/dc_j, where

$$y_i = \begin{cases} +1 & \text{if } u \text{ is increasing in } c_i \\ -1 & \text{if } u \text{ is decreasing in } c_i. \end{cases}$$

Once you have done this for every j other than your chosen i, you will have determined a vector $\mathbf{y}$ with $y_i \in \{+1, -1\}$ such that $u(\mathbf{c}) = u(\mathbf{c}^\dagger)$ if and only if $\mathbf{y}'\mathbf{c} = \mathbf{y}'\mathbf{c}^\dagger$. Naturally, any j such that u is constant in c_j has $y_j = 0$.

■ To prove that $y_j = y_i/dc_j$, it suffices to note that you are being asked to find dc_j such that $y_jc_j + y_ic_i = y_j(c_j + dc_j) + y_i(c_i - 1)$. Solving for y_j yields the asserted equality. ■

Note that no lottery comparisons are required in step M-1.

Step M-2 requires that you determine $\mathbf{z} = z^*\mathbf{y}$ and also the form of your utility function. If $\mathbf{E}\{\tilde{\mathbf{c}}|\ell\} = (E\{\tilde{c}_1|\ell\}, \ldots, E\{\tilde{c}_n|\ell\})$ is a certainty equivalent of ℓ for every ℓ, then your utility function is given by (5.7.3) and you can set $z^* = 1$ and $\mathbf{z} = \mathbf{y}$ without further ado, as here z^* is an unessential parameter of your utility function. If you are risk averse in the sense that $\mathbf{E}(\tilde{\mathbf{c}}|\ell) \gtrsim \ell$ for all lotteries ℓ, then a certainty equivalent of *one* lottery ℓ suffices to determine z^*. For convenience, take $(0, \ldots, 0, c_i^r, 0, \ldots, 0) = \mathbf{c}^r$ for $r \in \{0, 1\}$ such that $\mathbf{c}^1 > \mathbf{c}^0$. Then determine $c_i^{1/2}$ such that

$$(0, \ldots, 0, c_i^{1/2}, 0, \ldots, 0) \sim \ell^*(1/2; \mathbf{c}^1, \mathbf{c}^0).$$

Now z_i may be found by solving the nonlinear equation

$$e^{-z_ic_i^{1/2}} = .5e^{-z_ic_i^0} + .5e^{-z_ic_i^1},$$

just as in Example 5.4.2b. Then $z^* = z_i/y_i$, and $z_j = z^*y_j = z_iy_j/y_i$ for every $j \neq i$. You have now fitted all parameters of (5.7.2).

*■ More convenient for those who know about Normal distributions is to find Q_i such that $(0, \ldots, 0, Q_i, 0, \ldots, 0) \sim \ell$ for ℓ the lottery yielding outcomes $(0, \ldots, 0, \tilde{c}_i, 0, \ldots, 0)$ with $\tilde{c}_i$ having the Normal density function $f_N(\cdot|m, V)$, for in this case $z_i = 2[m - Q_i]/V$. See Equation (5.4.10). ■

Similar comparisons suffice to determine z_i, and thus z^* and all z_j's for $j \neq i$, when u is *convex*, in that $\ell \gtrsim \mathbf{E}\{\tilde{\mathbf{c}}|\ell\}$ for all lotteries ℓ, and thus when your utility function is of the form (5.7.4).

Finally, note that the *only* constant-risk-aversion utility function on R^n which is *separable* is (5.7.3). Popular assumptions in the context of

[t] If u were constant in some c_i, then most likely C_i would have been omitted from $C_1 \times \cdots \times C_n$.

Example 5.7.1 are utility functions of the form

$$u(c_1, \ldots, c_n) = \sum_{i=1}^{n} d^i u_i(c_i),$$

where d is a single-period discount factor and $u_i(c_i) = -e^{-z_i c_i}$ for every i, with all $z_i > 0$. Such utility functions do not have constant risk aversion on R^n.

***Exercise 5.7.1**

Show that u has constant risk aversion if and only if Equation (5.7.5) obtains.

Exercise 5.7.2

Show that $N_u(t_1) \cap N_u(t_2) = \emptyset$ if $t_1 \neq t_2$; that is, different indifference sets cannot intersect.

***Exercise 5.7.3**

Suppose that u is given by Equation (5.7.2) and that ℓ is a lottery with outcomes in R^n such that $\tilde{\mathbf{c}}$ has the Normal distribution with n-variate Normal density function $f_N^{(n)}(\cdot|\mathbf{m}, \mathbf{V})$, defined in *Section 3.11.

(A) Show that

$$E\{u(\tilde{\mathbf{c}})|\ell\} = -e^{-\mathbf{z}'\mathbf{m}+(1/2)\mathbf{z}'\mathbf{V}\mathbf{z}}.$$

(B) Show that $\mathbf{Q}\{\ell\}$ is any n-tuple such that

$$\mathbf{z}'\mathbf{Q}\{\ell\} = \mathbf{z}'[\mathbf{m} - (1/2)\mathbf{V}\mathbf{z}].$$

[In particular, $\mathbf{m} - (1/2)\mathbf{V}\mathbf{z}$ is a certainty equivalent of ℓ.]

■ Recently, Rothblum [**123**] has independently derived Theorem 5.7.1, together with a number of useful generalizations and variants. See also Keeney [**70**] and Pollak [**110**]. ■

6

QUANTIFICATION OF JUDGMENTS

Accident, hazard, chance, call it what you may, a mystery to ordinary minds, becomes a reality to superior men.

Napoleon I

6.1 INTRODUCTION

This chapter concerns convenient methods for quantifying judgments in the form of probabilities. All such methods described here involve combining $\mathscr{D}$'s probabilities of *some* events with assumed relationships among the probabilities of *all* events in order to determine all probabilities of interest.

The methods to be discussed may be classified as more or less *direct* and as more or less *indirect*. Suppose that $\tilde{\omega}$ is the random variable of interest to $\mathscr{D}$; that is, $\mathscr{D}$'s task is to quantify his judgments about events of the form $\tilde{\omega} \in \Omega'$. Then for our purposes a more or less direct method is one in which $\mathscr{D}$ (1) assesses directly his probabilities $P(\Omega')$ of *some* $\tilde{\omega}$-events Ω'; and (2) either invokes the laws of probability or applies some simple technique like graph smoothing in order to determine his probabilities of *all* $\tilde{\omega}$-events. Some such methods are described in Sections 6.3 and 6.4.

More or less indirect methods encompass all others, such as those discussed below.

(1) $\mathscr{D}$'s assessing his conditional probabilities of $\tilde{\omega}$-events, given $\tilde{\xi}$-events and his unconditional probabilities of $\tilde{\xi}$-events, where $\tilde{\xi}$ is some random variable of only indirect interest to $\mathscr{D}$, and then calculating the *unconditional* probabilities of the $\tilde{\omega}$-events via the easily verified fact that $P(\tilde{\omega} \in \Omega') = E\{P(\tilde{\omega} \in \Omega' | \tilde{\xi})\}$.

(2) If $\tilde{\omega}$ is functionally determined by $\tilde{\xi}$, in the sense that $P(\tilde{\omega} = y(\xi) | \tilde{\xi} = \xi) = 1$ for every $\xi \in \Xi$ and some function $y: \Xi \to \Omega$, $\mathscr{D}$'s assessing directly his probabilities of $\tilde{\xi}$-events and then calculating

the logically implied probabilities of $\tilde{\omega}$-events via the easily verified fact that $P(\tilde{\omega} \in \Omega') = P(\tilde{\xi} \in \{\xi: y(\xi) \in \Omega'\})$ for every $\tilde{\omega}$-event Ω'.

(3) $\mathcal{D}$'s acquiring direct experience with $\tilde{\omega}$, either through historical data on numerous instances under indistinguishable conditions or through repeated experimentation, the separate trials again being performed under indistinguishable conditions.

(4) $\mathcal{D}$'s delegation of the probability assessment task to someone else, this delegation being valid because, after all, a probability assessment is a decision, and the practical necessity of delegating some decisions is well established in the annals of management.

Indirect method (1) has been tacitly but extensively discussed in Chapter 3; see also Exercises *6.4.6 and *6.4.7. Indirect method (3) is introduced in Section 6.5, which constitutes a survey of the principal foundations of sampling theory, a topic revisited in Chapter 9. Section 6.6 synthesizes indirect method (2) and the results of Section 6.5 into an introduction to stochastic[a] simulation, which, despite popular views to the contrary, is simply a frequently useful computational methodology rather than a mysteriously omnipotent touchstone to the eternal verities.

Readers pressed for time may proceed to subsequent chapters without reading Sections 6.5 and 6.6, although Section 6.5 will be of interest in conjunction with Chapter 9.

Important complementary considerations can be found in Schlaifer [**128**], the book and papers [**145**]–[**148**] of Winkler, and the paper (with ensuing discussions) [**61**] of Hogarth, where further references may be found. Topics considered in these references include the calibration of probability assessors and the motivation of experts to report honestly their actual probability assessments rather than some slanted or biased version thereof.

As is our custom in Parts I and II of this book, we commence with a section of preliminaries.

6.2 PRELIMINARIES

First Preliminary: Definition of Probability

Let A and B be any two prizes whatsoever that are not equally desirable to $\mathcal{D}$; and let Ω' be an event in a sure event Ω. That is, Ω' is a subset of Ω. Then $\mathcal{D}$'s probability $P(\Omega')$ of Ω', or probability $P(\tilde{\omega} \in \Omega')$ that the random variable (= as-yet undetermined elementary event) $\tilde{\omega}$ falls in Ω', is by definition that canonical probability p such that $\ell^{\wedge}(\Omega'; A, B)$ and $\ell^*(p; A, B)$ are equally desirable.

■ Recall that $\ell^*(p; A, B)$ is a canonical lottery which results in A if an

[a] "Stochastic" is a popular synonym for "probabilistic."

event Λ' of canonical probability p occurs, and results in B otherwise (that is, results in B given the complementary event $\overline{\Lambda'}$, of canonical probability $1-p$). Also recall that $\ell^{\wedge}(\Omega'; A, B)$ is the lottery which results in A if event Ω' occurs and in B otherwise (that is, results in B if the complementary event $\overline{\Omega'}$ occurs). See Figure 6.1. ■

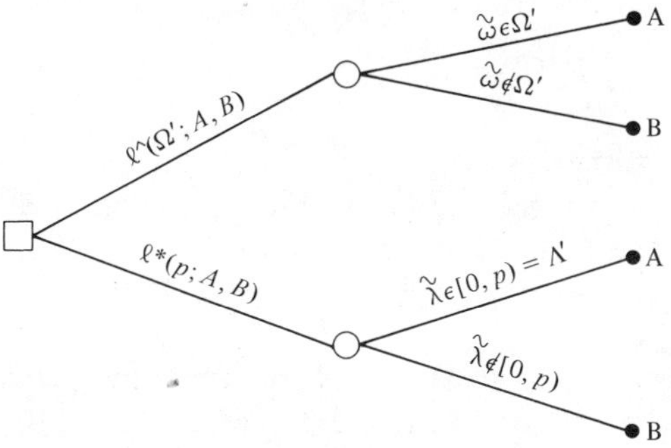

Figure 6.1 Lotteries for assessing $P(\Omega')$.

In principle, $\mathscr{D}$'s probability $P(\Omega')$ of Ω' is a *unique* nonnegative number, not exceeding one, although we do not recommend splitting hairs when one assesses probabilities of events unless extreme accuracy is essential.

Second Preliminary: Choice of Reference Prizes

We also showed in Chapter 3 that $P(\Omega')$ as defined above should not depend upon the reference prizes A and B, provided that they are not equally desirable. But as a practical matter, relatively simple and easily conceptualizable prizes should be chosen: say, $A =$ "receive \$1,000 with no strings attached" and $B =$ "receive \$0 with no strings attached." In this regard, it is often convenient to use reference prizes that do not appear as consequences of one's dpuu, because it could be quite unnatural and confusing to use consequences of the actual dpuu to assess probabilities of events in $\mathscr{E}$-moves of stages other than the last.

Third Preliminary: Relationships among Probabilities

Of the many relationships among probabilities of events discussed in Sections 3.6 and 3.7, a few are of particular importance in the judgment-quantification methods discussed in this chapter. We state them as observations.

Observation 6.2.1. If $\Omega^1, \ldots, \Omega^n$ are mutually exclusive events, then

$$P(\cup_{i=1}^{n} \Omega^i) = \sum_{i=1}^{n} P(\Omega^i).$$

Less formally, Observation 6.2.1 asserts that if no two of the events can *both* obtain, then $\mathscr{D}$'s probability that some one or another of them will obtain is just the sum of their individual probabilities. As a corollary, if these events are collectively exhaustive—some one of them is bound to obtain—then the sum of their individual probabilities is 1.

Observation 6.2.2. If $\Omega^1, \ldots, \Omega^n$ are mutually exclusive and collectively exhaustive, then

$$\sum_{i=1}^{n} P(\Omega^i) = 1.$$

Next, if one event *implies* another, then the obtaining of the former necessitates the latter's obtaining, so that the probability of the latter must be at least as great as that of the former. More formally:

Observation 6.2.3. If $\Omega' \subset \Omega''$, then $P(\Omega') \leq P(\Omega'')$.

Next, $\mathscr{D}$'s probability $P(\Omega' \cap \Omega'')$ that both Ω' and Ω'' obtain is the *product* (1) of his probability $P(\Omega'')$ that Ω'' obtains, and (2) of his conditional probability $P(\Omega'|\Omega'')$ that Ω' obtains *given* that Ω'' obtains.

Observation 6.2.4. If Ω' and Ω'' are any two events, then

$$P(\Omega' \cap \Omega'') = P(\Omega'|\Omega'') \cdot P(\Omega'').$$

■ The "product rule" as stated in Observation 6.2.4 differs from that in Theorem 3.7.1 only superficially. Both rules express the same consistency requirement. See Exercise 3.7.13, where Ξ' and Ξ^* correspond to Ω' and Ω'' respectively. ■

Fourth Preliminary: Assessment of Conditional Probabilities

A useful device for $\mathscr{D}$'s assessing directly his conditional probability $P(\Omega'|\Omega'')$ without constant reference to the "given" (or conditioning) event Ω'' is that of conditional lotteries. Let $\ell^*(p|\Omega''; A, B)$ and $\ell^\wedge(\Omega'|\Omega''; A, B)$ be defined as in Figure 6.2. If $\mathscr{D}$'s probability of Ω'' is greater than zero, then there is a unique value of p in $[0, 1]$ for which[b] $\ell^*(p|\Omega''; A, B) \sim \ell^\wedge(\Omega'|\Omega''; A, B)$, provided that A and B are not equally desirable to $\mathscr{D}$. This unique p is just $P(\Omega'|\Omega'')$.

■ *Proof.* $U(\ell^*(p|\Omega''; A, B)) = U(\ell^\wedge(\Omega'|\Omega''; A, B))$ when p is such that these lotteries are equally desirable. Suppose for notational convenience that $u(\cdot)$ has been chosen so that $u(B) = 0$. Then

(a) $$U(\ell^*(p|\Omega''; A, B)) = pP(\Omega'')u(A)$$

while

(b) $$U(\ell^\wedge(\Omega'|\Omega''; A, B)) = P(\Omega'|\Omega'')P(\Omega'')u(A).$$

Now equate the right-hand sides of (a) and (b) and cancel $P(\Omega'')u(A)$. ■

[b] Recall that "$\sim$" means "is as desirable as."

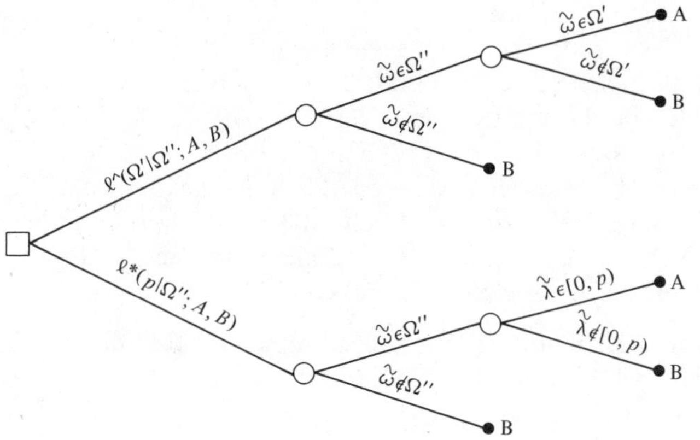

Figure 6.2 Lotteries for directly assessing $P(\Omega'|\Omega'')$.

Fifth Preliminary: Independence

Sweeping qualitative properties are of as much importance in quantifying judgments as they are in quantifying preferences. The most important such property in probability assessments is *independence*, the generally used synonym for our more suggestive "unrelatedness." Two events are said to be independent if learning about the occurrence or nonoccurrence of one of them does not lead $\mathscr{D}$ to revise his judgments about the occurrence of the other. The outcomes of successive tosses of a fair coin are generally regarded as independent events.[c] Similarly, two random variables are said to be independent if learning the value of one *cannot* induce $\mathscr{D}$ to revise his judgments about the other. We formalize slightly more general versions of the two preceding sentences as Definition 6.2.1.

Definition 6.2.1: Independence

(A) The events $\Omega^1, \ldots, \Omega^n$ are said to be (mutually) independent if

$$P(\cap_{i=1}^n \Omega^i) = \prod_{i=1}^n P(\Omega^i).$$

(B) The random variables $\tilde{\omega}_1, \ldots, \tilde{\omega}_n$ are said to be (mutually) independent if

$$P(\tilde{\omega}_1 \in \Omega_1', \ldots, \tilde{\omega}_n \in \Omega_n') = \prod_{k=1}^n P_k(\tilde{\omega}_k \in \Omega_k')$$

for *every* choice of $\Omega_1' \subset \Omega_1$, of $\Omega_2' \subset \Omega_2, \ldots,$ and of $\Omega_n' \subset \Omega_n$, where $P_k(\cdot)$ denotes $\mathscr{D}$'s probability function on $\tilde{\omega}_k$-events and $P(\cdot, \ldots, \cdot)$ denotes $\mathscr{D}$'s joint probability function.

[c] Similar arguments for chance devices in general establish what is generally called the "impossibility of gambling systems." The dice have no memory!

■ $\mathscr{D}$'s deeming random variables $\tilde{\omega}_1, \ldots, \tilde{\omega}_n$ to be independent is a *very* powerful qualitative property of his judgments. It implies that his probabilities of all *joint* events—subsets of the $\mathscr{E}$-move $\Omega_1 \times \cdots \times \Omega_n$ of all possible values of the joint random variable $(\tilde{\omega}_1, \ldots, \tilde{\omega}_n)$—are completely determined by his marginal probabilities of all $\tilde{\omega}_k$-events, for all k. Thus, if $\tilde{\omega}_1$, $\tilde{\omega}_2$, and $\tilde{\omega}_3$ are deemed independent and have 5, 20, and 100 possible values respectively, then the $5 \cdot 20 \cdot 100 = 10{,}000$ probabilities of joint elementary events are completely determined by the $5 + 20 + 100 = 125$ probabilities of marginal elementary events! ■

■ That Definition 6.2.1 manifests the preceding, behavioral definition of independence is easily verified by deducing that

$$P(\Omega^i | \cap_{j \neq i} \Omega^j) = P(\Omega^i) \tag{6.2.1}$$

for every i in Definition 6.2.1(A), and

$$P(\tilde{\omega}_k \in \Omega_k' | \tilde{\omega}_j \in \Omega_j' \quad \text{for every} \quad j \neq k) = P_k(\tilde{\omega}_k \in \Omega_k') \tag{6.2.2}$$

for every $\Omega_1', \ldots, \Omega_n'$ and every k in Definition 6.2.1(B). Now, Equation (6.2.1) says that $\mathscr{D}$'s probability of Ω^i is unchanged from $P(\Omega^i)$ by ascertaining that the other events obtain, whereas (6.2.2) says that $\mathscr{D}$'s judgments about $\tilde{\omega}_k$ cannot be swayed by *any* information he could acquire about the other random variables $\tilde{\omega}_j$. ■

We hasten to caution that many *bad* decision analyses have resulted from *blithely assuming* that random variables were independent when any reasonably intelligent person would strongly deny the behavioral implications of independence, and hence independence itself. Sometimes simplification of $\mathscr{D}$'s assessment task is bought at too high a price!

Exercise 6.2.1: Testing Independence in Your Judgments

In each of the following cases, state whether you deem the random variables therein to be independent, and explain your reasoning on this point.

(A) $\tilde{\omega}_1$ = price per share of General Motors Corp. common stock at the end of *next* year,
$\tilde{\omega}_2$ = price per share of Ford Motor Co. common stock at the end of *next* year.

(B) $\tilde{\omega}_1$ = result of your flipping a coin now,
$\tilde{\omega}_2$ = number of offshore drilling rigs in the Gulf of Mexico on December 22, year $x + 5$ if x is this year.

(C) $\tilde{\omega}_1$ = dollar sales of Texaco, Inc., this year,
$\tilde{\omega}_2$ = earnings per share of Texaco, Inc., this year.

(D) $\tilde{\omega}_1$ = whether or not U.S. combat troops are sent into action in a new theater next year,

$\tilde{\omega}_2$ = number of campus protest demonstrations next year.

(E) $\tilde{\omega}_1$ = next year-end exchange rate for marks (in \$/mark),
$\tilde{\omega}_2$ = next year-end exchange rate for yen (in \$/yen),
$\tilde{\omega}_3$ = next year-end exchange rate for krone (in \$/krone).

(F) $\tilde{\omega}_1$ = whether or not you have an auto accident tomorrow,
$\tilde{\omega}_2$ = whether or not your next-door neighbor has an auto accident tomorrow.

Exercise 6.2.2

Prove that

(A) Definition 6.2.1(A) implies Equation (6.2.1).
(B) Definition 6.2.1(B) implies Equation (6.2.2).

***Exercise 6.2.3**

Suppose that a function $y_k: \Xi_k \to \Omega_k$ is given for each of $k = 1, \ldots, n$; and write $\tilde{\omega}_k = y_k(\tilde{\xi}_k)$. Show that $\tilde{\omega}_1, \ldots,$ and $\tilde{\omega}_n$ are independent if $\tilde{\xi}_1, \ldots,$ and $\tilde{\xi}_n$ are independent.
[*Hint*: Because the value of $\tilde{\omega}_k$ is functionally determined from the value of $\tilde{\xi}_k$, it follows that

$$P(\tilde{\omega}_k \in \Omega'_k) = P(\tilde{\xi}_k \in \{\xi_k: y_k(\xi_k) \in \Omega'_k\}) \tag{6.2.3}$$

must obtain for every k.]

6.3 ASSESSING A PROBABILITY FUNCTION ON A (SMALL-) FINITE SET Ω

In this section we discuss briefly two direct methods for quantifying your judgments regarding events Ω' in an $\mathscr{E}$-move Ω which contains only a small finite[d] number m of elementary events ω.

Method f1 involves only the direct assessment of unconditional probabilities of events, and it entails repeated applications of Observations 6.2.3 and 6.2.1.

STEP f1-1 requires that you label and arrange the m elementary events in some convenient order, say ω^1 before ω^2, ω^2 before $\omega^3, \ldots,$ and ω^{m-1} before ω^m. Then you define $\Omega^0 = \emptyset$ (= the impossible event) and $\Omega^i = \{\omega^1, \ldots, \omega^i\}$ for $i = 1, \ldots, m$. (Note that $\Omega^m = \Omega$.)

STEP f1-2 requires that you assess your probability $P(\Omega^i)$ of each of the $m-1$ events $\Omega^1, \ldots, \Omega^{m-1}$ defined in step f1-1, according to the lottery comparisons described in the first preliminary in Section 6.2. [*Caution*:

[d] Both methods also apply *verbatim* to the case in which $\Delta^1, \ldots, \Delta^m$ are mutually exclusive and collectively exhaustive events in an $\mathscr{E}$-move Δ containing an *arbitrary* number of elementary events. Simply write ω^i for Δ^i and Ω^i for $\cup_{j=1}^{i} \Delta^j$.

Since $\Omega^i \subset \Omega^{i+1}$ for every i, Observation 6.2.3 requires that your probabilities must satisfy the inequalities

$$P(\Omega^0) = P(\emptyset) = 0 \leq P(\Omega^1) \leq P(\Omega^2) \leq \cdots \leq P(\Omega^{m-1}) \leq 1 = P(\Omega^m). \tag{6.3.1}$$

STEP f1-3 consists in finding your probability $P(\omega^i)$ of each elementary event ω^i by subtraction

$$P(\omega^i) = P(\Omega^i) - P(\Omega^{i-1}) \tag{6.3.2}$$

for $i = 1, 2, \ldots, m$.

■ *Proof of (6.3.2).* $\{\omega^i\}$ and $\Omega^{i-1} = \{\omega^1, \ldots, \omega^{i-1}\}$ are mutually exclusive and their union is Ω^i. Hence $P(\Omega^i) = P(\{\omega^i\}) + P(\Omega^{i-1})$ by Observation 6.2.1, and from this representation Equation (6.3.2) follows immediately once we simplify notation by writing $P(\omega^i)$ instead of $P(\{\omega^i\})$. ■

Example 6.3.1

Let ω denote the number of days next week on which the Dow Jones Industrial Average closes *above* its previous closing value. Then $\Omega = \{0, 1, 2, 3, 4, 5\} = \{\omega^1, \omega^2, \omega^3, \omega^4, \omega^5, \omega^6\}$. Suppose that you have assessed $P(\Omega^1) = .05$, $P(\Omega^2) = .20$, $P(\Omega^3) = .50$, $P(\Omega^4) = .75$, and $P(\Omega^5) = .95$; here, the labeling implies that Ω^i = "the DJIA advances on no more than $i - 1$ days next week," for $i = 1, \ldots, 5$, with $\Omega^0 = \emptyset$ and $\Omega^6 = \Omega$. Then

$$\begin{aligned}
P(0) &= P(\omega^1) = P(\Omega^1) - \quad 0 \quad = .05 - \ 0 \ = .05,\\
P(1) &= P(\omega^2) = P(\Omega^2) - P(\Omega^1) = .20 - .05 = .15,\\
P(2) &= P(\omega^3) = P(\Omega^3) - P(\Omega^2) = .50 - .20 = .30,\\
P(3) &= P(\omega^4) = P(\Omega^4) - P(\Omega^3) = .75 - .50 = .25,\\
P(4) &= P(\omega^5) = P(\Omega^5) - P(\Omega^4) = .95 - .75 = .20,
\end{aligned}$$

and

$$P(5) = P(\omega^6) = P(\Omega^6) - P(\Omega^5) = \ 1 \ - .95 = .05.$$

But, just as with the assessment of utility, most responsible decision analysts would advise you to *check* your assessed probabilities for *consistency* with your judgments about events Ω' whose probabilities you did not assess directly.

Example 6.3.2

In Example 6.3.1, you might ask yourself whether you would really put the odds at (.75/.25 =) three-to-one against the DJIA's closing higher on exactly three days next week, or at 19-to-one against five straight advances in the average.

Method f2 is a sequential approach rather like Method F2 for assessing a utility function on a finite set, in that it leads to some rather tedious calculations to obtain ultimately your probabilities $P(\omega^1), \ldots, P(\omega^m)$ of the elementary events.

STEP f2-1 requires that you assess $P(\omega^1)$ directly.

STEP f2-2 requires that, for each i in $\{1, \ldots, m-2\}$, you assess your *conditional* probability $P(\omega^{i+1}|\overline{\Omega^i}) = P(\tilde{\omega} = \omega^{i+1}|\tilde{\omega} \in \{\omega^{i+1}, \ldots, \omega^m\})$, which can be done by finding that p such that $\ell\hat{}(\omega^{i+1}|\overline{\Omega^i}; A, B) \sim \ell^*(p|\overline{\Omega^i}; A, B)$, where these lotteries are as defined in Figure 6.2 and the fourth preliminary in Section 6.2.

STEP f2-3 requires that you calculate the $2(m-2)$ quantities $P(\overline{\Omega^i})$ and $P(\omega^{i+1})$ *successively*—first for $i = 1$, then $i = 2$, then ..., and then $i = m-2$, according to the formulas

$$P(\overline{\Omega^i}) = P(\overline{\Omega^{i-1}}) - P(\omega^i) \tag{6.3.3}$$

and

$$P(\omega^{i+1}) = P(\omega^{i+1}|\overline{\Omega^i}) \cdot P(\overline{\Omega^i}). \tag{6.3.4}$$

■ *Proof.* (6.3.3) is immediate from Observation 6.2.1 and the obvious fact that $\{\omega^i\}$ and $\overline{\Omega^i}$ are mutually exclusive, and that $\overline{\Omega^{i-1}} = \{\omega^i\} \cup \overline{\Omega^i}$: Apply Observation 6.2.1 here by setting $n = 2$, $\Omega^1 = \{\omega^i\}$, and $\Omega^2 = \overline{\Omega^i}$. Equation (6.3.4) is immediate from Observation 6.2.4 and the fact that $\{\omega^{i+1}\} = \{\omega^{i+1}\} \cap \overline{\Omega^i}$: Apply Observation 6.2.4 here by setting $\Omega' = \{\omega^{i+1}\}$ and $\Omega'' = \overline{\Omega^i}$. ■

In applying Equations (6.3.3) and (6.3.4), recall that $P(\overline{\Omega^0}) = P(\overline{\emptyset}) = P(\Omega) = 1$. Also recall that $P(\omega^m)$ can be obtained by subtracting $\Sigma_{i=1}^{m-1} P(\omega^i)$ from 1.

Example 6.3.3

Suppose that you use Method f2 instead of Method f1 for the task in Example 6.3.1. If you assess $P(0) = P(\omega^1) = .05$, $P(\tilde{\omega} = 1|\tilde{\omega} > 0) = P(\omega^2|\overline{\Omega^1}) = 15/95$, $P(\tilde{\omega} = 2|\tilde{\omega} > 1) = P(\omega^3|\overline{\Omega^2}) = 30/80$, $P(\tilde{\omega} = 3|\tilde{\omega} > 2) = P(\omega^4|\overline{\Omega^3}) = 25/50$, and $P(\tilde{\omega} = 4|\tilde{\omega} > 3) = 20/25$, then successive applications of (6.3.3) and (6.3.4) yield

$$\begin{aligned}
P(\overline{\Omega^1}) &= P(\overline{\Omega^0}) - P(\omega^1) &&= 1 - .05 &&= .95,\\
P(\omega^2) &= P(\omega^2|\overline{\Omega^1}) \cdot P(\overline{\Omega^1}) &&= (15/95)(.95) &&= .15,\\
P(\overline{\Omega^2}) &= P(\overline{\Omega^1}) - P(\omega^2) &&= .95 - .15 &&= .80,\\
P(\omega^3) &= P(\omega^3|\overline{\Omega^2}) \cdot P(\overline{\Omega^2}) &&= (30/80)(.80) &&= .30\\
P(\overline{\Omega^3}) &= P(\overline{\Omega^2}) - P(\omega^3) &&= .80 - .30 &&= .50,\\
P(\omega^4) &= P(\omega^4|\overline{\Omega^3}) \cdot P(\overline{\Omega^3}) &&= (25/50)(.50) &&= .25\\
P(\overline{\Omega^4}) &= P(\overline{\Omega^3}) - P(\omega^4) &&= .50 - .25 &&= .25,\\
P(\omega^5) &= P(\omega^5|\overline{\Omega^4}) \cdot P(\overline{\Omega^4}) &&= (20/25)(.25) &&= .20,
\end{aligned}$$

and finally,

$$P(\omega^6) = 1 - \sum_{i=1}^{5} P(\omega^i) = 1 - (.05 + .15 + .30 + .25 + .20) = .05.$$

In this example we have *assumed* that your conditional-probability assessments were consistent with the probabilities you assessed in Example 6.3.1. Human fallibility makes such precise consistency rare in practice.

Exercise 6.3.1

Let ω^1 denote the (elementary) event that the Democratic candidate wins the next Presidential election; ω^2, the event that the Republican candidate wins; and ω^3, the event that some third-party candidate wins. Use Method f1 to assess your probabilities of ω^1, ω^2, and ω^3.

Exercise 6.3.2 (Continuation of Exercise 6.3.1.)

Let δ^i be the event that candidate i wins in the election *after* next, where $i = 1, 2, 3$ for Democrat, Republican, and some third party respectively.

(A) Assess your conditional probabilities $P(\delta^j|\omega^i)$, given each ω^i, for $j = 1, 2, 3$.
(B) Using the $P(\omega^i)$'s from Exercise 6.3.1 and the $P(\delta^j|\omega^i)$'s from (A) of this exercise, *calculate* $P(\delta^j) = \Sigma_{i=1}^3 P(\delta^j|\omega^i)P(\omega^i)$ for $j = 1, 2, 3$. Do these values accord with your basic judgments about the election after next? If not, then you are being inconsistent somewhere.

Exercise 6.3.3

Let $\tilde{\delta}$ denote your as-yet undetermined grade average (on a 0–100 scale) as of the end of the current semester. Let

$$\Delta^1 = \{\delta: 95 \le \delta \le 100\}, \quad \Delta^2 = \{\delta: 90 \le \delta < 95\}, \quad \Delta^3 = \{\delta: 85 \le \delta < 90\},$$
$$\Delta^4 = \{\delta: 80 \le \delta < 85\}, \quad \Delta^5 = \{\delta: 75 \le \delta < 80\}, \quad \Delta^6 = \{\delta: 65 \le \delta < 75\},$$
$$\Delta^7 = \{\delta: 60 \le \delta < 65\}, \quad \text{and} \quad \Delta^8 = \{\delta: \delta < 60\}.$$

Assess your probability $P(\Delta^i)$ of the event Δ^i for each $i \in \{1, \ldots, 8\}$.

6.4 ASSESSING A PROBABILITY FUNCTION ON A LARGE-FINITE OR INFINITE SET OF NUMBERS

This section is to Section 6.3 as Section 5.4 was to 5.3. We are concerned here with expressing your judgments about an as-yet undetermined elementary event $\tilde{\omega}$ which will assume a numerical value.[e] If the set Ω of possible values of $\tilde{\omega}$ contains only a few elements, then the methods of Section 6.3 are practical. But if Ω consists of, say, all temperature readings from a good thermometer between −30°F and +115°F, then you would be

[e] Such $\tilde{\omega}$'s are called *real random variables*. See Sections *3.9 and *3.10.

ill advised to split the hairs required by the methods in Equation 6.3. More expeditious methods will be developed here.

We shall be concerned solely with the case in which the actual possible values of $\tilde{\omega}$ are (1) evenly spaced and (2) individually very improbable. These conditions are frequently satisfied in practice when $\tilde{\omega}$ represents a measurement accurate to a given number of decimal places, a monetary sum, etc. In fact, the improbability of each specific value is one of the main deterrents to applying the assessment techniques of Section 6.3.

Both methods presented here result in your determining judgmentally the probability $P(\tilde{\omega} \in (-\infty, \omega])$ that $\tilde{\omega} \leq \omega$ for every real number ω. For notational convenience,[f] we denote $P(\tilde{\omega} \in (-\infty, \omega])$ by $F(\omega)$. Hence both methods result in your determining the *function* $F(\cdot)$. In Method R1, you determine a number of $(\omega, F(\omega))$ pairs and then graph a function through them; in Method R2 you determine as many $(\omega, F(\omega))$ pairs as there are parameters z of a chosen family $F(\omega|z)$ to fit.

After describing these assessment techniques, we then show (1) how to find probabilities of events Ω' from the graph of $F(\cdot)$, and (2) how to calculate expectations $E\{x(\tilde{\omega})\}$ of functions of $\tilde{\omega}$.

In both methods, you are asked as a preliminary to choose reference prizes A and B such that $A > B$, and then to vary "something" until

$$\ell^{\wedge}((-\infty, \omega]; A, B) \sim \ell^*(p; A, B), \tag{6.4.1}$$

since this equidesirability implies that $p = P(\tilde{\omega} \in (-\infty, \omega]) = P(\tilde{\omega} \leq \omega) = F(\omega)$, by definition of probability. But you *need not fix* ω *and vary* p until equidesirability (6.4.1) obtains. Instead, you *may fix* p *and vary* ω. This is much in the spirit of Section 5.4, where we fixed p and varied c until $c \sim \ell^*(p; A, B)$.

By virtue of our assuming that the possible values of $\tilde{\omega}$ are equally spaced and individually improbable, it also makes sense as an idealization to assume that if p is *strictly* between 0 and 1, then there is *exactly one number* ω^p such that $\ell^{\wedge}((-\infty, \omega^p)]; A, B) \sim \ell^*(p; A, B)$. This number ω^p is called the p*th fractile* of $\tilde{\omega}$. Our assumption means that there is exactly one number ω^p such that

$$F(\omega^p) = p. \tag{6.4.2}$$

Moreover, we know from Observation 6.2.3 that $F(\cdot)$ must be a *nondecreasing* function of ω, in that $F(\omega'') \geq F(\omega')$ whenever $\omega'' > \omega'$.

■ Let $\Omega' = (-\infty, \omega']$ and $\Omega'' = (-\infty, \omega'']$. If $\omega'' > \omega'$, then $\Omega' \subset \Omega''$. But $F(\omega') = P(\Omega')$ and $F(\omega'') = P(\Omega'')$. Hence $F(\omega') \leq F(\omega'')$ because $P(\Omega') \leq P(\Omega'')$ by Observation 6.2.3. ■

But nondecreasingness of $F(\cdot)$, existence and uniqueness of pth fractiles ω^p for $0 < p < 1$ and Equation (6.4.2) together imply that $F(\cdot)$ is a *strictly*

[f] This notation is consistent with the definition of cumulative distribution functions in Exercises *3.9.9 and *3.10.8.

increasing and continuous function on $\{\omega: 0 < F(\omega) < 1\}$; that $\{\omega: 0 < F(\omega) < 1\}$ is an *open* interval (ω^0, ω^1); that ω^1 might be $+\infty$ but if not, then $F(\omega) = 1$ for all $\omega \geq \omega^1$; and that ω^0 might be $-\infty$ but if not, then $F(\omega) = 0$ for all $\omega \leq \omega^0$.

These assumptions about $F(\cdot)$—and hence, by definition, also about $P(\tilde{\omega} \in (-\infty, \cdot])$—are as much guides to your assessments and curve graphings as the assumptions about $u(\cdot)$ in Section 5.4. Qualitative properties of functions to be assessed judgmentally are powerful tools for making the requisite assessments.

Method R1 requires that you select a number of values of p wisely, and then determine the corresponding values ω^p by lottery comparisons. Then you graph these (ω^p, p) pairs in the $(\omega, F(\omega))$ plane and draw a smooth and strictly increasing curve through these graphed points, very much in the spirit of the graphing in Method \$1 and Exercise 3.6.4.

■ In the following steps of this method, you may wish to refer to Figure 6.3, which depicts the result of the author's applying this method on August 23, 1971, in quantifying his judgments regarding the 1971-year-end value of the Dow-Jones Industrial Average. ■

We shall assume that you have chosen reference prizes A and B such that $A \succ B$.

STEP R1-1 requires that you determine the *largest* number ω^0 (if any) such that $\ell\hat{}((-\infty, \omega^0]; A, B) \sim \ell^*(0; A, B)$, and the *smallest* number ω^1 (if any) such that $\ell\hat{}((-\infty, \omega^1]; A, B) \sim \ell^*(1; A, B)$. But[g] if $\ell^*(1; A, B) \succ$

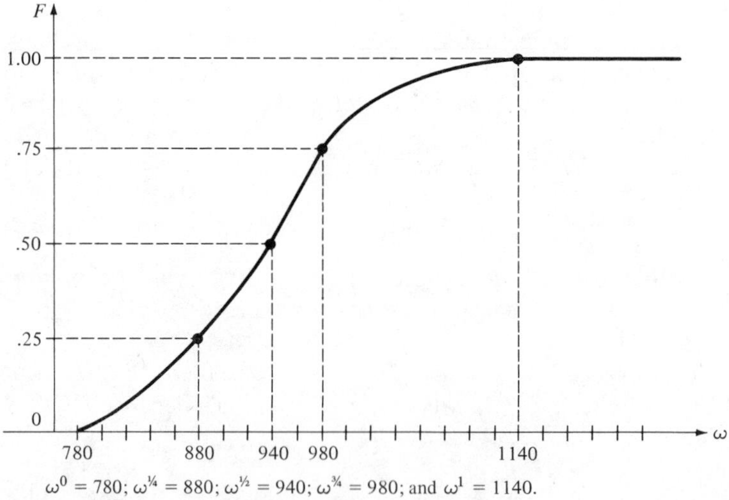

Figure 6.3 $F(\cdot)$ for $\tilde{\omega}$ = 1971-year-end DJIA, assessed on August 23, 1971.

[g] This sentence is superfluous if you are being brutally honest about life!

$\ell^\wedge((-\infty, \omega^1]; A, B)$ for every real number ω^1, then set $\omega^1 = +\infty$; and similarly, if $\ell^\wedge((-\infty, \omega^0]; A, B) > \ell^*(0; A, B)$ for every real number ω^0, then set $\omega^0 = -\infty$. In one sense, candor requires that $\omega^0 > -\infty$ and that $\omega^1 < +\infty$; on the other hand, if $-\infty < \omega^0 < \omega^1 < +\infty$, then it is easy to see that $P(\tilde{\omega} > \omega^1) = P(\tilde{\omega} < \omega^0) = 0$, and *no amount of evidence* could convince anyone that $\tilde{\omega} > \omega^1$ or $\tilde{\omega} < \omega^0$. Hence you should be careful to choose an ω^0 for which you are absolutely sure that $\tilde{\omega} \geq \omega^0$, and an ω^1 for which you are absolutely sure that $\tilde{\omega} \leq \omega^1$.

STEP R1-2[1/2] requires that you determine $\omega^{1/2}$ by Equation (6.4.1) for $p = 1/2$. Now graph the points $(\omega^0, 0)$, $(\omega^{1/2}, 1/2)$, and $(\omega^1, 1)$ in the $\omega - F(\omega)$ plane. Clearly, $\omega^0 < \omega^{1/2} < \omega^1$.

STEP R1-2[i/4] requires that you determine $\omega^{1/4}$ and $\omega^{3/4}$ by (6.4.1) for $p = 1/4$ and $p = 3/4$ respectively. Then add the points $(\omega^{1/4}, 1/4)$ and $(\omega^{3/4}, 3/4)$ to your graph. Clearly, $\omega^0 < \omega^{1/4} < \omega^{1/2} < \omega^{3/4} < \omega^1$. If more points are needed before confident graphing, the general step can be performed for $n = 3$, then $n = 4$, then . . . , etc.

STEP R1-2[$i/2^n$] requires that you determine $\omega^{i/2^n}$ by (6.4.1) for $i = 1, 3, 5, \ldots, 2^n - 1$; previous assessments have determined $\omega^{i/2^n}$ for $i = 0, 2, 4, \ldots, 2^n$. Then add the new points $(\omega^{i/2^n}, i/2^n)$ to your graph.

STEP R1-3, finally, requires you to draw a smooth, strictly increasing curve $\{(\omega^p, p)\colon 0 < p < 1\} = \{(\omega, F(\omega))\colon \omega^0 < \omega < \omega^1\}$ through the points $(\omega^{i/2^n}, F(\omega^{i/2^n})) = (\omega^{i/2^n}, i/2^n)$; this curve can be extended to the right of ω^1 by setting $F(\omega) = 1$ for all $\omega \geq \omega^1$, and to the left of ω^0 by setting $F(\omega) = 0$ for all $\omega \leq \omega^0$. Such has been done in Figure 6.3, in which all $\omega^{i/4}$'s were assessed.

Before pushing on to Method R2, we pause to mention an alternative general step R1-2[$i/2^n$] for $n = 1, 2, \ldots$ which requires you to ponder lotteries involving only $p = 1/2$.

STEP R1-2[$i/2^n$] (alternate) for $n = 1, 2, \ldots$ assumes that you have already determined $\omega^{i/2^n}$ for $i = 0, 2, \ldots, 2^n$ and hence have only $\omega^{i/2^n}$ for $i = 1, 3, \ldots, 2^n - 1$ left to assess. Let $\Omega^{i/2^n}$ be defined by

$$\Omega^{i/2^n} = \{\omega\colon \omega^{(i-1)/2^n} < \omega \leq \omega^{(i+1)/2^n}\} \tag{6.4.3}$$

for $i = 1, 3, \ldots, 2^n - 1$. Let $\ell^*(1/2|\Omega^{i/2^n}; A, B)$ and $\ell^\wedge((-\infty, \omega]|\Omega^{i/2^n}; A, B)$ be conditional lotteries as defined in Figure 6.2 and the fourth preliminary in Section 6.2. Then $\omega^{i/2^n}$ must satisfy

$$\ell^\wedge((-\infty, \omega^{i/2^n}]|\Omega^{i/2^n}; A, B) \sim \ell^*(1/2|\Omega^{i/2^n}; A, B). \tag{6.4.4}$$

Without notation, $\omega^{i/2^n}$ is your answer to the question, "If you were told infallibly that $\tilde{\omega}$ was between $\omega^{(i-1)/2^n}$ and $\omega^{(i+1)/2^n}$ (whose values you have already determined), then which number would you as soon wager that $\tilde{\omega}$ did not exceed as you would wager on the occurrence of an event of canonical probability 1/2?"

Method R2 is applicable when you have determined the functional form of $F(\cdot)$ up to some parameters, say, $z_1, \ldots, z_n$. Such is often the case in certain, special models. For these models, Method R2 is entirely analogous to Method \$2 for fitting parameters of a utility function for monetary returns. As step R2-1, you must choose, by qualitative considerations, the mathematical form $F(\cdot|z_1, \ldots, z_n)$ of $F(\cdot)$.

■ The qualitative considerations influencing your choice of mathematical form are not as easy to describe as those concerning utility for monetary returns. A balance must be struck (1) between flexibility of the form and hence capability of representing your judgments accurately for some values of its parameters and (2) between analytical tractability and hence ease of subsequent computations. Moreover, considerations of the physical and/or behavioral process which determines $\tilde{\omega}$ may also influence the choice of form. See *Examples 3.9.2–3.9.4 and *Exercise 6.4.6, for example. Often, considerations of process and/or tractability suggest a mathematical form, and then that form is tested for adequate flexibility in representing $\mathcal{D}$'s judgments. Further consideration of this subject is deferred to *Exercises 6.4.6 and 6.4.7 and to Chapter 9. ■

After having chosen the mathematical form $F(\cdot|z_1, \ldots, z_n)$ of $F(\cdot)$, you must as step R2-2 assess ω^p for as many $(= n)$ different values of p as there are parameters to be fitted. Then you either solve a system of equations for $z_1, \ldots, z_n$ in terms of $\omega^{p_1}, \ldots, \omega^{p_n}$, or you find $z_1, \ldots, z_n$ by looking up values in tables. Chapter 16 and the tables in LaValle [**78**] are much more explicit about Method R2; we shall content ourselves here with two examples.

Example 6.4.1

Suppose that, by some means or other, you have decided to express your judgments about $\tilde{\omega}$ in the form of $F(\cdot|z)$ as defined by

$$F(\omega|z) = \begin{cases} 0, & \omega \leq 0 \\ 1 - e^{-z\omega}, & \omega > 0, \end{cases} \tag{6.4.5}$$

where z is a *positive* parameter. If you assess $\omega^{1/2}$ judgmentally, then that single fractile suffices to determine the single parameter z, since, by virtue of Equations (6.4.2) and (6.4.5), we have

$$1/2 = F(\omega^{1/2}|z) = 1 - e^{-z\omega^{1/2}},$$

which implies, by rearranging, taking logs to the base e, and again rearranging, that

$$z = \log_e (2)/\omega^{1/2}.$$

***Example 6.4.2**

This example is quite important and is intelligible if you have followed our comments about Normal density functions. If you decide to express your judgments about $\tilde{\omega}$ in the form of a Normal density function with parameters $m \in (-\infty, +\infty)$ and $V > 0$, then

$$F(\omega|m, V) = \int_{-\infty}^{\omega} f_N(\tau|m, V)d\tau,$$

where $f_N(\cdot|m, V)$ is as defined in Example 3.10.7. Now, symmetry of $f_N(\cdot|m, V)$ about m implies that $m = \omega^{1/2}$, and hence determining m is easy. It can also be shown that $V \doteq [(\omega^{3/4} - \omega^{1/2})/.6745]^2 = [(\omega^{1/2} - \omega^{1/4})/.6745]^2$. Hence exactly two fractiles suffice to determine the two parameters of $F(\cdot|m, V)$; namely, $\omega^{1/2}$ and either $\omega^{3/4}$ or $\omega^{1/4}$.

So much for assessing $\{(\omega, F(\omega)): -\infty < \omega < +\infty\}$, or equivalently, for assessing your probability $P(\tilde{\omega} \in (-\infty])$ of every event of the form $(-\infty, \omega]$. We must now show (1) how to find $P(\tilde{\omega} \in \Omega')$ for events Ω' *not* of the form $(-\infty, \omega]$ and (2) how to calculate expectations $E\{x(\tilde{\omega})\}$ of functions $x(\cdot)$ of $\tilde{\omega}$ given all probabilities $P(\tilde{\omega} \in (-\infty, \omega])$.

If Ω' is an interval of the form $(a, b]$, then $(-\infty, b] = (a, b] \cup (-\infty, a]$, and hence $P(\tilde{\omega} \in (-\infty, b]) = F(b) = P(\tilde{\omega} \in (a, b]) + F(a)$, by mutual exclusivity of $(a, b]$ and $(-\infty, a]$. Hence

$$P(\tilde{\omega} \in (a, b]) = F(b) - F(a) \tag{6.4.6}$$

if $b \geq a$. Moreover, the assumed continuity of $F(\cdot)$ and an argument involving taking limits suggest that $P(\tilde{\omega} = a) = P(\tilde{\omega} \in [a, a]) = 0$ for every[h] real number a. Hence if $b > a$ and Ω' is *any* one of the intervals (a, b), $[a, b)$, $(a, b]$, and $[a, b]$, then

$$P(\tilde{\omega} \in \Omega') = F(b) - F(a). \tag{6.4.7}$$

Now Equation (6.4.7) shows how your assessed $F(\cdot)$ can be used to determine the probability of any interval. The probability of a union $\cup_{i=1}^{n} \Omega^i$ of mutually exclusive intervals is obtained by adding their individual probabilities $P(\tilde{\omega} \in \Omega^i)$, by virtue of Observation 6.2.1.

Example 6.4.3

We obtained the probabilities of the possible numbers of *Bon Vivants* demanded in Example 1.5.5, the newsboy problem, by drawing a reasonable-looking $F(\cdot)$ and then setting $P(\theta) = F(\theta) - F(\theta - 1)$ for every $\theta \in \{20, 21, \ldots, 40\}$. Naturally, if we were the newsboy, we would have been much less perfunctory in our determination of $F(\cdot)$.

Example 6.4.3 also suggests how one can find the expectation $E\{x(\tilde{\omega})\}$ of a numerical function $x(\cdot)$ of $\tilde{\omega}$, by what might be termed impressively the *rediscretization* of $\tilde{\omega}$. Our Methods R1 and R2 involved idealizing an

[h] See Exercise 3.4.3 on canonical probability, and *Section 3.10. That *every* real number has zero probability is a result of our idealizing the assumption that every possible value has *very small* probability.

originally discrete set Ω of possible values, and now we can return to that discrete set by applying (6.4.6).

Specifically, suppose that $\Omega = \{\omega^0 + ti: i = 1, \ldots, m\}$ with $\omega^1 = \omega^0 + tm$, where t is the constant (and positive) distance between any two successive possible values of $\tilde{\omega}$. Then, by definition of $F(\cdot)$, it follows clearly that

$$P(\tilde{\omega} = \omega^0 + ti) = F(\omega^0 + ti) - F(\omega^0 + t[i-1]) \tag{6.4.8}$$

for $i = 1, \ldots, m$.

■ Recall that $F(\omega^1) = F(\omega^0 + tm) = 1$ and that $F(\omega^0) = F(\omega^0 + t \cdot 0) = 0$. In practice, set ω^0 exactly t below the smallest possible value $\omega^0 + t$ of $\tilde{\omega}$, and designate as ω^1 the largest possible value $\omega^0 + tm$ of $\tilde{\omega}$. ■

Now, once we have determined the probabilities of the possible values $\omega^0 + ti$ of $\tilde{\omega}$, it is easy to calculate $E\{x(\tilde{\omega})\}$, since clearly

$$E\{x(\tilde{\omega})\} = \sum_{i=1}^{m} x(\omega^0 + ti)P(\tilde{\omega} = \omega^0 + ti). \tag{6.4.9}$$

That is all there is to the rediscretization of $\tilde{\omega}$, except for noting that small visual errors in determining the probabilities by Equation (6.4.8) do not generally make much difference in the computation of $E\{x(\tilde{\omega})\}$ by (6.4.9), and hence should not be cause for concern.

*■ An alternative to rediscretization is available, and involves the density function $f(\cdot)$ of $\tilde{\omega}$, as defined in *Section 3.10. Where $F(\cdot)$ is differentiable, $f(\omega) = dF(\omega)/d\omega$, the slope of $F(\cdot)$ at ω. As shown in *Section 3.10,

$$E\{x(\tilde{\omega})\} = \int_{\Omega} x(\omega) f(\omega) d\omega \tag{6.4.10}$$

where the integral is well defined. Finding $f(\cdot)$ from $F(\cdot)$ graphically may be quite a chore, but it is an easy task in the context of Method R2, since that technique produces a formula for $F(\cdot)$ which can be differentiated easily by analytical methods. In fact (6.4.9) may be regarded as an approximation to (6.4.10). But in view of the point that infinite cases are idealizations of finite reality, it is equally appropriate to call (6.4.10) an approximation to (6.4.9). ■

■ If the number m of possible values of $\tilde{\omega}$ is large, then the rediscretization procedure necessitates many graph readings and calculations. At the cost of a little accuracy, $E\{x(\tilde{\omega})\}$ may be approximated by a more crude discretization, involving three steps.

(1) Choose a positive integer k, usually much smaller than m for practical reasons; laboriousness as well as accuracy increases with k.

(2) For $i = 1, \ldots, k$, find the $[(2i-1)/(2k)]$th fractile $\omega^{(2i-1)/(2k)}$ of $\tilde{\omega}$ from the graph of $F(\cdot)$, by reading across from $(2i-1)/(2k)$ on the ordinate to $F(\cdot)$ and then down to the abscissa.

(3) The approximation is then to set

$$E\{x(\tilde{\omega})\} \doteq (1/k) \sum_{i=1}^{k} x(\omega^{(2i-1)/(2k)}). \tag{6.4.11}$$

The idea behind this technique is to imagine that the entire probability $1/k$ of the interval $\Omega^i = (\omega^{(i-1)/k}, \omega^{i/k}]$ is concentrated at the intermediate fractile $\omega^{(2i-1)/(2k)}$, which is the "probabilistic midpoint" of this interval, in the sense that $P(\tilde{\omega} \leq \omega^{(2i-1)/(2k)} | \tilde{\omega} \in \Omega^i) = 1/2$.

Example 6.4.4

Suppose that $F(\cdot)$ is given by the formula

$$F(\omega) = \begin{cases} 0, & \omega \leq 0 \\ 1-(1-\omega/10)^2, & 0 < \omega < 10 \\ 1, & \omega \geq 10. \end{cases}$$

In this case, it is easy to verify that $\omega^p = 10[1-(1-p)^{1/2}]$, for every $p \in (0, 1)$. If you have chosen $k = 4$, then you can calculate [instead of reading from the graph of $F(\cdot)$] $\omega^{1/8} \doteq .65$, $\omega^{3/8} \doteq 2.09$, $\omega^{5/8} \doteq 3.88$, and $\omega^{7/8} \doteq 6.46$. Now suppose that $x(\omega) = \omega^2$ for every $\omega \in (-\infty, +\infty)$. Then Equation (6.4.11) implies that $E\{x(\tilde{\omega})\} \doteq (1/4)[(.65)^2 + (2.09)^2 + (3.88)^2 + (6.46)^2] \doteq 15.39$. *How accurate this approximation is may be judged from the fact that we obtain $E\{x(\tilde{\omega})\} = 16\frac{2}{3}$ from (6.4.10) with $f(\omega) = dF(\omega)/d\omega = (1/50)(10-\omega)$ for $0 < \omega < 10$.

*■ In general, a very conservative bound on the accuracy of this approximation may be determined as follows. For every $j \in \{0, 1, \ldots, 2k-1\}$, let D_k^j be the maximum difference $x(\omega') - x(\omega'')$ as ω' and ω'' both range over the interval $[\omega^{j/(2k)}, \omega^{(j+1)/(2k)}]$. Clearly, every D_k^j is nonnegative. Now let $D_k = \max_j [D_k^j]$. Then the approximation to $E\{x(\tilde{\omega})\}$ must be within D_k of the actual value (except for rounding-off errors). In Example 6.4.4, we readily conclude that $D = x(10) - x(6.46) \doteq 100 - 41.73 = 58.27$, and hence that our approximation was *well* within D of $16\frac{2}{3}$. Moreover, it is easily verified that D_k is generally a decreasing function of k. ■

Exercise 6.4.1

Let $\tilde{\omega}$ denote the (naturally) as-yet undetermined year-end value of the Dow-Jones Industrial Average *two years* hence, that is, in year $x+2$ if "now" is year x. Quantify your judgments about $\tilde{\omega}$ in the form of $F(\cdot) = P(\tilde{\omega} \in (-\infty, \cdot])$ by using Method R1.

***Exercise 6.4.2**

Use $\omega^{1/2}$ and $\omega^{3/4}$ as determined in Exercise 6.4.1 to calculate the parameters m and V of a Normal density function for $\tilde{\omega}$, as discussed in *Example 6.4.2.

Exercise 6.4.3

Choose a street running past your house, and quantify your judgments regarding the total number $\tilde{\omega}$ of people who will ride past your home on that street within the 24-hour period starting one hour from now.

Exercise 6.4.4

Let $\tilde{\omega}$ denote the average age (rounded to the nearest year) of the people with whom you will speak during the week beginning tomorrow morning.

(A) Quantify your judgments about $\tilde{\omega}$ by using Method R1.
(B) Let $x(\omega) = \omega$ for every ω. Calculate $E\{x(\tilde{\omega})\}$.

Exercise 6.4.5

Now be more precise about your as-yet undetermined grade average $\tilde{\delta}$ as of the end of the current semester. (See Exercise 6.3.3.) Use method R1 to quantify your judgments about $\tilde{\delta}$.

■ The next two exercises constitute an introduction to the conditional-unconditional assessment techniques mentioned in Section 6.1 as indirect method (1). They also illustrate the tractability-enhancing nature of choosing compatible forms for representing your judgments about the random variables in question. ■

***Exercise 6.4.6: Conditional-Unconditional Assessments (I)**

Suppose that you must quantify your judgments about $\tilde{\omega}$, where $\tilde{\omega}$ denotes the as-yet unobserved number of declared Republicans out of a random sample with replacement of (predetermined) n Manhattanites. Let $\tilde{\xi}$ denote the as-yet unknown proportion of Manhattanites who would declare that they are Republicans if asked.

(A) Argue that, *given* ξ, your judgments about $\tilde{\omega}$ *should* be expressed by the binomial mass function $f_b(\cdot|n, \xi)$, defined in Example 3.9.2.
(B) Suppose that your judgments about $\tilde{\xi}$ can be expressed by a beta density function $f_\beta(\cdot|r', n')$ from some $r' > 0$ and some $n' > r'$, defined in Example 3.10.6. Show that your judgments about $\tilde{\omega}$, *unconditional* as regards $\tilde{\xi}$, are then necessarily expressed by the mass function $f_{hmu}^{(1)}(\cdot|r', n', n)$ defined in Example 3.11.4.

■ In the preceding exercise Method R2 would be required in (B) to quantify your judgments about $\tilde{\xi}$. As step R2-1, your choice of the beta-density-function form would have been strongly influenced by the convenience of the ensuing analysis which leads to the (believe me) relatively simple expression $f_{hmu}^{(1)}(\cdot|r', n', n)$ of your unconditional judgments about $\tilde{\omega}$. The same tractability considerations pertain in the following exercise as well. ■

■ There is a consistency moral in the preceding (and ensuing) exercise. You *cannot* (1) believe the sample will be chosen randomly and with replacement, (2) quantify your judgments about $\tilde{\xi}$ in the form of $f_{\beta}(\cdot|r', n')$, and (3) reject $f_{hmu}^{(1)}(\cdot|r', n', n)$ as an adequate expression of your judgments about $\tilde{\omega}$. To do so would be inconsistent. ■

***Exercise 6.4.7: Conditional-Unconditional Assessments (II)**

Suppose that $\tilde{\boldsymbol{\omega}}$ is a t-tuple of as-yet undetermined, year-end security prices, and that $\tilde{\boldsymbol{\xi}}$ is a k-tuple of as-yet undetermined economic and market indicators. (Generally, k is much smaller than t.) Suppose (1) that you have quantified your judgments about $\tilde{\boldsymbol{\xi}}$ in the form of a k-variate Normal density function $f_N^{(k)}(\cdot|\mathbf{m}_1, \mathbf{V}_{11})$ for some $\mathbf{m}_1 \in R^k$ [the ith element of $\mathbf{m}_1$ being $E\{\tilde{\xi}_i\}$] and some positive-definite, symmetric, k-by-k matrix $\mathbf{V}_{11}$ [the (i, j)th element of $\mathbf{V}_{11}$ being $V\{\tilde{\xi}_i, \tilde{\xi}_j\}$]; and (2) that you have quantified your judgments about $\tilde{\boldsymbol{\omega}}$ *conditional* upon the various possible values $\boldsymbol{\xi}$ in the form of a t-variate Normal density function $f_N^{(t)}(\cdot|\mathbf{m}_2 + \mathbf{Z}(\boldsymbol{\xi} - \mathbf{m}_1), \mathbf{V}^*)$, where $\mathbf{m}_2 \in R^t$, $\mathbf{Z}$ is a t-by-k matrix, and $\mathbf{V}^*$ is a positive-definite, symmetric, t-by-t matrix (the elements of which do *not* depend upon $\boldsymbol{\xi}$). Show that your judgments about $\tilde{\boldsymbol{\omega}}$, *unconditional* as regards $\tilde{\boldsymbol{\xi}}$, are necessarily expressed by the t-variate Normal density function $f_N^{(t)}(\cdot|\mathbf{m}_2, \mathbf{V}^* + \mathbf{Z}\mathbf{V}_{11}\mathbf{Z}')$, where $\mathbf{Z}'$ denotes the transpose of $\mathbf{Z}$. [*Hint*: The $(t + k)$-tuple $(\tilde{\boldsymbol{\xi}}', \tilde{\boldsymbol{\omega}}')'$ has the density function $f_N^{(t+k)}(\cdot|\mathbf{m}, \mathbf{V})$ for $\mathbf{m} = (\mathbf{m}_1', \mathbf{m}_2')'$ and

$$\mathbf{V} = \begin{bmatrix} \mathbf{V}_{11} & \mathbf{V}_{21}' \\ \mathbf{V}_{21} & \mathbf{V}_{22} \end{bmatrix},$$

where $\mathbf{V}_{21}\mathbf{V}_{11}^{-1} = \mathbf{Z}$ and $\mathbf{V}_{22} - \mathbf{V}_{21}\mathbf{V}_{11}^{-1}\mathbf{V}_{21}' = \mathbf{V}^*$.]

*■ The practical importance of this exercise lies in the fact that there may be choices of the economic and market indicators ξ_i such that you regard the different stock prices as unrelated, or independent, conditional on any value $\boldsymbol{\xi}$ of $\tilde{\boldsymbol{\xi}}$. In this case, $\mathbf{V}^*$ is a *diagonal* matrix [$V\{\tilde{\omega}_i, \tilde{\omega}_j|\boldsymbol{\xi}\} = 0$ unless $i = j$] and hence fully specified by its t diagonal elements, the variances of the individual $\tilde{\omega}_i$'s. On the other hand, the matrix $\mathbf{V}_{22}$ of unconditional covariances of the $\tilde{\omega}_i$'s requires $t(t + 1)/2$ specifications. Determining $\mathbf{m}_2$ and $\mathbf{Z}$ is not particularly difficult. Hence, if k is much smaller than t, the conditional-unconditional assessment

technique suggested by *Exercise 6.4.7 is attractive vis-à-vis alternatives. ■

6.5 INDEPENDENCE (I): INTRODUCTION TO SAMPLING THEORY

We noted in Section 6.1 that one way of assessing the probability $P(\Omega')$ of an event Ω' in an $\mathscr{E}$-move Ω is to observe what happened in previous instances, or in experimentally contrived instances, of that $\mathscr{E}$-move. For example, if Ω' occurred in 335 out of 1,000 observed instances, then we could estimate $P(\Omega')$ to be .335. Why, and under what conditions, is this procedure reasonable?

Similarly, people often use historical or experimentally contrived repetitions of an $\mathscr{E}$-move Ω to estimate the expectation $E\{x(\tilde{\omega})\}$ of some real-valued function $x(\cdot)$ of $\tilde{\omega}$. Their procedure is to take the function values $x(\omega_1), x(\omega_2), \ldots, x(\omega_n)$ of the respective outcomes $\omega_1, \omega_2, \ldots, \omega_n$ observed in the n repetitions to calculate the simple average or *sample mean* $\bar{x}_n$ of these numbers, defined by

$$\bar{x}_n = (1/n) \sum_{j=1}^{n} x(\omega_j), \tag{6.5.1}$$

and to use $\bar{x}_n$ as an estimate of $E\{x(\tilde{\omega})\}$.

■ In fact, statisticians calculate averages or means of the form (6.5.1) so frequently in applications that certain atrocious statements have been made, to the effect that statisticians are "average people" or "mean people." ■

Again, why and under what conditions is using $\bar{x}_n$ as an estimate of $E\{x(\tilde{\omega})\}$ a reasonable procedure?

We shall make one line of reasoning do the work of two by observing that the first problem, estimating $P(\Omega')$, is actually a special case of the second, estimating $E\{x(\tilde{\omega})\}$. To see this, simply define $x(\cdot)$ by

$$x(\omega) = \begin{cases} 1, & \omega \in \Omega' \\ 0, & \omega \notin \Omega'. \end{cases} \tag{6.5.2}$$

It follows readily that $E\{x(\tilde{\omega})\} = (1)P(\Omega') + (0)[1 - P(\Omega')] = P(\Omega')$ and that $\bar{x}_n = $ [number of occurrences of Ω' in the n repetitions]/[number of repetitions, $= n$], when $x(\cdot)$ and $\bar{x}_n$ are defined by Equations (6.5.2) and (6.5.1) respectively. Hence everything we want to know about estimating $P(\Omega')$ can be gotten from more general results about estimating $E\{x(\tilde{\omega})\}$.

Now $\mathscr{D}$ may have gotten started on quantifying his judgments about $\tilde{\omega}$, but the problem of this section does not arise unless he has not yet finished doing so; otherwise, why should he bother to look up or to contrive experimentally repetitions of $\tilde{\omega}$ in order to obtain information about some attribute of its probability function $P(\cdot)$? Let us call this as-yet unquantified something about the probability function of $\tilde{\omega}$ the (true) *state* (of affairs), denote it by θ, and let Θ denote the set of all conceivable states θ. We shall then suppose that $\mathcal{D}$ *has* assessed all the *conditional* probability functions $P(\cdot|\theta)$ on $\tilde{\omega}$-events given the state θ, for every θ in Θ. It follows that we may suppose $\mathcal{D}$ to have calculated $E\{x(\tilde{\omega})|\theta\}$ for every θ in Θ.

■ The sentence immediately preceding the one above can always be "forced" by appropriately defining θ, even when $\mathcal{D}$ has not even begun to assess his probability function on $\tilde{\omega}$-events. For in this case, θ is just a label for the entire probability function itself, and Θ is then the set of all conceivable probability functions on $\tilde{\omega}$-events. Thus if θ^* denotes the probability function $P^*(\cdot)$ on $\tilde{\omega}$-events, then for $\theta = \theta^*$ we have $P(\Omega'|\theta^*) = P^*(\Omega')$ for every $\tilde{\omega}$-event Ω'. The study of inferences in general when Θ is the set of all probability functions on $\tilde{\omega}$-events is called *nonparametric statistics*, the adjective "nonparametric" seemingly having arisen from an editor's Bowdlerizing a paper's title, "Problems Containing Too D___ Many Parameters." ■

Example 6.5.1

You wish to quantify your judgments about the event Ω' that an individual picked completely at random from the population holds an opinion i. Let θ denote the proportion of people in the population who hold opinion i. Since θ is a proportion, we have $\theta \in [0, 1] = \Theta$. Furthermore, your picking a person *at random* from the population and ascertaining his opinion means that $P(\Omega'|\theta) = \theta$ for every $\theta \in [0,1]$. Here, your *qualitative assumption* about picking someone at random *suffices to determine* $P(\Omega'|\theta)$.

Example 6.5.2

You have a capricious weighing scale (haven't we all?) and an object of unknown weight θ. Through vast past experience with this scale, you may know or at least feel comfortable in assessing the probability functions $P(\cdot|\theta)$ on the $\mathscr{E}$-move Ω of all possible readings' events for all possible true weights θ.

The justification for treating all the values $x(\omega_1), x(\omega_2), \ldots, x(\omega_n)$ observed on n repetitions of the $\tilde{\omega}$-$\mathscr{E}$-move with equal respect in (6.5.1) lies in supposing that *all repetitions, or trials, occur under conditions indistinguishable to* $\mathcal{D}$.

■ Many authors use "identical" in place of "indistinguishable." The latter term is apparently an innovation in Schlaifer [**128**]; it underscores admirably the essential subjectivity of the notion. Whether or not the trial conditions are exactly identical cannot be verified by $\mathcal{D}$; indeed, sufficiently fine description of the trial conditions can usually establish that no two trials are *ever* performed under *identical* conditions! ■

Before the trials are performed,[i] the assumption that their conditions are indistinguishable really means two things about $\mathscr{D}$'s judgments concerning their respective outcomes $\tilde{\omega}_1, \tilde{\omega}_2, \ldots$; namely,

(1) Conditional on θ, the random variables $\tilde{\omega}_1, \tilde{\omega}_2, \ldots$ have *identical* probability functions $P_1(\cdot|\theta)$, $P_2(\cdot|\theta), \ldots$ respectively—since nonidentical probability functions can arise only from $\mathscr{D}$'s being able to distinguish (at least tacitly) among the trial conditions.

(2) Conditional on θ, the random variables $\tilde{\omega}_1, \tilde{\omega}_2, \ldots$ are *independent*—or else the outcomes of previous trials are influencing $\mathscr{D}$'s judgments about subsequent trials *beyond* simple revision of his judgments about θ, as can happen only if trial conditions depend upon outcomes of the previous trials.

Example 6.5.3

In Example 6.5.1 on opinion polling, if you define $\tilde{\omega}_j$ to be the opinion of the jth person polled, and if you select a person at random on each trial j from the entire population (i.e., sample "with replacement"), then $\tilde{\omega}, \tilde{\omega}_2, \ldots$ satisfy (1) and (2), given the true proportion θ of people in the population who hold opinion i. Similarly in Example 6.5.2, if the object of true weight θ is weighed a number of times and the object does not damage the scale in the process, then the weighings $\tilde{\omega}_1, \tilde{\omega}_2, \ldots$ satisfy (1) and (2) in most people's judgment.

So much of current statistical theory and methodology is based on the concept of *experiments* consisting of trials whose outcomes satisfy these two conditions that a special terminology incorporating them has evolved and is popular in statistical circles.

Definition 6.5.1. Suppose that there is an $\mathscr{E}$-move Ω and a set $\{P(\cdot|\theta): \theta \in \Theta\}$ of probability functions on events in Ω such that $\mathscr{D}$'s judgments about the random variables $\tilde{\omega}_1, \ldots, \tilde{\omega}_n$ conditional on θ are expressible in the form

$$P(\tilde{\omega}_1 \in \Omega_1', \ldots, \tilde{\omega}_n \in \Omega_n' | \theta) = \prod_{j=1}^{n} P(\Omega_j'|\theta)$$

for every θ in Θ and every choice of n events $\Omega_1', \ldots, \Omega_n'$ in Ω. Then:

(A) $\tilde{\omega}_1, \ldots, \tilde{\omega}_n$ are said to be *independent and identically distributed* (abbreviated *iid*) *given* θ, or to be a *random sample* (of size n) *from* $P(\cdot|\theta)$.

[i] That is, before observing the historical data or performing the experiment.

(B) The probability function $P(\cdot|\theta)$ on Ω, unknown to $\mathscr{D}$ because of his uncertainty about θ, is called the *population distribution.*

■ The population distribution $P(\cdot|\theta)$ in Definition 6.5.1 is simply any one of the identical probability functions in assumption (1) about $\mathscr{D}$'s prior judgment that the trials are under indistinguishable conditions. ■

*■ If $\tilde{\omega}_1, \tilde{\omega}_2, \ldots$ are independent and identically distributed discrete random real k-tuples, we say that $\tilde{\omega}_1, \tilde{\omega}_2, \ldots$ are a random sample from the population mass function $f(\cdot|\theta)$; if they are continuous random real k-tuples, we say that they are a random sample from the population density function $f(\cdot|\theta)$. In each case, "iid-ness" implies that the joint mass or density function of $(\tilde{\omega}_1, \ldots, \tilde{\omega}_n)$, given θ, must satisfy

$$f(\omega_1, \ldots, \omega_n|\theta) = \prod_{j=1}^{n} f(\omega_j|\theta) \tag{6.5.3}$$

for every $\theta \in \Theta$ and every $(\omega_1, \ldots, \omega_n) \in \Omega \times \cdots \times \Omega$. Equation (6.5.3) is essentially equivalent to the probability-product equation in Definition 6.5.1. ■

To summarize briefly this section to date. Suppose that we are going to observe the outcomes $\omega_1, \omega_2, \ldots, \omega_n$ of n trials of an experiment, that before the trials are performed the random variables $\tilde{\omega}_1, \tilde{\omega}_2, \ldots, \tilde{\omega}_n$ constitute a random sample of size n from $P(\cdot|\theta)$, and that we will use $\bar{x}_n = (1/n)\sum_{j=1}^{n} x(\omega_j)$ as an estimate of $E\{x(\tilde{\omega})|\theta\}$, which we do not know because of our ignorance of the state θ. What justification is there for supposing that $\bar{x}_n$ is a *good* estimate of $E\{x(\tilde{\omega})|\theta\}$?

Before conducting the experiment, its trial outcomes $\tilde{\omega}_1, \ldots, \tilde{\omega}_n$ are random variables, and hence[j], so is $\tilde{\bar{x}}_n = (1/n)\sum_{j=1}^{n} x(\tilde{\omega}_j)$. We shall now show that $\tilde{\bar{x}}_n$ is likely to fall close to $E\{x(\tilde{\omega})|\theta\}$ if n is large. First, however, we deduce the expectation and the variance[k] of the real-valued random variable $\tilde{\bar{x}}_n$.

Lemma 6.5.1. If $\tilde{\bar{x}}_n = (1/n)\sum_{j=1}^{n} x(\tilde{\omega}_j)$, where $\tilde{\omega}_1, \ldots, \tilde{\omega}_n$ are a random sample from $P(\cdot|\theta)$, then

(A) $$E\{\tilde{\bar{x}}_n|\theta\} = E\{x(\tilde{\omega})|\theta\},$$

and

(B) $$V\{\tilde{\bar{x}}_n|\theta\} = (1/n)V\{x(\tilde{\omega})|\theta\}.$$

*■ *Proof.* Since $\tilde{\omega}_1, \ldots, \tilde{\omega}_n$ are independent and identically distributed, given θ,

(a) $$E\{x(\tilde{\omega}_j)|\theta\} = E\{x(\tilde{\omega})|\theta\},$$

[j] Note that this sentence necessitates another minor violation of our notational convention about Greek letters for random variables. At this point, however, your mathematical sophistication should be up to it.

[k] The variance $V\{x(\tilde{\omega})\}$ of a real-valued function of a random variable was defined in Exercise 3.7.11.

and

(b) $$V\{x(\tilde{\omega}_j)|\theta\} = V\{x(\tilde{\omega})|\theta\}$$

for every j and every $\theta \in \Theta$. By (a) and Exercises 3.7.4 and 3.7.5,

$$E\{\bar{\tilde{x}}_n|\theta\} = E\left\{(1/n)\sum_{j=1}^{n} x(\tilde{\omega}_j)|\theta\right\} = (1/n)\sum_{j=1}^{n} E\{x(\tilde{\omega}_j)|\theta\}$$
$$= (1/n)\cdot nE\{x(\tilde{\omega})|\theta) = E\{x(\tilde{\omega})|\theta\},$$

establishing (A). Similarly, (b) and Exercise 3.7.11 imply that

$$V\{\bar{\tilde{x}}_n|\theta\} = V\left\{(1/n)\sum_{j=1}^{n} x(\tilde{\omega}_j)|\theta\right\} = (1/n^2)\sum_{j=1}^{n} V\{x(\tilde{\omega}_j)|\theta\}$$
$$= (1/n^2)\cdot nV\{x(\tilde{\omega})|\theta\} = (1/n)V\{x(\tilde{\omega})|\theta\},$$

in which

$$V\left\{\sum_{j=1}^{n} x(\tilde{\omega}_j)|\theta\right\} = \sum_{j=1}^{n} V\{x(\tilde{\omega}_j)|\theta\}$$

because independence of $\tilde{\omega}_1, \ldots, \tilde{\omega}_n$ and *Exercise 6.2.3 imply independence of $x(\tilde{\omega}_1), \ldots, x(\tilde{\omega}_n)$, so that the conclusion of Exercise 3.7.11(*C) applies. ■

We are now in a position to assess the accuracy with which $\bar{x}_n$ estimates $E\{x(\tilde{\omega})|\theta\}$. Theorem 6.5.1(A) says that, *before* the experiment is conducted and *given* θ, the probability that $\bar{\tilde{x}}_n$ falls within a given distance b of $E\{x(\tilde{\omega})|\theta\}$ is at least $1 - V\{x(\tilde{\omega})|\theta\}/(nb^2)$. Naturally enough, this lower bound on the "probability of sufficient accuracy" is (1) an *increasing* function of the "tolerable error" b; (2) an *increasing* function of the number n of trials; and (3) a *decreasing* function of the variance $V\{x(\tilde{\omega})|\theta\}$ of a single measurement $x(\tilde{\omega})$. Theorem 6.5.1(B) gives a lower bound on the "probability of sufficient accuracy" which does not depend upon the unknown state θ. Theorem 6.5.1(C), an easy consequence of Theorem 6.5.1(A), is a famous theorem in mathematical probability theory called the *weak law of large numbers*; roughly, it states that an infinite number of trials should yield a sample mean right on the button. Theorems 6.5.1(A) and 6.5.1(B) are often called (versions of) *Chebyshev's Inequality*, a name also given to the result in *Exercise 3.7.14.

Theorem 6.5.1. Suppose that $\tilde{\omega}_1, \ldots, \tilde{\omega}_n$ are a random sample of size n from $P(\cdot|\theta)$, and that $x: \Omega \to R^1$. Let $\mu(\theta) = E\{x(\tilde{\omega})|\theta\}$ and $\upsilon(\theta) = V\{x(\tilde{\omega})|\theta\}$ for every $\theta \in \Theta$, and recall that $\bar{\tilde{x}}_n = (1/n)\Sigma_{j=1}^{n} x(\tilde{\omega}_j)$. Then for every $\theta \in \Theta$ and every $b > 0$:

(A) $P(-b \leq \bar{\tilde{x}}_n - \mu(\theta) \leq b|\theta) \geq 1 - \upsilon(\theta)/[nb^2]$;

(B) If $\upsilon(\theta) \leq \upsilon^*$ for every $\theta \in \Theta$, then $P(-b \leq \bar{\tilde{x}}_n - \mu(\theta) \leq +b|\theta) \geq 1 - \upsilon^*/[nb^2]$; and

(C) $\lim_{n\to\infty} [P(-b \leq \bar{\tilde{x}}_n - \mu(\theta) \leq +b|\theta)] = 1.$

*■ *Proof*. From *Exercise 3.7.14, applied to $P(\cdot|\theta)$ instead of simply to $P(\cdot)$, it follows that

$$P(-b \leq \tilde{\bar{x}}_n - E\{\tilde{\bar{x}}_n|\theta\} \leq +b|\theta) \geq 1 - V\{\tilde{\bar{x}}_n|\theta\}/b^2,$$

from which (A) follows readily by substituting $E\{\tilde{\bar{x}}_n|\theta\} = E\{x(\tilde{\omega})|\theta\} = \mu(\theta)$ and $V\{\tilde{\bar{x}}_n|\theta\} = V\{x(\tilde{\omega})|\theta\}/n = \upsilon(\theta)/n$, in each case the first equality following from Lemma 6.5.1. (B) follows readily from (A) and the fact that if $\upsilon(\theta) \leq \upsilon^*$ for every $\theta \in \Theta$, then $1 - \upsilon(\theta)/[nb^2] \geq 1 - \upsilon^*/[nb^2]$ for every $\theta \in \Theta$. (C) follows readily from (A) by taking $\lim_{n\to\infty}$ of each side of (A) and noting that $\lim_{n\to\infty} [1 - \upsilon(\theta)/[nb^2]] = 1 - (\lim_{n\to\infty} [1/n])\upsilon(\theta)/b^2 = 1 - 0 = 1$. ■

Example 6.5.4

In Examples 6.5.1 and 6.5.3 on opinion polling, it is easy to see that if each person is drawn (with replacement) at random from the population, if θ is the proportion of people in the population holding opinion i, and if $x(\omega_j)$ is defined by Equation (6.5.2) for every j, then $P(x(\tilde{\omega}) = 1|\theta) = \theta$ and $P(x(\tilde{\omega}) = 0|\theta) = 1 - \theta$. Hence $\mu(\theta) = E\{x(\tilde{\omega})|\theta\} = (1)\theta + (0)(1-\theta) = \theta$, and [see Exercise 3.7.11(*C)] $\upsilon(\theta) = V\{x(\tilde{\omega})|\theta\} = E\{[x(\tilde{\omega})]^2|\theta\} - [E\{x(\tilde{\omega})|\theta\}]^2 = (1)^2\theta + (0)^2(1-\theta) - \theta^2 = \theta - \theta^2 = \theta(1-\theta)$. Now, $\upsilon(\theta) = \theta(1-\theta)$ has a (least) upper bound of $1/4 = \upsilon^*$ as θ ranges between 0 and 1. Therefore, the probability of finding $\tilde{\bar{x}}_n$ = *sample* proportion holding opinion i to be within b of θ, whatever θ is, is at least $1 - [4nb^2]^{-1}$. More precisely, the *procedure* (1) of sampling the opinions of n people (randomly chosen, with replacement after each is polled) and (2) of estimating the proportion θ in the population holding opinion i by $\bar{x}_n$ is a procedure which, before the sample is conducted, has probability at least $1 - [4nb^2]^{-1}$ of producing an estimate that is within b of θ.

■ The involved language of the last sentence is unfortunately necessary, as we shall see in Chapter 9. The probability bound $1 - [4nb^2]^{-1}$ pertains to the as-yet unobserved estimate $\tilde{\bar{x}}_n$ and is *conditional on* θ. Most people misinterpret such probability statements as being conditional on $\bar{x}_n$ and pertaining to θ. Such misinterpretations are perfectly understandable, in view of the fact that it is θ rather than $\bar{x}_n$ which is the unknown of presumably decision-relevant interest. ■

Theorem 6.5.1(B) yields an easy corollary useful in determining the number of trials one must conduct in order to have sufficiently high probability of sufficiently accurate estimation.

Corollary 6.5.1. If $\upsilon(\theta) \leq \upsilon^*$ for every $\theta \in \Theta$, then

$$P(-b \leq \tilde{\bar{x}}_n - \mu(\theta) \leq +b|\theta) \geq p$$

for every $\theta \in \Theta$ if

$$n \geq [1-p]^{-1} \upsilon^*/b^2.$$

■ *Proof*. Exercise 6.5.3. ■

Example 6.5.5

In Example 6.5.4, if you want the number of trials to be large enough so that the preexperimenting probability of $\bar{\bar{x}}_n$'s being within .01 of θ is at least .95 whatever the value of θ in [0, 1], then Corollary 6.5.1 shows that any n not less than 50,000 trials will do.

As Example 6.5.5 indicates, the number of trials indicated by Corollary 6.5.1 may be discouragingly large. Remember, however, that the right-hand side of the inequality $P(\ldots) \geq 1 - \upsilon^*/(nb^2]$ in Theorem 6.5.1(B) is *only* a *lower* bound for the left-hand side; generally it is a *very* low bound, as when $1 - \upsilon^*/[nb^2]$ is negative. At the expense of some exactitude, much more reasonable statements about the accuracy of $\bar{\bar{x}}_n$ as an estimate of $E\{x(\tilde{\omega})|\theta\}$ can be obtained by using the famous Central Limit Theorem.

■ The proof of the Central Limit Theorem is well beyond the mathematical scope of this book, and even its statement and application in determining requisite sample sizes require familiarity with *Section 3.10. In *Exercise 6.5.7 the C.L.T. is stated so that you may derive certain of its consequences. An optional, further commentary follows that exercise. ■

Now, the preceding analysis dealt with only *one* measurement $x(\cdot)$ of the trial outcomes. There are easy generalizations of Theorem 6.5.1 and Corollary 6.5.1 which pertain to the case of *several* measurements, say $x_1(\cdot), \ldots, x_t(\cdot)$.

■ Concerning the special case in the first paragraph of this section, we would now have t distinct events $\Omega^1, \ldots, \Omega^t$ in Ω and would be observing repetitions of $\tilde{\omega}$ in order to obtain information about *all* t probabilities $P(\Omega^1), \ldots, P(\Omega^t)$. In this case, we would define $x_i(\omega_j) = 1$ if $\omega_j \in \Omega^i$ and $= 0$ if $\omega_j \notin \Omega^i$, for $i = 1, \ldots, t$ and $j = 1, \ldots, n$. ■

Each of the numerical measurements $x_i(\cdot)$ on Ω has a "population mean"

$$\mu_i(\theta) = E\{x_i(\tilde{\omega})|\theta\}, \tag{6.5.4}$$

a "population variance"

$$\upsilon_i(\theta) = V\{x_i(\tilde{\omega})|\theta\}, \tag{6.5.5}$$

and, given that a random sample $\omega_1, \ldots, \omega_n$ of size n is taken from $P(\cdot|\theta)$, a "sample mean"

$$\bar{x}_{in} = (1/n) \sum_{j=1}^{n} x_i(\omega_j) \tag{6.5.6}$$

may be computed.

■ In the preceding remark, we would use $\bar{x}_{in}$ as an estimate of $P(\Omega^i)$ or, in notation which properly isolates the reason for our not knowing that probability, we would use $\bar{\bar{x}}_{in}$ as our estimate of $P(\Omega^i|\theta)$, for $i = 1, \ldots, t$. ■

Now, Theorem 6.5.1 shows how to make preexperimentation probability statements about how good each $\bar{\bar{x}}_{in}$ should be as an estimate of $\mu_i(\theta)$; but can we make *joint* probability statements about the prospective *simultaneous* goodness of $\bar{\bar{x}}_{1n}, \ldots, \bar{\bar{x}}_{tn}$ as respective estimates of $\mu_1(\theta), \ldots, \mu_t(\theta)$? Yes. Theorem 6.5.2 is the direct generalization of Theorem 6.5.1.

Theorem 6.5.2. For every $\theta \in \Theta$ and every $b_1 > 0, \ldots, b_t > 0$:

(A) $P(-b_i \leq \bar{\bar{x}}_{in} - \mu_i(\theta) \leq +b_i$ for $i = 1, \ldots, t|\theta) \geq 1 - \Sigma_{i=1}^t (\upsilon_i(\theta)/[nb_i^2])$;

(B) If $\upsilon_i(\theta) \leq \upsilon_i^*$ for every $\theta \in \Theta$ and $i = 1, \ldots, t$, then $P(-b_i \leq \bar{\bar{x}}_{in} - \mu_i(\theta) \leq + b_i$ for $i = 1, \ldots, t|\theta) \geq 1 - \Sigma_{i=1}^t (\upsilon_i^*/[nb_i^2])$; and

(C) $\lim_{n\to\infty} [P(-b_i \leq \bar{\bar{x}}_{in} - \mu_i(\theta) \leq +b_i$ for $i = 1, \ldots, t|\theta)] = 1$.

*■ *Proof.* Let $\Delta^i = \{(\omega_1, \ldots, \omega_n): -b_i \leq \bar{x}_{in} - \mu_i(\theta) \leq +b_i\}$. Then the left-hand side of (A) is simply $P(\cap_{i=1}^t \Delta^i)$ and hence, by Theorem 3.6.3(10),

$$P(\cap_{i=1}^t \Delta^i) \geq 1 - t + \sum_{i=1}^t P(\Delta^i).$$

But $P(\Delta^i) \geq 1 - \upsilon_i(\theta)/[nb_i^2]$, by Theorem 6.5.1(A). Hence

$$P(\cap_{i=1}^t \Delta^i) \geq 1 - t + \sum_{i=1}^t (1 - \upsilon_i(\theta)/[nb_i^2]) = 1 - t + t - \sum_{i=1}^t (\upsilon_i(\theta)/[nb_i^2]),$$

as asserted. (B) and (C) follow from (A) just as (B) and (C) of Theorem 6.5.1 followed from Theorem 6.5.1(A). ■

The number of trials sufficient for adequately high probability of simultaneous requisite accuracy is given by the readily derivable Corollary 6.5.2.

Corollary 6.5.2. If $\upsilon_i(\theta) \leq \upsilon_i^*$ for every $\theta \in \Theta$ and $i = 1, \ldots, t$, then

$$P(-b_i \leq \bar{\bar{x}}_{in} - \mu_i(\theta) \leq +b_i \text{ for } i = 1, \ldots, t|\theta) \geq p$$

for every $\theta \in \Theta$ if

$$n \geq [1 - p]^{-1} \sum_{i=1}^t (\upsilon_i^*/b_i^2).$$

■ *Proof.* Exercise 6.5.4. ■

Example 6.5.6: Opinion Polls

Suppose that individuals are to be selected at random and with replacement from a given population for the purpose of estimating the proportions $\pi_1, \ldots, \pi_t$ of people who hold opinions $1, \ldots, t$ respectively. (These opinions need not be mutually exclusive and/or collectively exhaustive.) With each $x_i(\cdot)$ defined as in Equation (6.5.2) for $\Omega' =$ "holds opinion i," it follows that $\bar{x}_{in} =$ (number in sample holding opinion i)/(number in sample) = *sample* proportion holding opinion i, whereas $E\{x_i(\tilde{\omega})|\pi_1, \ldots, \pi_t\} = \pi_i =$ *population* proportion holding opinion i. What all the pollsters do (after more or less sophisticated modifications) is to use $\bar{x}_{in}$ as an estimate of π_i for every i. Theorem 6.5.2(C) states, roughly, that these estimates would be right on the button if the sample were infinite. Practical mortals must be satisfied with the hope that the estimates based on a finite sample n will not be too far off. Theorem 6.5.2(B) and Corollary 6.5.2 furnish quantitative renditions of that hope; in this case, each $\upsilon_i^* = 1/4$.

The preceding analysis furnishes much of the justification necessary for the popular practice of *smoothing cumulative relative frequency functions* to express judgments about $\tilde{\omega}$, which we now sketch.

Suppose that $\tilde{\omega}$ is a real-valued random variable with population distribution $P(\cdot|\theta)$, and also suppose that a random sample of size n from $P(\cdot|\theta)$ *has been* taken, the resulting observations being denoted $\omega_1^*, \ldots, \omega_n^*$. We define the *cumulative relative frequency function*, or *empirical cumulative distribution function*, of $\tilde{\omega}$ as the (step) function $\hat{F}_n(\cdot)$ satisfying

$$\hat{F}_n(\omega) = (1/n) \cdot (\text{number of } \omega_j^* \text{ which do not exceed } \omega) \tag{6.5.7}$$

for every $\omega \in (-\infty, \infty)$.

Example 6.5.7

Suppose that the numbers ω_j^* of calls for telephone repairmen on the past 20 "normal," indistinguishable days j in a given town were 16, 12, 17, 20, 18, 17, 15, 18, 13, 15, 14, 17, 17, 16, 15, 16, 14, 18, 16, and 17 on days $1, \ldots, 20$ respectively. Then $\hat{F}_{20}(\cdot)$ may be tabulated.

ω in	$\hat{F}_{20}(\omega)$	ω in	$\hat{F}_{20}(\omega)$
$(-\infty, 12)$	0	$[16, 17)$	11/20
$[12, 13)$	1/20	$[17, 18)$	16/20
$[13, 14)$	2/20	$[18, 19)$	19/20
$[14, 15)$	4/20	$[19, 20)$	19/20
$[15, 16)$	7/20	$[20, +\infty)$	1

For instance, exactly 16 of the ω_j^*'s were less than or equal to 17.4. We may also graph $\hat{F}_{20}(\cdot)$ as in Figure 6.4 and smooth that graph judgmentally to account for the apparent vagaries ascribable to the finiteness of the sample size. For example, if $\mathcal{D}$ knows no reason why there *could not* be 19 calls, then he should not assign probability zero to 19 calls! Similarly, some probability should be assigned to $\tilde{\omega} < 12$ and $\tilde{\omega} > 20$ unless there is an airtight reason for believing that fewer than 12 or more than 20 calls are impossible.

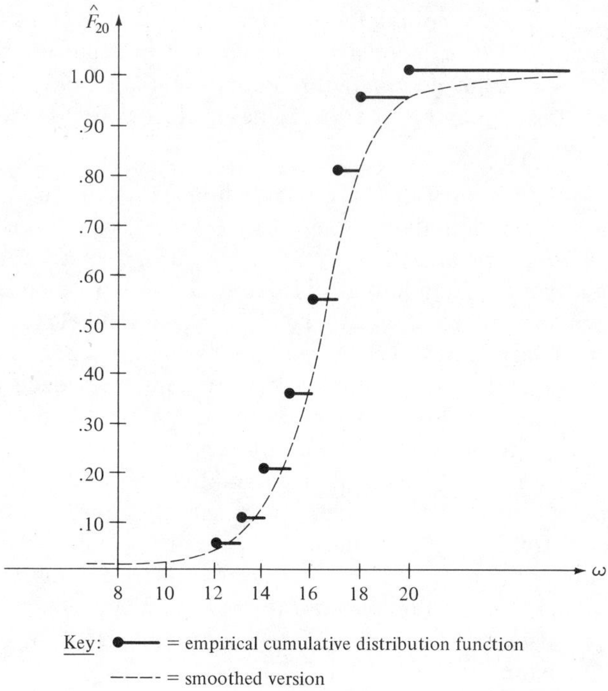

Figure 6.4 Smoothing an empirical cumulative distribution function.

The justification for the smoothing in Example 6.5.7 pertains generally, and hence this technique is simply an elaboration of Method R1 in Section 6.4, using the graph of $\hat{F}_n(\cdot)$ as an informal input to your assessment of $F(\cdot)$.

■ *If* you find that the "trials" producing the observed values $\omega_1^*, \ldots, \omega_n^*$ of $\tilde{\omega}$ are indistinguishable as to their underlying features, and if n is large, then $\hat{F}_n(\omega)$ should be close to $F(\omega|\theta) = P(\tilde{\omega} \leq \omega|\theta)$ for every ω, whatever the (fixed but unknown!) θ characterizing $P(\cdot|\theta)$. You may show that this is true in *Exercise 6.5.5. ■

■ Using observed relative frequencies as (bases for) estimates of corresponding probabilities has a long history in statistics and, from one philosophical point of view (obviously not ours), is the only way of assigning probabilities. The relative-frequency, or "frequentist," school of scientific philosophy holds that the *only* events to which probabilities may be assigned meaningfully are those whose occurrence or nonoccurrence can be observed repeatedly on trials conducted under identical conditions. They proceed from there to *define* the probability of such an event Ω' as the limit of the relative frequency $\bar{x}_n = (1/n) \cdot$ (number of occurrences of Ω' in n trials) as $n \to \infty$. There are several difficulties with this viewpoint, chief among which from our orientation is that it rules out of complete consistency analysis according to Chapters 1–4 the vast majority of decisions under uncertainty containing nonrepeatable $\mathscr{E}$-moves. Another difficulty is the operational inconsistency involved in requiring the demanding judgment that trials be conducted under identical conditions, while at the same time denying the usefulness of less demanding probability judgments. ■

■ In this section we have shown how the trial outcomes $\omega_1, \ldots, \omega_n$ can be used in gleaning information about functions $\mu(\theta) = E\{x(\tilde{\omega})|\theta\}$ of the unknown state θ *without* regarding Θ as an $\mathscr{E}$-move and quantifying judgments about its events. The method discussed above is to estimate $\mu(\theta)$ by $\bar{x}_n$; as justification, we showed that $\bar{x}_n$ is a priori likely to be close to $\mu(\theta)$ if n is large. But this estimation procedure cannot be completely satisfactory from the standpoint of fully consistent decision analysis, because a finite number of observations cannot dispel all uncertainty about $\mu(\theta)$ and because a single number $\bar{x}_n$ cannot furnish a complete description of $\mathscr{D}$'s judgments about $\tilde{\theta}$ and hence about $\mu(\tilde{\theta})$. It is more burdensome, but more consistent, to use the observations $\omega_1, \ldots, \omega_n$ in order to revise initial opinions $P_\theta(\cdot)$ about $\tilde{\theta}$ according to Bayes' Theorem—obtaining $P_{\theta|1\cdots n}(\cdot|\omega_1, \ldots, \omega_n)$ from the functions $P_\theta(\cdot)$ and $P(\omega_1, \ldots, \omega_n|\cdot) = \Pi_{j=1}^n P(\omega_j|\cdot)$ on Θ—and then to calculate $\mathscr{D}$'s probability function $P_{\mu|1\cdots n}(\cdot|\omega_1, \ldots, \omega_n)$ on all $\tilde{\mu} = \mu(\tilde{\theta})$ events M according to the evident relationship[1]

$$P_{\mu|1\cdots n}(\mu(\tilde{\theta}) \in M|\omega_1, \ldots, \omega_n) = P_{\theta|1\cdots n}(\{\theta: \mu(\theta) \in M\}|\omega_1, \ldots, \omega_n)$$

for every $\tilde{\mu}$-event M. Further discussion of these distinct approaches to the utilization of observations in making inferences is deferred to Chapter 9. ■

Exercise 6.5.1

Suppose that $\Omega = \{\omega^1, \ldots, \omega^4\}$, $\Theta = \{\theta^1, \theta^2, \theta^3\}$, and the population distribution $P(\cdot|\theta)$ is that determined by the following table.

	ω^1	ω^2	ω^3	ω^4
θ^1	.90	.05	.03	.02
θ^2	.02	.90	.05	.03
θ^3	.03	.02	.90	.05

Suppose that repeated observations may be made to obtain a random sample $\tilde{\omega}_1, \tilde{\omega}_2, \ldots, \tilde{\omega}_n$ from $P(\cdot|\theta)$.

(A) Suppose $x(\omega^m) = 5 + (m-1)^2$ for $m = 1, 2, 3, 4$.

(i) How many observations are required in order to have probability at least .95 that $\tilde{\bar{x}}_n$ will be within .5 of $E\{x(\tilde{\omega})|\theta\}$?

[1] Evident because $\mu(\theta) = \mu$ belongs to M if and only if $\theta \in \{\theta: \mu(\theta) \in M\}$. Hence these events are really the same event and thus must have equal probabilities. See also Equation (6.2.3) in the hint for *Exercise 6.2.3.

(ii) If three observations are to be obtained, what is the minimum probability of $\tilde{\bar{x}}_3$ falling within 2 of $E\{x(\tilde{\omega})|\theta\}$?

(B) Same questions as (A), but for the function $x(\omega^m) = m$ for $m = 1, 2, 3, 4$.

(C) Let $x_1(\omega^m) = 5 + (m-1)^2$ as in (A) and $x_2(\omega^m) = m$ as in (B).

(i) How many observations are required in order to have probability at least .95 that *each* $\tilde{\bar{x}}_{in}$ will be within .5 of its respective $E\{x_i(\tilde{\omega})|\theta\}$?

(ii) If three observations are to be obtained, what is the minimum probability of *each* $\tilde{\bar{x}}_{i3}$ falling within 2 of its respective $E\{x_i(\tilde{\omega})|\theta\}$?

***Exercise 6.5.2**

Suppose that $\tilde{\omega}_1, \tilde{\omega}_2, \ldots$ are independent and identically distributed with common density function

$$f(\omega|\theta) = \begin{cases} \theta^{-1} e^{-\omega/\theta}, & \omega > 0 \\ 0, & \omega \leq 0, \end{cases}$$

for some $\theta \in \Theta = (0, T]$, where $T > 0$. Let $x(\omega) = \omega$ for every $\omega \in (-\infty, \infty)$.

(A) Calculate $\mu(\theta) = E\{x(\tilde{\omega})|\theta\}$ and $\upsilon(\theta) = V\{x(\tilde{\omega})|\theta\}$.

(B) How many trials n suffice for $P(-1 \leq \tilde{\bar{x}}_n - \mu(\theta) \leq +1|\theta) \geq .75$ for every $\theta \in \Theta$?

Exercise 6.5.3

Deduce Corollary 6.5.1 from Theorem 6.5.1(B).

Exercise 6.5.4

Deduce Corollary 6.5.2 from Theorem 6.5.2(B).

***Exercise 6.5.5: Empirical Cumulative Distribution Functions**

Let ω' be a fixed real number in $\Omega = (-\infty, \infty)$ and let $x(\omega) = 1$ if $\omega \leq \omega'$ and $= 0$ if $\omega > \omega'$.

(A) Show that $\bar{x}_n = \hat{F}_n(\omega')$ as defined by Equation (6.5.7).

■ Hence, *before* any trials are performed, every *ordinate* at $\omega = \omega'$ of the empirical cumulative distribution function is a random variable, $\tilde{\hat{F}}_n(\omega')$. ■

(B) Show that

$$\lim_{n\to\infty} [P(-b \leq \tilde{\hat{F}}_n(\omega') - F(\omega'|\theta) \leq +b|\theta)] = 1$$

for every $b > 0$, every $\theta \in \Theta$, *and* every $\omega' \in (-\infty, \infty)$.

(C) Let $\omega^1, \ldots, \omega^t$ be *any* fixed real numbers. Show that

$$\lim_{n\to\infty} [P(-b \leq \tilde{\hat{F}}_n(\omega^i) - F(\omega^i|\theta) \leq +b \text{ for } i = 1, \ldots, t|\theta)] = 1$$

for every $b > 0$ and every $\theta \in \Theta$.

■ (C) obtains for any finite number t of distinct points $\omega^1, \ldots, \omega^t$ on the real line. There is an even stronger result, called the *weak Glivenko-Cantelli Theorem*, to the effect that (C) holds for *all* real numbers simultaneously:

$$\lim_{n\to\infty} [P(-b \leq \tilde{\hat{F}}_n(\omega') - F(\omega'|\theta) \leq +b \text{ for every } \omega' \in (-\infty, \infty)|\theta)] = 1.$$

What this really means is that ultimately the observed relative frequencies of *all* events should coincide with their respective probabilities. ■

*■ The adjectives "weak" in the above remark and the "weak law of large numbers," Theorem 6.5.1(C), arise because there are even stronger results. To introduce them, we must think of the as-yet unobserved outcome $\tilde{\boldsymbol{\omega}}^\infty = (\tilde{\omega}_1, \tilde{\omega}_2, \ldots$ *ad inf*) of an *infinite* random sample from $P(\cdot|\theta)$, with probabilities of $\tilde{\boldsymbol{\omega}}^\infty$-events being determined from infinite products $\Pi_{j=1}^\infty P(\Omega_j'|\theta)$ analogous to the finite products in Definition 6.5.1. Now think of observing the outcome $(\omega_1, \ldots, \omega_n)$ of a random sample of n observations from $P(\cdot|\theta)$ as being allowed to look at just the first n components of the outcome $\boldsymbol{\omega}^\infty$ of an infinite sample. Clearly, $\bar{x}_n$ is fully determined by $\boldsymbol{\omega}^\infty$, so let us write it more explicitly as $\bar{x}_n(\boldsymbol{\omega}^\infty)$. The *strong law of large numbers* is the assertion that

$$P(\tilde{\boldsymbol{\omega}}^\infty \in \{\boldsymbol{\omega}^\infty: \lim_{n\to\infty} [|\bar{x}_n(\boldsymbol{\omega}^\infty) - \mu(\theta)|] = 0\}|\theta) = 1$$

for every $\theta \in \Theta$. Similarly, be more explicit about the dependence of $\hat{F}_n(\cdot)$ upon $\boldsymbol{\omega}^\infty$ by writing it as $\hat{F}_n(\cdot; \boldsymbol{\omega}^\infty)$. Then the *strong Glivenko-Cantelli Theorem* is the assertion that

$$P(\tilde{\boldsymbol{\omega}}^\infty \in \{\boldsymbol{\omega}^\infty: \lim_{n\to\infty} [\sup_{\omega' \in R^1} [|\hat{F}_n(\omega'; \boldsymbol{\omega}^\infty) - F(\omega'|\theta)|]] = 0\}|\theta) = 1$$

for every $\theta \in \Theta$. The strong Glivenko-Cantelli Theorem implies the strong law of large numbers, hence the weak law of large numbers and the weak Glivenko-Cantelli Theorem. It states that, before sampling, there is probability one of obtaining an infinite sample $\boldsymbol{\omega}^\infty$ whose associated sequence $F_1(\cdot; \boldsymbol{\omega}^\infty)$, $F_2(\cdot; \boldsymbol{\omega}^\infty), \ldots$ *ad inf* of empirical cumulative distribution functions converges uniformly to the population cumulative distribution function. In other words, there is probability one that $\hat{F}_\infty(\cdot; \tilde{\boldsymbol{\omega}}^\infty)$ will coincide with $F(\cdot|\theta)$. The (strong) Glivenko-Cantelli Theorem furnishes the mathematical basis for the frequentist, sampling-theoretic approach to statistics, and for that reason is often called the *fundamental theorem of statistics*. See Tucker [**136**, pp. 127–128] for its derivation. ■

Exercise 6.5.6: Preview of Confidence Intervals

Let notation and concepts be as defined in Theorem 6.5.1 and the discussion preceding it. For every $\bar{x}_n \in (-\infty, \infty)$, define an interval $\iota(\bar{x}_n)$ by

$$\iota(\bar{x}_n) = [\bar{x}_n - (1-p)^{-1/2}\sqrt{\upsilon^*/n}, \bar{x}_n + (1-p)^{-1/2}\sqrt{\upsilon^*/n}],$$

where p is a given number in (0, 1). Show that

$$P(\iota(\tilde{\bar{x}}_n) \text{ contains } \mu(\theta)|\theta) \geq p$$

for every $\theta \in \Theta$.

■ If a random sample is taken, $\bar{x}_n$ is calculated, then $\iota(\bar{x}_n)$ is calculated, and the result is reported by statisticians in the language, "a 100p percent confidence interval for μ is $\iota(\bar{x}_n)$." This does *not* mean that the statistician's (revised by applying Bayes' Theorem to the observations) probability that $\mu(\tilde{\theta}) \in \iota(\bar{x}_n)$ is at least p; rather, it means that before the observations were gathered, there was probability at least p of the *uncertain interval's* containing $\mu(\theta)$. The probability p refers to the *procedure* for calculating intervals ι, and *not* to judgments about μ. See the remark preceding Exercise 6.5.1 and the remark following Example 6.5.4. More generally, a 100p percent confidence interval for a real-valued function $\tau(\theta)$ of θ is the output $\iota(\omega_1, \ldots, \omega_n)$ of *any* procedure for calculating intervals on the basis of the trial outcomes with the property that

$$P(\iota(\tilde{\omega}_1, \ldots, \tilde{\omega}_n) \text{ contains } \tau(\theta)|\theta) \geq p$$

for every $\theta \in \Theta$. ■

***Exercise 6.5.7: Ramifications of the Central Limit Theorem**

Let all assumptions and notation undefined here be as in Theorem 6.5.1. The (simple version of the) *Central Limit Theorem* concludes that

$$\lim_{n\to\infty}\left[P\left(d_1 \leq \frac{\tilde{\bar{x}}_n - \mu(\theta)}{\sqrt{\upsilon(\theta)/n}} \leq d_2\middle|\theta\right)\right] = F_N(d_2|0, 1) - F_N(d_1|0, 1)$$

for every θ such that $0 < \upsilon(\theta) < \infty$, every $d_1 \in (-\infty, \infty)$, and every $d_2 > d_1$, where $F_N(\cdot|0, 1)$ is the cumulative distribution function of a random variable with the "standardized Normal" density function $f_N(\cdot|0, 1)$, defined in Example 3.10.7. [See also Exercise 3.10.9(C).] The Central Limit Theorem is often expressed roughly by saying that "if the sample size n is large, then the 'standardized sample mean'

$$\tilde{z}_n = [\tilde{\bar{x}}_n - \mu(\theta)]/\sqrt{\upsilon(\theta)/n}$$

is approximately a standardized Normal random variable."[m]

[m] $\tilde{\zeta}$ is a "standardized Normal" random variable if its density function is $f_N(\cdot|0, 1)$.

(A) Show that the CLT implies, for large n, that

$$P(-b \leq \bar{\tilde{x}}_n - \mu(\theta) \leq +b|\theta) \doteq 2F_N(b\sqrt{n/\upsilon(\theta)}|0, 1) - 1$$

if $b > 0$ and $0 < \upsilon(\theta) < \infty$.

(B) Show that the CLT implies, for large n, that

$$P\left(d_1 \leq \sum_{j=1}^{n} x(\tilde{\omega}_j) \leq d_2|\theta\right) \doteq F_N(d_2|n\mu(\theta), n\upsilon(\theta)) - F_N(d_1|n\mu(\theta), n\upsilon(\theta)),$$

provided that $0 < \upsilon(\theta) < \infty$.

[*Hint*: Exercise 3.10.9(C) is indispensible for (A) and (B).]

■ Since $\tilde{\omega}_1, \tilde{\omega}_2, \ldots$ are independent and identically distributed, so are $x(\tilde{\omega}_1), x(\tilde{\omega}_2), \ldots$. Furthermore, $n\mu(\theta) = E\{\Sigma_{j=1}^{n} x(\tilde{\omega}_j)|\theta\}$ and $n\upsilon(\theta) = V\{\Sigma_{j=1}^{n} x(\tilde{\omega}_j)|\theta\}$. Let $\tilde{\psi}_n = \Sigma_{j=1}^{n} x(\tilde{\omega}_j)$. Then (B) shows that the probability function (given θ) of the sum $\tilde{\psi}_n$ of a large number n of independent and identically distributed random variables is approximately $P_N(\cdot|E\{\tilde{\psi}_n|\theta\}, V\{\tilde{\psi}_n|\theta\})$, i.e., approximately Normal with the appropriate expectation and variance. This indicates why many real-world measurements appear to have been generated by a Normal density function: If the overall measurement is the sum of (1) the quantity being measured and (2) the sum total of many "approximately iid" disturbances each having expectation zero, then that overall measurement should have a probability function well approximated by that of a Normal random variable with expectation equal to the quantity being measured. ■

*■ A proof of the Central Limit Theorem may be found in Chapter 13 of LaValle [**18**]; much deeper treatments of the CLT and its generalizations can be found in Loève [**87**] and in Tucker [**136**], for example. ■

***Exercise 6.5.8**

Suppose that $\tilde{\omega}_1, \ldots, \tilde{\omega}_n$ are a random sample from some population and that the typical measurement $x(\tilde{\omega})$ has the Normal density function $f_N(\cdot|\mu, \upsilon)$.

(A) Show that $\tilde{\boldsymbol{\xi}} = (x(\tilde{\omega}_1), \ldots, x(\tilde{\omega}_n))'$ has the k-variate Normal density function $f_N^{(k)}(\cdot|\mathbf{m}, \mathbf{V})$ for $\mathbf{m} = (\mu, \ldots, \mu)'$ and $\mathbf{V} = \upsilon\mathbf{I}$, where $\mathbf{I}$ is the nth order identity matrix.

(B) Applying Exercise 3.10.12 to (A), show that $\bar{\tilde{x}}_n$ has the Normal density function $f_N(\cdot|\mu, \upsilon/n)$.

■ Exercise 6.5.8(B) complements the Central Limit Theorem by showing that if each measurement is *exactly* Normal, then the sample mean is *exactly* what the Central Limit Theorem indicates as only an approximation when the individual measurements are non-Normal. ■

6.6 INDEPENDENCE (II): INTRODUCTION TO STOCHASTIC SIMULATION

For our purposes, *stochastic simulation* is adequately defined as the science and craft concerned with constructing, and statistically experimenting on, mathematical models of phenomena or systems for the purpose of obtaining information about the relevant attributes of these phenomena or systems.

This definition rules out "gaming," or "role playing," in which human beings are placed in hypothetical situations and asked to play their respective situational roles. Gaming is often an extremely valuable technique for acquiring information about how $\mathscr{D}$'s competitors would react to his acts, for example, but we shall not discuss it further here.

Stochastic simulation is nothing other than one of many branches of numerical analysis, as Example 6.6.1 indicates.

Example 6.6.1: Simulating the Area of a Blob

Consider the shaded blob in Figure 6.5; we wish to ascertain its area, which happens to coincide with the proportion θ of the unit square it covers. The irregularity of the blob implies that it would be very difficult, if even possible, to calculate θ via the calculus. *One* way of approximating θ is (1) to superimpose a very fine grid on the unit square by, say, subdividing each side into 10,000 equal subintervals, (2) to underestimate θ by adding up the areas of all squares lying wholly within the blob, (3) to overestimate θ by adding up the areas of all squares lying within or just overlapping the blob. Then we know that θ lies between its underestimate and overestimate. Now, this is a laborious approach. Another approach is to *construct a statistical experiment for estimating* θ. You will recall from Section 3.4 that one can use various devices for generating equally probable m-place random decimals; using such a device twice produces a pair $(\tilde{\lambda}, \tilde{\psi})$ of such decimals which may be

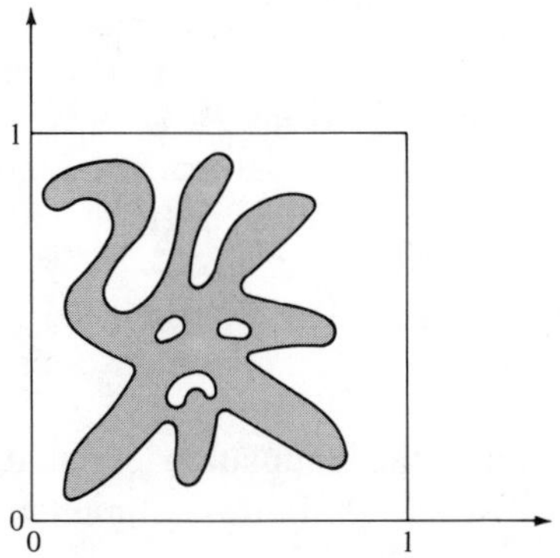

Figure 6.5 A blob in the unit square $[0, 1] \times [0, 1]$.

regarded as the coordinates of a single point in the unit square. Let $\tilde{\alpha} = (\tilde{\lambda}, \tilde{\psi})$. Abstracting from the finite number m of decimal places, the canonical nature of the randomizations $\tilde{\lambda}$ and $\tilde{\psi}$ implies that $\tilde{\alpha} = (\tilde{\lambda}, \tilde{\psi})$ may be regarded as a point to be picked completely at random from the unit square. Then the probability that $\tilde{\alpha}$ falls in the blob is just the blob's area θ. Now let $\Omega = \{$"α in blob," "α outside blob"$\}$. An inkling of how to construct an experiment for estimating now emerges: (1) generate n independent and identically distributed points $\tilde{\alpha}_1, \ldots, \tilde{\alpha}_n$ in the unit square; i.e., perform $2n$ basic canonical randomizations[n] and collect the results in consecutive pairs; (2) let $x(\omega_j) = 1$ if $\omega_j =$ "α_j in blob" and $x(\omega_j) = 0$ if $\omega_j =$ "α_j outside blob"; (3) calculate the proportion $\bar{x}_n = (1/n)\Sigma_{j=1}^{n} x(\omega_j)$ of "hits"; and (4) use $\bar{x}_n$ as an estimate of θ. Since $P(x(\tilde{\omega}) = 1|\theta) = \theta$ for every $\theta \in [0, 1]$, *all the reasoning in Example 6.5.4 for opinion polling continues to obtain here,* where θ is area rather than population proportion, and equal likelihood of points α in the plane corresponds to equal likelihood of selecting each member of the population.

Note carefully that what we did in Example 6.6.1 was to devise a means of *artificially experimenting* on θ, by utilizing some randomization device to produce outcomes with $P(x(\tilde{\omega}) = 1|\theta) = \theta$. Such *artificial* experimentation is what occasions the name "simulation" and differentiates this activity from statistical sampling.

In Example 6.6.1 we were fortunate in having a simple and obvious way of artificially obtaining independent and identically distributed observations $\tilde{\omega}_1, \tilde{\omega}_2, \ldots$ relevant for estimating θ, through repeated use of some (approximation to a) basic canonical randomization.

Actually, though, the method used in Example 6.6.1 may be extended. We shall now show how one can use the independent outcomes $\lambda_1, \lambda_2, \ldots$ of repetitions of a basic canonical randomization so as to produce a sequence $\xi_1, \xi_2, \ldots$ of real numbers which appear to be the outcome of a random sample from a specified population probability function $P(\cdot)$ on events in R^1. Hence $\xi_1, \xi_2, \ldots$ can be treated *as though they were* a random sample from $P(\cdot)$, even though they are produced by artificial means.

It will suffice to show how to use the outcome λ_j of a basic canonical randomization in producing a single observation ξ_j from $P(\cdot)$, since it is clear that repeated operation of the basic canonical randomization generates independent outcomes $\lambda_1, \lambda_2, \ldots$ producing independent and identically distributed observations $\xi_1, \xi_2, \ldots$ from $P(\cdot)$.

Method A for artificially producing one observation ξ_j from $P(\cdot)$ proceeds as follows:

STEP A1: Graph, or functionally describe, the population cumulative distribution function $F(\cdot) = P(\tilde{\xi} \in (-\infty, \cdot])$ of the "population random variable" $\tilde{\xi}$.

---

[n] See Section 3.4.

STEP A2: Perform a basic canonical randomization $\tilde{\lambda}_j$ according to some (necessarily approximate) method such as those described in Section 3.4.

STEP A3: Define the observation ξ_j from $P(\cdot)$ as

$$\xi_j = \min\{\xi\colon F(\xi) \geq \lambda_j\}, \tag{6.6.1}$$

where λ_j is the outcome of the basic canonical randomization performed in step A2.

Method A works, in that before using it the as-yet unobserved $\tilde{\xi}_j$ has a probability function coinciding with $P(\cdot)$. Here, $\tilde{\xi}_j$ is a "random sample of size 1" from $P(\cdot)$, albeit an artificially produced sample.

Theorem 6.6.1. Let $\tilde{\xi}_j$ denote the as-yet unobserved result of a prospective application of Method A. Then $\tilde{\xi}_j$ has the same probability function as has $\tilde{\xi}$.

*■ *Proof.* It suffices to show that the respective cumulative distribution functions $F_j(\cdot)$ and $F(\cdot)$ of $\tilde{\xi}_j$ and $\tilde{\xi}$ coincide. Consult Figure 6.6, which depicts the most complicated sort of cumulative distribution function of a real-valued random variable $\tilde{\xi}$. Let ξ be any real number. A moment's reflection in conjunction with Equation (6.6.1) and Figure 6.6 verifies that the event $\tilde{\xi}_j \leq \xi$ obtains if and only if the event $\tilde{\lambda}_j \leq \max\{\lambda\colon F(\xi) \geq \lambda\}$ obtains. Hence these events have the same probability, establishing the second equality in the following chain.

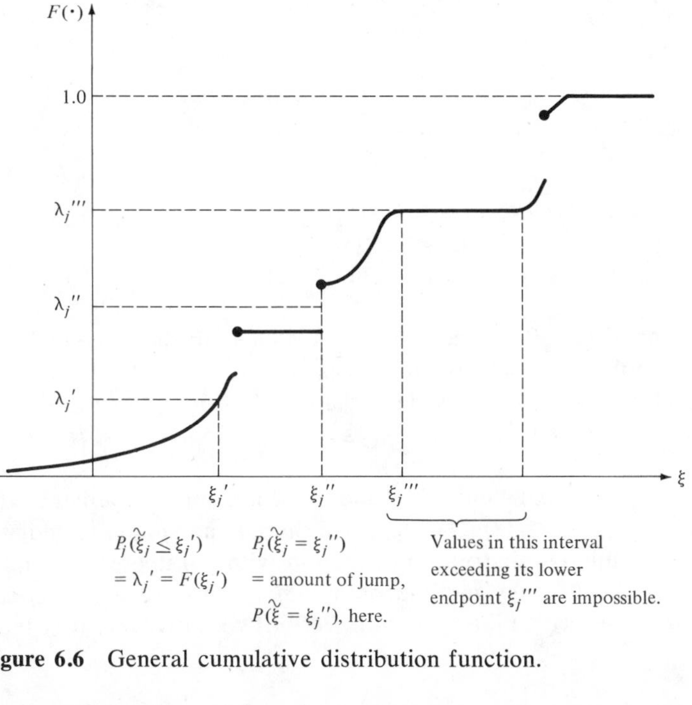

Figure 6.6 General cumulative distribution function.

$$\begin{aligned} F_j(\xi) &= P_j(\tilde{\xi}_j \le \xi) && \text{[definition of } F_j(\cdot)] \\ &= P(\tilde{\lambda}_j \le \max\{\lambda : F(\xi) \ge \lambda\}) && \text{[preceding discussion]} \\ &= \max\{\lambda : F(\xi) \ge \lambda\} && [\tilde{\lambda}_j \text{ basic canonical randomization}] \\ &= F(\xi), \end{aligned}$$

so that $F_j(\xi) = F(\xi)$ for every ξ, as asserted. ■

Example 6.6.2

Suppose that:

$$F(\xi) = \begin{cases} 0, & \xi < 0 \\ 1 - 10^{-\xi}, & \xi \ge 0. \end{cases}$$

This cumulative distribution function implies (1) that negative values of $\tilde{\xi}$ are impossible and (2) that small positive values of $\tilde{\xi}$ are more likely than large positive values. Here, Equation (6.6.1) implies that $\xi_j = -\log_{10}[1 - \lambda_j]$ for all $\lambda_j \in [0, 1)$. We "draw a random sample of $n = 30$ observations" ξ_j from $P(\cdot)$ by using a random-number table, thus obtaining the following:

j	λ_j	ξ_j	j	λ_j	ξ_j
1	.766	.631	16	.452	.261
2	.040	.018	17	.433	.246
3	.334	.177	18	.345	.184
4	.533	.331	19	.979	1.678
5	.804	.708	20	.670	.481
6	.085	.038	21	.398	.220
7	.914	1.066	22	.853	.833
8	.365	.197	23	.493	.295
9	.509	.309	24	.601	.399
10	.208	.101	25	.537	.334
11	.867	.876	26	.491	.293
12	.961	1.409	27	.670	.481
13	.398	.220	28	.784	.666
14	.969	1.509	29	.790	.678
15	.144	.068	30	.121	.056

Note that four of the 30 ξ_j's exceed 1; this is close to the three out of 30 which one would anticipate on the basis of $P(\tilde{\xi} > 1) = 1 - F(1) = 1/10$. Similarly, 14 of the 30 ξ_j's are less than $.301 \doteq \log_{10}[2]$, whereas one would have anticipated 15 on the basis of $P(\tilde{\xi} < \log_{10}[2]) = 1/2$.

*■ Applying (6.6.1) and hence Method A is not too convenient when $F(\cdot) = F_N(\cdot|m, V) = \int_{-\infty}^{\cdot} f_N(\zeta|m, V)d\zeta$, since there is no simple formula for the Normal cumulative distribution function with parameters m and V. There is a quick and easy method of obtaining an observation $\tilde{\xi}_j$ with approximately the probability function $P_N(\cdot|m, V)$, however; it is based

on the Central Limit Theorem (see *Exercise 6.5.7). Perform a basic canonical randomization (independently) a goodly number T (say at least 12) times and record the sample mean $\bar{\lambda}_{jT} = (1/T)\Sigma_{t=1}^{T} \lambda_{jt}$ of the outcomes $\lambda_{j1}, \ldots, \lambda_{jT}$. Then set

$$\xi_j = m + \sqrt{12TV}\,[\bar{\lambda}_{jT} - 1/2]. \tag{6.6.2}$$

The Central Limit Theorem, together with Exercises 3.10.9(C), 3.10.5(A), and 3.10.6(A), implies that the probability function of $\tilde{\xi}_j$ (before $\lambda_{j1}, \ldots, \lambda_{jT}$ were obtained) is approximately $P_N(\cdot|m, V)$. Showing this constitutes *Exercise 6.6.3. ■

*■ Suppose that you want to obtain independent and identically distributed observations $\tilde{\xi}_1, \ldots, \tilde{\xi}_n$ from a probability function $P(\cdot)$ such that the population random variable $\tilde{\xi}$ is expressible as a function $\tilde{\xi} = q(\tilde{\beta})$ of a random variable $\tilde{\beta}$ with a Normal density function $f_N(\cdot|m, V)$. Now, the Central Limit Theorem approximation in the previous remark may be used to produce $\beta_1, \ldots, \beta_n$, and then these can be transformed by $q(\cdot)$ to produce $\xi_1 = q(\beta_1), \ldots, \xi_n = q(\beta_n)$. That is,

$$\xi_j = q(m + \sqrt{12TV}[\bar{\lambda}_{jT} - 1/2]) \tag{6.6.3}$$

for $j = 1, \ldots, n$, where now you would have had to generate a total of nT independent random variables λ_{jt}. This is a convenient way of obtaining observations $\tilde{\xi}_1, \tilde{\xi}_2, \ldots$ from the so-called *lognormal* density function

$$f_{LN}(\xi|m, V) = \begin{cases} 0, & \xi \le 0 \\ (2\pi V\xi^2)^{-1/2}\, e^{-(1/2)(\log_e[\xi]-m)^2/V}, & \xi > 0. \end{cases}$$

Lognormal density functions are often convenient ways of quantifying judgments about surely nonnegative quantities. Now, it is easy to verify from *Exercise 3.10.9 that $\tilde{\xi}$ has the lognormal density function $f_{LN}(\cdot|m, V)$ if and only if $\tilde{\beta} = \log_e[\tilde{\xi}]$ has the Normal density function $f_N(\cdot|\mathrm{m}, \mathrm{V})$. Hence observations $\xi_1, \ldots, \xi_n$ from $f_{LN}(\cdot|m, V)$ may be obtained by Equation (6.6.3) for $\xi = q(\beta) = e^{\beta}$. ■

We now know a procedure for generating real numbers $\xi_1, \xi_2, \ldots$ which, before the procedure is conducted, promises to produce numbers behaving exactly as a random sample from $P(\cdot)$ would behave if repeated real observations were possible. Such procedures constitute the technology of stochastic simulation. They can become quite sophisticated, and hence a detailed discussion in this introduction would be out of place. See Naylor, Balintfy, Burdick, and Chu [**103**] and Naylor [**102**] for further discussion and references on this topic; also, Halton's survey [**50**].

How about procedures for generating real k-tuples $\boldsymbol{\xi}_1, \boldsymbol{\xi}_2, \ldots$ which beforehand promise to be just like a random sample from some probability function $P(\cdot)$ of a random real k-tuple $\tilde{\boldsymbol{\xi}}$, *when* k *exceeds one*? Doing so is easy in the special case wherein the components $\tilde{\xi}_1, \ldots, \tilde{\xi}_k$ of the population random variable $\tilde{\boldsymbol{\xi}}$ are independent, real-valued random variables, since each observation $\boldsymbol{\xi}_j = (\xi_{j1}, \ldots, \xi_{jk})$ may be obtained by applying Method A to each component $\tilde{\xi}_i$ separately, using its marginal cumulative distribution function $F_i(\cdot)$ in (6.6.1). But when the components $\tilde{\xi}_1, \ldots, \tilde{\xi}_i, \ldots, \tilde{\xi}_k$ of the population random variable $\tilde{\boldsymbol{\xi}}$ are dependent (= not independent), very much more sophisticated methods than independent applications of Method A are called for. Indeed, a large part of the art and craft in stochastic simulation involves *defining* problems in such a way that the observations which must be generated will all be independent.

*■ Suppose that the population random variable $\tilde{\boldsymbol{\xi}}$ has a k-variate Normal density function $f_N^{(k)}(\cdot|\mathbf{m}, \mathbf{V})$ as defined in Example 3.10.10. By Exercise 3.10.10(B), one can find a k-by-k matrix $\mathbf{Q}$ with the property that $\tilde{\boldsymbol{\epsilon}} = \mathbf{Q}(\tilde{\boldsymbol{\xi}} - \mathbf{m})$ has the k-variate Normal density function $f_N^{(k)}(\cdot|\mathbf{0}, \mathbf{I})$, where $\mathbf{0}$ is the zero vector in R^k and $\mathbf{I}$ is the k-by-k identity matrix. But it is easy to show that $f_N^{(k)}(\epsilon_1, \ldots, \epsilon_k|\mathbf{0}, \mathbf{I}) = \Pi_{i=1}^k f_N(\epsilon_i|0, 1)$, for every $(\epsilon_1, \ldots, \epsilon_k)$ in R^k. This means that $f_N^{(k)}(\cdot|\mathbf{0}, \mathbf{I})$ is just the joint density function of k independent and identically distributed "standardized Normal" random variables. Hence, we can apply (6.6.2) with $m = 0$ and $V = 1$ k times—using independent random decimals—to obtain (not ξ_j's as denoted there) $\epsilon_{j1}, \ldots, \epsilon_{jk}$, aggregate them in the order obtained into a vector $\boldsymbol{\epsilon}_j = (\epsilon_{j1}, \ldots, \epsilon_{jk})$, and know that $\boldsymbol{\epsilon}_j$ behaves like an observation from $f_N^{(k)}(\cdot|\mathbf{0}, \mathbf{I})$. Now we can invert the transformation $\boldsymbol{\epsilon} = \mathbf{Q}(\boldsymbol{\xi} - \mathbf{m})$ so as to obtain

$$\boldsymbol{\xi}_j = \mathbf{Q}^{-1}\boldsymbol{\epsilon}_j + \mathbf{m}. \tag{6.6.4}$$

Then $\boldsymbol{\xi}_j$ behaves like an observation from $f_N^{(k)}(\cdot|\mathbf{m}, \mathbf{V})$—as desired. [Note that generating n *independent* and identically distributed observations from $f_N^{(k)}(\cdot|\mathbf{m}, \mathbf{V})$ by using Equation (6.6.2) for each ϵ_{ji} necessitates $n \cdot k \cdot T$ independent performances of a basic canonical randomization.] ■

*■ Occasionally, one can establish that $\mathscr{D}$'s judgments about the population random vector $\tilde{\boldsymbol{\xi}}$ are expressible in the form $\tilde{\boldsymbol{\xi}} = \mathbf{q}(\tilde{\boldsymbol{\beta}})$ for some function $\mathbf{q}$: $R^k \to R^k$, where $\tilde{\boldsymbol{\beta}}$ has a k-variable Normal density function $f_N^{(k)}(\cdot|\mathbf{m}, \mathbf{V})$. Then the obvious analogue of Equation (6.6.3) obtains; that is, the k-tuple

$$\boldsymbol{\xi}_j = \mathbf{q}(\mathbf{Q}^{-1}[\boldsymbol{\epsilon}_j + \mathbf{m}]) \tag{6.6.5}$$

behaves for all the world like an observation from the probability function $P(\cdot)$ of $\tilde{\boldsymbol{\xi}}$ if $\boldsymbol{\epsilon}_j$ was generated according to the procedure described in the preceding remark. By defining $\mathbf{q}(\beta_1, \ldots, \beta_k) = (e^{\beta_1}, \ldots, e^{\beta_k})$, we obtain the so-called k-*variate lognormal* random variable, with the density function $f_{LN}^{(k)}(\cdot|\mathbf{m}, \mathbf{V})$ determined by applying Exercise 3.10.10 to the density function $f_N^{(k)}(\cdot|\mathbf{m}, \mathbf{V})$ of $\tilde{\boldsymbol{\beta}}$. The k-variate lognormal density functions are reasonable vehicles for expressing judgments about k nonindependent quantities $\tilde{\xi}_1, \ldots, \tilde{\xi}_k$ which, however, must all be nonnegative. ■

We now know how to simulate observations from a given probability function $P(\cdot)$ under certain rather general circumstances. Thus, we are now acquainted with the basic building block of stochastic simulation. The time has come to examine some applications in addition to the blob-area-estimation task in Example 6.6.1.

Stochastic simulation is often invoked as a computational technique in the analysis of how large, complex systems behave. In order to simulate the behavior of such systems, $\mathscr{D}$ must generally specify:

(A) the *description* of the system, in terms of a mathematical model which specifies:

(1) the *sub*systems $h = 1, \ldots, H$ of the system;
(2) pertinent measure(s) ξ_h, generally vectorial, of the behavior of subsystem h, for $h = 1, \ldots, H$;
(3) pertinent measure(s) ω, again generally vectorial, of the behavior of the system itself; and
(4) an at-least constructively definable° function $y: \Xi_1 \times \cdots \times \Xi_H \to \Omega$ which specifies system behavior ω as a deterministic function $y(\xi_1, \ldots, \xi_H)$ of the behaviors of all subsystems; and

(B) his probability function $P(\cdot)$ on the $\mathscr{E}$-move $\Xi = \Xi_1 \times \cdots \times \Xi_H$ of all possible subsystems' joint behaviors $\boldsymbol{\xi} = (\xi_1, \ldots, \xi_H)$.

Much art and ingenuity are often involved in specification (A), and the same goes for (B) if this task is to be performed in a responsible fashion.

■ Quite a bit of statistical and other empirical technology is involved in the so-called "validation" of the model, i.e., testing to see if your system description in (A) and your probability function in (B) are realistic characterizations of the actual system. A detailed discussion is not possible here; see Chapter 5 of Naylor [**102**], where further references may be found. ■

At this point, the reader with a long memory for notational nuances will recall Indirect Method (2) for assessing probability functions and say, "Aha! specifications (A) and (B) suffice for $\mathscr{D}$ to *compute* his logically induced probabilities $P(\tilde{\omega} \in \Omega')$ of all *system* events Ω' according to the necessary relationship in Section 6.1; namely,

$$P(\tilde{\omega} \in \Omega') = P(\tilde{\boldsymbol{\xi}} \in \{\boldsymbol{\xi}: y(\boldsymbol{\xi}) \in \Omega'\}) \tag{6.6.6}$$

for every $\tilde{\omega}$-event Ω'."

This is true, *in principle*, though *rarely in practice* with complex systems,

° By "at least constructively definable," we mean that the complexity of the system may rule out writing a short formula for $\omega = y(\xi_1, \ldots, \xi_H)$, but it does *not* rule out computing the *specific* ω which results from any *specific* $(\xi_1, \ldots, \xi_H)$ in $\Xi_1 \times \cdots \times \Xi_H$.

because $y(\cdot)$ is usually so complicated. In such cases it may be far simpler to simulate the system's behavior.

It is quite easy to describe how such a simulation proceeds in principle. *Each trial j* consists of:

(1) simulating one observation $\boldsymbol{\xi}_j = (\xi_{j1}, \ldots, \xi_{jH})$ from the probability function $P(\cdot)$ characterizing the joint behavior of the subsystems; and
(2) calculating $\omega_j = y(\boldsymbol{\xi}_j) = y(\xi_{j1}, \ldots, \xi_{jH})$.

We now know how to do (1), given that $P(\cdot)$ has some tractable form, and we know in principle how to do (2), because what is basically involved is tracing through some calculations. The *entire stimulation* consists of a large number n of independently simulated trials.

Example 6.6.3

Consider a project which consists of nine subprojects whose inceptions may have to await completions of other subprojects, where such interrelatedness is indicated in Figure 6.7 by writing

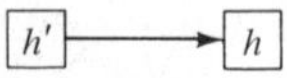

if subproject h cannot begin until h' is completed, and

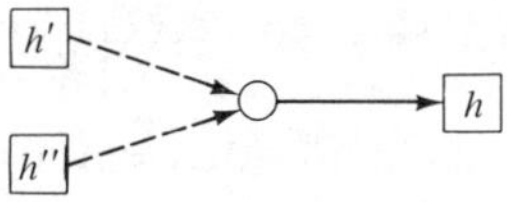

means that h cannot begin until *one* of h' and h'' is completed. Now you may wish to quantify your judgments about the *total* time $\tilde{\omega}$ required to perform the entire project, given some contemplated decision g regarding commitment and allocation of resources to the subprojects. But you would find this hard to do if many of the subproject durations ξ_h are uncertain, because of the complex fashion in which $(\tilde{\xi}_1, \ldots, \tilde{\xi}_9)$ determines $\tilde{\omega}$. Let us determine the "synthesis function" $y(\cdot)$, inductively, and assuming that every subproject $h > 1$ commences at the earliest possible opportunity. Subprojects 2, 3, and 4 begin at time $\tilde{\xi}_1$ after 1 begins (= time zero) and end at respective times $\tilde{\xi}_1 + \tilde{\xi}_2$, $\tilde{\xi}_1 + \tilde{\xi}_3$, and $\tilde{\xi}_1 + \tilde{\xi}_4$. Subproject 5 begins when 2 and 3 are both finished, which is at time $\max[\tilde{\xi}_1 + \tilde{\xi}_2, \tilde{\xi}_1 + \tilde{\xi}_3] = \tilde{\xi}_1 + \max[\tilde{\xi}_2, \tilde{\xi}_3]$; and 5 is completed at time $\tilde{\xi}_1 + \tilde{\xi}_5 + \max[\tilde{\xi}_2, \tilde{\xi}_3]$. Subproject 8 begins when the earlier of 2 and 4 is completed, or at time $\min[\tilde{\xi}_1 + \tilde{\xi}_2, \tilde{\xi}_1 + \tilde{\xi}_4] = \tilde{\xi}_1 + \min[\tilde{\xi}_2, \tilde{\xi}_4]$; and 8 is finished at time $\tilde{\xi}_1 + \tilde{\xi}_8 + \min[\tilde{\xi}_2, \tilde{\xi}_4]$. Subproject 6 begins when 5 is completed, and hence 6 is finished at time $\tilde{\xi}_1 + \tilde{\xi}_5 + \tilde{\xi}_6 + \max[\tilde{\xi}_2, \tilde{\xi}_3]$. Subproject 7 begins when 3 and 4 are both completed; i.e., at time $\max[\tilde{\xi}_1 + \tilde{\xi}_3, \tilde{\xi}_1 + \tilde{\xi}_4] = \tilde{\xi}_1 + \max[\tilde{\xi}_3, \tilde{\xi}_4]$, and hence 7 is finished at time $\tilde{\xi}_1 + \tilde{\xi}_7 + \max[\tilde{\xi}_3, \tilde{\xi}_4]$. Now, subproject 9 commences at the maximum of

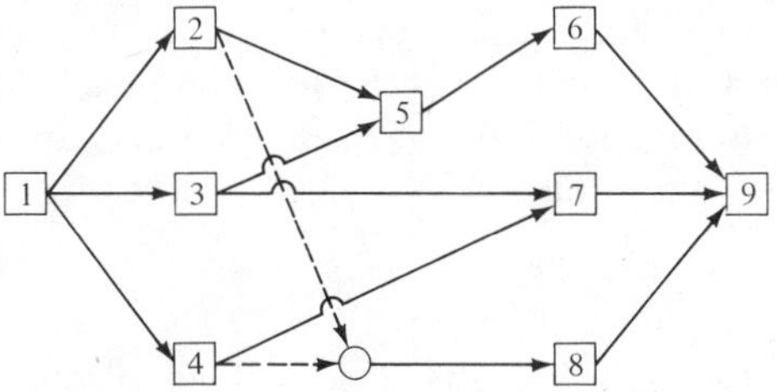

Figure 6.7 Project consisting of interdependent subprojects.

the completion times of subprojects 6, 7, and 8, and ends, together with the entire project, at time $\tilde{\omega} = \tilde{\xi}_9$ plus maximum of completion times of 6, 7, and 8. All written out, this is

$$\begin{aligned}\tilde{\omega} &= y(\tilde{\xi}_1, \ldots, \tilde{\xi}_9)\\ &= \tilde{\xi}_1 + \tilde{\xi}_9 + \max [\tilde{\xi}_5 + \tilde{\xi}_6 + \max [\tilde{\xi}_2, \tilde{\xi}_3],\\ &\qquad \tilde{\xi}_7 + \max [\tilde{\xi}_3, \tilde{\xi}_4],\\ &\qquad \tilde{\xi}_8 + \min [\tilde{\xi}_2, \tilde{\xi}_4]].\end{aligned} \tag{6.6.7}$$

This complicated function $y(\cdot)$ illustrates nicely the value of simulation, because it is clearly *impossible* to determine probabilities $P(\tilde{\omega} \in \Omega'|g)$ according to Equation (6.6.6) for this $y(\cdot)$, whereas it is *easy* to plug nine numbers $\xi_{j1}, \ldots, \xi_{j9}$ into (6.6.7) as substitutes for $\tilde{\xi}_1, \ldots, \tilde{\xi}_9$ respectively, and to calculate the resulting number ω_j. If these nine numbers constituted one simulated observation from $\mathscr{D}$'s joint probability function $P(\cdot|g)$ of $\tilde{\boldsymbol{\xi}}$-events, given his resources decision g, then the preceding sentence describes how the *second* step of performing trial j is completed. If a goodly number n of trials are independently simulated, the resulting random sample $\omega_1, \ldots, \omega_n$ of project durations should furnish a reasonably accurate estimate of $\mathscr{D}$'s induced probability $P(\tilde{\omega} \leq T|g)$ of meeting some contract deadline (say) T; from Section 6.4, $P(\tilde{\omega} \leq T|g)$ is generally estimated by (number of ω_j's $\leq T)/n$.

Example 6.6.4

Other attributes of the project might also be of interest; for instance, identifying which subprojects may deserve special attention because it is highly probable that their completions will be what is holding up subsequent subprojects and, hence, delaying the completion of the project. More specifically, for each possible $(\xi_1, \ldots, \xi_9)$ one or more of the six "paths" ① = $1 \to 2 \to 5 \to 6 \to 9$, ② = $1 \to 2 \to 8 \to 9$, ③ = $1 \to 3 \to 5 \to 6 \to 9$, ④ = $1 \to 3 \to 7 \to 9$, ⑤ = $1 \to 4 \to 7 \to 9$, and ⑥ = $1 \to 4 \to 8 \to 9$ is a *critical path*, meaning that the sum of the durations of its constituent subprojects equals the total project's duration $y(\xi_1, \ldots, \xi_9)$. On each trial j we can record along with $\omega_j = y(\xi_{j1}, \ldots, \xi_{j9})$ the subprojects on the critical path for that trial. After all n trials have been simulated, we can readily estimate the probability that subproject h will be on the critical path by the relative frequency (number of trials with h critical)$/n$. From

Figure 6.7 it is clear that subprojects 1 and 9 are bound to be on the critical path. On the other hand, if $\tilde{\xi}_3$ will surely be less than each of $\tilde{\xi}_2$ and $\tilde{\xi}_4$, then subproject 3 cannot be on the critical path, because it cannot hold up any subsequent subproject.

Example 6.6.5

In the following table, we record the results of a very small number $n = 15$ of independent trials, under the assumption that $P(\cdot|g)$ is such that $\tilde{\xi}_1, \ldots, \tilde{\xi}_9$ are independent and each is a basic canonical randomization.[p] Thus $\xi_{jh} = \lambda_{jh}$ for every h and j.

Trial	ξ_1	ξ_2	ξ_3	ξ_4	ξ_5	ξ_6	ξ_7	ξ_8	ξ_9
1	.14	.21	.34	.04	.99	.44	.69	.55	.72
	implying $\omega_1 = 2.63$, critical path $= 1 \to 3 \to 5 \to 6 \to 9$								
2	.01	.04	.83	.85	.48	.09	.37	.08	.94
	implying $\omega_2 = 2.35$, critical path $= 1 \to 3 \to 5 \to 6 \to 9$								
3	.37	.01	.22	.07	.99	.12	.78	.15	.53
	implying $\omega_3 = 2.23$, critical path $= 1 \to 3 \to 5 \to 6 \to 9$								
4	.32	.15	.02	.08	.50	.16	.28	.99	.56
	implying $\omega_4 = 1.95$, critical path $= 1 \to 4 \to 8 \to 9$								
5	.38	.95	.58	.49	.59	.63	.31	.31	.78
	implying $\omega_5 = 3.33$, critical path $= 1 \to 2 \to 5 \to 6 \to 9$								
6	.89	.93	.83	.89	.59	.04	.45	.31	.15
	implying $\omega_6 = 2.60$, critical path $= 1 \to 2 \to 5 \to 6 \to 9$								
7	.26	.02	.29	.60	.58	.52	.70	.17	.82
	implying $\omega_7 = 2.47$, critical path $= 1 \to 3 \to 5 \to 6 \to 9$								
8	.82	.58	.40	.93	.09	.14	.24	.04	.58
	implying $\omega_8 = 2.57$, critical path $= 1 \to 4 \to 7 \to 9$								
9	.51	.27	.63	.40	.73	.28	.10	.69	.92
	implying $\omega_9 = 3.07$, critical path $= 1 \to 3 \to 5 \to 6 \to 9$								
10	.52	.11	.04	.51	.29	.06	.06	.42	.78
	implying $\omega_{10} = 1.87$, critical path $= 1 \to 4 \to 7 \to 9$								
11	.89	.56	.73	.50	.73	.01	.80	.46	.22
	implying $\omega_{11} = 2.64$, critical path $= 1 \to 3 \to 7 \to 9$								
12	.02	.20	.66	.62	.21	.16	.33	.94	.01
	implying $\omega_{12} = 1.17$, critical path $= 1 \to 2 \to 8 \to 9$								
13	.64	.84	.95	.70	.46	.42	.47	.12	.20
	implying $\omega_{13} = 2.67$, critical path $= 1 \to 3 \to 5 \to 6 \to 9$								
14	.83	.46	.42	.48	.30	.03	.96	.69	.68
	implying $\omega_{14} = 2.95$, critical path $= 1 \to 4 \to 7 \to 9$								
15	.74	.82	.33	.48	.13	.93	.49	.66	.02
	implying $\omega_{15} = 2.64$, critical path $= 1 \to 2 \to 5 \to 6 \to 9$								

[p] Naturally, much more complex forms of $P(\cdot|g)$ are amenable to simulation, as we have already shown. This assumed form of $P(\cdot|g)$ is only for expositional convenience.

Many more trials should be simulated for confidence in the resulting estimates, but on the basis of this small sample we obtain the impression that additional resources are better expended on improving the efficiency of subprojects 1, 9, 5, 6, and 3 than on the efficiency of subprojects 8, 2, 4, and 7.

Our second large class of applications concerns the behavior over time of a "queueing system," in which people arrive at sporadic times, await service, and take varying lengths of time to be served. It illustrates the fact that $y(\cdot)$ need not be spelled out explicitly if $\mathcal{D}$ is certain of the uses to which he will put the simulated, make-believe trial histories of the queueing system.

Example 6.6.6

Dr. Marius Illby is interested in determining whether he can physically handle by himself all patients who come to his soon-to-be-opened Free Clinic for Minor Ailments. The policy of this clinic is to be (1) first come first served; (2) opening at 4:00 P.M. and closing at 6:00 P.M. Monday through Friday; and (3) patients who are waiting at 6:00 P.M. shall be treated rather than turned away. Dr. Illby sees no reason to distinguish among the weekdays as far as the inflow of patients is concerned. His waiting room seats seven, but others can stand, sit on the floor, or wait in the doorway. The number he treats in any given day is determined wholly by the number who arrive and join the waiting line (or queue), but the amount of time he is busy and how long after 6:00 P.M. he has to stay depend (1) upon how the arrivals are spaced throughout the open period and (2) upon how long it takes to treat each case. Dr. Illby is willing to quantify his judgments about how patients will arrive during the open period and about the time needed to treat each; what he wants in return is information about such things (1) as the probability that more than seven people will be waiting at some point during the open period; (2) as the probability of having a free ten minutes (no one being treated or waiting) beginning between 4:30 P.M. and 5:00 P.M., to call his broker; and (3) as the time at which the last patient of the day finishes treatment, or 6:00 P.M., whichever is later, representing Illby's departure for home. Regarding *treatment times* $\tilde{\tau}_1, \tilde{\tau}_2, \ldots$ of patients $1, 2, \ldots$ on any given day, Illby judges them to be independent and identically distributed, with common cumulative distribution function $F_\tau(\cdot)$ defined[q] by

$$F_\tau(\tau) = \begin{cases} 0, & \tau \leq 1 \\ (\tau - 1)^2/70, & 1 \leq \tau \leq 6 \\ 1 - (15 - \tau)^2/126, & 6 \leq \tau \leq 15 \\ 1, & \tau \geq 15. \end{cases}$$

[q] This is the *triangular* cumulative distribution function $F_\Delta(\cdot|1, 6, 15)$; see *Exercises 3.10.2 and 3.18.8(Cii).

Illby has effected a tripartite quantification of judgments about arrivals of patients. (I) He believes there will be $\tilde{\nu}$ waiting at 4:00 P.M. in an orderly line outside the door, and his judgments about $\tilde{\nu}$ are expressed by the following, tabulated cumulative distribution function

$\nu \in$	$F_\nu(\nu)$	$\nu \in$	$F_\nu(\nu)$
$(-\infty, 0)$	.00	$[5, 6)$	.86
$[0, 1)$	.15	$[6, 7)$	.91
$[1, 2)$	.35	$[7, 8)$	.95
$[2, 3)$	.60	$[8, 9)$	.98
$[3, 4)$	.70	$[9, 10)$	.99
$[4, 5)$	.80	$[10, +\infty)$	1.00

(II) Illby has also quantified his judgments about the "interarrival" times $\tilde{\psi}_1, \tilde{\psi}_2, \ldots$, where $\tilde{\psi}_1$ is the lapse (in minutes) between 4:00 P.M. and the first subsequent arrival and, for $i > 1$, $\tilde{\psi}_i$ is the time (in minutes) between the arrivals of the $(i-1)$st and ith patients. His quantified judgments about $\tilde{\psi}_1$ are given by the following cumulative distribution function[r] $F_1(\cdot)$.

$$F_1(\psi_1) = \begin{cases} 0, & \psi_1 \le 0 \\ (\psi_1)^2/162, & 0 < \psi_1 \le 9 \\ 1 - (18 - \psi_1)^2/162, & 9 < \psi_1 \le 18 \\ 1, & 18 \le \psi_1; \end{cases}$$

and his judgments about $\tilde{\psi}_i$ conditional on $\tilde{\psi}_1, \ldots, \tilde{\psi}_{i-1}$ are expressed by the following cumulative distribution function[s] $F_i(\cdot|\zeta)$ where $\zeta = \Sigma_{j=1}^{i-1} \psi_j$.

$$F_i(\psi_i|\zeta) = \begin{cases} 0, & \psi_i \le 0 \\ (\psi_i)^2/(x(\zeta)z(\zeta)), & 0 < \psi_i \le x(\zeta) \\ 1 - [z(\zeta) - \psi_i]^2/(z(\zeta)[z(\zeta) - x(\zeta)]), & x(\zeta) < \psi_i \le z(\zeta) \\ 1, & \psi_i \ge z(\zeta), \end{cases}$$

where $x(\cdot)$ and $z(\cdot)$ are the following functions.

$$x(\zeta) = \begin{cases} 12, & \zeta \in [0, 60] \cup [90, 120] \\ 6, & \zeta \in (60, 90), \end{cases}$$

and

$$z(\zeta) = \begin{cases} 20, & \zeta \in [0, 60] \cup [90, 120] \\ 7, & \zeta \in (60, 90). \end{cases}$$

[These definitions of $x(\cdot)$ and $z(\cdot)$ reflect Illby's judgment that arrivals become more intense between 5:00 P.M. and 5:30 P.M. as people get off from work.]

[r] Triangular, $F_\Delta(\cdot|0, 9, 18)$.

[s] Also triangular, $F_\Delta(\cdot|0, x(\zeta), z(\zeta))$.

(III) Illby's remaining task is to assess his probability $p(\phi)$ that an arrival will wait if there are ϕ people already waiting for treatment when he arrives, and hence also, his probability $1-p(\phi)$ that the arrival will be discouraged and go away. (This is called "balking" in queueing theory.) Suppose that Illby assesses the following probabilities.

ϕ	$p(\phi)$	ϕ	$p(\phi)$	ϕ	$p(\phi)$
0	1.00	6	.80	12	.20
1	1.00	7	.50	13	.10
2	.95	8	.35	14	.05
3	.90	9	.35	15	.02
4	.85	10	.30	16	.01
5	.80	11	.30	≥ 17	.00

All the above probability assessments suffice jointly for the simulation of a day in the life of the clinic. Because the arrival times, patients' "balking" decisions about joining the line or leaving, and treatment times are so interactive, there is little hope of determining the quantities of interest by straightforward mathematical analysis. Instead, the following example shows how we can obtain *one* simulated day's record ω_j of the clinic.

Example 6.6.7

Our simulation will proceed largely in chronological time.

(1) Simulate ν_j. We obtained random decimal .73, implying by Method A and the table $F_\nu(\cdot)$ that $\nu_j = 4$ people waiting at 4:00 P.M.

(2) Simulate $\tau_{j1}, \ldots, \tau_{j\nu_j} = \tau_{j4}$ by Method A and $F_\tau(\cdot)$.

i	Random decimal	τ_{ji}
1	.17283	$1+\sqrt{70(.17283)} = 4.48$
2	.85376	$15-\sqrt{126(.14624)} = 10.71$
3	.55892	$15-\sqrt{126(.44108)} = 7.55$
4	.13517	$1+\sqrt{70(.13517)} = 4.08$

(3) Simulate ψ_{j1} by Method A and $F_1(\cdot)$. Random decimal $= 0.30019$, implying $\psi_{j1} = \sqrt{162(.30019)} = 6.97$. Patient 1 has left; patient 2 is being treated; thus 2 are waiting when this first "after-four" arrives.

(4) Simulate whether the first "after-four" patient patiently stays [with probability $p(2) = .95$] or impatiently "balks" [with probability .05], by saying that he stays if a two-place random decimal is $\leq .94 = p(2) - .01$. We get random decimal .28, so he *stays*.

(5) Simulate the treatment time τ_{j5} for the first "after-four" arrival: Random decimal = .90543, implying by $F_\tau(\cdot)$ and Method A that $\tau_{j5} = 15 - \sqrt{126(1 - .90543)} = 11.55$.

(6) Simulate ψ_{j2} by Method A and $F_2(\cdot|\zeta = 6.97)$: Random decimal = .47146, implying $\psi_{j2} = \sqrt{240(.47146)} = 10.64$; so the second "after-four" arrival appears at four plus $6.97 + 10.64$, or $\zeta = 17.61$ minutes past four. By (2), Illby has finished treating two of the "before-fours" and is working on the third, so that one "before-four" and one "after-four" are waiting when the second "after-four" arrives. Hence $\phi = 2$ for:

(7) Simulating whether the second after-four joins the line. Random decimal = .83, less than $.95 = p(2)$, and hence he *stays.*

(8) Simulate the treatment time τ_{j6} for the second after-four arrival. Random decimal = .32611, implying by $F_\tau(\cdot)$ and Method A that $\tau_{j6} = 1 + \sqrt{70(.32611)} = 5.78$ minutes.

(9) Simulate ψ_{j3} by Method A and $F_3(\cdot|\zeta = 17.61)$. Random decimal = .56781, implying $\psi_{j3} = \sqrt{240(.56781)} = 11.67$; so the third after-four appears at $17.61 + 11.67 = 29.28 =$ (new) ζ minutes past four. Since $\Sigma_{i=1}^{4} \tau_{ji} = 26.82$ and $\Sigma_{i=1}^{5} \tau_{ji} = 38.37$, the first "after-four" is being treated and the second is waiting. Hence $\phi = 1$ for this third after-four arrival.

(10) Since $p(1) = 1$, it is obvious without generating a random decimal that the third after-four *stays.*

(11) Simulate the treatment time τ_{j7} for the third after-four. Random decimal = .90701, implying by $F_\tau(\cdot)$ and Method A that $\tau_{j7} = 15 - \sqrt{126(1 - .90701)} = 11.58$.

Steps (8)–(11) are repeated over and over until one has gotten N_j after-fours and generates $\psi_{j(N_j+1)}$ with the property that $\Sigma_{i=1}^{N_j} \psi_{ji} \leq 120 \leq \Sigma_{i=1}^{N_j+1} \psi_{ji}$. This means that the $(N_j + 1)$st after-four does not arrive before closing time. Without going through the details, we record the results of doing this in a Summary Table A; obviously, $N_j = 16$. Note that *this table is our single observation* ω_j, and hence that writing a formula of the form $\omega_j = y(\xi_{j1}, \ldots, \xi_{jH})$ would be clumsy to say the least.

■ It is impossible to overstress that Table A is only one observation. In this observation, Illby has *no* "idle time" between 4:00 P.M. and closing to call his broker or to do anything other than treat patients. But estimating the probability of having time for a ten-minute call as zero, by Section 6.5, would be ridiculous because such an estimate would be based on $n = 1$ observation! Also ridiculous would be placing great confidence in the estimate 6:17.86 P.M. of Illby's departure time, and in the estimate zero of the probability that more than seven people will be waiting at some point during the open period. ■

Table A Summary Table = ω_j

(1) Time	(2) Number arriving	(3) Stay?	(4) Number departing	(5) Number in line after (2)–(4)	(6) Patient being treated
4:00	4	yes	0	3	pre-four #1
4:04.48	0	—	1	2	pre-four #2
4:06.97	1	yes	0	3	pre-four #2
4:15.19	0	—	1	2	pre-four #3
4:17.61	1	yes	0	3	pre-four #3
4:22.74	0	—	1	2	pre-four #4
4:26.82	0	—	1	1	after-four #1
4:29.28	1	yes	0	2	after-four #1
4:38.37	0	—	1	1	after-four #2
4:40.71	1	yes	0	2	after-four #2
4:44.15	0	—	1	1	after-four #3
4:51.69	1	yes	0	2	after-four #3
4:55.73	0	—	1	1	after-four #4
4:61.87	1	yes	0	2	after-four #4
4:66.47	1	yes	0	3	after-four #4
4:67.03	0	—	1	2	after-four #5
4:70.01	1	yes	0	3	after-four #5
4:71.98	0	—	1	2	after-four #6
4:73.05	1	yes	0	3	after-four #6
4:78.60	1	yes	0	4	after-four #6
4:78.84	0	—	1	3	after-four #7
4:80.37	1	no	0	3	after-four #7
4:84.02	1	yes	0	4	after-four #7
4:90.53	1	yes	0	5	after-four #7
4:91.97	0	—	1	4	after-four #8
4:92.96	1	yes	0	5	after-four #8
4:94.53	0	—	1	4	after-four #9
4:106.06	0	—	1	3	after-four #10
4:110.63	0	—	1	2	after-four #12*
4:111.50	1	yes	0	3	after-four #12
4:112.79	0	—	1	2	after-four #13
4:118.09	1	yes	0	3	after-four #13
**4:120.00	0	—	0	3	after-four #13
4:120.71	0	—	1	2	after-four #14
4:126.67	0	—	1	1	after-four #15
4:133.20	0	—	1	0	after-four #16
4:137.86 = 6:17.86	0	—	1	0	Illby finished

* After-four No. 11 departed without waiting.

** Closed to new arrivals.

The clumsiness of simulating one observation in Example 6.6.7 by hand with a table of random digits[t] or with a canonical icosahedral die would be discouraging were it not for the existence of extremely rapid and convenient digital computer simulation techniques. The books of Naylor, Balintfy, Burdick, and Chu [**103**] and Naylor [**102**] constitute jointly a good introduction to these techniques. They also present certain important applications of simulation not introduced here.

We have said little in this section about the specific inferences which one may draw from the output of a stochastic simulation. The reason for this omission is that such inferences do not differ in any fundamental respect from statistical inferences in general, and hence nothing need be added to the discussion of this subject in Section 6.5 and Chapter 9. Also see Exercises 6.6.6 and 6.6.7, as well as the remarks following Exercise 6.6.7.

Exercise 6.6.1

Verify all entries in Table A, using the following simulated data. *Treatment times* τ_{j1} through τ_{j19} were 4.48, 10.71, 7.55, 4.08, 11.55, 5.78, 11.58, 11.30, 4.95, 6.86, 13.13, 2.56, 11.53, 4.57, 2.16, 7.92, 5.96, 6.53, and 4.66 minutes *respectively. Interarrival times* ψ_{j1} through ψ_{j17} were 6.97, 10.64, 11.67, 11.43, 10.98, 10.18, 4.60, 3.54, 3.04, 5.55, 1.77, 3.65, 6.51, 2.43, 18.54, 6.59, and 9.73 minutes *respectively*. The first 16 two-digit random decimals for determining whether after-four patients 1 through 16 wait or leave without treatment were .73, .83, .96, .62, .99, .55, .46, .85, .85, .33, .98, .20, .42, .12, .91, and .64.

Exercise 6.6.2

Simulate another day's table ω_j, for Illby's clinic, using the random decimals (rds) shown on page 289: (1) rd for simulating $\nu_{j'} = .05$.

(A) by using the tabulated decimals *in the order listed* (e.g., the decimal for simulating the first treatment time is .12917); and
(B) by using the tabulated decimals in the *reverse* of the order listed (e.g., the decimal for simulating the first treatment time is .16335).

[*Note*: We have tabulated *more* decimals than you will need for either part of this exercise, taking into consideration that no arrivals are accepted after 6:00 = 4:120.00 P.M.]

***Exercise 6.6.3**

Show that if T is large, then

$$\tilde{\xi}_j = m + \sqrt{12TV}\left[(1/T)\sum_{t=1}^{T}\tilde{\lambda}_{jt} - 1/2\right]$$

[t] Such as *A Million Random Digits with 100,000 Normal Deviates* by the RAND Corp. [**116**]. My colleague Professor Steven A. Zeff has repeatedly warned of the inconsistency in the title of this work.

(2) rds for $\psi_{j'}$'s	(3) rds for $p(\phi)$'s	(4) rds for $\tau_{j'}$'s
.06185	.03	.12917
.97251	.50	.41189
.57314	.87	.79811
.86207	.10	.52407
.74092	.79	.00275
.29603	.37	.42224
.68141	.69	.02776
.97382	.50	.44141
.39780	.79	.59860
.49479	.17	.64529
.89806	.81	.59703
.68731	.52	.78369
.90965	.99	.79858
.52359	.97	.08576
.81057	.23	.63394
.87102	.29	.05762
.84389	.98	.55093
.60164	.63	.35856
.28270	.82	.36410
.93778	.51	.08703
.15046	.11	.50242
.79348	.53	.50197
.11031	.90	.69051
.69691	.24	.20731
.83346	.71	.16335

has approximate probability function $P_N(\cdot|m, V)$, provided that $\tilde{\lambda}_{j1}, \ldots, \tilde{\lambda}_{jT}$ are independent (large m-place) random decimals.

***Exercise 6.6.4: Facts about Lognormals**

We say that the continuous random real k-tuple $\tilde{\boldsymbol{\xi}} = (\tilde{\xi}_1, \ldots, \tilde{\xi}_k)$ has the k-variate *lognormal* density function $f_{\mathrm{LN}}^{(k)}(\cdot|\mathbf{m}, \mathbf{V})$ if the density function of $\tilde{\boldsymbol{\beta}} = (\tilde{\beta}_1, \ldots, \tilde{\beta}_k)$ is $f_{\mathrm{N}}^{(k)}(\cdot|\mathbf{m}, \mathbf{V})$ and $\tilde{\beta}_i = \log_e [\tilde{\xi}_i]$ for $i = 1, \ldots, k$. Let $\boldsymbol{\beta}(\boldsymbol{\xi}) = (\log_e [\xi_1], \ldots, \log_e [\xi_k])$ for every $\boldsymbol{\xi}$ with all components positive.
(A) Show that

$$f_{\mathrm{LN}}^{(k)}(\cdot|\mathbf{m}, \mathbf{V}) = \begin{cases} \left(\prod_{i=1}^{k} \xi_i\right)^{-1} (2\pi)^{-k/2}|\mathbf{V}|^{-1/2} e^{-(1/2)(\boldsymbol{\beta}(\boldsymbol{\xi})-\mathbf{m})'\mathbf{V}^{-1}(\boldsymbol{\beta}(\boldsymbol{\xi})-\mathbf{m})}, & \text{each } \xi_i > 0 \\ 0, & \text{some } \xi_i \leq 0. \end{cases}$$

(B) Show that

$$E_{\mathrm{LN}}^{(k)}\{\tilde{\xi}_i|\mathbf{m}, \mathbf{V}\} = e^{m_i+(1/2)V_{ii}}$$

for $i = 1, \ldots, k$.

(C) Show that

$$V_{\mathrm{LN}}^{(k)}\{\tilde{\xi}_i, \tilde{\xi}_j|\mathbf{m}, \mathbf{V}\} = e^{m_i+m_j+(1/2)(V_{ii}+V_{jj})}[e^{V_{ij}} - 1]$$

for $i = 1, \ldots, k$ and $j = 1, \ldots, k$.

Exercise 6.6.5: Discussion Question

Is there such a thing as "objective randomness"? The following points are pertinent.

(1) If one knew enough physics, could measure sufficiently accurately, and had a sufficiently sensitive touch, then he could control the outcome of a die cast. Casinos take apparent cognizance of such faculties by imposing a rule that dice must rebound from a backboard.

(2) Once computed, there is nothing random about the million random digits published in the RAND book [**116**]; they are simply "patternless" as judged by certain statistical tests and, hence, "behave for all the world *as if*" they were a random sample from the mass function assigning probability 1/10 to each of the ten digits $0, 1, \ldots, 9$.

(3) Computers produce and use so-called "pseudo-random" decimals δ_j by an *algebraic formula* such as δ_j = decimal point followed by the rightmost ten digits of the integer $(9.9997 \times 10^{14})\delta_{j-1}$ for $j = 1, 2, \ldots,$ where δ_0 can be chosen somewhat arbitrarily as 10^{-10} times any ten-digit integer not divisible by 5 (e.g., .1279674813). What is so random about the output of a mechanistic procedure such as this?

■ It can be shown that all procedures such as the one just described produce "periodic" sequences of decimals. For the numerical instance cited, the maximum number of δ_j's that can be obtained without repeating the cycle is 500 million. ■

Exercise 6.6.6: Inferences from Simulation Experiments (I)

(A) Suppose that $n = 100$ observations ω_j were simulated on the blob of Example 6.6.1 and that 63 of them were "hits" on the blob. Hence, Section 6.5 suggests estimating the blob's area θ to be .63. Before running this simulation, what was a lower bound for the probability of obtaining an estimate $\bar{x}_{100}$ of θ within .05 of θ?

(B) Using the simulated behavior ($n = 15$) of the project system in Examples 6.6.3–6.6.5, Section 6.5 suggests estimating the probability of subproject 5 being on the critical path to be 9/15, or .60. (Why?) With at least what pre-15-trial-simulation probability would the resulting *estimate* of P(5 on critical path$|g$) be within .20 of P(5 on critical path$|g$)?

(C) Suppose that on 14 out of 100 simulated days' histories of the clinic, Dr. Illby had the ten minutes for his call to his broker. With at least what presimulation probability would his relative-frequency estimate (number of days had time for call)/100 be within 1/4 of the actual probability of having time for the call?

***Exercise 6.6.7: Inferences from Simulation Experiments (II)**

This exercise is greatly facilitated by a review of *Example 3.11.1 and the final remark in Section 6.5 on alternatives to sample-mean-estimate-based inferences, alternatives that incorporate $\mathscr{D}$'s prior judgments about the quantity of interest.

(A) In Example 6.6.1, suppose that prior to the simulation $\mathscr{D}$ had quantified his judgments about the area $\tilde{\theta}$ of the blob in the form of some beta density function $f_\beta(\cdot|r', n')$. Show that *after* the 100 trials with 63 "hits" cited in Exercise 6.6.6(A), $\mathscr{D}$'s judgments about $\tilde{\theta}$ are expressed by the beta density function $f_\beta(\cdot|r' + 63, n' + 100)$.

(B) Similarly, suppose in the context of Exercise 6.6.6(B) that prior to the simulation, $\mathscr{D}$ had quantified his judgments about the as-yet unknown (because uncalculated) probability $\pi = P(5 \text{ on critical path}|g)$ in the form of a beta density function $f_\beta(\cdot|r', n')$. What density function expresses $\mathscr{D}$'s judgments about $\tilde{\pi}$ as revised by the simulation output cited in Example 6.6.5?

(C) In the context of Exercise 6.6.6(C), suppose that Illby had (optimistically) quantified his judgments about the probability $\tilde{\pi}$ of his having the ten minutes free in the form of the beta density function $f_\beta(\cdot|3, 4)$.

   (i) What was Illby's expectation $E\{\tilde{\pi}\}$ of $\tilde{\pi}$ prior to conducting the simulation?

   (ii) What density function expresses Illby's judgments about $\tilde{\pi}$ as revised by finding that on 14 out of 100 simulated days he did have the time?

   (iii) What is Illby's revised expectation $E\{\tilde{\pi}|\text{simulation output}\}$?

■ It might appear that the specter of logically infinite regress materializes from Illby's assessing a probability function on his *deducible* (from other assessed probabilities) and (in principle) *computable* probability $\tilde{\pi}$ of having time for the call to his broker. But such an appearance is illusory because it amounts to failure to recognize the two senses in which the notion of probability is used in this line of reasoning. The following detailed remarks should help to dispel such unhealthy notions about infinite regress.

(1) Illby's other probability assessments *suffice to determine uniquely* the number π, which can therefore be calculated in principle, but only with great difficulty, difficulty which simulation seeks to alleviate.

(2) *Before* calculating or estimating by simulation the number π, it is, tautologically, as yet undetermined, and hence a random variable, written $\tilde{\pi}$.

(3) Illby is called on to quantify his judgments about this random variable $\tilde{\pi}$ in the form of a probability function (which we assumed to be representable by a beta density function).
(4) The random variable $\tilde{\pi}$ *just happens* to denote an as-yet undetermined probability. That is why in (1)–(3) above we have referred to "the number π" and to "the random variable $\tilde{\pi}$." ■

■ We can generalize from the preceding remark to observe that one *can* quantify his judgments about the number θ to be produced by any deterministic calculation, e.g., the as-yet undetermined (by me) integer $\theta = (146)(215)(313) - (113)(223)(247)$. Then Bayes' Theorem preserves ordinary arithmetical common sense in the following respect. Let the result of the arithmetical calculation be denoted by ζ, and assume that your arithmetic is *error-free.* This assumption means that

$$(*) \qquad P(\zeta|\theta) = \begin{cases} 1, & \zeta = \theta \\ 0, & \zeta \neq \theta, \end{cases}$$

and hence that obtaining ζ is really obtaining *perfect information* about θ, in the sense of Example 3.7.7. That example went on to show that your judgments $P(\theta|\zeta)$ about θ as revised by ascertaining ζ are now in conformity with arithmetical common sense:

$$(**) \qquad P(\theta|\zeta) = \begin{cases} 1, & \theta = \zeta \\ 0, & \theta \neq \zeta. \end{cases}$$

Now, the role of stochastic simulation is put in proper perspective by observing (once again) that the perfect-information calculation ζ may be very expensive, and therefore that one might do better on balance by trading off perfection of information for drastically lower cost information in the form of the output of a simple simulation. ■

■ How many people really trust their arithmetic in the sense of (honestly!) assessing (*) for all ζ and θ? I for one do not trust my arithmetic nearly that well. Machines also produce errors occasionally. ■

III

FURTHER TOPICS IN INDIVIDUAL DECISION MAKING

The time has come, the walrus said
To speak of many things
Of ships and shoes and sealing wax
And cabbages and kings.
Lewis Carroll

7

DECISIONS IN NORMAL FORM AND SENSITIVITY ANALYSIS

Why, sir, a cow is a very good animal in the field, but we turn her out of the garden.
Dr. Samuel Johnson, opposing broadened suffrage

7.1 INTRODUCTION

Rarely is a decision analysis conducted by $\mathscr{D}$ in total isolation from colleagues who may differ with him regarding (1) availability of some courses of action, (2) preferences for the potential consequences, and (3) judgments about the possible events. With the problem represented in extensive form, $\mathscr{D}$ can ascertain the ramifications of such disagreements only by performing a separate, complete recursion analysis for each set of assumptions, according to the methodology of Chapter 4.

The extensive form of a decision problem under uncertainty is thus not well suited to the garden in which (1) we eliminate all consensually inferior courses of action and (2) we study the importance of remaining disagreements. These activities will be referred to somewhat loosely as *sensitivity analysis.*

Another representation of decision problems under uncertainty, the so-called *normal form,* does facilitate sensitivity analysis. With the normal form, it is conceptually and often practicably simple to determine what changes are and/or should be wrought by given changes in the assumptions.

Definition 7.1.1. A *dpuu in normal form* consists of the following.

(1) A nonempty set S of elements s, called *pure strategies for* $\mathscr{D}$.
(2) A nonempty set Σ of elements σ, called *pure strategies for* $\mathscr{E}$.
(3) A function $c[\cdot,\cdot]$ mapping $S\times\Sigma$ into a set C of consequences, with $c[s,\sigma]$ denoting the consequence resulting from $\mathscr{D}$'s and $\mathscr{E}$'s choices of s and σ respectively.

■ There is no relationship between the "normal form" of a dpuu (= decision problem under uncertainty) and the Normal density functions defined in *Section 3.10. ■

■ The generic symbol s is intended to suggest strategy; σ is Greek s and is also intended to suggest strategy. Its use for $\mathscr{E}$'s strategy is consistent with our notational convention, for our parlor-game model of a dpuu implies that $\mathscr{E}$'s choice of σ is the same thing as the occurrence of elementary event σ. ■

■ To appreciate the conceptual simplicity of the normal form, you may think of S as a set of *lotteries* s, all on the same $\mathscr{E}$-move Σ, with consequences denoted by $c[s, \sigma]$ instead of by the $s(\sigma)$ which would be in perfect conformity with the $\ell(\omega)$ in Chapters 3 through 5 for lotteries ℓ on Ω. ■

Suppose that a dpuu in normal form has a finite number $\#(S)$ of pure strategies s for $\mathscr{D}$ and a finite number $\#(\Sigma)$ of pure strategies σ for $\mathscr{E}$. Then it can be arranged in *tabular* format, with rows $s^1, s^2, \ldots, s^{\#(S)}$, columns $\sigma^1, \sigma^2, \ldots, \sigma^{\#(\Sigma)}$, and table entries $c[s^i, \sigma^j]$.

We have intimated that *any* dpuu can be expressed in normal form. Section 7.2 is devoted to showing that this is so, by describing useful methods for converting to normal form a dpuu originally expressed in extensive form. But the techniques in Section 7.2 actually establish more; namely, that *in the resulting normal form, $\mathscr{D}$'s judgments about events Σ' in Σ do not depend upon his choice of* s.

■ That any dpuu can be put in a normal form in which $\mathscr{D}$'s judgments are independent of his choice of pure strategy s is in pleasant conformity with the model of a dpuu in normal form as a set of lotteries s all on the same $\mathscr{E}$-move domain Σ. ■

If we think of a dpuu in normal form as a set of lotteries s all on the same $\mathscr{E}$-move Σ, it is clear that $\mathscr{D}$ should choose a pure strategy s so as to maximize the expectation of his utility $u(c[s, \tilde{\sigma}])$ with respect to his probability function on Σ. The pure strategy so chosen is optimal in the same sense, and for exactly the same reasons, as given in Theorem 3.2.1. Moreover, this optimal strategy in the *normal* form of a dpuu will agree precisely with an optimal entire course of action, in the sense of Chapter 4, for the same dpuu in extensive form, provided that $\mathscr{D}$ uses the same utility function in both formulations and that his probability function on Σ is consistent with his probability functions on the $\mathscr{E}$-moves in the dpuu's extensive form.

■ Section 7.2 shows that consistent probability functions can always be obtained, by calculating $\mathscr{D}$'s probability function on Σ from his probability functions on the $\mathscr{E}$-moves of his dpuu in extensive form—or vice versa. ■

The extensive form of a dpuu is both intuitively natural and, usually, very amenable to rapid computation of an optimal course of action. In these respects it is superior to the normal form, which often consists of a huge table with many columns σ and many rows s; in fact, each pure strategy s for $\mathscr{D}$ is a complete course of action in the extensive form of his dpuu. Hence the extensive form is preferable unless one wishes to undertake sensitivity analyses or to deviate otherwise from the basic analysis as set forth in Chapters 2–4.

As noted, the normal form is better suited to sensitivity analysis. It also provides a conceptually clean framework for other digressions from the straight and narrow, such as (1) introducing *other* approaches to the analysis of dpuu's and (2) generalizing the way in which $\mathscr{D}$ chooses a strategy so as to permit his making this choice by operating a randomization device.

The other approaches to decision analysis differ from ours as developed in Chapters 3 and 4, and hence they may violate in any given instance one or more of the basic consistency principles as applied to $\mathscr{D}$'s quantified judgments and preferences. But they may enable $\mathscr{D}$ to avoid some quantification efforts, and they are interesting in their own right. We shall examine one such approach, the *maximin* criterion, in the exercises in subsequent sections.

Generalizing $\mathscr{D}$'s choice of strategy from a simple act of will to the operation of a randomization device is easy. If $\mathscr{D}$ is free to choose between pure strategies s' and s'', then he is also free to choose between s' and s'' on the basis of the flip of a coin if he so desires. More generally, $\mathscr{D}$ may decide to use a randomization device by which his (prerandomization!) probability of ultimately choosing a pure strategy in a given subset S' of S is $P_S(S')$, for every S' in S, where P_S is a probability function on S.

■ $\mathscr{D}$ can always use (an approximation to) a basic canonical randomization Λ, defined in Section 3.4, as the actual randomization device. Its outcomes λ can be relabeled appropriately, via a function $t: \Lambda \to S$, so that $P_S(S') = P_\Lambda(\tilde{\lambda} \in \{\lambda : t(\lambda) \in S'\})$. For instance, if S contains $\#(S) = 4$ pure strategies, and if $\mathscr{D}$ wants to choose s^1, s^2, s^3, and s^4 with respective probabilities .14, .23, .57, and .06, then he can define $t(\cdot)$ so as to choose $t(\lambda) = s^1$ if $\lambda \in [0, .14)$, $t(\lambda) = s^2$ if $\lambda \in [.14, .37)$, $t(\lambda) = s^3$ if $\lambda \in [.37, .94)$, and $t(\lambda) = s^4$ if $\lambda \in [.94, 1.00]$. In fact, he needs only a two-place random decimal (e.g., two rolls of an icosahedral die) to conduct this randomization. *Fancier methods of "simulating an observation from a probability function" are introduced in Section 6.6. ■

Note that $\mathcal{D}$'s choice of a randomization device has the effect of converting S from a $\mathcal{D}$-move to an $\mathcal{E}$-move,[a] but one with the important distinction that $\mathcal{D}$ *chooses* rather than *assesses* his probability function on its events.

By the same token, $\mathcal{D}$'s probability function P_Σ on events in Σ, representing his judgments about $\mathcal{E}$'s choice of pure strategy σ, is technically equivalent to $\mathcal{D}$'s knowing for sure that $\mathcal{E}$ is irrevocably committed to the operation of some randomization device which will result in his choosing a pure strategy in Σ' with probability $P_\Sigma(\Sigma')$, for every event Σ' in Σ. This alternative interpretation of P_Σ facilitates a player-symmetric representation of the notion of choosing a pure strategy via a randomization device.

Definition 7.1.2: Randomized Strategies

(A) A *randomized strategy*, or *mixed strategy*, for $\mathcal{E}$ is a probability function P_Σ on events Σ' in the set Σ of all of $\mathcal{E}$'s pure strategies σ.

(B) A *randomized strategy*, or *mixed strategy*, for $\mathcal{D}$ is a probability function P_S on events S' in the set S of all of $\mathcal{D}$'s pure strategies s.

(C) We denote by $\mathcal{Q}_\Sigma$ and $\mathcal{Q}_S$ the sets of *all* randomized strategies for $\mathcal{E}$ and for $\mathcal{D}$ respectively.

Note that a pure strategy s^* (say) for $\mathcal{D}$ *is* a randomized strategy for $\mathcal{D}$;[b] s^* is, in fact, the randomized strategy P_S in which s^* is chosen with probability one.

*■ Technically, s^* is the randomized strategy (= probability function) P_S such that $P_S(S') = 1$ if $s^* \in S'$ and $P_S(S') = 0$ if $s^* \notin S'$. ■

Similarly, a pure strategy σ^* (say) for $\mathcal{E}$ *is* a randomized strategy for $\mathcal{E}$.

Now, at this point it is natural to question the desirability of widening $\mathcal{D}$'s range of choice from the set S of all his pure strategies to the set $\mathcal{Q}_S$ of all his randomized strategies. Can $\mathcal{D}$ do *better* by choosing some genuinely randomized strategy[c] than he can by choosing any pure strategy? In our approach to decision analysis, the answer is "no," but in other approaches such as "maximin," it may be "yes."

For the ensuing analysis it is essential to have a concise notation system for $\mathcal{D}$'s utility of the various pairs of strategies, one for him and one for $\mathcal{E}$, as well as for $\mathcal{D}$'s utility of choosing a strategy optimally from a given subset of $\mathcal{Q}_S$.

Definition 7.1.3: Concise Utility Notation. Let $u: C \to R^1$ denote $\mathcal{D}$'s utility function on the set C of consequences of the dpuu. Then:

(A) For every $s \in S$ and $\sigma \in \Sigma$, we define $U(s, \sigma)$ to be $\mathcal{D}$'s utility $u(c[s, \sigma])$ of $c[s, \sigma]$.

---

[a] This means a violation of the Latin-versus-Greek notation conventions. By now you should have little need for such crutches.

[b] This clause is at least as true as the billboards which allege "(Actor) x *is* (role) y"!

[c] Since a pure strategy is a randomized strategy, we must phrase carefully references to "impure" strategies!

(B) For every $P_S \in \mathcal{Q}_S$ and every $P_\Sigma \in \mathcal{Q}_\Sigma$, we define $U(P_S, P_\Sigma)$ to be $\mathcal{D}$'s prerandomizations expectation (with respect to P_S and P_Σ) of his utility $U(\tilde{s}, \tilde{\sigma}) = u(c[\tilde{s}, \tilde{\sigma}])$ of the uncertain consequence $c[\tilde{s}, \tilde{\sigma}]$ which will result from $\mathcal{D}$'s and $\mathcal{E}$'s implementing the randomized strategies P_S and P_Σ respectively.

(C) For every $P_\Sigma \in \mathcal{Q}_\Sigma$ and for every closed[d] and nonempty subset $\mathcal{Q}'_S$ of the set $\mathcal{Q}_S$ of all randomized strategies for $\mathcal{D}$, we define $U(\mathcal{Q}'_S, P_\Sigma)$ to be $\mathcal{D}$'s prerandomizations expected utility of choosing P_S *optimally* from $\mathcal{Q}'_S$; that is,

$$U(\mathcal{Q}'_S, P_\Sigma) = \max_{P_S \in \mathcal{Q}'_S} [U(P_S, P_\Sigma)].$$

■ When S and Σ are finite sets, with $S = \{s^1, \ldots, s^{\#(S)}\}$ and $\Sigma = \{\sigma^1, \ldots, \sigma^{\#(\Sigma)}\}$, it is clear that

$$U(P_S, P_\Sigma) = \sum_{i=1}^{\#(S)} \sum_{j=1}^{\#(\Sigma)} U(s^i, \sigma^j) P_\Sigma(\sigma^j) P_S(s^i). \tag{7.1.1}$$

Equation (7.1.1) is indicative of the notational economy furnished by Definition 7.1.3(B). When P_S is the pure strategy s^*, then we write $U(s^*, P_\Sigma)$ in place of $U(P_S, P_\Sigma)$; and similarly, we write $U(P_S, \sigma^*)$ in place of $U(P_S, P_\Sigma)$ when P_Σ is the pure strategy σ^* for $\mathcal{E}$. If Σ is finite, then

$$U(s^*, P_\Sigma) = \sum_{j=1}^{\#(\Sigma)} U(s^*, \sigma^j) P_\Sigma(\sigma^j); \tag{7.1.2}$$

and if S is finite, then

$$U(P_S, \sigma^*) = \sum_{i=1}^{\#(S)} U(s^i, \sigma^*) P_S(s^i). \tag{7.1.3}$$

Equations (7.1.2) and (7.1.3) follow immediately from (7.1.1), together with the respective assumptions that $P_S(s^*) = 1$ and that $P_\Sigma(\sigma^*) = 1$. ■

So much for now about randomized strategies and utility notation. You may test your understanding of these matters by working Exercises 7.1.1 and 7.1.2.

Our primary motivation in studying the normal form of dpuu's is to provide a flexible model for analyzing potential controversy regarding

(1) judgments and preferences, and
(2) availabilities of certain courses of action in the dpuu.

Potential controversy regarding judgments and preferences is the subject of Sections 7.3 through 7.5; potential controversy regarding available courses of action is considered at the end of this section and further in Section 7.6.

In Sections 7.3 and 7.4 we assume that there is no controversy whatsoever regarding preferences; all concerned with the dpuu agree com-

[d] Closed sets are defined in Appendix 3. We assume that $\mathcal{Q}'_S$ is closed in order to ensure that there *is* an optimal choice of P_S from $\mathcal{Q}'_S$.

pletely concerning the relative desirabilities of any two lotteries with outcomes in $C = \{c[s, \sigma]: s \in S, \sigma \in \Sigma\}$, and hence all have the same utility function[e] $u: C \to R^1$. In these sections we focus on the ramifications of greater or lesser potential controversy regarding judgments, i.e., regarding P_Σ.

Section 7.3 concerns what happens if one can assume no consensus whatsoever about any feature of P_Σ. We develop a geometric model of the dpuu in normal form and then relate our usual concept of optimality to a different notion, *cardinal admissibility*, which arises as follows. We say that a randomized strategy P''_S *cardinally dominates* another randomized strategy P'_S, *given* ($\mathscr{D}$'s utility function) u (1) if P''_S is at least as good as P'_S, given every pure strategy σ of $\mathscr{E}$ [and hence $U(P''_S, \sigma) \geq U(P'_S, \sigma)$ for every $\sigma \in \Sigma$]; and (2) if P''_S is actually better than P'_S, given some pure strategy σ of $\mathscr{E}$ [hence $U(P''_S, \sigma) > U(P'_S, \sigma)$ for at least one $\sigma \in \Sigma$]. A moment's reflection convinces one that any people who agree on the utility function u will also agree that P'_S can be eliminated from further consideration if there is some strategy P''_S which cardinally dominates it, given u. Hence these people can agree without further ado that a strategy P^*_S (say) is worthy of additional attention and analysis only if it is *not* cardinally dominated, given u. Such a strategy is called *cardinally admissible*, given u.

We recommend that these people confine further attention and controversy to the set $\mathscr{A}_{\mathscr{C}}[\mathscr{Q}_S|u]$ of all cardinally admissible strategies in $\mathscr{Q}_S$ given the agreed-upon utility function u, because so doing amounts to full exploitation of the hypothesized agreement about u in the task of arriving at an optimal choice of strategy. The main point of Section 7.3 is that we cannot narrow this choice down any further than to the set $\mathscr{A}_{\mathscr{C}}[\mathscr{Q}_S|u]$ without either eliminating some strategy which is optimal, given some P_Σ, or hypothesizing some agreed-upon aspects of P_Σ. The final result in Section 7.3 addresses the issue of how serious the remaining controversy over the strategies in $\mathscr{A}_{\mathscr{C}}[\mathscr{Q}_S|u]$ can be, with the rather feeble statement that relatively small disagreements about P_Σ are less serious in their $U(P_S, P_\Sigma)$ ramifications than are relatively large disagreements about P_Σ.

Section 7.4 concerns what occurs if there is not only complete consensus on u, as in Section 7.3, but also complete consensus on some aspects of P_Σ. As intimated in the preceding paragraph, one can now narrow controversy down quite a bit more in most dpuu's.

Section 7.5 moves in the direction opposite to that of Section 7.4 from Section 7.3, by examining how far one can go if the concerned parties do not share a consensus on the utility function u, but rather only about the relative magnitudes of the utility numbers. More specifically, we assume in Section 7.5 that all concerned agree only about the relative desirabilities of the consequences and that there is potentially total divergence as to risk

---

[e] To be precise, all must have *essentially* the same utility function, in that their individual utility functions must all be related to one another in the manner described in Theorem 3.5.2.

attitude. If C were monetary returns, for instance, they all might agree only that more return is preferable to less, and hence all that can be assumed about u is that it is a strictly increasing function of returns c. For the general case, we assume only (1) that $u(c'') \geq u(c')$ whenever all agree that $c'' \gtrsim c'$; and (2) that for every pair c', c'' of consequences, all must agree that $c'' > c'$, or all must agree that $c' > c''$, or all must agree that $c' \sim c''$. We show that, in general, controversy can still be focused on some (not cardinally!) admissible set, but it cannot be focused as sharply as in Section 7.3, not to mention Section 7.4. This is as intuition suggests: the greater the consensus, the fewer the strategies which are not consensually inferior.

As far as pure strategies are concerned, we can introduce briefly at this point the useful concept of ordinal admissibility. The set $\mathscr{A}_{\mathscr{O}}[S|\gtrsim]$ of all ordinally admissible *pure* strategies S is as far as we can narrow the optimal choice down to when only the basic preference ranking "$\gtrsim$" of C is agreed upon. To determine $\mathscr{A}_{\mathscr{O}}[S|\gtrsim]$ we proceed as follows. First, we say that the pure strategy s'' *ordinally dominates* the pure strategy s', given $\gtrsim$ (1) if, given every pure strategy σ for $\mathscr{E}$, s'' yields at least as desirable a consequence as does s' (that is, $c[s'', \sigma] \gtrsim c[s', \sigma]$ for every $\sigma \in \Sigma$); and (2) if, given at least one pure strategy σ for $\mathscr{E}$, s'' yields a more desirable consequence than does s' (that is, $c[s'', \sigma] > c[s', \sigma]$ for at least one $\sigma \in \Sigma$). Second, we eliminate from further consideration every pure strategy that is ordinally dominated by another pure strategy. This leaves the set $\mathscr{A}_{\mathscr{O}}[S|\gtrsim]$ of all ordinally admissible pure strategies, given $\gtrsim$.

Frequently $\mathscr{A}_{\mathscr{O}}[S|\gtrsim]$ is a very small subset of S, and hence a great economy is effected by confining further analysis and argument to the strategies in $\mathscr{A}_{\mathscr{O}}[S|\gtrsim]$. The main point, however, is that *nobody* with the given desirability ranking "$\gtrsim$" of the consequences need look outside of $\mathscr{A}_{\mathscr{O}}[S|\gtrsim]$ for a strategy optimal with respect to his judgments P_Σ and his *completely* specified, numerical utility function u—whatever P_Σ and u may be!

Example 7.1.1

As noted after Definition 7.1.1, dpuu's with finite numbers $\#(S)$ and $\#(\Sigma)$ of pure strategies for $\mathscr{D}$ and $\mathscr{E}$ respectively are often arranged in tabular format, with $c[s^i, \sigma^j]$ in the row-i and column-j cell. In the following dpuu, $\#(S) = 7$ and $\#(\Sigma) = 2$.

	σ^1	σ^2
s^1	c^0	c^2
s^2	c^1	c^1
s^3	c^2	c^1
s^4	c^2	c^0
s^5	c^3	c^0
s^6	c^1	c^2
s^7	c^2	c^2

Suppose that the consequences have been labeled in conformity with the given desirability ranking "$\succsim$" in such a way that $c^3 > c^2 > c^1 > c^0$. Then

s^1 is ordinally dominated by s^6 and by s^7;
s^2 is ordinally dominated by s^3, s^6, and s^7;
s^3 is ordinally dominated by s^7;
s^4 is ordinally dominated by s^5; and
s^6 is ordinally dominated by s^7.

Neither s^5 nor s^7 is ordinally dominated by any pure strategy, and hence $\mathscr{A}_O[S|\succsim] = \{s^5, s^7\}$. Further analysis and argument may be confined to s^5 and s^7 without any fear of having rejected "good" candidates for optimality.

You may test your understanding of ordinal admissibility by working Exercise 7.1.3.

Ordinal admissibility is reintroduced more generally, for genuinely randomized strategies as well as for pure strategies, in Section 7.5. We shall see (1) that *all* the sets of admissible strategies considered in Sections 7.3 through 7.5 are subsets of the set $\mathscr{A}_O[\mathscr{Q}_S|\succsim]$ of all ordinally admissible randomized strategies; and (2) that $S \cap \mathscr{A}_O[\mathscr{Q}_S|\succsim] = \mathscr{A}_O[S|\succsim]$, which means that in the ensuing analysis we will never have reason to regret having said *here*, in this introduction, that attention may be confined to $\mathscr{A}_O[S|\succsim]$.

The foregoing introduction to various sorts of admissibility and their roles in the determination of an optimal strategy in the presence of controversy about judgments and/or preferences has been silent about controversy concerning the availability of certain courses of action in the dpuu, or equivalently, certain strategies in the dpuu in normal form. There are (at least) two levels at which such controversies can arise.

The first level contains almost all cases which can conceivably arise in applications. In a certain, fundamental sense, almost all practical disagreements about the feasibility of pure strategies arise from disagreements about the feasibility (availability) of acts in non-initial $\mathscr{D}$-moves in the decision in extensive form.[f] All such disagreements can be reformulated as disagreements about probabilities. Suppose, for instance, that in your original decision tree you and a colleague differ about whether an act g^* (say) will be feasible in a given $\mathscr{D}$-move G, with you saying that g^* *is* feasible and your colleague holding otherwise. You can then embellish your tree (1) by replacing G with an $\mathscr{E}$-move {"g^* feasible," "g^* infeasible"}; (2) by attaching *your* version of G and all ensuing parts of the decision tree to the "g^* feasible" branch of your new $\mathscr{E}$-move; and (3) by attaching your colleague's abbreviated version of G and all ensuing parts of the decision tree[g] to the "g^* infeasible" branch of your new $\mathscr{E}$-move. *Now* your disagreement about the feasibility of g^* amounts to a dis-

[f] You may verify the converse of this assertion as Exercise 7.2.10.

[g] Namely, G with "g^* feasible" and everything emanating to the right of g^* pruned.

agreement about probabilities; you say P ("g^* feasible") = 1, whereas your colleague says P ("g^* infeasible") = 1.

Hence practically arising disagreements about strategy feasibilities can be handled by embellishing the decision tree and then using the methodology of Sections 7.3 through 7.5. Such an approach, however, tends to becloud the analysis of "what if" certain strategies were declared infeasible, by fiat or tentative assumption. Therefore, we shall examine in Section 7.6 the ramifications of potentially more general feasibility disagreements, ones which we shall term simply second-level feasibility disagreements.

We shall assume that these fully general, second-level disagreements take the form of different individuals' believing that different subsets $\mathcal{Q}'_S$ of the set $\mathcal{Q}_S$ of all randomized strategies for $\mathcal{D}$ are feasible. One very important sort of subset $\mathcal{Q}'_S$ is that which arises from first-level disagreements. If, for example, you believe that the only *pure* strategies ultimately implementable are those in some subset S^* (say) of S, then you must also believe that the only surely feasible randomized strategies are those which assign probability one to S^* and probability zero to $\overline{S^*}$. We shall call this set of randomized strategies $\mathcal{Q}_S^{S^*}$. Since we shall need to cite such sets frequently from Section 7.3 on, we give a formal definition of the notation here.

Definition 7.1.4. If S^* is any given nonempty subset of S, then $\mathcal{Q}_S^{S^*}$ is defined by

$$\mathcal{Q}_S^{S^*} = \{P_S: P_S(S^*) = 1\}.$$

Example 7.1.2

In the dpuu of Example 7.1.1, if you believed that s^6 and s^7 could not be implemented, then your $S^* = \{s^1, \ldots, s^5\}$ and your $\mathcal{Q}_S^{S^*} = \{P_S: P_S(s^6) = P_S(s^7) = 0\}$, because you should not take a chance on choosing a pure strategy that you regard as infeasible.

Throughout Sections 7.3 through 7.6, we shall assume, not at all heroically, that all concerned with the dpuu agree on two structural matters.

(1) All concerned agree in principle on some basic canonical randomization mechanism, so that a given randomized strategy P_S denotes the same probabilities $P_S(S')$ to all.
(2) All concerned agree on largest potential sets S and Σ of pure strategies, and hence on largest potential sets $\mathcal{Q}_S$ and $\mathcal{Q}_\Sigma$ of randomized strategies, as well as on the descriptions $c[s, \sigma]$ of all consequences obtaining, given $s \in S$ and $\sigma \in \Sigma$.

■ The first structural assumption is a rather mild one in view of the rather innocuous mechanisms discussed in Section 3.4. The second structural assumption can always be guaranteed in practice by putting into S all pure strategies which someone concerned regards as feasible for $\mathscr{D}$; by putting into Σ all pure strategies which someone regards as feasible for $\mathscr{E}$; and by including in the description of $c[s, \sigma]$ all attributes which someone regards as relevant. ■

Sections 7.3 through 7.6 are not essential prerequisites for any of the ensuing chapters, but Section 7.3 conveys so much of the flavor of the entire analysis, and so many fundamental results, that it would be a shame if any reader of this book passed over Section 7.3 completely and never returned to it. The exercises in Section 7.3 introduce a widely cited alternative approach to the choice of strategy, namely, the "maximin" criterion.

Exercise 7.1.1: Randomized Strategies and Vectors

For every positive integer n, let $V^n = \{(x_1, \ldots, x_n): \Sigma_{i=1}^n x_i = 1,\ x_i \geq 0$ for every $i\}$. Clearly, $V^N \subset R^n$, and every n-tuple in V^n can be regarded as a probability vector because its components are nonnegative and add up to one.

(A) If S consists of $\#(S)$ pure strategies, labeled $s^1, s^2, \ldots, s^{\#(S)}$, show that every randomized strategy P_S corresponds to exactly one vector in $V^{\#(S)}$, and conversely, that every vector in $V^{\#(S)}$ corresponds to exactly one randomized strategy P_S.

(B) To what $\#(S)$-tuple does the pure strategy s^1 correspond? How about $s^{\#(S)}$?

■ Exactly the same correspondence obtains for randomized strategies P_Σ for $\mathscr{E}$. If Σ consists of $\#(\Sigma)$ pure strategies labeled $\sigma^1, \sigma^2, \ldots, \sigma^{\#(\Sigma)}$, then every P_Σ corresponds to exactly one vector in $\mathbf{V}^{\#(\Sigma)}$, and every vector in $V^{\#(\Sigma)}$ corresponds to exactly one randomized strategy P_Σ for $\mathscr{E}$. ■

■ The conclusion to be drawn from the preceding remark and from Exercise 7.1.1 is that *$\mathscr{Q}_S$ may be identified with* $\mathbf{V}^{\#(S)}$, and *$\mathscr{Q}_\Sigma$ may be identified with $V^{\#(\Sigma)}$*, so that you always think of a randomized strategy P_S for $\mathscr{D}$ as the $\#(S)$-tuple $(P_S(s^1), \ldots, P_S(s^{\#(S)}))$ in $V^{\#(S)}$, and you may always think of the randomized strategy P_Σ as the $\#(\Sigma)$-tuple $(P_\Sigma(\sigma^1), \ldots, P_\Sigma(\sigma^{\#(\Sigma)}))$ in $V^{\#(\Sigma)}$. In Section 7.5 we shall develop a similar concrete representation of utility functions. ■

Exercise 7.1.2

Consider the dpuu in normal form given by the table

	σ^1	σ^2	σ^3	σ^4
s^1	c^0	c^2	c^1	c^2
s^2	c^3	c^0	c^0	c^0
s^3	c^2	c^2	c^0	c^1
s^4	c^1	c^0	c^3	c^1
s^5	c^1	c^1	c^1	c^0

and assume that $u(c^0) = 0$, $u(c^1) = 2$, $u(c^2) = 5$, and $u(c^3) = 6$. If you have not already done so, read Exercise 7.1.1 and the remarks following it, since for convenience we shall identify randomized strategies P_S for $\mathscr{D}$ with 5-tuples and randomized strategies P_Σ for $\mathscr{E}$ with 4-tuples.

(A) Find $U(s^2, \sigma^3)$.
(B) Find $U(P_S, \sigma^3)$ for $P_S = (0, 1, 0, 0, 0)$.
(C) Find $U(P_S, \sigma^4)$ for $P_S = (.2, .2, 0, 0, .6)$.
(D) Find $U(s^2, P_\Sigma)$ for $P_\Sigma = (0, 0, 1, 0)$.
(E) Find $U(s^3, P_\Sigma)$ for $P_\Sigma = (1/4, 1/4, 1/4, 1/4)$.
(F) Find $U(P_S, P_\Sigma)$ for $P_S = (1/3, 0, 1/3, 1/3, 0)$ and $P_\Sigma = (1/2, 0, 1/4, 1/4)$.
(G) Show that $U(P_S, \sigma^1) = 4.00$ if $P_S = (.1, .4, .2, .2, .1)$.
(H) Show that $U(P_S, \sigma^2) = 1.70$ if $P_S = (.1, .4, .2, .2, .1)$.
(I) Show that $U(P_S, \sigma^3) = 1.60$ if $P_S = (.1, .4, .2, .2, .1)$.
(J) Show that $U(P_S, \sigma^4) = 1.30$ if $P_S = (.1, .4, .2, .2, .1)$.
(K) Show that $U(P_S, P_\Sigma) = 2.51$ if $P_S = (.1, .4, .2, .2, .1)$ and $P_\Sigma = (.4, .1, .3, .2)$.
(L) Show that $U(\mathscr{Q}'_S, P_\Sigma) = 2.90$ if $P_\Sigma = (.4, .1, .3, .2)$ and $\mathscr{Q}'_S = \{s^2, s^3, s^5\}$.

Exercise 7.1.3

(A) If $c^3 > c^2 > c^1 > c^0$, show that $\mathscr{A}_O[S|\gtrsim] = S$ for the dpuu tabulated in Exercise 7.1.2.
(B) Find $\mathscr{A}_O[S|\gtrsim]$ if $c^2 > c^0 > c^3 > c^1$.

[*Hint*: Start with s^1, seeing if s^2 ordinally dominates s^1, then if s^3 ordinally dominates s^1, and so on, to determine whether $s^1 \in \mathscr{A}_O[S|\gtrsim]$. Then determine in a similar fashion whether $s^2 \in \mathscr{A}_O[S|\gtrsim]$, and so on.]

***Exercise 7.1.4: Physical-Averaging Context for Genuinely Randomized Strategies**

Suppose:

(I) That there are n identical dpuu's confronting $\mathscr{D}$ and $\mathscr{E}$, each called a *trial*.
(II) That $\mathscr{D}$ chooses strategies for the n trials at random, but in such a way that he uses s^1 in proportion $P_S(s^1)$ of the trials, s^2 in proportion $P_S(s^2)$ of the trials, . . . , and $s^{\#(S)}$ in proportion $P_S(s^{\#(S)})$ of the trials; and similarly,
(III) That $\mathscr{E}$ chooses strategies for the n trials at random, but in such a way that he uses σ^1 in proportion $P_\Sigma(\sigma^1)$ of the trials, σ^2 in proportion $P_\Sigma(\sigma^2)$ of the trials, . . . , and $\sigma^{\#(\Sigma)}$ in proportion $P_\Sigma(\sigma^{\#(\Sigma)})$ of the trials.

This exercise leads you to the conclusion that Equation (7.1.1) is an appropriate, approximate guide to the choice of the proportion vector P_S under certain conditions. Specifically, show that as the number n of

trials increases toward $+\infty$, the pregame probability that $\mathscr{D}$'s actual postgame *average utility per trial* will be within $e > 0$ of $U(P_S, P_\Sigma)$ increases toward $+1$.
[*Hint*: Weak law of large numbers.]

■ In competitive situations at the policy-making level, $\mathscr{D}$ may be faced with just such a large number n of identical dpuu's. The conclusion of this exercise implies that $\mathscr{D}$'s decision to use different pure strategies in the proportions implied by P_S, given his anticipation that $\mathscr{E}$ will behave similarly with respect to P_Σ, can be evaluated by $U(P_S, P_\Sigma)$, under the assumption that no further information about $\mathscr{E}$'s specific choices in specific "trials" will become available. ■

Exercise 7.1.5

If you delete s^6 and s^7 in the dpuu of Example 7.1.1, you obtain the diminished set S^* of pure strategies discussed in Example 7.1.2. Find $\mathscr{A}_{\mathscr{E}}[S^*|\gtrsim]$ if $c^3 > c^2 > c^1 > c^0$ specifies "$\gtrsim$."

7.2 ANY dpuu CAN BE REPRESENTED IN NORMAL FORM

In this section we show that any dpuu in extensive form may be reexpressed in normal form, in such a way that $\mathscr{D}$'s judgments about $\tilde{\sigma}$ are independent of his choice of s. Throughout, it will be assumed that the given dpuu in extensive form consists of a finite number of moves, each of which has a finite number of elements. This finiteness assumption can be removed in certain classes of idealized dpuu's having special structures.

The basic procedure for converting to normal form a given dpuu in extensive form, the *successive partial normalization* procedure, consists of three steps.

(1) Defining pure strategies s for $\mathscr{D}$ by moving all $\mathscr{D}$'s choices to the beginning of the tree in a certain, permissible way.
(2) Defining pure strategies σ for $\mathscr{E}$ by moving all $\mathscr{E}$'s choices to the beginning of the tree in a certain, permissible way.
(3) Determining the consequence $c[s, \sigma]$ of each (s, σ) pair by tracing through the action implications of s and σ in the original, given tree.

While describing these steps, we show how $\mathscr{D}$'s original probabilities on the $\mathscr{E}$-moves of the dpuu in extensive form serve to determine a logically consistent probability function P_Σ on the set Σ of $\mathscr{E}$'s pure strategies. Then, finally, we note that in certain dpuu's it is possible to replace step 2 with a more economical means of defining Σ.

The "certain, permissible way" of moving all $\mathscr{D}$'s choices to the beginning of the tree so as to define pure strategies s entails repeated applications of a technique for *reversing the order of an $\mathscr{E}$-move with attached, contingent $\mathscr{D}$ moves,* in such a way that $\mathscr{D}$'s resulting, pre-$\mathscr{E}$-move choice reflects the information that he will have obtained from that $\mathscr{E}$-move in the original tree. This technique may be regarded as a supplement to similar tree manipulations discussed in Chapter 2.

Consider the partial tree in Figure 7.1, in which $\mathscr{D}$-moves G_1 and G_2 are attached to branches γ^1 and γ^2 of an $\mathscr{E}$-move Γ, a consequence c is attached to the lowest branch γ^3 of Γ, the bracketed decimals are $\mathscr{D}$'s probabilities, and b^1–b^5 denote ensuing parts of the decision tree.

At the right-hand endpoint of branch a, the decision maker can exercise foresight by planning his future choices from G_1 and G_2 and hence can be prepared if γ^1 or γ^2 obtains. By virtue of Exercise 3.1.7, he can adopt six possible plans: (1) g^1 if γ^1 and g^4 if γ^2, (2) g^1 if γ^1 and g^5 if γ^2, (3) g^2 if γ^1 and g^4 if γ^2, (4) g^2 if γ^1 and g^5 if γ^2, (5) g^3 if γ^1 and g^4 if γ^2, and (6) g^3 if γ^1 and g^5 if γ^2.

Now consider the following sequence. (1) $\mathscr{D}$ chooses one of his six plans; (2) $\mathscr{E}$ chooses a γ from Γ; and (3) attention passes to whichever of c and b^1–b^5 obtains, given γ and the act dictated by $\mathscr{D}$'s plan. Since $\mathscr{D}$'s plans reflect precisely the information on which his choices from G_1 and G_2 are based in Figure 7.1, it follows that *this plan-event sequence, depicted in Figure 7.2, is strategically equivalent for $\mathscr{D}$ to the original sequence in Figure 7.1.* In Figure 7.2, copies of $\mathscr{E}$ are attached to the plan branches, and the implications of plan *cum* γ are traced through from Figure 7.1 to determine what to attach to each γ-endpoint.

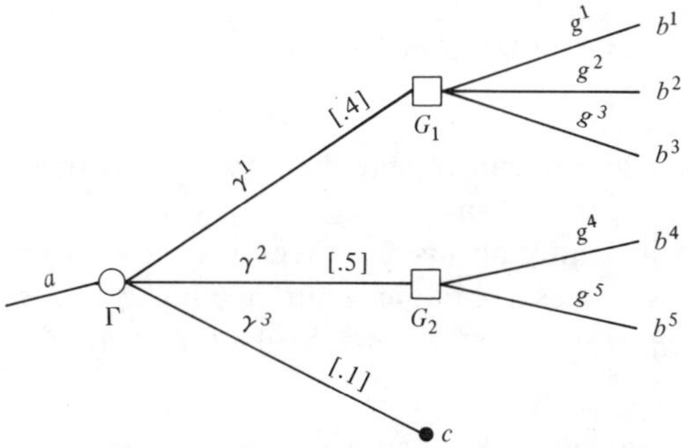

Figure 7.1 Partial decision tree.

■ The tree in Figure 7.2 is *not* strategically equivalent *for* $\mathscr{E}$ to that in Figure 7.1, since here it appears that $\mathscr{E}$ can choose γ after learning $\mathscr{D}$'s plan, whereas $\mathscr{E}$'s choice of γ in Figure 7.1 precedes $\mathscr{D}$'s choices. From a less personified view of $\mathscr{E}$, however, all we have done is attach copies of Γ to the plan branches. Furthermore, $\mathscr{D}$'s probabilities on the copies of $\mathscr{E}$ should certainly be unaffected by his permissibly reformulating his dpuu. ■

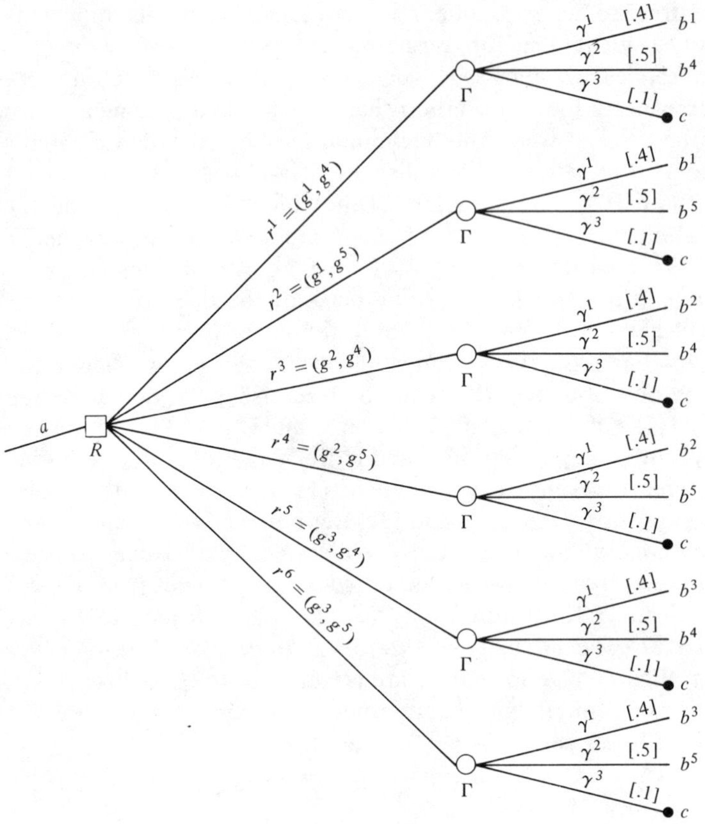

Figure 7.2 Partial decision tree equivalent for $\mathscr{D}$ of Figure 7.1.

It should be clear at this point that $\mathscr{D}$ can repeat this reversal of order any number of times, by recognizing that R in Figure 7.2 is just a $\mathscr{D}$-move. Ultimately, he will obtain a tree in which all his $\mathscr{D}$-moves precede all the $\mathscr{E}$-moves. Then he can combine his successive choices into one big $\mathscr{D}$-move, as shown in Section 2.11. *This big $\mathscr{D}$-move is $\mathscr{D}$'s set* S *of pure strategies.*

■ The plans, or *partial pure strategies*, obtained after every reversal of order of an $\mathscr{E}$-move with attached $\mathscr{D}$-moves, are complete specifications of what $\mathscr{D}$ will do after obtaining information from $\mathscr{E}$. Hence the ultimate, big $\mathscr{D}$-move S consists of pure strategies s that are complete prescriptions for action in every contingency. Each specifies an initial act and, subsequently, an act in every $\mathscr{D}$-move that can obtain, given $\mathscr{D}$'s earlier choices. ■

Step 2, obtaining $\mathscr{E}$'s set Σ of pure strategies by moving all of $\mathscr{E}$'s choices to the beginning of the tree in a certain, permissible way, is carried out by reversing the roles of $\mathscr{D}$ and $\mathscr{E}$ and by applying the partial normalization technique described above for $\mathscr{D}$. This means that each step involves reversing the order of a $\mathscr{D}$-move with attached, contingent $\mathscr{E}$-moves and hence the creation of plans, or partial pure strategies, for $\mathscr{E}$. Repeating such order reversals to the fullest extent possible produces a tree with a big, initial $\mathscr{E}$-move, to the branches of which are attached $\mathscr{D}$-moves. This tree is strategically equivalent to the original one for $\mathscr{E}$, though not for $\mathscr{D}$. The big initial $\mathscr{E}$-move in this tree is $\mathscr{E}$'s set Σ of pure strategies.

■ Since this procedure mimics step 1, further discussion is unnecessary, except for noting that $\mathscr{D}$'s probability of any pure strategy for $\mathscr{E}$ of the form (γ', γ'') can be defined as the *product* of his probabilities of γ' and of γ''. To see this, note that the basic probability-assessment technique of Section 3.6 is tantamount to converting the $\mathscr{E}$-moves $\Gamma_1, \ldots, \Gamma_M$ (say) of the decision tree into basic canonical randomizations, and also that these basic canonical randomizations may be regarded as completely unrelated (or independent; see Section 6.5). Hence $\mathscr{E}$ may set $P(\gamma', \gamma'') = P(\gamma')P(\gamma'')$. ■

Now that we know how to define S and Σ, it remains only to describe how to find the consequence $c[s, \sigma]$ resulting from any (s, σ) pair. First, note by generalizing from Figure 7.2 that every pure strategy s for $\mathscr{D}$ is a sequence $(g', g'', \ldots, g^{[n]})$ of acts from $\mathscr{D}$-moves in the original tree; and similarly, that every pure strategy σ for $\mathscr{E}$ is a sequence $(\gamma', \gamma'', \ldots, \gamma^{[m]})$ of elementary events from $\mathscr{E}$-moves in the original tree. Thus (s, σ) is a pair of such sequences. Now, $c[s, \sigma]$ is readily found by *tracing forward* in the original tree: (1) See what $\mathscr{D}$ chose in the initial $\mathscr{D}$-move. (2) Given $\mathscr{D}$'s initial choice, find the choice in the move attached to $\mathscr{D}$'s initial choice branch, and so on, ultimately arriving at a consequence. One is never stymied in this procedure, because pure strategies for each player are *complete* prescriptions for action in *every* contingency that can arise.

Example 7.2.1: Medical Treatment

See Figure 7.3, representing a medical treatment problem, in which g_1^1 = "treat with drug A," g_1^2 = "treat with drug B," $\gamma_1^1 = \gamma_2^1$ = "temperature drops to normal within 24 hours," $\gamma_1^2 = \gamma_2^2$ = "temperature does not drop to normal within 24 hours," $g_2^1 = g_3^1 = g_4^1$ = "give massive dose of drug D," $g_2^2 = g_3^2 = g_4^2$ = "terminate medication," γ_3^1 = "patient is cured," and γ_3^2 = "patient suffers harmful side effects of medication overdosage." Suppose that giving the patient the massive dose of D is out of the question if he had initially been given B and his temperature dropped to normal. We also suppose that side effects are possible *only* when D is administered to a patient who had taken A and whose temperature is normal. In step 1, successive partial normalization for $\mathscr{D}$,

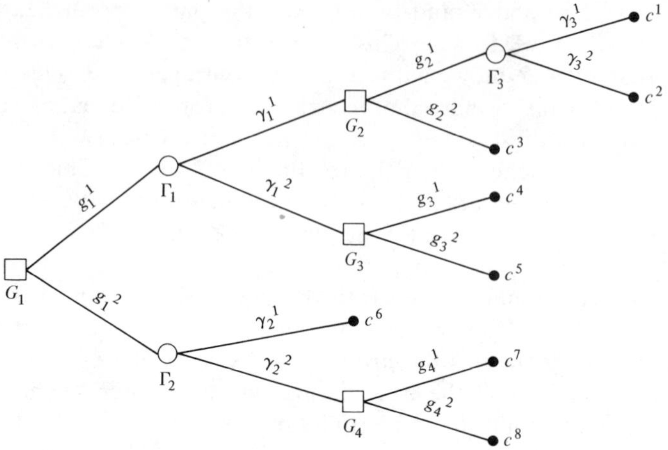

Figure 7.3 Dpuu in Example 7.2.1.

reversing the order of Γ_1 with G_2 and G_3 attached results in Γ_1 being attached to each of four plan branches of the form (g_2^i, g_3^j), and the $\mathscr{D}$-move consisting of these four plans is attached to g_1^1. Similarly, reversing the order of Γ_2 with G_4 attached results in Γ_2 being attached to each of the two branches of G_4, which is attached to g_1^2; here, G_4 is itself the plan $\mathscr{D}$-move because it is the only $\mathscr{D}$-move in the tree attached to any of the branches of Γ_2. Now, combining the successive choices by $\mathscr{D}$ gives rise to the tree in Figure 7.4. Hence, $\mathscr{D}$ has six pure strategies: $s^1 = (g_1^1, g_2^1, g_3^1)$, $s^2 = (g_1^1, g_2^1, g_3^2)$, $s^3 = (g_1^1, g_2^2, g_3^1)$, $s^4 = (g_1^1, g_2^2, g_3^2)$, $s^5 = (g_1^2, g_4^1)$, and $s^6 = (g_1^2, g_4^2)$. (The order in which the strategies are labeled $s^1, \ldots, s^6$ is arbitrary, of course.) For step 2, you may verify as Exercise 7.2.1 that $\mathscr{E}$ also has six pure strategies; namely, $\sigma^1 = (\gamma_1^1, \gamma_3^1, \gamma_2^1)$, $\sigma^2 = (\gamma_1^1, \gamma_3^1, \gamma_2^2)$, $\sigma^3 = (\gamma_1^1, \gamma_3^2, \gamma_2^1)$, $\sigma^4 = (\gamma_1^1, \gamma_3^2, \gamma_2^2)$, $\sigma^5 = (\gamma_1^2, \gamma_2^1)$, and $\sigma^6 = (\gamma_1^2, \gamma_2^2)$. In step 3, the consequence $c[s^2, \sigma^4]$ of $(s^2, \sigma^4) = ((g_1^1, g_2^1, g_3^2), (\gamma_1^1, \gamma_3^2, \gamma_2^2))$ is found by tracing down branch g_1^1, then down branch γ_1^1, then down branch g_2^1, and then down branch γ_3^2 to find c^2. Similar tracings through Figure 7.3 establish the 35 other $c[s, \sigma]$'s in the following consequence table of the dpuu in normal form.

	σ^1	σ^2	σ^3	σ^4	σ^5	σ^6
s^1	c^1	c^1	c^2	c^2	c^4	c^4
s^2	c^1	c^1	c^2	c^2	c^5	c^5
s^3	c^3	c^3	c^3	c^3	c^4	c^4
s^4	c^3	c^3	c^3	c^3	c^5	c^5
s^5	c^6	c^7	c^6	c^7	c^6	c^7
s^6	c^6	c^8	c^6	c^8	c^6	c^8

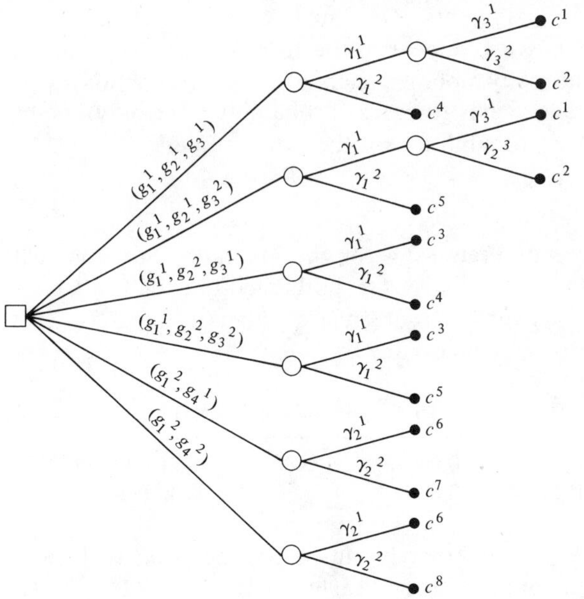

Figure 7.4 Complete successive partial normalization for $\mathscr{D}$ of Figure 7.3.

Finally, as we have noted, $\mathscr{D}$'s probability $P_\Sigma(\sigma)$ of any given pure strategy σ for $\mathscr{E}$ may be defined consistently to be the product of his probabilities of the elementary events comprising Σ. Thus if $\mathscr{D}$ had assessed $P(\gamma_1^{\,1}) = .3$, $P(\gamma_2^{\,1}) = .3$, and $P(\gamma_3^{\,1}) = .9$, it follows that $\mathscr{D}$ may take

$$P_\Sigma(\sigma^1) = P_\Sigma(\gamma_1^{\,1}, \gamma_3^{\,1}, \gamma_2^{\,1}) = (.3)(.9)(.3) = .081,$$
$$P_\Sigma(\sigma^2) = P_\Sigma(\gamma_1^{\,1}, \gamma_3^{\,1}, \gamma_2^{\,2}) = (.3)(.9)(.7) = .189,$$
$$P_\Sigma(\sigma^3) = P_\Sigma(\gamma_1^{\,1}, \gamma_3^{\,2}, \gamma_2^{\,1}) = (.3)(.1)(.3) = .009,$$
$$P_\Sigma(\sigma^4) = P_\Sigma(\gamma_1^{\,1}, \gamma_3^{\,2}, \gamma_2^{\,2}) = (.3)(.1)(.7) = .021,$$
$$P_\Sigma(\sigma^5) = P_\Sigma(\gamma_1^{\,2}, \gamma_2^{\,1}) = (.7)(.3) = .210,$$

and

$$P_\Sigma(\sigma^6) = P_\Sigma(\gamma_1^{\,2}, \gamma_2^{\,2}) = (.7)(.7) = .490.$$

■ Note that in Example 7.2.1 the $P_\Sigma(\sigma^j)$'s sum to 1, as required for P_Σ to be a valid probability function. It is not hard to show that whenever Σ is defined by step 2, the product formula

$$P_\Sigma(\gamma'_{i_1}, \ldots, \gamma'_{i_M}) = \prod_{m=1}^{M} P(\gamma'_{i_m}) \tag{7.2.1}$$

for every $\sigma = (\gamma'_{i_1}, \ldots, \gamma'_{i_M})$ in Σ yields $P_\Sigma(\sigma)$'s which sum to 1. ■

An *economical alternative to step 2* for defining Σ is available for certain dpuu's in which $\mathscr{D}$'s probabilities in his original decision tree were derived from other, more basic judgments. *One* such case is that in which (1) several copies of an $\mathscr{E}$-move Γ appear in the tree, and (2) $\mathscr{D}$ is convinced that if branch γ' in one copy obtains, then γ' must also obtain in every other copy as well. Hence $\mathscr{D}$ has to assess probabilities for only one copy, and then duplicate those probabilities on the other copies.

Example 7.2.2

In the decision tree of Figure 2.4 for the insurance problem, with a copy of $\Gamma = \{$"*D*isaster," "*N*o *D*isaster"$\}$ attached to each branch of the initial $\mathscr{D}$-move $G = \{$"*I*nsure" "*N*ot *I*nsure"$\}$, it is easy to verify that steps 1–3 yield the consequence table

	(D|I, D|NI)	(D|I, ND|NI)	(ND|I, D|NI)	(ND|I, ND|NI)
I	10,000	10,000	10,000	10,000
NI	0	12,000	0	12,000

Now suppose that $\mathscr{D}$ is sure that his insurance decision will have no influence whatsoever on whether disaster befalls the *Pride of Old Sarum*. This means (1) that each copy of Γ in Figure 2.4 should have the same probability of disaster, and (2) even more fundamentally, that there is really only *one* $\mathscr{E}$-move rather than two. Hence, $\mathscr{D}$ can use the abbreviated consequence table

	D	ND
I	10,000	10,000
NI	0	12,000

The economizing effect of such judgments on the normal form of a dpuu is evident.

Another case in which probabilities for the decision tree arise from more basic judgments is Example 1.5.1c and similar problems. You will recall that the majority of the work in Section 4.4 required obtaining probabilities for the tree from $\mathscr{D}$'s readily obtainable joint probabilities of the eight events specifying (1) the arrival state AD or AN of the subsystem, (2) the outcome PD or PN of the pretest, and (3) the outcome LD or LN of the less reliable test. Since in $\mathscr{D}$'s estimation $\mathscr{E}$ really must choose precisely one of these eight events in order to specify a choice in each $\mathscr{E}$-move of Figure 4.3, it follows that these eight events constitute a sufficient description of Σ.

Example 7.2.3

In the notation of Section 4.4, the eight sufficiently defined pure strategies for $\mathscr{E}$ are: $\sigma^1 = (\text{PD, LD, AD})$, $\sigma^2 = (\text{PD, LD, AN})$, $\sigma^3 =$

(PD, LN, AD), $\sigma^4 =$ (PD, LN, AN), $\sigma^5 =$ (PN, LD, AD), $\sigma^6 =$ (PN, LD, AN), $\sigma^7 =$ (PN, LN, AD), and $\sigma^8 =$ (PN, LN, AN). Their joint probabilities are given in Table B of Section 4.4. [*Caution*: Our definition of the pure strategies σ differs from that obtained via step 2, and hence the product rule (7.2.1) does *not* pertain.] If step 2 had been applied to Figure 4.3 in defining Σ, there would have been 34 pure strategies for $\mathscr{E}$, as you may verify in Exercise 7.2.2. Now, applying step 1 to Figure 4.3 ultimately results in 74 pure strategies for $\mathscr{D}$. Let P and L stand for "pretest" and "less reliable test" respectively; and let R and NR stand for "rebuild before installing" and "do not rebuild before installing" respectively. With this notation, we define the 74 pure strategies s for $\mathscr{D}$ in Table A. Table B is the consequence table for this dpuu in normal form; there are 26 ordinally admissible pure strategies for $\mathscr{D}$, listed before the ordinally dominated pure strategies, and the last column

Table A Pure Strategies for $\mathscr{D}$

$s^1 =$ don't test, NR.
$s^2 =$ don't test, R.
$s^3 =$ P, R only if PD.
$s^4 =$ P, R regardless of outcome.
$s^5 =$ P, R only if PN.
$s^6 =$ P, NR regardless of outcome.
$s^7 =$ P, R if PD; L if PN, and R only if LD.
$s^8 =$ P, R if PD; L if PN, and R regardless of outcome of L.
$s^9 =$ P, R if PD; L if PN, and R only if LN.
$s^{10} =$ P, R if PD; L if PN, and NR regardless of outcome of L.
$s^{11} - s^{14}$: like $s^7 - s^{10}$ respectively but with NR if PD.
$s^{15} - s^{18}$: like $s^7 - s^{10}$ respectively but with roles of PD and PN reversed.
$s^{19} - s^{22}$: like $s^{11} - s^{14}$ respectively but with roles of PD and PN reversed.
$s^{23} =$ P and then L; then R regardless of outcomes of P and L.
$s^{24} =$ P and then L; then R except if (PN, LN).
$s^{25} =$ P and then L; then R except if (PN, LD).
$s^{26} =$ P and then L; then R if PD regardless of outcome of L.
$s^{27} =$ P and then L; then R except if (PD, LN).
$s^{28} =$ P and then L; then R if LD regardless of outcome of P.
$s^{29} =$ P and then L; then R if (PD, LD) or (PN, LN).
$s^{30} =$ P and then L; then R only if (PD, LD).
$s^{31} =$ P and then L; then R except if (PD, LD).
$s^{32} =$ P and then L; then R if (PD, LN) or (PN, LD).
$s^{33} =$ P and then L; then R if LN regardless of outcome of P.
$s^{34} =$ P and then L; then R only if (PD, LN).
$s^{35} =$ P and then L; then R if PN regardless of outcome of L.
$s^{36} =$ P and then L; then R only if (PN, LD).
$s^{37} =$ P and then L; then R only if (PN, LN).
$s^{38} =$ P and then L; then NR regardless of outcomes of P and L.
$s^{39} - s^{74}$: like $s^3 - s^{38}$ respectively but with roles of P and L reversed everywhere.

Table B Consequence Table of Example 1.5.1c in Normal Form

	PD $\sigma^1 = $ LD AD	PD $\sigma^2 = $ LD AN	PD $\sigma^3 = $ LN AD	PD $\sigma^4 = $ LN AN	PN $\sigma^5 = $ LD AD	PN $\sigma^6 = $ LD AN	PN $\sigma^7 = $ LN AD	PN $\sigma^8 = $ LN AN	ORD DOM BY
s^1	−250	+250	−250	+250	−250	+250	−250	+250	—
s^2	+150	+150	+150	+150	+150	+150	+150	+150	—
s^3	+130	+130	+130	+130	−270	+230	−270	+230	—
s^{43}	+145	+145	+125	+125	+145	+145	−275	+225	—
s^{55}	+125	+125	−255	+245	−275	+225	−255	+245	—
s^5	−270	+230	−270	+230	+130	+130	+130	+130	—
s^7	+130	+130	+130	+130	+125	+125	−275	+225	—
s^9	+130	+130	+130	+130	−275	+225	+125	+125	—
s^{11}	−270	+230	−270	+230	+125	+125	−275	+225	—
s^{13}	−270	+230	−270	+230	−275	+225	+125	+125	—
s^{15}	+125	+125	−275	+225	+130	+130	+130	+130	—
s^{17}	−275	+225	+125	+125	+130	+130	+130	+130	—
s^{19}	+125	+125	−275	+225	−270	+230	−270	+230	—
s^{21}	−275	+225	+125	+125	−270	+230	−270	+230	—
s^{29}	+125	+125	−275	+225	−275	+225	+125	+125	—
s^{32}	−275	+225	+125	+125	+125	+125	−275	+225	—
s^{39}	+145	+145	−255	+245	+145	+145	−255	+245	—
s^{41}	−255	+245	+145	+145	−255	+245	+145	+145	—
s^{45}	+145	+145	−275	+225	+145	+145	+125	+125	—
s^{47}	−255	+245	+125	+125	−255	+245	−275	+225	—
s^{49}	−255	+245	−275	+225	−255	+245	+125	+125	—
s^{51}	+125	+125	+145	+145	−275	+225	+145	+145	—
s^{53}	−275	+225	+145	+145	+125	+125	+145	+145	—
s^{57}	−275	+225	−255	+245	+125	+125	−255	+245	—
s^{65}	+125	+125	−275	+225	−275	+225	+125	+125	—
s^{68}	−275	+225	+125	+125	+125	+125	−275	+225	—

s^4	+130	+130	+130	+130	+130	+130	+130	+130	s^2
s^6	−270	+230	−270	+230	−270	+230	−270	+230	s^1
s^8	+130	+130	+130	+130	+125	+125	+125	+125	s^2
s^{10}	+130	+130	+130	+130	−275	+225	−275	+225	s^3
s^{12}	−270	+230	−270	+230	+125	+125	+125	+125	s^5
s^{14}	−270	+230	−270	+230	−275	+225	−275	+225	s^1
s^{16}	+125	+125	+125	+125	+130	+130	+130	+130	s^2
s^{18}	−275	+225	−275	+225	+130	+130	+130	+130	s^5
s^{20}	+125	+125	+125	+125	−270	+230	−270	+230	s^3
s^{22}	−275	+225	−275	+225	−270	+230	−270	+230	s^1
s^{23}	+125	+125	+125	+125	+125	+125	+125	+125	s^2
s^{24}	+125	+125	+125	+125	+125	+125	−275	+225	s^7
s^{25}	+125	+125	+125	+125	−275	+225	+125	+125	s^9
s^{26}	+125	+125	+125	+125	−275	+225	−275	+225	s^3
s^{27}	+125	+125	−275	+225	+125	+125	+125	+125	s^{15}
s^{28}	+125	+125	−275	+225	+125	+125	−275	+225	s^{39}
s^{30}	+125	+125	−275	+225	−275	+225	−275	+225	s^{19}
s^{31}	−275	+225	+125	+125	+125	+125	+125	+125	s^{17}
s^{33}	−275	+225	+125	+125	−275	+225	+125	+125	s^{41}
s^{34}	−275	+225	+125	+125	−275	+225	−275	+225	s^{21}
s^{35}	−275	+225	−275	+225	+125	+125	+125	+125	s^5
s^{36}	−275	+225	−275	+225	+125	+125	−275	+225	s^{11}
s^{37}	−275	+225	−275	+225	−275	+225	+125	+125	s^{13}
s^{38}	−275	+225	−275	+225	−275	+225	−275	+225	s^1
s^{40}	+145	+145	+145	+145	+145	+145	+145	+145	s^2
s^{42}	−255	+245	−255	+245	−255	+245	−255	+245	s^1
s^{44}	+145	+145	+125	+125	+145	+145	+125	+125	s^2
s^{46}	+145	+145	−275	+225	+145	+145	−275	+225	s^{39}
s^{48}	−255	+245	+125	+125	−255	+245	+125	+125	s^{41}
s^{50}	−255	+245	−275	+225	−255	+245	−275	+225	s^1
s^{52}	+125	+125	+145	+145	+125	+125	+145	+145	s^2

Table B (*Concl.*)

	PD $\sigma^1 = $ LD AD	PD $\sigma^2 = $ LD AN	PD $\sigma^3 = $ LN AD	PD $\sigma^4 = $ LN AN	PN $\sigma^5 = $ LD AD	PN $\sigma^6 = $ LD AN	PN $\sigma^7 = $ LN AD	PN $\sigma^8 = $ LN AN	ORD DOM BY
s^{54}	−275	+225	+145	+145	−275	+225	+145	+145	s^{41}
s^{56}	+125	+125	−255	+245	+125	+125	−255	+245	s^{39}
s^{58}	−275	+225	−255	+245	−275	+225	−255	+245	s^{1}
s^{59}	+125	+125	+125	+125	+125	+125	+125	+125	s^{2}
s^{60}	+125	+125	+125	+125	+125	+125	−275	+225	s^{43}
s^{61}	+125	+125	−275	+225	+125	+125	+125	+125	s^{45}
s^{62}	+125	+125	−275	+225	+125	+125	−275	+225	s^{39}
s^{63}	+125	+125	+125	+125	−275	+225	+125	+125	s^{51}
s^{64}	+125	+125	+125	+125	−275	+225	−275	+225	s^{3}
s^{66}	+125	+125	−275	+225	−275	+225	−275	+225	s^{55}
s^{67}	−275	+225	+125	+125	+125	+125	+125	+125	s^{53}
s^{69}	−275	+225	−275	+225	+125	+125	+125	+125	s^{5}
s^{70}	−275	+225	−275	+225	+125	+125	−275	+225	s^{57}
s^{71}	−275	+225	+125	+125	−275	+225	+125	+125	s^{41}
s^{72}	−275	+225	+125	+125	−275	+225	−275	+225	s^{47}
s^{73}	−275	+225	−275	+225	−275	+225	+125	+125	s^{49}
s^{74}	−275	+225	−275	+225	−275	+225	−275	+225	s^{1}

of Table B enables you to check that the last listed 48 pure strategies are indeed ordinally dominated. Since negative returns are always accompanied by "LP" and positive returns by "MP" in this problem, we have abbreviated "\$-x; LP" and "\$ + y; MP" to "$-x$" and "$+y$" respectively.

■ Although Table B appears formidable at first glance, its formulation required only a degree of diligent plodding and therefore can be delegated. As in Section 4.4, the importance of this example is that it illustrates that a reasonably complicated problem such as Example 1.5.1c *can* be handled! Furthermore, our discussion of ordinal admissibility in Section 7.1 establishes that only the part of Table B above the dashed line need be given any further attention; all rows below the dashed line may be deleted without strategic disability to $\mathscr{D}$, provided that he always prefers more profit to less. ■

*■ For many analytical purposes, the normal form of a dpuu is generally more flexible than its extensive form. In this concluding remark we argue briefly that if P_Σ is consistent with $\mathscr{D}$'s probability functions on the $\mathscr{E}$-moves of his decision tree, then s'' is preferable to s' in the normal form if and only if s'' is preferable to s' in the extensive form. *First*, "pruning" all but one act in each $\mathscr{D}$-move of the decision tree (the extensive form) produces a pure strategy for $\mathscr{D}$, and hence a lottery on C, with probabilities determined by his probability functions on the $\mathscr{E}$-moves of the tree. The recursive pruning of all but one optimal act in each $\mathscr{D}$-move, according to Chapter 4, is designed to preserve optimality throughout the recursion, which therefore ultimately determines an optimal strategy for $\mathscr{D}$(= an optimal lottery on C). *Second*, P_Σ is *consistent* with $\mathscr{D}$'s probability functions on the $\mathscr{E}$-moves of his tree if and only if, for every pure strategy s for $\mathscr{D}$, P_Σ produces the *same* lottery on C as does the set of his probability functions on the $\mathscr{E}$-moves of his tree. When Σ is defined according to step 2, the product rule (Equation 7.2.1) defining P_Σ is consistent by virtue of regarding the $\mathscr{E}$-moves of the tree as independent canonical randomizations; and when the "economical alternative" to step 2 is used to define Σ, consistency obtains because P_Σ is used to determine the probability functions on the $\mathscr{E}$-moves of the tree, just as in Section 4.4. *Third*, by virtue of the first two points, any pure strategy s for $\mathscr{D}$ determines the same lottery on C in the normal form as in the extensive form if P_Σ and the probability functions on the $\mathscr{E}$-moves of the tree are consistent. Hence the extensive and normal forms of a given dpuu lead to the same action implications as far as pure strategies are concerned. For randomized strategies, it is clear on reflection that a randomized strategy P_S for $\mathscr{D}$ induces a set of probability functions on the *$\mathscr{D}$-moves* of the decision tree which, together with the probability functions on the $\mathscr{E}$-moves (perhaps induced by P_Σ, but assumed consistent with P_Σ at any rate), determine a lottery on C identical to the lottery determined by P_S and

P_Σ. Hence relative desirabilities of randomized strategies P_S for $\mathscr{D}$, given their extensive-form action implications, coincide in the extensive and normal forms. But how analytically convenient it is to have a *compact* terminology such as $U(P_S, P_\Sigma)$ vis-à-vis something even more complex than the extensive-form terminology of Chapter 4! ■

Exercise 7.2.1

Apply the successive partial normalization technique of step 2 to Figure 7.3 to determine the set Σ of $\mathscr{E}$'s pure strategies for that dpuu.

Exercise 7.2.2

Apply step 2 to Figure 4.3 to determine all 34 pure strategies for $\mathscr{E}$.

Exercise 7.2.3

Suppose that in the dpuu of Example 7.2.1 all concerned are sure that the initial dosage will have no effect on the patient's temperature, in the sense that $\gamma_1^{\ 1}$ will obtain if and only if $\gamma_2^{\ 1}$ will obtain. Show that Σ may be defined so as to contain only three pure strategies, and determine the consequence table under this assumption.

Exercise 7.2.4

Example 1.5.1b is depicted in Figure 2.12 in extensive form. Reexpress it in normal form by performing step 1, the economical alternative to step 2, and step 3.

Exercise 7.2.5

In Figure 2.15, how many pure strategies are there for (A) $\mathscr{D}$, according to step 1, and (B) $\mathscr{E}$, according to step 2?

Exercise 7.2.6

In Example 7.2.2, suppose that $\mathrm{P(D|I) = P(D|NI) = P(D)} = p$ (say). Show that, for every $p \in [0, 1]$, $\mathscr{D}$'s pure strategies I and NI produce the same respective lotteries on C in the first consequence table as they do in the second, where the product rule (Equation 7.2.1) is used to produce a consistent P_Σ for the first table.

Exercise 7.2.7

Verify that only the 26 pure strategies above the dashed line in Table B are ordinally admissible given the natural desirability ranking.

Exercise 7.2.8

Verify that disagreements about feasibility of an act in a $\mathscr{D}$-move in a decision tree give rise to disagreements about the feasibility of one or more pure strategies in the normal form of the given dpuu.

***Exercise 7.2.9: Complete Pure Strategies**

A wasteful but easily described way of obtaining S and Σ is to define S as the Cartesian product of all the $\mathscr{D}$-moves in the decision tree, and Σ as the Cartesian product of all the $\mathscr{E}$-moves in the tree. Thus a pure strategy s for $\mathscr{D}$ consists of one act from *every* $\mathscr{D}$-move—not just from the $\mathscr{D}$-moves which can arise given $\mathscr{D}$'s earlier choices. Pure strategies s and σ so defined are called *complete* pure strategies for $\mathscr{D}$ and for $\mathscr{E}$ respectively.

(A) Show that in Example 1.5.1c $\mathscr{D}$ has 393,216 complete pure strategies and $\mathscr{E}$ has 524,288 complete pure strategies. [*Hint*: See Exercise 3.1.7.]

(B) Define the sets S and Σ of complete pure strategies for $\mathscr{D}$ and $\mathscr{E}$ respectively in the dpuu of Example 7.2.1 and Figure 7.3, and determine the resulting consequence table.

■ Comparison of the consequence table you obtain for Exercise 7.2.9(B) with that in Example 7.2.1 or in Exercise 7.2.3 shows how wastefulness arises with complete pure strategies. The complete impracticality of determining a 393,216-by-524,288 consequence table for Example 1.5.1c should be evident. That is why we stress the harder-to-describe successive partial normalization procedure. ■

***Exercise 7.2.10: More on Partial Normalization**

The process of determining partial pure strategies for $\mathscr{D}$, or for $\mathscr{E}$, can be stopped at any point. In Figure 2.16, pretend that h_2^1–h_2^{13} are consequences, and "combine stages one and two":

(1) by defining 9 partial pure strategies r for $\mathscr{D}$ which determine his behavior in the first two stages in Figure 2.16;

(2) by defining 16 partial pure strategies ρ for $\mathscr{E}$ which describe $\mathscr{E}$'s behavior in the first two stages in Figure 2.16;

(3) by determining h_2 as a function of r and ρ in the form of a 9-by-16 "partial history" table.

Also *sketch* the two-stage tree reflecting: in stage one, $\mathscr{D}$'s choice of r followed by $\mathscr{E}$'s choice of ρ; and in stage two, $\mathscr{D}$'s choice of rebuilding action, followed by $\mathscr{E}$'s choice of subsystem state. Record the appropriate probability on each branch of each $\mathscr{E}$-move of this tree. [How can you find $P(\mathrm{AD}|\rho, r, \mathrm{NR})$, for example?]

■ This exercise has important ramifications in Chapter 8, because it implies that every "sequential-information-acquisition" problem is reexpressible as a "one-shot-information-acquisition" problem, i.e., as a two-stage dpuu in which stage one is the acquisition of information and stage two is ultimate action. Some scholars have intimated that there is a *fundamental* distinction between the one-shot and the sequential problems; but, as we have just shown, this distinction does not exist—at least at the conceptual level. ■

7.3 SENSITIVITY ANALYSIS WHEN u IS AGREED ON

As promised in the introduction, this section concerns (1) how to eliminate all consensually inferior strategies P_S for $\mathscr{D}$ and (2) how to analyze the gravity of potentially remaining disagreements, both under the assumption that all concerned with the dpuu agree fully upon the utility function u.

Most of the section concerns the methodology for eliminating consensually inferior strategies; only at the end do we comment on the analysis of remaining disagreement. We shall proceed as follows.

(1) Introduce a convenient, geometrically based model for the dpuu in normal form with the utility function u fixed.
(2) Define and deduce basic properties of cardinal dominance and cardinal admissibility.
(3) Define and deduce basic properties of optimal strategies, called Bayes strategies in the present analytical setting.
(4) Deduce in Theorem 7.3.4 the relationship between cardinally admissible and Bayes strategies.
(5) Discuss briefly how one can determine the set of all cardinally admissible strategies.
(6) Discuss how one can go about analyzing the ramifications of remaining controversy regarding P_Σ.

To introduce the geometrically based model, we note first that when the utility function u is agreed upon, the *consequence* table of a dpuu, such as Table B in Section 7.2, may be replaced by the *utility* table, in which $U(s, \sigma)$ replaces $c[s, \sigma]$ for every row s and column σ.

Example 7.3.1

Suppose $u(c^0) = 0$, $u(c^1) = 1$, $u(c^2) = 2$, $u(c^3) = 3$, $u(c^4) = 6$, $u(c^5) = 7$, and $u(c^6) = 10$. Then

Consequence Table implies Utility Table

	σ^1	σ^2
s^1	c^6	c^1
s^2	c^4	c^4
s^3	c^2	c^5
s^4	c^3	c^2
s^5	c^0	c^6
s^6	c^6	c^0

	σ^1	σ^2
s^1	10	1
s^2	6	6
s^3	2	7
s^4	3	2
s^5	0	10
s^6	10	0

The geometric model builds on the concept that each *row* of the utility table is a *point* in $R^{\#(\Sigma)}$. Generalized to randomized strategies, this concept motivates the definition of the utility characteristic of a strategy for $\mathscr{D}$.

Definition 7.3.1. Let P_S be any randomized strategy for $\mathscr{D}$. Then $\mathscr{D}$'s utility characteristic $\mathbf{x}(P_S)$ of P_S is the $\#(\Sigma)$-tuple of utilities $U(P_S, \sigma^j)$ of choosing P_S given that $\tilde{\sigma} = \sigma^j$. That is,

$$\mathbf{x}(P_S) = (U(P_S, \sigma^1), U(P_S, \sigma^2), \ldots, U(P_S, \sigma^{\#(\Sigma)})).$$

■ Recall that a pure strategy is a randomized strategy. The utility characteristic of pure strategy s^i is $\mathbf{x}(s^i) = (U(s^i, \sigma^1), \ldots, U(s^i, \sigma^{\#(\Sigma)}))$; that is, $\mathbf{x}(s^i)$ is simply the ith row of the utility table of the dpuu in normal form. ■

Example 7.3.2

For the dpuu of Example 7.3.1, $\mathbf{x}(s^1) = (10, 1)$, $\mathbf{x}(s^2) = (6, 6)$, $\mathbf{x}(s^3) = (2, 7)$, $\mathbf{x}(s^4) = (3, 2)$, $\mathbf{x}(s^5) = (0, 10)$, $\mathbf{x}(s^6) = (10, 0)$; and if $P_S = (.4, .3, 0, .1, .2, 0)$, then $\mathbf{x}(P_S) = (6.1, 4.4)$.

Definition 7.3.2. The attainable utility set of a dpuu (in normal form) is defined to be the set $X(\mathscr{Q}_S)$ of all utility characteristics. Symbolically,

$$X(\mathscr{Q}_S) = \{\mathbf{x}(P_S): P_S \in \mathscr{Q}_S\}.$$

■ The adjective "attainable" is meant to be suggestive, but it is well to be careful about the sense in which utility characteristics in $X(\mathscr{Q}_S)$ are attainable. If P_S is a genuinely randomized (= impure) strategy, then $U(P_S, \sigma^j)$ may not be in $\{u(c): c \in C\}$ for any j in $\{1, \ldots, \#(\Sigma)\}$, so that $\mathscr{D}$'s ultimate consequence could not possibly have utility equal to one of the components of $\mathbf{x}(P_S)$. The "attainability" is in the "before-conducting-the-randomization" sense. Equivalently, by choosing a randomized strategy P_S, $\mathscr{D}$ attains the utility-characteristic vector $\mathbf{x}(P_S)$, then $\mathscr{E}$ chooses a component j of that vector, and this component is $\mathscr{D}$'s prerandomizing (expected) utility of the as-yet-to-be determined consequence $c[\tilde{s}, \sigma^j]$. ■

Now, from Example 7.3.2 it is clear that the utility characteristic of *any* randomized strategy P_S for $\mathcal{D}$ is the weighted average, or convex combination, of the utility characteristics of $\mathcal{D}$'s *pure* strategies, with $\mathbf{x}(s^i)$ having weight $P_S(s^i)$. That is,

$$\mathbf{x}(P_S) = \sum_{i=1}^{\#(S)} \mathbf{x}(s^i) P_S(s^i) \tag{7.3.1}$$

for every P_S in $\mathcal{Q}_S$. This implies that the attainable utility set $X(\mathcal{Q}_S)$ is just the convex hull of the utility characteristics of $\mathcal{D}$'s pure strategies!

Theorem 7.3.1. Let $X(\mathcal{Q}_S)$ be the attainable utility set of a dpuu in which $\mathcal{D}$ has a finite number of pure strategies. Then

$$X(\mathcal{Q}_S) = \text{CONV}[\{\mathbf{x}(s): s \in S\}].$$

*■ *Proof*. Easy from the definition of CONV[·] in Appendix 3 and from Equation (7.3.1): If $\mathbf{x} \in X(\mathcal{Q}_S)$, then $\mathbf{x} = \mathbf{x}(P_S)$ for some P_S in $\mathcal{Q}_S$. Therefore $\mathbf{x} = \Sigma_{i=1}^{\#(S)} \mathbf{x}(s^i) P_S(s^i)$, implying that $\mathbf{x} \in \text{CONV}[\{\mathbf{x}(s): s \in S\}] = \text{CONV}[\{\mathbf{x}(s^i): i = 1, \ldots, \#(S)\}]$, and hence $X(\mathcal{Q}_S) \subset \text{CONV}[\{\mathbf{x}(s): s \in S\}]$. The reverse inclusion is obvious. ■

Example 7.3.3

Figure 7.5 depicts $X(\mathcal{Q}_S)$ for the dpuu in Examples 7.3.1 and 7.3.2. (The dashed lines are explained later.) Note that $X(\mathcal{Q}_S)$ is the *entire*, shaded convex polyhedron, not just the vertices, which are the utility characteristics of s^1, s^2, s^5, s^4, and s^6.

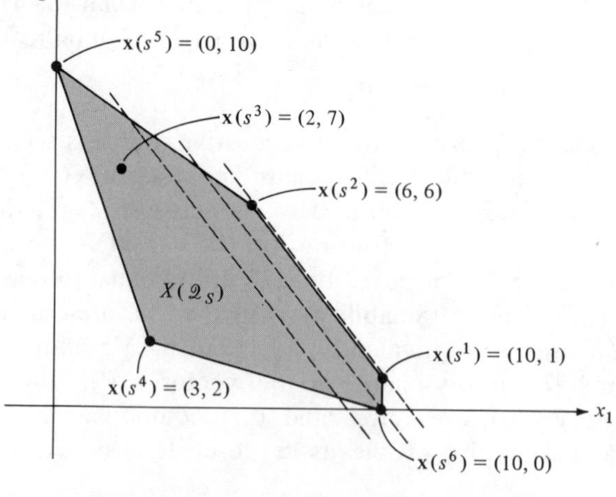

Figure 7.5 Attainable utility set in Examples 7.3.1–7.3.3.

Example 7.3.4

Suppose that a dpuu has the following normal-form utility table.

	σ^1	σ^2	σ^3
s^1	10	0	0
s^2	0	0	0
s^3	0	10	0
s^4	0	0	10
s^5	4	4	4
s^6	3	1	2

Then the attainable utility set is the solid polyhedron, appearing in Figure 7.6, with vertices (0, 0, 0), (10, 0, 0), (0, 10, 0), (0, 0, 10), and (4, 4, 4); (3, 1, 2) is in the interior.

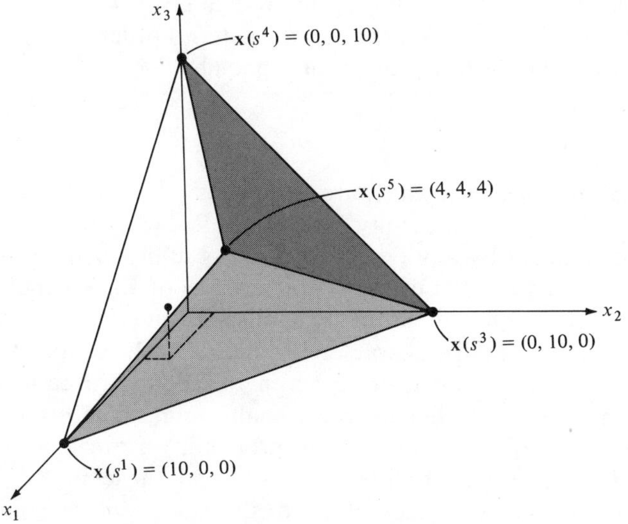

Figure 7.6 Attainable utility set for Example 7.3.4.

So much for the basic ideas of utility characteristics and attainable utility sets. The geometric intuition these concepts can evoke is useful in what follows.

Our next task is to define and study the concepts of cardinal dominance and cardinal admissibility.

Definition 7.3.3: Cardinal Dominance and Cardinal Admissibility

(A) Given two strategies P'_S and P''_S in $\mathcal{Q}_S$, we say that P''_S *cardinally dominates* P'_S *given* u, or that P'_S is cardinally dominated by P''_S

given u, if:

(1) $U(P''_S, \sigma) \geq U(P'_S, \sigma)$ for every σ in Σ, and
(2) $U(P''_S, \sigma) > U(P'_S, \sigma)$ for at least one σ in Σ,

or equivalently, P''_S cardinally dominates P'_S given u, if $\mathbf{x}(P''_S) \geq \mathbf{x}(P'_S)$ but $\mathbf{x}(P''_S) \neq \mathbf{x}(P'_S)$.

(B) We say that a strategy P'_S is *cardinally admissible given u* if no strategy P''_S in $\mathcal{Q}_S$ cardinally dominates P'_S given u.

(C) We denote by $\mathcal{A}_{\mathscr{C}}[\mathcal{Q}_S|u]$ the set of all strategies in $\mathcal{Q}_S$ which are cardinally admissible given u.

Definition 7.3.3 simply formalizes the definitions of cardinal dominance and cardinal admissibility given in Section 7.1, except for the equivalent, second definition of cardinal dominance in terms of vector inequalities between utility characteristics. That definition implies, for the case in which $\#(\Sigma) = 2$, that P''_S cardinally dominates P'_S given u if its utility characteristic $\mathbf{x}(P''_S)$ is "north and/or east" of the utility characteristic $\mathbf{x}(P'_S)$ of P'_S. By the same token, P'_S is cardinally admissible given u if its utility characteristic $\mathbf{x}(P'_S)$ is somewhere on the "northeastern border" of $X(\mathcal{Q}_S)$, or is *efficient in* $X(\mathcal{Q}_S)$, in the terminology of Appendix 3.

Example 7.3.5

In Figure 7.5, we see that the pure strategies s^1, s^2, and s^5 are cardinally admissible, given the utility function u. But even though s^3 is not cardinally dominated by any pure strategy,[h] its utility characteristic "lies south and/or west of" the utility characteristics of some genuinely randomized strategies. s^6 is not cardinally admissible given u, because it is dominated by s^1; its utility characteristic lies directly south of the utility characteristic of s^1. Most strategies in this dpuu are cardinally dominated given u. In fact, the only cardinally admissible strategies given u are those in $\mathcal{Q}_S^{\{s1, s2\}}$—which assign probability zero to all pure strategies *not* in $\{s^1, s^2\}$ (see Definition 7.1.4)—and those in $\mathcal{Q}_S^{\{s2, s5\}}$—which assign probability zero to all pure strategies *not* in $\{s^2, s^5\}$. Strategies in $\mathcal{Q}_S^{\{s1, s2\}}$ are those which have utility characteristics on the line joining $\mathbf{x}(s^1)$ and $\mathbf{x}(s^2)$; strategies in $\mathcal{Q}_S^{\{s2, s5\}}$ are those which have utility characteristics on the line joining $\mathbf{x}(s^2)$ and $\mathbf{x}(s^5)$. These two lines collectively constitute the northeastern border of $X(\mathcal{Q}_S)$. Hence

$$\mathcal{A}_{\mathscr{C}}[\mathcal{Q}_S|u] = \mathcal{Q}_S^{\{s1, s2\}} \cup \mathcal{Q}_S^{\{s2, s5\}}.$$

Example 7.3.6

In Figure 7.6, the three "visible" faces of the polyhedron consist of the utility characteristics of the cardinally admissible strategies given u

[h] s^3 is also *ordinally* undominated, given the desirability ranking consistent with u.

in this dpuu. Again recalling Definition 7.1.4 of $\mathcal{Q}_S^{S^*}$, we see that

$$\mathcal{A}_{\mathcal{C}}[\mathcal{Q}_S|u] = \mathcal{Q}_S^{\{s1, s4, s5\}} \cup \mathcal{Q}_S^{\{s3, s4, s5\}} \cup \mathcal{Q}_S^{\{s1, s3, s5\}}.$$

The practical significance of $\mathcal{A}_{\mathcal{C}}[\mathcal{Q}_S|u]$ should be clear from the introduction; $\mathcal{D}$ has no incentive whatsoever to consider choosing a strategy *not* in $\mathcal{A}_{\mathcal{C}}[\mathcal{Q}_S|u]$, because any such strategy *can be inferior* to some other strategy and *could never be superior* to that same strategy. A more formal statement of this fact constitutes part (A) of Theorem 6.3.2; part (B) says that one need never go outside $\mathcal{A}_{\mathcal{C}}[\mathcal{Q}_S|u]$ to find cardinally dominat*ing* strategies given u.

Theorem 7.3.2: Cardinal Dominance, Strategy Evaluation, and Cardinal Admissibility

(A) If P''_S cardinally dominates P'_S given u, then:

(1) $U(P''_S, P_\Sigma) \geq U(P'_S, P_\Sigma)$ for every P_Σ in $\mathcal{Q}_\Sigma$; and
(2) $U(P''_S, P_\Sigma) > U(P'_S, P_\Sigma)$ for every P_Σ in $\mathcal{Q}_\Sigma$ which assigns positive probability to $\{\sigma: U(P''_S, \sigma) > U(P'_S, \sigma)\}$.

(B) If S and Σ are finite sets of pure strategies and if P'_S is cardinally dominated given u, then it is cardinally dominated by a strategy P''_S *in* $\mathcal{A}_{\mathcal{C}}[\mathcal{Q}_S|\mathrm{u}]$.

*■ *Proof.* Part (A) is straightforward and is left to you as Exercise 7.3.5. Part (B) follows readily from (A) of the Efficiency Theorem in Appendix 3, because finiteness of S and Σ imply that $x(\mathcal{Q}_S)$ is a closed, bounded, and convex subset of the finite-dimensional real vector space $R^{\#(\Sigma)}$, so that $X(\mathcal{Q}_S)$ is a fortiori closed above and locally bounded above. To apply the Efficiency Theorem, simply identify P_Σ here with $\mathbf{w}$ in Appendix 3, $\mathbf{x}(P_S)$ here with $\mathbf{x}$ there, and $X(\mathcal{Q}_S)$ here with X there. ■

*■ If S were allowed to be countably infinite, $X(\mathcal{Q}_S)$ could fail to be closed above, thus jeopardizing the conclusion of Theorem 7.3.2(B). For example, let $S = \{1, 2, \ldots, \textit{ad inf}\}$, let $\#(\Sigma) = 1$ with $\Sigma = \{\sigma^1\}$, and let $U(s, \sigma^1) = s$ for every positive integer s. Every s' is cardinally dominated by every $s'' > s'$, and there is no cardinally undominated s. Hence $\mathcal{A}_{\mathcal{C}}[\mathcal{Q}_S|u] = \emptyset$ and the conclusion of Theorem 7.3.2(B) is dramatically false. ■

*■ A subset $\mathcal{K}$ of $\mathcal{Q}_S$ is said to be a *complete class* if every strategy P'_S *not* in $\mathcal{K}$ is cardinally dominated by some strategy P''_S *in* $\mathcal{K}$. With this definition, Theorem 7.3.2(B) says that $\mathcal{A}_{\mathcal{C}}[\mathcal{Q}_S|u]$ is a complete class. When $\#(S) \geq 2$, there are many complete classes; indeed, $\mathcal{Q}_S$ itself is always a complete class. See Ferguson [**34**] for more information on complete classes, which are important in the infinite idealizations of advanced statistical decision theory. ■

Our *third* task is to discuss optimal strategies and their manifestations in the geometric model. From Definition 7.1.3(B), which establishes $U(P_S, P_\Sigma)$ as $\mathscr{D}$'s a priori utility evaluation of his strategy P_S, given his judgments P_Σ about $\tilde{\sigma}$ (or given $\mathscr{E}$'s strategy P_Σ); it is clear that a strategy P_S° (say) is optimal for $\mathscr{D}$ if P_S° maximizes $U(P_S, P_\Sigma)$ as P_S varies throughout $\mathscr{Q}_S$. That is, from Definition 7.3.1(C), P_S° is *optimal* if

$$U(P_S^\circ, P_\Sigma) = U(\mathscr{Q}_S, P_\Sigma).$$

Being a little more explicit about the role of P_Σ results in Definition 7.3.4(A), which also introduces the synonym "Bayes" for "optimal," a synonym we shall use frequently in the sequel and which is very popular in decision analysis, particularly in connection with statistical decision problems. Definitions 7.3.4(B) and 7.3.4(C) arise from 7.3.4(A) by supposing that P_Σ is controversial and hence may be varied throughout $\mathscr{Q}_\Sigma$.

Definition 7.3.4: Bayes Strategies for $\mathscr{D}$

(A) A (randomized) strategy P_S° for $\mathscr{D}$ is *Bayes given* u *against*, or *optimal given* u *with respect to*, a strategy P_Σ for $\mathscr{E}$ if

$$U(P_S^\circ, P_\Sigma) = U(\mathscr{Q}_S, P_\Sigma) = \max_{P_S \in \mathscr{Q}_S} [U(P_S, P_\Sigma)].$$

(B) A (randomized) strategy P_S° for $\mathscr{D}$ is simply called *Bayes* given u if there is at least one P_Σ in $\mathscr{Q}_\Sigma$ such that P_S° is Bayes given u against P_Σ.

(C) We denote by $\mathscr{B}[\mathscr{Q}_S|u]$ the set of all strategies P_S° in $\mathscr{Q}_S$ which are Bayes given u.

■ In set-theoretic terminology, it follows succinctly from Definitions 7.3.4(B) and (C) that

$$\mathscr{B}[\mathscr{Q}_S|u] = \cup_{P_\Sigma \in \mathscr{Q}_\Sigma} \{P_S^\circ\colon P_S^\circ \text{ is Bayes against } P_\Sigma \text{ given } u\}. \blacksquare$$

Now let us see how Bayes strategies appear in our geometric model. We begin by fixing the utility function u—as always in this section—, fixing P_Σ, choosing some number z—soon to be varied—, and then asking the question, "Does $\mathscr{D}$ have a strategy or strategies P_S with a priori evaluation $U(P_S, P_\Sigma) = z$?" To answer this question within our geometric model amounts to seeing if there is a utility characteristic $\mathbf{x} = (x_1, \ldots, x_{\#(\Sigma)})$ in $X(\mathscr{Q}_S)$ with the property that

$$\sum_{j=1}^{\#(\Sigma)} x_j P_\Sigma(\sigma^j) = z, \tag{7.3.2}$$

because if so, then there is some strategy P_S with $\mathbf{x} = \mathbf{x}(P_S)$ and $x_j = U(P_S, \sigma^j)$ for every j, so that the left-hand side of Equation (7.3.2) is really

$$\sum_{j=1}^{\#(\Sigma)} U(P_S, \sigma^j) P_\Sigma(\sigma^j) = \sum_{j=1}^{\#(\Sigma)} \sum_{i=1}^{\#(S)} U(s^i, \sigma^j) P_S(s^i) P_\Sigma(\sigma^j)$$

$$= \sum_{i=1}^{\#(S)} \sum_{j=1}^{\#(\Sigma)} U(s^i, \sigma^j) P_\Sigma(\sigma^j) P_S(s^i) = U(P_S, P_\Sigma).$$

To see if (7.3.2) obtains for some utility characteristic, it suffices to examine the relationship of the line (when $\#(\Sigma) = 2$), plane (when $\#(\Sigma) = 3$), or hyperplane

$$H(P_\Sigma, z) = \left\{\mathbf{x}: \mathbf{x} \in R^{\#(\Sigma)}, \sum_{j=1}^{\#(\Sigma)} x_j P_\Sigma(\sigma^j) = z\right\} \qquad (7.3.3)$$

to $X(\mathcal{Q}_S)$. If $H(P_\Sigma, z)$ has points in common with $X(\mathcal{Q}_S)$, then the answer to our direct question is "yes," but if $H(P_\Sigma, z) \cap X(\mathcal{Q}_S) = \emptyset$, then the answer is "no."

Now, for a fixed P_Σ, varying z moves $H(P_\Sigma, z)$ parallel to itself; and, since $P_\Sigma(\sigma^j) \geq 0$ for every j, it follows that if $z'' > z'$, then strategies P_S'' with utility characteristics in $H(P_\Sigma, z'')$ are "better" against P_Σ—have higher a priori evaluations $U(P_S, P_\Sigma)$—than are strategies P_S' with utility characteristics in $H(P_S, z')$.

It follows immediately that the points common to $X(\mathcal{Q}_S)$ and the "highest and/or rightmost" member $H(P_\Sigma, z^*)$ of the family $\{H(P_\Sigma, z): -\infty < z < +\infty\}$ of parallel hyperplanes which has points in common with $X(\mathcal{Q}_S)$ are precisely the utility characteristics of strategies P_S° Bayes given u against P_Σ.

Example 7.3.7

For the dpuu in Figure 7.5 and Examples 7.3.1–7.3.3 and 7.3.5, let us take $P_\Sigma = (5/9, 4/9)$ and find strategies which are Bayes against (= optimal with respect to) this P_Σ. For any real number z, $H(P_\Sigma, z)$ is the locus of all (x_1, x_2) such that $(5/9)x_1 + (4/9)x_2 = z$, or equivalently, $H(P_\Sigma, z)$ is the line with slope $-(5/9)/(4/9) = -5/4$ which passes through (z, z). Several members of this family of parallel lines are depicted as dashed lines in Figure 7.5. The largest z for which $H((5/9, 4/9), z) \cap X(\mathcal{Q}_S) \neq \emptyset$ is $z^* = 6$; and it is clear that $\mathbf{x}(P_S)$ belongs to $H((5/9, 4/9), 6) \cap X(\mathcal{Q}_S)$ if and only if P_S is a randomized strategy in $\mathcal{Q}_S^{\{s1, s2\}}$, that is, if and only if P_S assigns positive probability only to s^1 and/or s^2.

Example 7.3.8

To continue the preceding example, let us now vary P_Σ and ascertain which strategies P_S are Bayes against every possible P_Σ. This task is not hard; you may perform it as Exercise 7.3.2(A), thus verifying all entries in the following summary table, called an *optimal response* (to judgments P_Σ) *table.*

If P_Σ is such that:	Then P_S is Bayes against P_Σ, where:	
$0 \leq P_\Sigma(\sigma^1) < .40$	$P_S = (0, 0, 0, 0, 1, 0)$	(i.e., s^5)
$P_\Sigma(\sigma^1) = .40$	$P_S \in \mathcal{Q}_S^{\{s2, s5\}}$	
$.40 < P_\Sigma(\sigma^1) < 5/9$	$P_S = (0, 1, 0, 0, 0, 0)$	(i.e., s^2)
$P_\Sigma(\sigma^1) = 5/9$	$P_S \in \mathcal{Q}_S^{\{s1, s2\}}$	
$5/9 < P(\sigma^1) < 1.00$	$P_S = (1, 0, 0, 0, 0, 0)$	(i.e., s^1)
$P_\Sigma(\sigma^1) = 1.00$	$P_S \in \mathcal{Q}_S^{\{s1, s6\}}$	

From the optimal response table, it follows immediately that

$$\mathscr{B}[\mathscr{Q}_S|u] = \mathscr{Q}_S^{\{s1, s6\}} \cup \mathscr{Q}_S^{\{s1, s2\}} \cup \mathscr{Q}_S^{\{s2, s5\}}.$$

A brief comparison of Examples 7.3.5 and 7.3.8 shows that $\mathscr{A}_{\mathscr{C}}[\mathscr{Q}_S|u]$ is a proper subset of $\mathscr{B}[\mathscr{Q}_S|u]$, the difference being that $\mathscr{B}[\mathscr{Q}_S|u]$ includes all of $\mathscr{Q}_S^{\{s1, s6\}}$, whereas $\mathscr{A}_{\mathscr{C}}[\mathscr{Q}_S|u]$ includes only the pure strategy s^1 from $\mathscr{Q}_S^{\{s1, s6\}}$. *No* P_S of the form $(1-p, 0, 0, 0, 0, p)$ with $0 < p \leq 1$ is cardinally admissible, whereas *all* such strategies P_S are *optimal* with respect to $P_\Sigma = (1, 0) = \sigma^1$. Yet, it is equally clear that $\mathscr{D}$ has no positive incentive to choose any P_S in $\mathscr{Q}_S^{\{s1, s6\}}$ other than s^1, because[i] he can do just as well and maybe better with s^1.

Before proceeding to our fourth task, which is to show that $\mathscr{A}_{\mathscr{C}}[\mathscr{Q}_S|u]$ is a subset of $\mathscr{B}[\mathscr{Q}_S|u]$ under general conditions and to deduce other facts about the relationship between $\mathscr{A}_{\mathscr{C}}[\mathscr{Q}_S|u]$ and $\mathscr{B}[\mathscr{Q}_S|u]$, we shall pause to state as Theorem 7.3.3 some almost obvious facts about Bayes strategies.

Theorem 7.3.3(A) simply means that two strategies, each optimal with respect to P_Σ, must have the same a priori utility. This is intuitively obvious because it is the a priori utility by which optimality is determined! Theorem 7.3.3(B) means that, given u, the only way a genuinely randomized strategy can be optimal with respect to P_Σ is for it to allocate positive probabilities *only* to pure strategies which are themselves optimal with respect to P_Σ.

Theorem 7.3.3: Structure of $\mathscr{B}[\mathscr{Q}_S|u]$

(A) If both P'_S and P''_S are Bayes given u against P_Σ, then $U(P'_S, P_\Sigma) = U(P''_S, P_\Sigma)$.

(B) Let $S[P_\Sigma] = \{s: s \text{ is Bayes given } u \text{ against } P_\Sigma\}$. Then P_S is Bayes given u against P_Σ if and only if $P_S \in \mathscr{Q}_S^{S[P_\Sigma]}$.

■ *Proof.* *Exercise 7.3.6. ■

Example 7.3.9

In Example 7.3.8, assume that $P_\Sigma = (5/9, 4/9)$. Then each of s^1 and s^2 is Bayes against P_Σ, with $U(s^1, P_\Sigma) = U(s^2, P_\Sigma) = 6$. So is any P_S of the form $(p, 1-p, 0, 0, 0, 0)$ for $0 < p < 1$. Conversely, any strategy P_S^* (say) for which $P_S^*(\{s^3, s^4, s^5, s^6\}) > 0$ *cannot* be Bayes against $P_\Sigma = (5/9, 4/9)$.

It is important to note that Theorem 7.3.3 answers "no" to the question we raised in Section 7.1 regarding $\mathscr{D}$'s need to consider adopting a genuinely randomized strategy, given that $\mathscr{D}$ is following our approach to decision analysis, "our approach" being that $\mathscr{D}$ quantifies his preferences, u, and his judgments, P_Σ, and then chooses P_S so as to maximize his a priori utility $U(P_S, P_\Sigma)$. When we introduce the maximin choice criterion in Exercises 7.3.8–7.3.10, it will be shown that under this criterion, $\mathscr{D}$ may be

---

[i] This follows from Theorem 7.3.2(A) with $P''_S = s^1$ and $P'_S =$ any strategy of the form $(1-p, 0, 0, 0, 0, p)$ with $p > 0$.

able to do *better* with a genuinely randomized strategy than he can do with any pure strategy—better, that is, in terms of the strategy evaluation appropriate for the maximin criterion.

We come now to our *fourth* task, which is to elucidate the relationship between $\mathscr{A}_{\mathscr{C}}[\mathscr{Q}_S|u]$ and $\mathscr{B}[\mathscr{Q}_S|u]$. Theorem 7.3.4, which presents the salient facts, is really the keystone of this section, because it not only provides information of great conceptual interest but also hints at how we can go about determining $\mathscr{A}_{\mathscr{C}}[\mathscr{Q}_S|u]$ in specific cases, a subject we take as our fifth task.

We have argued previously that the only strategies worthy of $\mathscr{D}$'s consideration are cardinally admissible strategies given u. Theorem 7.3.4(A) shows, therefore, that *every strategy worthy of consideration is actually optimal with respect to some set of judgments quantitatively expressed by* P_Σ. In fact, such a strategy must be optimal with respect to some P_Σ which assigns *positive* probability to every σ. Conversely, part (B) of Theorem 7.3.4 says that any strategy optimal with respect to such a P_Σ is cardinally admissible. Part (C) completes the picture by showing that we need not worry about $\mathscr{D}$'s having to use a Bayes strategy which is cardinally inadmissible, because any Bayes strategy that is cardinally dominated must also be "tied" for optimality with a cardinally admissible strategy. Hence cardinally *in*admissible Bayes strategies need never be chosen, and $\mathscr{D}$'s search for a good strategy may be confined to the set $\mathscr{A}_{\mathscr{C}}[\mathscr{Q}_S|u]$ of cardinally admissible strategies.

Theorem 7.3.4: Relationships between Cardinally Admissible and Bayes Strategies, Given u. Assume throughout that S and Σ are both finite sets.

(A) If P_S is cardinally admissible given u, then P_S is Bayes given u against at least one P_Σ such that $P_\Sigma(\sigma)>0$ for every $\sigma\in\Sigma$. Hence

$$\mathscr{A}_{\mathscr{C}}[\mathscr{Q}_S|u]\subset\mathscr{B}[\mathscr{Q}_S|u].$$

(B) Conversely, if P_S is Bayes given u against some P_Σ such that $P_\Sigma(\sigma)>0$ for every $\sigma\in\Sigma$, then P_S is cardinally admissible given u.

(C) If a cardinally dominated strategy P'_S given u is Bayes given u against some P_Σ, then there is a cardinally admissible strategy P''_S given u which is also Bayes given u against that same P_Σ, with (naturally) $U(P''_S, P_\Sigma)=U(P'_S, P_\Sigma)$.

*■ *Proof.* Since S and Σ are finite sets, $X(\mathscr{Q}_S)$ is closed, bounded, and convex, and hence a fortiori closed above, convex above, and locally bounded above. Part (A) follows immediately from part (D) of the Efficiency Theorem by virtue of Theorem 7.3.1, given the notational identifications in the proof of Theorem 7.3.2(B). Part (B) is immediate from part (B2) of the Efficiency Theorem; part (C) is a similarly immediate consequence of part (C) of the Efficiency Theorem. ■

*■ In Figure 7.7 we depict a number of possible relationships among $Y(P_S)$, $X(\mathcal{Q}_S)$, and $\mathbf{x}(P_S)$—for P_S cardinally admissible given u in Figure 7.7(a), (b), and (d), and for P_S Bayes given u but cardinally dominated given u in Figure 7.7(c). The sets $Y(P_S)$ are analogous to the sets $Y(\mathbf{x})$ in the Efficiency Theorem in Appendix 3; they are defined by

$$Y(P_S) = Y(\mathbf{x}(P_S)) = \{\mathbf{x}: \mathbf{x} \in R^{\#(\Sigma)}, \ \mathbf{x} \geq \mathbf{x}(P_S)\}.$$

In Figure 7.7(d), P_S is cardinally admissible given u, but the conclusion of Theorem 7.3.4(A) is false because we must have $P_\Sigma = (1, 0)$ to separate $X(\mathcal{Q}_S)$ and $Y(P_S)$ at $\mathbf{x}(P_S)$. Such a nonpolyhedral attainable utility set $X(\mathcal{Q}_S)$ can arise *only* when S is an *infinite* set of pure strategies, or when $\mathcal{D}$'s choice is constrained to a nonpolyhedral subset of $\mathcal{Q}_S$ (as envisioned in Section 7.6). ■

Our *fifth* task is to discuss how to determine $\mathcal{A}_{\mathcal{C}}[\mathcal{Q}_S|u]$. When $\#(\Sigma) = 2$, the graphical approach, typified by Figure 7.5 and Example 7.3.5 or by the graphs in Figure 7.7, works very well. When $\#(\Sigma) = 3$, "eyeballing" a graph is satisfactory, as we have seen with Figure 7.6 and Example 7.3.6, but drawing the graph itself can be difficult. In fact, the art work rather

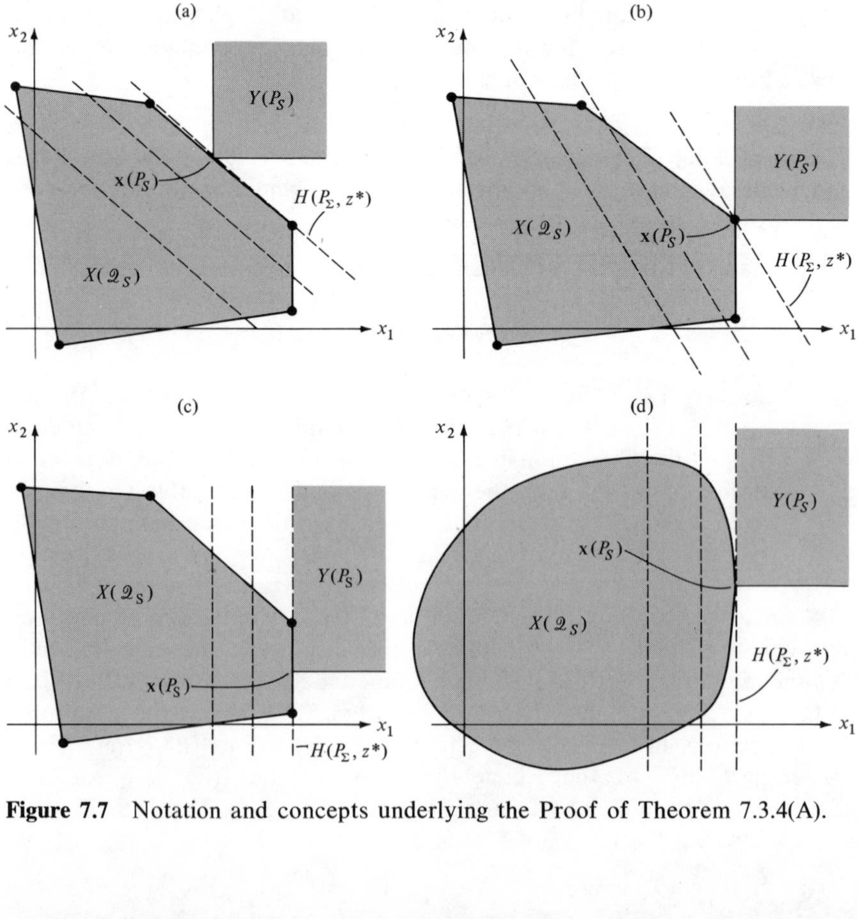

Figure 7.7 Notation and concepts underlying the Proof of Theorem 7.3.4(A).

begs the question about $\mathscr{A}_{\mathscr{C}}[\mathscr{Q}_S|u]$, because conscientiously accurate graphing when $\#(\Sigma) = 3$ virtually requires one's first determining $\mathscr{A}_{\mathscr{C}}[\mathscr{Q}_S|u]$!

Graphical methods are completely out of the question when $\#(\Sigma)$ is four or more. In such cases, the *analytical* technique of maximizing $U(P_S, P_\Sigma)$ for various strategies P_Σ for $\mathscr{D}$ with all $P_\Sigma(\sigma)$'s positive, a technique suggested by Theorem 7.3.4(A), assumes a dominant role.

Here is an informal, "groping" way of searching for $\mathscr{A}_{\mathscr{C}}[\mathscr{Q}_S|u]$. *First*, it follows from Theorems 7.3.3(B) and 7.3.4(A) that[j]

$$\mathscr{A}_{\mathscr{C}}[\mathscr{Q}_S|u] = \cup_{t=1}^{T} \mathscr{Q}_S^{S[P_\Sigma{}^t]} \tag{7.3.4}$$

for a finite number T of strategies $P_\Sigma{}^t$ for $\mathscr{E}$, each of which assigns positive probability $P_\Sigma{}^t(\sigma)$ to every pure strategy σ for $\mathscr{E}$. *Second*, it is easy to see from Definition 7.1.4 that $\mathscr{Q}_S{}^{S'} \subset \mathscr{Q}_S{}^{S''}$ whenever $S' \subset S''$. *Third*, and consequently, we can go about finding $\mathscr{A}_{\mathscr{C}}[\mathscr{Q}_S|u]$ by fiddling around with P_Σ's all components $P_\Sigma(\sigma)$ of which are positive, until we find a $P_\Sigma{}^t$ producing a *maximal* number of pure strategies tied for optimality given u with respect to $P_\Sigma{}^t$. That is, find a P_Σ^* and an associated $S^* = \{s°: s° \text{ is optimal given } u$ with respect to $P_\Sigma^*\}$ with the property that varying P_Σ^* so as to add *any* element to S^* would cause one or more of the elements now in S^* to drop out. Each such P_Σ^* is one of the $P_\Sigma{}^t$'s in (7.3.4). We repeat the process until we are sure that we have found all $P_\Sigma{}^t$'s. Clearly, this approach is very informal.

Example 7.3.10

Consider the dpuu in normal form with utility table[k]

	$\sigma^1 = (\sigma_I^1, \sigma_{II}^1)$	$\sigma^2 = (\sigma_I^1, \sigma_{II}^2)$	$\sigma^3 = (\sigma_I^2, \sigma_{II}^1)$	$\sigma^4 = (\sigma_I^2, \sigma_{II}^2)$
s^1	10	0	0	0
s^2	0	10	0	0
s^3	0	0	10	0
s^4	0	0	0	10
s^5	3	3	3	3
s^6	0	6	6	0

Although we cannot *graph* $X(\mathscr{Q}_S)$ in four-dimensional space, it is not difficult to show that all six *pure* strategies are cardinally admissible. Clearly, s^1 through s^4 are cardinally admissible given u, because $\mathscr{D}$ does better with s^i given σ^i than with any $s^{i'} \neq s^i$, for $i = 1, 2, 3, 4$. By a rather laborious fiddling around with P_Σ's, it can be determined that $\mathscr{A}_{\mathscr{C}}[\mathscr{Q}_S|u]$ is the union of the $\mathscr{Q}_S^{S[P_\Sigma{}^t]}$'s for the following sets $S[P_\Sigma{}^t]$.

[j] Recall from Theorem 7.3.3(B) that $S[P_{\Sigma^t}] = \{s°: U(s°, P_{\Sigma^t}) = \max_{s \in S}[U(s, P_{\Sigma^t})]\}$.

[k] The alternative notation for the σ^j's will be explained and used in Section 7.4.

t	"Tieing" $P_\Sigma{}^t$	$S[P_\Sigma{}^t]$
1	(.3, .3, .1, .3)	$\{s^1, s^2, s^4, s^5\}$
2	(.3, .1, .3, .3)	$\{s^1, s^3, s^4, s^5\}$
3	(.3, .3, .2, .2)	$\{s^1, s^2, s^5, s^6\}$
4	(.3, .2, .3, .2)	$\{s^1, s^3, s^5, s^6\}$
5	(.2, .3, .2, .3)	$\{s^2, s^4, s^5, s^6\}$
6	(.2, .2, .3, .3)	$\{s^3, s^4, s^5, s^6\}$

No other set of four pure strategies is tied for optimality given u against any P_Σ.

Systematic methods for determining $\mathscr{A}_{\mathscr{C}}[\mathscr{Q}_S|u]$ given any finite numbers $\#(S)$ and $\#(\Sigma)$ of pure strategies for $\mathscr{D}$ and $\mathscr{E}$, respectively, have recently been developed by Shachtman [**130**], Walker [**137**], and Cabot [**16**]. Shachtman's method "works up to" the efficient boundary of $X(\mathscr{Q}_S)$ by creating a sequence of hyperplanes; Walker's and Cabot's methods use linear-programming techniques[1] to flip-flop a hyperplane around the efficient boundary. Computational experience to date with problems of significant size is scanty and not too encouraging.

*■ Actually, the problem of finding cardinally admissible strategies, given u, is usually studied in mathematical optimization theory as the problem of finding efficient (utility-characteristic) points $\mathbf{x}$ in a convex, compact set X [$= X(\mathscr{Q}_S)$] (see Appendix 3). In general, X is defined only implicitly, as the range of a given function $\mathbf{F}\colon Y \to R^{\#(\Sigma)}$ on a closed, convex domain Y which itself may be defined only implicitly, as the set of all $\mathbf{y}$ in R^k such that $\mathbf{G}(\mathbf{y}) \geq \mathbf{0}$ for a given function $\mathbf{G}\colon R^k \to R^m$. The *vector maximum problem* may now be posed in the following general form: Given the functions $\mathbf{G}$ and $\mathbf{F}$ and certain assumptions about them which imply that Y and X have "nice" properties, find a $\mathbf{y}^*$ such that $\mathbf{G}(\mathbf{y}^*) \geq \mathbf{0}$ and $\mathbf{F}(\mathbf{y}^*)$ if efficient in $X = \{\mathbf{F}(\mathbf{y})\colon \mathbf{G}(\mathbf{y}) \geq \mathbf{0}\}$; and, more generally, find the set of *all* such $\mathbf{y}^*$'s. For some recent results and further references on the vector maximum problem, see Geoffrion [**47**]. It is important to note that if $\#(\Sigma) = 1$, then $F(\mathbf{y}^*)$ is efficient in $X = \{F(\mathbf{y})\colon \mathbf{G}(\mathbf{y}) \geq \mathbf{0}\}$ if and only if $\mathbf{y}^*$ maximizes $F(\mathbf{y})$ over $Y = \{\mathbf{y}\colon \mathbf{G}(\mathbf{y}) \geq \mathbf{0}\}$. Hence for $\#(\Sigma) = 1$ the vector maximum problem specializes to the general "mathematical programming problem," or "constrained optimization problem." But a more general version of Theorem 7.3.4(A) obtains for $\#(\Sigma) > 1$, and with it the *general* vector maximum problem can be reduced to a constrained optimization problem under certain circumstances. The typical theorem of this sort says that, under certain assumptions about $\mathbf{F}(\cdot)$ and $\mathbf{G}(\cdot)$, the efficiency of $\mathbf{F}(\mathbf{y}^*)$ in $\{\mathbf{F}(\mathbf{y})\colon \mathbf{G}(\mathbf{y}) \geq \mathbf{0}\}$ implies and/or is implied by $\mathbf{y}^*$ being a maximizer of $f(\mathbf{y}) = \Sigma_{j=1}^{\#(\Sigma)} F_j(\mathbf{y}) w_j$

[1] See the second of the following remarks.

in $\{\mathbf{y}: \mathbf{G}(\mathbf{y}) \geq \mathbf{0}\}$ for some vector $\mathbf{w}$ of nonnegative relative weights w_j (corresponding to our $P_\Sigma(\sigma^j)$'s. The literature on constrained optimization is voluminous; see Mangasarian [**93**] and Zangwill [**150**] for a running start. ■

*■ The most intensively studied and widely applied class of problems within constrained optimization theory is the class of *linear programming* problems, which concern finding a maximizer $\mathbf{y}^*$ of a real-valued ($\#(\Sigma) = 1$) *and linear* function $f(\mathbf{y}) = \sum_{h=1}^{k} t_h y_h$, where $t_1, \ldots, t_k$ are given constants and where $\mathbf{y} = (y_1, \ldots, y_k)$ must satisfy m constraints of the form $\sum_{h=1}^{k} a_{\ell h} y_h \leq b_\ell$ for $\ell = 1, \ldots, m$, in which all $a_{\ell h}$ and all b_ℓ are given constants. Define $G_\ell(\mathbf{y}) = b_\ell - \sum_{h=1}^{k} a_{\ell h} y_h$; then $\sum_{h=1}^{k} a_{\ell h} y_h \leq b_\ell$ if and only if $G_\ell(\mathbf{y}) \geq 0$. Then define $\mathbf{G}(\mathbf{y}) = (G_1(\mathbf{y}), \ldots, G_m(\mathbf{y}))$. Then the linear inequality constraints are equivalent to the constraint $\mathbf{G}(\mathbf{y}) \geq \mathbf{0}$. Now, $\{\mathbf{y}: \mathbf{G}(\mathbf{y}) \geq \mathbf{0}\}$ is a convex polyhedron, and $\{\mathbf{y}: f(\mathbf{y}) = z\}$ is a hyperplane, and so the linear programming problem amounts to the same thing as finding Bayes strategies [= cardinally admissible strategies if every $t_h > 0$, by virtue of Theorem 7.3.4, parts (A) and (B).] Moreover, sensitivity analysis of the effects of varying $(t_1, \ldots, t_k)$ is highly developed in linear programming literature. The reason why we cannot apply this work directly to the task of finding all cardinally admissible strategies, given the geometric similarity of the problems, is the following. In linear programming, the convex polyhedron is *defined by the linear inequalities*, whereas in the normal-form analysis the convex polyhedron is *defined as the convex hull of a set of points*. Actually, the thrust of the Shachtman [**16**], Walker [**130**], and Cabot [**137**] procedures is to determine linear inequalities which produce, precisely, the given convex hull. ■

Our *sixth* and final task in this section is to explore the ramifications of the potential disagreements which remain after all persons concerned with $\mathscr{D}$'s choice have agreed to confine their attention to $\mathscr{A}_{\mathscr{C}}[\mathscr{Q}_S | u]$. Such disagreements can crop up at two levels. First, people may quibble about judgmental differences of opinion so minor as to be irrelevant, in that all concerned agree on the optimality of some particular pure strategy $s°$. Is this not what is really meant by the expression "splitting hairs"?

Example 7.3.11

Suppose that in the dpuu of Example 7.3.8 $\mathscr{D}$ and all his colleagues have judgments P_Σ which assign probability .80 or higher to σ^1. Then their quibbles about P_Σ are irrelevant, because all should agree that $s° = s^1$.

The second, and more substantive, level of disagreements involves those in which different parties advocate different cardinally admissible strategies as optimal. The optimal response tables [for $\#(\Sigma) = 2$] introduced in Example 7.3.8 enable rapid determination of the materiality of disagreements regarding P_Σ, but these tables or their equivalents become very clumsy when $\#(\Sigma) \geq 3$. Of course, you can always derive the set $S[P_\Sigma]$ of optimal pure strategies, given each participant's judgments P_Σ, and then see if there is any pure strategy common to all these sets. Given many cooks each with his own "recipe" P_Σ, there probably will not be any such commonly optimal pure strategy. In its absence, the only statements we are prepared to make on the subject are those in Theorem 7.3.5.

Part (A) of Theorem 7.3.5 addresses part of this issue; its continuity assertion means that if two people with the same utility function u, confronting the same dpuu, have judgments $P_\Sigma = (P_\Sigma(\sigma^1), \ldots, P_\Sigma(\sigma^{\#(\Sigma)})$ close together, then their a priori utility evaluations $U(\mathcal{Q}_S, P_\Sigma)$ of their optimal strategies will be close together. The convexity assertion is much more important from a practical point of view, because it implies that "more information is always at least as desirable as less," a conclusion you may deduce in Exercise 7.3.7.

Theorem 7.3.5(B) addresses another part of the same issue. Suppose that $\mathcal{D}$ is forced to use a strategy optimal with respect to a compromise $P_\Sigma{}^t = tP_\Sigma{}^1 + (1-t)P_\Sigma{}^0$ between *his* judgments $P_\Sigma{}^0 = (P_\Sigma{}^0(\sigma^1), \ldots, P_\Sigma{}^0(\sigma^{\#(\Sigma)}))$ and someone *else*'s judgments $P_\Sigma{}^1 = (P_\Sigma{}^1(\sigma^1), \ldots, P_\Sigma{}^1(\sigma^{\#(\Sigma)}))$, where t, between 0 and 1, denotes the extent to which $\mathcal{D}$ has to give in. Then Theorem 7.3.5(B) says that $\mathcal{D}$'s *own* utility (evaluated with respect to *his* judgments $P_\Sigma{}^0$) of having to compromise is decreasing in the extent to which he has to give in.

Theorem 7.3.5: Analysis of Irresolvable Controversies. Assume that both S and Σ are finite sets.

(A) $U(\mathcal{Q}_S, P_\Sigma)$ is a continuous and convex function of P_Σ.

(B) Let $P_\Sigma{}^0$ and $P_\Sigma{}^1$ be any strategies in Σ such that there is a unique optimal strategy for $\mathcal{D}$ with respect to $P_\Sigma{}^0$. For every t in $[0, 1]$, let $P_\Sigma{}^t = tP_\Sigma{}^1 + (1-t)P_\Sigma{}^0$, and let $P_S^*(P_\Sigma{}^t)$ be not only optimal with respect to $P_\Sigma{}^t$ but also at least as good against $P_\Sigma{}^0$ as is any other P_S also optimal with respect to $P_\Sigma{}^t$. Then $U(P_S^*(P_\Sigma{}^t), P_\Sigma{}^0)$ is a *nonincreasing* and *upper semicontinuous* function of t.

*■ *Proof*: *Of (A), Continuity.* For every s,

$$U(s, P_\Sigma) = \sum_{j=1}^{\#(\Sigma)} U(s, \sigma^j) P_\Sigma(\sigma^j)$$

is a linear and therefore continuous function of P_Σ. By Theorem 7.3.3(B), for every P_Σ there is at least one *pure* strategy optimal with respect to P_Σ, so that

$$U(\mathcal{Q}_S, P_\Sigma) = U(S, P_\Sigma) \qquad \text{for every } P_\Sigma \text{ in } \mathcal{Q}_\Sigma. \tag{7.3.5}$$

Since S is a finite set, it follows that $U(\mathcal{Q}_S, \cdot)$ is the pointwise maximum of a *finite* number of continuous functions $U(s, \cdot)$, and is therefore continuous.

Of (A), Convexity. We shall show that

$$(*)\qquad U(\mathcal{Q}_S, tP_\Sigma{}^1 + (1-t)P_\Sigma{}^0) \le tU(\mathcal{Q}_S, P_\Sigma{}^1) + (1-t)U(\mathcal{Q}_S, P_\Sigma{}^0).$$

We have:

$$\begin{aligned}
U(\mathcal{Q}_S, tP_\Sigma{}^1 + (1-t)P_\Sigma{}^0) &=^1 U(S, tP_\Sigma{}^1 + (1-t)P_\Sigma{}^0)\\
&=^2 \max_{s\in S}\left[\sum_{j=1}^{\#(\Sigma)} U(s,\sigma^j)\{tP_\Sigma{}^1(\sigma^j) + (1-t)P_\Sigma{}^0(\sigma^j)\}\right]\\
&=^3 \max_{s\in S}\left[t\sum_{j=1}^{\#(\Sigma)} U(s,\sigma^j)P_\Sigma{}^1(\sigma^j)\right.\\
&\qquad\left. + (1-t)\sum_{j=1}^{\#(\Sigma)} U(s,\sigma^j)P_\Sigma{}^0(\sigma^j)\right]\\
&=^4 \max_{s\in S}[tU(s, P_\Sigma{}^1) + (1-t)U(s, P_\Sigma{}^0)]\\
&\le^5 t\cdot \max_{s\in S}[U(s, P_\Sigma{}^1)]\\
&\qquad + (1-t)\cdot \max_{s\in S}[U(s, P_\Sigma{}^0)]\\
&=^6 tU(S, P_\Sigma{}^1) + (1-t)U(S, P_\Sigma{}^0)\\
&=^7 tU(\mathcal{Q}_S, P_\Sigma{}^1) + (1-t)U(\mathcal{Q}_S, P_\Sigma{}^0),
\end{aligned}$$

equalities 1 and 7 following from Equation (7.3.5); 2 and 6, from Definition 7.1.3(C); 3, from rearranging the summation and factoring; 4, from (7.1.2); and inequality 5, from Exercise 3.7.5(C). The resulting inequality between the first and last members of this chain of relations is (*).

Of (B). By (A), $U(\mathcal{Q}_S, P_\Sigma{}^t)$ is a convex function of t in $[0, 1]$; and, on a little reflection, it follows that $P_S^*(P_\Sigma{}^t)$ can always be chosen to be a pure strategy. Moreover, it is easy to verify that

$$(\dagger)\qquad U(s, P_\Sigma{}^t) = U(s, P_\Sigma{}^0) - [U(s, P_\Sigma{}^0) - U(s, P_\Sigma{}^1)]t$$

for every s in S and every t in $[0, 1]$. These facts, and finiteness of S, imply that there is a finite set $s^{*1}, s^{*2}, \ldots, s^{*N}$ of pure strategies and a corresponding, increasing set of real numbers $0 = t_0 < t_1 < \cdots < t_{N-1} < t_N = 1$ such that:

(i) $U(\mathcal{Q}_S, P_\Sigma{}^t) = U(s^{*n}, P_\Sigma{}^t) = U(P_S^*(P_\Sigma{}^t), P_\Sigma{}^t)$ for every t in $(t_{n-1}, t_n]$; and

(ii) if $n'' > n'$, then $U(S^{*n''}, P_\Sigma{}^0) < U(s^{*n'}, P_\Sigma{}^0)$ and

$$[U(s^{*n''}, P_\Sigma{}^0) - U(s^{*n''}, P_\Sigma{}^1] < [U(s^{*n'}, P_\Sigma{}^0) - U(s^{*n'}, P_\Sigma{}^1)].$$

[(i) says that the t_n's for $2 \le n \le N-1$ are changeover points of the optimal strategies s^{*n}; and (ii) synthesizes (†) and convexity of $U(\mathcal{Q}_S, P_\Sigma{}^t)$ in t.] But the first inequality in (ii) *is* the nonincreasingness assertion, because $P_S^*(P^t) = s^{*n}$ for all t in $(t_{n-1}, t_n]$. The upper semicontinuity assertion follows from our choice, among strategies tied for optimality with respect to $P_\Sigma{}^t$, that strategy most desirable given $\mathcal{D}$'s judgments $P_\Sigma{}^0$. ■

■ Even though P_Σ may be thought of as a vector with $\#(\Sigma)$ components $P_\Sigma(\sigma^j)$, the requirement that $\Sigma_{j=1}^{\#(\Sigma)} P_\Sigma(\sigma^j) = 1$ implies that one of them is determined by the others. Hence $U(\mathcal{Q}_S, \cdot)$ may be regarded as a function of $[\#(\Sigma)-1]$-tuples rather than of $\#(\Sigma)$-tuples: omit $P_\Sigma(\sigma^{\#(\Sigma)})$, for example. This change of domain does not impair continuity or convexity in the slightest; it is merely a change of viewpoint. ■

Example 7.3.12

We apply Theorem 7.3.5 to the dpuu of Figure 7.5 and Examples 7.3.8–7.3.11. From the optimal response table in Example 7.3.8 we readily see that

$$U(\mathcal{Q}_S, P_\Sigma) = \begin{cases} U(s^5, P_\Sigma) & \text{if} \quad 0 \le P_\Sigma(\sigma^1) \le .40 \\ U(s^2, P_\Sigma) & \text{if} \quad .40 \le P_\Sigma(\sigma^1) \le 5/9 \\ U(s^1, P_\Sigma) & \text{if} \quad 5/9 \le P_\Sigma(\sigma^1) \le 1.00. \end{cases}$$

But $U(s^5, P_\Sigma) = 10P_\Sigma(\sigma^2) = 10[1 - P_\Sigma(\sigma^1)]$, $U(s^2, P_\Sigma) = 6$, and $U(s^1, P_\Sigma) = 10P_\Sigma(\sigma^1) + 1P_\Sigma(\sigma^2) = 10P_\Sigma(\sigma^1) + 1[1 - P_\Sigma(\sigma^1)] = 9P_\Sigma(\sigma^1) + 1$, for every $P_\Sigma = (P_\Sigma(\sigma^1), P_\Sigma(\sigma^2)) = (P_\Sigma(\sigma^1), 1 - P_\Sigma(\sigma^1))$. These facts are summarized in Figure 7.8.

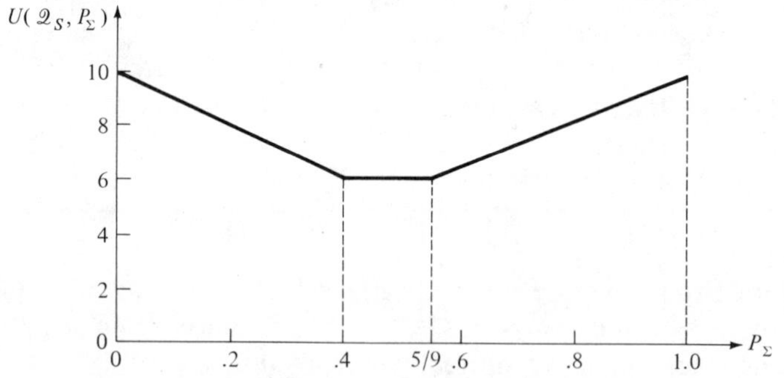

Figure 7.8 $U(\mathcal{Q}_S, \cdot)$ for the dpuu in Figure 7.5.

Example 7.3.13

For the dpuu in Example 7.3.12, we apply Theorem 7.3.5(B) with $P_\Sigma{}^0 = (.2, .8)$ and $P_\Sigma{}^1 = (.9, .1)$. Therefore $P_\Sigma{}^t(\sigma^1) = tP_\Sigma{}^1(\sigma^1) + (1-t)P_\Sigma{}^0(\sigma^1) = .9t + .2(1-t) = .2 + .7t$. By again using the optimal response table in Example 7.3.8, we readily deduce that

$$P_S^*(P_\Sigma{}^t) = \begin{cases} s^5 & \text{if} \quad .2 \le P_\Sigma{}^t(\sigma^1) \le .4; & \text{i.e.,} \quad 0 \le t \le 18/63 \\ s^2 & \text{if} \quad .4 < P_\Sigma{}^t(\sigma^1) \le 5/9; & \text{i.e.,} \quad 18/63 < t \le 32/63 \\ s^1 & \text{if} \quad 5/9 < P_\Sigma{}^t(\sigma^1) \le .9; & \text{i.e.,} \quad 32/63 < t \le 1. \end{cases}$$

Hence

$$U(P_S^*(P_\Sigma{}^t), P_\Sigma{}^0) = \begin{cases} U(s^5, P_\Sigma{}^0) = 8 & \text{if} \quad 0 \le t \le 18/63 \\ U(s^2, P_\Sigma{}^0) = 6 & \text{if} \quad 18/63 < t \le 32/63 \\ U(s^1, P_\Sigma{}^0) = 2.8 & \text{if} \quad 32/63 < t \le 1. \end{cases}$$

This function is depicted in Figure 7.9.

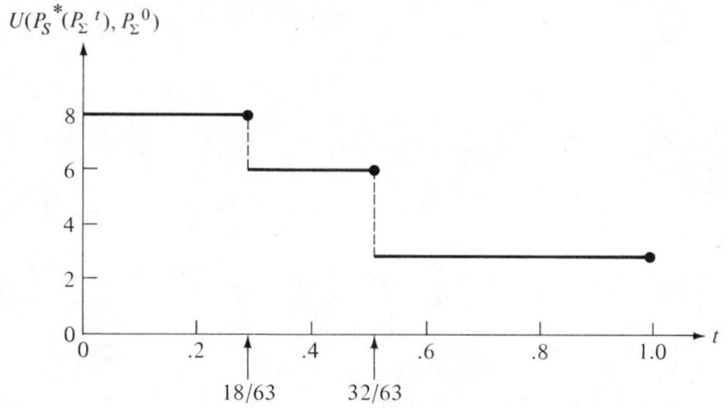

Figure 7.9 $U(P_S^*(P_\Sigma{}^t), P_\Sigma{}^0)$ in Example 7.3.13.

Example 7.3.14

Our final example concerns the subsystem problem, Example 1.5.1c, whose consequence table is Table B in Section 7.2. Let us suppose that all agree on the utility function u given as Table A in Section 4.4. Since we can ignore ordinally dominated strategies, given the desirability ranking "$\succsim$" agreeing with u, it follows that the relevant utility table is the 26-by-8 Table C. An application of a crudely programmed version of Walker's algorithm [**137**] had not yielded all of $\mathscr{A}_{\mathscr{C}}[\mathscr{Q}_S|u]$ at the time this book goes to press, but preliminary evidence suggests that it is the union of a very great many sets of the form $\mathscr{Q}_S^{S'}$.

Exercise 7.3.1: Finding $\mathscr{A}_{\mathscr{C}}[\mathscr{Q}_S|u]$ and $\mathscr{B}[\mathscr{Q}_S|u]$ when $\#(\Sigma) = 2$

Consider the following dpuu's in normal form, summarized by their utility tables.

dpuu (I)

	σ^1	σ^2
s^1	80	0
s^2	40	25
s^3	50	10
s^4	−25	60
s^5	60	30

dpuu (II)

	σ^1	σ^2
s^1	31	−51
s^2	−145	35
s^3	18	9
s^4	0	11
s^5	−191	90

Table C Utility Table for Example 1.5.1c, with Ordinally Dominated Pure Strategies Deleted

	PD $\sigma^1 = $ LD AD	PD $\sigma^2 = $ LD AN	PD $\sigma^3 = $ LN AD	PD $\sigma^4 = $ LN AN	PN $\sigma^5 = $ LD AD	PN $\sigma^6 = $ LD AN	PN $\sigma^7 = $ LN AD	PN $\sigma^8 = $ LN AN
s^1	.140	1.000	.140	1.000	.140	1.000	.140	1.000
s^2	.970	.970	.970	.970	.970	.970	.970	.970
s^3	.960	.960	.960	.960	.030	.993	.030	.993
s^{43}	.965	.965	.950	.950	.965	.965	0	.990
s^{55}	.950	.950	.110	.999	0	.990	.110	.999
s^5	.030	.993	.030	.993	.960	.960	.960	.960
s^7	.960	.960	.960	.960	.950	.950	0	.990
s^9	.960	.960	.960	.960	0	.990	.950	.950
s^{11}	.030	.993	.030	.993	.950	.950	0	.990
s^{13}	.030	.993	.030	.993	0	.990	.950	.950
s^{15}	.950	.950	0	.990	.960	.960	.960	.960
s^{17}	0	.990	.950	.950	.960	.960	.960	.960
s^{19}	.950	.950	0	.990	.030	.993	.030	.993
s^{21}	0	.990	.950	.950	.030	.993	.030	.993
s^{29}	.950	.950	0	.990	0	.990	.950	.950
s^{32}	0	.990	.950	.950	.950	.950	0	.990
s^{39}	.965	.965	.110	.999	.965	.965	.110	.999
s^{41}	.110	.999	.965	.965	.110	.999	.965	.965
s^{45}	.965	.965	0	.990	.965	.965	.950	.950
s^{47}	.110	.999	.950	.950	.110	.999	0	.990
s^{49}	.110	.999	0	.990	.110	.999	.950	.950
s^{51}	.950	.950	.965	.965	0	.990	.965	.965
s^{53}	0	.990	.965	.965	.950	.950	.965	.965
s^{57}	0	.990	.110	.999	.950	.950	.110	.999
s^{65}	.950	.950	0	.990	0	.990	.950	.950
s^{68}	0	.990	.950	.950	.950	.950	0	.990

dpuu (III)

	σ^1	σ^2
s^1	−5	20
s^2	15	−2
s^3	4	12
s^4	3	19
s^5	18	−6
s^6	8	11

dpuu (IV)

	σ^1	σ^2
s^1	10	1
s^2	4	4
s^3	2	9
s^4	3	2
s^5	0	10
s^6	10	0

(A) Graph the attainable utility set $X(\mathcal{Q}_S)$, clearly labeling the utility characteristics of the pure strategies s^i, for:

(i) dpuu (I) (iii) dpuu (III)
(ii) dpuu (II) (iv) dpuu (IV)

(B) Find $\mathcal{A}_{\mathscr{C}}[\mathcal{Q}_S|u]$ and express it as a union of strategy sets of the form $\mathcal{Q}_S^{S^*}$, for:

(i) dpuu (I) (iii) dpuu (III)
(ii) dpuu (II) (iv) dpuu (IV)

(C) Find $\mathcal{B}[\mathcal{Q}_S|u]$ and express it as a union of strategy sets of the form $\mathcal{Q}_S^{S^*}$, for:

(i) dpuu (I) (iii) dpuu (III)
(ii) dpuu (II) (iv) dpuu (IV)

Exercise 7.3.2: Optimal Response Tables

(A) Verify all aspects of the optimal response table in Example 7.3.8.
(B) Deduce the optimal response table for:

(i) dpuu (I) (iii) dpuu (III)
(ii) dpuu (II) (iv) dpuu (IV)

all these dpuu's being defined as in Exercise 7.3.1.

Exercise 7.3.3: Theorem 7.3.5(A) Graphs

Graph $U(\mathcal{Q}_S, P_\Sigma)$ as a function of $P_\Sigma(\sigma^1)$ for:

(A) dpuu (I) of Exercise 7.3.1 (C) dpuu (III) of Exercise 7.3.1
(B) dpuu (II) of Exercise 7.3.1 (D) dpuu (IV) of Exercise 7.3.1

***Exercise 7.3.4: Optimal Response Tables and Theorem 7.3.5(A) Graphs When $\#(\Sigma) = 3$**

(A) Deduce the optimal response table for the dpuu of Figure 7.6 and Example 7.3.4. [*Hint*: Identify each P_Σ with a point in the triangle $\{(p_1, p_2)$: $p_1 + p_2 \leq 1$, each $p_j \geq 0\}$ by setting $p_j = \{P_\Sigma(\sigma^j)\}$ for $j = 1, 2$

and $P_\Sigma(\sigma^3) = 1 - p_1 - p_2$. Partition this triangle into regions within each of which a given pure strategy is optimal and randomizations are optimal along borders.]

(B) Try to graph $U(\mathcal{Q}_S, (p_1, p_2))$ as a function of (p_1, p_2) over the triangle defined in (A).

(C) Now suppose that in Example 7.3.4 the utility characteristic $\mathbf{x}(s^5)$ was changed from (4, 4, 4) to (5, 5, 5). Determine $\mathcal{A}_{\mathscr{C}}[\mathcal{Q}_S|u]$ and $\mathcal{B}[\mathcal{Q}_S|u]$ for this modified dpuu.

(D) Deduce the optimal response table for the dpuu modified as in (C).

***Exercise 7.3.5**

Prove Theorem 7.3.2(A).

***Exercise 7.3.6**

Prove Theorem 7.3.3.

***Exercise 7.3.7: "More Costless Information Is Always at Least As Desirable As Less"**

Suppose that in a given dpuu $\mathscr{D}$ has the option (1) of deferring his choice of P_S until he has ascertained whether an "informational" event Ω' or its complement $\overline{\Omega'}$ has occurred or (2) of choosing P_S now, before learning if Ω' occurs. Also suppose that waiting to learn if Ω' occurs has no bearing whatsoever on the feasibility of strategies or on the consequences. Let P'_Σ and P''_Σ denote his judgments about $\tilde{\sigma}$ conditional upon Ω' and upon $\overline{\Omega'}$ respectively, and let t denote $\mathscr{D}$'s probability that Ω' will occur. Verify the following statements.

(A) $\mathscr{D}$'s judgments about $\tilde{\sigma}$, unconditional as regards Ω' or $\overline{\Omega'}$, are expressed by the randomized strategy $tP'_\Sigma + (1-t)P''_\Sigma$ for $\mathscr{E}$, where by definition $[tP'_\Sigma + (1-t)P''_\Sigma](\Sigma^*) = tP'_\Sigma(\Sigma^*) + (1-t)P''_\Sigma(\Sigma^*)$ for every event Σ^* in Σ.

(B) $\mathscr{D}$'s utility of optimally choosing *now*, i.e., his utility of option (2), is $U(\mathcal{Q}_S, tP'_\Sigma + (1-t)P''_\Sigma)$.

(C) $\mathscr{D}$'s utility of optimally choosing after learning that Ω' [respectively: $\overline{\Omega'}$] has occurred is $U(\mathcal{Q}_S, P'_\Sigma)$ [respectively: $U(\mathcal{Q}_S, P''_\Sigma)$].

(D) $\mathscr{D}$'s utility *now* of optimally choosing *after* learning whether Ω' or $\overline{\Omega'}$ has occurred is $tU(\mathcal{Q}_S, P'_\Sigma) + (1-t)U(\mathcal{Q}_S, P''_\Sigma)$.

(E) From (D), (B), and Theorem 7.3.5(A), option 1 is at least as desirable as option 2.

■ The conclusion of Exercise 7.3.7 is *not* that one should wait forever to do anything, because in most significant decisions the consequences and the feasibility of strategies are highly time-dependent. ■

Exercise 7.3.8: Maximin Criterion (I): "Semipessimism"

When "the environment" $\mathscr{E}$ is a competitor who is very interested in $\mathscr{D}$'s choice of pure strategy s, it is argued in the theory of games that $\mathscr{D}$ might well consider being semipessimistic, by assuming: (1) that $\mathscr{E}$ will ascertain $\mathscr{D}$'s utility function u and his chosen randomized strategy P_S before $\mathscr{E}$ must choose P_Σ; (2) that $\mathscr{E}$ will *not* ascertain the pure strategy s resulting from the randomization P_S (hence only semipessimism); and (3) that $\mathscr{E}$ will choose P_Σ so as to do $\mathscr{D}$ the greatest amount of damage—that is, so as to minimize $\mathscr{D}$'s utility $U(P_S, P_\Sigma)$. This implies that $\mathscr{D}$ should evaluate each of his strategies P_S by the *minimum* of his utility $U(P_S, P_\Sigma)$ as P_Σ ranges over $\mathscr{Q}_\Sigma$. Denote this minimum by $U(P_S, \mathscr{Q}_\Sigma)$; more generally, if $\mathscr{Q}'_\Sigma$ is any closed and nonempty subset of $\mathscr{Q}_\Sigma$, define $U(P_S, \mathscr{Q}'_\Sigma)$ by

$$U(P_S, \mathscr{Q}'_\Sigma) = \min_{P_\Sigma \in \mathscr{Q}'_\Sigma} [U(P_S, P_\Sigma)]. \tag{7.3.6}$$

Now $\mathscr{D}$ should select P_S so as to maximize his utility evaluation $U(P_S, \mathscr{Q}_\Sigma)$ over *his* set $\mathscr{Q}_S$ of strategies. Such a maximizer $P_S^\dagger$ is called a *maximin* strategy for $\mathscr{D}$. Suppose, more generally, that $\mathscr{D}$ has a closed and nonempty subset $\mathscr{Q}'_S$ of feasible strategies; then we define $U(\mathscr{Q}'_S, \mathscr{Q}'_\Sigma)$ to be $\mathscr{D}$'s maximum attainable utility of choosing via the maximin criterion when $\mathscr{E}$'s set of minimizing strategies is $\mathscr{Q}'_\Sigma$. Clearly,

$$U(\mathscr{Q}'_S, \mathscr{Q}'_\Sigma) = \max_{P_S \in \mathscr{Q}'_S} [U(P_S, \mathscr{Q}'_\Sigma)] = \max_{P_S \in \mathscr{Q}'_S} [\min_{P_\Sigma \in \mathscr{Q}'_\Sigma} [U(P_S, P_\Sigma)]]. \tag{7.3.7}$$

[Note that Equations (7.3.6) and (7.3.7) extend Definition 7.1.3(C), and that $U(P_S, P_\Sigma)$ and $U(\mathscr{Q}'_S, P_\Sigma)$ there coincide with $U(P_S, \{P_\Sigma\})$ and $U(\mathscr{Q}'_S, \{P_\Sigma\})$ here, so that there is no conflict if we as practical people refuse to distinguish between elements P_Σ and their corresponding one-element sets $\{P_\Sigma\}$.]

(A) Show that

$$U(P_S, \mathscr{Q}_\Sigma) = \min [U(P_S, \sigma^1), \ldots, U(P_S, \sigma^{\#(\Sigma)})] \tag{7.3.8}$$

for every P_S in $\mathscr{Q}_S$. [*Hint*: Suppose σ^* minimizes the right-hand side of (7.3.6) for the given P_S. Show that $U(P_S, P_\Sigma) \geq U(P_S, \sigma^*)$ for every P_Σ in $\mathscr{Q}_\Sigma$.]

(B) Show that

$$U(\mathscr{Q}_S, P_\Sigma) = \max [U(s^1, P_\Sigma), \ldots, U(s^{\#(S)}, P_\Sigma)] \tag{7.3.9}$$

for every P_Σ in $\mathscr{Q}_\Sigma$.

(C) We cannot combine Equations (7.3.8) and (7.3.9) in any simple fashion; for example, by asserting that

$$U(\mathscr{Q}_S, \mathscr{Q}_\Sigma) = \max [U(s^1, \mathscr{Q}_\Sigma), \ldots, U(s^{\#(S)}, \mathscr{Q}_\Sigma)]$$

or that

$$U(\mathscr{Q}_S, \mathscr{Q}_\Sigma) = \min [U(\mathscr{Q}_S, \sigma^1), \ldots, U(\mathscr{Q}_S, \sigma^{\#(\Sigma)})],$$

since these equations hold only under added restrictions. To show that this is so, consider the normal-form utility table

	σ^1	σ^2
s^1	10	0
s^2	0	10

(i) Show that $\mathscr{E}$'s minimizer σ^* in (7.3.7), given P_S, is σ^1 if $P_S(s^1) < .5$, is σ^2 if $P_S(s^1) > .5$, and is either σ^1 or σ^2 if $P_S(s^1) = .5$.

(ii) Show that $U(s^1, \mathscr{Q}_\Sigma) = U(s^2, \mathscr{Q}_\Sigma) = 0$ and that $U((.5, .5), \mathscr{Q}_\Sigma) = 5$. [Therefore $U(\mathscr{Q}_S, \mathscr{Q}_\Sigma) \neq \max_i [U(s^i, \mathscr{Q}_\Sigma)] = U(S, \mathscr{Q}_\Sigma)$.]

(iii) Show that $\mathscr{D}$'s maximizer s^* in (7.3.8), given P_Σ, is s^1 if $P_\Sigma(\sigma^1) > .5$, is s^2 if $P_\Sigma(\sigma^1) < .5$, and is either s^1 or s^2 if $P_\Sigma(\sigma^1) = .5$.

(iv) Show that $U(\mathscr{Q}_S, \sigma^1) = U(\mathscr{Q}_S, \sigma^2) = 10$ and that $U(\mathscr{Q}_S, (.5, .5)) = 5$. [Therefore $U(\mathscr{Q}_S, \mathscr{Q}_\Sigma) \neq \min_j [U(\mathscr{Q}_S, \sigma^j)]$.]

(D) Show that if $\mathscr{Q}_S' \subset \mathscr{Q}_S''$, then $U(\mathscr{Q}_S', \mathscr{Q}_\Sigma') \leq U(\mathscr{Q}_S'', \mathscr{Q}_\Sigma')$ for any available strategy set $\mathscr{Q}_\Sigma'$ for $\mathscr{E}$.

(E) Show that if $\mathscr{Q}_\Sigma' \subset \mathscr{Q}_\Sigma''$, then $U(\mathscr{Q}_S', \mathscr{Q}_\Sigma') \geq U(\mathscr{Q}_S', \mathscr{Q}_\Sigma'')$ for any available strategy set $\mathscr{Q}_S'$ for $\mathscr{D}$.

[*Hints*: In (D), more strategies in $\mathscr{Q}_S''$ are available for $\mathscr{D}$'s choice of maximizer; in (E), more strategies in $\mathscr{Q}_\Sigma''$ are available for each $\mathscr{E}$'s choices of minimizer.]

(F) Verify the appropriateness of the decision-tree model for the maximin criterion sketched in Figure 7.10.

■ We subtitled Exercise 7.3.8 a *semi*pessimistic approach, because the rationale for $\mathscr{D}$'s choosing P_S so as to maximize $U(P_S, \mathscr{Q}_\Sigma')$ for some given $\mathscr{Q}_\Sigma'$ assumes that $\mathscr{E}$ will discover $\mathscr{D}$'s choice of P_S ***but not the result s of $\mathscr{D}$'s conducting the randomization***! If $\mathscr{D}$ assumed, ***fully*** pessimistically, that $\mathscr{E}$ would discover his ultimate pure strategy s, then $\mathscr{D}$ should choose s so as to maximize $U(s, \mathscr{Q}_\Sigma')$ and should dispense with

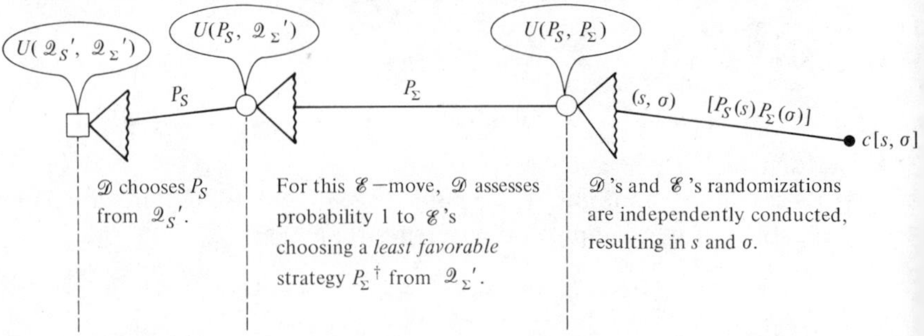

Figure 7.10 Decision-tree model for the maximin criterion.

randomized strategies (provided, of course, that he is free to do so). In part (Cii) of Exercise 7.3.8, this would mean that $\mathcal{D}$ should choose either s^1 or s^2 and plan on a utility of 0 rather than 5. ■

■ The *order* in which the maximization and minimization are taken in (7.3.7)—minimization and then maximization—is very important in general. In fact, little of generality in game theory existed until 1928, when J. von Neumann [**105**] derived conditions under which the order becomes immaterial. A version of his famous *minimax theorem* follows: If each of $\mathcal{Q}'_S$ and $\mathcal{Q}'_\Sigma$ is a nonempty, convex, and compact set of $\#(S)$- and $\#(\Sigma)$-tuples of strategies respectively, then

$$\max_{P_S \in \mathcal{Q}'_S} [\min_{P_\Sigma \in \mathcal{Q}'_\Sigma} [U(P_S, P_\Sigma)]] = \min_{P_\Sigma \in \mathcal{Q}'_\Sigma} [\max_{P_S \in \mathcal{Q}'_S} [U(P_S, P_\Sigma)]] \tag{7.3.10}$$

so that $U(\mathcal{Q}'_S, \mathcal{Q}'_\Sigma)$ could equally as well be evaluated by the right-hand side of (7.3.10). The sets $\mathcal{Q}_S$ and $\mathcal{Q}_\Sigma$ of all randomized strategies of $\mathcal{D}$ and $\mathcal{E}$ respectively are convex and compact, and hence the minimax theorem (7.3.10) holds for unrestricted choices of randomized strategies by $\mathcal{D}$ and $\mathcal{E}$. ■

Exercise 7.3.9: Maximin Criterion (II)

In this exercise you may relate maximin, Bayes, and cardinally admissible strategies via the geometric model. Let z be a real number, and define $\mathrm{GE}(z) = \{\mathbf{x}: \mathbf{x} \in R^{\#(\Sigma)}, x_j \geq z \text{ for every } j\}$.

(A) Show that if $U(P_S, \sigma^j) = z$ for every j, then $\mathrm{GE}(z) = Y(P_S)$, the right-hand side being defined as in Figure 7.7.

(B) Since $\mathcal{D}$'s semipessimistically conservative evaluation of every strategy P_S is $\min[U(P_S, \sigma^1), \ldots, U(P_S, \sigma^{\#(\Sigma)})]$—see Exercise 7.3.8(A)—argue that $\mathcal{D}$ can "attain" a utility (before randomizing) of z with P_S if and only if $U(P_S, \sigma^j) \geq z$ for every j.

(C) Argue geometrically from (B) that $\mathcal{D}$ can attain a prerandomizing utility of z if and only if $X(\mathcal{Q}_S) \cap \mathrm{GE}(z) \neq \emptyset$.

(D) There is a largest number z such that $X(\mathcal{Q}_S) \cap \mathrm{GE}(z) \neq \emptyset$, because $X(\mathcal{Q}_S)$ is closed and bounded (because compact). Call that largest number $z^\dagger$. Argue that $P_S^\dagger$ is a maximin strategy for $\mathcal{D}$ if and only if $\mathbf{x}(P_S^\dagger) \in X(\mathcal{Q}_S) \cap \mathrm{GE}(z^\dagger)$.

(E) Analogize the proof of Theorem 7.3.4(A) to show that if $P_S^\dagger$ is a maximin strategy for $\mathcal{D}$, then $P_S^\dagger$ is optimal with respect to some randomized strategy $P_\Sigma^\dagger$ (say) of $\mathcal{E}$. This $P_\Sigma^\dagger$ is called a *least favorable* strategy for $\mathcal{E}$—obviously from $\mathcal{D}$'s viewpoint.

(F) Show that if a maximin strategy $P_S^\dagger$ is cardinally dominated, then $P_S^\dagger$ is cardinally dominated by a cardinally *admissible* maximin strategy $P_S^{\dagger\prime}$. [*Hint*: Figure 7.11 depicts some relationships between $X(\mathcal{Q}_S)$ and $\mathrm{GE}(z^\dagger)$ for $\#(\Sigma) = 2$.]

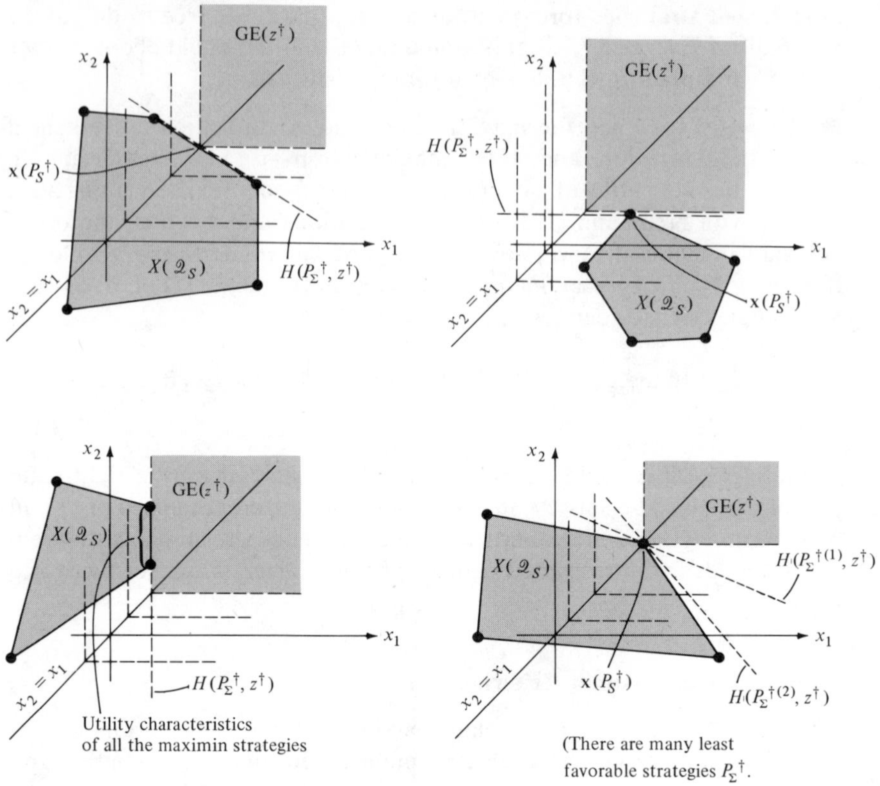

Figure 7.11 Relationships between $X(\mathcal{Q}_S)$ and GE($z^\dagger$).

Exercise 7.3.10: Maximin Criterion (III)

Find a maximin strategy $P_S^\dagger$ for:

(A) dpuu (I) in Exercise 7.3.1
(B) dpuu (II) in Exercise 7.3.1
(C) dpuu (III) in Exercise 7.3.1
(D) dpuu (IV) in Exercise 7.3.1
(E) the dpuu in Figure 7.5

■ The maximin philosophy, if perhaps not precisely the maximin criterion itself, is incorporated in military doctrine, which holds that strategy should be based on enemy capabilities, not enemy intentions. Moreover, a close reading of Chandler's *Campaigns of Napoleon* [**17**] and von Wartenburg's *Napoleon as a General* [**139**] indicates that Napoleon used both our approach and the maximin approach in choosing his strategic postures for different campaigns, his selection of approach seeming to depend upon his confidence in his judgments about the personality of the opposing commander. ■

7.4 SENSITIVITY ANALYSIS WHEN u AND SOME JUDGMENTS ARE AGREED ON

In this section we show that the range of meaningful controversy can be narrowed still further than in Section 7.3 when all participants agree on some judgments about $\tilde{\sigma}$ as well as on the utility function u.

More specifically, suppose that Σ can be expressed as a Cartesian product, $\Sigma_I \times \Sigma_{II}$, and that all concerned agree on all conditional probabilities $P_{II|I}(\Sigma'_{II}|\sigma_I)$ of events Σ'_{II} in Σ_{II}, given each σ_I in Σ_I. We will abbreviate the entire set $\{P_{II|I}(\cdot|\sigma_I)\colon \sigma_I \in \Sigma_I\}$ of conditional probability functions on $\tilde{\sigma}_{II}$-events by $P_{II|I}$.

Example 7.4.1

The consequence table of our subsystem problem, Example 1.5.1c, is given as Table B in Section 7.2, and its utility table as Table C in Section 7.3 for the utility function given in Table A of Section 4.4. Let $\Sigma_{II} = \{(\mathrm{PD}, \mathrm{LD}), (\mathrm{PD}, \mathrm{LN}), (\mathrm{PN}, \mathrm{LD}), (\mathrm{PN}, \mathrm{LN})\}$ and $\Sigma_I = \{\mathrm{AD}, \mathrm{AN}\}$. If all participants agree on the properties of these tests, as embodied in the conditional probabilities of the joint test outcomes given each of the arrival states, this agreement can be exploited to reduce the effective number of pure strategies for $\mathscr{E}$ from eight to two, as we shall see in Example 7.4.4.

We now show that $\mathscr{D}$ and his colleagues should base all further discussion, not on the original, $\#(S)$-by-$\#(\Sigma)$ utility table of $U(s, \sigma)$'s, but rather on a "conditional-expected" utility table consisting of $\#(S)$ rows s (unchanged), $\#(\Sigma_I)$ columns σ_I (down by an order of magnitude), with entries $U_I(s, \sigma_I)$ given by

$$U_I(s, \sigma_I) = \mathcal{E}_{II|I}\{U(s, (\sigma_I, \tilde{\sigma}_{II}))|\sigma_I\} \tag{7.4.1}$$

for every s in S and every σ_I in Σ_I.

■ All spelled out,

$$U_I(s, \sigma_I) = \sum_{k=1}^{\#(\Sigma_{II})} U(s, (\sigma_I, \sigma_{II}^k))P_{II|I}(\sigma_{II}^k|\sigma_I) \tag{7.4.2}$$

for every s in S and every σ_I in Σ_I. ■

■ The justification for the assertion in the sentence containing Equation (7.4.1) is very easily obtained. The original dpuu in normal form is depicted in Figure 7.12(a) as a one-stage decision tree with an initial $\mathscr{D}$-move S and terminal $\mathscr{E}$-moves all copies of Σ. Figure 7.12(b) is obtained from Figure 7.12(a) by representing the $\Sigma = \Sigma_I \times \Sigma_{II}$ $\mathscr{E}$-moves as "Σ_I before Σ_{II}" in each case.[m] The bubbled evaluations follow readily from the fundamental theory in Chapter 3. Since $P_{II|I}$ and u are agreed

[m] See Figure 3.17 and the related discussion for an explanation of the expression in quotes.

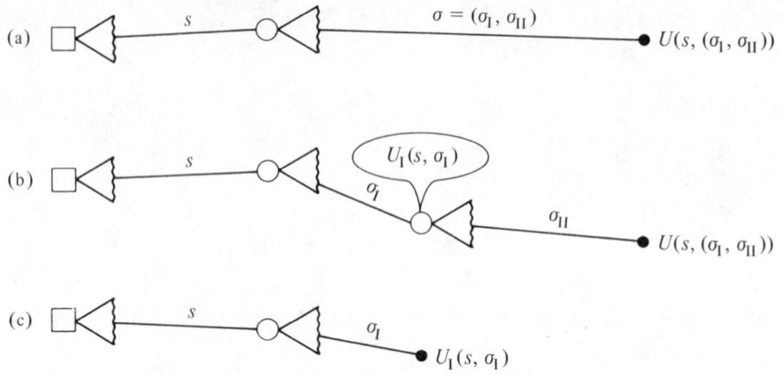

Figure 7.12 Equivalent representations of a dpuu with u and $P_{II|I}$ agreed on.

on, no one can take issue with these evaluations. But the equivalent tree in Figure 7.12(c) is qualitatively identical to Figure 7.12(a); and the $\#(S)$-by-$\#(\Sigma_I)$ table of $U_I(s, \sigma_I)$'s is as equivalent to Figure 7.12(c) as the $\#(S)$-by-$\#(\Sigma)$ table of $U(s, \sigma)$'s is to Figure 7.12(a). ■

Example 7.4.2

Consider the dpuu in Example 7.3.10. Suppose that, in addition to the utilities in that table, all participants agree that $P_{II|I}(\sigma_{II}^1|\sigma_I^1) = .3$ and $P_{II|I}(\sigma_{II}^1|\sigma_I^2) = .9$. Then (7.4.1) readily implies the following utility table.

	σ_I^1	σ_I^2
s^1	3.0	0
s^2	7.0	0
s^3	0	9.0
s^4	0	1.0
s^5	3.0	3.0
s^6	4.2	5.4

Note that in *this* table the only cardinally admissible pure strategies are s^2, s^3, and s^6, whereas all six pure strategies were cardinally admissible in Example 7.3.10. The difference is that *there* all six pure strategies were cardinally admissible given u, whereas *here* only s^2, s^3, and s^6 are cardinally admissible given u *and* given $P_{II|I}$.

It is important to note that once one has passed to the abbreviated, $\#(S)$-by-$\#(\Sigma_I)$ table of $U_I(s, \sigma_I)$'s, all the analysis of Section 7.3 can be brought to bear, with Σ_I replacing Σ and U_I replacing U. Furthermore, letting $\mathcal{Q}_I$ denote the set of all probability functions P_I on Σ_I (= "abbreviated randomized strategies for $\mathcal{E}$"), it is clear that $\mathcal{Q}_I$ and P_I here play the roles of $\mathcal{Q}_\Sigma$ and P_Σ respectively in Section 7.3.

Hence it only remains to examine the relationships (1) between the cardinally admissible strategies $\mathcal{A}_{\mathcal{C}}[\mathcal{Q}_S|u, P_{II|I}]$ given both u and $P_{II|I}$ with the cardinally admissible strategies $\mathscr{A}_{\mathscr{C}}[\mathscr{Q}_S|u]$ given only u, and (2) between the Bayes strategies $\mathscr{B}[\mathscr{Q}_S|u, P_{\text{II}|\text{I}}]$ given both u and $P_{\text{II}|\text{I}}$ with the Bayes strategies $\mathscr{B}[\mathscr{Q}_S|u]$ given only u. That is, we explore the effect of obtaining consensus on $P_{\text{II}|\text{I}}$ on the range of potentially remaining controversy. Obviously, that range can only narrow as consensus increases; this is the gist of Theorem 7.4.1.

■ Because the $\#(S)$-by-$\#(\Sigma_\text{I})$ table of $U_\text{I}(s, \sigma_\text{I})$'s performs exactly the same function here as the $\#(S)$-by-$\#(\Sigma)$ table of $U(s, \sigma)$'s performed in Section 7.3, it follows readily (1) that we should define $U_\text{I}(P_S, P_\text{I})$ by

$$U_\text{I}(P_S, P_\text{I}) = \sum_{i=1}^{\#(S)} \sum_{r=1}^{\#(\Sigma_\text{I})} U_\text{I}(s^i, \sigma_\text{I}{}^r) P_\text{I}(\sigma_\text{I}{}^r) P_S(s^i)$$

$$= \sum_{i=1}^{\#(S)} U_I(s^i, P_\text{I}) P_S(s^i) = \sum_{r=1}^{\#(\Sigma_\text{I})} U(P_S, \sigma_\text{I}{}^r) P_\text{I}(\sigma_\text{I}{}^r)$$

for every P_S in $\mathscr{Q}_S$ and every P_I in $\mathscr{Q}_\text{I}$; (2) that $P_S^\circ \in \mathscr{B}[\mathscr{Q}_S|u, P_{\text{II}|\text{I}}]$ if and only if

$$U_\text{I}[P_S^\circ, P_\text{I}) = \max_{P_S \in \mathscr{Q}_S} [U_\text{I}(P_S, P_\text{I})] = U_\text{I}(\mathscr{Q}_S, P_\text{I})$$

for at least one P_I in Σ_I; (3a) that P_S'' cardinally dominates P_S' given u and $P_{\text{II}|\text{I}}$ if $U_\text{I}(P_S'', \sigma_\text{I}) \geq U_\text{I}(P_S', \sigma_\text{I})$ for all σ_I in Σ_I and if $U_\text{I}(P_S'', \sigma_\text{I}) > U_\text{I}(P_S', \sigma_\text{I})$ for at least one σ_I in Σ_I; and (3b) that $P_S \in \mathscr{A}_{\mathscr{C}}[\mathscr{Q}_S|u, P_{\text{II}|\text{I}}]$ if and only if P_S is not cardinally dominated given u and $P_{\text{II}|\text{I}}$ by any strategy in $\mathscr{Q}_S$. These definitions simply mimic their counterparts in Section 7.3, but in the context of the table of $U_\text{I}(s, \sigma_\text{I})$'s rather than the table of $U(s, \sigma)$'s. ■

While Theorem 7.4.1 gives the results of greatest practical importance, Lemma 7.4.1 furnishes the key to the derivation of Theorem 7.4.1 via invocation of Theorem 7.3.4. As a preliminary, we recall that $P_{\text{II}|\text{I}}$ is fixed (because consensus obtains about it), whereas P_I is potentially completely controversial—by assumption. Therefore, P_Σ is partly constrained and partly unconstrained. For every P_I in $\mathscr{Q}_\text{I}$, we shall denote by $P_{\text{II}|\text{I}} \cdot P_\text{I}$ the unique strategy in $\mathscr{Q}_\Sigma$ induced by P_I and $P_{\text{II}|\text{I}}$; that is,

$$P_{\text{II}|\text{I}} \cdot P_\text{I}(\sigma_\text{I}, \sigma_\text{II}) = P_{\text{II}|\text{I}}(\sigma_\text{II}|\sigma_\text{I}) \cdot P_\text{I}(\sigma_\text{I}) \tag{7.4.3}$$

for every $(\sigma_\text{I}, \sigma_\text{II})$ in $\Sigma_\text{I} \times \Sigma_\text{II} = \Sigma$.

■ $P_{\text{II}|\text{I}} \cdot P_\text{I}$ as defined by Equation (7.5.3) is simply the unique "joint probability function" on the combined $\mathscr{E}$-move $\Sigma = \Sigma_\text{I} \times \Sigma_\text{II}$ which is consistent (in the sense of obeying Theorem 3.7.1) with P_I on Σ_I and the conditional probability functions $P_{\text{II}|\text{I}}(\cdot|\sigma_\text{I})$ on the copies of Σ_II attached to the branches σ_I of Σ_I. ■

Lemma 7.4.1(A) shows that $\mathscr{D}$'s utility of P_S given P_I in the present context is just his utility of P_S given P_I *and* $P_{\text{II}|\text{I}}$ in the context of Section

7.3. This should be obvious once stated! Lemma 7.4.1(B) is just a restatement of Theorem 7.3.4(A) and (B) in the context of this section.

Lemma 7.4.1. Useful Facts. Assume that both S and Σ are finite sets.

(A) $U_I(P_S, P_I) = U(P_S, P_{II|I} \cdot P_I)$ for every P_S in $\mathcal{Q}_S$ and every P_I in $\mathcal{Q}_I$.
(B) $P_S^* \in \mathcal{A}_{\mathscr{C}}[\mathcal{Q}_S|u, P_{II|I}]$ if and only if $U_I(P_S^*, P_I) = U_I(\mathcal{Q}_S, P_I)$ for some P_I such that $P_I(\sigma_I) > 0$ for every σ_I in Σ_I.

■ *Proof.* Part (A) is left to you as Exercise 7.4.3. Part (B) follows from the definitions of $\mathcal{A}_{\mathscr{C}}[\mathcal{Q}_S|u, P_{II|I}]$ and $\mathcal{B}[\mathcal{Q}_S|u, P_{II|I}]$ in the preceding remark, by transliterating the proofs of parts (A) and (B) of Theorem 7.3.4. ■

Now for the principal result of this section, Theorem 7.4.1. Part (A) says that the strategies which are Bayes given both u and $P_{II|I}$ must also be Bayes given only the utility function u. Similarly, part (B) says that the strategies which are cardinally admissible given both u and $P_{II|I}$ must also be cardinally admissible given only u, *provided* that $P_{II|I}(\sigma_{II}|\sigma_I)$ is *positive* for every $(\sigma_I, \sigma_{II}) = \sigma$ in $\Sigma = \Sigma_I \times \Sigma_{II}$. It is too bad that there is no decent way to replace "$\subset$" in (A) and (B) with some symbol meaning "is a *usually drastically proper* subset of," because such a symbol would fit our experience in applying Theorem 7.4.1 to a "T"!

Theorem 7.4.1. Effects on Bayesness and Admissibility of Attaining Consensus on $P_{II|I}$. Assume that S and Σ are both finite sets.

(A) $$\mathcal{B}[\mathcal{Q}_S|u, P_{II|I}] \subset \mathcal{B}[\mathcal{Q}_S|u].$$

(B) If $P_{II|I}(\sigma_{II}|\sigma_I) > 0$ for every (σ_I, σ_{II}) in $\Sigma_I \times \Sigma_{II}$, then

$$\mathcal{A}_{\mathscr{C}}[\mathcal{Q}_S|u, P_{II|I}] \subset \mathcal{A}_{\mathscr{C}}[\mathcal{Q}_S|u].$$

■ *Proof*: *Of (A)*. If $P_S^\circ \in \mathcal{B}[\mathcal{Q}_S|u, P_{II|I}]$, then there is a P_I^* (say) in $\mathcal{Q}_I$ such that $U_I(P_S^\circ, P_I^*) = U_I(\mathcal{Q}_S, P_I^*)$. By Lemma 7.4.1(A), $U_I(P_S^\circ, P_I^*) = U(P_S^\circ, P_{II|I} \cdot P_I^*)$ and $U_I(\mathcal{Q}_S, P_I^*) = U(\mathcal{Q}_S, P_{II|I} \cdot P_I^*)$. Hence P_S° is Bayes, given u against $P_{II|I} \cdot P_I^*$ in the context of Section 7.3, and hence $P_S^\circ \in \mathcal{B}[\mathcal{Q}_S|u]$. [That $\mathcal{B}[\mathcal{Q}_S|u, P_{II|I}]$ is a usually drastically proper subset of $\mathcal{B}[\mathcal{Q}_S|u]$ follows from the now obvious fact that the only strategies in $\mathcal{B}[\mathcal{Q}_S|u]$ which belong to $\mathcal{B}[\mathcal{Q}_S|u, P_{II|I}]$ are those which are Bayes given u against P_Σ's of the very special form $P_{II|I} \cdot P_I$ for the agreed-upon $P_{II|I}$.]

Of (B). Suppose that $P_S \in \mathcal{A}_{\mathscr{C}}[\mathcal{Q}_S|u, P_{II|I}]$. By Lemma 7.4.1(B), it follows that P_S is Bayes given u and $P_{II|I}$ against some P_I with $P_I(\sigma_I) > 0$ for every σ_I in Σ_I. If also $P_{II|I}(\sigma_{II}|\sigma_I) > 0$ for every (σ_I, σ_{II}) in $\Sigma_I \times \Sigma_{II}$, then $P_{II|I} \cdot P_I(\sigma_I, \sigma_{II}) = P_{II|I}(\sigma_{II}|\sigma_I)P_I(\sigma_I) > 0$, so that P_S is Bayes given u against a P_Σ—namely, some $P_{II|I} \cdot P_I$—such that $P_\Sigma(\sigma) = P_\Sigma(\sigma_I, \sigma_{II}) > 0$ for every σ in Σ, and hence $P_S \in \mathcal{A}_{\mathscr{C}}[\mathcal{Q}_S|u]$, by virtue of Theorem 7.3.4(B). ■

Example 7.4.3

For the dpuu introduced in Example 7.3.10, whose table of $U_I(s, \sigma_I)$'s appears in Example 7.4.2 for the $P_{II|I}$ defined there, it follows readily by graphical inspection that

$$\mathscr{A}_{\mathscr{C}}[\mathscr{Q}_S|u, P_{II|I}] = \mathscr{Q}_S^{\{s^2,s^6\}} \cup \mathscr{Q}_S^{\{s^6,s^3\}}.$$

(See Figure 7.13.) Now refer back to Example 7.3.10 to see just how tiny a subset $\mathscr{A}_{\mathscr{C}}[\mathscr{Q}_S|u, P_{II|I}]$ is of $\mathscr{A}_{\mathscr{C}}[\mathscr{Q}_S|u]$! This is what attaining consensus on $P_{II|I}$ has wrought. Clearly, $\mathscr{A}_{\mathscr{C}}[\mathscr{Q}_S|u, P_{II|I}] = \mathscr{B}[\mathscr{Q}_S|u, P_{II|I}]$ here.

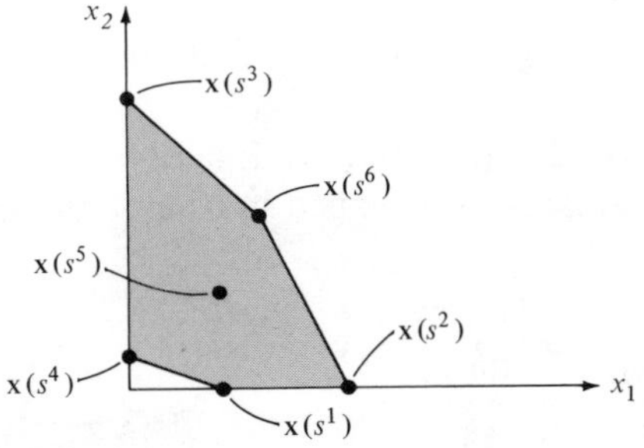

Figure 7.13 Attainable "conditional-expected" utility set $X_I(\mathscr{Q}_S)$ for the dpuu in Examples 7.3.10, 7.4.2, and 7.4.3.

Example 7.4.4

Assume that in Example 1.5.1c consensus obtains both on the utility function u given in Table A of Section 4.4 and on the conditional probabilities given in Example 1.5.1 of the test outcomes given the arrival states; they imply the following table.

	$\sigma_{II}^1 =$ (PD, LD)	$\sigma_{II}^2 =$ (PD, LN)	$\sigma_{II}^3 =$ (PN, LD)	$\sigma_{II}^4 =$ (PN, LN)
$\sigma_I^1 =$ AD	.52	.28	.13	.07
$\sigma_I^2 =$ AN	0	0	.25	.75

Using these probabilities in conjunction with Equation (7.4.2) and Table C in Section 7.3 yields the 26-by-2 Table D of $U_I(s, \sigma_I)$'s. Again by graphical inspection, it is easy to verify that

$$\mathscr{A}_{\mathscr{C}}[\mathscr{Q}_S|u, P_{II|I}] = \mathscr{Q}_S^{\{s^1,s^{55}\}} \cup \mathscr{Q}_S^{\{s^{55},s^3\}} \cup \mathscr{Q}_S^{\{s^3,s^{43}\}} \cup \mathscr{Q}_S^{\{s^{43},s^2\}},$$

a presumably small subset of $\mathscr{A}_{\mathscr{C}}[\mathscr{Q}_S|u]$.

Table D Conditional-Expected Utility Table for Example 1.5.1c

	$\sigma_I^1 = \text{AD}$	$\sigma_I^2 = \text{AN}$		$\sigma_I^1 = \text{AD}$	$\sigma_I^2 = \text{AN}$		$\sigma_I^1 = \text{AD}$	$\sigma_I^2 = \text{AN}$
s^1	.14000	1.00000	s^{13}	.09050	.96000	s^{45}	.69375	.95375
s^2	.97000	.97000	s^{15}	.68600	.96000	s^{47}	.33750	.99225
s^3	.77400	.99300	s^{17}	.45800	.96000	s^{49}	.13800	.96225
s^{43}	.89325	.98375	s^{19}	.50000	.99300	s^{51}	.83175	.97125
s^{55}	.53250	.99675	s^{21}	.27200	.99300	s^{53}	.46125	.96125
s^5	.21600	.96000	s^{29}	.56050	.96000	s^{57}	.16200	.98675
s^7	.89150	.98000	s^{32}	.38950	.98000	s^{65}	.56050	.96000
s^9	.83450	.96000	s^{39}	.66575	.99050	s^{68}	.38950	.98000
s^{11}	.14750	.98000	s^{41}	.40925	.97350			

Exercise 7.4.1

Find $\mathscr{A}_{\mathscr{C}}[\mathscr{Q}_S|u, P_{\text{II}|\text{I}}]$ and $\mathscr{B}[\mathscr{Q}_S|u, P_{\text{II}|\text{I}}]$ for the dpuu in Example 7.4.3, assuming that the utility function u remains as specified in (the original) Example 7.3.10, but that $P_{\text{II}|\text{I}}$:

(A) is given by

	σ_{II}^1	σ_{II}^2
σ_I^1	.7	.3
σ_I^2	.1	.9

(B) is given by

	σ_{II}^1	σ_{II}^2
σ_I^1	.8	.2
σ_I^2	.8	.2

(C) is given by

	σ_{II}^1	σ_{II}^2
σ_I^1	.9	.1
σ_I^2	.5	.5

(D) is given by

	σ_{II}^1	σ_{II}^2
σ_I^1	0	1
σ_I^2	0	1

Exercise 7.4.2

Find $\mathscr{A}_{\mathscr{C}}[\mathscr{Q}_S|u, P_{\text{II}|\text{I}}]$ for the subsystem problem, Example 1.5.1c, using the utility function u given by Table A′ in Exercise 4.4.2 and $P_{\text{II}|\text{I}}$ as given in Example 7.4.4.

Exercise 7.4.3

Prove Lemma 7.4.1(A).

Exercise 7.4.4: Maximin Criterion (IV)

If all participants agree on both u and $P_{\text{II}|\text{I}}$, then in effect $\mathscr{E}$ is regarded as being able to choose only P_{I} so as to do $\mathscr{D}$ the greatest harm, and $\mathscr{D}$ should choose a strategy $P_{\hat{S}}$ so as to maximize

$$U_{\text{I}}(P_{\hat{S}}, \mathscr{Q}_{\text{I}}) = \min_{P_{\text{I}} \in \mathscr{Q}_{\text{I}}} [U_{\text{I}}(P_S, P_{\text{I}})];$$

that is, let $P_{\hat{S}}$ satisfy

$$U_{\text{I}}(P_{\hat{S}}, \mathscr{Q}_{\text{I}}) = \max_{P_S \in \mathscr{Q}_S} [U_{\text{I}}(P_S, \mathscr{Q}_{\text{I}})] = U_{\text{I}}(\mathscr{Q}_S, \mathscr{Q}_{\text{I}}).$$

Such a strategy is called "maximin," but it is maximin for $\mathscr{D}$ usually only in the context of the consensus on $P_{\text{II}|\text{I}}$, as this exercise shows.

(A) Show that

$$U_{\text{I}}(P_{\hat{S}}, \mathscr{Q}_{\text{I}}) \geq U(P_S^{\dagger}, \mathscr{Q}_{\Sigma}), \tag{7.4.4}$$

where $P_S^{\dagger}$ is maximin for $\mathscr{D}$ in the absence of agreement on $P_{\text{II}|\text{I}}$. [*Hint*: Use Lemma 7.4.1(A) and apply Exercise 7.3.8(D) and (E) for $\mathscr{Q}'_S = \mathscr{Q}''_S = \mathscr{Q}_S$, $\mathscr{Q}''_{\Sigma} = \mathscr{Q}_{\Sigma}$, and $\mathscr{Q}'_{\Sigma} = \{P_{\text{II}|\text{I}} \cdot P_{\text{I}}: P_{\text{I}} \in \mathscr{Q}_{\text{I}}\}$.]

(B) Show that equality obtains in (7.4.4) if and only if there is some P_{I} in $\mathscr{Q}_{\text{I}}$ such that $P_{\text{II}|\text{I}} \cdot P_{\text{I}}$ is a least favorable strategy $P_{\Sigma}^{\dagger}$ for $\mathscr{E}$ from $\mathscr{D}$'s viewpoint [as defined in Exercise 7.3.9(E)].

Exercise 7.4.5: Maximin Criterion (V)

(A) Find a maximin strategy P_S for $\mathscr{D}$ in the dpuu with u and $P_{\text{II}|\text{I}}$ agreed upon, in:

(i) Exercise 7.4.1(A)
(ii) Exercise 7.4.1(B)
(iii) Exercise 7.4.1(C)
(iv) Exercise 7.4.1(D)

(B) Find a maximin strategy for $\mathscr{D}$ in Example 1.5.1c, with assumptions about u and $P_{\text{II}|\text{I}}$ as given in:
(i) Example 7.4.4
(ii) Exercise 7.4.2
(iii) Example 7.3.14—that is, given no consensus on $P_{\text{II}|\text{I}}$.
[*Hint*: Apply common sense to the question of how $\mathscr{E}$ could choose the arrival state and the test outcomes so as to cause $\mathscr{D}$ the greatest amount of harm.]

*7.5 SENSITIVITY ANALYSIS WHEN "$\gtrsim$" IS AGREED ON

This section concerns situations in which full consensus does not obtain as regards u, but all concerned with the dpuu do agree on a desirability ranking "$\gtrsim$" of the set C of potential consequences. Hence disagreement regarding u centers on attitudes toward risk.

The parties to the decision may also disagree on P_{Σ}, but we shall initially suppose that P_{Σ} is noncontroversial. For the case in which both P_{Σ} and

"$\gtrsim$" are agreed on, the notion analogous to cardinal admissibility is *stochastic*[n] *admissibility*. We shall show that stochastic admissibility is just as closely related to optimality, or Bayesness, as is cardinal admissibility. After examining stochastic admissibility, we drop the assumption that consensus obtains concerning P_Σ, and hence suppose that only "$\gtrsim$" is noncontroversial. For this case, as noted in the introduction, the appropriate notion is *ordinal admissibility*.

The pace in this section is somewhat faster than that in Section 7.3, because the close parallels with concepts from Section 7.3 reduce the novelty of the material introduced here.

■ Further information on stochastic and ordinal admissibility may be found in LaValle [**76**], Fishburn [**36**], and the references cited in these works. ■

Throughout this section we assume that "$\gtrsim$" is agreed on. By the Substitutability Principle, we can assume that the set $C = \{c^0, c^1, \ldots, c^t\}$ of potential consequences of the dpuu contains no two equally desirable consequences, because if it originally contained c' and $c'' \sim c'$, we could substitute c' for c'' wherever c'' appeared in the $\#(S)$-by-$\#(\Sigma)$ consequence table of $c[s, \sigma]$'s. Furthermore, by relabeling consequences if necessary, we can assume that

$$c^t > c^{t-1} > \cdots > c^1 > c^0. \tag{7.5.1}$$

Since "$\gtrsim$" and the Substitutability Principle are noncontroversial, no party to the decision should disagree with (7.5.1), once the consequences have been labeled appropriately. Finally, whatever the parties' various attitudes toward risk may be, each should be willing to let $u(c^0) = 0$ and $u(c^t) = 1$, because each party's utility function should satisfy Theorem 3.5.2 (or Observation 5.2.4) concerning the arbitrariness and inessentiality of changes in origin and/or scale.

Hence we shall agree that

$$0 = u(c^0) < u(c^1) < \cdots < u(c^t) = 1. \tag{7.5.2}$$

Disagreement about risk attitudes is reflected in the exact placement of $u(c^1), \ldots, u(c^{t-1})$. Now let d^k denote the kth *utility* (*first*) *difference*, defined by

$$d^k = u(c^k) - u(c^{k-1}) \tag{7.5.3}$$

for $k = 1, \ldots, t$. Clearly, d^k must be *strictly greater than* zero because $u(c^k) > u(c^{k-1})$. Moreover, you may easily verify that

$$u(c^k) = \sum_{h=1}^{k} d^h \tag{7.5.4}$$

for $k = 1, \ldots, t$. Hence $\Sigma_{k=1}^{t} d^k = 1$. Thus the utility differences $d^1, \ldots, d^t$

---

[n] "Stochastic" means "random," "uncertain," or "probabilistic."

remind one of a set of strictly positive probabilities. To preview coming attractions, d^k is analogous to $P_\Sigma(\sigma^j)$ in Section 7.3. Analogous to the entire probability function P_Σ in Section 7.3, we shall consider

$$d = (d^1, \ldots, d^t).$$

Additional preliminary notions concern the probability $P(\tilde{c} \in C'|P_S, P_\Sigma)$ of obtaining a consequence in the subset C' of C given $\mathscr{D}$'s choice of P_S and his judgments P_Σ, prior to $\mathscr{D}$'s conducting the randomization or observing σ. Clearly,

$$P(\tilde{c} \in C'|s, \sigma) = \begin{cases} 1, & c[s, \sigma] \in C' \\ 0, & c[s, \sigma] \notin C'. \end{cases} \tag{7.5.5}$$

You may verify, as Exercise 7.5.1, that

$$P(\tilde{c} \in C'|s^i, P_\Sigma) = \sum_{j=1}^{\#(\Sigma)} P(\tilde{c} \in C'|s^i, \sigma^j) P_\Sigma(\sigma^j), \tag{7.5.6a}$$

$$P(\tilde{c} \in C'|P_S, \sigma^j) = \sum_{i=1}^{\#(S)} P(\tilde{c} \in C'|s^i, \sigma^j) P_S(s^i), \tag{7.5.6b}$$

$$P(\tilde{c} \in C'|P_S, P_\Sigma) = \sum_{i=1}^{\#(S)} P(\tilde{c} \in C'|s^i, P_\Sigma) P_S(s^i) \tag{7.5.6c}$$

$$= \sum_{j=1}^{\#(\Sigma)} P(\tilde{c} \in C'|P_S, \sigma^j) P_\Sigma(\sigma^j) \tag{7.5.6d}$$

$$= \sum_{i=1}^{\#(S)} \sum_{j=1}^{\#(\Sigma)} P(\tilde{c} \in C'|s^i, \sigma^j) P_S(s^i) P_\Sigma(\sigma^j). \tag{7.5.6e}$$

Introducing the probability function on C induced by P_Σ and P_S, reminiscent of Chapter 5, enables us to reexpress $U(P_S, P_\Sigma)$ in a manner more amenable to our present task.

Lemma 7.5.1 gives three expressions for $U(P_S, P_\Sigma)$, two of which are in terms of $P(\cdot|P_S, P_\Sigma)$ and d, with the third being familiar to you and serving to make the appropriate analogies with Section 7.3.

Lemma 7.5.1. Let U and d be obtained from a given utility function $u(\cdot)$ on C. Then:

(A)
$$U(P_S, P_\Sigma) = \sum_{j=1}^{\#(\Sigma)} \sum_{h=1}^{t} P(\tilde{c} \gtrsim c^h|P_S, \sigma^j) d^h P_\Sigma(\sigma^j)$$
$$= \sum_{j=1}^{\#(\Sigma)} \sum_{h=1}^{t} \left[\sum_{i=1}^{\#(S)} P(\tilde{c} \gtrsim c^h|s^i, \sigma^j) P_S(s^i) \right] d^h P_\Sigma(\sigma^j);$$

(B)
$$U(P_S, P_\Sigma) = \sum_{h=1}^{t} P(\tilde{c} \gtrsim c^h|P_S, P_\Sigma) d^h$$
$$= \sum_{h=1}^{t} \left[\sum_{i=1}^{\#(S)} P(\tilde{c} \gtrsim c^h|s^i, P_\Sigma) P_S(s^i) \right] d^h;$$

and

$$\text{(C)} \qquad U(P_S, P_\Sigma) = \sum_{j=1}^{\#(\Sigma)} U(P_S, \sigma^j) P_\Sigma(\sigma^j)$$

$$= \sum_{j=1}^{\#(\Sigma)} \left[\sum_{i=1}^{\#(S)} U(s^i, \sigma^j) P_S(s^i) \right] P_\Sigma(\sigma^j).$$

*■ *Proof*. Part (C) is immediate from Equations (7.1.1)–(7.1.3). The second equalities in (A) and (B) are immediate from (7.5.6); and the first equality in (A) follows via (7.5.6d) and interchange of summation order from the first equality in (B), to prove which we have:

$$\begin{aligned} U(P_S, P_\Sigma) &=^1 \sum_{k=1}^{t} u(c^k) P(c^k | P_S, P_\Sigma) \\ &=^2 \sum_{k=1}^{t} \left(\sum_{h=1}^{k} d^h \right) P(c^k | P_S, P_\Sigma) \\ &=^3 \sum_{h=1}^{t} \sum_{k=h}^{t} P(c^k | P_S, P_\Sigma) d^h \\ &=^4 \sum_{h=1}^{t} P(\tilde{c} \in \{c^h, \ldots, c^t\} | P_S, P_\Sigma) d^h \\ &=^5 \sum_{h=1}^{t} P(\tilde{c} \gtrsim c^h | P_S, P_\Sigma) d^h, \end{aligned}$$

in which equality (1) is immediate from the definition of $P(\cdot | P_S, P_\Sigma)$; (2) follows from (7.5.3); (3) is interchanging summation order; (4) is factoring out d^h and summing probabilities of mutually exclusive events; and (5) follows from the assumption that $c^t > c^{t-1} > \cdots > c^{j+1} > c^h$, which implies that $\tilde{c} \gtrsim c^h$ if and only if $\tilde{c} \in \{c^h, \ldots, c^t\}$. ■

Table 7.1 is the key to the analogies with Section 7.3 and will enable us to move rapidly through stochastic and part of ordinal admissibility analysis. You may wish to refer back to it occasionally.

Recall our modus operandi in Section 7.3. (1) We defined the utility characteristic $\mathbf{x}(P_S)$ of P_S; (2) we defined the attainable utility set $X(\mathcal{Q}_S)$; (3) we noted that $X(\mathcal{Q}_S)$ is the convex hull of the utility characteristics $\mathbf{x}(s^i)$ of $\mathcal{D}$'s pure strategies s^i; (4) we defined cardinal dominance and cardinal admissibility; (5) we deduced as Theorem 7.3.2 some elementary facts about cardinal dominance and admissibility and evaluations of P_S; (6) we defined Bayes strategies; (7) we deduced as Theorem 7.3.3 some elementary facts about Bayes strategies; and (8) we gave as Theorem 7.3.4 the key results about the relationship between cardinal admissibility and Bayesness.[o] We shall sketch the counterparts of these steps for the case in which

[o] There was more to the program of Section 7.3, but we shall not analogize it here.

Table 7.1 Analogous Terms and Concepts

Cardinal Admissibility	Stochastic Admissibility	Ordinal Admissibility
Consensus on $u(\cdot)$	Consensus on $\gtrsim$ and P_Σ	Consensus on $\gtrsim$
j, σ^j	h, c^h	$(j, h), (\sigma^j, c^h)$
$\#(\Sigma)$-tuple P_Σ	t-tuple d	$t \cdot \#(\Sigma)$-tuple w
$P_\Sigma(\sigma^j)$	d^h	g^{jh}
$U(P_S, \sigma^j)$	$P(\tilde{c} \gtrsim c^h \mid P_S, P_\Sigma)$	$P(\tilde{c} \gtrsim c^h \mid P_S, \sigma^j)$
$\mathbf{x}(P_S)$	$\mathbf{z}(P_S)$	$\mathbf{w}(P_S)$
$X(\mathcal{Q}_S)$	$Z(\mathcal{Q}_S)$	$W(\mathcal{Q}_S)$
given u	given $\gtrsim$ and P_Σ	given $\gtrsim$
$\mathcal{A}_{\mathcal{C}}[\mathcal{Q}_S \mid u]$	$\mathcal{A}_{\mathcal{S}}[\mathcal{Q}_S \mid \gtrsim, P_\Sigma]$	$\mathcal{A}_{\mathcal{O}}[\mathcal{Q}_S \mid \gtrsim]$
cardinally	stochastically	ordinally

"$\gtrsim$" and P_Σ are agreed upon. Proofs are largely left to you as exercises, as they mimic their counterparts in Section 7.3.

Definitions and Theorems 7.5.1–7.5.4 are counterparts of like-numbered definitions and theorems in Section 7.3.

Definition 7.5.1. Let $\gtrsim$ and P_Σ be given. Let P_S be any randomized strategy for $\mathcal{D}$. Then $\mathcal{D}$'s *risk characteristic* $\mathbf{z}(P_S)$ of P_S is the t-tuple whose hth component is $P(\tilde{c} \gtrsim c^h \mid P_S, P_\Sigma)$.

Definition 7.5.2. The *attainable risk set* of a dpuu is the set $Z(\mathcal{Q}_S)$ of all risk characteristics; i.e.,

$$Z(\mathcal{Q}_S) = \{\mathbf{z}(P_S)\colon P_S \in \mathcal{Q}_S\}.$$

Theorem 7.5.1. Let $Z(\mathcal{Q}_S)$ be the attainable risk set of a dpuu in which $\mathcal{D}$ has a finite number of pure strategies. Then

$$Z(\mathcal{Q}_S) = \text{CONV}[\{\mathbf{z}(s)\colon s \in S\}].$$

■ *Proof.* Exercise 7.5.2. ■

Example 7.5.1

Consider the dpuu with consequence table

	σ^1	σ^2	σ^3	σ^4
s^1	\$10	\$10	\$0	\$2
s^2	\$2	\$2	\$2	\$2
s^3	\$10	\$0	\$10	\$2

and suppose that $\$c'' \gtrsim \$c'$ whenever $c'' > c'$. Let $P_\Sigma(\sigma^j) = p_j$ for $j = 1, 2, 3, 4$. Then $c^0 = \$0$, $c^1 = \$2$, and $c^2 = \$10$, with $t = 2$ because there are three unequally desirable consequences. Now, $P(\tilde{c} \gtrsim c^1 \mid s^1, P_\Sigma) = P(\tilde{c} \in \{\$2, \$10\} \mid s^1, P_\Sigma) = P_\Sigma(\tilde{\sigma} \in \{\sigma^1, \sigma^2, \sigma^4\}) = p_1 + p_2 + p_4$. Moreover,

$P(\tilde{c} \gtrsim c^2|s^1, P_\Sigma) = P(\tilde{c} = \$10|s^1, P_\Sigma) = P_\Sigma(\tilde{\sigma} \in \{\sigma^1, \sigma^2\}) = p_1 + p_2$. For s^2, we have $P(\tilde{c} \gtrsim c^1|s^2, P_\Sigma) = P(\tilde{c} \in \{\$2, \$10\}|s^2, P_\Sigma) = 1$ because $\tilde{c} = \$2$ is certain, whereas $P(\tilde{c} \gtrsim c^2|s^2, P_\Sigma) = P(\tilde{c} = \$10|s^2, P_\Sigma) = 0$ because $c = \$10$ is impossible. For s^3, we have $P(\tilde{c} \gtrsim c^1|s^3, P_\Sigma) = P(\tilde{c} \in \{\$2, \$10\}|s^3, P_\Sigma) = P_\Sigma(\tilde{\sigma} \in \{\sigma^1, \sigma^3, \sigma^4\}) = p_1 + p_3 + p_4$, while $P(\tilde{c} \gtrsim c^2|s^3, P_\Sigma) = P(\tilde{c} = \$10|s^3, P_\Sigma) = P_\Sigma(\tilde{\sigma} \in \{\sigma^1, \sigma^3\}) = p_1 + p_3$. Hence $\mathbf{z}(s^1) = (p_1 + p_2 + p_4, p_1 + p_2)$, $\mathbf{z}(s^2) = (1, 0)$, and $\mathbf{z}(s^3) = (p_1 + p_3 + p_4, p_1 + p_3)$. If these three $\mathbf{z}$'s were plotted on a graph in the plane as points, $Z(\mathcal{Q}_S)$ would be the convex hull of these points. See Figure 7.14.

■ As Example 7.5.1 suggests and as checking the definition verifies, the components of any given $\mathbf{z}(P_S)$ are nonincreasing, because $h > k$ implies that $P(\tilde{c} \gtrsim c^h|P_S, P_\Sigma) = P(\tilde{c} \in \{c^h, \ldots, c^t\}|P_S, P_\Sigma) \leq P(\tilde{c} \in \{c^k, \ldots, c^t\}|P_S, P_\Sigma) = P(\tilde{c} \gtrsim c^k|P_S, P_\Sigma)$. ■

Definition 7.5.3: Stochastic Dominance and Stochastic Admissibility

(A) Given two strategies P'_S and P''_S in $\mathcal{Q}_S$, we say that P''_S *stochastically dominates* P'_S, given $\gtrsim$ and P_Σ, or that P'_S is stochastically dominated by P''_S, given $\gtrsim$ and P_Σ, if

(i) $P(\tilde{c} \gtrsim c^h|P''_S, P_\Sigma) \geq P(\tilde{c} \gtrsim c^h|P'_S, P_\Sigma)$ for every $h \in \{1, \ldots, t\}$,

and

(ii) $P(\tilde{c} \gtrsim c^h|P''_S, P_\Sigma) > P(\tilde{c} \gtrsim c^h|P'_S, P_\Sigma)$ for at least one such h;

or equivalently, P''_S stochastically dominates P'_S, given $\gtrsim$ and P_Σ if $\mathbf{z}(P''_S) \geq \mathbf{z}(P'_S)$ but $\mathbf{z}(P''_S) \neq \mathbf{z}(P'_S)$.

(B) We say that P'_S is *stochastically admissible given* $\gtrsim$ and P_Σ if no strategy P''_S in $\mathcal{Q}_S$ stochastically dominates P'_S given $\gtrsim$ and P_Σ.

(C) We denote by $\mathcal{A}_{\mathcal{S}}[\mathcal{Q}_S|\gtrsim, P_\Sigma]$ the set of all strategies in $\mathcal{Q}_S$ which are stochastically admissible given $\gtrsim$ and P_Σ.

■ Definition 7.5.3(B) says that P'_S is stochastically admissible given $\gtrsim$ and P_Σ if $\mathbf{z}(P'_S)$ is on the "northeastern border" of the set $Z(\mathcal{Q}_S)$ of all risk characteristics. ■

Example 7.5.2

In Example 7.5.1, suppose that $p_2 > p_3$. Then s^1 stochastically dominates s^3, because in this case we have $\mathbf{z}(s^1) > \mathbf{z}(s^3)$. As long as $p_1 + p_2 + p_4 < 1$, s^1 does not stochastically dominate s^2. In fact, as Figure 7.14 indicates, $\mathcal{A}_{\mathcal{S}}[\mathcal{Q}_S|\gtrsim, P_\Sigma] = \mathcal{Q}_S^{\{s^1, s^2\}}$ when $p_2 > p_3$ and $p_1 + p_2 + p_4 < 1$.

Theorem 7.5.2: Stochastic Dominance, Strategy Evaluation, and Stochastic Admissibility

(A) If P''_S stochastically dominates P'_S given $\gtrsim$ and P_Σ, then $U(P''_S, P_\Sigma) > U(P'_S, P_\Sigma)$ for every U which is consistent with (i.e., reflects) "$\gtrsim$."

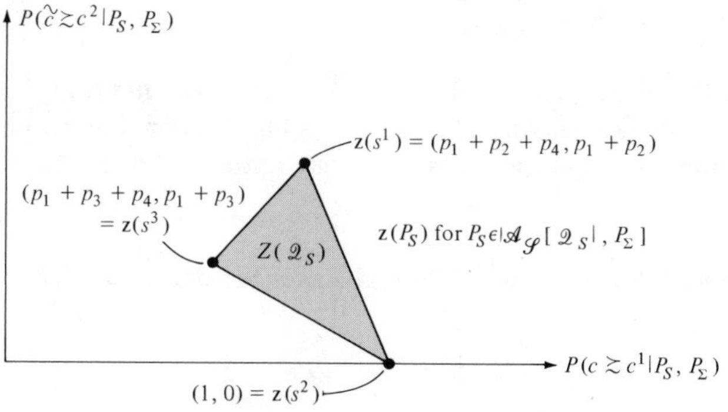

Figure 7.14 Attainable risk set for Examples 7.5.1 and 7.5.2, $p_2 > p_3$ and $p_1 + p_2 + p_4 < 1$.

(B) If S and C are finite sets of pure strategies and if P'_S is stochastically dominated given $\succsim$ and P_Σ, then P'_S is stochastically dominated by a strategy in $\mathcal{A}_{\mathcal{S}}[\mathcal{Q}_S|\succsim, P_\Sigma]$.

■ *Proof.* Exercise 7.5.3. Note that part (A) is *less* complicated than Theorem 7.3.2(A), because here *every* d^h must be *positive* in order for the associated $u(\cdot)$ and $U(\cdot)$ to be consistent with "$\succsim$." ■

No essential change in the definition (Definition 7.3.4) of Bayes strategies is needed, except to note that in Section 7.3 we were given u (or equivalently, both $\succsim$ and d) and had to "fill in" P_Σ, whereas here we are given $\succsim$ and P_Σ and have to "fill in" d.

Definition 7.5.4: Bayes Strategies for $\mathcal{D}$

(A) A strategy P_S° for $\mathcal{D}$ is *Bayes given* $\succsim$ *and* P_Σ *for*, or *optimal given* $\succsim$ *and* P_Σ *with respect to*, a first-difference vector d if

$$\sum_{h=1}^{t} P(\tilde{c} \succsim c^h | P_S^\circ, P_\Sigma) d^h = \max_{P_S \in \mathcal{Q}_S} \left[\sum_{h=1}^{t} P(\tilde{c} \succsim c^h | P_S, P_\Sigma) d^h \right].$$

(B) A strategy P_S° for $\mathcal{D}$ is called Bayes given $\succsim$ and P_Σ if there is at least one d (strictly positive, with components summing to one) such that P_S° is Bayes given $\succsim$ and P_Σ for d.

(C) We denote by $\mathcal{B}[\mathcal{Q}_S|\succsim, P_\Sigma]$ the set of all strategies P_S° in $\mathcal{Q}_S$ which are Bayes given $\succsim$ and P_Σ.

Theorem 7.5.3: Structure of $\mathcal{B}[\mathcal{Q}_S|\succsim, P_\Sigma]$

(A) If both P'_S and P''_S are Bayes given $\succsim$ and P_Σ for d, then

$$\sum_{h=1}^{t} P(\tilde{c} \succsim c^h | P'_S, P_\Sigma) d^h = \sum_{h=1}^{t} P(\tilde{c} \succsim c^h | P''_S, P_\Sigma) d^h.$$

(B) Let $S[d] = \{s: s \text{ is Bayes given } \succsim \text{ and } P_\Sigma \text{ for } d\}$. Then P_S is Bayes given $\succsim$ and P_Σ for d if and only if $P_S \in \mathcal{Q}_S^{S[d]}$.

■ *Proof.* Exercise 7.5.4. ■

Theorem 7.5.4 is less complicated than its counterpart, Theorem 7.3.4, because only d's with strictly positive components can be considered here, whereas P_Σ's wth some zero components can arise naturally in the context of Section 7.3.

Theorem 7.5.4: Stochastically Admissible and Bayes Strategies. If S and C are both finite sets, then

$$\mathscr{A}_{\mathscr{S}}[\mathscr{Q}_S|\gtrsim, P_\Sigma] = \mathscr{B}[\mathscr{Q}_S|\gtrsim, P_\Sigma].$$

■ *Proof.* Exercise 7.5.5. ■

Theorem 7.5.4 suggests that exactly the same approaches can be used to find stochastically admissible strategies as can be used to find cardinally admissible strategies. The discussion in Section 7.3 can be translated intact to this context with only minor notational amendments, chief among which is that d here plays the role of P_Σ in Section 7.3. In the present context, the graphical approach, according to Figure 7.5 (say), really amounts to saying that P_S is stochastically admissible, given $\gtrsim$ and P_Σ if and only if P_S maximizes $U(P_S, P_\Sigma)$, where $U(\cdot, \cdot)$ and its underlying $u(\cdot)$ derive from $\gtrsim$ and some strictly positive d.

To carry this reasoning one step further, suppose that $P_\Sigma(\sigma) > 0$ for every σ. Then $P_S \in \mathscr{A}_{\mathscr{S}}[\mathscr{Q}_S|\gtrsim, P_\Sigma]$ if and only if P_S satisfies the hypotheses of Theorem 7.3.4(B) for some u consistent with $\gtrsim$, and hence $P_S \in \mathscr{A}_{\mathscr{C}}[\mathscr{Q}_S|u]$ for any such u. Conversely, if $P_S \in \mathscr{A}_{\mathscr{C}}[\mathscr{Q}_S|u]$, then, by Theorem 7.3.4(A), P_S maximizes $U(P_S, P_\Sigma)$ for some P_Σ with every $P_\Sigma(\sigma)$ positive, so that, by Theorem 7.5.4, $P_S \in \mathscr{A}_{\mathscr{S}}[\mathscr{Q}_S|\gtrsim, P_\Sigma]$ for any such P_Σ. Hence we have shown what might be called the equal statūs of stochastic admissibility (for fixed, strictly positive P_Σ) and cardinal admissibility (for fixed, necessarily strictly positive, first-difference vector d representing risk attitudes). A precise statement of this parity is Corollary 7.5.1.

Corollary 7.5.1: Parity of Cardinal and Stochastic Admissibilities

$$\cup\{\mathscr{A}_{\mathscr{C}}[\mathscr{Q}_S|u]\colon u \text{ reflects } \gtrsim\} = \cup\{\mathscr{A}_{\mathscr{S}}[\mathscr{Q}_S|\gtrsim, P_\Sigma]\colon P_\Sigma \text{ strictly positive}\}.$$

In a pure sense, Corollary 7.5.1 solves the problem of the parties to the decision who can agree only on "$\gtrsim$", because it states that they can narrow down further controversy only to the common set described in the corollary. But in most realistic problems involving several σ's and several c's, finding this set by attempting to evaluate either side of the set equation in the corollary is likely to be very difficult. A conceptually cleaner, but less economical approach is to find a still reasonably small superset of this set. Such a superset is the set $\mathscr{A}_{\mathscr{O}}[\mathscr{Q}_S|\gtrsim]$ of all *ordinally* admissible strategies given $\gtrsim$.

In Definitions 7.5.5 and 7.5.6, we define the ordinal characteristic of a strategy P_S, the attainable ordinal set of a dpuu, ordinal dominance, and ordinal admissibility.

Definition 7.5.5: Ordinal Concepts

(A) Suppose that some ordering of the $t \cdot \#(\Sigma)$ pairs (h, j) for $h \in \{1, \ldots, t\}$ and $j \in \{1, \ldots, \#(\Sigma)\}$ is chosen and fixed; say, for $j = 1$, let h run from 1 through t, then for $j = 2$, let h run from 1 through t, and so on. Then $\mathcal{D}$'s *ordinal characteristic* of any strategy P_S is the $t \cdot \#(\Sigma)$-tuple $\mathbf{w}(P_S)$ whose kth component is $P(\tilde{c} \gtrsim c^h | P_S, \sigma^j)$ for the kth (h, j) in the chosen ordering.

(B) The *attainable ordinal set* of a dpuu is the set $W(\mathcal{Q}_S)$ of all ordinal characteristics $\mathbf{w}(P_S)$.

■ From Lemma 7.5.1(A), it is clear that $W(\mathcal{Q}_S) = \text{CONV}[\{\mathbf{w}(s)\colon s \in S\}]$. Indeed, $W(\mathcal{Q}_S)$ performs the same function in ordinal admissibility analysis as $X(\mathcal{Q}_S)$ and $Z(\mathcal{Q}_S)$ do in cardinal and stochastic admissibility analyses respectively. ■

Definition 7.5.6: Ordinal Dominance and Ordinal Admissibility. Let $\gtrsim$ be fixed.

(A) Given two strategies P'_S and P''_S in $\mathcal{Q}_S$, we say that P''_S *ordinally dominates* P'_S *given* $\gtrsim$, or that P'_S is ordinally dominated by P''_S given $\gtrsim$, if

(i) $P(\tilde{c} \gtrsim c^h | P''_S, \sigma^j) \geq P(\tilde{c} \gtrsim c^h | P'_S, \sigma^j)$ for *every* (h, j),

and

(ii) $P(\tilde{c} \gtrsim c^h | P''_S, \sigma^j) > P(\tilde{c} \gtrsim c^h | P'_S, \sigma^j)$ for at least one (h, j);

or equivalently, P''_S ordinally dominates P'_S given $\gtrsim$ if $\mathbf{w}(P''_S) \geq \mathbf{w}(P'_S)$ but $\mathbf{w}(P''_S) \neq \mathbf{w}(P'_S)$.

(B) We say that P'_S is *ordinally admissible* given $\gtrsim$ if no strategy P''_S in $\mathcal{Q}_S$ ordinally dominates P'_S given $\gtrsim$.

(C) We denote by $\mathcal{A}_O[\mathcal{Q}_S|\gtrsim]$ the set of all strategies in $\mathcal{Q}_S$ which are ordinally admissible given $\gtrsim$.

Theorem 7.5.5 is the counterpart of Theorems 7.3.4 and 7.5.4 for the ordinal-admissibility case. Here, it does not make sense to talk about Bayes strategies, for the reasons noted after the statement of the theorem.

Theorem 7.5.5. If S, C, and Σ are all finite sets, then $P^*_S \in \mathcal{A}_O[\mathcal{Q}_S|\gtrsim]$ if and only if

$$\sum_{k=1}^{t \cdot \#(\Sigma)} w_k(P^*_S) g^k = \max_{P_S \in \mathcal{Q}_S} \left[\sum_{k=1}^{t \cdot \#(\Sigma)} w_k(P_S) g^k \right]$$

for some set $\{g^1, \ldots, g^k, \ldots, g^{t \cdot \#(\Sigma)}\}$ of *positive* weights which sum to one.

■ *Proof.* Exercise 7.5.6. ■

■ In less compact notation, Theorem 7.5.5 says that P_S^* is ordinally admissible if and only if P_S^* maximizes

$$\sum_{j=1}^{\#(\Sigma)} \sum_{h=1}^{t} P(\tilde{c} \gtrsim c^h | P_S, \sigma^j) g^{hj} \tag{7.5.7}$$

for some set of positive weights g^{hj} which sum to one. Now, (7.5.7) would be the right-hand side of Lemma 7.5.1(A) if the weights g^{hj} were necessarily all of the "factorable" form

$$g^{hj} = P_\Sigma(\sigma^j) \cdot d^h$$

for some strictly positive P_Σ and some strictly positive d. But the g^{hj} need not be of this form. If they were, then $\mathcal{A}_\mathcal{O}[\mathcal{Q}_S|\gtrsim]$ would be the set described in Corollary 7.5.1. (Why? Exercise 7.5.7.) In general, the set in Corollary 7.5.1 is a proper subset of $\mathcal{A}_\mathcal{O}[\mathcal{Q}_S|\gtrsim]$. ■

The remainder of this section focuses on ordinally admissible *pure* strategies, that is, on the elements of $\mathcal{A}_\mathcal{O}[S|\gtrsim]$. In **[76]**, it is shown that if a pure strategy s is ordinally dominated given $\gtrsim$ by some randomized strategy, then it must also be ordinally dominated by a pure strategy. Hence the pure strategies in $\mathcal{A}_\mathcal{O}[\mathcal{Q}_S|\gtrsim]$ are precisely those in $\mathcal{A}_\mathcal{O}[S|\gtrsim]$.

Because even very small dpuu's can have $t \cdot \#(\Sigma)$ rather large and because $W(\mathcal{Q}_S)$ is a (convex) set of $t \cdot \#(\Sigma)$-tuples, it follows that finding $\mathcal{A}_\mathcal{O}[\mathcal{Q}_S|\gtrsim]$ may be a difficult task even though it is one designed to circumvent finding the union described in Corollary 7.5.1. But, as noted in the Introduction, finding $\mathcal{A}_\mathcal{O}[S|\gtrsim]$ is a very simple matter, and we can use $\mathcal{A}_\mathcal{O}[S|\gtrsim]$ to find a superset of $\mathcal{A}_\mathcal{O}[\mathcal{Q}_S|\gtrsim]$.

Theorem 7.5.6.

$$\mathcal{A}_\mathcal{O}[\mathcal{Q}_S|\gtrsim] \subset \mathcal{Q}_S^{\mathcal{A}_\mathcal{O}[S|\gtrsim]}.$$

■ *Proof.* If $P_S' \notin \mathcal{Q}_S^{\mathcal{A}_\mathcal{O}[S|\gtrsim]}$, then, by the definitions of $\mathcal{Q}_S^{S'}$ and $\mathcal{A}_\mathcal{O}[S|\gtrsim]$, the strategy P_S' places positive probability on some ordinally dominated pure strategy s^*. Suppose s^{**} ordinally dominates s^*, and let P_S'' coincide with P_S' except that the probability that P_S' accords to s^* is shifted to s^{**}. It is easily verified that P_S'' ordinally dominates P_S'. Hence $P_S' \notin \mathcal{A}_\mathcal{O}[\mathcal{Q}_S|\gtrsim]$. ■

Theorem 7.5.6 thus furnishes a criterion by which it can be shown that some randomized strategies are not ordinally admissible.

One question, however, remains hitherto unanswered and is highly important. Since pure strategies are ordinarily of much greater practical interest than are genuinely randomized strategies, we need to ascertain whether every ordinally admissible pure strategy can be an optimal strategy given some set of risk attitudes (reflected by d) and some set of judgments (reflected by P_Σ). The answer is yes, and we prove it by showing that every ordinally admissible pure strategy is cardinally admissible for some utility function representing $\gtrsim$. Theorem 7.3.4(A) then completes the argument.

Theorem 7.5.7. Suppose that S and Σ are finite sets. If s^* is ordinally admissible given $\succsim$, then s^* is cardinally admissible given some u which reflects $\succsim$.

*■ *Proof.* (From Fishburn and LaValle [**43**]) For simplicity of exposition, let $\mathcal{D}$'s pure strategies be $S = \{i: i = 1, \ldots, m\}$, and $\mathcal{E}$'s be $\Sigma = \{j: j = 1, \ldots, n\}$; and suppose that $\mathcal{D}$'s preference ranking has been numerically represented by an m-by-n matrix of (ordinal) utilities u_{ij}. Since the admissibility status—whether cardinal, ordinal, or stochastic—of a strategy is independent of the labeling of the rows and of the columns, we shall suppose that the s^* in question is the first row of the matrix of u_{ij}'s. We shall also use the clear-on-a-little-reflection facts (1) that duplicating columns of the u_{ij}-matrix affects no strategy's admissibility status and (2) that strategies for $\mathcal{D}$ with identical rows in the u_{ij}-matrix are effectively the same thing and may be coalesced into a single strategy. Because $s^* =$ row 1 is ordinally admissible, it follows that, for every $h \neq 1$, there is a column $j(h)$ such that $u_{1j(h)} > u_{hj(h)}$, because otherwise row h would either coincide with or ordinally dominate row 1. For each $h > 1$, pick exactly one such $j(h)$ and rearrange—and duplicate as necessary—the columns of the u_{ij}-matrix so that $j(h)$ becomes h for $h = 2, \ldots, m$. By the preceding remarks, row 1 has exactly the same ordinal and cardinal admissibility statūs in the resulting matrix as it had before the column rearrangements and duplications. Thus our rearranged table has the property that $u_{1j} > u_{jj}$ for $j = 2, \ldots, m$. Subsequent attention is confined to columns 2 through m of this table. Some of the u_{1j}'s may be equal; denote their distinct values by x_a, with $x_1 < x_2 < \cdots < x_p (p \leq m)$, and define $x_0 = -\infty$ and $x_{p+1} = +\infty$. We now indicate how to transform ordinally all u_{ij}'s so that row 1 of the matrix of transformed utilities $v_{ij} = t(u_{ij})$ is cardinally admissible. Let $\delta = 1/m^p - 1/m^{p+1}$, and define $y_a = 1 - 1/m^a$ for $a = 0, 1, \ldots, p$. Now define the ordinally transformed utilities $v_{ij} = t(u_{ij})$ so that:

$$v_{ij} = y_a \quad \text{if} \quad u_{ij} = x_a, \tag{7.5.8}$$

$$y_a < v_{ij} < y_a + \delta \quad \text{if} \quad x_a < u_{ij} < x_{a+1}, \tag{7.5.9}$$

and

$$v_{ij} > v_{hk} \quad \text{whenever} \quad u_{ij} > u_{hk}. \tag{7.5.10}$$

Such a transformation of utilities is always possible given the hypotheses, and the matrix of v_{ij}'s is obviously ordinally equivalent to the original matrix of u_{ij}'s. We now show that row 1 is *cardinally* admissible in the matrix of v_{ij}'s. Suppose to the contrary. Then row 1 would be cardinally dominated by a randomized strategy placing probability q_i (say) on row i, for $i = 1, \ldots, m$, with $q_1 = 0$, so that

$$\sum_{i=2}^{m} q_i v_{ij} \geq v_{1j} \quad \text{for} \quad j = 2, \ldots, m. \tag{7.5.11}$$

Since $v_{ij} < y_p + \delta$ for all (i, j) and $u_{jj} < u_{1j} = x_a$ for some $a \in \{1, \ldots, p\}$

implies that $v_{ij} < y_{a-1} + \delta$, it follows from (7.5.11) that

$$q_j[y_{a-1} + \delta] + (1 - q_j)[y_p + \delta] > \sum_{i=2}^{m} q_i v_{ij} \geq v_{1j} = y_a$$

for some $a = a(j)$ in $\{1, \ldots, p\}$ and for every j, which implies that

$$q_j < \frac{y_p + \delta - y_a}{y_p - y_{a-1}} = \frac{1}{m}$$

for every j, the equality following for any $a \in \{1, \ldots, p\}$ on substituting definitions of δ and the y's. But $q_j < 1/m$ for $j = 2, \ldots, m$ and $q_1 = 0$ imply that $\Sigma_{i=1}^{m} q_i < 1$, contradicting the fact that probabilities in a randomized strategy must sum to one. Thus no randomized strategy cardinally dominates row 1. ■

Finally, you should note that finding cardinally, stochastically, or ordinally admissible strategies is tantamount to finding the northeastern boundary of a set: $X(\mathscr{Q}_S)$, $Z(\mathscr{Q}_S)$, and $W(\mathscr{Q}_S)$ for cardinal, stochastic, and ordinal admissibility, respectively. Since all such forms of admissibility produce the same mathematical problem, the development of efficient algorithms for obtaining concrete solutions to large-scale, realistic problems should enjoy high priority.

The following, and final, section concerns some of the ramifications of restricting $\mathscr{D}$'s choice to some proper subset $\mathscr{Q}'_S$ of $\mathscr{Q}_S$, so as to study the issue of which strategies are worth considering if $\mathscr{D}$ were barred from choosing any $P_S \notin \mathscr{Q}'_S$.

■ Stochastic admissibility as defined in this section is often termed *first-degree* stochastic admissibility, to distinguish it from second- and third-degree stochastic admissibilities, defined respectively in Hadar and Russell [**49**] and Whitmore [**142**]. Second-degree stochastic admissibility is the appropriate notion when you know that $u(\cdot)$ is "concave," in the sense that $d^1 > d^2 > \cdots > d^k$; third-degree stochastic admissibility is appropriate when you know that $u(\cdot)$ has "decreasing risk aversion," in the sense that $d^1 - d^2 > d^2 - d^3 > \cdots > d^{k-1} - d^k$. (With finite sets of consequences, concavity and decreasing risk aversion are primarily suggestive terms, and hence have been placed in quotes.) See also Fishburn [**35**] and [**36**] for important recent contributions in this area. ■

■ Recent work by Yu, introduced in [**149**], generalizes significantly the entire concept of dominance and admissibility. Yu's definition of domination structures allows, in effect, the criterion to depend upon the (point representing the) provisionally chosen strategy. ■

Exercise 7.5.1

Derive Equation (7.5.6).

Exercise 7.5.2

Prove Theorem 7.5.1.

Exercise 7.5.3

Prove Theorem 7.5.2.

Exercise 7.5.4

Prove Theorem 7.5.3.

Exercise 7.5.5

Prove Theorem 7.5.4.

Exercise 7.5.6

Prove Theorem 7.5.5.

Exercise 7.5.7

In the context of Equation (7.5.7), in the remark following Theorem 7.5.5: Show that if g^{hj} were necessarily of the form $P_\Sigma(\sigma^j) \cdot d^h$ for some strictly positive P_Σ and some strictly positive d, then $\mathscr{A}_\mathscr{O}[\mathscr{Q}_S|\succsim]$ would be the set described in Corollary 7.5.1.

Exercise 7.5.8: Maximin Again

It is possible to define maximin strategies in the contexts of stochastic admissibility (with minimization being performed on d) and of ordinal admissibility (with minimization being performed on g). But is there a realistic behavioral scenario to justify using maximin strategies in these contexts? If so, describe it.

Exercise 7.5.9

Find $\mathscr{A}_\mathscr{S}[\mathscr{Q}_S|\succsim, P_\Sigma]$ for the dpuu

	σ^1	σ^2	σ^3	σ^4
s^1	\$3	\$1	\$0	\$1
s^2	\$1	\$1	\$1	\$1
s^3	\$0	\$0	\$3	\$1
s^4	\$1	\$0	\$1	\$3
s^5	\$3	\$0	\$1	\$3

if $\$x > \$y$ whenever $x > y$ and $[P_\Sigma(\sigma^1), P_\Sigma(\sigma^2), P_\Sigma(\sigma^3), P_\Sigma(\sigma^4)] =$

(A) [.4, .1, .2, .3];
(B) [.3, .4, .1, .2]; and

(C) [.1, .2, .3, .4].

(D) Is s^4 *always* stochastically dominated? If so, why? If not, for what P_Σ does $s^4 \in \mathscr{A}_\mathscr{S}[\mathscr{Q}_S|\gtrsim, P_\Sigma]$?

Exercise 7.5.10

In the dpuu of Exercise 7.5.9, show that $\mathscr{A}_\mathscr{O}[\mathscr{Q}_S|\gtrsim] = \mathscr{Q}_S^{\{s^1, s^2, s^3, s^5\}}$. [*Hint*: For each (s, σ), first tabulate two numbers: the first being 1 if $\$c \geq \$1$ (zero otherwise), and the second being 1 if $\$c \geq \$3$ (zero otherwise). Thus $\mathbf{w}(s^1) = (1, 1; 1, 0; 0, 0; 1, 0)$. Complete this table of **w**'s. Then see if you can find $(g^1, \ldots, g^8)$, nonnegative and summing to one, such that the numbers $\Sigma_{k=1}^{8} w_k(s^i)g^k$ are equal for $i = 1, 2, 3, 5$. (This involves finding relationships among the g^k's.)]

***Exercise 7.5.11**

Find a utility function and a set of probabilities for the σ's in the dpuu of the preceding exercise, for which every P_S in $\mathscr{Q}_S^{\{s^1, s^2, s^3, s^5\}}$ is optimal. [*Hint*: Utilize the machinery of the preceding exercise, together with the fact that *if* every P_S in the asserted set is optimal for some d and some P_Σ, *then* $g^1/g^2 = g^3/g^4 = g^5/g^6 = g^7/g^8 = d^1/(1 - d^1)$. (Why?)]

7.6 RAMIFICATIONS OF CONFINING $\mathscr{D}$'S CHOICE TO A PROPER SUBSET $\mathscr{Q}_S'$ OF $\mathscr{Q}_S$.

In the Introduction we noted that controversies about the feasibility of certain courses of action could be handled in (at least) two ways, one which transformed the disagreement to one over probabilities, and another which leads to a study of the consequences of restricting $\mathscr{D}$'s choice to a subset $\mathscr{Q}_S'$ of the set $\mathscr{Q}_S$ of all conceivable randomized strategies. The latter situation is examined in this section.

Everything here proceeds under the assumption that $\mathscr{D}$'s preferences are fully quantified, risk attitudes included. Hence cardinal admissibility is the theme. But all results have perfect analogues in other situations, where stochastic or ordinal admissibility is appropriate.

For a given $\mathscr{Q}_S'$, the attainable utility set $X(\mathscr{Q}_S)$, and the sets $\mathscr{B}[\mathscr{Q}_S'|u]$ and $\mathscr{A}_C[\mathscr{Q}_S'|u]$ of Bayes and cardinally admissible strategies given u and given the necessity of choosing P_S from $\mathscr{Q}_S'$, are all defined much as in Section 7.3. Recall from Definition 7.1.3(C) that $U(\mathscr{Q}_S', P_\Sigma)$ is the maximum of $U(P_S, P_\Sigma)$ as P_S ranges over $\mathscr{Q}_S'$.

Definition 7.6.1. Let $\mathscr{Q}_S'$ be a nonempty, closed subset of $\mathscr{Q}_S$ (so that the following will exist). Then:

(A) $X(\mathscr{Q}_S') = \{\mathbf{x}(P_S): P_S \in \mathscr{Q}_S'\}$;

(B) $\mathscr{B}[\mathscr{Q}_S'|u] = \{P_S^*: P_S^* \in \mathscr{Q}_S', U(P_S^*, P_\Sigma) = U(\mathscr{Q}_S', P_\Sigma)$ for some $P_\Sigma \in \mathscr{Q}_\Sigma\}$;

and

(C) $\mathcal{A}_{\mathcal{C}}[\mathcal{Q}'_S|u] = \{P^*_S\colon P^*_S \in \mathcal{Q}'_S$, no strategy in $\mathcal{Q}'_S$ cardinally dominates $P^*_S\}$.

Example 7.6.1

Suppose that in the dpuu of Figure 7.5 and Example 7.3.1 $\mathcal{Q}'_S = \mathcal{Q}_S^{\{s^1, s^3, s^4, s^5\}}$; that is, we make s^2, s^6, and all randomized strategies with $P_S(\{s^2, s^6\}) > 0$ unavailable to $\mathcal{D}$. Then $X(\mathcal{Q}'_S)$ is the triangle with vertices $\mathbf{x}(s^4)$, $\mathbf{x}(s^1)$, and $\mathbf{x}(s^5)$. Clearly, $\mathcal{A}_{\mathcal{C}}[\mathcal{Q}'_S|u] = \mathcal{B}[\mathcal{Q}'_S|u] = \mathcal{Q}_S^{\{s^1, s^5\}}$. Note that all genuinely randomized strategies in $\mathcal{A}_{\mathcal{C}}[\mathcal{Q}'_S|u]$ are dominated if in fact $\mathcal{D}$ were not forbidden to choose s^2 or s^6.

Some rather obvious but useful facts are gathered together as Theorem 7.6.1.

Theorem 7.6.1

(A) If $\mathcal{Q}'_S \subset \mathcal{Q}''_S \subset \mathcal{Q}_S$, then $X(\mathcal{Q}'_S) \subset X(\mathcal{Q}''_S) \subset X(\mathcal{Q}_S)$.
(B) If $\mathcal{Q}'_S$ is nonempty and closed, then $X(\mathcal{Q}'_S)$ is nonempty and closed.
(C) If $\mathcal{Q}'_S$ is convex, then $X(\mathcal{Q}'_S)$ is convex.

■ *Proof*. (A) is obvious from Definition 7.6.1(A), as is nonemptiness of $X(\mathcal{Q}'_S)$. The closedness part of (B) is a less elementary consequence of the linearity (and therefore continuity) of the mapping $\mathbf{x}(\cdot)\colon \mathcal{Q}'_S \to R^{\#(\Sigma)}$; we omit the details. For (C): If $\mathbf{x}^i \in X(\mathcal{Q}'_S)$ for $i = 1, 2$, then $\mathbf{x}^i = \mathbf{x}(P_S^i)$ for $i = 1, 2$ and some P_S^1 and P_S^2 belonging to $\mathcal{Q}'_S$. But convexity of $\mathcal{Q}'_S$ implies that $pP_S^1 + (1-p)P_S^2 \in \mathcal{Q}'_S$, and it is easily verified that $\mathbf{x}(pP_S^1 + (1-p)P_S^2) = p\mathbf{x}(P_S^1) + (1-p)\mathbf{x}(P_S^2) = p\mathbf{x}^1 + (1-p)\mathbf{x}^2$, which therefore belongs to $X(\mathcal{Q}'_S)$. Hence the line segment $L(\mathbf{x}^1, \mathbf{x}^2)$ joining any two points $\mathbf{x}^1$ and $\mathbf{x}^2$ of $X(\mathcal{Q}'_S)$ is a subset of $X(\mathcal{Q}'_S)$, thus verifying convexity of $X(\mathcal{Q}'_S)$. ■

Example 7.6.2

In the dpuu of Examples 7.3.1, 7.6.1, and Figure 7.5, suppose that, in order to placate a regulatory agency, you must assign probability at least 1/2 to s^4. Thus your choice is restricted to $\mathcal{Q}'_S = \{P_S\colon P_S(s^4) \geq 1/2\}$. This is a nonempty, closed, convex set. It is not hard to verify that $X(\mathcal{Q}'_S)$ is as graphed in Figure 7.15. [Each "corner" of $X(\mathcal{Q}'_S)$ is halfway between $\mathbf{x}(s^4) = (3, 2)$ and the corresponding "corner" of $X(\mathcal{Q}_S)$.] Clearly, $\mathcal{A}_{\mathcal{C}}[\mathcal{Q}'_S|u] = \{P_S\colon P_S = (1/2)P^*_S + (1/2)(0, 0, 0, 1, 0, 0)$ for some P^*_S in $\mathcal{A}_{\mathcal{C}}[\mathcal{Q}_S|u]\}$; that is, P_S is cardinally admissible if, in 6-tuple form, it is either $(0, p, 0, 1/2, 1/2-p, 0)$ for some $p \in [0, 1/2]$ or $(p, 1/2-p, 0, 1/2, 0, 0)$ for some $p \in [0, 1/2]$. Bayes strategies in $\mathcal{Q}'_S$ include the cardinally admissible strategies, and also the strategies $(p, 0, 0, 1/2, 0, 1/2-p)$ for some $p \in [0, 1/2]$. (Why?)

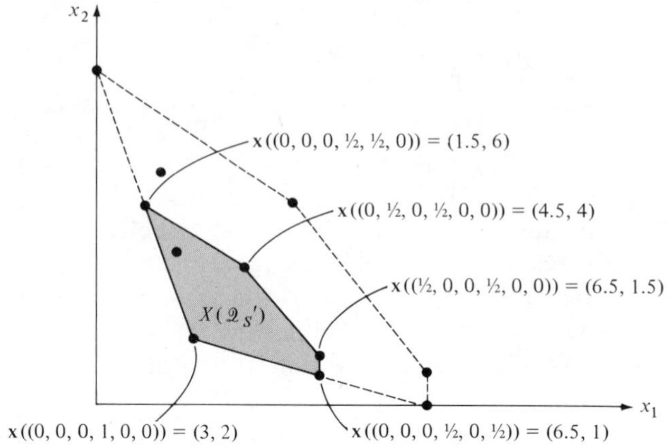

Key: Shaded region is $X(\mathcal{Q}_S')$; dashed lines encompass the rest of $X(\mathcal{Q}_S)$ [See Figure 7.5]. Strategies written in vector form.

Figure 7.15 $X(\mathcal{Q}_S')$ for Example 7.6.2.

Now for the three principal results concerning imposing a constraint on $\mathcal{D}$'s choice, by restricting that choice to the strategies in $\mathcal{Q}_S'$. Theorem 7.6.2(A) shows that such a constraint never improves the quality of the constrained-admissible strategies, in that $\mathcal{D}$ could get at least as good a utility characteristic if he could make his choice from $\mathcal{A}_{\mathcal{C}}[\mathcal{Q}_S|u]$ instead of from $\mathcal{A}_{\mathcal{C}}[\mathcal{Q}_S'|u]$. Theorem 7.6.2(B) embodies the same spirit and is, in fact, *the most fundamental fact of life in all decision analysis*: *a constraint cannot help and may hurt.* A converse to (B), Theorem 7.6.2(C), is sometimes taken as an axiom for solution procedures in game theory and abstract decision theory. A more generally phrased version of (C) was first used by J. F. Nash as an axiom, called "independence of irrelevant alternatives." It says that, conversely to (B), if a constraint does not rule out $\mathcal{D}$'s optimal strategy, then that constraint is harmless.

Theorem 7.6.2: Principal Facts. Let $\mathcal{Q}_S'$ and $\mathcal{Q}_S''$ be nonempty, closed subsets of $\mathcal{Q}_S$ such that $\mathcal{Q}_S' \subset \mathcal{Q}_S''$. Then:

(A) For every P_S' in $\mathcal{A}_{\mathcal{C}}[\mathcal{Q}_S'|u]$ there is a P_S'' in $\mathcal{A}_{\mathcal{C}}[\mathcal{Q}_S''|u]$ such that $\mathbf{x}(P_S'') \geq \mathbf{x}(P_S')$.

(B) $U(\mathcal{Q}_S', P_\Sigma) \leq U(\mathcal{Q}_S'', P_\Sigma)$ for every P_Σ in $\mathcal{Q}_\Sigma$—where $U(\mathcal{Q}_S^*, P_\Sigma) = \max_{P_S \in \mathcal{Q}_S^*}[U(P_S, P_\Sigma)]$ for every nonempty, closed subset $\mathcal{Q}_S^*$ of $\mathcal{Q}_S$.

(C) If a strategy P_S'' optimal against P_Σ in $\mathcal{B}[\mathcal{Q}_S''|u]$ belongs to $\mathcal{B}[\mathcal{Q}_S'|u]$, then $U(\mathcal{Q}_S', P_\Sigma) = U(\mathcal{Q}_S'', P_\Sigma)$.

■ *Proof*: *Of* (A). Since $\mathcal{A}_{\mathcal{C}}[\mathcal{Q}_S'|u] \subset \mathcal{Q}_S' \subset \mathcal{Q}_S''$, there is a P_S^* in $\mathcal{Q}_S''$ such that $\mathbf{x}(P_S^*) \geq \mathbf{x}(P_S')$; take $P_S^* = P_S'$, as a matter of fact. Now, if $P_S^* \in \mathcal{A}_{\mathcal{C}}[\mathcal{Q}_S''|u]$, we set $P_S'' = P_S^*$ and are done; if not, then, by Theorem 7.3.2(B), there is a P_S'' in $\mathcal{A}_{\mathcal{C}}[\mathcal{Q}_S''|u]$ which dominates P_S^* and therefore also P_S'. In any event, $\mathbf{x}(P_S'') \geq \mathbf{x}(P_S')$, as required. [The proof of Theorem

7.3.2(B) made no use of convexity of $X(\mathscr{Q}_S)$, only closedness above and local boundedness above.]

Of (B). Suppose P_S° in $\mathscr{B}[\mathscr{Q}'_S|u]$ maximizes $U(P_S, P_\Sigma)$ in $\mathscr{Q}'_S$; that is, $U(P_S^\circ, P_\Sigma) = U(\mathscr{Q}'_S, P_\Sigma)$. Since $\mathscr{Q}'_S$ is a subset of $\mathscr{Q}''_S$, clearly P_S° belongs to $\mathscr{Q}''_S$ and either maximizes $U(P_S, P_\Sigma)$ in $\mathscr{Q}''_S$—in which case $U(\mathscr{Q}'_S, P_\Sigma) = U(\mathscr{Q}''_S, P_\Sigma)$—or does not maximize, in which case $U(\mathscr{Q}'_S, P_\Sigma) < U(\mathscr{Q}''_S, P_\Sigma)$. Assertion (B) is true in either case.

Proof *of (C)* requires an obvious modification of the reasoning for (B). ■

As noted in Section 7.3, $\mathscr{D}$ may throw away, or constrain away, cardinally *inadmissible* strategies with impunity, because there is always an admissible strategy which is optimal against any given P_Σ.

■ See Theorem 7.3.4(C). Furthermore, there is always an admissible maximin strategy; see *Exercise 7.3.9(F). ■

In fact, you may show as Exercise 7.6.3 that

$$\mathscr{A}_{\mathscr{C}}[\mathscr{A}_{\mathscr{C}}[\mathscr{Q}'_S|u]|u] = \mathscr{A}_{\mathscr{C}}[\mathscr{Q}'_S|u] \tag{7.6.1}$$

whenever $\mathscr{Q}'_S$ is a closed and nonempty subset of $\mathscr{Q}_S$.

It is also worth noting that the inclusion relationship

$$\mathscr{A}_{\mathscr{C}}[\mathscr{Q}'_S|u] \subset \mathscr{B}[\mathscr{Q}'_S|u] \tag{7.6.2}$$

between the cardinally admissible and Bayes strategy sets continues to obtain *when $\mathscr{Q}'_S$ is convex* as well as closed and nonempty. But Figure 7.7(d) shows that some strategies in $\mathscr{A}_{\mathscr{C}}[\mathscr{Q}'_S|u]$ may be optimal only against some P_Σ with zero components. That is not bad. What is quite horrid, however, is that if $\mathscr{Q}'_S$ is *nonconvex*, then (7.6.2) may not be true at all; see Figure 7.16.

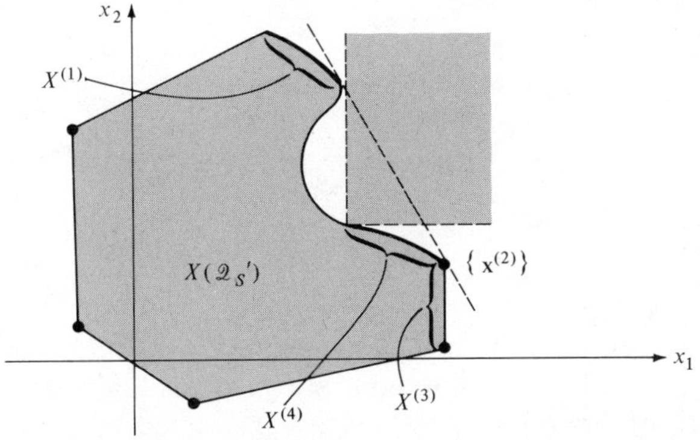

$P_S \in \mathscr{A}_{\mathscr{C}}[\mathscr{Q}'_S|u]$ if and only if $\mathbf{x}(P_S) \in X^{(1)} \cup \{\mathbf{x}^{(2)}\} \cup X^{(4)}$
$P_S \in \mathscr{B}[\mathscr{Q}'_S|u]$ if and only if $\mathbf{x}(P_S) \in X^{(1)} \cup \{\mathbf{x}^{(2)}\} \cup X^{(3)}$
$X^{(4)} = \{\mathbf{x}(P_S)$: P_S cardinally admissible but not Bayes$\}$.

Figure 7.16 Discomforts of nonconvexity of $\mathscr{Q}'_S$.

In this section we have noted some of the effects of constraints on $\mathscr{D}$'s choice of strategy, particularly as regards $\mathscr{D}$'s utility of choosing a "good" strategy. Most of the results should have been obvious to the educated intuition of anyone who has assiduously worked through Section 7.3, but their importance warrants special emphasis. They are significantly applied in Chapter 8, on the monetary evaluation of decision opportunities and of information, and again in Chapter 11 on risk sharing.

Exercise 7.6.1

Find $\mathscr{A}_{\mathscr{C}}[\mathscr{Q}'_S|u]$ and $\mathscr{B}[\mathscr{Q}'_S|u]$ for the dpuu in Examples 7.3.1, 7.6.1, and 7.6.2, and in Figures 7.5 and 7.15, for:

(A) $\mathscr{Q}'_S = \{P_S: P_S \in \mathscr{Q}_S, P_S(s^3) = 0\}$;
(B) $\mathscr{Q}'_S = \{P_S: P_S \in \mathscr{Q}_S, P_S(s^5) = 0\}$;
(C) $\mathscr{Q}'_S = \{P_S: P_S \in \mathscr{Q}_S, P_S(s^3) = P_S(s^5) = 0\}$;
(D) $\mathscr{Q}'_S = \{P_S: P_S \in \mathscr{Q}_S, P_S(s^1) = P_S(s^5) = 0\}$;
(E) $\mathscr{Q}'_S = \{P_S: P_S \in \mathscr{Q}_S, P_S(s^2) \leq .40\}$; and
* (F) $\mathscr{Q}'_S = \{P_S: P_S \in \mathscr{Q}_S, P_S(s^1) + P_S(s^5) \leq \frac{1}{2}\}$.

[*Hint* for (F): For each $s^i \notin \{s^1, s^5\}$, find the line segment $L(s^i, \frac{1}{2}) = \{(1-\frac{1}{2})\mathbf{x}(s^i) + p\mathbf{x}(s^1) + (\frac{1}{2} - p)\mathbf{x}(s^5): 0 \leq p \leq \frac{1}{2}\}$ and note on which side of this segment the utility characteristics will lie if the $\frac{1}{2}$'s in this definition are replaced by something less than $\frac{1}{2}$. Doing this for s^2, s^3, s^4, and s^6 will demarcate $X(\mathscr{Q}'_S)$.]

Exercise 7.6.2

Find $\mathscr{A}_{\mathscr{C}}[\mathscr{Q}'_S|u]$ and $\mathscr{B}[\mathscr{Q}'_S|u]$ for the dpuu in Example 7.3.4 and Figure 7.6 for:

(A) $\mathscr{Q}'_S = \{P_S: P_S \in \mathscr{Q}_S, P_S(s^5) = 0\}$; and
(B) $\mathscr{Q}'_S = \{P_S: P_S \in \mathscr{Q}_S, P_S(s^1) = 0\}$.

Exercise 7.6.3

Derive (7.6.1).

***Exercise 7.6.4: Maximin Again**

The maxmin approach was introduced in *Exercises 7.3.8–7.3.10. By a *constrained maximin strategy in* $\mathscr{Q}'_S$ we mean a strategy $P_S^\dagger$ in $\mathscr{Q}'_S$ such that $U(P_S^\dagger, \mathscr{Q}_\Sigma) = U(\mathscr{Q}'_S, \mathscr{Q}_\Sigma)$, where these U-notations are as defined in *Exercise 7.3.8. For the dpuu of Exercise 7.6.1, find a constrained maximin strategy in $\mathscr{Q}'_S$ if $\mathscr{Q}'_S$ is as specified in:

(A) Exercise 7.6.1(A)
(B) Exercise 7.6.1(B)
(C) Exercise 7.6.1(C)
(D) Exercise 7.6.1(D)
(E) Exercise 7.6.1(E)
(F) Exercise 7.6.1(F)

***Exercise 7.6.5: More Maximin**

Suppose that $\mathcal{Q}'_S$ and $\mathcal{Q}''_S$ are nonempty and closed subsets of $\mathcal{Q}_S$, with $\mathcal{Q}'_S \subset \mathcal{Q}''_S$. Show that:

(A) $U(\mathcal{Q}'_S, \mathcal{Q}_\Sigma) \leq U(\mathcal{Q}''_S, \mathcal{Q}_\Sigma)$.
(B) There is a constrained maximin strategy in $\mathcal{Q}'_S$ which is in $\mathscr{A}_{\mathscr{E}}[\mathcal{Q}'_S|u]$.
(C) If $\mathcal{Q}'_S$ is convex as well as nonempty and closed, then a constrained maximin strategy in $\mathcal{Q}'_S$ is also a Bayes strategy, in $\mathscr{B}[\mathcal{Q}'_S|u]$; that is, $\mathscr{E}$ has a "least favorable" strategy $P^\dagger_\Sigma$ such that $U(\mathcal{Q}'_S, \mathcal{Q}_\Sigma) = U(\mathcal{Q}'_S, P^\dagger_\Sigma)$.

*■ That (C) may not be true if $\mathcal{Q}'_S$ is nonconvex is demonstrated graphically in Figure 7.17. ■

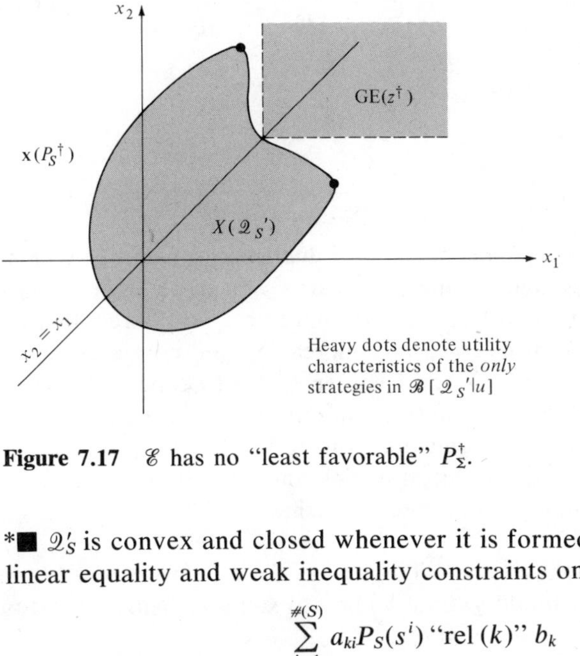

Figure 7.17 $\mathscr{E}$ has no "least favorable" $P^\dagger_\Sigma$.

*■ $\mathcal{Q}'_S$ is convex and closed whenever it is formed from $\mathcal{Q}_S$ by imposing linear equality and weak inequality constraints on P_S, of the form

$$\sum_{i=1}^{\#(S)} a_{ki} P_S(s^i) \text{ "rel}\,(k)\text{" } b_k$$

for $k = 1, \ldots, K$; where the b_k's and all a_{ki}'s are real numbers, and, for each k, "rel (k)" is one of "$\geq$", "$=$", and "$\leq$". [In Example 7.6.2, $K = 1$, "rel (1)" = "$\geq$", $b_1 = \frac{1}{2}$, $a_{14} = 1$, and $a_{11} = a_{12} = a_{13} = a_{15} = a_{16} = 0$.] But $\mathcal{Q}'_S$ will be empty if there is no P_S which simultaneously satisfies all K constraints, and hence nonemptiness of $\mathcal{Q}'_S$ always must be verified by producing a P_S which does satisfy all K constraints. ■

8

MONETARY EVALUATIONS OF OPPORTUNITIES AND INFORMATION

Invoke then the master spirit of the earth, to come into your midst and sanctify the scales and the reckoning that weighs value against value.

Kahlil Gibran

8.1 INTRODUCTION

This chapter concerns the special class of decision problems in which (1) all potential consequences are monetary returns[a] or can be made so by use of a substitute function, and (2) $\mathscr{D}$'s utility function u on the real line R^1 of all potential monetary returns is strictly increasing and continuous.[b]

We already know from Part I and Chapter 7 how to cope with the most general dpuu's, and therefore our attention here will be focused on problems different from the preceding ones. In this chapter, we shall develop *monetary* evaluations for strategies and for sets of strategies, in order to answer questions such as the following.

(1) How much should you be willing to pay for a decision opportunity?
(2) For how much should you be willing to sell a decision opportunity you own?
(3) How much should you be willing to pay in order to effect some changes in a decision opportunity?
(4) Less generally, how much should you be willing to pay for information which promises to reduce your uncertainty about $\mathscr{E}$'s choices (in $\mathscr{E}$-moves or of strategies)?

In order to address these questions properly, it is necessary first to define

[a] More precisely, monetary returns as of the same date and accompanied by the same nonmonetary attribute.

[b] These assumptions about u are justified in Section 5.4.

what we mean by a decision opportunity, or simply *opportunity*. Roughly, we construe an opportunity as consisting of two things: a set of strategies for $\mathscr{D}$ (= you), together with their corresponding, uncertain consequences (= monetary returns).

For convenience, we shall suppose that all opportunities derive from some basic, given dpuu. This supposition entails no loss of generality in view of the fact that in practice all opportunities are analyzed in the context of the set of all alternative opportunities available at the moment of analysis. Hence the given, basic dpuu may be formed by aggregating all opportunities.

In Chapter 7 we showed how such a basic dpuu can be expressed in normal form, in terms of

(1) a set $S=\{s^1,\ldots,s^{\#(S)}\}$ of all pure strategies for $\mathscr{D}$;
(2) a set $\Sigma=\{\sigma^1,\ldots,\sigma^{\#(\Sigma)}\}$ of all pure strategies for $\mathscr{E}$; and
(3) a function $c\colon S\times\Sigma\to R^1$ such that $c[s,\sigma]$ denotes $\mathscr{D}$'s monetary return (as of the given date, with the given nonmonetary concomitant) from choosing his pure strategy s and $\mathscr{E}$'s choosing his pure strategy σ.

■ In this chapter we shall denote the *function* $c\colon S\times\Sigma\to R^1$ by $\mathbf{c}$ rather than by the more standard but space-consuming $c[\cdot,\cdot]$. ■

In addition, we assume that $\mathscr{D}$ has quantified his preferences and judgments by specifying

(4) a strictly increasing and continuous utility function $u\colon R^1\to R^1$ for monetary returns c (as of the given date, etc.); and
(5) a randomized strategy $P_\Sigma=(P_\Sigma(\sigma^1),\ldots,P_\Sigma(\sigma^{\#(\Sigma)}))$ for $\mathscr{E}$, with $P_\Sigma(\sigma^j)$ representing $\mathscr{D}$'s judgments regarding the event that $\tilde{\sigma}=\sigma^j$.

■ As in Chapter 7, we are assuming that the dpuu is finite: $\#(S)<+\infty$ and $\#(\Sigma)<+\infty$. This assumption is brutally realistic but can be relaxed in many classes of problems without changing the qualitative results from the ensuing analysis. But here, in contrast to Chapter 7, we shall keep P_Σ fixed, and hence we shall not study the ramifications of varying it. ■

Our precise Definition 8.1.1 of an opportunity may seem rather strange at first, even with the preceding discussion, since it ignores specifications (4) and (5) above, which are specific to $\mathscr{D}$ and thus subjective rather than natural parts of an "opportunity," but it also ignores the given consequence function (3) of the basic dpuu! Ignoring the given consequence function $\mathbf{c}$ makes sense once we build a consequence function into the definition of an opportunity. Flexibility in redefining consequence functions will prove useful in defining opportunities whose evaluations have real economic significance to $\mathscr{D}$. One such instance constitutes Example 8.1.2.

Definition 8.1.1. Given the pure-strategy sets S and Σ of the basic dpuu, an *opportunity* is defined to be a *pair* $(\mathcal{Q}'_S, \mathbf{c}')$ consisting of (1) a nonempty and closed subset $\mathcal{Q}'_S$ of the set $\mathcal{Q}_S$ of all randomized strategies for $\mathcal{D}$ and (2) a monetary-return consequence function $\mathbf{c}' = c'$: $S \times \Sigma \to R^1$.

■ Assuming that $\mathcal{Q}'_S$ is a closed set if convenient in the ensuing analysis and not at all restrictive in practice, all the important sets of the form

$$\mathcal{Q}_S^{S'} = \{P_S: P_S(S') = 1\}$$

are closed, and convex to boot. ■

Example 8.1.1

In Example 1.5.1c, the sets of pure strategies of the basic dpuu may be taken as $S = \{s^1, \ldots, s^{74}\}$ and $\Sigma = \{\sigma^1, \ldots, \sigma^8\}$—see Example 7.2.3. Moreover, we take the consequences to be monetary returns accompanied by MP in every case; we assumed that $(\$c_1; \text{LP}) \sim (\$c_1 - 50(000); \text{MP})$ for every c_1, as indicated in Example 5.2.1. One opportunity here is to accept a consideration of \$30(000) for not performing any test. This opportunity is representable by $(\mathcal{Q}'_S, \mathbf{c}')$ for $\mathcal{Q}'_S = \mathcal{Q}_S^{\{s^1, s^2\}}$ and $c'[s, \sigma] = c[s, \sigma] + 30(000)$ for every (s, σ) in $S \times \Sigma$.

Example 8.1.2

Another opportunity in the context of Example 1.5.1c might be to get out of the whole dpuu. If work has yet to commence and no costs have been irrevocably sustained to date, getting out means forfeiting all chances to make decisions and to reap gains or losses. Now, it is not clear how to reconcile this opportunity with Definition 8.1.1, because none of s^1 through s^{74} lets $\mathcal{D}$ off the hook of having to make delivery of the completed computer. But since $\mathcal{D}$'s utility of this opportunity is clearly $u(\$0)$, it follows that its *desirability* is identical to that of an opportunity $(\mathcal{Q}'_S, \mathbf{c}')$ in which $\mathcal{Q}'_S$ is *any* nonempty subset of $\mathcal{Q}_S$ and $c'[s, \sigma] = \$0$ for every (s, σ) in $S \times \Sigma$, since $\mathcal{D}$'s utility of any such opportunity is obviously $u(\$0)$. Because any such $(\mathcal{Q}'_S, \mathbf{c}')$ is equally as desirable as the opportunity of interest, we can substitute $(\mathcal{Q}'_S, \mathbf{c}')$ for the opportunity of interest throughout the ensuing analysis.

■ Example 8.1.2 illustrates the desirability of using a little ingenuity in defining $\mathbf{c}'$ in order to avoid having to enrich the definition of the basic dpuu. ■

Since we shall vary $\mathbf{c}'$ and $\mathcal{Q}'_S$ but not P_Σ in what follows, we shall introduce some convenient notation for $\mathcal{D}$'s utility of an opportunity.

Definition 8.1.2. $\mathcal{D}$'s utility of any opportunity $(\mathcal{Q}'_S, \mathbf{c}')$ is denoted by $\mathcal{U}(\mathcal{Q}'_S, \mathbf{c}')$; that is,

$$\mathcal{U}(\mathcal{Q}'_S, \mathbf{c}) = \max_{P_S \in \mathcal{Q}'_S} \left[\sum_{i=1}^{\#(S)} \sum_{j=1}^{\#(\Sigma)} u(c'[s^i, \sigma^j]) P_\Sigma(\sigma^j) P_S(s^i) \right].$$

■ The assumption that $\mathcal{Q}'_S$ is a *closed* subset of $\mathcal{Q}_S$ ensures that the maximum in Definition 8.1.2 exists (and is therefore attained by some strategy P°_S in $\mathcal{Q}'_S$). ■

One more bit of notation. In many of the following cases we shall vary an opportunity $(\mathcal{Q}'_S, \mathbf{c}')$ by adding a nonrandom return \$$x$ to every possible return $c'[s, \sigma]$ in the opportunity. It is then logical to denote the resulting opportunity by $(\mathcal{Q}'_S, \mathbf{c}' + x)$, so as to avoid the pedantry of "$(\mathcal{Q}''_S, \mathbf{c}'')$ where $\mathcal{Q}''_S = \mathcal{Q}'_S$ and $c''[s, \sigma] = c'[s, \sigma] + x$ for every (s, σ) in $S \times \Sigma$."

Definition 8.1.3. For any $x \in R^1$, we denote by $\mathbf{c}' + x$ the return function $\mathbf{c}''$ such that $c''[s, \sigma] = c'[s, \sigma] + x$ for every (s, σ) in $S \times \Sigma$.

In Section 8.2 we derive some general properties of $\mathcal{D}$'s utility $\mathcal{U}(\mathcal{Q}'_S, \mathbf{c}')$ of an opportunity $(\mathcal{Q}'_S, \mathbf{c}')$. These properties are of some academic interest in and of themselves, but their major role is to act as stepping-stones to the operationally more important results in later sections. Readers willing to take on faith certain assertions of a technical nature later on may skim or even omit Section 8.2 altogether.

Section 8.3 concerns the *evaluation* of opportunities and hence provides the analytical framework for answering the sorts of questions posed at the beginning of this section.

Sections 8.4 and 8.5 constitute a brief introduction to statistical decision theory and the evaluation of information. Readers who wish to pursue this important subject should consult works devoted (more or less) exclusively to it; e.g., Pratt, Raiffa, and Schlaifer [**112**], LaValle [**78**], Ferguson [**34**], and Weiss [**140**]. These books make much use of the probability models developed in Part C of Chapter 3 (and many other models as well), and they list many additional references.

Section 8.4 is an introduction to the statistical decision problem. It indicates clearly the sense in which statistical decision problems constitute a small (but important) subset of the set of all dpuu's. Section 8.5 introduces the theory of information evaluation in statistical decision problems.

Sections 8.4 and 8.5 may be read independently of Sections 8.2 and 8.3 by those willing to accept on faith some proofs and other indented remarks of a technical nature.

8.2 PROPERTIES OF $\mathcal{U}(\mathcal{Q}'_S, \mathbf{c}')$

This section consists of a number of results demonstrating the reasonable behavior of $\mathcal{D}$'s utility $\mathcal{U}(\mathcal{Q}'_S, \mathbf{c}')$ of an opportunity $(\mathcal{Q}'_S, \mathbf{c}')$ as a function of that opportunity.

Theorem 8.2.1 shows that $\mathcal{U}(\mathcal{Q}'_S, \mathbf{c}')$ never decreases when $\mathbf{c}'$ is "improved on."

Theorem 8.2.1. If $c'[s, \sigma] \leq c^*[s, \sigma]$ for every (s, σ) such that

$P_\Sigma(\sigma)P_S(s) > 0$ for some P_S in $\mathcal{Q}'_S$, then

$$\mathcal{U}(\mathcal{Q}'_S, \mathbf{c}') \leq \mathcal{U}(\mathcal{Q}'_S, \mathbf{c}^*).$$

■ *Proof.* Let P_S° be any optimal strategy in $(\mathcal{Q}'_S, \mathbf{c}')$. Then

$$\begin{aligned}\mathcal{U}(\mathcal{Q}'_S, \mathbf{c}') &=^1 \sum_{i=1}^{\#(S)} \sum_{j=1}^{\#(\Sigma)} u(c'[s^i, \sigma^j]) P_\Sigma(\sigma^j) P_S^\circ(s^i) \\ &\leq^2 \sum_{i=1}^{\#(S)} \sum_{j=1}^{\#(\Sigma)} u(c^*[s^i, \sigma^j]) P_\Sigma(\sigma^j) P_S^\circ(s^i) \\ &\leq^3 \mathcal{U}(\mathcal{Q}'_S, \mathbf{c}^*),\end{aligned}$$

from which Theorem 8.2.1 follows readily. Equality 1 is immediate from Definition 8.1.2 and the assumed optimality of P_S° in $(\mathcal{Q}'_S, \mathbf{c}')$. Inequality 2 follows because the hypothesis implies that $c^*[s^i, \sigma^j] \geq c'[s^i, \sigma^j]$ for every i and j such that $P_\Sigma(\sigma^j)P_S^\circ(s^i) > 0$, and hence that $u(c^*[s^i, \sigma^j]) \geq u(c'[s^i, \sigma^j])$ for every such i and j because $u(\cdot)$ is strictly increasing. Inequality 3 is an immediate consequence of the observation that P_S° is available, but not necessarily optimal, in the opportunity $(\mathcal{Q}'_S, \mathbf{c}^*)$. ■

Various corollaries follow readily from Theorem 8.2.1 and its proof. The most obvious is Corollary 8.2.1, to the effect that how one defines $\mathbf{c}'$ for pure-strategy pairs (s, σ) which *cannot* arise from $(\mathcal{Q}'_S, \mathbf{c}')$—i.e., pairs such that $P_\Sigma(\sigma)P_S(s) = 0$ for every P_S in $\mathcal{Q}'_S$—is immaterial.

Corollary 8.2.1. If $c'[s, \sigma] = c^*[s, \sigma]$ for every (s, σ) such that $P_\Sigma(\sigma)P_S(s) > 0$ for some P_S in $\mathcal{Q}'_S$, then

$$\mathcal{U}(\mathcal{Q}'_S, \mathbf{c}') = \mathcal{U}(\mathcal{Q}'_S, \mathbf{c}^*).$$

■ *Proof.* Obvious. ■

Corollary 8.2.2 specifies about the weakest possible set of assumptions under which a strict inequality along these lines can be deduced.

Corollary 8.2.2. Suppose that there is a strategy P_S° optimal for $\mathcal{D}$ in the opportunity $(\mathcal{Q}'_S, \mathbf{c}')$ with the properties: (1) that $c'[s, \sigma] < c^*[s, \sigma]$ for *some* (s, σ) such that $P_\Sigma(\sigma)P_S^\circ(s) > 0$; and (2) that $c'[s, \sigma] \leq c^*[s, \sigma]$ for *all* (s, σ) such that $P_\Sigma(\sigma)P_S^\circ(s) > 0$. Then

$$\mathcal{U}(\mathcal{Q}'_S, \mathbf{c}') < \mathcal{U}(\mathcal{Q}'_S, \mathbf{c}^*).$$

■ *Proof.* It is easy to check that the hypothesis minimally suffices to make inequality 2 in the proof of Theorem 8.2.1 strict, when applied in conjunction with strict increasingness of the utility function u. ■

Theorem 8.2.2 furnishes information about the behavior of $\mathcal{U}(\mathcal{Q}'_S, \mathbf{c}')$ as $\mathcal{Q}'_S$ is "enriched."

Theorem 8.2.2. If $\mathcal{Q}'_S \subset \mathcal{Q}''_S$, then

(A) $\mathcal{U}(\mathcal{Q}'_S, \mathbf{c}') \leq \mathcal{U}(\mathcal{Q}''_S, \mathbf{c}')$; and

(B) $\mathscr{U}(\mathscr{Q}'_S, \mathbf{c}') < \mathscr{U}(\mathscr{Q}''_S, \mathbf{c}')$ if and only if no strategy P°_S optimal for $\mathscr{D}$ in the opportunity $(\mathscr{Q}'_S, \mathbf{c}')$ is also optimal for $\mathscr{D}$ in the opportunity $(\mathscr{Q}''_S, \mathbf{c}')$.

■ *Proof.* Both assertions follow readily from Definition 8.1.2 and the fact that $\max_{x \in X}[t(x)] \le \max_{x \in X'}[t(x)]$ whenever $X \subset X'$, with strict inequality obtaining whenever no maximizer in X of $t(\cdot)$ is also a maximizer in X' of $t(\cdot)$. ■

Various corollaries, such as Corollary 8.2.3, are readily obtained by combining the hypotheses of Theorem 8.2.2 with those of Theorem 8.2.1 or its corollaries.

Corollary 8.2.3. If $\mathscr{Q}'_S \subset \mathscr{Q}''_S$ and if there is a strategy P°_S optimal for $\mathscr{D}$ in $(\mathscr{Q}'_S, \mathbf{c}')$ with the property that $c'[s, \sigma] \le c''[s, \sigma]$ for every (s, σ) such that $P_\Sigma(\sigma)P^\circ_S(s) > 0$, then

$$\mathscr{U}(\mathscr{Q}'_S, \mathbf{c}') \le \mathscr{U}(\mathscr{Q}''_S, \mathbf{c}'').$$

■ *Proof.* Exercise 8.2.3. ■

Theorem 8.2.3 is the first of two important results showing that $\mathscr{U}(\mathscr{Q}'_S, \mathbf{c}' + x)$ behaves very much in accord with raw intuition as a function of the "sure increment" x to $\mathscr{D}$'s net monetary return function $\mathbf{c}'$.

Theorem 8.2.3. As a function of x, $\mathscr{D}$'s utility $\mathscr{U}(\mathscr{Q}'_S, \mathbf{c}' + x)$ of the variable opportunity $(\mathscr{Q}'_S, \mathbf{c}' + x)$ is
(A) strictly increasing, and
(B) continuous.

■ *Proof*: *Of (A).* Let P_S^1 be any strategy optimal for $\mathscr{D}$ in the opportunity $(\mathscr{Q}'_S, \mathbf{c}' + x^1)$, and let $x^2 > x^1$. Then

$$\begin{aligned}\mathscr{U}(\mathscr{Q}'_S, \mathbf{c}' + x^1) &= \sum_{i=1}^{\#(S)} \sum_{j=1}^{\#(\Sigma)} u(c'[s^i, \sigma^j] + x^1) P_\Sigma(\sigma^j) P_S^1(s^i) \\ &< \sum_{i=1}^{\#(S)} \sum_{j=1}^{\#(\Sigma)} u(c'[s^i, \sigma^j] + x^2) P_\Sigma(\sigma^j) P_S^1(s^i) \\ &\le \mathscr{U}(\mathscr{Q}'_S, \mathbf{c}' + x^2),\end{aligned}$$

which implies that $\mathscr{U}(\mathscr{Q}'_S, \mathbf{c}' + x^1) < \mathscr{U}(\mathscr{Q}'_S, \mathbf{c}' + x^2)$ if $x^1 < x^2$, as asserted. The weak inequality "$\le$" obtains above because P_S^1 need not be optimal for $\mathscr{D}$ in the opportunity $(\mathscr{Q}'_S, \mathbf{c}' + x^2)$. The strict inequality "$<$" is an easy consequence of strict increasingness of $u(\cdot)$ and the fact that $c'[s, \sigma] + x^1 < c'[s, \sigma] + x^2$ for every s and σ whenever $x^1 < x^2$.

Of (B). For any real number x and any $e > 0$, continuity of $u(\cdot)$ implies the existence of a number $d_{ij}(e, x) > 0$ such that $u(c'[s^i, \sigma^j] + x^*)$ is within e of $u(c'[s^i, \sigma^j] + x)$ whenever x^* is within $d_{ij}(e, x)$ of x. Let $d(e, x) = \min_{ij}[d_{ij}(e, x)]$. Then $u(c'[s^i, \sigma^j] + x^*)$ is within e of its (i, j)th counterpart $u(c'[s, \sigma^j] + x)$ whenever x^* is within $d(e, x)$ of x; and hence it is clear that, for any given P_Σ,

$$\sum_{i=1}^{\#(S)} \sum_{j=1}^{\#(\Sigma)} u(c'[s^i, \sigma^j] + x^*) P_\Sigma(\sigma^j) P_S(s^i)$$

is within e of

$$\sum_{i=1}^{\#(S)} \sum_{j=1}^{\#(\Sigma)} u(c'[s^i, \sigma^j] + x) P_\Sigma(\sigma^j) P_S(s^i)$$

whenever x^* is within $d(e, x)$ of x, *for every* P_S *in* $\mathcal{Q}'_S$. Although (B) should now be obvious, we proceed rigorously by assuming that $x - d(e, x) < x^* < x$, and denoting by P_S^x any strategy optimal for $\mathcal{D}$ in $(\mathcal{Q}'_S, \mathbf{c}' + x)$. We have shown that

$$\begin{aligned}\mathcal{U}(\mathcal{Q}'_S, \mathbf{c}' + x) - e &= \sum_{i=1}^{\#(S)} \sum_{j=1}^{\#(\Sigma)} u(c'[s^i, \sigma^j] + x) P_\Sigma(\sigma^j) P_S^x(s^i) - e \\ &<^1 \sum_{i=1}^{\#(S)} \sum_{j=1}^{\#(\Sigma)} u(c'[s^i, \sigma^j] + x^*) P_\Sigma(\sigma^j) P_S^x(s^i) \\ &\leq^2 \mathcal{U}(\mathcal{Q}'_S, \mathbf{c}' + x^*) \\ &<^3 \mathcal{U}(\mathcal{Q}'_S, \mathbf{c}' + x),\end{aligned}$$

inequality 1 following by the preceding reasoning; 2, because P_S^x may not be optimal for $\mathcal{D}$ in the opportunity $(\mathcal{Q}'_S, \mathbf{c}' + x^*)$; and 3, by assertion (A). Hence $\mathcal{U}(\mathcal{Q}'_S, \mathbf{c}' + x^*)$ is less than e below $\mathcal{U}(\mathcal{Q}'_S, \mathbf{c}' + x)$ whenever x^* is less than $d(e, x)$ below x. Hence decreasing x never causes a precipitous tumble of $\mathcal{U}(\mathcal{Q}'_S, \mathbf{c}' + x)$. Since this is true for every x, assertion (B) obtains. ■

*■ In infinite-Σ cases, some restrictions are necessary for (A) and (B). For (A), it suffices to assume existence of optimal strategies P_S^x in $(\mathcal{Q}'_S, \mathbf{c}' + x)$ for all real numbers x. For (B), it suffices (and is not unrealistic) to assume that c': $S \times \Sigma \to R^1$ is bounded, so that a proof invoking uniform continuity of $u(\cdot)$ on compact intervals can be constructed. ■

Although Theorem 8.2.3 tells us that $\mathcal{U}(\mathcal{Q}'_S, \mathbf{c}' + x)$ is a strictly increasing and continuous function of x, it does *not* tell us that the range $\{\mathcal{U}(\mathcal{Q}'_S, \mathbf{c}' + x): -\infty < x < +\infty\}$ of this function coincides with the range $\{u(y): -\infty < y < +\infty\}$ of $u(\cdot)$. That this coincidence does obtain is not only of academic interest, but also of great importance in our subsequent developments. Theorem 8.2.4 phrases this fact in a useful manner.

Theorem 8.2.4. Let $(\mathcal{Q}'_S, \mathbf{c}')$ be fixed. Then

(A) for every real number y there exists a unique real number x_y such that $\mathcal{U}(\mathcal{Q}'_S, \mathbf{c}' + x_y) = u(y)$; and

(B) for every real number x there exists a unique real number y_x such that $\mathcal{U}(\mathcal{Q}'_S, \mathbf{c}' + x) = u(y_x)$.

■ *Proof.* (B) is easy. Because $\mathcal{U}(\mathcal{Q}'_S, \mathbf{c}' + x)$ is a weighted average of values $u(c'[s^i, \sigma^j] + x)$, it is bracketed by the u-values

$u(\min_{ij}[c'[s^i, \sigma^j]] + x) = u_*$ and $u(\max_{ij}[c'[s^i, \sigma^j]] + x) = u^*$, below and above respectively. But since $u(\cdot)$ is strictly increasing and continuous, $u(y)$ assumes each value in $[u_*, u^*]$ *exactly once* as y goes up from $\min_{ij}[c'[s^i, \sigma^j]] + x$ to $\max_{ij}[c'[s^i, \sigma^j]] + x$. Thus the asserted y_x exists and is unique. The only complication in proving (A) is showing that x's can be chosen so that $u(y)$ is bracketed in the range of $\mathcal{U}$; from there on, proof of (A) is similar to that of (B) by virtue of the strict increasingness and continuity of $\mathcal{U}(\mathcal{Q}'_S, \mathbf{c}' + x)$ in x, given in Theorem 8.2.3. Let x^* be any number exceeding $y\text{-}\min_{ij}[c'[s^i, \sigma^j]]$. Then $c'[s^i, \sigma^j] + x^* \geq \min_{ij}[c'[s^i, \sigma^j]] + x^* > y$ for every (s^i, σ^j), so that $u(c'[s^i, \sigma^j] + x^*) > u(y)$ for every (s^i, σ^j), which implies that $\mathcal{U}(\mathcal{Q}'_S, \mathbf{c}' + x^*) > u(y)$, obviously. Similarly, let x_* be any number less than $y\text{-}\max_{ij}[c'[s^i, \sigma^j]]$. Then $c'[s^i, \sigma^j] + x_* \leq \max_{ij}[c'[s^i, \sigma^j]] + x_* < y$ for every (s^i, σ^j), which implies that $\mathcal{U}(\mathcal{Q}'_S, \mathbf{c}' + x_*) < u(y)$, obviously. Hence $u(y)$ is indeed bracketed in the range of the function $\mathcal{U}(\mathcal{Q}'_S, \mathbf{c}' + x)$ of x. ■

The final theorem of this section furnishes some very useful results about $(\mathcal{Q}'_S, \mathbf{c}')$ for the two constant-risk-aversion utility functions for monetary returns introduced in Examples 5.4.1 and 5.4.2; namely, the *linear* utility function $u(c) = c$ for every c, and the *concave-exponential* utility function $u(c) = -e^{-zc}$ for every c and some $z > 0$.

Theorem 8.2.5. Let $(\mathcal{Q}'_S, \mathbf{c}')$ be fixed. If x is any real number, and if $\mathbf{y} = y\colon S \times \Sigma \to R^1$ is a real-valued function on $S \times \Sigma$ which does not depend upon s, then

(A) if $u(c) = c$ for every c in R^1, then

$$\mathcal{U}(\mathcal{Q}'_S, \mathbf{c}' + \mathbf{y}) = \mathcal{U}(\mathcal{Q}'_S, \mathbf{c}') + \sum_{j=1}^{\#(\Sigma)} y[*, \sigma^j] P_\Sigma(\sigma^j);$$

and

(B) if $u(c) = -e^{-zc}$ for every c in R^1 and some $z > 0$, then

$$\mathcal{U}(\mathcal{Q}'_S, \mathbf{c}' + x) = e^{-zx}\mathcal{U}(\mathcal{Q}'_S, \mathbf{c}').$$

■ *Proof.* Exercise 8.2.4. ■

■ Note that we replaced s in $y[s, \sigma^j]$ by $*$ to emphasize the hypothesis that $y[\cdot, \cdot] = \mathbf{y}$ does not depend upon its first argument; that is, that $y[s', \sigma] = y[s'', \sigma]$ for every s' and s'' in S and every σ in Σ. In that vein, a "sure increment" x can be regarded as a function $\mathbf{x} = x\colon S \times \Sigma \to R^1$ which depends upon neither s nor σ. (Why?) ■

■ If we call an increment x or $\mathbf{y}$ to $\mathbf{c}'$ "strategically neutral" provided that it does not depend upon s, then assertion (A) of Theorem 8.2.5 shows how *any* strategically neutral increment $\mathbf{y}$ to $\mathbf{c}'$ may be "isolated," or separated out, of $\mathcal{D}$'s utility of optimally acting in the opportunity $(\mathcal{Q}'_S, \mathbf{c}')$, when his utility is linear. By contrast, assertion (B) shows only how *sure* strategically neutral increments x may be isolated when $\mathcal{D}$'s utility is concave exponential. ■

Exercise 8.2.1: Computational Opportunity

Consider the following dpuu in normal form, in which the table entries denote monetary returns $c[s, \sigma]$.

	σ^1	σ^2
s^1	10	0
s^2	0	10
s^3	9	2
s^4	4	7

(A) Suppose that $P_\Sigma = (.4, .6)$ and that $u(c) = -(25-c)^2$ for every $c \leq 25$. Recall Definition 7.1.4 of $\mathcal{Q}_S{}^{S'}$; and find x_0 such that $\mathcal{U}(\mathcal{Q}_S{}^{S'}, \mathbf{c} + x_0) = u(0)$ for

(i) $S' = \{s^1\}$
(ii) $S' = \{s^2\}$
(iii) $S' = \{s^3\}$
(iv) $S' = \{s^4\}$
(v) $S' = \{s^1, s^2\}$
(vi) $S' = \{s^1, s^3\}$
(vii) $S' = \{s^1, s^4\}$
(viii) $S' = \{s^2, s^3\}$
(ix) $S' = \{s^2, s^4\}$
(x) $S' = \{s^3, s^4\}$
(xi) $S' = \{s^1, s^2, s^3\}$
(xii) $S' = \{s^1, s^2, s^4\}$
(xiii) $S' = \{s^1, s^3, s^4\}$
(xiv) $S' = \{s^2, s^3, s^4\}$
(xv) $S' = S$

(B) Redo (A) under the assumption that $P_\Sigma = (.4, .6)$, as before, but that $u(c) = c$ for every c in R^1.

(C) Redo (A) under the assumption that $P_\Sigma = (.4, .6)$, as before, but that $u(c) = -e^{-c}$ for every c in R^1.

(D) Suppose that $P_\Sigma = (.4, .6)$ and that $u(c) = -(25-c)^2$ for every $c \leq 25$, as in (A). Find y_0 such that $\mathcal{U}(\mathcal{Q}_S{}^{S'}, \mathbf{c}) = u(y_0)$ for

(i) $S' = \{s^1\}$
(ii) $S' = \{s^2\}$
(iii) $S' = \{s^3\}$
(iv) $S' = \{s^4\}$
(v) $S' = \{s^1, s^2\}$
(vi) $S' = \{s^1, s^3\}$
(vii) $S' = \{s^1, s^4\}$
(viii) $S' = \{s^2, s^3\}$
(ix) $S' = \{s^2, s^4\}$
(x) $S' = \{s^3, s^4\}$
(xi) $S' = \{s^1, s^2, s^3\}$
(xii) $S' = \{s^1, s^2, s^4\}$
(xiii) $S' = \{s^1, s^3, s^4\}$
(xiv) $S' = \{s^2, s^3, s^4\}$
(xv) $S' = S$

(E) Redo (D) under the assumptions that $P_\Sigma = (.4, .6)$, as before, but that $u(c) = c$ for every c in R^1.

(F) Redo (D) under the assumptions that $P_\Sigma = (.4, .6)$, as before, but that $u(c) = -e^{-c}$ for every c in R^1.

(G) Can you infer anything about a relationship between x_0 and y_0?

Exercise 8.2.2: Computational Opportunity

Suppose that $u(c) = -e^{-c} + .15c$ for every c in R^1; this utility function has decreasing risk aversion (defined in Example 5.4.3). Consider the following dpuu in normal form with table entries being $c[s, \sigma]$'s.

	σ^1	σ^2	σ^3	σ^4
s^1	10	0	0	0
s^2	0	12	0	0
s^3	0	0	9	0
s^4	0	0	2	15
s^5	8	4	1	−3
s^6	−10	25	0	10

Assume that $P_\Sigma = (.25, .10, .15, .50)$.

(A) Find y_0, correct to two decimal places, such that $\mathcal{U}(\mathcal{Q}_S, \mathbf{c}) = u(y_0)$.
(B) Find x_0, correct to two decimal places, such that $\mathcal{U}(\mathcal{Q}_S, \mathbf{c} + x_0) = u(0)$.
[*Hints*: (1) In (A), write $r(y) = u(y) - \mathcal{U}(\mathcal{Q}_S, \mathbf{c})$, so that $r(y) = 0$ for $y = y_0$. Use Appendix 4 to find y_0. Note that y_0 must belong to $[0, 25]$. (Why?)
(2) In (B), write $r(x) = \mathcal{U}(\mathcal{Q}_S, \mathbf{c} + x) - u(0)$, so that $r(x) = 0$ for $x = x_0$. Use Appendix 4. Why does $x_0 \in [-25, 0]$?]
(C) Redo (A) and (B) under the assumption that $u(c) = c$ for every c in R^1.
(D) Redo (A) and (B) under the assumption that $u(c) = -e^{-c}$ for every c in R^1.

Exercise 8.2.3

Prove Corollary 8.2.3.

Exercise 8.2.4

Prove (A) Assertion (A) of Theorem 8.2.5 and (B) Assertion (B) of Theorem 8.2.5.

Exercise 8.2.5: Maximin Revisited (I)

The semipessimistically conservative, maximin choice criterion was introduced in various exercises in Chapter 7. In this framework, we generalize Definition 8.1.2 by assuming that $\mathcal{Q}'_S$ and $\mathcal{Q}'_\Sigma$ are nonempty and closed subsets of $\mathcal{Q}_S$ and $\mathcal{Q}_\Sigma$ respectively, and then by defining $\mathcal{D}$'s maximin utility evaluation of the "generalized opportunity" $\mathcal{U}(\mathcal{Q}'_S, \mathcal{Q}'_\Sigma, \mathbf{c}')$ by

$$\mathcal{U}(\mathcal{Q}'_S, \mathcal{Q}'_\Sigma, \mathbf{c}') = \max_{P_S \in \mathcal{Q}'_S}\left[\min_{P_\Sigma \in \mathcal{Q}'_\Sigma}\left[\sum_{i=1}^{\#(S)} \sum_{j=1}^{\#(\Sigma)} u(c'[s^i, \sigma^j]) P_\Sigma(\sigma^j) P_S(s^i)\right]\right].$$

In this exercise you will show, in effect, that all the principal results in this section continue to obtain for $\mathcal{U}(\mathcal{Q}'_S, \mathcal{Q}'_\Sigma, \mathbf{c}')$ with suitable modifications. Indeed, $\mathcal{U}(\mathcal{Q}'_S, \mathbf{c}')$ is the special case $\mathcal{U}(\mathcal{Q}'_S, \{P_\Sigma\}, \mathbf{c}')$ of $\mathcal{U}(\mathcal{Q}'_S, \mathcal{Q}'_\Sigma, \mathbf{c}')$.

(A) Why is this last sentence true?

(B) **Theorem 8.2.1M.** Prove that if $c'[s,\sigma]\begin{Bmatrix}<\\=\\\le\end{Bmatrix}c^*[s,\sigma]$ for every (s,σ) in $S\times\Sigma$, then

$$\mathcal{U}(\mathcal{Q}'_S, \mathcal{Q}'_\Sigma, \mathbf{c}')\begin{Bmatrix}<\\=\\\le\end{Bmatrix}\mathcal{U}(\mathcal{Q}'_S, \mathcal{Q}'_\Sigma, \mathbf{c}').$$

(C) **Theorem 8.2.2M.** Prove that if $\mathcal{Q}'_S \subset \mathcal{Q}''_S$ and $\mathcal{Q}'_\Sigma \subset \mathcal{Q}''_\Sigma$, then

(i) $$\mathcal{U}(\mathcal{Q}'_S, \mathcal{Q}'_\Sigma, \mathbf{c}') \le \mathcal{U}(\mathcal{Q}''_S, \mathcal{Q}'_\Sigma, \mathbf{c}')$$

and

(ii) $$\mathcal{U}(\mathcal{Q}'_S, \mathcal{Q}'_\Sigma, \mathbf{c}') \ge \mathcal{U}(\mathcal{Q}'_S, \mathcal{Q}''_\Sigma, \mathbf{c}').$$

(D) **Theorem 8.2.3M.** Prove that $\mathcal{U}(\mathcal{Q}'_S, \mathcal{Q}'_\Sigma, \mathbf{c}'+x)$ is a strictly increasing and continuous function of x.

(E) **Theorem 8.2.4M.** Prove that

(i) for every y in R^1 there exists a unique real number x_y such that $\mathcal{U}(\mathcal{Q}'_S, \mathcal{Q}'_\Sigma, \mathbf{c}'+x_y) = u(y)$; and

(ii) for every x in R^1 there exists a unique real number y_x such that $\mathcal{U}(\mathcal{Q}'_S, \mathcal{Q}'_\Sigma, \mathbf{c}'+x) = u(y_x)$.

(F) **Theorem 8.2.5M.** Prove that if x is any real number, then

(i) if $u(c) = c$ for every c in R^1, then

$$\mathcal{U}(\mathcal{Q}'_S, \mathcal{Q}'_\Sigma, \mathbf{c}'+x) = \mathcal{U}(\mathcal{Q}'_S, \mathcal{Q}'_\Sigma, \mathbf{c}') + x;$$

and

(ii) if $u(c) = -e^{-zc}$ for every c in R^1 and some $z > 0$, then

$$\mathcal{U}(\mathcal{Q}'_S, \mathcal{Q}'_\Sigma, \mathbf{c}'+x) = e^{zx}\mathcal{U}(\mathcal{Q}'_S, \mathcal{Q}'_\Sigma, \mathbf{c}').$$

(G) Why does the straightforward generalization of Theorem 8.2.5(A) not obtain under the maximin criterion?

***Exercise 8.2.6**

Suppose that the set C of potential consequences of the dpuu in normal form is representable as a Cartesian product $C_1 \times C_2$, so that $c = (c_1, c_2)$, in which c_1 denotes net monetary return (as of a given date) and c_2 denotes all nonmonetary attributes. Continue to denote the consequence functions c: $S\times\Sigma \to C$, c': $S\times\Sigma\to C$, etc., by $\mathbf{c}$, $\mathbf{c}'$, etc.; clearly, $c'[s,\sigma] = (c'_1[s,\sigma], c'_2[s,\sigma])$. Assume that, for every c_2 in C_2, the decision maker's utility function $u((\cdot, c_2))$ for net monetary return given c_2 is strictly increasing and continuous.

(A) Show that the conclusions of Theorems 8.2.1 and 8.2.2, and of Corollaries 8.2.1–8.2.3, continue to obtain, provided that the *numerical* relations $c'[s, \sigma] \leq c^*[s, \sigma]$, $c'[s, \sigma] = c^*[s, \sigma]$, and $c'[s, \sigma] < c^*[s, \sigma]$ are replaced by their corresponding *desirability* relations $c'[s, \sigma] \lesssim c^*[s, \sigma]$, $c'[s, \sigma] \sim c^*[s, \sigma]$, and $c'[s, \sigma] \prec c^*[s, \sigma]$ respectively in the hypotheses.

(B) For any real number x and any consequence function $\mathbf{c}' = (c_1'[\cdot, \cdot], c_2'[\cdot, \cdot])$, let $c' + x$ denote the consequence function $(c_1'[\cdot, \cdot] + x, c_2'[\cdot, \cdot])$. Show that the conclusion of Theorem 8.2.3 continues to obtain.

(C) With notation as in (B), assume that $c_2^\ddagger$ is a fixed, nonmonetary reference attribute. Show that the conclusion of Theorem 8.2.4 continues to obtain, once $u(y)$ and $u(y_x)$ are replaced by $u((y, c_2^\ddagger))$ and $u((y_x, c_2^\ddagger))$ respectively.

(D) Generalize all parts of Exercise 8.2.5 in the context of this exercise.

■ The first systematic study of the subject of allowing utility functions for monetary returns to depend upon other attributes of the real problem, without using substitute functions, appears to be that of Hirschleifer [**60**]. ■

8.3 EXCHANGE VALUES AND THEIR APPLICATIONS

Suppose that you currently possess the opportunity $(\mathcal{Q}_S', \mathbf{c}')$ but are eyeing another opportunity, $(\mathcal{Q}_S'', \mathbf{c}'')$. How much should you be willing to pay for the privilege of exchanging $(\mathcal{Q}_S', \mathbf{c}')$ for $(\mathcal{Q}_S'', \mathbf{c}'')$? This question is hardly academic, because you might sustain breach-of-contract fines, intelligence-effort-redirection costs, etc., which are not embodied in $\mathbf{c}''$ if you decide to make the exchange.

If you agree to pay the sure amount \$$x$ to make the exchange, then your new opportunity is really not the $(\mathcal{Q}_S'', \mathbf{c}'')$ you have been eyeing, but rather the opportunity $(\mathcal{Q}_S'', \mathbf{c}'' - x)$, which reflects your cost of exchanging. The exchange cost must be figured into any evaluation of the desirability of exchanging.

If you *do not* exchange, then your utility evaluation of making the best of your opportunity $(\mathcal{Q}_S', \mathbf{c}')$ is, naturally,

$$\mathcal{U}(\mathcal{Q}_S', \mathbf{c}'), \tag{8.3.1a}$$

as given by Definition 8.1.2. But if you *do* exchange, at cost \$$x$, then your utility evaluation of making the best of your resulting opportunity $(\mathcal{Q}_S'', \mathbf{c}'' - x)$ is

$$\mathcal{U}(\mathcal{Q}_S'', \mathbf{c}'' - x), \tag{8.3.1b}$$

again as given by Definition 8.1.2.

Now, it is clear that you *should exchange at cost* \$x if

$$\mathcal{U}(\mathcal{Q}''_S, \mathbf{c}'' - x) > \mathcal{U}(\mathcal{Q}'_S, \mathbf{c}'), \tag{8.3.2a}$$

while you *should not exchange at cost* \$x if

$$\mathcal{U}(\mathcal{Q}''_S, \mathbf{c}'' - x) < \mathcal{U}(\mathcal{Q}'_S, \mathbf{c}'). \tag{8.3.2b}$$

If (8.3.1a) and (8.3.1b) are equal, then exchanging and not exchanging at cost $\$x$ are equally desirable to you, and hence your decision can be based on pure whimsy.

But Theorems 8.2.3 and 8.2.4(A) imply that there is a *unique* potential exchange value $\$x^*$ such that exchanging is preferable if its cost $\$x$ is less than $\$x^*$, and not exchanging is preferable if $\$x$ exceeds $\$x^*$. This $\$x^*$ is therefore the *maximum justifiable* cost of exchanging $(\mathcal{Q}'_S, \mathbf{c}')$ for $(\mathcal{Q}''_S, \mathbf{c}'')$. We call it the *exchange value* and denote it by $V(\mathcal{Q}'_S, \mathbf{c}' \rightarrow \mathcal{Q}''_S, \mathbf{c}'')$ for explicitness.

Definition 8.3.1: Exchange Values. The *value* $V(\mathcal{Q}'_S, \mathbf{c}' \rightarrow \mathcal{Q}''_S, \mathbf{c}'')$ *of exchanging* $(\mathcal{Q}'_S, \mathbf{c}')$ *for* $(\mathcal{Q}''_S, \mathbf{c}'')$ is defined implicitly by

$$\mathcal{U}(\mathcal{Q}''_S, \mathbf{c}'' - V(\mathcal{Q}'_S, \mathbf{c}' \rightarrow \mathcal{Q}''_S, \mathbf{c}'')) = \mathcal{U}(\mathcal{Q}'_S, \mathbf{c}').$$

■ *By Theorem 8.2.4(B) there is a unique real number y'_0 such that

$$\mathcal{U}(\mathcal{Q}'_S, \mathbf{c}') = u(y'_0);$$

moreover, by Theorem 8.2.4(A), there is a unique real number x''_0 such that

$$\mathcal{U}(\mathcal{Q}''_S, \mathbf{c}'' + x''_0) = u(y'_0).$$

Hence

$$\mathcal{U}(\mathcal{Q}''_S, \mathbf{c}'' + x''_0) = \mathcal{U}(\mathcal{Q}'_S, \mathbf{c}')$$

for exactly one real number x''_0. Put $V(\mathcal{Q}'_S, \mathbf{c}' \rightarrow \mathcal{Q}''_S, \mathbf{c}'') = -x''_0$ to obtain existence and uniqueness of the exchange value. ■

We have noted that the exchange value $V(\mathcal{Q}'_S, \mathbf{c}' \rightarrow \mathcal{Q}''_S, \mathbf{c}'')$ represents the maximum justifiable cost of exchanging. This quantity has a number of intuitively natural properties when viewed as a function of the current and/or the contemplated opportunity. These properties constitute Theorems 8.3.1 and 8.3.2.

Theorem 8.3.1(A) indicates that if $\mathcal{Q}''_S$ is enlarged to a superset $\mathcal{Q}^*_S$, then the contemplated opportunity (naturally) becomes at least as attractive (maybe more so), and hence the exchange value will not decrease; indeed, you should be willing to pay more for a more attractive opportunity. By contrast, Theorem 8.3.1(B) indicates that if $\mathcal{Q}'_S$ is enlarged to some superset, then the current opportunity becomes at least as attractive (maybe more so), and hence exchanging is *not* made more attractive, so that the exchange value will not increase; indeed, you should be willing to pay less to get out of a more attractive situation!

Theorem 8.3.1

(A) If $\mathcal{Q}''_S \subset \mathcal{Q}^*_S$, then $V(\mathcal{Q}'_S, \mathbf{c}' \to \mathcal{Q}''_S, \mathbf{c}'') \leq V(\mathcal{Q}'_S, \mathbf{c}' \to \mathcal{Q}^*_S, \mathbf{c}'')$.
(B) If $\mathcal{Q}'_S \subset \mathcal{Q}^*_S$, then $V(\mathcal{Q}'_S, \mathbf{c}' \to \mathcal{Q}''_S, \mathbf{c}'') \geq V(\mathcal{Q}^*_S, \mathbf{c}' \to \mathcal{Q}''_S, \mathbf{c}'')$.

■ **Proof*: *Of* (*A*). In the relation chain

$$\begin{aligned}
&\mathcal{U}(\mathcal{Q}^*_S, \mathbf{c}'' - V(\mathcal{Q}'_S, \mathbf{c}' \to \mathcal{Q}^*_S, \mathbf{c}'')) \\
&= \mathcal{U}(\mathcal{Q}'_S, \mathbf{c}') && \text{[Definition 8.3.1]} \\
&= \mathcal{U}(\mathcal{Q}''_S, \mathbf{c}'' - V(\mathcal{Q}'_S, \mathbf{c}' \to \mathcal{Q}''_S, \mathbf{c}'')) && \text{[Definition 8.3.1]} \\
&\leq \mathcal{U}(\mathcal{Q}^*_S, \mathbf{c}'' - V(\mathcal{Q}'_S, \mathbf{c}' \to \mathcal{Q}''_S, \mathbf{c}'')), && \text{[Theorem 8.2.2(A)]}
\end{aligned}$$

it is clear that the first member does not exceed the last. By Theorem 8.2.3, this can be true if and only if $-V(\mathcal{Q}'_S, \mathbf{c}' \to \mathcal{Q}^*_S, \mathbf{c}'') \leq -V(\mathcal{Q}'_S, \mathbf{c}' \to \mathcal{Q}''_S, \mathbf{c}'')$, which obviously implies (A).

Of (*B*). A similar invocation of Theorem 8.2.3 to the inequality between the first and last members of the chain

$$\begin{aligned}
&\mathcal{U}(\mathcal{Q}''_S, \mathbf{c}'' - V(\mathcal{Q}'_S, \mathbf{c}' \to \mathcal{Q}''_S, \mathbf{c}'')) \\
&= \mathcal{U}(\mathcal{Q}'_S, \mathbf{c}') && \text{[Definition 8.3.1]} \\
&\leq \mathcal{U}(\mathcal{Q}^*_S, \mathbf{c}') && \text{[Theorem 8.2.2(A)]} \\
&= \mathcal{U}(\mathcal{Q}''_S, \mathbf{c}'' - V(\mathcal{Q}^*_S, \mathbf{c}' \to \mathcal{Q}''_S, \mathbf{c}'')) && \text{[Definition 8.3.1]}
\end{aligned}$$

suffices. ■

■ Strict inequalities in Theorem 8.3.1(A) result when the equivalence in Theorem 8.2.2(B) obtains for the relationship of $\mathcal{Q}''_S$ to $\mathcal{Q}^*_S$; and similarly in Theorem 8.3.1(B) for the relationship of $\mathcal{Q}'_S$ to $\mathcal{Q}^*_S$. ■

Whereas Theorem 8.3.1 concerns the relationship of exchange values to the "richness" of current and contemplated opportunity sets, Theorem 8.3.2 concerns the relationship of exchange values to the monetary-return consequences in the current and contemplated opportunities. Theorem 8.3.2(A) indicates the intuitive truism that "improving" the return in the contemplated opportunity does not lessen the value of exchanging into it; Theorem 8.3.2(B) states the similarly appealing fact that "improving" the current opportunity cannot increase the value of exchanging *out* of it.

Theorem 8.3.2

(A) If $c''[s, \sigma] \leq c^*[s, \sigma]$ for every (s, σ) in $S \times \Sigma$, then

$$V(\mathcal{Q}'_S, \mathbf{c}' \to \mathcal{Q}''_S, \mathbf{c}'') \leq V(\mathcal{Q}'_S, \mathbf{c}' \to \mathcal{Q}''_S, \mathbf{c}^*).$$

(B) If $c'[s, \sigma] \leq c^*[s, \sigma]$ for every (s, σ) in $S \times \Sigma$, then

$$V(\mathcal{Q}'_S, \mathbf{c}' \to \mathcal{Q}''_S, \mathbf{c}'') \geq V(\mathcal{Q}'_S, \mathbf{c}^* \to \mathcal{Q}''_S, \mathbf{c}'').$$

■ **Proof*: *Of* (*A*). Abbreviate to V'' and V^* the two exchange values in question, and consider the following relation chain.

$$\begin{aligned}
&\mathcal{U}(\mathcal{Q}''_S, \mathbf{c}'' - V^*) \\
&\le \mathcal{U}(\mathcal{Q}''_S, \mathbf{c}^* - V^*) && \text{[Theorem 8.2.1]} \\
&= \mathcal{U}(\mathcal{Q}'_S, \mathbf{c}') && \text{[Definition 8.3.1]} \\
&= \mathcal{U}(\mathcal{Q}''_S, \mathbf{c}'' - V''), && \text{[Definition 8.3.1]}
\end{aligned}$$

in which the weak inequality obtains because:

(1) The hypothesis of (A) implies that $c''[s, \sigma] - V^* \le c^*[s, \sigma] - V^*$ for every (s, σ) in $S \times \Sigma$ and every conceivable V^* in R^1.
(2) Therefore the hypothesis of Theorem 8.2.1 obtains, with $\mathbf{c}'' - V^*$ in place of $\mathbf{c}'$ and $\mathbf{c}'' - V^*$ in place of $\mathbf{c}^*$.
(3) Hence, the conclusion of Theorem 8.2.1 obtains, with (of course) $\mathcal{Q}''_S$ in place of $\mathcal{Q}'_S$.

Now, this chain of relations implies that

$$\mathcal{U}(\mathcal{Q}''_S, \mathbf{c}'' - V^*) \le \mathcal{U}(\mathcal{Q}''_S, \mathbf{c}'' - V''),$$

so that by applying Theorem 8.2.3 we readily conclude that $-V^* \le -V''$, from which (A) is obvious.

Of (B). Abbreviate to V' and V^* the exchange values in this part, and similarly conclude that $-V' \le -V^*$ from examining the relation chain

$$\begin{aligned}
&\mathcal{U}(\mathcal{Q}''_S, \mathbf{c}'' - V') \\
&= \mathcal{U}(\mathcal{Q}'_S, \mathbf{c}') && \text{[Definition 8.3.1]} \\
&\le \mathcal{U}(\mathcal{Q}'_S, \mathbf{c}^*) && \text{[Theorem 8.2.1]} \\
&= \mathcal{U}(\mathcal{Q}''_S, \mathbf{c}'' - V^*)), && \text{[Definition 8.3.1]}
\end{aligned}$$

in which the weak inequality obtains because the hypothesis of (B) implies the hypothesis of Theorem 8.2.1. ■

Before going on to important special cases of exchange values, it is desirable that we devote some attention to the practical problem of how to *calculate* $V(\mathcal{Q}'_S, \mathbf{c}' \to \mathcal{Q}''_S, \mathbf{c}'')$ when an actual number in an actual problem is required.

In two special cases, the linear utility function and the concave-exponential utility functions, we can write down *explicit* formulae for the exchange value in terms of the (more or less) easily calculated utility evaluations $\mathcal{U}(\mathcal{Q}'_S, \mathbf{c}')$ and $\mathcal{U}(\mathcal{Q}''_S, \mathbf{c}'')$.

Theorem 8.3.3

(A) If $u(c) = c$ for every c in R^1, then

$$V(\mathcal{Q}'_S, \mathbf{c}' \to \mathcal{Q}''_S, \mathbf{c}'') = \mathcal{U}(\mathcal{Q}''_S, \mathbf{c}'') - \mathcal{U}(\mathcal{Q}'_S, \mathbf{c}').$$

(B) If $u(c) = -e^{-zc}$ for every c in R^1 and some $z > 0$, then

$$V(\mathcal{Q}'_S, \mathbf{c}' \to \mathcal{Q}''_S, \mathbf{c}'') = (1/z) \log_e [\mathcal{U}(\mathcal{Q}'_S, \mathbf{c}')/\mathcal{U}(\mathcal{Q}''_S, \mathbf{c}'')].$$

■ *Proof.* An easy application of Theorem 8.2.5 to Definition 8.3.1; left to you as Exercise 8.3.5. ■

Unfortunately, the linear and concave-exponential utility functions are the *only* useful utility functions for which exchange values may be written explicitly. Therefore, other utility functions require the application of approximation techniques to the definition of exchange value.

*■ The difficulty with most utility functions is that $\mathcal{D}$'s optimal strategy P_S^x in $(\mathcal{Q}_S'', \mathbf{c}'' - x)$ depends upon x, and hence need not be the same in $(\mathcal{Q}_S'', \mathbf{c}'' - V)$ as in $(\mathcal{Q}_S'', \mathbf{c}'')$. Hence the relationship of V to $\mathcal{U}(\mathcal{Q}_S'', \mathbf{c}'')$ is at best tenuous. But the linear and concave-exponential utility functions have the property of *constant risk aversion*, defined in Example 5.4.2, and this property implies that any strategy P_S^0 optimal for $\mathcal{D}$ in $(\mathcal{Q}_S'', \mathbf{c}'')$ is also optimal in every opportunity $(\mathcal{Q}_S'', \mathbf{c}'' - x)$ as x traverses the real line. ■

One practical way of approximating an exchange value is to use the bisection technique, presented in Appendix 4. First, we define a function $r(\cdot)$ by

$$r(x) = \mathcal{U}(\mathcal{Q}_S', \mathbf{c}') - \mathcal{U}(\mathcal{Q}_S'', \mathbf{c}'' - x) \tag{8.3.3}$$

for every x in R^1. Now $r(\cdot)$ has the following properties.

(1) $r(\cdot)$ is strictly increasing and continuous, by virtue of Theorem 8.2.3;
(2) $r(x)$ is negative for very small x and positive for very large x, by virtue of Theorem 8.2.4 and its proof; hence
(3) $r(\cdot)$ has exactly one root x^*; that is, exactly one number x^* such that $r(x^*) = 0$; and
(4) the root x^* of $r(\cdot)$ is $V(\mathcal{Q}_S', \mathbf{c}' \to \mathcal{Q}_S'', \mathbf{c}'')$, since $r(x^*) = 0$ if and only if $\mathcal{U}(\mathcal{Q}_S'', \mathbf{c}'' - x^*) = \mathcal{U}(\mathcal{Q}_S', \mathbf{c}')$.

Thus the exchange value may be approximated by applying the bisection technique in Appendix 4 to the function $r(\cdot)$ defined by (8.3.3). To do so, we must specify the endpoints A and B of the interval (A, B) within which x^* is sure to be. One very conservative specification is to set

$$A = \min_{ij} [c''[s^i, \sigma^j]] - \max_{ij} [c'[s^i, \sigma^j]] \tag{8.3.4a}$$

and

$$B = \max_{ij} [c''[s^i, \sigma^j]] - \min_{ij} [c'[s^i, \sigma^j]]. \tag{8.3.4b}$$

■ The rationale for (8.3.4b) is that $V(\mathcal{Q}_S', \mathbf{c}' \to \mathcal{Q}_S'', \mathbf{c}'')$ cannot exceed what it would be if the current opportunity were *sure* to result in the *smallest* conceivable return, $\min_{ij} [c'[s^i, \sigma^j]]$, and the *contemplated* opportunity were *sure* to result in the *largest* conceivable return, $\max_{ij} [c''[s^i, \sigma^j]]$. Under these assumptions, it is clear that the exchange value should equal the improvement, B. By reversing the roles of the current and contemplated opportunities, it follows similarly that the exchange value should be A. ■

But to minimize the number of burdensome computations required to approximate the exchange value to a given degree of accuracy, we should pick as small an interval (A, B) as possible, consistent with ensuring that it contains the exchange value. A number of techniques exist for being less conservative about A and B than (8.3.4), but we shall not consider them explicitly.

Example 8.3.1

Consider the following table of monetary returns $c[s, \sigma]$ of a dpuu in normal form.

	σ^1	σ^2
s^1	20	0
s^2	0	10
s^3	20	10
s^4	15	8

Suppose that $P_\Sigma = (.2, .8)$, that $\mathcal{Q}'_S = \mathcal{Q}_S^{\{s^1, s^4\}}$, and that $\mathcal{Q}''_S = \mathcal{Q}_S$. If $\mathcal{D}$'s utility is the linear function $u(c) = c$, we may apply Theorem 8.3.3(A) to see that

$$V(\mathcal{Q}'_S, \mathbf{c} \to \mathcal{Q}''_S, \mathbf{c}) = \mathcal{U}(\mathcal{Q}''_S, \mathbf{c}) - \mathcal{U}(\mathcal{Q}'_S, \mathbf{c}).$$

Now, each of $\mathcal{Q}'_S$ and $\mathcal{Q}''_S$ is a set of the form $\mathcal{Q}_S^{S^*}$ for some nonempty subset S^* of S, and in all such cases it follows from Theorem 7.3.3 that

$$\mathcal{U}(\mathcal{Q}_S^{S^*}, \mathbf{c}) = \max_{s \in S^*} [\mathcal{U}(s, \mathbf{c})] = \max_{s \in S^*} [(.2)u(c[s, \sigma^1]) + (.8)u(c[s, \sigma^2])].$$

Here,

$$\mathcal{U}(s^1, \mathbf{c}) = 20(.2) + 0(.8) = 4.0,$$
$$\mathcal{U}(s^2, \mathbf{c}) = 0(.2) + 10(.8) = 8.0,$$
$$\mathcal{U}(s^3, \mathbf{c}) = 20(.2) + 10(.8) = 12.0,$$

and

$$\mathcal{U}(s^4, \mathbf{c}) = 15(.2) + 8(.8) = 9.4,$$

so that

$$\mathcal{U}(\mathcal{Q}'_S, \mathbf{c}) = \max [4.0, 9.4] = 9.4$$

and

$$\mathcal{U}(\mathcal{Q}''_S, \mathbf{c}) = \max [4.0, 8.0, 12.0, 9.4] = 12.0.$$

Therefore

$$V(\mathcal{Q}'_S, \mathbf{c} \to \mathcal{Q}''_S, \mathbf{c}) = 12.0 - 9.4 = 2.6.$$

Example 8.3.2

Keep all assumptions as in Example 8.3.1, except to assume now that $u(c) = -e^{-(.05)c}$ for every c in R^1. Here,

$$\mathcal{U}(s^1, \mathbf{c}) = -e^{-(.05)(20)}(.2) - e^{-(.05)(0)}(.8) = -.873574,$$
$$\mathcal{U}(s^2, \mathbf{c}) = -e^{-(.05)(0)}(.2) - e^{-(.05)(10)}(.8) = -.685224,$$
$$\mathcal{U}(s^3, \mathbf{c}) = -e^{-(.05)(20)}(.2) - e^{-(.05)(10)}(.8) = -.558798,$$

and

$$\mathcal{U}(s^4, \mathbf{c}) = -e^{-(.05)(15)}(.2) - e^{-(.05)(8)}(.8) = -.630730,$$

so that

$$\mathcal{U}(\mathcal{Q}'_S, \mathbf{c}) = \max[-.873574, -.630730] = -.630730$$

and

$$\mathcal{U}(\mathcal{Q}''_S, \mathbf{c}) = \max[-.873574, -.685224, -.558798, -.630730] = -.558798.$$

Now, according to Theorem 8.3.3(B),

$$V(\mathcal{Q}'_S, \mathbf{c} \to \mathcal{Q}''_S, \mathbf{c}) = (1/.05)\log_e[-.630730/-.558798]$$
$$= 20\log_e[1.1287] \doteq 2.42,$$

which differs significantly from the exchange value 2.60 when utility was assumed linear in Example 8.3.1.

■ In order to use tables of logarithms to the base 10, it is convenient to reexpress $\mathcal{D}$'s utility function as

$$u(c) = -10^{-(.4342944819)zc},$$

and then to use the following modification of Theorem 8.3.3(B).

$$V = \frac{2.3025850930}{z}\log_{10}\left[\frac{\mathcal{U}(\mathcal{Q}'_S, \mathbf{c}')}{\mathcal{U}(\mathcal{Q}''_S, \mathbf{c}'')}\right]. \blacksquare$$

Example 8.3.3

Keep all assumptions as in Example 8.3.1, except to assume now that $u(c) = -(20-c)^4$ for every c in $(-\infty, 20]$. This utility function has *increasing risk aversion* (see Example 5.4.5). We shall use the bisection technique to find V, after applying a bit of common sense to simplify the ensuing calculations. *First*, in the defining equation (8.3.3) of $r(\cdot)$, it is clear that $\mathcal{U}(\mathcal{Q}'_S, \mathbf{c}')$ must be calculated only once. Since

$$\mathcal{U}(s^1, \mathbf{c}) = -(20-20)^4(.2) - (20-0)^4(.8) = -128{,}000$$

and

$$\mathcal{U}(s^4, \mathbf{c}) = -(20-15)^4(.2) - (20-8)^4(.8) = -16{,}713.8,$$

it follows that

$$\mathcal{U}(\mathcal{Q}'_S, \mathbf{c}) = \max[-128{,}000, -16{,}713.8] = -16{,}713.8.$$

Second, we observe that, in this particular dpuu,[c] s^3 will be optimal in $(\mathcal{Q}''_S, \mathbf{c}-x)$ for every x, since each of s^1, s^2, and s^4 is ordinally dominated by s^3 for any x and any strictly increasing utility function. Hence

$$\mathcal{U}(\mathcal{Q}''_S, \mathbf{c}-x) = \mathcal{U}(s^3, \mathbf{c}-x),$$

which implies that

$$\begin{aligned} r(x) &= (20-[20-x])^4(.2)+(20-[10-x])^4(.8)-16{,}713.8 \\ &= x^4(.2)+(10+x)^4(.8)-16{,}713.8. \end{aligned}$$

We have a strong prior hunch that $x^* = V(\mathcal{Q}'_S, \mathbf{c} \to \mathcal{Q}''_S, \mathbf{c})$ is between 2 and 3, but we shall set $A=-20$ and $B=+20$ in accordance with Equation (8.3.4). For two-decimal-place accuracy, Equation (A4.2) in Appendix 4 implies that we shall have to calculate $r(x)$ for eleven values of x. Here goes.

(1) Set $x=(-20+20)/2=0$; calculate $r(0)=(0)^4(.2)+(10)^4(.8)-16{,}713.8=-8{,}713.8$. Hence new $A=0$ and B remains 20.

(2) Set $x=(0+20)/2=10$; calculate $r(10)=(10)^4(.2)+(20)^4(.8)-16{,}713.8=+113{,}286.2$. Hence new $B=10$ and A remains 0.

(3) Set $x=(0+10)/2=5$; calculate $r(5)=(5)^4(.2)+(15)^4(.8)-16{,}713.8=+23{,}911.2$. Hence new $B=5$ and A remains 0.

(4) Set $x=(0+5)/2=2.5$; calculate $r(2.5)=(2.5)^4(.2)+(12.5)^4(.8)-16{,}713.8=+2{,}825.2625$. Hence new $B=2.5$ and A remains 0.

(5) Set $x=(0+2.5)/2=1.25$; calculate $r(1.25)=(1.25)^4(.2)+(11.25)^4(.8)-16{,}713.8=-3{,}898.858594$. Hence new $A=1.25$ and B remains 2.50.

(6) Set $x=(1.25+2.50)/2=1.875$; calculate $r(1.875)=(1.875)^4(.2)+(11.875)^4(.8)-16{,}713.8=-803.002879$. Hence new $A=1.875$ and B remains 2.500.

(7) Set $x=(1.875+2.500)/2=2.1875$; calculate $r(2.1875)=(2.1875)^4(.2)+(12.1875)^4(.8)-16{,}713.8=+940.93360$. Hence new $B=2.1875$ and A remains 1.875.

(8) Set $x=(1.8750+2.1875)/2=2.03125$; calculate $r(2.03125)=(2.03125)^4(.2)+(12.03125)^4(.8)-16{,}713.8=+51.880960$. Hence new $B=2.03125$ and A remains 1.87500.

(9) Set $x=(1.87500+2.03125)/2=1.953125$; calculate $r(1.953125)=(1.953125)^4(.2)+(11.953125)^4(.8)-16{,}713.8=-379.774852$. Hence new $A=1.953125$ and B remains 2.031250.

(10) Set $x=(1.953125+2.031250)/2=1.992187$; calculate $r(1.992187)=(1.992187)^4(.2)+(11.992187)^4(.8)-16{,}713.8=-165.010303$. Hence new $A=1.992187$ and B remains 2.031250.

(11) Set $x=(1.992187+2.031250)/2=2.011718$; calculate $r(2.011718)=(2.011718)^4(.2)+(12.011718)^4(.8)-16{,}713.8=-56.833555$. Hence new $A=2.011718$ and B remains 2.031250. *STOP*: $B-A<.02$; $V=$ root x^* is estimated at approximately 2.02.

[c] In general, the optimal strategy depends upon x and hence $\mathcal{U}(s, \mathbf{c}'')$ ordinarily has to be calculated for every s at each trial value of x, as a preliminary to determining $\mathcal{U}(\mathcal{Q}''_S, \mathbf{c}''-x)$.

Two very important special cases of exchange values are obtained by making special assumptions about the current and contemplated opportunities. The first of these cases is $\mathcal{D}$'s *bid ceiling* $B(\mathcal{Q}'_S, \mathbf{c}')$ of an opportunity $(\mathcal{Q}'_S, \mathbf{c}')$; it is the answer to the question as to the maximum amount which $\mathcal{D}$ should be willing to pay for the opportunity $(\mathcal{Q}'_S, \mathbf{c}')$ if he does not now own any of the opportunities under consideration.

Definition 8.3.2. $\mathcal{D}$'s *bid ceiling* $B(\mathcal{Q}'_S, \mathbf{c}')$ of an opportunity $(\mathcal{Q}'_S, \mathbf{c}')$ is the unique real number which satisfies

$$\mathcal{U}(\mathcal{Q}'_S, \mathbf{c}' - B(\mathcal{Q}'_S, \mathbf{c}')) = u(0).$$

As noted in Example 8.1.2, not owning any opportunity is for all intents and purposes the same as owning the opportunity $(\mathcal{Q}'_S, \mathbf{0})$, where $\mathbf{0}$ is defined by

$$0[s, \sigma] = 0 \quad \text{for every } (s, \sigma) \text{ in } S \times \Sigma.$$

Clearly,

$$u(0) = \mathcal{U}(\mathcal{Q}'_S, \mathbf{0}),$$

so that, by Definition 8.3.2,

$$\mathcal{U}(\mathcal{Q}'_S, \mathbf{c}' - B(\mathcal{Q}'_S, \mathbf{c}')) = \mathcal{U}(\mathcal{Q}'_S, \mathbf{0}),$$

which readily implies that

$$B(\mathcal{Q}'_S, \mathbf{c}') = V(\mathcal{Q}'_S, \mathbf{0} \to \mathcal{Q}'_S, \mathbf{c}'), \tag{8.3.5}$$

by virtue of Definition 8.3.1 and Theorem 8.2.3. Thus $B(\mathcal{Q}'_S, \mathbf{c}')$ is the maximum amount which $\mathcal{D}$ should be willing to pay to exchange the opportunity $(\mathcal{Q}'_S, \mathbf{0})$ to receive nothing for the opportunity $(\mathcal{Q}'_S, \mathbf{c}')$.

Example 8.3.4

For the dpuu in Example 8.3.1, it is easy to verify that $B(\mathcal{Q}'_S, \mathbf{c}) = 9.40$, since $\mathcal{U}(\mathcal{Q}'_S, \mathbf{0}) = 0$. For the dpuu in Example 8.3.2, we note that $\mathcal{U}(\mathcal{Q}'_S, \mathbf{0}) = -1$, so that

$$\begin{aligned} B(\mathcal{Q}'_S, \mathbf{c}) &= V(\mathcal{Q}'_S, \mathbf{0} \to \mathcal{Q}'_S, \mathbf{c}) = \left(\frac{1}{z}\right) \log_e \left[\frac{\mathcal{U}(\mathcal{Q}'_S, \mathbf{0})}{\mathcal{U}(\mathcal{Q}'_S, \mathbf{c})}\right] \\ &= \left(\frac{1}{z}\right) \log_e [|\mathcal{U}(\mathcal{Q}'_S, \mathbf{c})|^{-1}] = -(2.302585/.05) \log_{10} [|\mathcal{U}'_S, \mathbf{c})|] \\ &= (-46.0517) \log_{10} (.630730) = (-46.0517)(-.20016) \doteq 9.22, \end{aligned}$$

somewhat less than the 9.40 for the linear utility function. An application of the bisection technique in Appendix 4 to find $B(\mathcal{Q}'_S, \mathbf{c})$ for the dpuu in Example 8.3.3 results in $B(\mathcal{Q}'_S, \mathbf{c}) \doteq 8.89$ (to two decimal places).

In Example 8.3.4 we needed the explicit formulae for $B(\mathcal{Q}'_S, \mathbf{c}')$ in Corollary 8.3.1.

Corollary 8.3.1

(A) If $u(c) = c$ for every c in R^1, then

$$B(\mathcal{Q}'_S, \mathbf{c}') = \mathcal{U}(\mathcal{Q}'_S, \mathbf{c}').$$

(B) If $u(c) = -e^{-zc}$ for every c in R^1 and some $z > 0$, then

$$B(\mathcal{Q}'_S, \mathbf{c}') = -(1/z) \log_e [|\mathcal{U}(\mathcal{Q}'_S, \mathbf{c}')|].$$

■ *Proof*. Exercise 8.3.6. ■

One must not fall into the trap of believing that $B(\mathcal{Q}'_S, \mathbf{c}')$ also represents the smallest amount which $\mathcal{D}$ should be willing to accept in exchange for the opportunity $(\mathcal{Q}'_S, \mathbf{c}')$ if he owns it; that quantity is called $\mathcal{D}$'s *reservation price* for $(\mathcal{Q}'_S, \mathbf{c}')$ and is denoted by $R(\mathcal{Q}'_S, \mathbf{c}')$. Since $u(\cdot)$ is strictly increasing and continuous (by assumption), it follows that reservation prices are as given by Definition 8.3.3.

Definition 8.3.3. $\mathcal{D}$'s *reservation price* $R(\mathcal{Q}'_S, \mathbf{c}')$ of an opportunity $(\mathcal{Q}'_S, \mathbf{c}')$ is the unique real number which satisfies

$$u(R(\mathcal{Q}'_S, \mathbf{c}')) = \mathcal{U}(\mathcal{Q}'_S, \mathbf{c}').$$

Since $\mathcal{D}$ is sure to end up richer by $R(\mathcal{Q}'_S, \mathbf{c}')$ in the opportunity $(\mathcal{Q}'_S, \mathbf{0} + R(\mathcal{Q}'_S, \mathbf{c}'))$, it is clear that

$$\mathcal{U}(\mathcal{Q}'_S, \mathbf{0} + R(\mathcal{Q}'_S, \mathbf{c}')) = u(R(\mathcal{Q}'_S, \mathbf{c}')),$$

so that

$$\mathcal{U}(\mathcal{Q}'_S, \mathbf{0} + R(\mathcal{Q}'_S, \mathbf{c}')) = \mathcal{U}(\mathcal{Q}'_S, \mathbf{c}'),$$

by Definition 8.3.3. Hence

$$R(\mathcal{Q}'_S, \mathbf{c}') = -V(\mathcal{Q}'_S, \mathbf{c}' \to \mathcal{Q}'_S, \mathbf{0}), \tag{8.3.6}$$

by virtue of Definition 8.3.1 and Theorem 8.2.3. But (8.3.6) is quite natural: a person's minimum selling price (or reservation price) for $(\mathcal{Q}'_S, \mathbf{c}')$ is the negative of the maximum he would be willing to pay to get out of this opportunity.

We have said that $R(\mathcal{Q}'_S, \mathbf{c}')$ and $B(\mathcal{Q}'_S, \mathbf{c}')$ are generally distinct numbers. There are important special cases in which these quantities must coincide—the usual special cases.

Theorem 8.3.4. $R(\mathcal{Q}'_S, \mathbf{c}') = B(\mathcal{Q}'_S, \mathbf{c}')$ for every opportunity $(\mathcal{Q}'_S, \mathbf{c}')$ if either

(A) $u(c) = c$ for every c in R^1, or
(B) $u(c) = -e^{-zc}$ for every c in R^1 and some $z > 0$.

■ *Proof*. Exercise 8.3.7. ■

Example 8.3.5

From Example 8.3.4, $R(\mathcal{Q}'_S, \mathbf{c}) = 9.40$ for the linear utility function and $R(\mathcal{Q}'_S, \mathbf{c}) = 9.22$ for the concave-exponential utility function. For the quartic utility function introduced for this dpuu in Example 8.3.3, we can solve explicitly for $R(\mathcal{Q}'_S, \mathbf{c}') = R$ by recalling that $\mathcal{U}(\mathcal{Q}'_S, \mathbf{c}) = -16{,}713.8$ and hence that $-(20 - R)^4 = -16{,}713.8$, by virtue of Definition 8.3.3, so that $R = 20 - (16{,}713.8)^{1/4} \doteq 20 - 11.37 = 8.63$, which is quite different from the $B(\mathcal{Q}'_S, \mathbf{c}) = 8.89$ in Example 8.3.4.

The second[d] important case of exchange values is best introduced by observing that $V(\mathcal{Q}'_S, \mathbf{c}' \to \mathcal{Q}''_S, \mathbf{c}'')$ is the maximum *certain* cost of exchanging $(\mathcal{Q}'_S, \mathbf{c}')$ for $(\mathcal{Q}''_S, \mathbf{c}'')$ which $\mathcal{D}$ should be willing to sustain. But as yet it provides no indication as to how $\mathcal{D}$ should determine whether such an exchange is worthwhile if the cost of exchanging is *uncertain*, or if it depends upon his choice of strategy. In practice, such situations are more the rule than the exception! To make exchange values germane in such situations, we will have to replace uncertain exchange costs $\mathbf{v}$ with certain ones, x, which are a priori equally desirable. Such x's are called *cash equivalents* and are denoted by the dollar sign.[e]

Definition 8.3.4. Let $(\mathcal{Q}''_S, \mathbf{c}'')$ be any opportunity, and let $\mathbf{y} = y: S \times \Sigma \to R^1$ be any "exchange-cost function" whose values are to be subtracted from the corresponding values of $\mathbf{c}''$ in order to determine $\mathcal{D}$'s net monetary return. The *cash equivalent* $\$(\mathbf{y}; \mathcal{Q}''_S, \mathbf{c}'')$ of $\mathbf{y}$ in relation to $(\mathcal{Q}''_S, \mathbf{c}'')$ is defined to be the unique real number which satisfies

$$\mathcal{U}(\mathcal{Q}''_S, \mathbf{c}'' - \$(\mathbf{y}; \mathcal{Q}''_S, \mathbf{c}'')) = \mathcal{U}(\mathcal{Q}''_S, \mathbf{c}'' - \mathbf{y}).$$

A glance at the equations in Definitions 8.3.1 and 8.3.4 suffices to establish that

$$\$(\mathbf{y}; \mathcal{Q}''_S, \mathbf{c}'') = V(\mathcal{Q}''_S, \mathbf{c}'' - \mathbf{y} \to \mathcal{Q}''_S, \mathbf{c}'') \tag{8.3.7}$$

for every $\mathbf{y} = y: S \times \Sigma \to R^1$ and every opportunity $(\mathcal{Q}''_S, \mathbf{c}'')$. Thus existence and uniqueness of $\$(\mathbf{y}; \mathcal{Q}''_S, \mathbf{c}'')$ follow from the same properties of V.

If $y[s, \sigma] = x$ for every (s, σ) in $S \times \Sigma$, so that in effect $\mathbf{y}$ is a simple constant, x, it is easy to show that the cash equivalent of $\mathbf{y}$ is x.

Theorem 8.3.5. If $y[s, \sigma] = x$ for every (s, σ) in $S \times \Sigma$, then

$$\$(\mathbf{y}; \mathcal{Q}''_S, \mathbf{c}'') = x$$

for every opportunity $(\mathcal{Q}''_S, \mathbf{c}'')$.

■ *Proof.* Exercise 8.3.8(A). ■

In view of Theorem 8.3.5, it follows that $\mathcal{D}$ should compare $\$(\mathbf{y}; \mathcal{Q}''_S, \mathbf{c}'')$ with $V(\mathcal{Q}'_S, \mathbf{c}' \to \mathcal{Q}''_S, \mathbf{c}'')$ in order to determine whether exchanging his current opportunity $(\mathcal{Q}'_S, \mathbf{c}')$ for $(\mathcal{Q}''_S, \mathbf{c}'')$ is economically desirable.

Theorem 8.3.6: Economics of Exchanging One Opportunity for Another. $\mathcal{D}$ should be willing to exchange $(\mathcal{Q}'_S, \mathbf{c}')$ for $(\mathcal{Q}''_S, \mathbf{c}'')$ at cost $\mathbf{y}$ only if

$$\$(\mathbf{y}; \mathcal{Q}''_S, \mathbf{c}'') \leq V(\mathcal{Q}'_S, \mathbf{c}' \to \mathcal{Q}''_S, \mathbf{c}'').$$

■ *Proof.* Exercise 8.3.8(B). ■

---

[d] Reservation price, though intimately related to exchange value via (8.3.6), is not a special case if one wants to be technical with minus-sign hairsplitting.

[e] No offense to other currencies is intended; we are simply running out of Latin, Greek, and script symbols.

Bid ceilings and reservation prices have their uses, as we have seen; but neither indicates how *much better off* $\mathscr{D}$ would be by "paying" **y** to exchange $(\mathscr{Q}'_S, \mathbf{c}')$ for $(\mathscr{Q}''_S, \mathbf{c}'')$. Such a question is rather natural, but the indications are that it is ill-posed in that several answers are possible, depending upon what is meant by "better off."

We shall suggest three possible answers, called *net gains* of exchanging, together with their accompanying interpretations of "better off." The *first* interpretation of "better off" is "increase in reservation price"; therefore, we should define the *net* R-*gain* $G_R(\mathbf{y};\ \mathscr{Q}'_S,\ \mathbf{c}' \to \mathscr{Q}''_S,\ \mathbf{c}'')$ of paying **y** to exchange $(\mathscr{Q}'_S, \mathbf{c}')$ for $(\mathscr{Q}''_S, \mathbf{c}'')$ by

$$G_R(\mathbf{y};\ \mathscr{Q}'_S, \mathbf{c}' \to \mathscr{Q}''_S, \mathbf{c}'') = R(\mathscr{Q}''_S, \mathbf{c}'' - \mathbf{y}) - R(\mathscr{Q}'_S, \mathbf{c}'). \tag{8.3.8}$$

■ A moment's reflection on Definition 8.3.3 shows that $R(\mathscr{Q}'_S, \mathbf{c}')$ is $\mathscr{D}$'s *certainty equivalent* of the net-return lottery determined: (1) by the net-return function c': $S \times \Sigma \to R^1$; (2) by the probability P_Σ on Σ; and (3) by the probability P°_S on S, where P°_S is an optimal strategy for $\mathscr{D}$ in $(\mathscr{Q}'_S, \mathbf{c}')$. Similarly, $R(\mathscr{Q}''_S, \mathbf{c}'' - \mathbf{y})$ is $\mathscr{D}$'s certainty equivalent of his "optimal net-return lottery" in $(\mathscr{Q}''_S, \mathbf{c}'' - \mathbf{y})$. (See Chapter 5 for a discussion of certainty equivalents.) Hence G_R indicates how much $\mathscr{D}$ increases his certainty equivalent by paying **y** to effect the exchange. ■

A *second* interpretation of "better off" is "the minimum bribe $\mathscr{D}$ should accept for *not* making the exchange." The corresponding net gain is called the *net* B-*gain* $G_B(\mathbf{y}; \mathscr{Q}'_S, \mathbf{c}' \to \mathscr{Q}''_S, \mathbf{c}'')$ of paying **y** to exchange $(\mathscr{Q}'_S, \mathbf{c}')$ for $(\mathscr{Q}''_S, \mathbf{c}'')$. It is defined implicitly as the unique real number G_B which satisfies

$$\mathscr{U}(\mathscr{Q}'_S, \mathbf{c}' + G_B) = \mathscr{U}(\mathscr{Q}''_S, \mathbf{c}'' - \mathbf{y}). \tag{8.3.9}$$

A *third* interpretation of "better off" is "the maximum constant amount by which **y** could be increased and still have the exchange rationally justifiable." The corresponding net gain is the *net O-gain* $G_O(\mathbf{y};\ \mathscr{Q}'_S,\ \mathbf{c}' \to \mathscr{Q}''_S, \mathbf{c}'')$; it is defined as the unique real number G_O which satisfies

$$\mathscr{U}(\mathscr{Q}''_S, \mathbf{c}'' - \mathbf{y} - G_O) = \mathscr{U}(\mathscr{Q}'_S, \mathbf{c}'). \tag{8.3.10}$$

These three definitions of net gain generally yield different monetary amounts; the usual exception pertains, however.

Theorem 8.3.7. All three numbers $G_i(\mathbf{y}; \mathscr{Q}'_S, \mathbf{c}' \to \mathscr{Q}''_S, \mathbf{c}'')$ coincide if either

(A) $u(c) = c$ for every c in R^1 or
(B) $u(c) = -e^{-zc}$ for every c in R^1 and some $z > 0$.

■ *Proof.* Exercise 8.3.9. ■

In addition to providing answers to various evaluation questions, the three types of net gains have a more important role in certain applications, by determining for which one of *a number of* contemplated opportunities of the form $(\mathcal{Q}_S^m, \mathbf{c}^m)$, if any, $\mathcal{D}$ should exchange his current opportunity $(\mathcal{Q}'_S, \mathbf{c}')$, given that it would cost $\mathbf{y}^m$ to exchange $(\mathcal{Q}'_S, \mathbf{c}')$ for $(\mathcal{Q}_S^m, \mathbf{c}^m)$. Remember, Theorem 8.3.6 envisioned only *one* contemplated opportunity.

Theorem 8.3.8. Let i be any one of R, B, and O. If the cost of exchanging $(\mathcal{Q}'_S, \mathbf{c}')$ for $(\mathcal{Q}_S^m, \mathbf{c}^m)$ is $\mathbf{y}^m$ for $m = 1, 2, \ldots, M$, then:

(A) No exchange should be made if

$$\max_m [G_i(\mathbf{y}^m; \mathcal{Q}'_S, \mathbf{c}' \to \mathcal{Q}_S^m, \mathbf{c}^m)] < 0.$$

(B) And if

$$\max_m [G_i(\mathbf{y}^m; \mathcal{Q}'_S, \mathbf{c}' \to \mathcal{Q}_S^m, \mathbf{c}^m)] > 0,$$

then an optimal opportunity $(\mathcal{Q}^\circ_S, \mathbf{c}^\circ)$ is any maximizer (as $m = 1, 2, \ldots, M$) of $G_i(\mathbf{y}^m; \mathcal{Q}'_S, \mathbf{c}' \to \mathcal{Q}_S^m, \mathbf{c}^m)$.

■ *Proof.* It is easy to verify from Equations (8.3.8)–(8.3.10), together with Theorem 8.2.3 and strict increasingness and continuity of u, that

$$G_i(\mathbf{y}^{m'}; \mathcal{Q}'_S, \mathbf{c}' \to \mathcal{Q}_S^{m'}, \mathbf{c}^{m'}) < G_i(\mathbf{y}^{m''}; \mathcal{Q}'_S, \mathbf{c}' \to \mathcal{Q}_S^{m''}, \mathbf{c}^{m''})$$

if and only if

$$\mathcal{U}(\mathcal{Q}_S^{m'}, \mathbf{c}^{m'} - \mathbf{y}^{m'}) < \mathcal{U}(\mathcal{Q}_S^{m''}, \mathbf{c}^{m''} - \mathbf{y}^{m''}).$$

This, together with the (easily verified) fact that

$$G_i(\mathbf{0}; \mathcal{Q}'_S, \mathbf{c}' \to \mathcal{Q}'_S, \mathbf{c}') = 0,$$

suffices for both (A) and (B). ■

Exercise 8.3.1: Computational Opportunity

Consider the dpuu in normal form tabulated in Exercise 8.2.1, with $P_\Sigma = (.4, .6)$.

(A) If $u(c) = c$ for every c in R^1 and $c'[s, \sigma] = c''[s, \sigma] = c[s, \sigma]$ for every (s, σ) in $S \times \Sigma$, find $R(\mathcal{Q}_S^{S'}, \mathbf{c}')$ and $V(\mathcal{Q}_S^{S'}, \mathbf{c}' \to \mathcal{Q}_S^{S''}, \mathbf{c}'')$ for:

- (i) $S' = \{s^1\}$ and $S'' = \{s^3, s^4\}$
- (ii) $S' = \{s^4\}$ and $S'' = \{s^1, s^3\}$
- (iii) $S' = \{s^2\}$ and $S'' = \{s^1, s^4\}$
- (iv) $S' = \{s^3, s^4\}$ and $S'' = \{s^2, s^4\}$
- (v) $S' = \{s^3\}$ and $S'' = S = \{s^1, s^2, s^3, s^4\}$
- (vi) $S' = S'' = S$

(B) Under the assumptions of (A), find $\$(\mathbf{y}; \mathcal{Q}_S^{S''}, \mathbf{c}'')$ and $G_R(\mathbf{y}; \mathcal{Q}_S^{S'}, \mathbf{c}' \to \mathcal{Q}_S^{S''}, \mathbf{c}'')$ for:

- (i) part (i) of (A)
- (ii) part (ii) of (A)
- (iii) part (iii) of (A)

(iv) part (iv) of (A)
(v) part (v) of (A)
(vi) part (vi) of (A)

given that **y** is defined by the following table.

	σ^1	σ^2
s^1	2	−1
s^2	0	5
s^3	0	1
s^4	2	0

(C) Repeat (A) under the assumption that $u(c) = -e^{-(.10)c}$ for every c in R^1.
(D) Repeat (B) under the assumption that $u(c) = -e^{-(.10)c}$ for every c in R^1.
(E) Repeat (A) under the assumption that $u(c) = -(15-c)^2$ for every $c \leq 15$.
(F) Repeat (B) under the assumption that $u(c) = -(15-c)^2$ for every $c \leq 15$.

Exercise 8.3.2: Computational Opportunity

(A) Redo Example 8.3.3 under the assumption that $u(c) = -(20-c)^2$ for every $c \leq 20$ rather than $u(c) = -(20-c)^4$ for every $c \leq 20$.
(B) Find $B(\mathcal{Q}'_S, \mathbf{c})$ under these assumptions.
(C) Find $R(\mathcal{Q}'_S, \mathbf{c})$ under these assumptions.

Exercise 8.3.3: Computational Opportunity

Consider the dpuu in Exercise 8.2.2, and let $\mathbf{y}^1$ and $\mathbf{y}^2$ be given by the following tables.

$\mathbf{y}^1$

	σ^1	σ^2	σ^3	σ^4
s^1	0	1	0	0
s^2	1	0	0	0
s^3	.50	0	.50	.50
s^4	1	−1	1	12
s^5	.50	.50	.50	0
s^6	.50	0	0	1

$\mathbf{y}^2$

	σ^1	σ^2	σ^3	σ^4
s^1	.25	.25	.25	.25
s^2	1	0	0	0
s^3	1	2	1	−1
s^4	1	−1	1	12
s^5	.50	.50	.50	0
s^6	−1	2	1	0

(A) Let $u(c) = c$ for every c in R^1 and $P_\Sigma = (.25, .10, .15, .50)$. Suppose that $S' = \{s^1, s^2\}$ and $S'' = \{s^2, s^4, s^5\}$. Compute:

(i) $\$(\mathbf{y}^1; \mathcal{Q}_S^{S'}, \mathbf{c} \to \mathcal{Q}_S^{S''}, \mathbf{c})$
(ii) $\$(\mathbf{y}^2; \mathcal{Q}_S^{S'}, \mathbf{c} \to \mathcal{Q}_S^{S''}, \mathbf{c})$
(iii) $V(\mathcal{Q}_S^{S'}, \mathbf{c} \to \mathcal{Q}_S^{S''}, \mathbf{c})$
(iv) $G_R(\mathbf{y}^1; \mathcal{Q}_S^{S'}, \mathbf{c} \to \mathcal{Q}_S^{S''}, \mathbf{c})$

(v) $G_R(\mathbf{y}^2; \mathscr{Q}_S^{S'}, \mathbf{c} \to \mathscr{Q}_S^{S''}, \mathbf{c})$

(vi) $G_B(\mathbf{y}^1; \mathscr{Q}_S^{S'}, \mathbf{c} \to \mathscr{Q}_S^{S''}, \mathbf{c})$

(vii) $G_B(\mathbf{y}^2; \mathscr{Q}_S^{S'}, \mathbf{c} \to \mathscr{Q}_S^{S''}, \mathbf{c})$

(viii) $G_0(\mathbf{y}^1; \mathscr{Q}_S^{S'}, \mathbf{c} \to \mathscr{Q}_S^{S''}, \mathbf{c})$

(ix) $G_0(\mathbf{y}^2; \mathscr{Q}_S^{S'}, \mathbf{c} \to \mathscr{Q}_S^{S''}, \mathbf{c})$

(B) Repeat (A) under the assumption that $u(c) = -e^{-(.10)c}$ for every c in R^1.

(C) Repeat (A) under the assumption that $u(c) = -(30-c)^2$ for every $c \leq 30$.

Exercise 8.3.4: Computational Opportunity

Consider the following tabulated net-return function $\mathbf{c}''$ for the dpuu in Exercise 8.3.3.

	σ^1	σ^2	σ^3	σ^4
s^1	0	1	10	3
s^2	12	5	3	18
s^3	0	15	8	5
s^4	9	1	4	3
s^5	6	5	5	0
s^6	20	0	18	6

(A) Replace $(\mathscr{Q}_S^{S''}, \mathbf{c})$ with $(\mathscr{Q}_S^{S''}, \mathbf{c}'')$ and redo Exercise 8.3.3(A).

(B) Replace $(\mathscr{Q}_S^{S''}, \mathbf{c})$ with $(\mathscr{Q}_S^{S''}, \mathbf{c}'')$ and redo Exercise 8.3.3(B).

(C) Replace $(\mathscr{Q}_S^{S''}, \mathbf{c})$ with $(\mathscr{Q}_S^{S''}, \mathbf{c}'')$ and redo Exercise 8.3.3(C).

Exercise 8.3.5

Prove (A) Theorem 8.3.3(A) and (B) Theorem 8.3.3(B).

Exercise 8.3.6

Prove (A) Corollary 8.3.1(A) and (B) Corollary 8.3.1(B).

Exercise 8.3.7

Prove (A) Theorem 8.3.4(A) and (B) Theorem 8.3.4(B).

Exercise 8.3.8

Prove (A) Theorem 8.3.5 and (B) Theorem 8.3.6.

Exercise 8.3.9

Prove Theorem 8.3.7.

***Exercise 8.3.10**

Prove that for every $(\mathscr{Q}'_S, \mathbf{c}')$ and every $(\mathscr{Q}''_S, \mathbf{c}'')$,

$$V(\mathscr{Q}'_S, \mathbf{c}' \to \mathscr{Q}''_S, \mathbf{c}'' + x) = x + V(\mathscr{Q}'_S, \mathbf{c}' \to \mathscr{Q}''_S, \mathbf{c}'')$$

for every x in R^1.

■ How about adding x to $\mathbf{c}'$? Don't jump to a hasty conclusion! ■

***Exercise 8.3.11: Implications of Concavity of $u(\cdot)$**

This is the first of three exercises which establish bounds on V, R, and B, given less powerful properties than linearity and concave exponentiality. For any opportunity $(\mathcal{Q}'_S, \mathbf{c}')$ let $\mathcal{U}^{\text{lin}}(\mathcal{Q}'_S, \mathbf{c}')$ denote what $\mathcal{D}$'s utility of $(\mathcal{Q}'_S, \mathbf{c}')$ would be *if* $u(c)$ equaled c for every c in R^1. Continue to let $\mathcal{U}(\mathcal{Q}'_S, \mathbf{c}')$ denote $\mathcal{D}$'s utility of $(\mathcal{Q}'_S, \mathbf{c}')$, given his *actual* utility function $u(\cdot)$.

(A) Show that if $u(\cdot)$ is concave, then

$$V(\mathcal{Q}'_S, \mathbf{c}' \to \mathcal{Q}''_S, \mathbf{c}'') \leq \mathcal{U}^{\text{lin}}(\mathcal{Q}''_S, \mathbf{c}'') - R(\mathcal{Q}'_S, \mathbf{c}'). \tag{8.3.11}$$

[*Hints*: (1) If $u(\cdot)$ is concave and V is the left-hand side of (8.3.11), then $R(\mathcal{Q}''_S, \mathbf{c}'' - V) \leq E\{\ell^V - V\}$, where $\ell^V - V$ is any return lottery $c''[\tilde{s}, \tilde{\sigma}] - V$ which is optimal in $(\mathcal{Q}''_S, \mathbf{c}'' - V)$ with respect to $\mathcal{D}$'s actual utility function $u(\cdot)$. (Why?)

(2) $E\{\ell^V - V\} \leq \mathcal{U}^{\text{lin}}(\mathcal{Q}''_S, \mathbf{c}'') - V$. (Why?)

(3) $\mathcal{U}^{\text{lin}}(\mathcal{Q}''_S, \mathbf{c}'') - V < R(\mathcal{Q}'_S, \mathbf{c}')$ if (8.3.11) were false.

(4) Prove that falsity of (8.3.11) implies that $R(\mathcal{Q}''_S, \mathbf{c}'' - V) < R(\mathcal{Q}'_S, \mathbf{c}')$ and that this inequality is a contradiction.]

(B) Show that if $u(\cdot)$ is concave, then

$$B(\mathcal{Q}'_S, \mathbf{c}') \leq \mathcal{U}^{\text{lin}}(\mathcal{Q}'_S, \mathbf{c}'). \tag{8.3.12}$$

(C) Show that if $u(\cdot)$ is concave, then

$$R(\mathcal{Q}'_S, \mathbf{c}') \leq \mathcal{U}^{\text{lin}}(\mathcal{Q}'_S, \mathbf{c}'). \tag{8.3.13}$$

■ The importance of (8.3.11)–(8.3.13) lies in the facts that $\mathcal{U}^{\text{lin}}(\mathcal{Q}''_S, \mathbf{c}'')$ is usually quite rapid to compute and that $R(\mathcal{Q}'_S, \mathbf{c}')$ is usually much easier to compute than an exchange value or a bid ceiling. Hence using these upper bounds in place of B as defined by (8.3.4b) may result in saving vexatious bisection-technique computations of $\mathcal{U}(\mathcal{Q}''_S, \mathbf{c}'' - x)$ at small cost, by reducing the number of these computations required for approximation to a given degree of accuracy. ■

***Exercise 8.3.12: Implications of $u(\cdot)$ Having Decreasing Risk Aversion**

Show that if $u(\cdot)$ has decreasing risk aversion, then for any opportunity $(\mathcal{Q}'_S, \mathbf{c}')$ either

$$0 \leq B(\mathcal{Q}'_S, \mathbf{c}') \leq R(\mathcal{Q}'_S, \mathbf{c}') \tag{8.3.14a}$$

or

$$0 \geq B(\mathcal{Q}'_S, \mathbf{c}') \geq R(\mathcal{Q}'_S, \mathbf{c}'). \tag{8.3.14b}$$

[*Hints*: (1) If $u(\cdot)$ has decreasing risk aversion, then $x - Q\{\ell + x\}$ is a nonincreasing function of x for any lottery ℓ. (Why?)

(2) If $u(\cdot)$ has decreasing risk aversion, then $x - R(\mathcal{Q}'_S, \mathbf{c}' + x)$ is a nonincreasing function of x for any opportunity $(\mathcal{Q}'_S, \mathbf{c}')$. (Why?)

(3) $B - R(\mathcal{Q}'_S, \mathbf{c}' - B + B) \leq -R(\mathcal{Q}'_S, \mathbf{c}' - B)$ if $B \geq 0$.

(4) $R(\mathcal{Q}'_S, \mathbf{c}' - B(\mathcal{Q}'_S, \mathbf{c}')) = 0$. (Why?)]

■ Since $R(\mathcal{Q}'_S, \mathbf{c}')$ is usually easier to calculate than $B(\mathcal{Q}'_S, \mathbf{c}')$, calculating the latter may be expedited by first calculating the former and then using it as a bound for the latter. ■

■ If $u(\cdot)$ has *increasing* risk aversion, then the roles of $R(\mathcal{Q}'_S, \mathbf{c}')$ and $B(\mathcal{Q}'_S, \mathbf{c}')$ in Equations (8.3.14a) and (8.3.14b) are reversed: either

$$0 \leq R(\mathcal{Q}'_S, \mathbf{c}') \leq B(\mathcal{Q}'_S, \mathbf{c}') \tag{8.3.15a}$$

or

$$0 \geq R(\mathcal{Q}'_S, \mathbf{c}') \geq B(\mathcal{Q}'_S, \mathbf{c}'). \tag{8.3.15b}$$

Again, reservation price can be calculated as a bound on bid ceiling. ■

Exercise 8.3.13

That reservation price and bid ceiling both lie on the same side of zero in (8.3.14) and (8.3.15) is no mere happenstance, and it is not even a special attribute of monotone risk aversion. Show that strict increasingness and continuity of $u(\cdot)$ imply that $R(\mathcal{Q}'_S, \mathbf{c}') \geq 0$ if and only if $B(\mathcal{Q}'_S, \mathbf{c}') \geq 0$. [*Hint*: $\mathcal{U}(\mathcal{Q}'_S, \mathbf{c}') \geq \mathcal{U}(\mathcal{Q}'_S, \mathbf{c}' - B)$ if and only if $B \geq 0$.]

Exercise 8.3.14: Iterated Exchanges

Let $(\mathcal{Q}'_S, \mathbf{c}')$, $(\mathcal{Q}''_S, \mathbf{c}'')$, and $(\mathcal{Q}'''_S, \mathbf{c}''')$ by three opportunities, and suppose that the cost of exchanging $(\mathcal{Q}'_S, \mathbf{c}')$ for $(\mathcal{Q}''_S, \mathbf{c}'')$ is $\mathbf{y}$.

(A) Show that

$$V(\mathcal{Q}''_S, \mathbf{c}'' - \mathbf{y} \to \mathcal{Q}'''_S, \mathbf{c}''') \leq V(\mathcal{Q}'_S, \mathbf{c}' \to \mathcal{Q}'''_S, \mathbf{c}''')$$

if and only if

$$\$(\mathbf{y}; \mathcal{Q}''_S, \mathbf{c}'') \leq V(\mathcal{Q}'_S, \mathbf{c}' \to \mathcal{Q}''_S, \mathbf{c}'').$$

(B) What is the intuitive meaning of (A)?

(C) Use (A) to show that

$$V(\mathcal{Q}''_S, \mathbf{c}'' - V(\mathcal{Q}'_S, \mathbf{c}' \to \mathcal{Q}''_S, \mathbf{c}'') \to \mathcal{Q}'_S, \mathbf{c}') = 0.$$

(D) What is the intuitive meaning of (C)?

(E) Use (C) and *Exercise 8.3.10 to show that

$$V(\mathcal{Q}'_S, \mathbf{c}' \to \mathcal{Q}''_S, \mathbf{c}'') = -V(\mathcal{Q}''_S, \mathbf{c}'' - V'' \to \mathcal{Q}'_S, \mathbf{c}' - V''),$$

where

$$V'' = V(\mathcal{Q}'_S, \mathbf{c}' \to \mathcal{Q}''_S, \mathbf{c}'').$$

(F) What is the intuitive meaning of (E)?

***Exercise 8.3.15: Maximin Revisited (II)**

Virtually everything in Section 8.3 remains true for the maximin approach, developed partly in Exercise 8.2.5. Given the notation of Exercise 8.2.5 and given the more general definition $(\mathcal{Q}'_S, \mathcal{Q}'_\Sigma, \mathbf{c}')$ of an opportunity in the maximin approach, the principal concepts of this section are defined as follows.

(i) $\mathcal{U}(\mathcal{Q}''_S, \mathcal{Q}''_\Sigma, \mathbf{c}'' - V(\mathcal{Q}'_S, \mathcal{Q}'_\Sigma, \mathbf{c}' \to \mathcal{Q}''_S, \mathcal{Q}''_\Sigma, \mathbf{c}'')) = \mathcal{U}(\mathcal{Q}'_S, \mathcal{Q}'_\Sigma, \mathbf{c}')$, to define $V(\mathcal{Q}'_S, \mathcal{Q}'_\Sigma, \mathbf{c}' \to \mathcal{Q}''_S, \mathcal{Q}''_\Sigma, \mathbf{c}'')$

(ii) $\mathcal{U}(\mathcal{Q}'_S, \mathcal{Q}'_\Sigma, \mathbf{c}' - B(\mathcal{Q}'_S, \mathcal{Q}'_\Sigma, \mathbf{c}')) = u(0)$, to define $B(\mathcal{Q}'_S, \mathcal{Q}'_\Sigma, \mathbf{c}')$

(iii) $u(R(\mathcal{Q}'_S, \mathcal{Q}'_\Sigma, \mathbf{c}')) = \mathcal{U}(\mathcal{Q}'_S, \mathcal{Q}'_\Sigma, \mathbf{c}')$, to define $R(\mathcal{Q}'_S, \mathcal{Q}'_\Sigma, \mathbf{c}')$

(iv) $\mathcal{U}(\mathcal{Q}''_S, \mathcal{Q}''_\Sigma, \mathbf{c}'' - \$(\mathbf{y}; \mathcal{Q}''_S, \mathcal{Q}''_\Sigma, \mathbf{c}'')) = \mathcal{U}(\mathcal{Q}''_S, \mathcal{Q}''_\Sigma, \mathbf{c}'' - \mathbf{y})$, to define $\$(\mathbf{y}; \mathcal{Q}''_S, \mathcal{Q}''_\Sigma, \mathbf{c}'')$

(v) $G_R(\mathbf{y}; \mathcal{Q}'_S, \mathcal{Q}'_\Sigma, \mathbf{c}' \to \mathcal{Q}''_S, \mathcal{Q}''_\Sigma, \mathbf{c}'') = R(\mathcal{Q}''_S, \mathcal{Q}''_\Sigma, \mathbf{c}'' - \mathbf{y}) - R(\mathcal{Q}'_S, \mathcal{Q}'_\Sigma, \mathbf{c}')$

(vi) $\mathcal{U}(\mathcal{Q}'_S, \mathcal{Q}'_\Sigma, \mathbf{c}' + G_B) = \mathcal{U}(\mathcal{Q}''_S, \mathcal{Q}''_\Sigma, \mathbf{c}'' - \mathbf{y})$, to define $G_B = G_B(\mathbf{y}; \mathcal{Q}'_S, \mathcal{Q}'_\Sigma, \mathbf{c}' \to \mathcal{Q}''_S, \mathcal{Q}''_\Sigma, \mathbf{c}'')$

(vii) $\mathcal{U}(\mathcal{Q}''_S, \mathcal{Q}''_\Sigma, \mathbf{c}'' - \mathbf{y} - G_O) = \mathcal{U}(\mathcal{Q}'_S, \mathcal{Q}'_\Sigma, \mathbf{c}')$, to define $G_O = G_O(\mathbf{y}; \mathcal{Q}'_S, \mathcal{Q}'_\Sigma, \mathbf{c}' \to \mathcal{Q}''_S, \mathcal{Q}''_\Sigma, \mathbf{c}'')$

(A) **Theorem 8.3.1M.** Prove that:

(i) If $\mathcal{Q}''_S \subset \mathcal{Q}^*_S$, then

$$V(\mathcal{Q}'_S, \mathcal{Q}'_\Sigma, \mathbf{c}' \to \mathcal{Q}''_S, \mathcal{Q}''_\Sigma, \mathbf{c}'') \leq V(\mathcal{Q}'_S, \mathcal{Q}'_\Sigma, \mathbf{c}' \to \mathcal{Q}^*_S, \mathcal{Q}''_\Sigma, \mathbf{c}'').$$

(ii) If $\mathcal{Q}'_S \subset \mathcal{Q}^*_S$, then

$$V(\mathcal{Q}'_S, \mathcal{Q}'_\Sigma, \mathbf{c}' \to \mathcal{Q}''_S, \mathcal{Q}''_\Sigma, \mathbf{c}'') \geq V(\mathcal{Q}^*_S, \mathcal{Q}'_\Sigma, \mathbf{c}' \to \mathcal{Q}''_S, \mathcal{Q}''_\Sigma, \mathbf{c}'').$$

(i*) If $\mathcal{Q}''_\Sigma \subset \mathcal{Q}^*_\Sigma$, then

$$V(\mathcal{Q}'_S, \mathcal{Q}'_\Sigma, \mathbf{c}' \to \mathcal{Q}''_S, \mathcal{Q}''_\Sigma, \mathbf{c}'') \geq V(\mathcal{Q}'_S, \mathcal{Q}'_\Sigma, \mathbf{c}' \to \mathcal{Q}''_S, \mathcal{Q}^*_\Sigma, \mathbf{c}'').$$

(ii*) If $\mathcal{Q}'_\Sigma \subset \mathcal{Q}^*_\Sigma$, then

$$V(\mathcal{Q}'_S, \mathcal{Q}'_\Sigma, \mathbf{c}' \to \mathcal{Q}''_S, \mathcal{Q}''_\Sigma, \mathbf{c}'') \leq V(\mathcal{Q}'_S, \mathcal{Q}^*_\Sigma, \mathbf{c}' \to \mathcal{Q}''_S, \mathcal{Q}''_\Sigma, \mathbf{c}'').$$

(B) **Theorem 8.3.2M.** Prove that:

(i) If $c''[s, \sigma] \leq c^*[s, \sigma]$ for every (s, σ) in $S \times \Sigma$, then

$$V(\mathcal{Q}'_S, \mathcal{Q}'_\Sigma, \mathbf{c}' \to \mathcal{Q}''_S, \mathcal{Q}''_\Sigma, \mathbf{c}'') \leq V(\mathcal{Q}'_S, \mathcal{Q}'_\Sigma, \mathbf{c}' \to \mathcal{Q}''_S, \mathcal{Q}''_\Sigma, \mathbf{c}^*).$$

(ii) If $c'[s, \sigma] \leq c^*[s, \sigma]$ for every (s, σ) in $S \times \Sigma$, then

$$V(\mathcal{Q}'_S, \mathcal{Q}'_\Sigma, \mathbf{c}' \to \mathcal{Q}''_S, \mathcal{Q}''_\Sigma, \mathbf{c}'') \geq V(\mathcal{Q}'_S, \mathcal{Q}'_\Sigma, \mathbf{c}^* \to \mathcal{Q}''_S, \mathcal{Q}''_\Sigma, \mathbf{c}'').$$

(C) **Theorem 8.3.3M.** Prove that:

(i) If $u(c) = c$ for every c in R^1, then

$$V(\mathcal{Q}'_S, \mathcal{Q}'_\Sigma, \mathbf{c}' \to \mathcal{Q}''_S, \mathcal{Q}''_\Sigma, \mathbf{c}'') = \mathcal{U}(\mathcal{Q}''_S, \mathcal{Q}''_\Sigma, \mathbf{c}'') - \mathcal{U}(\mathcal{Q}'_S, \mathcal{Q}'_\Sigma, \mathbf{c}').$$

(ii) If $u(c) = -e^{-zc}$ for every c in R^1 and some $z > 0$, then

$$V(\mathcal{Q}'_S, Q'_\Sigma, \mathbf{c}' \to \mathcal{Q}''_S, \mathcal{Q}''_\Sigma, \mathbf{c}'') = \frac{1}{z}\log_e\left[\frac{\mathcal{U}(\mathcal{Q}'_S, \mathcal{Q}'_\Sigma, \mathbf{c}')}{\mathcal{U}(\mathcal{Q}''_S, \mathcal{Q}''_\Sigma, \mathbf{c}'')}\right].$$

(D) **Theorem 8.3.4M.** Show that $R(\mathcal{Q}'_S, \mathcal{Q}'_\Sigma, \mathbf{c}') = B(\mathcal{Q}'_S, \mathcal{Q}'_\Sigma, \mathbf{c}')$ if either $u(c) = c$ for every c in R^1 or $u(c) = -e^{-zc}$ for every c in R^1 and some $z > 0$.

(E) **Theorem 8.3.5M.** Show that if $y[s, \sigma] = x$ for every (s, σ) in $S \times \Sigma$, then

$$\$(\mathbf{y}; \mathcal{Q}''_S, \mathcal{Q}''_\Sigma, \mathbf{c}'') = x$$

for every opportunity $(\mathcal{Q}''_S, \mathcal{Q}''_\Sigma, \mathbf{c}'')$.

(F) **Theorem 8.3.6M.** Show that $\mathcal{D}$ should be willing to exchange $(\mathcal{Q}'_S, \mathcal{Q}'_\Sigma, \mathbf{c}')$ for $(\mathcal{Q}''_S, \mathcal{Q}''_\Sigma, \mathbf{c}'')$ at cost $\mathbf{y}$ only if

$$\$(\mathbf{y}; \mathcal{Q}''_S, \mathcal{Q}''_\Sigma, \mathbf{c}'') \le V(\mathcal{Q}'_S, \mathcal{Q}'_\Sigma, \mathbf{c}' \to \mathcal{Q}''_S, \mathcal{Q}''_\Sigma, \mathbf{c}'').$$

(G) **Theorem 8.3.8M.** Prove the (by now) obvious generalization of Theorem 8.3.8.

8.4 STATISTICAL DECISION PROBLEMS

Example 1.5.2b (see Figure 2.12) is a statistical decision problem, and so is Example 1.5.2c (see Figure 2.16). In these dpuu's the nature of all stages except the Nth (in standard extensive form) is the gleaning of information to enable $\mathcal{D}$ to make a wiser choice of g_N.

Figure 8.1 depicts schematically the simplest sort of statistical decision problem, involving one round of experimentation (= testing, or information gathering) before making the choice of terminal action (= the operationally important choice of g_N). In the figure we introduce popular notation for statistical decision problems: e for experiments, ζ for the outcomes of experiments, a for terminal acts, and θ for the "state" of affairs about which information is desired. We denote "do not experiment" by e^-, and the "null outcome" of e^- by ζ^-. We shall always assume that e^- belongs to any set E of all available experiments under discussion.

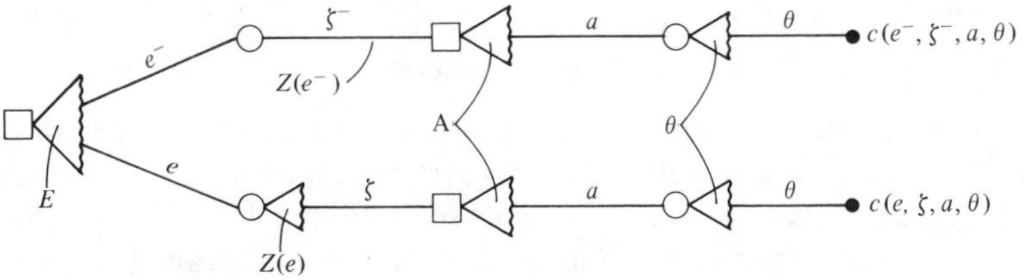

Figure 8.1 Single-stage-of-experimentation problem.

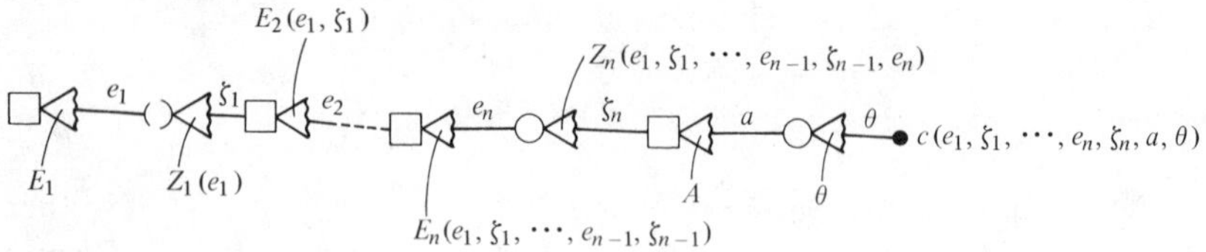

Figure 8.2 n-Stages-of-experimentation problem.

More generally, Figure 8.2 depicts schematically a statistical decision problem in which there are $n+1$ $(=N)$ stages in standard extensive form, the first n of which develop information about θ. Again, we shall assume that e^- belongs to every set of available experiments.

Actually, the greater generality depicted in Figure 8.2 vis-à-vis Figure 8.1 is largely illusory. As Exercise 8.4.3, you may show that any n-stages-of-experimentation problem may be reformulated as a single-stage-of-experimentation problem by partial normalization.

Now, if statistical decision problems were nothing more than general decision problems in standard extensive form with different notation, we would be wasting space here. But there is more to it than that. For our purposes, a dpuu in standard extensive form is not a *statistical* decision problem unless it satisfies certain restrictive assumptions about its structure and about $\mathscr{D}$'s probability functions:

I. The set A of available terminal acts a does not depend upon the preceding program of experimentation, and neither does the set Θ of all possible states θ;

and, more importantly,

II. $\mathscr{D}$'s contemplated program of experimentation and terminal action does not affect his judgments about $\tilde{\theta}$ prior to commencing experimentation.

■ The second part of I is easy to "engineer" in most cases. In Examples 1.5.2b and 1.5.2c, simply replace "defective in use" and "nondefective in use" with "arrived defective" and "arrived nondefective" respectively, and replace the dummy $\mathscr{E}$-moves "nondefective in use" in Figures 2.12 and 2.16 with the nondummy $\mathscr{E}$-moves describing the arrival state, duplicating the consequences at the endpoints appropriately. Thus the $\$+150$, MP″ attached to "nondefective in use" in Figure 2.12 would be attached to each of the arrival states' branches in the modified tree. ■

■ There is an artificial way to "engineer" the first part of I, by making A consist of all conceivable terminal acts and by attaching horrible but artificial consequences to any act not *actually* available, given the previous course of experimentation, so that no actually *un*available will be chosen. This is an illustration of what is called a "penalty method" in mathematical programming (or optimization)—a method of converting a problem with constraints to one with fewer or no constraints in such a way that an optimal solution to the transformed problem is also necessarily an optimal solution to the original problem. ■

The second assumption can be reexpressed more lucidly with some additional notation. II amounts to assuming that $\mathscr{D}$'s probabilities on all branches of the tree are derivable by Bayes' Theorem from

IIa. his *prior* probability function $P_\theta(\cdot)$ on events in Θ;

IIb. for every θ in Θ and e_1 in E_1, his *conditional sampling* probability function $P_{\zeta_1|e_1,\theta}(\cdot|e_1, \theta)$ on events in $Z_1(e_1)$;

and, if $n > 1$,

IIc. for every θ in Θ, every $i \in \{2, \ldots, n\}$, and every possible $(e_1, \zeta_1, \ldots, e_{i-1}, \zeta_{i-1}, e_i)$, his *conditional sampling* probability function $P_{\zeta_i|e_1,\zeta_1,\ldots,\zeta_{i-1},e_i,\theta}(\cdot|e_1, \zeta_1, \ldots, \zeta_{i-1}, e_i, \theta)$ on events in $Z_i(e_1, \zeta_1, \ldots, e_{i-1}, \zeta_{i-1}, e_i)$.

In Section 4.4 we went through just such a derivation of the probabilities needed for the tree and analysis of Example 1.5.1c.

■ IIa–IIc are restrictive and not always satisfied even in those dpuu's in standard extensive form which naturally satisfy I. In particular, they imply that the corresponding branches of all copies of Θ attached to acts in a single copy of A have the same probability; that is, the choice of terminal act a does not affect $\mathscr{D}$'s judgments about the state θ. Only the previous course of experimentation, $(e_1, \zeta_1, \ldots, e_n, \zeta_n)$, affects these judgments. ■

By Bayes' Theorem, $\mathscr{D}$ may derive from IIa–IIc the probabilities he needs for the branches of his decision tree; namely, $P_{\zeta_1|e_1}(\cdot|e_1)$, $P_{\zeta_i|e_1,\zeta_1,\ldots,e_{i-1},\zeta_{i-1},e_i}(\cdot|e_1, \zeta_1, \ldots, e_{i-1}, \zeta_{i-1}, e_i)$ for $i = 2, \ldots, n$, and $P_{\theta|e_1,\zeta_1,\ldots,e_n,\zeta_n}(\cdot|e_1, \zeta_1, \ldots, e_n, \zeta_n)$. This last is called $\mathscr{D}$'s *posterior* probability function on events in Θ; it represents his judgments about $\tilde{\theta}$ posterior to—or as revised by—the previous course of experimentation. The mechanics of obtaining these probabilities should be obvious from Chapter 3 and Section 4.4.

The general procedure of Chapter 4 suffices for determining an optimal program of experimentation and, ultimately, terminal action given the experimentation. Hence no more need be said in general. But a number of interesting and useful results are obtainable if we make the following assumptions about the consequences.

III. All consequences are net monetary returns as of a given date, accompanied by some fixed, nonmonetary attribute (such as the *status quo*);

IV. $\$c(e_1, \zeta_1, \ldots, e_n, \zeta_n, a, \theta) = \$v(a, \theta) - \$k(e_1, \zeta_1, \ldots, e_n, \zeta_n)$;

and

V. $\mathscr{D}$'s utility function $u(\cdot)$ on the set R^1 of all conceivable net monetary returns is strictly increasing and continuous.

*■ In [**115**] Raiffa and Schlaifer introduce an alternative set of assumptions, to the effect that $\mathscr{D}$'s utility $u(c(e_1, \zeta_1, \ldots, e_n, \zeta_n, a, \theta))$ is separable (see Section 5.5), so that it can be written in the form

$$v(a, \theta) - k(e_1, \zeta_1, \ldots, e_n, \zeta_n),$$

with v and $-k$ playing the roles of u_1 and u_2 in Section 5.5. This does *not* assume that the consequences are monetary. Under this assumption, V is irrelevant, or trivial, in that setting $u(v-k) = v-k$ for every $v-k$ in R^1 works. The assumption of monetary consequences and fairly general utility was introduced and studied in [**81**]. ■

Exercise 8.4.1: Discussion Question

The term "experiment" has very specialized connotations, but it is entrenched in the literature. Discuss: "A company commander performs an experiment when he sends out a reconnaissance patrol." (Is it necessary to assume that the enemy will surely be unaware of the existence of the patrol?)

Exercise 8.4.2: Discussion Question

Political opinion polls may or may not be experiments in statistical decision problems as defined in this section, according to whether one believes that polls can influence opinion. Discuss this statement.

Exercise 8.4.3

Show that *any* statistical decision problem can be expressed in the single-stage-of-experimentation form of Figure 8.1 by using partial normalization.

8.5 SINGLE-STAGE-OF-EXPERIMENTATION PROBLEMS

In this section we discuss the *value of information* conveyed by an experiment e in the single-stage-of-experimentation model depicted in Figure 8.1. The notation for this model is much simpler than for the n-stages-of-experimentation model, and the latter can be reformulated in terms of the former (see Exercise 8.4.3), so we are actually not restricting unduly the generality of our analysis. This section is a close counterpart to, and a specialization of, Section 8.3; the same sorts of topics and ideas are discussed in both 8.3 and 8.5.

We shall omit here stage subscripts for e and ζ, because there is only one stage of experimentation. Furthermore, we shall assume without loss of generality that the cost of no experimentation is zero; that is, that $k(e^-, \zeta^-) = 0$.

Roughly, the value $I(e)$ of an experiment e is the maximum constant

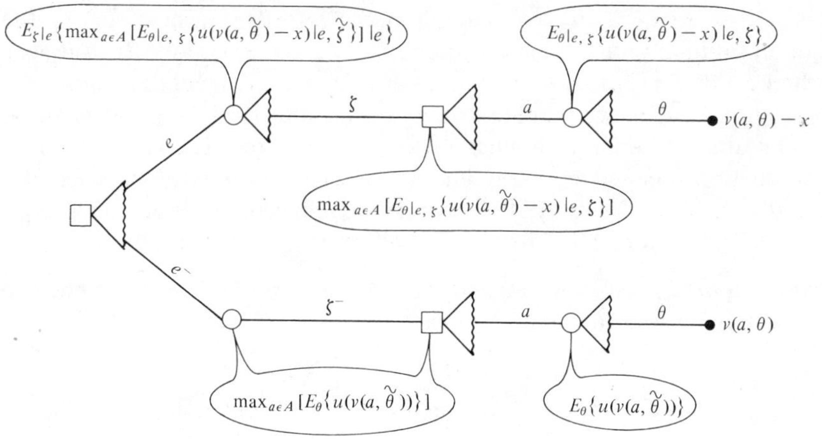

Figure 8.3 Implications of paying x for e-information.

amount which $\mathscr{D}$ should be willing to pay for the privilege of choosing his terminal act a after observing the outcome ζ of e, instead of choosing his terminal act a without experimentation. Figure 8.3 depicts the implications of his contemplated payment of \$$x$ for the e-information; naturally, his utility of paying \$$x$ is a strictly decreasing and continuous function of x. Hence there exists a unique value $I(e)$ such that e-information at cost $I(e)$ is equally as desirable as no information e^- at cost zero.

■ Clearly, $P_\theta(\cdot) = P_{\theta|e^-,\zeta^-}(\cdot|e^-, \zeta^-)$, and similarly for the corresponding expectations. ■

Definition 8.5.1. The *value* I(e) *of* e-*information* is defined to be the unique real number x^* such that

$$E_{\zeta|e}\{\max_{a\in A}[E_{\theta|e,\zeta}\{u(v(a, \tilde{\theta}) - x^*)|e, \tilde{\zeta}\}]|e\} = \max_{a\in A}[E_\theta\{u(v(a, \tilde{\theta}))\}].$$

■ $I(e)$ is a special case of exchange value V. To see this, we first let S' be the set of all pure strategies in the normal form of the statistical decision problem which commence with $\mathscr{D}$'s choosing e^-, and we let S'' be the set of all pure strategies with $\mathscr{D}$'s initially choosing e. Second, we note that every pair (s, σ) of pure strategies determines a unique pair (a, θ). Third, we define $\mathbf{c}'$ and $\mathbf{c}''$ by setting $c'[s, \sigma]$ and $c''[s, \sigma]$ each equal to $v(a, \theta)$ for the (a, θ) determined by (s, σ), for every (s, σ) in $S \times \Sigma$. Then

$$I(e) = V(\mathscr{Q}_S{}^{S'}, \mathbf{c}' \to \mathscr{Q}_S{}^{S''}, \mathbf{c}'').$$

Hence existence and uniqueness of $I(e)$ follow from existence and uniqueness of exchange values. ■

Because $E_{\zeta|e}\{\max_{a\in A}[E_{\theta|e,\zeta}\{u(v(a, \tilde{\theta}) - x)|e, \tilde{\zeta}\}]|e\}$ is strictly decreasing and continuous in x, it is clear that $\mathscr{D}$ should be willing to pay x for

e-information if $x < I(e)$, and should *not* if $x > I(e)$. Thus $I(e)$ is the maximum rationally justifiable constant cost of e-information. But suppose that the cost of e-information is uncertain, with $k(e, \tilde{\zeta})$ depending upon $\tilde{\zeta}$ as well as upon e. Then we cannot compare $I(e)$ and $k(e, \tilde{\zeta})$ except in those unusually drastic situations where $k(e, \tilde{\zeta})$ is *certain* to exceed $I(e)$ or *certain* to be exceeded by $I(e)$. Therefore what we must do is to find a *constant* cost $\$(k(e, \tilde{\zeta}); e)$ which is a priori equally as undesirable as $k(e, \tilde{\zeta})$, given that experiment e is chosen. Definition 8.5.2 fills this bill.

Definition 8.5.2. The *cash equivalent* $\$(k(e, \tilde{\zeta}); e)$ of $k(e, \tilde{\zeta})$ is defined to be the unique real number x^* such that

$$E_{\zeta|e}\{\max_{a\in A}[E_{\theta|e,\zeta}\{u(v(a, \tilde{\theta}) - x^*)|e, \tilde{\zeta}\}]|e\}$$
$$= E_{\zeta|e}\{\max_{a\in A}[E_{\theta|e,\zeta}\{u(v(a, \tilde{\theta}) - k(e, \tilde{\zeta}))|e, \tilde{\zeta}\}]|e\}.$$

■ Let S'' and $\mathbf{c}''$ be as in the preceding remark, and let $\mathbf{y} = y[\cdot, \cdot]$ be defined by setting $y[s, \sigma]$ equal to $k(e, \zeta)$ for the unique (e, ζ) determined by (s, σ), for every (s, σ) in $S \times \Sigma$. Then $\$(k(e, \tilde{\zeta}); e)$ is simply $\$(\mathbf{y}; \mathcal{Q}_S^{S''}, \mathbf{c}'')$ as defined in Section 8.3. Hence existence and uniqueness of $\$(k(e, \tilde{\zeta}); e)$ follow from the same properties of cash equivalents as defined in Section 8.3. ■

Now we can state the fundamental result on the evaluation of *an* experiment.

Theorem 8.5.1. Experiment e at cost $k(e, \tilde{\zeta})$ is at least as desirable as no information whatsoever if and only if

$$\$(k(e, \tilde{\zeta}); e) \leq I(e).$$

■ *Proof.* Follows readily from Theorem 8.3.6, given the preceding definitions of $\mathcal{Q}'_S$, $\mathcal{Q}''_S$, $\mathbf{c}'$, $\mathbf{c}''$, and $\mathbf{y}$. ■

Unless $u(\cdot)$ is of particularly decent form, finding $I(e)$ and $\$(k(e, \tilde{\zeta}); e)$ is a task in numerical approximation. The bisection technique discussed in Appendix 4 is pertinent in both cases. But $I(e)$ and $\$(k(e, \tilde{\zeta}); e)$ can be expressed *explicitly* when either $u(c) = c$ for every c in R^1 or $u(c) = -e^{-zc}$ for every c in R^1 and some $z > 0$.

Theorem 8.5.2.

(A) If $u(c) = c$ for every c in R^1, then

(i) $I(e) = E_{\zeta|e}\{\max_{a\in A}[E_{\theta|e,\zeta}\{v(a, \tilde{\theta})|e, \tilde{\zeta}\}]|e\} - \max_{a\in A}[E_\theta\{v(a, \tilde{\theta})\}]$,

and

(ii) $$\$(k(e, \tilde{\zeta}); e) = E_{\zeta|e}\{k(e, \tilde{\zeta})|e\}.$$

(B) If $u(c) = -e^{-zc}$ for every c in R^1 and some $z > 0$, then

(i) $$I(e) = \left(\frac{1}{z}\right)\log_e\left[\frac{\min_{a\in A}[E_\theta\{e^{-zv(a,\tilde{\theta})}\}]}{E_{\zeta|e}\{\min_{a\in A}[E_{\theta|e,\zeta}\{e^{-zv(a,\theta)}|e, \tilde{\zeta}\}]|e\}}\right],$$

and

$$\text{(ii)}\quad \$(k(e, \tilde{\zeta}); e) = \left(\frac{1}{z}\right) \log_e \left[\frac{E_{\zeta|e}\{\min_{a\in A}[E_{\theta|e,\zeta}\{e^{-z[v(a,\tilde{\theta})-k(e,\tilde{\zeta})]}|e, \tilde{\zeta}\}]|e\}}{E_{\zeta|e}\{\min_{a\in A}[E_{\theta|e,\zeta}\{e^{-zv(a,\tilde{\theta})}|e, \tilde{\zeta}\}]|e\}}\right]$$

■ *Proof.* Exercise 8.5.1. ■

Example 8.5.1

Suppose $A = \{a^1, a^2\}$, $\Theta = \{\theta^1, \theta^2\}$, $Z(e) = \{\zeta^1, \zeta^2\}$, $v(a^1, \theta^1) = v(a^2, \theta^2) = 10$, $v(a^1, \theta^2) = v(a^2, \theta^1) = 0$, $P_\theta(\theta^1) = .7$, $P_\theta(\theta^2) = .3$, $P_{\zeta|e,\theta}(\zeta^1|e, \theta^1) = P_{\zeta|e,\theta}(\zeta^2|e, \theta^2) = .8$, and $P_{\zeta|e,\theta}(\zeta^2|e, \theta^1) = P_{\zeta|e,\theta_1}(\zeta^1|e, \theta^2) = .2$. The terminal-action problem as represented by $v\colon A \times \Theta \to R^1$ might be regarded as an estimation problem, with a^i denoting the estimate that $\tilde{\theta} = \theta^i$, and with a reward of \$10 for correctness. Experiment e might be regarded as a prediction of $\tilde{\theta}$, with ζ^i predicting that $\tilde{\theta} = \theta^i$, and with predictions having probability .8 of correctness. Now suppose that $u(c) = c$ for every c in R^1. *Without* information,

$$E_\theta\{v(a^1, \tilde{\theta})\} = (10)(.7) + (0)(.3) = 7$$

and

$$E_\theta\{v(a^2, \tilde{\theta})\} = (0)(.7) + (10)(.3) = 3,$$

so that the optimal act without information is a^1 and $\max_{a\in A}[E_\theta\{v(a, \tilde{\theta})\}] = 7$. *With* e-information, we must first determine $P_{\theta|e,\zeta}(\cdot|e, \zeta^1)$ and $P_{\theta|e,\zeta}(\cdot|e, \zeta^2)$ by Bayes' Theorem. This is easily done and yields the following table.

$P_{\theta|e,\zeta}(\theta^j|e, \zeta^i)$

	θ^1	θ^2	$P_{\zeta|e}(\zeta^i|e)$
ζ^1	56/62	6/62	.62
ζ^2	14/38	24/38	.38

Hence, given that $\tilde{\zeta} = \zeta^1$,

$$E_{\theta|e,\zeta}\{v(a^1, \tilde{\theta})|e, \zeta^1\} = (10)(56/62) + (0)(6/62) = 560/62$$

and

$$E_{\theta|e,\zeta}\{v(a^2, \tilde{\theta})|e, \zeta^1\} = (0)(56/62) + (10)(6/62) = 60/62,$$

implying that the optimal act given $\tilde{\zeta} = \zeta^1$ is a^1, and that

$$\max_{a\in A}[E_{\theta|e,\zeta}\{v(a, \tilde{\theta})|e, \zeta^1\}] = 560/62.$$

But given $\tilde{\zeta} = \zeta^2$,

$$E_{\theta|e,\zeta}\{v(a^1, \tilde{\theta})|e, \zeta^2\} = (10)(14/38) + (0)(24/38) = 140/38$$

and

$$E_{\theta|e,\zeta}\{v(a^2, \tilde{\theta})|e, \zeta^2\} = (0)(14/38) + (10)(24/38) = 240/38,$$

implying that the optimal act given $\tilde{\zeta} = \zeta^2$ is a^2, and that

$$\max_{a\in A}[E_{\theta|e,\zeta}\{v(a, \tilde{\theta})|e, \zeta^2\}] = 240/38.$$

Hence

$$E_{\zeta|e}\{\max_{a\in A}[E_{\theta|e,\zeta}\{v(a, \tilde{\theta})|e, \tilde{\zeta}\}]|e\} = (560/62)(.62) + (240/38)(.38) = 8.00.$$

Hence Theorem 8.5.2(Ai) implies that $I(e) = 8.00 - 7.00 = \$1.00$. By Theorem 8.5.2(Aii) and Theorem 8.5.1, $\mathcal{D}$ should want to purchase e vis-à-vis nothing if $.62k(e, \zeta^1) + .38k(e, \zeta^2) < 1.00$.

Example 8.5.2

If everything remains as in Example 8.5.1 except that now $P_{\zeta|e,\theta}(\zeta^1|e, \theta^1) = P_{\zeta|e,\theta}(\zeta^2|e, \theta^2) = .6$ instead of .8, it is easy to verify, in Exercise 8.5.2, that the optimal act given each of ζ^1 and ζ^2 is a^1, that

$$\max_{a\in A}[E_{\theta|e,\zeta}\{v(a, \tilde{\theta})|e, \zeta^1\}] = 420/54,$$

that

$$\max_{a\in A}[E_{\theta|e,\zeta}\{v(a, \tilde{\theta})|e, \zeta^2] = 280/46,$$

and hence that

$$E_{\zeta|e}\{\max_{a\in A}[E_{\theta|e,\zeta}\{v(a, \tilde{\theta})|e, \tilde{\zeta}\}]|e\} = 7.00,$$

so $I(e) = 0$. Now, with the probabilities of correct prediction lowered, the prediction is not sufficiently reliable to be able to influence $\mathcal{D}$'s terminal act a; and because he knows for sure what he will do (namely, choose a^1) *after* observing the outcome $\tilde{\zeta}$ of e, why bother to observe $\tilde{\zeta}$ or, indeed, to conduct e? Experiment e is therefore worthless.

It should be obvious that noninformation is valueless, but this fact is of some use later.

Corollary 8.5.1. $I(e^-) = 0$.

■ *Proof.* Immediate from the fact that Definition 8.5.1 requires that

$$\max_{a\in A}[E_\theta\{u(v(a, \tilde{\theta}) - I(e^-))\}] = \max_{a\in A}[E_\theta\{u(v(a, \tilde{\theta}))\}]. \blacksquare$$

It turns out that $I(e^-) = 0$ is a *lower bound* on $I(e)$ for every conceivable e. That is, we always have $I(e) \geq 0$. An *upper* bound on $I(e)$ is the value $I(e^+)$ of *perfect* information. We defined perfect information in Example 3.7.7, but an obviously equivalent definition is that perfect information consists in allowing $\mathcal{D}$ to choose a after observing θ itself. Therefore, $I(e^+)$ is the unique real number x^* such that

$$E_\theta\{\max_{a\in A}[u(v(a, \tilde{\theta}) - x^*)]\} = \max_{a\in A}[E_\theta\{u(v(a, \tilde{\theta}))\}]. \quad (8.5.1)$$

With $I(e^-) = 0$ and the definition of $I(e^+)$ in appropriate form, we are prepared to state the principal bounding inequalities on $I(e)$.

Theorem 8.5.3. If e is any experiment in a statistical decision problem, then

$$0 \leq I(e) \leq I(e^+).$$

■ *Proof.* $0 \leq I(e)$ follows readily by comparing the first and last terms of the following inequality chain and noting that $E_{\zeta|e}\{\max_{a \in A} [E_{\theta|e,\zeta}\{u(v(a, \tilde{\theta}) - x)|e, \tilde{\zeta}\}]|e\}$ is strictly decreasing and continuous in x.

$$\begin{aligned}
&E_{\zeta|e}\{\max_{a \in A} [E_{\theta|e,\zeta}\{u(v(a, \tilde{\theta}) - I(e))|e, \tilde{\zeta}\}]|e\} \\
&= \max_{a \in A} [E_\theta\{u(v(a, \tilde{\theta}))\}] && \text{[Definition 8.5.1]} \\
&= \max_{a \in A} [E_\theta\{E_{\zeta|e,\theta}\{u(v(a, \tilde{\theta}))|e, \tilde{\theta}\}\}] && \text{[Exercise 3.7.7]} \\
&= \max_{a \in A} [E_{\zeta|e}\{E_{\theta|e,\zeta}\{u(v(a, \tilde{\theta}))|e, \tilde{\zeta}\}|e\}] && \text{[Exercise 3.7.6 twice]} \\
&\leq E_{\zeta|e}\{\max_{a \in A} [E_{\theta|e,\zeta}\{u(v(a, \hat{\theta}))|e, \tilde{\zeta}\}]|e\}. && \text{[Exercise 3.7.5(C)]}
\end{aligned}$$

To prove that $I(e) \leq I(e^+)$ involves a similar comparison of the first and last terms in the chain

$$\begin{aligned}
&E_\theta\{\max_{a \in A} [u(v(a, \tilde{\theta}) - I(e))]\} \\
&= E_{\zeta|e}\{E_{\theta|e,\zeta}\{\max_{a \in A} [u(v(a, \tilde{\theta}) - I(e))]|e, \tilde{\zeta}\}|e\} && \text{[Exercise 3.7.7]} \\
&\geq E_{\zeta|e}\{\max_{a \in A} [E_{\theta|e,\zeta}\{u(v(a, \tilde{\theta}) - I(e))|e, \tilde{\zeta}\}]|e\} && \text{[Exercise 3.7.5(C)]} \\
&= \max_{a \in A} [E_\theta\{u(v(a, \tilde{\theta}))\}] && \text{[Definition 8.5.1]} \\
&= E_\theta\{\max_{a \in A} [u(v(a, \tilde{\theta}) - I(e^+))]\}. && \text{[Definition 8.5.1]}
\end{aligned}$$

In the first chain, the reference to Exercise 3.7.6 twice means that

$$E_1\{E_{2|1}\{x(\tilde{\omega}_1, \tilde{\omega}_2)|\tilde{\omega}_1\}\} = E_2\{E_{1|2}\{x(\tilde{\omega}_1, \tilde{\omega}_2)|\tilde{\omega}_2\}\}. \blacksquare$$

Example 8.5.3

In the statistical decision problem of Example 8.5.1,

$$\max_{a \in A} [v(a, \theta^1) - x] = 10 - x = \max_{a \in A} [v(a, \theta^2) - x],$$

so that

$$E_\theta\{\max_{a \in A} [v(a, \tilde{\theta}) - x]\} = 10 - x.$$

Hence

$$10 - I(e^+) = 7.00 = \max_{a \in A} [E_\theta\{v(a, \tilde{\theta})\}],$$

implying that $I(e^+) = 3.00$.

The value $I(e^+)$ of perfect information plays an important role in screening available experiments, even though *perfect information is usually unavailable at any cost.* Suppose, for example, that someone proposes a complicated experiment e with a constant cost $k(e)$ which happens to exceed $I(e^+)$. Then you can reject e without determining $I(e)$, since if $k(e) > I(e^+)$, then $k(e) > I(e)$ because $I(e^+) \geq I(e)$, and hence e is preferable to e at cost $k(e)$.

Corollary 8.5.2. If $\$(k(e, \tilde{\zeta}); e) > I(e^+)$, then e may be rejected as undesirable.

■ *Proof.* An obvious application of Theorems 8.5.1 and 8.5.3. ■

■ That Corollary 8.5.2 is a valid generalization of the preceding discussion follows from the fact that if $k(e)$ is nonrandom, then

$$\$(k(e); e) = k(e), \tag{8.5.2}$$

which you may prove in Exercise 8.5.3. ■

Up to this point this section has been concerned with the comparison between *one* experiment e at cost $k(e, \tilde{\zeta})$ and the no-information experiment e^-, at cost zero. Nothing has yet been said about comparing any two experiments with each other and about the determination of an optimal experiment. To do so necessitates defining the *net gain* of paying $k(e, \tilde{\zeta})$ for e-information, and this definition requires the preliminary definition of $\mathscr{D}$'s *reservation price* $R(e, k(e, \tilde{\zeta}))$ of the decision-*cum*-e-information at cost $k(e, \tilde{\zeta})$.

Definition 8.5.3. The smallest amount $R(e, k(e, \tilde{\zeta}))$ which $\mathscr{D}$ should be willing to accept for his right to take terminal action after e-information acquired at cost $k(e, \tilde{\zeta})$ is called his *reservation price*, and is defined implicitly by

$$u(R(e, k(e, \tilde{\zeta}))) = E_{\zeta|e}\{\max_{a\in A}[E_{\theta|e,\zeta}\{u(v(a, \tilde{\theta}) - k(e, \tilde{\zeta}))|e, \zeta\}]|e\}.$$

Now one natural measure of $\mathscr{D}$'s net gain of acquiring e-information at its cost is the amount by which his reservation price $R(e, k(e, \tilde{\zeta}))$ of the decision *cum* information and cost exceeds his reservation price $R(e^-, 0)$ of the decision *cum* no information and no cost.

Definition 8.5.4. $\mathscr{D}$'s *net* R-*gain* $G_R(e)$ of experiment e is defined by

$$G_R(e) = R(e, k(e, \tilde{\zeta})) - R(e^-, 0).$$

■ Let S'', $\mathbf{c}''$, and $\mathbf{y}$ be as defined in preceding remarks. Then $R(e, k(e, \tilde{\zeta}))$ is simply $R(\mathscr{Q}_S{}^{S''}, \mathbf{c}'' - \mathbf{y})$ as defined in Section 8.3. Furthermore, if S' and $\mathbf{c}$ are also as defined in the preceding remarks, then $G_R(e)$ is simply $G_R(\mathbf{y}; \mathscr{Q}_S{}^{S'}, \mathbf{c}' \to \mathscr{Q}_S{}^{S''}, \mathbf{c}'')$ as defined by Equation (8.3.8). ■

It is possible to define other types of net gain, as should be clear to any alumnus of Section 8.3, but we shall refrain from doing so here.

Net R-gains can be expressed in an intuitively natural form when $u(\cdot)$ has one of the two "nice" forms.

Theorem 8.5.4. If either $u(c) = c$ for every c in R^1 or $u(c) = -e^{-zc}$ for every c in R^1 and some $z > 0$, then

$$G_R(e) = I(e) - \$(k(e, \tilde{\zeta}); e).$$

■ *Proof.* Exercise 8.5.4. ■

The usefulness of net R-gain is manifested in our final theorem, 8.5.5, which shows that choosing e so as to maximize $G_R(e)$ is equivalent to choosing e so as to maximize $\mathcal{D}$'s utility $E_{\zeta|e}\{\max_{a\in A}[E_{\theta|e,\zeta}\{u(v(a,\tilde{\theta})-k(e,\tilde{\zeta}))|e,\tilde{\zeta}\}]|e\}$.

Theorem 8.5.5. An *optimal* experiment e° is any maximizer of $G_R(e)$.

■ *Proof.* Note that $G_R(e'')\geq G_R(e')$ if and only if $R(e'',k(e'',\tilde{\zeta}))\geq R(e',k(e',\tilde{\zeta}))$, which [by strict increasingness of $u(\cdot)$ and Definition 8.5.3] is true if and only if

$$E_{\zeta|e}\{\max_{a\in A}[E_{\theta|e,\zeta}\{u(v(a,\tilde{\theta})-k(e'',\tilde{\zeta}))|e'',\tilde{\zeta}\}]|e''\}$$
$$\geq' E_{\zeta|e}\{\max_{a\in A}[E_{\theta|e,\zeta}\{u(v(a,\tilde{\theta})-k(e',\tilde{\xi}))|e',\tilde{\zeta}\}]|e'\}.$$

Hence $G_R(e'')\geq G_R(e')$ if and only if paying $k(e'',\tilde{\zeta})$ for e''-information is a priori at least as desirable as paying $k(e',\tilde{\zeta})$ for e'-information. ■

Example 8.5.4

In Example 8.5.1, suppose that $k(e,\zeta^1)=.25$ and $k(e,\zeta^2)=1.20$. Then $\$(k(e,\tilde{\zeta});e)=E_{\zeta|e}\{k(e,\tilde{\zeta})|e\}=(.25)(.62)+(1.20)(.38)=.611$, so that $G_R(e)=1.000-.611=.389$. Next, suppose that the revised experiment in Example 8.5.2 is denoted by e^*, and that $k(e^*,\zeta^1)=-2.00$ and $k(e^*,\zeta^2)=2.00$. Then $\$(k(e^*,\tilde{\zeta});e^*)=E_{\zeta|e}\{k(e^*,\tilde{\zeta})|e^*\}=(-2.00)(.62)+(2.00)(.38)=-.480$. Hence $G_R(e^*)=0.000+.480=.480$, and $\mathcal{D}$ should choose e^* over e and e^-. Clearly, the desirability of e^* here is due solely to the negative cost given $\tilde{\zeta}=\zeta^1$—a subsidy for experimentation!

■ A great deal of literature exists on the statistical decision problem. Most of it—for example, Chapters 17–19 of LaValle [**78**]—consists of detailed studies of sampling problems and makes extensive use of models such as those in Part C of Chapter 3. See the references in Section 8.1. You may try your hand at one such problem as *Exercise 8.5.9. ■

Exercise 8.5.1

Prove Theorem 8.5.2.

Exercise 8.5.2

(A) In Example 8.5.2, verify that the optimal act given each of ζ^1 and ζ^2 is a^1, and verify the asserted values of

$$\max_{a\in A}[E_{\theta|e,\zeta}\{v(a,\tilde{\theta})|e,\zeta^i\}].$$

(B) Prove that $I(e)=0$ whenever there is a terminal act a^* which is optimal for every ζ in $Z(e)$.

Exercise 8.5.3: Properties of $\$(\cdot; e)$

Prove that

(A) $\$(x; e) = x$ for every x in R^1, and
(B) $\$(k''(e, \tilde{\zeta}); e) \geq \$(k'(e, \tilde{\zeta}); e)$ if $k''(e, \zeta) \geq k'(e, \zeta)$ for every ζ in $Z(e)$.

Exercise 8.5.4

Prove Theorem 8.5.4.

Exercise 8.5.5: Computational Opportunity

Keep all assumptions as in Example 8.5.1, except now assume that $P_{\zeta|e,\theta}(\zeta^i|e, \theta^i) = p$ for $i = 1, 2$, and for p variable in $[0, 1]$.

(A) Determine $I(e)$ as a function of p.
(B) Graph $I(e)$ as a function of p.

Exercise 8.5.6: Computational Opportunity

Let $\Theta = \{\theta^1, \theta^2, \theta^3\}$, $A = \{a^1, a^2, a^3, a^4\}$, and suppose that $v(a, \theta)$ is given by the following table.

	θ^1	θ^2	θ^3
a^1	20	0	0
a^2	0	15	0
a^3	0	0	10
a^4	10	10	5

Suppose that $P_\theta(\theta^1) = .30$ and $P_\theta(\theta^2) = .40$. Also suppose that $u(c) = c$ for every c in R^1. Finally, suppose that there are two available experiments besides e^-, namely, e^1 and e^2, characterized by their tables of conditional sampling probabilities.

$P_{\zeta|e,\theta}(\cdot|e^1, \cdot)$

	θ^1	θ^2	θ^3
ζ^1	.50	.20	.10
ζ^2	.25	.60	.10
ζ^3	.25	.20	.80

$P_{\zeta|e,\theta}(\cdot|e^2, \cdot)$

	θ^1	θ^2	θ^3
ζ^1	.60	.05	.00
ζ^2	.10	.65	.00
ζ^3	.05	.05	.75
ζ^4	.25	.25	.25

Compute:

(A) $I(e^+)$.
(B) $I(e^1)$.
(C) $I(e^2)$.
(D) Suppose that an experiment e^* yielding 163 possible outcomes is available, and its (skyscraperish) table of conditional sampling

probabilities indicates that it should convey very good information about $\tilde{\theta}$. Suppose that $k(e^*) = 7.83$. Do you wish more information about e^* before considering it seriously?

Exercise 8.5.7: Computational Opportunity

Assume that $u(c) = -(20 - c)^2$, and redo parts (A), (B), and (C) of Exercise 8.5.6 with this single altered assumption.

Exercise 8.5.8: Decision Functions and Admissibility

Decision functions are encountered in much of the statistical decision-theoretic literature. Given an experiment e with outcome $\mathscr{E}$-move $Z(e)$, a *decision function* is a function $d: Z(e) \to A$, specifying a terminal act $a = d(\zeta)$ for each potential outcome ζ of e. Let $D(e)$ be the set of all decision functions given e. A decision function d for e is said to be *admissible* if there is no other decision function d^* for e such that

$$E_{\zeta|e,\theta}\{u(v(d^*(\tilde{\zeta}), \theta) - k(e, \tilde{\zeta}))|e, \theta\} \geq E_{\zeta|e,\theta}\{u(v(d(\tilde{\zeta}), \theta) - k(e, \tilde{\zeta})|e, \theta\}$$

for every $\theta \in \Theta$ with strict inequality holding for some $\theta \in \Theta$.

(A) Show that if $d \in D(e)$, then (e, d) is a pure strategy for $\mathscr{D}$.
(B) Show that the definition of admissibility of a decision function is consistent with the terminology in Chapter 7.

***Exercise 8.5.9: Point Estimation with Quadratic Loss**

This exercise presents an opportunity to apply the probability models of part C in Chapter 3 to an important class of statistical decision problems.

(A) *Loss functions.* Show that if $q: \Theta \to R^1$ is any real-valued function on Θ and

$$\ell(a, \theta) = q(\theta) - v(a, \theta) \tag{8.5.3}$$

for every (a, θ) in $A \times \Theta$, then

(i) $a°$ maximizes $E_\theta\{v(a, \tilde{\theta})\}$ if and only if $a°$ minimizes $E_\theta\{\ell(a, \tilde{\theta})\}$; and
(ii) $e°$ maximizes $E_{\zeta|e}\{\max_{a\in A}[E_{\theta|e,\zeta}\{v(a, \tilde{\theta})|e, \tilde{\zeta}\}]|e\}$ if and only if $e°$ minimizes $E_{\zeta|e}\{\min_{a\in A}[E_{\theta|e,\zeta}\{\ell(a, \tilde{\theta})|e, \tilde{\zeta}\}]|e\}$.

■ Equation (8.5.3) defines a special case of an (*opportunity*) *loss function,* defined more generally in LaValle [**78**, p. 472], for example. The loss function (8.5.3) is useful in conjunction with the assumption that $u(c) = c$ for every c in R^1. Exercise 8.5.9(A) shows that $\ell(\cdot, \cdot)$ is just as useful as $v(\cdot, \cdot)$ for deriving solutions to statistical decision problems with linear utility, and this is important, because in many applied statistical decision problems it is easier to determine the values of $\ell(\cdot, \cdot)$ than the values of $v(\cdot, \cdot)$. ■

(B) *Point-estimation problems with quadratic loss.* Suppose that $A = \Theta = R^1$ and that there is a positive constant k_d such that

$$\ell(a, \theta) = k_d(a - \theta)^2 \tag{8.5.4}$$

for every (a, θ) in $A \times \Theta$. Show that:

(i) $a°$ minimizes $E_\theta\{\ell(a, \tilde{\theta})\}$ if and only if $a° = E_\theta\{\tilde{\theta}\}$; and
(ii) $\min_{a \in A} [E_\theta\{\ell(a, \tilde{\theta})\}] = k_d V_\theta\{\tilde{\theta}\}$.

■ The terminal acts a in Exercise 8.5.9(B) may be regarded as *estimates* of θ. The quadratic loss function (8.5.4) is zero only if $a = \theta$, that is, only if the estimate is precisely correct. Otherwise, the loss is positive and increases with the magnitude $|a - \theta|$ of the error. This is quite in accord with rough judgments to the effect that it is good to be right and bad to be wrong. ■

■ In the remainder of this exercise it will be assumed that $u(c) = c$ for every c in R^1, and that $\ell(\cdot, \cdot)$ is given by Equation (8.5.4). ■

(C) *Beta-binomial case.* Suppose that θ denotes the probability that a specified phenomenon occurs on any given instance, that $\mathscr{D}$'s prior judgments about $\tilde{\theta}$ are expressed by a *beta* density function $f_\beta(\cdot|r', n')$ as defined in Example 3.10.6, and that $\mathscr{D}$ can obtain information about $\tilde{\theta}$ by observing the occurrence or nonoccurrence of the phenomenon in each of n instances. Given that occurrences in different instances are unrelated (or independent), the total number $\tilde{\zeta}$ of occurrences in n observed instances had the *binomial* mass function $f_b(\cdot|n, \theta)$ as defined in Example 3.9.2, conditional on θ. Let e^n denote the experiment which consists in observing n instances. Use Example 3.11.1, Exercise 3.10.6, and whatever other results in Part C of Chapter 3 are pertinent to show that:

(i) $\min_{a \in A} [E_{\theta|e,\zeta}\{\ell(a, \tilde{\theta})|e^n, \zeta\}] = k_d V_\beta\{\tilde{\theta}|r' + \zeta, n' + n\}$;
(ii) $E_{\zeta|e}\{\min_{a \in A} [E_{\theta|e,\zeta}\{\ell(a, \tilde{\theta})|e^n, \hat{\zeta}\}]|e^n\} = [n'/(n' + n)]k_d V_\beta\{\tilde{\theta}|r', n'\}$;
(iii) $I(e^n) = [n/(n' + n)]k_d V_\beta\{\tilde{\theta}|r', n'\}$; and
(iv) if there is a *positive* constant k_e and a *nonnegative* constant K_e such that $k(e^n, \zeta) = K_e + k_e \cdot n$ for every (n, ζ) with $n > 0$, then

$$G_R(e^n) = \begin{cases} 0, & n = 0 \\ \left(\dfrac{n}{n' + n}\right) k_d V_\beta\{\tilde{\theta}|r', n'\} - K_e - k_e \cdot n, & n > 0. \end{cases}$$

■ Now an *optimal* number $n°$ of observations can be determined according to a two-step procedure. Step 1 consists in ignoring K_e and determining a maximizer n^* of $G_R^*(n) = [n/(n' + n)]k_d V_\beta\{\tilde{\theta}|r', n'\} - k_e \cdot n$ on $\{n: n = 0, 1, 2, \ldots\}$. By noting that n^* is either zero or must satisfy the inequalities $G_R^*(n^*) \geq G_R^*(n^* - 1)$ and $G_R^*(n^*) \geq G_R^*(n^* + 1)$, it follows by a little algebra that either $n^* = 0$ or

$$n' + n^* - 1 \leq \left(\frac{n'}{n' + n^*}\right)\left(\frac{k_d}{k_e}\right) V_\beta\{\tilde{\theta}|r', n'\} \leq n' + n^* + 1.$$

[Alternatively, one can approximate n^* by assuming that n can be any nonnegative real number (not necessarily an integer), set the first derivative $dG_R^*(n)/dn$ equal to zero, and ascertain that $n^* = [k_d n' V_\beta\{\tilde{\theta}|r', n'\}/k_e]^{1/2} - n'$ unless this is negative, in which case $n^* = 0$.] Step 2 consists in calculating $G_R(e^{n^*}) = G_R^*(n^*) - K_e$. If this is positive, then $n° = n^*$; otherwise, $n° = 0$. (Why?) ■

(D) *Gamma-Poisson case.* Suppose that θ denotes the intensity rate of a Poisson process, in that the number $\tilde{\zeta}$ of occurrences of a phenomenon during any time interval of length t has the *Poisson* mass function $f_{Po}(\cdot|\theta t)$ as defined in Example 3.9.6, conditional on θ. Suppose that $\mathscr{D}$'s prior judgments about $\tilde{\theta}$ are expressed by a *gamma* density function $f_\gamma(\cdot|r', t')$. Let e^t denote the experiment which consists in observing the process for a time interval of length t. Use Example 3.11.2, Exercise 3.10.6, and whatever other results in part C of Chapter 3 are pertinent to show that:

(i) $\min_{a\in A}[E_{\theta|e,\zeta}\{\ell(a, \tilde{\theta})|e^t, \zeta\}] = k_d V_\gamma\{\tilde{\theta}|r' + \zeta, t' + t\}$;
(ii) $E_{\tilde{\zeta}|e}\{\min_{a\in A}[E_{\theta|e,\zeta}\{\ell(a, \tilde{\theta})|e^t, \tilde{\zeta}\}]|e^t\} = [t'/(t' + t)]k_d V_\gamma\{\tilde{\theta}|r', t'\}$;
(iii) $I(e^t) = [t/(t' + t)]k_d V_\gamma\{\theta|r', t'\}$;
(iv) if there is a *positive* constant k_e and a *nonnegative* constant K_e such that $k(e^t, \zeta) = K_e + k_e \cdot t$ for every (t, ζ) with $t > 0$, then

$$G_R(e^t) = \begin{cases} 0, & t = 0 \\ \left(\dfrac{t}{t' + t}\right) k_d V_\gamma\{\tilde{\theta}|r', t'\} - K_e - k_e \cdot t, & t > 0; \end{cases}$$

(v) the optimal timespan $t°$ to observe the process is $t^* = [k_d t' V_\gamma\{\tilde{\theta}|r', t'\}/k_e]^{1/2} - t'$ provided that $t^* > 0$ and $G_R(e^{t^*}) > 0$; otherwise, $t° = 0$.

(E) *Normal-normal case.* Suppose that θ denotes some unknown quantity about which information can be obtained in the form of unrelated observations $\tilde{\xi}_1, \tilde{\xi}_2, \ldots$, each of which has the univariate *Normal* density function $f_N(\cdot|\theta, V)$ as defined in Example 3.10.7, conditional on θ. It can then be shown that if $\tilde{\zeta} = (1/n)\Sigma_{i=1}^n \tilde{\xi}_i$—the "sample mean"—then $\tilde{\zeta}$ has the Normal density function $f_N(\cdot|\theta, V/n)$ conditional on θ. Let e^n denote the experiment in which n unrelated observations $\tilde{\xi}_i$ are obtained and $\tilde{\zeta}$ is recorded. Suppose that $\mathscr{D}$'s prior judgments about $\tilde{\theta}$ are expressed by a Normal density function $f_N(\cdot|m', V')$, and write $V' = V/n'$ to define n'. Use Example 3.11.3, Exercise 3.10.6, and whatever other results in part C of Chapter 3 are pertinent to show that:

(i) $\min_{a\in A}[E_{\theta|e,\zeta}\{\ell(a, \tilde{\theta})|e^n, \zeta\}] = k_d[V/(n' + n)]$;
(ii) $E_{\tilde{\zeta}|e}\{\min_{a\in A}[E_{\theta|e,\zeta}\{\ell(a, \tilde{\theta})|e^n, \tilde{\zeta}\}]|e^n\} = k_d[V/(n' + n)]$;
(iii) $I(e^n) = [n/(n' + n)]k_d[V/n']$.

■ Defining $G_R(e^n)$ and finding $n°$ now proceeds exactly as in part (C), the beta-binomial case, with V/n' here replacing $V_\beta\{\tilde{\theta}|r', n'\}$ there. ■

9

APPROACHES TO STATISTICAL INFERENCE

> The whole art of war consists in getting at what
> lies on the other side of the hill, or, in other
> words, in deciding what we do not know from what we do.
>
> F.M. the Duke of Wellington

9.1 INTRODUCTION

Statistics is an immeasurable subject—and, broadly defined, ancient. Because the subject is so vast, it is useful to subdivide it into three, necessarily related, sub-subjects: statistical description, statistical inference, and statistical decision theory.

Statistical description, clearly the oldest branch of statistics, is concerned with the organization, summarization, and presentation of data and includes much if not all of accounting. We could stretch the point and assert that statistical description includes the work of the cave-dwelling artists who presented to their youths the essential data concerning the effective hunting of mammoths. In more recent times, this area of statistics is most often thought of as concerning effective graphing and tabulating of quantitative information.

Statistical inference is concerned with making conclusions and predictions about some aspect of reality on the basis of some imperfect information. Whenever we take a small sample from a big population and use the proportion *in the sample* favoring Brand X as our estimate of the proportion *in the population* favoring Brand X, we are making a statistical inference.

Statistical decision theory is concerned with making good decisions based on information. We have already encountered statistical decision theory, in Sections 8.4 and 8.5. What differentiates statistical decision theory from statistical inference is that statistical decision theory makes *explicit* use of (at least assumptions about) the *user's* (or decision maker's) available acts, his utility function, and the economics of his decision problem. In statistical inference these features are usually not considered formally.

There are no hard and fast boundaries between these three branches of statistics. Our trichotomy exists only for the purpose of placing the status of this chapter in an appropriate perspective.

■ Historically, these three branches of statistics developed roughly in the order listed as far as their central intellectual orientations are concerned. All are active areas for current research in view of the knowledge and data "explosion." Furthermore, general decision analysis as expounded in Chapters 1–4 and 7 of this book is a relatively immediate and painless outgrowth of statistical decision theory. ■

When the decision maker wishes to make inferences about a true state of affairs on the basis of less than perfect information, he needs a *model* which relates the unknown to the known. For our purposes in this chapter, the model will consist of the following.

(1) A nonempty set Θ of mutually exclusive and collectively exhaustive *states* θ.
(2) An *experiment* e which will furnish an *outcome* belonging to a nonempty set $Z(e)$ of mutually exclusive and collectively exhaustive outcomes ζ.
(3) All *conditional-sampling* probability functions $P_{\zeta|e,\theta}(\cdot|e, \theta)$ on events in $Z(e)$, one for each θ in Θ; and sometimes also
(4) A *prior* probability function $P_\theta(\cdot)$ on events in Θ.

■ We shall be only sporadically concerned in this chapter with the ramifications of varying e. By contrast, the main thrust of Sections 8.4 and 8.5 was determining the *optimal* e. Choosing e so that inferences possess desirable properties is known in statistical inference as the experimental (or statistical) *design* problem. ■

■ Experiments consisting of a predetermined number n of indistinguishable trials were discussed in Section 6.5 and constitute a special case of great importance in statistical practice, because it includes random sampling with replacement from a population. In such an experiment $\zeta = (\omega_1, \ldots, \omega_n)$ and hence

$$P_{\zeta|e,\theta}(\tilde{\zeta} \in \Omega_1' \times \cdots \times \Omega_n'|e, \theta) = \prod_{i=1}^{n} P(\tilde{\omega}_i \in \Omega_i'|\theta) \tag{9.1.1}$$

for every n events Ω_i' in Ω_i and every θ in Θ. When every Ω_i is a finite set, all $\tilde{\zeta}$-event probabilities can be obtained from

$$P_{\zeta|e,\theta}(\omega_1, \ldots, \omega_n|e, \theta) = \prod_{i=1}^{n} P(\omega_i|\theta), \tag{9.1.2}$$

the special case of (9.1.1) in which all Ω_i' are elementary events. ■

*■ In keeping with the analysis in Sections 3.9–3.11, we shall replace (3) and (4) in starred material to follow with their appropriate counterparts involving mass or density functions $f_{\zeta|e,\theta}(\cdot|e, \theta)$ and $f_\theta(\cdot)$ when $Z(e)$ and Θ are sets of real numbers or vectors. These starred items about statistical inference for infinite idealizations are of more than ordinary importance in relation to comparable passages elsewhere in this book, because they facilitate the reader's consultation of texts on statistical inference, most of which are almost exclusively concerned with infinite idealizations. ■

Example 9.1.1

In the subsystem problem, Example 1.5.1c, there are seven experiments besides the null experiment "no testing." In the notation of Sections 4.4 and 7.2, they are: $e^1 = \text{P}$, with $Z(e^1) = \{\text{PD}, \text{PN}\}$; $e^2 = \text{L}$, with $Z(e^2) = \{\text{LD}, \text{LN}\}$; $e^3 = \text{P}$ and L, with $Z(e^3) = \{(\text{PD}, \text{LD}), (\text{PD}, \text{LN}), (\text{PN}, \text{LD}), (\text{PN}, \text{LN})\}$ $e^4 = \text{P}$, and then L only if PD, with $Z(e^4) = \{\text{PN}, (\text{PD}, \text{LD}), (\text{PD}, \text{LN})\}$; $e^5 = \text{P}$, and then L only if PN, with $Z(e^5) = \{\text{PD}, (\text{PN}, \text{LD}), (\text{PN}, \text{LN})\}$; $e^6 = \text{L}$, and then P only if LD, with $Z(e^6) = \{\text{LN}, (\text{LD}, \text{PD}), (\text{LD}, \text{PN})\}$; and $e^7 = \text{L}$, and then P only if LN, with $Z(e^7) = \{\text{LD}, (\text{LN}, \text{PD}), (\text{LN}, \text{PN})\}$. In this problem, $\Theta = \{\text{AD}, \text{AN}\}$. For each of these experiments the conditional-sampling probabilities are readily deducible from the data in Examples 1.5.1b and 1.5.1c, which imply the following table of joint-test probabilities given the arrival states.

	(PD, LD)	(PD, LN)	(PN, LD)	(PN, LN)	*sum*
AD	.52	.28	.13	.07	1.00
AN	0	0	.25	.75	1.00

Thus the appropriate table for e^4 would be just this table with the first two columns summed.

	$P_{\zeta|e,\theta}(\zeta|e^4, \theta)$		
	$\zeta^1 = \text{PD}$	$\zeta^2 = (\text{PN}, \text{LD})$	$\zeta^3 = (\text{PN}, \text{LN})$
$\theta^1 = \text{AD}$	.80	.13	.07
$\theta^2 = \text{AN}$	0	.25	.75

Now, given whichever model, (1)–(3) or (1)–(4), of inference is adopted, the central question is: What inferences can be made about θ on the basis of having observed ζ?

■ In subsequent sections we shall occasionally consider a more general version of this question, one which assumes (1) that only the value δ of a *summarization function*, or *statistic*, $\delta : Z(e) \to \Delta$ of the outcome ζ will be observed; and/or (2) that inferences are desired, not about θ itself, but rather about the value ϕ of some *attribute* function $\phi : \Theta \to \Phi$ of the state θ. In the experiments e consisting of n indistinguishable trials, discussed in Section 6.5, $\zeta = (\omega_1, \ldots, \omega_n)$, but we used $\delta(\zeta) = \bar{x} = (1/n)\Sigma_{j=1}^{n} x(\omega_j)$ to make inferences about $\phi(\theta) = \mu(\theta) = E_{\omega|\theta}\{x(\tilde{\omega})|\theta\}$. Hence the more general version of the central question is: What inferences can be made about the value ϕ of a given function $\phi: \Theta \to \Phi$ of θ on the basis of having observed the value δ of a given function $\delta: Z(e) \to \Delta$ of ζ? ■

Our central question is still too nebulous, in that we have not yet specified the types of inference desired. Basically, there are two types: (1) *estimates* of θ, being either single values or sets of values; and (2) *tests of hypotheses* about θ. An *estimate* is a prediction of θ on the basis of the evidence ζ; a *test* of a hypothesis about θ is a statement, based on the evidence ζ, concerning whether that hypothesis is so inconsistent with the observed evidence as to be untenable.

Now, there are two basic *approaches to* statistical inference, one of which uses *only* the conditional-sampling probabilities $P_{\zeta|e,\theta}$ assumed given in the third part of the specification of the inference model. This approach is termed *sampling-theoretic.* The other, or *Bayesian* approach, uses the prior probability function P_θ in conjunction with $P_{\zeta|e,\theta}$ in order to determine, via Bayes' Theorem, the *posterior* probability function $P_{\theta|e,\zeta}$ on which all inferences are then based.

■ There appear to be no commonly accepted synonyms for "Bayesian inference," but there are several popular synonyms for "sampling-theoretic inference"; namely, "non-Bayesian inference," "orthodox inference," and "classical inference," the last two of which have pejorative connotations. "Classical inference" is downright misleading, because Bayesian inference preceded much of the development of sampling-theoretic inference, which itself was motivated by dissatisfaction with the consequences of careless assessments of P_θ. ■

Sampling-theoretic inferences are logically quite subtle and difficult to grasp at first—and sometimes at second and third as well! The reason for this subtlety ultimately devolves from the fact that sampling-theoretic methods endeavor to arrive at judgments posterior to experimentation while ignoring the existence of judgments prior to experimentation. Sections 9.2 and 9.3 concern sampling-theoretic estimation and sampling-theoretic testing of hypotheses respectively.

Section 9.4 concerns Bayesian inferences, always based on the posterior probability function $P_{\theta|e,\zeta}$ rather than on the conditional-sampling probability functions $P_{\zeta|e,\theta}$. The meanings of Bayesian inferences are not subtle; they mean what the layman mistakenly takes sampling-theoretic inferences to mean.

Section 9.5 concludes the chapter with a discussion of some basic behavioral principles for inference. These principles, or at least the central one, are intuitively compelling to many people, but the logical consequence is that many of the sampling-theoretic techniques are invalid! Why, then, should we spend time on this subject? For two reasons: first, to acquaint the reader with the meaning of sampling-theoretic inferences, which are widely used despite their often dubious justification; and second, because sampling-theoretic inferences are often adequate approximations to their Bayesian counterparts and thus their cost in terms of inaccuracy can be more than recaptured in the time saved by their not requiring the careful assessment of P_θ.

The principal objective of this chapter is to acquaint the reader with the basic approaches to statistical inference. In no sense can it supplant compendious tomes devoted exclusively to that subject. The reader interested in pursuing statistical inference may consult Brunk [**15**], Hays and Winkler [**56**], LaValle [**78**], Lindgren [**84**], and Mood and Graybill [**98**], among others. An exceptionally fine survey of the principles of Bayesian inference and how it compares with sampling-theoretic inference is that of Lindley [**85**]. Bayesian methods in econometric model building are impressively covered by Zellner [**151**].

Exercise 9.1.1: Discussion Question

In this Introduction we have presented less and more complete models, (1)–(3) and (1)–(4) respectively, which relate the known, ζ, to the unknown, θ, in a probabilistic fashion. Nonprobabilistic models, or at least not overtly probabilistic models, which attempt such a relation, have a long history. How about Druidical sacrifices for purposes of augury? How about principles of logic as the model by which one deduces the as-yet unknown consequences of known axioms? Can you think of other models?

Exercise 9.1.2

Can you cite an authority to support the statement that military science is a branch of statistical inference?

9.2 SAMPLING-THEORETIC ESTIMATION

As indicated in the Introduction, there are two basic types of estimates, *point* estimates and *set* estimates. The former calls for naming a single value as the estimate of θ, and we shall discuss it first, from the sampling-theoretic viewpoint.

Sampling-theoretic point estimation is a three-step procedure.

(1) Using $P_{\zeta|e,\theta}(\cdot|e,\cdot)$ to derive a function $\hat{\theta}[\cdot|e]$ on $Z(e)$, called an estima*tor* of θ.
(2) Performing e and observing its outcome ζ.
(3) Reporting $\hat{\theta}[\zeta|e]$ as the estimate of θ.

Clearly, (2) and (3) are purely mechanical; hence discussion will center on (1). It is in determining the estima*tor* $\hat{\theta}[\cdot|e]$ that controversy arises, because about all we can claim in the absence of a utility function $u(\cdot)$ for consequences of all estimate-state pairs $(\hat{\theta}[\cdot|e], \theta)$ and a prior probability function $P_\theta(\cdot)$ on Θ is that "it is good to be right and bad to be wrong." Thus the estimator should be chosen with a view to being right.

■ In *sampling-theoretic* inference, all inferences must be chosen on the basis of properties of the inferential *procedure* rather than properties of the inference per se, because the only basis on which to determine inferences is the preexperimentation, state-conditional sampling probability function $P_{\zeta|e,\theta}(\cdot|e, \cdot)$ on $Z(e) \times \Theta$. ■

We shall present and discuss two ways of choosing point estimates. The first is the method of maximum likelihood. As a preliminary, we define the likelihood function of e.

Definition 9.2.1: Likelihood Function. Given an experiment e with as-yet unobserved outcome ζ. The likelihood function $L(\cdot|e, \zeta)$ on Θ to R^1 is defined by

$$L(\theta|e, \zeta) = P_{\zeta|e,\theta}(\zeta|e, \theta)$$

for every θ in Θ and every ζ in $Z(e)$.

■ It is important to notice in Definition 9.2.1 that the likelihood function is simply the conditional-sampling probability function, but with the roles of ζ and θ reversed: θ is the "variable" in $L(\cdot|e, \cdot)$; it is the "parameter" in $P_{\zeta|e,\theta}(\cdot|e, \cdot)$. It is productive to interpret $L(\cdot|e, \zeta)$ as the function showing how probable the various possible states θ make the observation ζ. ■

*■ In the context of *Sections 3.9–3.11, the likelihood function $L(\cdot|e, \zeta)$ is defined by

$$L(\cdot|e, \zeta) = f_{\zeta|e,\theta}(\zeta|e, \cdot), \tag{9.2.1}$$

for every "parameter value" ζ. When $\{f_{\zeta|e,\theta}(\cdot|e, \theta): \theta \in \Theta\}$ is a set of density functions, the "how probable the various θ's make ζ" interpretation is preserved by replacing (9.2.1) with its probability-element equivalent $L(\cdot|e, \zeta)d\zeta = f_{\zeta|e,\theta}(\zeta|e, \cdot)d\zeta$, and then ignoring $d\zeta$ henceforth. ■

Example 9.2.1

When $P_{\zeta|e,\theta}(\cdot|e, \cdot)$ is expressed as a table with rows θ and columns ζ, it follows that $L(\cdot|e, \zeta)$ appears as column ζ of this table. For e^4 in Example 9.1.1, for instance, $L(\cdot|e^4, \zeta^2)$ is the second column of the $P_{\zeta|e,\theta}(\cdot|e^4, \cdot)$ table in Example 9.1.1. Hence $L(\theta^1|e^4, \zeta^2) = .13$ and $L(\theta^2|e^4, \zeta^2) = .25$.

Example 9.2.2

Consider an experiment e with the following table of conditional-sampling probabilities.

$P_{\zeta|e,\theta}(\zeta|e,\theta)$

	ζ^1	ζ^2	ζ^3	ζ^4
θ^1	.80	.10	.06	.04
θ^2	.10	.80	.04	.06
θ^3	.06	.10	.80	.04
θ^4	.10	.06	.04	.80

Then $L(\cdot|e, \zeta^3)$ is given by the third column, with $L(\theta^1|e, \zeta^3) = .06$ and $L(\theta^3|e, \zeta^3) = .80$, for instance.

Example 9.2.3

For subsequent purposes, we shall require a slightly less "regular" table, and hence consider the experiment e characterized by the following table.

$P_{\zeta|e,\theta}(\zeta|e,\theta)$

	ζ^1	ζ^2	ζ^3	ζ^4
θ^1	.90	.05	.03	.02
θ^2	.05	.90	.02	.03
θ^3	.02	.01	.90	.07
θ^4	.80	.13	.01	.06

Here, $L(\cdot|e, \zeta^1)$ is the first column of the table, meaning that $L(\theta^i|e, \zeta^1)$ for $i = 1, 2, 3, 4$ are .90, .05, .02, and .80 respectively.

***Example 9.2.4**

If an experiment e consists in performing n indistinguishable trials in each of which a phenomenon either occurs or does not occur, if it occurs with probability θ on any given trial, and if on trial j we set $\omega_j = 1$ or 0 according to whether the phenomenon did or did not occur, then $\zeta = (\omega_1, \ldots, \omega_n)$ and

$$L(\theta|e, \zeta) = \theta^{\delta}(1-\theta)^{n-\delta}, \tag{9.2.2a}$$

where $\delta = \Sigma_{j=1}^{n} \omega_j =$ the total number of occurrences. But from Example 3.10.6 we note that $L(\theta|e, \zeta)$ may be written instead as

$$L(\theta|e, \zeta) = k(\zeta) f_{\beta}(\theta|\delta + 1, n + 2), \tag{9.2.2b}$$

where $k(\zeta) = \delta!(n-\delta)!/(n+1)!$. Hence $L(\cdot|e, \zeta)$ has the same shape as the beta density function $f_{\beta}(\cdot|\delta + 1, n + 2)$.

*■ In Section 9.5 we shall see that there is good reason for asserting that *only the shape* of the likelihood function should be relevant for

inferences. That is, if outcomes ζ' of e' and ζ'' of e'' produce likelihood functions $L(\cdot|e', \zeta')$ and $L(\cdot|e'', \zeta'')$ which are proportional, in that

$$L(\theta|e', \zeta') = k(\zeta', \zeta'')L(\theta|e'', \zeta'')$$

for every θ in Θ, then a statistician who performed e' and obtained ζ' should make exactly the same inferences as if he had performed e'' and obtained ζ''. This is called the *likelihood principle.* ■

***Example 9.2.5**

In the notation of Chapter 6, suppose that e consists in performing n indistinguishable trials j and recording the value $x(\omega_j)$ of a measurement $x(\cdot)$ of the trial outcome ω_j. Suppose also that the population density function, of the typical measurement $x(\tilde{\omega})$, is the Normal density function $f_N(\cdot|\mu, v)$, where $v = V\{x(\tilde{\omega})\}$ is known but $\mu = E\{x(\tilde{\omega})\}$ is unknown. Then $\zeta = (x(\omega_1), \ldots, x(\omega_n))$ is the complete record of the measurements; and

$$L(\mu|e, \zeta) = k(\zeta)f_N|\mu(\bar{x}, v/n), \tag{9.2.3}$$

where

$$\bar{x} = (1/n)\sum_{j=1}^{n} x(\omega_j)$$

and

$$k(\zeta) = n^{-1/2}(2\pi v)^{-(n-1)/2}e^{-(1/2v)\sum_{j=1}^{n}[x(\omega_j)-\bar{x}]^2},$$

which you may verify as *Exercise 9.2.5. Now (9.2.3) and the functional nondependence of $k(\zeta)$ upon μ imply that $L(\cdot|e, \zeta)$ has the same shape as the Normal density function $f_N(\cdot|\bar{x}, v/n)$: It attains its maximum at $\mu = \bar{x}$ and falls away from that maximum symmetrically on both sides of $\bar{x}$.

We are now prepared to define the maximum-likelihood method of point estimation.

Definition 9.2.2

(A) An estimate $\hat{\theta}[\zeta|e]$ is called a *maximum-likelihood estimate* if

$$L(\hat{\theta}[\zeta|e]|e, \zeta) = \max_{\theta\in\Theta}[L(\theta|e, \zeta)].$$

(B) An estimator $\hat{\theta}[\cdot|e]$ is called a *maximum-likelihood estimator* if $\hat{\theta}[\zeta|e]$ is a maximum-likelihood estimate for every ζ in $Z(e)$.

Note that a maximum-likelihood estimate $\hat{\theta}[\zeta|e]$ makes the outcome ζ of e more likely than does any non-maximum-likelihood estimate. But that is by no means a guarantee of the goodness of maximum-likelihood estimators, since unlikely things occur all the time!

Example 9.2.6

For e^4 in the subsystem problem, whose likelihood function given the various ζ's is the various columns of the second table in Example 9.1.1, it is immediate that $\hat{\theta}[\zeta|e^4]$ is simply the row yielding the maximum in column ζ. Hence $\hat{\theta}[\zeta^1|e^4] = \theta^1 = \text{AD}$, $\hat{\theta}[\zeta^2|e^4] = \theta^2 = \text{AN}$, and $\hat{\theta}[\zeta^3|e^4] = \theta^2 = \text{AN}$. Having thus determined the maximum-likelihood estimator $\hat{\theta}[\cdot|e^4]$, it remains only for the experiment to be carried out and the outcome ζ plugged into $\hat{\theta}[\cdot|e^4]$.

Example 9.2.7

For e with $P_{\zeta|e,\theta}(\cdot|e, \cdot)$ tabulated in Example 9.2.2, it follows by reading down column ζ^1 that the maximum-likelihood estimate $\hat{\theta}[\zeta^1|e] = \theta^1$. In fact, it is clear that in this example $\hat{\theta}[\zeta^i|e] = \theta^i$ for $i = 1, 2, 3, 4$. In Example 9.2.3, however, θ^4 can never be the maximum-likelihood estimate, because $\hat{\theta}[\zeta^i|e] = \theta^i$ for $i = 1, 2, 3$, but $\hat{\theta}[\zeta^4|e] = \theta^3$.

Example 9.2.8

For e as defined in *Example 9.2.4, it is easy to verify that $\hat{\theta}[\zeta|e] = \delta/n$; that is, the maximum-likelihood estimate of the probability of an occurrence on any trial is the proportion δ/n of trials on which there were occurrences. A similar result obtains for e as defined in *Example 9.2.5, where (9.2.3) implies readily that the maximum-likelihood estimate $\hat{\mu}[\zeta|e]$ of μ should be the sample mean $\bar{x}$.

■ Although the intuitive justification of maximum-likelihood estimation may be somewhat dubious, this approach has a number of highly desirable properties as a method. First, it has the *invariance* property: If $t: \Theta \to T$ is a one-to-one function[a] of θ and if $\hat{\theta}[\cdot|e]$ is a maximum-likelihood estimator of θ, then $t(\hat{\theta}[\cdot|e])$ is a maximum-likelihood estimator of $t(\theta)$. Second, it has various good properties of an "asymptotic" sort as $n \to +\infty$ in experiments consisting of n indistinguishable trials. Description of these properties are beyond the level of this book. Third, it satisfies the likelihood principle, discussed in Section 9.5. ■

Another method for making point estimates of θ is applicable only when we desire a point estimate of a real-valued function $\phi(\theta)$ of θ, or when θ is itself real-valued.[b] For historical reasons, this method bears the overly connotive name of *unbiased* estimation.

Definition 9.2.3. An estimator $\hat{\phi}[\cdot|e]$ of a real-valued function $\phi: \Theta \to R^1$ of θ is called *unbiased* if

$$E_{\zeta|e,\theta}\{\hat{\phi}[\tilde{\zeta}|e]|e, \theta\} = \phi(\theta)$$

for every θ in Θ.

[a] A function $t: X \to Y$ is *one-to-one* if $t(x') \neq t(x'')$ whenever $x' \neq x''$; that is, distinct x's have distinct t-images.

[b] If θ is real-valued, define $\phi(\theta) = \theta$ to see that this case is a particular instance of assuming that we wish to estimate a real-valued function of θ.

In many instances, unbiased estimators can be found, but there exist situations in which no unbiased estimator exists. In other situations, a unique unbiased estimator exists but is ridiculous. In still other situations, infinitely many more or less reasonable unbiased estimators exist.

■ When Θ and $Z(e)$ are finite sets, so that $P_{\zeta|e,\theta}(\cdot|e,\cdot)$ can be tabulated as in Examples 9.1.1, 9.2.1–9.2.3, 9.2.6, and 9.2.7, the question of existence and uniqueness of an unbiased estimator reduces to a question of existence and uniqueness of a solution to a system of simultaneous linear equations. Let $p_{ij} = P_{\zeta|e,\theta}(\zeta^j|e, \theta^i)$ for all i and all j; let $\hat{\phi}_j = \hat{\phi}[\zeta^j|e]$; and let $\phi_i = \phi(\theta^i)$. Then the estimator $\hat{\phi}[\cdot|e]$ is just the set of numbers $\hat{\phi}_j$, and the definition of an unbiased estimator is equivalent to $\hat{\phi}_1, \ldots, \hat{\phi}_{\#(Z(e))}$ satisfying the $\#(\Theta)$ linear equations

$$\sum_{j=1}^{\#(Z(e))} p_{ij}\hat{\phi}_j = \phi_i \tag{9.2.4}$$

for $i = 1, \ldots, \#(\Theta)$. *[In linear-algebraic terms, (9.2.4) is the linear system $\mathbf{P}\hat{\boldsymbol{\phi}} = \boldsymbol{\phi}$ for $\mathbf{P}$ the matrix whose (i, j)th element is p_{ij}, $\hat{\boldsymbol{\phi}}$ the $\#(Z(e))$-tuple of $\hat{\phi}_j$'s, and $\boldsymbol{\phi}$ the $\#(\Theta)$-tuple of ϕ_i's. It is an elementary fact of linear-algebraic life that a solution $\hat{\boldsymbol{\phi}}$ *exists* if and only if the rank of $\mathbf{P}$ equals the rank of the augmented matrix $[\mathbf{P}, \boldsymbol{\phi}]$, in which $\boldsymbol{\phi}$ is tacked onto $\mathbf{P}$ as an extra column. If a solution $\hat{\boldsymbol{\phi}}$ exists, it is *unique* if and only if the rank of $\mathbf{P}$ is $\#(Z(e))$.] ■

Example 9.2.9

For e^4 in Example 9.1.1, suppose that $\theta^1 = \text{AD}$ has $\phi(\theta^1) = \phi_1 = 0$ and $\theta^2 = \text{AN}$ has $\phi(\theta^2) = 1$. Then the estimator $\hat{\phi}[\cdot|e^4]$ is unbiased if and only if the values $\hat{\phi}_j = \hat{\phi}[\zeta^j|e^4]$ satisfy the equations

$$.80\hat{\phi}_1 + .13\hat{\phi}_2 + .07\hat{\phi}_3 = 0,$$
$$.25\hat{\phi}_2 + .75\hat{\phi}_3 = 1.$$

It is easy to deduce that the solution exists but is not unique; indeed, *any* $(\hat{\phi}_1, \hat{\phi}_2, \hat{\phi}_3)$ of the form $(.40\hat{\phi}_3 - .65, 4 - 3\hat{\phi}_3, \hat{\phi}_3)$ solves this pair of equations. In particular, setting $\hat{\phi}_3 = 0$ yields the unbiased estimator with values $\hat{\phi}[\zeta^1|e^4] = -.65$, $\hat{\phi}[\zeta^2|e^4] = 4$, and $\hat{\phi}[\zeta^3|e^4] = 0$.

■ Note that in Example 9.2.9 unbiased estimates need not be possible values of $\phi(\theta)$. In that example, $-.65$ and 4 are not possible values. Furthermore, note that ζ^3 strongly indicates that $\theta^2 = \text{AN}$ obtains, but $\phi(\theta^2) = 1$, so that $\hat{\phi}_3 = 0$ is contraindicative. ■

Example 9.2.10

Ferguson [**34**] and Lindley [**85**] present the following example of a nonsensical but unique unbiased estimator. Suppose that the number $\tilde{\zeta}$ of calls which come into a switchboard during any timespan of t minutes is known to have a Poisson mass function $f_{Po}(\cdot|\theta t)$, but θ is unknown. Suppose also that the new telephone operator will observe the number ζ of calls coming in during the first minute on the job and use ζ to estimate the probability $e^{-2\theta}$ of no calls coming in to interrupt her subsequent, two-minute coffee break. Ferguson shows that the only unbiased estimator $\hat{\phi}[\cdot|e]$ of $\phi(\theta) = e^{-2\theta}$ is the function $(-1)^{\zeta}$, which says that the probability of not being interrupted is 1 if ζ is even and -1 if ζ is odd!!

Nonuniqueness of unbiased estimators is a more frequent problem than nonexistence. Usually, statisticians narrow down the choice in such cases by trying to find an unbiased estimator with minimum variance $V_{\zeta|e,\theta}\{\hat{\phi}[\tilde{\zeta}|e]\|e, \theta\}$ for all θ.

Example 9.2.11

For the case in Example 9.2.9, the statistician would choose $\hat{\phi}_3$ (and thus also $\hat{\phi}_1$ and $\hat{\phi}_2$) so as to minimize the variance of the resulting estimator with respect to the conditional-sampling probabilities. We readily calculate that

$$\begin{aligned} V_{\zeta|e,\theta}\{\hat{\phi}[\tilde{\zeta}|e^4]\|e^4, \theta^1\} &= .80[(.40\hat{\phi}_3 - .65) - 0]^2 + .13[(4 - 3\hat{\phi}_3) - 0]^2 \\ &\quad + .07[(\hat{\phi}_3) - 0]^2 \\ &= 1.368(\hat{\phi}_3)^2 - 3.536\hat{\phi}_3 + 2.418, \end{aligned}$$

minimized at $\hat{\phi}_3 = 3536/2736 \doteq 1.29$; and

$$\begin{aligned} V_{\zeta|e,\theta}\{\hat{\phi}[\tilde{\zeta}|e^4]\|e^4, \theta^2\} &= .25[(4 - 3\hat{\phi}_3) - 1]^2 + .75[(\hat{\phi}_3) - 1]^2 \\ &= 3(\hat{\phi}_3 - 1)^2, \end{aligned}$$

minimized at $\hat{\phi}_3 = 1.00$. Hence we cannot choose $\hat{\phi}_3$ so as to minimize $V_{\zeta|e,\theta}\{\hat{\phi}[\tilde{\zeta}|e^4]\|e^4, \theta^i\}$ for *both* $i = 1$ *and* $i = 2$. What the statistician might do in this situation is to choose $\hat{\phi}_3$ so as to minimize the larger of these two variances; the resulting choice is $\hat{\phi}_3 = 1.138$ (approximately), implying that $\hat{\phi}_2 = .586$ and $\hat{\phi}_1 = -.1948$.

■ In many situations, there is an unbiased estimate which minimizes the conditional-sampling variance for *all* θ, but when no such unbiased estimate exists, all that can be done is to introduce some notion of dominance; e.g., say that unbiased estimator $\hat{\phi}''[\cdot|e]$ dominates unbiased estimator $\hat{\phi}'[\cdot|e]$ if (1) $V_{\zeta|e,\theta}\{\hat{\phi}''[\tilde{\zeta}|e]\|e, \theta\} \leq V_{\zeta|e,\theta}\{\hat{\phi}'[\tilde{\zeta}|e]\|e, \theta\}$ for every θ in Θ, and (2) the inequality in (1) is *strict* for at least one θ in Θ. Then say that an unbiased estimator is *admissible* if it is not dominated by any other unbiased estimator. Then we can restrict our attention to the class of admissible unbiased estimators. In Example 9.2.11, the admissible unbiased estimators are precisely those for which $1 \leq \hat{\phi}_3 \leq 3536/2736$. ■

About the only generally applicable intuitive justification of the property of unbiasedness is that *if* e were to be repeated a large number n of times, each repetition j constituting a single trial of the "superexperiment," and if a given unbiased estimator $\hat{\phi}[\cdot|e]$ were used each time, then the simple average $(1/n)\sum_{j=1}^{n}\hat{\phi}[\zeta_j|e]$ of the individual estimates $\hat{\phi}[\zeta_j|e]$ should be close to $\phi(\theta)$.

■ This follows from the weak law of large numbers in Section 6.5; write $\omega_j = \zeta_j$, $x(\omega_j) = \hat{\phi}[\zeta_j|e]$, and $\mu(\theta) = \phi(\theta)$. ■

But this interpretation is of small comfort to the individual who for various reasons will be able to perform e only once.

If by now you gather that sampling-theoretic point-estimation procedures are rather ad hoc, you have gathered correctly. Much the same can be said of sampling-theoretic *set*-estimation procedures, to which we now turn.

Again given only the tabulated conditional-sampling probabilities $P_{\zeta|e,\theta}(\cdot|e,\cdot)$, our object now is to describe a more modest procedure for making estimates of θ, a procedure which results in asserting that θ belongs to some subset $\iota(\zeta|e)$ of Θ.

(1) We choose a probability level p in the interval $[0, 1]$.

(2) We then determine a set $\{\iota(\zeta|e): \zeta \in Z(e)\}$ of subsets of Θ with the property that

$$P_{\zeta|e,\theta}(\iota(\tilde{\zeta}|e) \text{ contains } \theta \,|e, \theta) \geq p \tag{9.2.5}$$

for every θ in Θ.

(3) We then perform e, observe ζ.

(4) Finally, we report "a $100p$ percent *confidence set* for θ is $\iota(\zeta|e)$."

Thus the final inference is the prediction that the true state θ belongs to a certain subset $\iota(\zeta|e)$ of Θ, that subset being determined by ζ and called a $100p$ percent confidence set. The entire set $\{\iota(\zeta|e): \zeta \in Z(e)\}$ of such subsets of Θ, determined as step (2) above, is called a $100p$ percent *confidence procedure*; the stipulated probability level p is called the *confidence coefficient*.

■ When Θ is an interval in R^1, $\iota(\zeta|e)$ is ordinally chosen to be an interval for every ζ, in which case we speak of *confidence intervals* rather than confidence sets. Similarly, when $\Theta \subset R^k$ for $k > 1$, we usually choose the $\iota(\zeta|e)$'s to be connected and speak of confidence *regions*. ■

It is advisable that you stare at (9.2.5) for a minute or so, noting carefully where the tilde is placed. Most people—even those who frequently apply confidence procedures, albeit in a necessarily mechanical way—believe that the statement "a 95 percent confidence interval for θ is $[a, b]$" means "the statistician assesses probability (at least) .95 to the event that $\tilde{\theta} \in [a, b]$." *It means no such thing*, because in sampling-theoretic inference probabilities cannot be assigned to subsets of Θ. Instead, "a 95 percent confidence interval for θ is $[a, b]$" means "before conducting our experiment we chose a formula for calculating intervals as a function of the outcome of the experiment; this procedure promised to yield with probability at least .95 an interval bracketing θ; then we performed the experiment, plugged its outcome into the prechosen formula, and got

$[a, b]$." In other words, "the interval $[a, b]$ was obtained by a procedure which has preexperimentation probability at least .95 of yielding an interval bracketing θ."

■ The interpretation of confidence intervals or, more generally, confidence sets, is very much removed from the common-sense way of making probabilistic inferences. This is necessarily so, because common sense dictates that the probability of some $\tilde{\theta}$-event should be stated, whereas no probabilities of such events can be considered in the sampling-theoretic framework. ■

It remains to show how step (2) can be accomplished, determining a set $\{\iota(\zeta|e): \zeta \in Z(e)\}$ of subsets of Θ satisfying (9.2.5). One procedure follows.

(2A) For every θ in Θ, determine a subset $Z^*(\theta|e)$ of $Z(e)$ with the properties that

(a)
$$P_{\zeta|e,\theta}(\tilde{\zeta} \in Z^*(\theta|e)|e, \theta) \geq p; \tag{9.2.6}$$

(b)
$$\text{if } \zeta'' \in Z^*(\theta|e) \text{ and } \zeta' \notin Z^*(\theta|e),$$

then

$$P_{\zeta|e,\theta}(\zeta''|e, \theta) \geq P_{\zeta|e,\theta}(\zeta'|e, \theta);$$

and

(c) no proper subset of $Z^*(\theta|e)$ satisfies (9.2.6);

and then

(2B) for every $\zeta \in Z(e)$, define $\iota(\zeta|e)$ by

$$\iota(\zeta|e) = \{\theta: \zeta \in Z^*(\theta|e)\}. \tag{9.2.7}$$

■ Property (b) of $Z^*(\theta|e)$ means that we do not put less likely ζ's into $Z^*(\theta|e)$ and leave more likely ζ's out. Property (c) means that we want to keep $Z^*(\theta|e)$ as small as possible in number of elements; this is often desirable as a means of keeping the $\iota(\zeta|e)$'s small. ■

■ This procedure does yield sets $\iota(\zeta|e)$ collectively satisfying (9.2.5), because, by (9.2.7), $\theta \in \iota(\zeta|e)$ if and only if $\zeta \in Z^*(\theta|e)$. Hence the events [in $Z(e)$!] $\theta \in \iota(\tilde{\zeta}|e)$ and $\tilde{\zeta} \in Z^*(\theta|e)$ have the same conditional probabilities given θ, for every θ in Θ. ■

Example 9.2.12

For the situation in Example 9.2.2, suppose we want a 90 percent confidence procedure. From the tabulated $P_{\zeta|e,\theta}(\cdot|e, \cdot)$, we readily determine by (2A) that $Z^*(\theta^1|e) = \{\zeta^1, \zeta^2\} = Z^*(\theta^2|e)$, $Z^*(\theta^3|e) = \{\zeta^2, \zeta^3\}$, and $Z^*(\theta^4|e) = \{\zeta^1, \zeta^4\}$; for each θ, we constructed $Z^*(\theta|e)$ by taking the successively most likely ζ's until the sum of their probabilities was at least .90. Now, from (9.2.7) it follows that $\iota(\zeta^1|e) = \{\theta^1, \theta^2, \theta^4\}$, $\iota(\zeta^2|e) = \{\theta^1, \theta^2, \theta^3\}$, $\iota(\zeta^3|e) = \{\theta^3\}$, and $\iota(\zeta^4|e) = \{\theta^4\}$. If we then performed e and its outcome were ζ^2, we would report "a 90 percent confidence set for θ is $\{\theta^1, \theta^2, \theta^3\}$."

Example 9.2.13

For the situation in Example 9.2.3, an 85 percent confidence procedure is obtained as follows. From (2A), $Z^*(\theta^i|e) = \{\zeta^i\}$ for $i = 1, 2, 3$, and $Z^*(\theta^4|e) = \{\zeta^1, \zeta^2\}$. Hence $\iota(\zeta^1|e) = \{\theta^1, \theta^4\}$, $\iota(\zeta^2|e) = \{\theta^2, \theta^4\}$, $\iota(\zeta^3|e) = \{\theta^3\}$, and $\iota(\zeta^4|e) = \emptyset$—yes, empty! This is *not* a mistake; because the probability of ζ^4 given every θ is less than .15, there is preexperimentation probability less than .15 of making a surely incorrect inference!

*■ In many important applications Θ is an interval of real numbers and $\tilde{\zeta}$ can be taken to be a continuous real random variable. In such applications the formulae for $f_{\zeta|e,\theta}(\cdot|e, \theta)$ are used to determine two functions $\alpha(\cdot)$ and $\beta(\cdot)$ of θ with the properties that

$$\int_{\alpha(\theta)}^{\beta(\theta)} f_{\zeta|e,\theta}(\zeta|e, \theta)d\zeta = P_{\zeta|e,\theta}(\alpha(\theta) \le \tilde{\zeta} \le \beta(\theta)|e, \theta) = p,$$

and then $\iota(\zeta|e) = \{\theta: \alpha(\theta) \le \zeta \le \beta(\theta)\}$ is defined. ■

***Example 9.2.14**

For the situation in *Example 9.2.5, it is an important fact of statistical life, discussed in Section 9.5, that all inferences about μ when υ is known can be based on $\bar{x}$ without any loss of information. From *Exercise 6.5.8, $\tilde{\bar{x}}$ has conditional-sampling density function $f_{\zeta|e,\mu}(\bar{x}|e, \mu) = f_N(\bar{x}|\mu, \upsilon/n)$. Now $P_N(\alpha(\mu) \le \tilde{\bar{x}} \le \beta(\mu)|\mu, \upsilon/n) = p$ if $\alpha(\mu) = \mu - N_q\sqrt{\upsilon/n}$ and $\beta(\mu) = \mu + N_q\sqrt{\upsilon/n}$, where $q = ½(1 + p)$ and N_q is the "qth fractile" of the standardized Normal random variable; i.e., N_q is the number which satisfies $F_N(N_q|0, 1) = q$. The following table approximates N_q for some of the more popular values of q.

q	N_q	q	N_q
.750	0.64	.975	1.96
.875	1.15	.990	2.33
.900	1.28	.995	2.58
.950	1.65	.999	3.09

Now, $\tilde{\bar{x}} \le \mu + N_q\sqrt{\upsilon/n}$ if and only if $\mu \ge \tilde{\bar{x}} - N_q\sqrt{\upsilon/n}$, and $\tilde{\bar{x}} \ge \mu - N_q\sqrt{\upsilon/n}$ if and only if $\mu \le \tilde{\bar{x}} + N_q\sqrt{\upsilon/n}$. Hence

$$P_N(\tilde{\bar{x}} - N_q\sqrt{\upsilon/n} \le \mu \le \tilde{\bar{x}} + N_q\sqrt{\upsilon/n}|\mu, \upsilon/n) = p,$$

meaning that

$$\iota(\bar{x}|e) = [\bar{x} - N_{(1+p)/2}\sqrt{\upsilon/n}, \bar{x} + N_{(1+p)/2}\sqrt{\upsilon/n}] \tag{9.2.8}$$

is a $100p$ percent confidence interval for μ. Since $N_{(1+p)/2}$ is an increasing function of p, it is clear that increasing the confidence coefficient p spreads out the confidence interval and makes the inference vaguer. This is to be expected: Greater probability p of correctness is purchased at the cost of greater vagueness about μ. Moreover, (9.2.8) shows clearly that it is the endpoints of the interval which, prior to the experiment, are random variables to be determined by the outcome of the experiment.

Exercise 9.2.1

(A) Determine the 95 percent confidence procedure for the situation in Example 9.2.2, according to (2A) and (2B).
(B) How do the resulting confidence sets compare with the 90 percent confidence sets determined in Example 9.2.12?

Exercise 9.2.2

Consider the experiment e with the following conditional-sampling probability table $P_{\zeta|e,\theta}(\cdot|e, \cdot)$.

	ζ^1	ζ^2
θ^1	.90	.10
θ^2	.40	.60

(A) Find a maximum-likelihood estimator of θ.
(B) If $t(\theta^1) = 15$ and $t(\theta^2) = -1.732$, find a maximum-likelihood estimator of $t(\theta)$.
(C) Let $\phi(\theta^i) = i$ for $i = 1, 2$, and find, if possible, an unbiased estimator of $\phi(\theta)$.
(D) Now let $\phi(\theta^i)$ = the square of i, for $i = 1, 2$, and find, if possible, an unbiased estimator of $\phi(\theta)$.

■ (C) and (D) imply that unbiased estimators do not possess the invariance property of maximum-likelihood estimators. ■

Exercise 9.2.3

Consider the experiment e with the following conditional-sampling probability table $P_{\zeta|e,\theta}(\cdot|e, \cdot)$.

	ζ^1	ζ^2
θ^1	.90	.10
θ^2	.90	.10
θ^3	.50	.50

Let $\phi(\theta^i) = i$ for $i = 1, 2, 3$. Find, if possible, an unbiased estimator of $\phi(\theta)$.

Exercise 9.2.4

Consider the experiment e with the following conditional-sampling probability table $P_{\zeta|e,\theta}(\cdot|e, \cdot)$.

	ζ^1	ζ^2	ζ^3	ζ^4	ζ^5
θ^1	.40	.30	.20	.05	.05
θ^2	.05	.40	.20	.05	.30
θ^3	.20	.20	.20	.20	.20
θ^4	.30	.05	.20	.40	.05

(A) Find a maximum-likelihood estimator of θ. How many maximum-likelihood estimators are there?
(B) Find a 40 percent confidence procedure for θ. What criterion or criteria did you use to "break any ties" that might have arisen?
(C) Find an 80 percent confidence procedure for θ.
(D) Find a 90 percent confidence procedure for θ.

***Exercise 9.2.5**

Derive (9.2.3).

Exercise 9.2.6: More Difficulty with Confidence Procedures

We saw in Example 9.2.13 that a valid confidence procedure may yield an inference of the form $\iota(\zeta|e) = \emptyset$, which is nonsensical. Consider now the problem of making a 500/6 percent confidence statement about next year's GNP, θ; and suppose that e consists in casting a fair, six-sided die once, declaring that $\iota(\zeta|e) = \{\theta: -\infty < \theta < +\infty\}$ if $\zeta \in \{1, \ldots, 5\}$, and declaring that $\iota(\zeta|e) = \emptyset$ if $\zeta = 6$. Verify that $\{\iota(\zeta|e): \zeta = 1, \ldots, 6\}$ is a valid 500/6 percent confidence procedure for estimating next year's GNP. Is it meaningful?

***Exercise 9.2.7**

Show that the maximum-likelihood estimate of $e^{-2\theta}$ in Example 9.2.10 is $e^{-2\zeta}$.
[*Hint*: Find the maximum-likelihood estimate of θ and use invariance.]

9.3 SAMPLING-THEORETIC TESTS OF HYPOTHESES

Often the purpose of performing an experiment is not to estimate θ but rather to glean evidence on the reasonableness of some preconceived hypothesis about θ. In Example 1.5.1c with e^4 as defined in Example 9.1.1, for instance, we might perform e^4 in order to shed light on whether it is reasonable to suppose that $\theta = \text{AD}$, and hence that the subsystem should be rebuilt.

Suppose the hypothesis H_0 about which information is desired is the hypothesis that θ belongs to a specified, nonempty subset Θ_0 of Θ. We denote this hypothesis by writing H_0: $\theta \in \Theta_0$. If the outcome ζ of e indicates that H_0: $\theta \in \Theta_0$ is too implausible, we say that we *reject* H_0 in favor of the alternative hypothesis H_A: $\theta \notin \Theta_0$.

Now, a *test* of H_0: $\theta \in \Theta_0$ is a procedure for deciding, on the basis of the outcome ζ of e, whether to reject H_0 in favor of H_A or not to reject H_0 in favor of H_A; this chapter will not consider "sequential tests," in which we have the third option of deciding to conduct another experiment. It follows readily that choosing a test of H_0 versus H_A is tantamount to choosing the subset R of $Z(e)$ consisting of all outcomes ζ for which H_0 is to be *r*ejected.

■ Since there are $2^{\#(Z(e))}$ subsets of $Z(e)$, including $Z(e)$ itself and $\emptyset$, it follows that there are $2^{\#(Z(e))}$ tests of H_0 based on e, each test being identified with its corresponding rejection set, or critical set, R. For brevity we shall say "test R" instead of "test with rejection set R." ■

Example 9.3.1

In each of Examples 9.2.2 and 9.2.3, there are $2^4 = 16$ tests, regardless of the nature of H_0.

Obviously, a test R cannot be chosen without reference to *some* criterion of goodness of tests. Without reference to a utility function, it is still clear that it is good to be right and bad to be wrong; and a test can be chosen so as to minimize (in a certain sense) the probability of being wrong.

Now, "being wrong" in testing H_0 can mean either erroneously rejecting H_0 or erroneously failing to reject H_0. Erroneous rejection of H_0, or rejecting H_0 given that θ does belong to Θ_0, is called a *type-I error*, or an error of the first kind. Erroneous failure to reject H_0, or not rejecting H_0 given that θ does not belong to Θ_0, is called a *type-II error*, or an error of the second kind.

■ Pundits have defined two other types of errors. An error of the third kind, or a type-III error, is committed far too often for professional complacency and consists in furnishing an elegant solution to the wrong problem, a solution based on woeful lack of apprehension of the realities of the situation. An error of the fourth kind, or a type-IV error, is committed far too often by students and consists in being wrong because of arithmetical carelessness. ■

Therefore, we seek a test R with the property that

$$P_{\zeta|e,\theta}(\tilde{\zeta} \in R \mid e, \theta) = \begin{cases} 0, & \theta \in \Theta_0 \\ 1, & \theta \notin \Theta_0. \end{cases}$$

Unfortunately, rarely do such tests exist. We must seek a more generally applicable criterion for choosing tests.

Since various (conditional-sampling!) probabilities of $\tilde{\zeta} \in R$ and $\tilde{\zeta} \notin R$ given θ are common in the hypothesis-testing literature, we introduce the conventional terminology in Definition 9.3.1.

Definition 9.3.1: Test Characteristics. Suppose that H_0: $\theta \in \Theta_0$, $P_{\zeta|e,\theta}(\cdot|e, \cdot)$, and a test R are given.

(A) The *power function* $\mathscr{P}(R|e, \cdot): \Theta \to [0, 1]$ of R is the probability of rejecting H_0, as a function of θ; that is,

$$\mathscr{P}(R|e, \theta) = P_{\zeta|e,\theta}(\tilde{\zeta} \in R|e, \theta)$$

for every θ in Θ.

(B) The *operating characteristic* $\mathscr{O}(R|e, \cdot): \Theta \to [0, 1]$ of R is the probability of not rejecting H_0, as a function of θ; that is,

$$\mathscr{O}(R|e, \theta) = 1 - \mathscr{P}(R|e, \theta)$$

for every θ in Θ.

(C) The *error characteristic* $\mathscr{E}(R|e, \cdot): \Theta \to [0, 1]$ of R is the probability of erring with R, as a function of θ; that is,

$$\mathscr{E}(R|e, \theta) = \begin{cases} \mathscr{P}(R|e, \theta), & \theta \in \Theta_0 \\ \mathscr{O}(R|e, \theta), & \theta \notin \Theta_0. \end{cases}$$

Because erring is bad, we would like to choose a test so as to minimize $\mathscr{E}(R|e, \theta)$ for every θ in Θ. But, again, this is usually impossible, for exactly the same reasons as in Chapter 7 we could not rely on finding a strategy that simultaneously maximized all components of the utility characteristic.

The sampling-theoretic criterion for choosing a test is as follows.

(1) Choose a probability level α in [0, 1], called the *significance level*, to define a constraint on the allowable probability of type-I error, and eliminate from further consideration any test R which violates the constraint

$$\mathscr{E}(R|e, \theta) \le \alpha \tag{9.3.1}$$

for every θ in Θ_0.

(2) Further eliminate from consideration any test R' which is "dominated by" another test, R'', in the sense that $\mathscr{E}(R''|e, \theta) \le \mathscr{E}(R'|e, \theta)$ for all $\theta \notin \Theta_0$ and $\mathscr{E}(R''|e, \theta) < \mathscr{E}(R'|e, \theta)$ for at least one $\theta \notin \Theta_0$.

It might be hoped that this procedure would isolate a single test, but this too is rarely true.

Example 9.3.2

Suppose that e in Example 9.2.2 is for the purpose of testing H_0: $\theta \in \{\theta^1, \theta^2\}$ *versus* H_A: $\theta \in \{\theta^3, \theta^4\}$. Also suppose that the significance level α is .07. Three tests R satisfy the step-(1) constraint $\mathscr{E}(R|e, \theta) \le .07$ for both $\theta \in \{\theta^1, \theta^2\}$; since for $\theta \in \Theta_0$ we have $\mathscr{E}(R|e, \theta) = P_{\zeta|e,\theta}(\tilde{\zeta} \in R|e, \theta)$, we readily verify that $R = \emptyset$, $R = \{\zeta^3\}$, and $R = \{\zeta^4\}$ are the only tests which "pass" step (1); the other thirteen tests are eliminated from further consideration. Now we apply step (2), first tabulating $\mathscr{E}(R|e, \theta^3)$ and $\mathscr{E}(R|e, \theta^4)$ for each of the three noneliminated tests.

R	$\mathscr{E}(R\|e, \theta^3)$	$\mathscr{E}(R\|e, \theta^4)$
$\emptyset$	1.00	1.00
$\{\zeta^3\}$	0.20	0.96
$\{\zeta^4\}$	0.96	0.20

Hence $\emptyset$ is dominated by each of the other tests, neither of which dominates the other. Therefore we would select either $\{\zeta^3\}$ or $\{\zeta^4\}$ as the test of H_0: $\theta \in \{\theta^1, \theta^2\}$.

Example 9.3.3

If the significance level α were chosen to be .10 rather than .07 in Example 9.3.2, then four tests would "pass" step (1); namely, $\{\zeta^3, \zeta^4\}$ and the three which passed in Example 9.3.2. Of these four, $\{\zeta^3, \zeta^4\}$ is clearly "best" in the sense of step (2), since $\mathscr{E}(\{\zeta^3, \zeta^4\}|e, \theta^i) = .16 < \min[.20, .96]$ for $i = 3, 4$, and hence it dominates the others. Here step (2) does yield a single test.

If step (2) yields a test R^* such that $\mathscr{E}(R^*|e, \theta) \leq \mathscr{E}(R'|e, \theta)$ for every $\theta \notin \Theta_0$ and every R' passing step (1), then it is clearly optimal within the sampling-theoretic framework, and it is called *uniformly most powerful* at significance level α.

■ Since $\mathscr{E}(R|e, \theta) = 1 - \mathscr{P}(R|e, \theta)$ for every R and every $\theta \notin \Theta_0$, the R^* described above can also be characterized by the power-function property that $\mathscr{P}(R^*|e, \theta) \geq \mathscr{P}(R'|e, \theta)$ for every $\theta \notin \Theta_0$ and every R' passing step (1). Hence the terminology "uniformly (for all $\theta \notin \Theta_0$) most powerful." ■

Uniformly most powerful tests do not exist at every significance level α, as Example 9.3.2 shows. Hence an additional criterion is required for choosing among the occasionally many tests which pass both steps (1) and (2). This criterion is best introduced by considering briefly what it means to conduct an experiment and find its outcome ζ in R. If R has been chosen to satisfy (9.3.1) for a given significance level α and if we call an event of probability not exceeding α a "virtual miracle," then finding ζ in R means that *either* a virtual miracle has occurred or that H_0: $\theta \in \Theta_0$ is not true. The choice of R means that the statistician does not believe in virtual miracles, because he rejects H_0: $\theta \in \Theta_0$ if $\zeta \in R$.

But the dichotomy posed by $\zeta \in R$, between believing in a virtual miracle having occurred and rejecting H_0, would not be so simple if there were a third possibility; namely, that *both* a virtual miracle has occurred *and* H_0 is false.

Example 9.3.4

In Example 9.2.3, suppose that we wish to test H_0: $\theta \in \{\theta^1, \theta^2\}$ at the $\alpha = .03$ level of significance, and that we call events of probability not exceeding .03 virtual miracles. Then steps (1) and (2) yield two tests, namely, $\{\zeta^3\}$ and $\{\zeta^4\}$. Suppose that we decide to use $\{\zeta^3\}$ and hence will reject H_0: $\theta \in \{\theta^1, \theta^2\}$ if $\tilde{\zeta} = \zeta^3$. Now suppose we conduct e and do obtain ζ^3. Then, given that $\theta = \theta^4$ and hence that H_0 is false, we find that a virtual miracle has occurred; indeed, getting $\tilde{\zeta} = \zeta^3$ is even *more* miraculous given θ^4 than it is given each of the two states for which H_0 is true! This is very embarrassing to the interpreter of a test result. Hence test $\{\zeta^4\}$ would be opted in preference to $\{\zeta^3\}$ in this problem.

To rule out this third possibility, statisticians often search for "unbiased" tests. A test passing step (1) for the given significance level α is called an *unbiased test* at level α if $\mathscr{P}(R|e, \theta) \geq \alpha$ for every $\theta \notin \Theta_0$, or equivalently, if $\mathscr{E}(R|e, \theta) \leq 1 - \alpha$ for every $\theta \notin \Theta_0$. If a test is unbiased at level α, then $\tilde{\zeta} \in R$ is not more miraculous when H_0 is false than when H_0 is true.

***Example 9.3.5**

Suppose that a social worker wishes to test, at significance level α, the hypothesis that the average annual income of a wage earner in a given locale does not exceed a given amount μ_0 per year. If the locale is at all populous, it is reasonable to suppose that the density function of the income $x(\tilde{\omega})$ of a randomly chosen wage earner is approximately Normal, $f_N(\cdot|\mu, \upsilon)$, around the true locale average μ. Now suppose that she randomly selects n wage earners and, after careful verification of their individual incomes, records the sample mean $\bar{x}$. We now show how $\bar{x}$ can be used to test the hypothesis H_0: $\mu \leq \mu_0$ against the alternative H_A: $\mu > \mu_0$ at significance level α, assuming that υ is known. (This assumption is easily obviated, as virtually every statistics text demonstrates.) From *Example 9.2.14, we know that the conditional-sampling density function of $\tilde{\bar{x}}$ is $f_N(\cdot|\mu, \upsilon/n)$, and hence $P_N(\tilde{\bar{x}} \geq \gamma(\mu)|\mu, \upsilon/n) = \alpha$ for $\gamma(\mu) = \mu + N_{1-\alpha}\sqrt{\upsilon/n}$, where $N_{1-\alpha}$ is as defined in *Example 9.2.14. Now, the social worker wishes to reject H_0: $\mu \leq \mu_0$ if $\bar{x}$ is so large as to be virtually miraculous given that H_0 is true. She may take $R = [\mu_0 + N_{1-\alpha}\sqrt{\upsilon/n}, +\infty)$—that is, reject H_0 if $\bar{x} \geq \mu_0 + N_{1-\alpha}\sqrt{\upsilon/n}$—because the preexperimentation probability of the event $\tilde{\bar{x}} \geq \mu_0 + N_{1-\alpha}\sqrt{\upsilon/n}$ is at most α for every $\mu \leq \mu_0$. It is not hard to verify that this test R is both uniformly most powerful and unbiased.

Finally, a word about researchers' strategy in empirical testing is in order. Because of the asymmetric way in which the sampling-theoretic criterion treats H_0: $\theta \in \Theta_0$ and H_A: $\theta \notin \Theta_0$, it is fair to say that hypothesis tests cannot "prove" H_0 but can only cast doubt on it. Therefore, it is better to say that the researcher has failed to reject H_0 than to say that she has accepted H_0, if $\zeta \notin R$.

Now consider a researcher with some notion that he would like others to accept. Suppose he can perform an experiment e, with possible outcomes ζ having conditional-sampling probability function $P_{\zeta|e,\theta}(\cdot|e, \cdot)$, such that his notion is true if and only if θ belongs to some subset Θ' of Θ. In order to get others to accept his notion, he might well test the *contrary* notion, playing devil's advocate with himself by choosing a low significance level α so as to guard against erroneously rejecting the contrary notion. Thus he would test the "null" hypothesis H_0: $\theta \in \Theta_0$ for $\Theta_0 = \overline{\Theta'}$ at a low significance level α, so that if he succeeds (as he hopes) in rejecting H_0 in favor of the alternative hypothesis H_A: $\theta \notin \Theta_0$, or H_A: $\theta \in \Theta'$, he has demonstrated good empirical support for his notion. The lower the level α at which he can reject H_0, the better the empirical support, and the more likely others are to concur on the validity of his notion.

■ Certain scholarly journals will not publish results of tests at significance levels exceeding .05, their reasoning being that at most 5 percent of the results published in their journals will be erroneous given this policy. But this editorial policy is itself based on fallacious reasoning, as Raiffa **[113]** has observed. Suppose M people have correct hypotheses, N people have incorrect hypotheses, and all test their hypotheses at the .05 level of significance. Then we expect $.95M$ "publishable" correct results and $.05N$ "publishable" incorrect results. The proportion of expected incorrect to expected total publishable results is $.05N/(.05N + .95M)$, which fails to exceed .05 only if $N \le M$. If all the researchers are wrong ($M = 0$), then all the published results will be wrong! ■

Exercise 9.3.1

Suppose that e is as given in Example 9.2.3. Then there are 16 possible tests of each conceivable hypothesis.

(A) List all tests R of H_0: $\theta \in \{\theta^1, \theta^4\}$ at significance level .20, and tabulate $\mathscr{E}(R|e, \theta)$, as in Example 9.3.2, for each such test. Which would you choose, and why?
(B) Same questions, but for significance level .19.
(C) Same questions, but for significance level .18.
(D) Same questions, but for significance level .13.
(E) Same questions, but for significance level .12.
(F) Same questions, but for significance level .06.
(G) Same questions, but for significance level .05.
(H) Same questions, but for significance level .02.
(I) Which of the tests R in (A) are unbiased?

Exercise 9.3.2: Discussion Question

There appear to be interesting correspondences between certain legal concepts and concepts in the sampling-theoretic approach to testing hypotheses. For instance, guilt or innocence versus H_0 or H_A, and "a man is innocent until proven guilty" versus "use a small α to guard against erroneous rejection of H_0." Can quizzing of prospective jurors be interpreted as an attempt to exclude from the jury persons whose prior probabilities of the defendant's guilt deviate from 1/2, or to exclude persons who even admit to possession of prior probabilities? Can one interpret the prosecution's efforts as directed toward creation of a ζ within a critical set R for testing H_0: "innocence" at a low significance level α?

***Exercise 9.3.3**

(A) Show that if z and υ are fixed, then $P_N(\tilde{\bar{x}} \geq z|\mu, \upsilon/n)$ is an increasing function of μ.

(B) Show that the test R described in Example 9.3.5 is unbiased.

(C) Show that if z and υ are fixed, then $P_N(\tilde{\bar{x}} \leq z|\mu, \upsilon/n)$ is a decreasing function of μ.

(D) Let everything be as in Example 9.3.5, except that we wish to test H_0: $\mu \geq \mu_0$ versus H_A: $\mu < \mu_0$ at significance level α. Show that $R = (-\infty, \mu_0 - N_{1-\alpha}\sqrt{\upsilon/n}]$ is such a test, and that it is unbiased.

[*Hint*: $N_\alpha = -N_{1-\alpha}$.]

9.4 BAYESIAN INFERENCE

None of the sampling-theoretic inferences discussed in Sections 9.2 and 9.3 depends in any way upon a prior probability function $P_\theta(\cdot)$ on events in Θ. By contrast, all Bayesian inferences are based directly on a posterior probability function $P_{\theta|e,\zeta}(\cdot|e, \zeta)$ and hence stem indirectly, but in an essential manner, from a prior probability function $P_\theta(\cdot)$. The relationship of posterior to prior probability and likelihood functions is specified by Bayes' Theorem, which can be written in several ways, among which are

$$P_{\theta|e,\zeta}(\theta|e, \zeta) = k(\zeta|e)L(\theta|e, \zeta)P_\theta(\theta) \tag{9.4.1}$$

for every θ in Θ, where

$$k(\zeta|e) = 1\Big/\sum_{i=1}^{\#(\Theta)} L(\theta^i|e, \zeta)P_\theta(\theta^i);$$

and, more succinctly,

$$P_{\theta|e,\zeta}(\cdot|e, \zeta) \propto L(\cdot|e, \zeta)P_\theta(\cdot). \tag{9.4.2}$$

■ Expression (9.4.2) conveys information about the relationship of the *shape* of the posterior probability function to the shapes of the prior probability and likelihood functions. Here, "is proportional to" means "equals $k(\zeta|e)$—not dependent upon θ—times." ■

The grounding of all Bayesian inferences on posterior, and hence also on prior, probability functions presents both drawbacks and advantages. An undeniable drawback to ready acceptance of Bayesian inferential methods by the scientific community is the fact that people confronted with the same empirical evidence (e, ζ) may arrive at different conclusions because their prior probability functions differ.

In principle, such objections are readily countered by observing that a prior probability function may be regarded as posterior—to its assessor's entire background and experience. Hence the addition of (e, ζ) to different backgrounds should yield different posterior [to (e, ζ)] probability functions and hence potentially different inferences as well.

In practice, such objections are somewhat blunted by the realization that if (e, ζ) is very conclusive in relation to people's differing prior experiences, then their individual posterior probability functions $P_{\theta|e,\zeta}(\cdot|e, \zeta)$ should all be very close, to the "scaled" likelihood function $L(\cdot|e, \zeta)/\Sigma_{i=1}^{\#(\Theta)} L(\theta^i|e, \zeta)$; and hence such objections really amount to a plaintive plea for more conclusive evidence. This assertion is a direct consequence of (9.4.2) and is elaborated lucidly in Edwards, Lindman, and Savage [**30**].

Example 9.4.1

Suppose that $\Theta = \{\theta^1, \ldots, \theta^5\}$ and that $L(\cdot|e, \zeta)$ and three prior probability functions $P_\theta^1(\cdot)$, $P_\theta^2(\cdot)$, and $P_\theta^3(\cdot)$ are as given by the following table.

	θ^1	θ^2	θ^3	θ^4	θ^5
$L(\theta\|e, \zeta)$	.01	.02	.44	.02	.01
$P_\theta^1(\theta)$	.15	.20	.30	.20	.15
$P_\theta^2(\theta)$	.40	.20	.10	.20	.10
$P_\theta^3(\theta)$	.20	.20	.20	.20	.20

As Exercise 9.4.4, you may verify from Equation (9.4.1) that the three posterior probability functions $P_{\theta|e,\zeta}^h(\cdot|e, \zeta)$ resulting from the evidence (e, ζ) and the three prior probability functions $P_\theta^h(\cdot)$ are, to two decimal places, as follows.

	θ^1	θ^2	θ^3	θ^4	θ^5
$P_{\theta\|e,\zeta}^1(\theta\|e, \zeta)$	.01	.03	.92	.03	.01
$P_{\theta\|e,\zeta}^2(\theta\|e, \zeta)$	.07	.07	.77	.07	.02
$P_{\theta\|e,\zeta}^3(\theta\|e, \zeta)$	.02	.04	.88	.04	.02

Posterior judgments still differ, but nowhere nearly as much as prior judgments differed. All three posterior probability functions assign a preponderance of probability to θ^3, even $h = 2$, which arose from prior judgments in which θ^3 was tied for least likely. Finally, note also that $P_{\theta|e,\zeta}^3(\cdot|e, \zeta)$ *is* the "scaled" likelihood function $L(\cdot|e, \zeta)/\Sigma_{i=1}^{\#(\Theta)} L(\theta^i|e, \zeta)$.

The practical implication of these elementary observations is that if the evidence (e, ζ) is overwhelmingly conclusive in relation to our as-yet incompletely quantified prior judgments, then we are spared the task of quantifying these judgments by using the "scaled" likelihood function $L(\cdot|e, \zeta)/\Sigma_{i=1}^{\#(\Theta)} L(\theta^i|e, \zeta)$ as a good approximation to our real posterior probability function. In other words, if the evidence (e, ζ) is very conclusive, then we can behave *as if* our prior probability function were "uniform"—$P_\theta(\theta) = 1/\#(\Theta)$ for every θ in Θ—because in that case the posterior probability function coincides with the scaled likelihood function. (Verifying this assertion is an easy task, left to you as Exercise 9.4.5.)

In statistical sampling problems, such as those in which e consists in observing the outcomes of n indistinguishable trials, the laws of large numbers introduced in Section 6.5 imply that, as n is increased, $L(\cdot|e, \zeta)$ may be expected to be more and more "peaked" about the true θ, and hence the evidence (e, ζ) may be expected to be more and more conclusive. Thus the popular expression "Large samples wash out prior opinion."

■ Except if prior opinion is very opinionated. Anyone who assesses $P_\theta(\theta^3) = 0$ in Example 9.4.1, for instance, would have $P_{\theta|e,\zeta}(\theta^3|e, \zeta) = 0$ as well. *No* amount of evidence can convince the completely opinionated! ■

***Example 9.4.2**

For the experiment e described in *Example 9.2.5, with likelihood function given by (9.2.3), suppose that prior judgments about $\tilde{\mu}$ are expressible (at least to a good approximation) by some Normal density function $f_N(\cdot|m', \upsilon/n')$ where υ is the known variance $V\{x(\tilde{\omega})\}$ of a single measurement. It then follows readily from Example 3.11.3 (with $\omega_1 = \mu$, $V = \upsilon/n$, and $V' = \upsilon/n'$) that the resulting posterior judgments about $\tilde{\mu}$ are expressed in the form of a Normal density function $f_N(\cdot|m'', V'')$, where

$$V'' = \frac{\upsilon}{n' + n} = \left[\frac{n}{n' + n}\right]\frac{\upsilon}{n}$$

and

$$m'' = \left[\frac{n'}{n' + n}\right]m' + \left[\frac{n}{n' + n}\right]\bar{x}.$$

As n increases, m'' gets closer and closer to $\bar{x}$ and V'' gets closer and closer to υ/n, so that if n is large in relation to n', then $f_N(\cdot|m'', V'')$ is close to the scaled likelihood function $f_N(\cdot|\bar{x}, \upsilon/n)$.

*■ The setting of *Example 9.4.2 is an infinite idealization, in which $\Theta = (-\infty, +\infty)$. Here, $f_N(\cdot|m'', V'')$ could coincide exactly with $f_N(\cdot|\bar{x}, \upsilon/n)$ only if n' were zero, that is, if V' were $+\infty$. But one cannot set $V' = +\infty$ even in the limit, $\lim_{V' \to +\infty} f_N(\mu|m', V') = 0$ for every μ, and the zero function is *not* a density function, because the area under it is zero. There is no such thing as a uniform density function on an unbounded

interval, because the area under it is zero if its height is zero, and $+\infty$ if its height is any positive number. Such functions are useful as limit concepts, however, just as in the example at hand. They are called *improper* prior density functions. ■

As a valedictory comment on the general topic of the relationship between prior experience and the conclusiveness of evidence, the author once, but only temporarily, convinced a colleague whose specialization is advertising that Bayes' Theorem is nothing but the precise, mathematical formalization of the accretion of wisdom. One wishes that the convictions of others in such cases could be less ephemeral.

We proceed now to a brief discussion of some Bayesian versions of the estimation and hypothesis-testing problems. Before jumping in, we once again stress that *all Bayesian inferences stem from the posterior probability function*, which alone and in its entirety embodies judgments as revised by evidence. Bayesian estimates, whether point or set, are crude summaries of the posterior probability function, and hypothesis tests are somewhat ad hoc decisions based thereon.

If there are many possible states θ, it is clear that no single point estimate $\hat{\theta}$ can provide an adequate summary of the entire posterior probability function $P_{\theta|e,\zeta}(\cdot|e, \zeta)$. Nevertheless, such estimates are frequently made because of their general intelligibility, and also because they suffice for certain, limited decision purposes. Bayesian point estimates are often made in the context of statistical decision problems,[c] but our attention here will be confined to Bayesian analogues of maximum-likelihood and unbiased estimators discussed in Section 9.2.

The natural Bayesian analogue of the maximum-likelihood estimate is the maximum-posterior-probability estimate, or posterior mode estimate.

Definition 9.4.1

(A) An estimate $\hat{\theta}[\zeta|e]$ is called a *posterior-mode estimate* if

$$P_{\theta|e,\zeta}(\hat{\theta}[\zeta|e]\|e, \zeta) = \max_{\theta\in\Theta}[P_{\theta|e,\zeta}(\theta|e, \zeta)].$$

(B) An estimator $\hat{\theta}[\cdot|e]$ is called a *posterior-mode estimator* if $\hat{\theta}[\zeta|e]$ is a posterior-mode estimate for every ζ in $Z(e)$.

Note that a posterior-mode estimate means what the ill-informed think a maximum-likelihood estimate means. When you quote your posterior-mode estimate of something, you are stating your "best bet," in a sense made precise in *Exercise 9.4.6(A).

Example 9.4.3

Consider the experiment e whose likelihood function is tabulated in Example 9.2.2, and suppose that the prior probability function is given

[c] See *Exercises 8.5.9 and 9.4.6.

by $P_\theta(\theta^1) = .80$, $P_\theta(\theta^2) = .15$, $P_\theta(\theta^3) = .04$, and $P_\theta(\theta^4) = .01$. The resulting posterior probability functions, for the four possible ζ's, are

	$P_{\theta\|e,\zeta}(\theta\|e,\zeta)$			
	ζ^1	ζ^2	ζ^3	ζ^4
θ^1	6400/6584	800/2046	480/864	320/506
θ^2	150/6584	1200/2046	60/864	90/506
θ^3	24/6584	40/2046	320/864	16/506
θ^4	10/6584	6/2046	4/864	80/506

It is clear that $\hat{\theta}[\zeta^1|e] = \theta^1$, $\hat{\theta}[\zeta^2|e] = \theta^2$, and $\hat{\theta}[\zeta^3|e] = \hat{\theta}[\zeta^4|e] = \theta^1$.

*■ In infinite idealizations where $\tilde{\theta}$ has a posterior density function $f_{\theta|e,\zeta}(\cdot|e,\zeta)$, the posterior-mode estimate may be defined as a maximizer of the posterior density function; that is, $\hat{\theta}[\zeta|e]$ satisfies

$$f_{\theta|e,\zeta}(\hat{\theta}[\zeta|e]|e,\zeta) = \max_{\theta\in\Theta}[f_{\theta|e,\zeta}(\theta|e,\zeta)].$$

Thus, for the situation in *Example 9.4.2, the posterior-mode estimate $\hat{\mu}[\zeta|e]$ is m'', which agrees with the maximum-likelihood estimate $\bar{x}$ generally only when $m' = \bar{x}$, which is rare luck, and when $n' = 0$, which implies the "improper" prior density function with infinite variance. ■

Since scale factors do not affect the location of maximizers, it is clear that $\hat{\theta}[\zeta|e]$ is a posterior-mode estimate if and only if

$$L(\hat{\theta}[\zeta|e]|e,\zeta)P_\theta(\hat{\theta}[\zeta|e]) = \max_{\theta\in\Theta}[L(\theta|e,\zeta)P_\theta(\theta)], \tag{9.4.3}$$

from which it is clear that posterior-mode estimates need not agree with maximum-likelihood estimates. If, however, the prior probability function is uniform, with $P_\theta(\theta) = 1/\#(\Theta)$ for every θ in Θ, then all $P_\theta(\cdot)$ terms in (9.4.3) are constant and can be canceled, and the posterior-mode estimates must agree with the maximum-likelihood estimates.

A Bayesian analogue of unbiased estimation is a posterior-mean estimation.

Definition 9.4.2. An estimator $\hat{\phi}[\cdot|e]$ of a real-valued function $\phi\colon \Theta \to R^1$ of $\tilde{\theta}$ is called a *posterior-mean* estimator if

$$\hat{\phi}[\zeta|e] = E_{\theta|e,\zeta}\{\phi(\tilde{\theta})|e,\zeta\}$$

for every ζ in $Z(e)$.

In other words, the estimate of $\phi(\tilde{\theta})$ is just the expectation of $\phi(\tilde{\theta})$ with respect to the posterior probability function of $\tilde{\theta}$.

In contrast to the embarrassing situation with unbiased estimators, posterior-mean estimators always exist and are unique.

***Example 9.4.4**

In Example 9.2.10 we saw that an unbiased estimate might exist and be unique, but that it might be nonsensical. We now see what Bayesian inference can contribute to the peace of mind of the telephone operator in that example. From Example 3.11.2, with $\omega_1 = \theta$, $\omega_2 = \zeta$, and $t = 1$, we see that *if* she had assessed a gamma prior density function $f_\gamma(\cdot|r', t')$ for $\tilde{\theta}$, then her posterior density function of $\tilde{\theta}$ would be $f_\gamma(\cdot|r' + \zeta, t' + 1)$. From Example 3.10.11, with $\omega = \theta$ and $z = -2$, we see that

$$E_{\theta|e,\zeta}\{\phi(\tilde{\theta})|e, \zeta\} = E_\gamma\{e^{-2\tilde{\theta}}|r' + \zeta, t' + 1\} = [(t' + 1)/(t' + 3)]^{r'+\zeta},$$

which is her posterior-mean estimate of $e^{-2\theta}$ in terms of the number ζ of calls coming in during the next minute and in terms of the parameters r' and t' of her prior density function of $\tilde{\theta}$. Note that the estimate is a *decreasing* function of ζ, which is intuitively reasonable because many calls coming in during the next minute imply small probability $e^{-2\theta}$ of *no* calls coming in during the next two minutes. It can be shown[d] that her posterior-mode estimate of $\tilde{\phi} = e^{-2\tilde{\theta}}$ is given by

$$\hat{\phi}[\zeta|e] = e^{-2(r'+\zeta-1)/(t'-1)},$$

again a decreasing function of ζ. It agrees with the maximum-likelihood estimate $e^{-2\zeta}$ (see Exercise 9.2.7) if $r' = 1$ and $t' = 2$.

So much for Bayesian point estimation. *Bayesian set estimation* is much more straightforward than sampling-theoretic set estimation. As in Section 9.2, our object is to determine a procedure resulting in one's asserting that $\tilde{\theta}$ belongs to some subset $\iota(\zeta|e)$ of Θ. Here is how we could go about the task if we had posterior probability functions $P_{\theta|e,\zeta}(\cdot|e, \zeta)$, one for each $\zeta \in Z(e)$, to work from.

(1) We choose a probability level p in the interval $[0, 1]$.
(2) We then determine a set $\{\iota(\zeta|e): \zeta \in Z(e)\}$ of subsets of Θ with the property that

$$P_{\theta|e,\zeta}(\hat{\theta} \in \iota(\zeta|e)|e, \zeta) \geq p \qquad (9.4.4)$$

for every ζ in $Z(e)$.
(3) We now perform e, observe ζ.
(4) Finally, we report "a $100p$ percent *Bayesian confidence set* for $\tilde{\theta}$ is $\iota(\zeta|e)$."

You should compare (9.4.4) and (9.2.5) very carefully, because they are radically different in meaning. (9.4.4) and the report in quotes mean what most people mistakenly think that sampling-theoretic confidence statements mean. The Bayesian confidence set statement means that the re-

[d] By using Exercise 3.10.9(B) with $\omega = \phi$ and $\xi = \theta$ to obtain the posterior density function of $\tilde{\phi} = e^{-2\tilde{\theta}}$ corresponding to her posterior density function $f_\gamma(\cdot|r' + \zeta, t' + 1)$ of $\tilde{\theta}$, and then differentiating with respect to ϕ and equating the derivative to zero.

porter's posterior probability of the random variable $\tilde{\theta}$'s belonging to the set $\iota(\zeta|e)$ is at least p.

We have described a four-step procedure only for purposes of comparability with the sampling-theoretic confidence procedure in Section 9.2. In practice, it is not necessary to go through step (2) for every ζ in $Z(e)$ or to perform step (2) before step (3). On the contrary, it is much simpler to choose p, then perform e and observe ζ, then determine the posterior probability function $P_{\theta|e,\zeta}(\cdot|e, \zeta)$ *only* for the ζ which occurred, and finally, perform step (2) only for this ζ and resulting $P_{\theta|e,\zeta}(\cdot|e, \zeta)$. Bayesian methods never draw on properties of outcomes ζ which might have occurred but did not.

■ In performing step (2) here, there are usually many possible subsets $\iota(\zeta|e)$ of Θ which satisfy (9.4.4). For purposes of informativeness, one usually wishes to minimize the number of elements in $\iota(\zeta|e)$, and hence he generally constructs $\iota(\zeta|e)$ by putting into it states θ of highest posterior probability, then next highest, and so forth, until the posterior probability that $\tilde{\theta} \in \iota(\zeta|e)$ just satisfies the constraint (9.4.4). The resulting set $\iota(\zeta|e)$ will then have the following properties.

(a′) $P_{\theta|e,\zeta}(\tilde{\theta} \in \iota(\zeta|e)|e, \zeta) \geq p$.

(b′) If $\theta'' \in \iota(\zeta|e)$ and $\theta' \notin \iota(\zeta|e)$, then $P_{\theta|e,\zeta}(\theta''|e, \zeta) \geq P_{\theta|e,\zeta}(\theta'|e, \zeta)$.

(c′) No proper subset of $\iota(\zeta|e)$ satisfies (9.4.4) [or (a′)].

Note that (a′)–(c′) are analogous to (a)–(c) of step (2A) in Section 9.2. Also note that (a)–(c) pertain only to the intermediately useful subsets $Z^*(\theta|e)$ of $Z(e)$ and thus only indirectly to the sets $\iota(\zeta|e)$ of interest. ■

*■ When, in infinite idealizations, $\tilde{\theta}$ has a density function, step (b′) is usually modified to

(b″) if $\theta'' \in \iota(\zeta|e)$ and $\theta' \notin \iota(\zeta|e)$,

then

$$f_{\theta|e,\zeta}(\theta''|e, \zeta) \geq f_{\theta|e,\zeta}(\theta'|e, \zeta).$$

The intent is to make $\iota(\zeta|e)$ of minimum length (if $\Theta = R^1$), area (if $\Theta = R^2$), volume (if $\Theta = R^3$), etc. Criterion (b″) prompts the term "highest posterior density" (HPD) region for an $\iota(\zeta|e)$ satisfying (b″). ■

Example 9.4.5

Consider the problem in Example 9.4.3, which concerns the experiment e defined in Example 9.2.2. Suppose we want 90 percent Bayesian confidence sets. From the posterior probability functions tabulated in Example 9.4.3, it follows readily that $\iota(\zeta^1|e) = \{\theta^1\}$, $\iota(\zeta^2|e) = \{\theta^1, \theta^2\}$, $\iota(\zeta^3|e) = \{\theta^1, \theta^3\}$, and $\iota(\zeta^4|e) = \{\theta^1, \theta^2, \theta^4\}$. These Bayesian confidence sets differ radically from the 90 percent (sampling-theoretic) confidence sets determined in Example 9.2.12.

Example 9.4.6

The prior probability function in Examples 9.4.5 and 9.4.3 is drastically nonuniform, and hence, it is reasonable to suppose that much of the difference between the 90 percent sampling-theoretic and Bayesian confidence sets is ascribable to the nonuniformity of the prior probability function. Now suppose that $P_\theta(\theta) = 1/4$ for each of the four possible states. Then it follows that the posterior probability functions are in the table (from Example 9.2.2).

$P_{\theta|e,\zeta}(\theta|e, \zeta)$

	ζ^1	ζ^2	ζ^3	ζ^4
θ^1	80/106	10/106	6/94	4/94
θ^2	10/106	80/106	4/94	6/94
θ^3	6/106	10/106	80/94	4/94
θ^4	10/106	6/106	4/94	80/94

Then 90 percent Bayesian confidence sets are $\iota(\zeta^1|e) = \{\theta^1, \theta^2, \theta^4\}$, $\iota(\zeta^2|e) = \{\theta^1, \theta^2, \theta^3\}$, $\iota(\zeta^3|e) = \{\theta^1, \theta^3\}$, and $\iota(\zeta^4|e) = \{\theta^2, \theta^4\}$, which agree for ζ^1 and ζ^2 with the sampling-theoretic confidence sets but disagree for ζ^3 and ζ^4.

***Example 9.4.7**

Consider the experiment e and prior density function discussed in *Example 9.4.2, now supposing that a $100p$ percent Bayesian confidence interval for $\tilde{\mu}$ is desired. Since the posterior density function of $\tilde{\mu}$ is $f_N(\cdot|m'', V'')$, where m'' and V'' are defined in *Example 9.4.2, it follows readily from the symmetry of Normal density functions that the shortest $100p$ percent Bayesian confidence interval for $\tilde{\mu}$ is of the form

$$\iota(\bar{x}|e) = [m'' - \mathrm{N}_{(1+p)/2}\sqrt{V''},\ m'' + \mathrm{N}_{(1+p)/2}\sqrt{V''}],$$

where N_q is defined as satisfying $F_N(\mathrm{N}_q|0, 1) = q$ and is tabulated for some values of q in *Example 9.2.14. Note that $\iota(\bar{x}|e)$ here differs from the $100p$ percent sampling-theoretic confidence interval (9.2.8) whenever $V'' \neq \upsilon/n$. Hence agreement between sampling-theoretic and Bayesian confidence intervals obtains in this problem only when the prior density function is improper, with $n' = 0$ and $V' = \upsilon/n' = +\infty$.

Finally, it is important to note that nonsensical inferences, such as $\iota(\zeta^4|e) = \emptyset$ in Example 9.2.13, are impossible with Bayesian confidence procedures. A $100p$ percent Bayesian confidence set can be empty only if $p = 0$.

There are several Bayesian approaches to hypothesis testing, but we shall introduce only the most straightforward of these; namely, the approach in which:

(1) One chooses a probability level α in $[0, 1]$, called the Bayesian significance level.
(2) H_0: $\theta \in \Theta_0$ is rejected in favor of H_A: $\theta \notin \Theta_0$ if $P_{\theta|e,\zeta}(\tilde{\theta} \in \Theta_0|e, \zeta) \leq \alpha$.

This procedure is crudely similar in spirit to the type-I-error-probability constraint in (9.3.1); namely, that $P_{\zeta|e,\theta}(\tilde{\zeta} \in R|e, \theta) \leq \alpha$ for every $\theta \in \Theta_0$. In both cases, one is guarding against the erroneous rejection of H_0.

But there is a noticeable difference between the sampling-theoretic approach and this Bayesian approach, concerning the means by which error is controlled. In the Bayesian approach, type-I error is a $\tilde{\theta}$-event with a single probability conditional on the evidence (e, ζ) actually obtained, whereas in the sampling-theoretic approach, type-I error is a $\tilde{\zeta}$-event with potentially many probabilities, conditional on the different states θ.

Example 9.4.8

Suppose one wishes to test H_0: $\theta \in \{\theta^1, \theta^2\}$ versus H_A: $\theta \in \{\theta^3, \theta^4\}$ at Bayesian significance level $\alpha = .07$ in the situation described in Examples 9.2.2 and 9.4.3. From the table of posterior probabilities in Example 9.4.3 it is clear that $P_{\theta|e,\zeta}(\tilde{\theta} \in \{\theta^1, \theta^2\}|e, \zeta) > .07$ for every ζ in $Z(e)$, and therefore, e cannot result in rejection of H_0 at the .07 level of significance, using our Bayesian procedure. This is in contrast to the sampling-theoretic tests $R = \{\zeta^3\}$ and $R = \{\zeta^4\}$ in Example 9.3.2. The difference is readily explained, but only in part, by noting that the prior probability function is "heavily biased" in favor of H_0, with $P_\theta(\tilde{\theta} \in \{\theta^1, \theta^2\}) = .80 + .15 = .95$. If $P_\theta(\cdot)$ were the uniform prior probability function, then the posterior-probability table would be that in Example 9.4.6, and we would *still* have $P_{\theta|e,\zeta}(\tilde{\theta} \in \{\theta^1, \theta^2\}|e, \zeta) > .07$ for every ζ, so that the "bias" of the prior probability function is only a partial explanation for the contrast between the sampling-theoretic and Bayesian tests.

■ Bayesian and sampling-theoretic approaches to hypothesis testing can yield persistently divergent conclusions, as Lindley [**85**] notes with an example of a situation where, as the number n of trials increases, the posterior probability of H_0 being true approaches one, whereas the sampling-theoretic test persistently rejects H_0. ■

***Example 9.4.9**

Consider testing H_0: $\mu \leq \mu_0$ versus H_A: $\mu > \mu_0$ at Bayesian significance level α in the problem discussed in *Examples 9.4.2 and 9.3.5. Using the results in *Example 9.4.2, it can be shown that the social worker should reject H_0 if $m'' \geq \mu_0 + N_{1-\alpha}\sqrt{V''}$, a prescription that agrees in general with the sampling-theoretic result in *Example 9.3.5 only if the improper prior density function, with $n' = 0$, where used.

There are other Bayesian approaches to hypothesis testing, and many interesting results for special models. But the main point, concerning estimation as well as testing, remains that the most complete Bayesian inference is the full posterior probability function itself.

■ Although the full posterior probability function is the most *complete* Bayesian inference, it is inadequate for scientific reporting purposes because of its dependence upon the prior probability function $P_\theta(\cdot)$, about which there may be radical disagreement. In many experimental contexts, the likelihood function is only minimally, if at all, controversial. Since anyone's prior probability function can be combined [via (9.4.1)] with the likelihood function to obtain his posterior probability function, it follows that the likelihood function is better for reporting purposes than the experimenter's own posterior probability function; his audience is not constrained to agree with his prior judgments. See Hildreth [**59**] for a much more complete discussion of this topic. ■

Exercise 9.4.1

Consider the experiment e with conditional-sampling probability table given in Exercise 9.2.3.

(A) Let $P_\theta(\theta^1) = .4$, $P_\theta(\theta^2) = P_\theta(\theta^3) = .3$. Find:
   (i) the posterior-mode estimator $\hat{\theta}[\cdot|e]$ of $\tilde{\theta}$.
   (ii) the posterior-mean estimator $\hat{\phi}[\cdot|e]$ of $\phi(\tilde{\theta})$, where $\phi(\theta^i) = i$ for $i = 1, 2, 3$.

(B) Let $P_\theta(\theta^i) = p_i$ for $i = 1, 2, 3$. Show that $P_{\theta|e,\zeta}(\theta^1|e, \zeta) \geq P_{\theta|e,\zeta}(\theta^2|e, \zeta)$ if and only if $p_1 \geq p_2$. Why is this so?

Exercise 9.4.2

Consider experiment e with conditional-sampling probability table given in Example 9.2.3.

(A) Let $P_\theta(\theta^1) = .2$, $P_\theta(\theta^2) = .4$, $P_\theta(\theta^3) = .3$, and $P_\theta(\theta^4) = .1$. Find:
   (i) the posterior-mode estimator $\hat{\theta}[\cdot|e]$ of $\tilde{\theta}$.
   (ii) the posterior-mean estimator $\hat{\phi}[\cdot|e]$ of $\phi(\tilde{\theta})$, where $\phi(\theta^1) = \phi(\theta^4) = 0$, $\phi(\theta^2) = 2$, and $\phi(\theta^3) = -4$.
   (iii) for each $\zeta \in Z(e)$, an 85 percent Bayesian confidence set $\iota(\zeta|e)$.

(B) Let $P_\theta(\theta^i) = 1/4$ for $i = 1, 2, 3, 4$. Under this assumption, do: (i) part (i) of (A); (ii) part (ii) of (A); and (iii) part (iii) of (A).

Exercise 9.4.3

(A) With all assumptions as given in Exercise 9.4.2(A), for what outcomes ζ of e can you reject H_0: $\theta \in \{\theta^1, \theta^2\}$ at Bayesian significance level $\alpha = .03$?

(B) Same question, but with assumptions as given in Exercise 9.4.2(B).

Exercise 9.4.4

Using the prior probability functions and the likelihood function given in Example 9.4.1, calculate the three posterior probability functions and show that they agree with those stated to two decimal places.

Exercise 9.4.5

Show that if $P_\theta(\theta) = 1/\#(\Theta)$ for every θ in Θ, then

$$P_{\theta|e,\zeta}(\theta|e,\zeta) = \frac{L(\theta|e,\zeta)}{\sum_{i=1}^{\#(\Theta)} L(\theta^i|e,\zeta)}$$

for every θ in Θ.

Exercise 9.4.6: Optimality of Some Bayesian Point Estimates

Posterior-mode and posterior-mean estimates are optimal in the statistical-decision-theoretic sense (Chapter 8) if the forecaster's set A of available terminal acts and his utility function $u(v(a, \theta))$ satisfy certain properties.

(A) Let $A = \Theta$ and suppose that

$$u(v(a,\theta)) = \begin{cases} K_1, & a = \theta \\ K_0, & a \neq \theta\,, \end{cases}$$

for every θ in Θ, where $K_1 > K_0$. ("A miss is as good as a mile.") Show that $a° = \hat{\theta}[\zeta|e]$, where $\hat{\theta}[\zeta|e]$ is the posterior-mode estimate of $\tilde{\theta}$.

(B) Let $A = (-\infty, +\infty)$ and suppose that

$$u(v(a,\theta)) = K - k[a - \phi(\theta)]^2$$

for every θ in Θ, where $k > 0$ and $\phi(\cdot)$ is a real-valued function on Θ. Show that $a° = \hat{\phi}[\zeta|e]$, where $\hat{\phi}[\zeta|e]$ is the posterior-mean estimate of $\phi(\tilde{\theta})$.

9.5 SOME AXIOMATIC PRINCIPLES OF INFERENCE

This section concerns several more or less equivalent principles that can serve as axiomatic constraints on the methodology of statistical inference in much the same way as the principles of consistent decision making under uncertainty served in Chapters 3 and 4 as constraints on the way in which quantified judgments and preferences were to be processed to determine optimal courses of action. Here, however, our aim is more modest; we shall merely show that a certain, intuitively compelling principle logically implies a less compelling but quite operationally useful principle, the gist of which is that all we should look at for inferential purposes is the likelihood function $L(\cdot|e', \zeta')$ given only the outcome ζ' of e' that *actually* occurred.

■ This section is based substantially on Birnbaum's important paper [**10**], with ensuing discussion, on the foundations of inference. See also Durbin [**28**], Savage [**124**], Birnbaum [**11**], and the references in these papers. ■

We shall focus attention on pairs (e, ζ) consisting of an experiment e and an outcome ζ of e. To shorten verbiage, we shall write

$$(e', \zeta') \equiv (e'', \zeta'')$$

to mean that (e', ζ') and (e'', ζ'') convey exactly the same information about $\tilde{\theta}$, and hence that the inferences, conclusions, or decisions about $\tilde{\theta}$ that one could make had he performed e' and obtained the outcome $\zeta' \in Z(e')$ should be exactly the same as if he had performed e'' and obtained the outcome $\zeta'' \in Z(e'')$.

The principle we seek to justify by appeal to a more intuitively compelling principle is the likelihood principle. It says that if (e', ζ') and (e'', ζ'') yield proportional likelihood functions, $L(\cdot|e', \zeta') \propto L(\cdot|e'', \zeta'')$, then $(e', \zeta') \equiv (e'', \zeta'')$.

Definition 9.5.1: Likelihood Principle. Given (e', ζ') and (e'', ζ''). If there is a number $k = k(\zeta', \zeta'') > 0$ such that

$$L(\theta|e', \zeta') = kL(\theta|e'', \zeta'')$$

for every θ in Θ, then

$$(e', \zeta') \equiv (e'', \zeta'').$$

Nothing in the Likelihood Principle prevents e' and e'' from being the *same* experiment, say e, in which case the Likelihood Principle implies that any two outcomes ζ' and ζ'' of an experiment e which have proportional columns in the conditional-sampling probability table $P_{\zeta|e,\theta}(\cdot|e, \cdot)$ should induce exactly the same inferences.

The Likelihood Principle essentially states that inferences should be based solely on the outcome ζ' of e' which actually occurred and not on outcomes which might have occurred but did not; this follows by noting that the Likelihood Principle makes no references to outcomes of e' and e'' other than ζ' and ζ'', and no reference to any relationships among such outcomes.

Our second principle has much intuitive appeal and is, in fact, equivalent to the Likelihood Principle. It relies upon the concept of a *randomized experiment* e^*, defined as consisting of two stages: in stage one, the value λ of a random variable $\tilde{\lambda}$ is observed; in stage two, an associated experiment e_λ is performed and its outcome ζ_λ is observed. Moreover, it is assumed that $\tilde{\lambda}$ is independent of, or entirely unrelated to, the state $\tilde{\theta}$ about which information is sought. The complete outcome of such a randomized experiment is thus describable as $(e_\lambda, \zeta_\lambda)$; and e^* together with an outcome of e^* is denoted by $(e^*, (e_\lambda, \zeta_\lambda))$.

Now, in such a randomized experiment, it does not seem reasonable to allow inferences to depend upon the probability $P_\lambda(\lambda)$ of performing the second-stage experiment e_λ that you actually performed, in view of the assumption that knowledge of $\tilde{\lambda}$ per se conveys nothing about $\tilde{\theta}$. For instance, if you chose between two test markets on the basis of a coin flip, then your final inference should depend solely upon the results in the chosen test market, and it should not be influenced by subsequent information that the coin was "unfair." The formalization of this idea is the Conditionality Principle.

Definition 9.5.2: Conditionality Principle. Suppose that e^* is a randomized experiment in which the probability function of $\tilde{\lambda}$, and hence of the second-stage experiment $e_{\tilde{\lambda}}$, is independent of $\tilde{\theta}$. Then

$$(e^*, (e_\lambda, \zeta_\lambda)) \equiv (e_\lambda, \zeta_\lambda)$$

for every λ and every $\zeta_\lambda \in Z(e_\lambda)$.

■ An informal but fully descriptive version of the Conditionality Principle is this: For inference purposes you can ignore the method by which you chose an experiment as long as θ had no influence on that choice. ■

We introduce another principle which, strictly speaking, is unnecessary because it follows from the Conditionality Principle, but which is of interest in its own right. First, however, some preliminary definitions are necessary. Recall that in the Introduction we defined a *statistic* as a function $\delta: Z(e) \to \Delta$ of the outcome of an experiment.

The role of a statistic is to summarize the outcome ζ of e; its interpretation is that one looks at δ and ignores the particular ζ which gave rise functionally to δ. [There may be many such ζ, just as there are many n-tuples $(x(\omega_1), \ldots, x(\omega_n))$ giving rise to the same sample mean $\delta = \bar{x}_{n}$.] Since in any given situation there are many statistics, the question arises as to just which statistics retain *all* the information about $\tilde{\theta}$ in ζ without losing any. Such statistics are called *sufficient* and are defined by requiring that the conditional probability function of $\tilde{\zeta}$ given θ *and* given the value δ of the statistic be independent of θ.

Definition 9.5.3: Sufficient Statistic. Given e and $Z(e)$, a statistic $\delta: Z(e) \to \Delta$ is called *sufficient* (for $\tilde{\theta}$ in e) if, for every ζ in $Z(e)$ and every δ in Δ, $P_{\zeta|e,\theta,\delta}(\zeta|e, \theta, \delta)$ is independent of θ.

■ Since δ is functionally determined by ζ, it is clear that δ and ζ must be such that $\delta = \delta(\zeta)$, although many different ζ's may possess this property for a given $\delta \in \Delta$. In this case, it is easy to see that $P_{\delta,\zeta|e,\theta}(\delta, \zeta|e, \theta) = P_{\zeta|e,\theta}(\zeta|e, \theta)$, and hence that

$$P_{\zeta|e,\theta}(\zeta|e, \theta) = P_{\delta|e,\theta}(\delta|e, \theta)P_{\zeta|e,\theta,\delta}(\zeta|e, \theta, \delta). \tag{9.5.1}$$

Now, (9.5.1) holds for *any* statistic, but if $\delta(\cdot)$ is *sufficient* for $\tilde{\theta}$ in e, then $P_{\zeta|e,\theta,\delta}(\zeta|e, \theta, \delta)$ is independent of θ. In fact, knowledge of the value δ of a sufficient statistic allows one to duplicate the probabilistic properties of the actual outcome by simulating an observation ζ from the *known* (because independent of θ) probability function of $\tilde{\zeta}$ given (θ and) δ. ■

Since $P_{\zeta|e,\theta,\delta}(\zeta|e, \theta, \delta)$ is independent of θ when δ is a sufficient statistic, common sense dictates that one who knows δ gains nothing by also learning the actual outcome ζ of e.

Definition 9.5.4: Sufficiency Principle. Let e' be an experiment with sufficient statistic $\delta: Z(e) \to \Delta$, and suppose that e'' is the experiment in which the outcome $\delta(\zeta)$ occurs if and only if ζ occurs in e'. Then

$$(e'', \delta) \equiv (e', \zeta')$$

for every ζ' such that $\delta = \delta(\zeta')$.

Example 9.5.1

Consider the conditional-sampling probability table for e'

	ζ^1	ζ^2	ζ^3
θ^1	.4	.4	.2
θ^2	.1	.6	.3
θ^3	.7	.2	.1

together with $\delta(\cdot)$ defined by $\delta(\zeta^1) = \delta^1$ and $\delta(\zeta^2) = \delta(\zeta^3) = \delta^2$, where $\delta^1 \neq \delta^2$. In e', you observe ζ^1, ζ^2, or ζ^3. In e'', you observe δ^1 or δ^2 with conditional-sampling probabilities

	δ^1	δ^2
θ^1	.4	.6
θ^2	.1	.9
θ^3	.7	.3

It is easy to see that $P_{\zeta|e,\theta,\delta}(\zeta^1|e, \theta, \delta^1) = 1$, and that

$$P_{\zeta|e,\theta,\delta}(\zeta^2|e, \theta, \delta^2) = 2P_{\zeta|e,\theta,\delta}(\zeta^3|e, \theta, \delta^2) = 2/3,$$

in all cases independent of θ. Hence $\delta(\cdot)$ is a sufficient statistic. The Sufficiency Principle contends that e'' is just as good as e' for making inferences about $\tilde{\theta}$.

■ From (9.5.1) follows a characterization of sufficient statistics which is very convenient in sampling applications. If $P_{\zeta|e,\theta,\delta}(\zeta|e, \theta, \delta)$ is independent of θ, then (9.5.1) implies that

$$P_{\zeta|e,\theta}(\zeta|e, \cdot) = P_{\delta|e,\theta}(\delta|e, \cdot)k(\zeta). \tag{9.5.2}$$

But the left-hand side of (9.5.2) is just $L(\cdot|e, \zeta)$. Hence δ is a sufficient statistic if the likelihood function $L(\cdot|e, \delta)$ can be factored (1) into a factor $h(\zeta)$ independent of θ and (2) into a factor dependent upon θ and upon ζ only via the statistic $\delta(\zeta)$. This is written frequently in the form

$$L(\theta|e, \zeta) = g(\theta, \delta(\zeta))h(\zeta). \tag{9.5.3}$$

Now, (9.5.3) suggests a very useful technique for identifying sufficient statistics: just see how $L(\theta|e, \zeta)$ factors. [In certain cases, some ingenuity is required in reexpressing $L(\theta|e, \zeta)$ so as to satisfy (9.5.3) for the statistic of interest; we had to do so in *Example 9.2.5, as discussed in *Example 9.5.3 below.] ■

***Example 9.5.2**

In *Example 9.2.4, we noted that if e consisted in observing a process on n indistinguishable instances and $\omega_j = 1$ for occurrence of a given phenomenon, $\omega_j = 0$ for nonoccurrence, then $\zeta = (\omega_1, \ldots, \omega_n)$ and $L(\theta|e, \zeta) = \theta^\delta(1-\theta)^{n-\delta}$, where θ is the probability of an occurrence in any given instance and $\delta = \Sigma_{j=1}^n \omega_j$ = total number of occurrences. From (9.5.3) it is immediate that δ is a sufficient statistic, with $h(\zeta) = 1$ for every ζ. By using (9.5.3), we are spared having to determine the binomial mass function of $\tilde{\delta}$.

***Example 9.5.3**

Suppose e consists in performing n indistinguishable trials and recording the n measurements $x(\omega_j)$, where each $x(\tilde{\omega})$ had the Normal density function $f_N(\cdot|\mu, \upsilon)$, as discussed in *Example 9.2.5, where υ is assumed known. Then $L(\mu|e, \zeta) = k(\zeta)f_N(\mu|\bar{x}, \upsilon/n)$, from (9.2.3), and hence $\bar{x}$ is a sufficient statistic for μ.

The recognition and exploitation of sufficient statistics is an important part of the practicing statistician's life. That the overwhelming majority of statisticians *are* willing to exploit sufficient statistics by ignoring other attributes of experimental outcomes is evidence of the acceptability of the Sufficiency Principle.

In the remainder of this section, we shall (1) present two useful lemmas; (2) show that the Conditionality Principle implies the Sufficiency Principle; (3) show that the Conditionality Principle implies the Likelihood Principle; and (4) discuss the momentous implications of the Likelihood Principle for statistical inference.

Our first lemma concerns the existence of sufficient statistics. It says that if one column, say ζ'', of the conditional-sampling probability table is proportional to another column, say ζ', then there is a sufficient statistic $\delta(\cdot)$ such that $\delta(\zeta') = \delta(\zeta'')$.

Lemma 9.5.1. If $L(\theta|e', \zeta') = kL(\theta|e', \zeta'')$ for every θ in Θ and some $k = k(\zeta', \zeta'') > 0$, then there is a sufficient statistic $\delta(\cdot)$ for $\tilde{\theta}$ in e' such that $\delta(\zeta') = \delta(\zeta'')$.

■ *Proof.* Exercise 9.5.1. ■

The second lemma is that the Sufficiency Principle implies the "weak likelihood principle."

Lemma 9.5.2. The Sufficiency Principle implies the Weak Likelihood Principle; namely: if $L(\theta|e', \zeta') = kL(\theta|e', \zeta'')$ for every θ in Θ and some $k = k(\zeta', \zeta'') > 0$, then $(e', \zeta') \equiv (e', \zeta'')$.

■ *Proof.* If $L(\theta|e', \zeta') = kL(\theta|e', \zeta'')$ for every θ in Θ and some $k > 0$, then by Lemma 9.5.1 there is a sufficient statistic $\delta(\cdot)$ such that $\delta(\zeta') = \delta(\zeta'')$. By the Sufficiency Principle, $(e', \zeta') \equiv (e'', \delta)$ and $(e', \zeta'') \equiv (e'', \delta)$, where $\delta = \delta(\zeta') = \delta(\zeta'')$ and e'' is the experiment in which $\delta(\zeta)$ occurs if and only if ζ occurs in e'. Hence $(e', \zeta') \equiv (e', \zeta'')$, both being $\equiv (e'', \delta)$. ■

■ The Weak Likelihood Principle is "weak" only because it does not encompass the case in which the experiment is varied; compare it with the "nonweak" Likelihood Principle in Definition 9.5.1. ■

Our first key result, Theorem 9.5.1, is that the Conditionality Principle implies the Sufficiency Principle.

Theorem 9.5.1. The Conditionality Principle implies the Sufficiency Principle.

■ *Proof.* Let e' be an experiment and suppose that $\delta(\cdot)$ is a sufficient statistic for $\tilde{\theta}$ in e'. Now, the complete outcome ζ of e' can be regarded as "unfolding" via a two-stage process: (1) The value δ of the sufficient statistic $\delta(\tilde{\zeta})$ is observed; (2) ζ itself is observed. See Figure 9.1, in which we also suppose for didactic purposes that ζ is necessarily followed by the "null experiment with null outcome," (e^-, ζ^-). Since $P_{\zeta|e,\theta,\delta}(\zeta|e, \theta, \delta)$ is independent of θ, because $\delta(\cdot)$ is assumed to be a sufficient statistic, it follows from the Conditionality Principle applied to e^*_δ that $(e^*_\delta, (e^-, \zeta^-)) \equiv (e^-, \zeta^-)$, which means that no additional information can be provided by e^*_δ. Letting e'' denote the truncation of e' after the first stage, i.e., after observing δ, it follows that $(e', \zeta) \equiv (e', \delta(\zeta))$, which is the Sufficiency Principle. ■

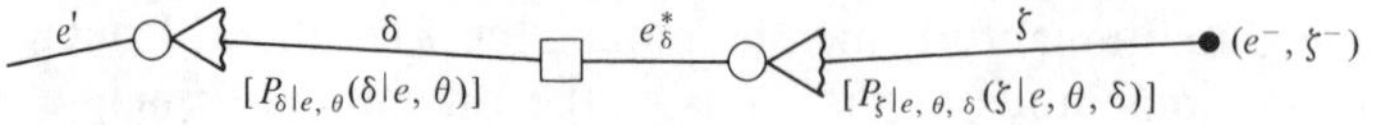

Figure 9.1 "Unfolding" model of e'.

Because the Conditionality Principle is much more intuitively compelling to many people than the Likelihood Principle, it is important to recognize that the Likelihood Principle follows logically from the Conditionality Principle. This fact, Theorem 9.5.2, is the central result of this section.

Theorem 9.5.2. The Conditionality Principle implies the Likelihood Principle.

■ *Proof.* Given (e', ζ') and (e'', ζ''), suppose that $L(\theta|e', \zeta') = kL(\theta|e'', \zeta'')$ for every θ in Θ and some $k = k(\zeta', \zeta'') > 0$; this is the hypothesis of the Likelihood Principle. Because we have already proved that the Conditionality Principle implies the Sufficiency Principle, we are free to use both of these to deduce the conclusion, $(e', \zeta') \equiv (e'', \zeta'')$, of the Likelihood Principle. Consider the randomized experiment e^* in which e' and e'' are each chosen with probability 1/2. By the Conditionality Principle,

(a′) $$(e^*, (e', \zeta')) \equiv (e', \zeta'),$$

and

(a″) $$(e^*, (e'', \zeta'')) \equiv (e'', \zeta'').$$

But e^* can also be regarded as a single-stage experiment in which $Z(e^*) = \{(e', \zeta)\colon \zeta \in Z(e')\} \cup \{(e'', \zeta)\colon \zeta \in Z(e'')\}$ and

$$P_{\zeta|e,\theta}(\zeta^*|e^*, \theta) = \begin{cases} \tfrac{1}{2}P_{\zeta|e,\theta}(\zeta|e', \theta) & \text{if } \zeta^* = (e', \zeta) \\ \tfrac{1}{2}P_{\zeta|e,\theta}(\zeta|e'', \theta) & \text{if } \zeta^* = (e'', \zeta). \end{cases}$$

By assumption, $L(\cdot|e', \zeta') = kL(\cdot|e'', \zeta'')$. Hence in the one-stage model of e^*, we have $P_{\zeta|e,\theta}((e', \zeta')|e^*, \theta) = kP_{\zeta|e,\theta}((e'', \zeta'')|e^*, \theta)$ for every θ in Θ, or $L(\theta|e^*, (e', \zeta')) = kL(\theta|e^*, (e'', \zeta''))$. This is the hypothesis of the "weak likelihood principle," and hence Lemma 9.5.2 implies that

(b) $$(e^*, (e', \zeta')) \equiv (e^*, (e'', \zeta'')),$$

given the Sufficiency Principle, which follows from the Conditionality Principle (Theorem 9.5.1). From (a′), (a″), and (b) we obtain $(e', \zeta') \equiv (e'', \zeta'')$, the desired conclusion. That this argument, requiring construction of a randomized experiment e^*, is a valid proof follows from the fact that one could always construct such an e^* and obtain a logical contradiction if the conclusion $(e', \zeta') \equiv (e'', \zeta'')$ were denied. ■

Now that we have given a strong intuitive justification for the Likelihood Principle, it remains for us to examine its implications for inferential methodology. Beginning on a positive note, it is clear that maximum-likelihood estimation satisfies the Likelihood Principle. So do all Bayesian inferences, because they all stem from the posterior probability function, which depends upon (e, ζ) only via $L(\cdot|e, \zeta)$. (You may furnish proofs of these assertions as Exercise 9.5.2.)

But sampling-theoretic techniques of set estimation and hypothesis testing and of unbiased estimation do *not* satisfy the Likelihood Principle; the estimator or critical set, as the case may be, is determined with reference to all the outcomes ζ which might occur, and not just to the outcome which in fact does occur.

Example 9.5.4

Consider e' and e'' as described by the following conditional-sampling probability tables $P_{\zeta|e,\theta}(\cdot|e, \cdot)$,

e'	ζ^1	ζ^2
$\theta^1 = 0$	.8	.2
$\theta^2 = 1$	.9	.1

e''	ζ^3	ζ^4
$\theta^1 = 0$	.2	.8
$\theta^2 = 1$	.6	.4

Since $L(\cdot|e'\zeta^2) = \frac{1}{4}L(\cdot|e'', \zeta^4)$, the Likelihood Principle implies that $(e', \zeta^2) \equiv (e'', \zeta^4)$. But it is easy to verify that the unique unbiased estimator of θ in e' is $\hat{\theta}[\zeta^1|e'] = 2$, $\hat{\theta}[\zeta^2|e'] = -8$, whereas the unique unbiased estimator of θ in e'' is $\hat{\theta}[\zeta^3|e''] = 2$, $\hat{\theta}[\zeta^4|e''] = -0.5$. Clearly, $\hat{\theta}[\zeta^2|e'] \neq \hat{\theta}[\zeta^4|e'']$, and the Likelihood Principle has been violated. Suppose we wanted 80 percent confidence sets; using the procedure in Section 9.2 yields $\iota(\zeta^2|e') = \emptyset$ and $\iota(\zeta^4|e'') = \{\theta^1, \theta^2\}$, a disparity again in violation of the Likelihood Principle. Similarly, the uniformly most powerful test of H_0: $\theta = \theta^1$ in e' at significance level $\alpha = .2$ calls for rejecting H_0 given ζ^2, whereas the uniformly most powerful test of H_0: $\theta = \theta^1$ in e'' at significance level $\alpha = .2$ calls for *not* rejecting H_0 given ζ^4.

All this is, and should be, very disconcerting to statisticians brought up on sampling-theoretic methods who find the Conditionality Principle compelling. Yet, not all sampling-theoretic statistics is lost, because much of the probabilistic model building in that area is applicable with only minor (if any) adaptation in inferential approaches which do satisfy the Likelihood Principle, and because sampling-theoretic inferences which violate the Likelihood Principle may, in "conclusive" experiments, violate it in so minor a way as to remain good approximations to, say, Bayesian methods which do not violate it.

Exercise 9.5.1

Prove Lemma 9.5.1. [*Hint*: Define $\delta(\zeta'') = \zeta'$ and $\delta(\zeta) = \zeta$ for every $\zeta \neq \zeta''$.]

Exercise 9.5.2

Suppose that $L(\theta|e', \zeta') = kL(\theta|e'', \zeta'')$ for every θ in Θ and some $k = k(\zeta', \zeta'') > 0$.

(A) Show that θ^* is a maximum-likelihood estimate of θ given (e', ζ') if and only if θ^* is a maximum-likelihood estimate of θ given (e'', ζ'').

(B) Given any prior probability function $P_\theta(\cdot)$ on events in Θ, show that $P_{\theta|e,\zeta}(\theta|e', \zeta') = P_{\theta|e,\zeta}(\theta|e'', \zeta'')$ for every θ in Θ.

■ Exercise 9.5.2 shows that maximum-likelihood estimation and Bayesian inferences satisfy the Likelihood Principle. ■

Exercise 9.5.3

Prove that the Likelihood Principle implies (A) the Sufficiency Principle; (B) the Conditionality Principle; and (C) the Weak Likelihood Principle.

■ Exercise 9.5.3 and Theorem 9.5.2 together show that the Likelihood Principle is equivalent to the Conditionality Principle; each implies the other. Since the Weak Likelihood Principle implies the Sufficiency Principle, these two principles are also equivalent, by virtue of Lemma 9.5.2. ■

10

INTRODUCTION TO MARKOVIAN DECISION PROCESSES

The first and greatest concern for the immense majority of every nation is the stability of the laws, and their uninterrupted action—never their change.

Prinz Klemens Wenzel Nepomunk Lothar von Metternich, secret memorandum to Tsar Alexander I, Troppau, 1820

10.1 INTRODUCTION

When you read Chapters 2 and 4, you must have noticed that decisions in standard extensive form tend to become unwieldy as the number N of "stages" increases above some very small number. A glance at Figure 2.16 confirms that the number of partial histories h_n of the decision through stage n tends to become huge as n increases, and it is this fact which tends to limit the practical application of decision analysis to problems with relatively short horizons in terms of number of stages explicitly modeled in tree form.

What then can be said for a commonly occurring problem like Example 1.5.12, concerning inventory policy? Can decision analysis even begin to cope with, say, Example 1.5.12b, in which the number of future time periods is assumed to be infinite?

Fortunately, the answer is a qualified yes. Certain multistage dpuu's possess an underlying structure sufficiently simple to permit our deriving optimal strategies for $\mathscr{D}$ even when N is very large or even infinite. The key attributes of such a simple-structured dpuu are twofold.

(1) At every stage n, only a *very succinct* summarization x_n of the history of the dpuu up to stage n is needed to specify fully the effect of that history on $\mathscr{D}$'s utility of the ultimate consequence, $\mathscr{D}$'s probabilities of events in stage n and beyond, and $\mathscr{D}$'s sets of available acts in stages n and beyond.

(2) The qualitative natures of the decisions made in the various stages are sufficiently similar to permit, for the infinite-N case, taking limits as $N \to +\infty$.

■ Attributes (1) and (2) are often intuitively evident. When (1) fails, for one reason or another, the dpuu is beset by what the father of dynamic programming, Richard Bellman, calls "the curse of dimensionality." Dynamic programming may be defined as multistage optimization methodology; it includes the extensive-form analysis of Part I of this book, as well as ingenious approaches to deterministic optimization problems. Some flavor for the general subject is suggested by the exercises for this section; the interested reader should consult a work devoted to the subject, such as Bellman and Dreyfus [**8**], White [**141**], or Nemhauser [**104**]. ■

In many problems of this type, the overall consequence is some function of N monetary returns earned in the N stages of the dpuu. Box diagrams are often used to depict the flow of the dpuu from one stage to the next; see Figure 10.1, for example, where, you will note, we have *reversed* the numbering of the stages. This reversal of stage numbering is occasionally used to facilitate taking limits as $N \to +\infty$, since in the reversed numbering N is the initial stage.

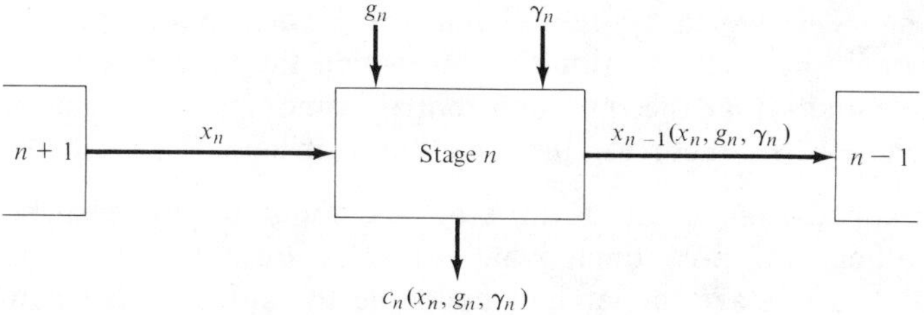

Figure 10.1 Box diagram of a dpuu.

■ Interpreting the box diagram is simple. "Inputs" to stage n are (1) the summary x_n of the preceding history; (2) $\mathscr{D}$'s chosen act g_n; and (3) $\mathscr{E}$'s chosen ee γ_n. "Outputs" from stage n are (1) the net monetary return c_n, a function of x_n, g_n, and γ_n; and (2) the summary x_{n-1} of the history up to stage $n-1$, the next stage. This summary is, naturally, dependent upon x_n, g_n, and γ_n. ■

The succinct summary x_n is usually called the (stage-n) *state*, or *state variable*. Be careful not to confuse this use of the word "state," which refers to the state of the system under consideration, with our uses of the term in Chapter 9, Chapter 7, and elsewhere.

Except for in the exercises for this section, in the remainder of this chapter we shall consider an even more specialized problem, the (*homogeneous, finite*) *Markovian*[a] *decision problem*. It is specified by:

(A) a finite set $X = \{x: x = 1, \ldots, I\}$ of states, to which every x_n must belong.

[a] "Markovian" is an adjective referring to the probabilist A. A. Markov, since the underlying probabilistic structure of the problems under consideration is the *Markov chain*, which Markov introduced and first studied.

(B) for each state x, a finite set $G(x) = \{g: g = 1, \ldots, K_x\}$ of available *acts* for $\mathscr{D}$ given x.

(C) for each state x and act g, a probability p_{xy}^g that the state at stage $n-1$ will be y, with $\Sigma_{y=1}^{I} p_{xy}^g = 1$, of course.

(D) an expected, immediate net return $c_n = q_x{}^g$ in stage n, attributable to being in state x and making decision g.

(E) a discount factor $\beta \in [0, 1]$ such that $\mathscr{D}$'s utility in stage N of any sequence $c_N, c_{N-1}, \ldots, c_1$ of current and future net returns is *linear* in the present value $\Sigma_{n=1}^{N} \beta^{N-n} c_n$ of that sequence.

■ Note that the stage number n has no effect on the set of stage-n states, the set of available acts given the state, the "transition (from state to state) probabilities" p_{xy}^g, the expected immediate return $q_x{}^g$, or the discount factor β. This independence of the problem parameters from the stage numbers makes all stages look alike and justifies the adjective "homogeneous." In Metternich's terms, the laws of the system are stable. ■

■ The effect of the discount factor is to express the judgment by $\mathscr{D}$ that \$1 one stage hence is worth \$$\beta$ now. Thus $\beta = 1$ corresponds to "no discounting," or, in practice, to situations in which the stages succeed one another so rapidly that the time differential in monetary evaluation is negligible, whereas $\beta = 0$ corresponds to a "now-or-never" attitude. ■

■ Often, the expected immediate returns $q_x{}^g$ are not given but must be derived from other data. Sometimes what is given are returns r_{xy}^g at the beginning of the next stage, dependent upon the transition to the new state y at that stage. The discounted present value, $\beta \, \Sigma_{y=1}^{I} r_{xy}^g p_{xy}^g$, would then be $q_x{}^g$. ■

Example 10.1.1

We show that Example 1.5.12 is a Markovian decision problem. At each stage n, the *state variable* x is the inventory (number) of widgets remaining from the preceding period, stage $n+1$ in reverse numbering. It must be nonnegative (because excess demands in preceding periods are not backordered), and it cannot exceed the maximum number, say M, of widgets that you can physically possess. The stage-n *act* g is the number of widgets that you decide to restock; g must be nonnegative, and it cannot exceed $M - x$, the number for which space could be made available. Hence

$$X = \{x: x = 0, 1, \ldots, M\}$$

and

$$G(x) = \{g: G = 0, 1, \ldots, M - x\}.$$

[Do not worry about the difference in labeling of states and acts from specifications (A) and (B), as the labeling in (A) and (B) is convenient but entirely arbitrary in the ensuing analysis.] The stage-n random variable $\tilde{y}_n$ is the number of widgets demanded by your customers. Recall that $\tilde{y}_1, \tilde{y}_2, \ldots$ are assumed to be completely unrelated, or independent. Now note that the next state y, the beginning inventory for period $n-1$, is related to g, x, and γ_n by $y = \max[0, x+g-\gamma_n]$, so that

$$p_{xy}^g = \begin{cases} P(\tilde{\gamma}_n \geq x+g \mid x, g), & y = 0 \\ P(\tilde{\gamma}_n = x+g-y \mid x, g), & y > 0 \end{cases}$$

for every x, g, and y. [Note (1) that for p_{xy}^g to be independent of n, the probability functions of $\tilde{\gamma}_1, \tilde{\gamma}_2, \ldots$ must all be the same, given x and g; and (2) that we permit dependence of the probability function of $\tilde{\gamma}_n$ on x and g—since a large stock $x+g$ of goods on hand, for example, may stimulate impulse purchasing.] Determining q_x^g, the expected immediate net return for stage n, is a bit involved. We shall determine $c_n(x, g, \gamma_n)$, since $q_x^g = E\{c_n(x, g, \tilde{\gamma}_n) \mid x, g\}$. For every γ_n, x, and g, the net return $c_n(x, g, \gamma_n)$ attributable to stage n is *proceeds* less *costs*. Proceeds amount to $\$P \cdot \min[\gamma_n, x+g]$, reflecting immediate sale of γ_n widgets if available, or of $x+g$ widgets if demand exceeds supply. Total *cost* is the sum of three kinds of cost: the purchase cost $\$C \cdot g$ of g widgets; the cost $\$h \cdot \max[0, x+g-\gamma_n]$ of holding inventory (if any) in period n; and the ordering cost $\$K + k \cdot g$ if $g > 0$, or $\$0$ if $g = 0$. The discount factor, β, is simply the r of Example 1.5.12, which we have not yet needed.

Example 10.1.2

For subsequent purposes, we shall work with a numerical version of Example 10.1.1, in which: for every n, $\tilde{\gamma}_n = 0, 1$, or 2 with respective probabilities .1, .7, and .2; the maximum inventory which can be carried is $M = 5$; the selling price per widget is $P = \$10$; the purchase cost per widget is $C = \$4$; the holding cost per widget period is $h = \$1$; the fixed ordering cost is $K = \$2$; and the variable ordering cost per widget is $k = \$.50$. As Exercise 10.1.1(A), you may verify that p_{xy}^g is as given in the following table.

	p_{xy}^g: $y = x+g$	$y = x+g-1$	$y = x+g-2$
$x+g=0$	1.0	0	0
$x+g=1$	.1	.9	0
$x+g=2$	.1	.7	.2

As Exercise 10.1.1(B), you may verify that q_x^g is as given in the following table.

	$g=0$	1	2	3	4	5
$x=0$	0.0	2.4	−0.9	−6.4	−11.9	−17.4
1	8.9	3.6	−1.9	−7.4	−12.9	*
2	10.1	2.6	−2.9	−8.4	*	*
3	9.1	1.6	−3.9	*	*	*
4	8.1	0.6	*	*	*	*
5	7.1	*	*	*	*	*

where $*$ denotes an (x, g) pair which is impossible because of the constraint that $x+g \leq 5$.

Now, we could use Theorem 4.3.1 to solve any Markovian decision problem in which N is finite, but to do so would amount to not exploiting any of the regularity features we have introduced. In fact, to *define* the overall utility of the dpuu at stage N of acting at stages $n, n-1, \ldots, 1$ in reverse numbering requires knowledge of the complete partial history up to stage n, even though the additive nature of $\mathscr{D}$'s utility—from specification (E)—implies that the returns on stages preceding n are irrelevant to choice of optimal action in stages $n, n-1, \ldots, 1$.

■ Because of specification (E), $\mathscr{D}$'s overall utility of an N-stage Markovian decision problem is expressible, in present value at the initial stage N, as

$$u(c_N, \ldots, c_1) = \sum_{i=n+1}^{N} \beta^{N-i} c_i + \sum_{i=1}^{n} \beta^{N-i} c_i.$$

But $c_j = c_j(x_j, g_j, \gamma_j)$ for $j = 1, \ldots, N$, and hence the maximum expected utility U_n of acting optimally on stages $n, n-1, \ldots, 1$ depends upon x_N, g_N, γ_N, g_{N-1}, $\gamma_{N-1}, \ldots, g_{n+1}$, γ_{n+1}, even though only the resulting stage-n state variable x_n is relevant to optimal action in stages n and following, because $\tilde{\gamma}_n, \tilde{\gamma}_{n-1}, \ldots, \tilde{\gamma}_1$ depend upon $(x_N, g_N, \gamma_N, \ldots, g_{n+1}, \gamma_{n+1})$ only via x_n and

$$E_{n,n-1,\ldots,1}\{u(c_N, \ldots, c_{n+1}, \tilde{c}_n, \ldots, \tilde{c}_1)|x_n\}$$
$$= \sum_{i=n+1}^{n} \beta^{N-i} c_i + E_{n,n-1,\ldots,1}\left\{\sum_{i=1}^{n} \beta^{N-i} \tilde{c}_i | x_n\right\}.$$

The fact that $\Sigma_{i=n+1}^{N} \beta^{N-i} c_i$ "separates out" from the remaining summands, unrevealed to $\mathscr{D}$ as of choosing action in stage n, implies that $c_N, \ldots, c_{n+1}$ are irrelevant to this and subsequent choices. Hence so is the partial history up to stage n, except for its summary x_n. ■

But there is an alternative, really a corollary, to Theorem 4.3.1, which does exploit the simplifying features of the Markovian decision problem in determining an optimal strategy for $\mathscr{D}$.

Theorem 10.1.1: Recursion for Markovian Decision Problems. For $n = 1, 2, \ldots$, let $v_n(x)$ denote the expected present value as of stage n of acting optimally in stages $n, n-1, \ldots, 1$, and let $v_0(x) = 0$ for every x. Then

$$v_n(x) = \max_{g \in G(x)} \left[q_x^g + \beta \sum_{y=1}^{I} v_{n-1}(y) p_{xy}^g \right] \tag{10.1.1}$$

for $n = 1, 2, \ldots, N$, and for every $N \geq 1$.

*■ *Proof.* By the previous remark we may refer curtly to $U_n(x)$ and $U_{n-1}(y)$ even though U_n and U_{n-1} depend upon their entire preceding

partial histories. By Theorem 4.3.1, with reversed numbering,

$$U_n(x) = \max_{g\in G(x)}\left[\sum_{y=1}^{I} U_{n-1}(y)p_{xy}^g\right]. \tag{a}$$

But, by definition of $v_n(\cdot)$ and specification (E),

$$U_n(x) = \sum_{i=n+1}^{N} \beta^{N-i}c_i + \beta^{N-n}v_n(x) \tag{b}$$

and, similarly,

$$U_{n-1}(y) = \sum_{i=n+1}^{N} \beta^{N-i}c_i + \beta^{N-n}c_n + \beta^{N-n+1}v_{n-1}(y). \tag{c}$$

Note, moreover, that

$$\beta^{N-n}q_x^g = \sum_{y=1}^{I} \beta^{N-n}c_n p_{xy}^g. \tag{d}$$

Substituting (b) and (c) into (a), and then (d) into the result, and performing some algebraic simplifications, yields

$$\sum_{i=n+1}^{N} \beta^{N-i}c_i + \beta^{N-n}v_n(x) = \sum_{i=n+1}^{N} \beta^{N-i}c_i + \beta^{N-n}\max_{g\in G(x)}\left[q_x^g + \sum_{y=1}^{I} v_{n-1}(y)p_{xy}^g\right],$$

from which (10.1.1) follows readily by subtracting $\Sigma_{i=n+1}^{N}\beta^{N-i}c_i$ from both sides and canceling β^{N-n}. ■

Naturally, g^* is an *optimal* act in stage n if and only if

$$q_x^{g^*} + \beta\sum_{y=1}^{I} v_{n-1}(y)p_{xy}^{g^*} = \max_{g\in G(x)}\left[q_x^g + \beta\sum_{y=1}^{I} v_{n-1}(y)p_{xy}^g\right]. \tag{10.1.2}$$

Optimal act(s) in stage n will be denoted by $g_n^\circ(x)$ to indicate dependence upon both x and n. You should note that $g_n^\circ(x)$ will depend, in principle, upon the number n of stages to go even though the problem parameters do not depend upon n. This dependence is a "horizon effect" and is one of the primary matters of interest in this subject area.

You should also note that $v_n(x)$ and thus also $g_n^\circ(x)$ do not depend in any way upon the total number N of stages under consideration, but rather only upon the number of stages to go. This suggests that we can readily study the effects of increasing N simply by increasing n, that is, by solving for all $v_1(x)$ and $g_1^\circ(x)$, then for all $v_2(x)$ and $g_2^\circ(x)$ using the previously obtained $v_1(x)$'s, and so on and so on.

In Example 10.1.3 we present the results of such an approach, called *value iteration* by Howard [**63**], for $n = 1, 2, 3, 4, 5$, and also for $n = +\infty$ by using results presented in Section 10.2.

Example 10.1.3

The following table was derived, except for the final two rows, by direct application of Theorem 10.1.1, assuming a discount factor of $\beta = .90$. Note in particular the difference between $g_1^\circ(0)$ and $g_n^\circ(0)$ for

$n>1$; at the last stage a more conservative stock level is called for by the tacit assumption embodied in $v_0(x)=0$ that any stock remaining after stage 1 is valueless. The optimal strategy is very easy to describe: "As long as this is not the last stage, replenish your inventory by two widgets whenever it has run out."

Value Iteration for Example 10.1.2

	$x=0$	1	2	3	4	5
$g_1^\circ(x)=$	1	0	0	0	0	0
$v_1(x)=$	2.40	8.90	10.10	9.10	8.10	7.10
$g_2^\circ(x)=$	2	0	0	0	0	0
$v_2(x)=$	6.05	11.64	16.95	17.88	16.38	14.48
$g_3^\circ(x)=$	2	0	0	0	0	0
$v_3(x)=$	9.05	14.85	20.05	23.48	23.89	21.94
$g_4^\circ(x)=$	2	0	0	0	0	0
$v_4(x)=$	11.89	17.57	22.89	26.52	28.65	28.35
$g_5^\circ(x)=$	2	0	0	0	0	0
$v_5(x)=$	14.37	20.11	25.37	29.07	31.51	32.47
$\vdots$			$\vdots$			
$g_\infty^\circ(x)=$	2	0	0	0	0	0
$v_\infty(x)=$	36.94	42.66	47.94	51.63	54.13	55.49

As suggested by Example 10.1.3, it is very important to discover whether the limits of the $v_n(x)$ exist as $n\to+\infty$ and whether these limits satisfy (10.1.1), so that they can be used for determining $\mathscr{D}$'s optimal strategy when the number n of remaining stages is so great as to warrant using results valid for the limiting case.

Existence of these limits depends critically upon whether $0\le\beta<1$ or $\beta=1$. Section 10.2 considers the simpler case where $0\le\beta<1$; Section 10.3 concerns the more difficult case where $\beta=1$, limits typically fail to exist, and different quantities must be introduced to determine $g_\infty^\circ(\cdot)$. Section 10.4 concerns generalizations and extensions of the material discussed in Sections 10.2 and 10.3.

Proofs of the central results in Sections 10.2 and 10.3 are often omitted, since they require a background in probability theory and mathematical analysis not only beyond the level of this book but also too extensive for a succinct summary in an appendix. Much more rigorous expositions are contained in Howard [**63**], White [**141**], Derman [**25**], and references in these texts.

Exercise 10.1.1

In Example 10.1.2, derive all tabulated values of (A) p_{xy}^g and (B) $q_x{}^g$.

Exercise 10.1.2

Redo Example 10.1.3 for $n = 1, \ldots, 5$ with the sole changed assumption that (A) $\beta = 0$ and (B) $\beta = .50$.
[*Hint*: (A) is very easy. Think before you calculate!]

Exercise 10.1.3

Redo Example 10.1.3 for $n = 1, \ldots, 5$ with the sole changed assumptions (1) that for every n, $\tilde{\gamma}_n = 0$, 1, 2, or 3 with respective probabilities .2, .4, .3, and .1; and (2), that (A) $\beta = .90$, (B) $\beta = 0$, and (C) $\beta = .50$.

Exercise 10.1.4: The Toymaker

(Adapted from Howard [**63**].) A toymaker may commence any stage with either a successful toy (= state 1) or an unsuccessful toy (= state 2). Given state 1, he may either advertise ($g = 1$) or not advertise ($g = 2$); given state 2, he may either undertake research ($g = 1$) or forgo research to make current income look better to his stockholders ($g = 2$). The p_{xy}^g and $q_x{}^g$ are as follows.

State x	Act g	p_{x1}^g	p_{x2}^g	$q_x{}^g$
1	1	.4	.6	8
1	2	.8	.2	5
2	1	.6	.4	−4
2	2	.3	.7	−1

Find $g_n^\circ(x)$ and $v_n(x)$ for $x = 1$, 2 and $n = 1, 2, 3, 4$, if (A) $\beta = .90$, (B) $\beta = .60$, and (C) $\beta = .20$.

■ The recursion-analytic philosophy of dynamic programming can be invoked productively in a wide variety of optimization problems even in the absence of a natural, sequential interpretation. Often an ingenious sequential interpretation of the problem can be formulated; equations similar to (10.1.1) relating optimal values between adjacent "stages" are deduced from the "principle of optimality"; and then an optimal strategy is deduced from these equations. The principle of optimality, actually a rather self-evident theorem, holds that *an optimal strategy is such that, regardless of how one arrived at a given stage, the remaining decisions must constitute an optimal strategy in the smaller problem which commences at the given stage.* The following two exercises enable you to deduce and apply dynamic programming models in some not atypical contexts. ■

Exercise 10.1.5: Separable Constrained Deterministic Optimizations

(A) Consider the problem of maximizing $\Sigma_{i=1}^{N} R_i(g_i)$ via choice of real numbers $g_1, \ldots, g_N$, subject to the constraint that $\Sigma_{i=1}^{N} t_i(g_i) \leq X$, where $X > 0$; the R_i are real valued functions, and the t_i are nonnegative real-valued functions. For $0 \leq x \leq X$, let $G_n(x) = \{g_n: x - t_n(g_n) \geq 0\}$. Let $U_n(x)$ denote the maximum of $\Sigma_{i=1}^{N} R_i(g_i)$ subject to the constraint that $\Sigma_{i=n}^{N} t_i(g_i) \leq x$. Show that

$$U_n(x) = \max_{g_n \in G_n(x)} [R_n(g_n) + U_{n+1}(x - t_n(g_n))], \qquad (*)$$

where $U_{N+1}(x) = 0$ for every x.

(B) What does x represent in (*)? How could one handle additional constraints of the form, say, $m_i \leq t_i^*(g_i) \leq y_i$ for $i = 1, \ldots, N$?

(C) Consider the problem of choosing real numbers $g_1, \ldots, g_N$ so as to maximize $\Sigma_{i=1}^{N} R_i(g_i)$ subject to W constraints of the form

$$\sum_{i=1}^{N} t_i^w(g_i) \leq X^w,$$

with $X^w > 0$ and t_i^w nonnegative functions. Define $G_n(x^1, \ldots, x^W)$ and $U_n(x^1, \ldots, x^W)$ so that

$$U_n(x^1, \ldots, x^W) = \max_{g_n \in G_n(x^1, \ldots, x^W)} [R_n(g_n) + U_{n+1}(y^1, \ldots, y^W)] \qquad (**)$$

with U_{N+1} always defined as zero, where

$$y^w = x^w - t_n^W(g_n)$$

for $w = 1, \ldots, W$.

(D) *A Knapsack Problem.* A smuggler wishes to pack his knapsack with the most remunerative contraband by choosing integers g_1, g_2, and g_3 so as to maximize $7g_1 + 6g_2 + 5g_3$, subject to the constraints that total weight not exceed 43 pounds and total volume not exceed 11 cubic feet. Items 1, 2, and 3 weigh 16, 3, and 7 pounds respectively and occupy 3, 4, and 1 cubic feet per unit, respectively. Find the optimal number g_i of units of item i to carry, for $i = 1$, 2, and 3. What difficulties do you encounter in arriving at the answer?

Exercise 10.1.6

The success of a system's mission depends upon whether all N of its subsystems do not fail. A subsystem, n, consists of one circuit, plus $g_n - 1$ redundant copies of that circuit, and the subsystem does not fail unless all g_n circuits fail. Let p_n denote the probability that a type-n circuit will fail, and assume that *all* failures in the system are independent events.

(A) Show that the reliability, or probability of nonfailure, of subsystem n is $1 - p_n^{g_n}$.

(B) Show that the reliability, or probability of nonfailure, of the overall system is $\Pi_{n=1}^N (1-p_n^{g_n})$.

(C) Suppose that the system is to be designed so as to maximize overall reliability, subject to a cost constraint of $\$C$ and a weight constraint of W pounds, and that each type-n circuit weighs w_n pounds and costs $\$c_n$. Show that an optimal design $(g_1^\circ, \ldots, g_N^\circ)$ can be derived by recursively solving the system of equations

$$U_n(c, w) = \max_{1 \leq g_n \leq \min[c/c_n, w/w_n]} [(1-p_n^{g_n}) \cdot U_{n+1}(c - c_n g_n, w - w_n g_n)],$$

where $U_{N+1}(c, w) = 1$ for all (c, w) and $U_n(c, w)$ is the maximum of $\Pi_{i=n}^N (1-p_i^{g_i})$ subject to the constraints that $\Sigma_{i=n}^N w_i g_i \leq w$ and $\Sigma_{i=n}^N c_i g_i \leq c$.

*(D) Find g_1°, g_2°, and g_3° for $N=3$, $C=20$, $W=25$, $c_1=1$, $c_2=3$, $c_3=4$, $w_1=5$, $w_2=8$, $w_3=2$, $p_1=.40$, $p_2=.20$, and $p_3=.10$. [*Hint*: For $g_2 \geq 1$ and $g_3 \geq 1$, the constraints imply that $g_1 \in \{1, 2, 3\}$. Similarly, $g_2 \in \{1, 2\}$ and $g_3 \in \{1, 2, 3, 4\}$. Would this numerical problem be only a little easier, or much easier, if there were only one constraint instead of two?]

10.2 INFINITE-HORIZON MARKOVIAN dpuu WHEN $0 \leq \beta < 1$

In general, not necessarily Markovian dpuu's, Chapters 4 and 7 have established that a *strategy* for $\mathscr{D}$ is a choice of act given every partial history which might occur given his earlier choices. In Markovian decisions, we have noted that in an optimal strategy, $\mathscr{D}$'s choice at each stage need not depend upon more of the partial history than x. Hence in such problems, with infinite horizons, attention may be confined to the sequences

$$\mathscr{g} = \{g_1(\cdot), g_2(\cdot), \ldots \textit{ ad inf}\},$$

in forward rather than backward numbering, where each $g_i(\cdot)$ is a function of the stage-i state x.

Now, a strategy $\mathscr{g}$ for the infinite-horizon problem has the property that the substrategy $\mathscr{g}'$ which prescribes what to do on stages n and thereafter is *also* a strategy for the infinite-horizon problem, but $\mathscr{g}$ and $\mathscr{g}'$ may in general appear very different. Only when $\mathscr{g}$ prescribes the same act $g(x)$ in *every* stage whose state is x, and hence is of the form $\mathscr{g} = \{g(\cdot), g(\cdot), \ldots \textit{ ad inf}\}$, will $\mathscr{g}$ and $\mathscr{g}'$ constitute the same strategy. It happens, as Theorem 10.2.1 asserts, that the search for an optimal strategy may be confined to just such strategies.

Definition 10.2.1: Policies; Stationary Strategies

(A) A *policy* is a function prescribing act $g(x) \in G(x)$ given state x, for every x.

(B) A *stationary strategy* $\mathscr{g}$ is an infinite sequence $\{g(\cdot), g(\cdot), \ldots \textit{ ad inf}\}$

which requires implementing the same policy $g(\cdot)$ at every stage of an infinite-horizon problem.

■ Howard [**63**], Derman [**25**], Blackwell [**12**], and others use "policy" where we use "strategy." Our usage of "policy" appears more in keeping with common usage. ■

Theorem 10.2.1: Optimality of Stationary Strategies. If $0 \leq \beta \leq 1$, and if the number I of states and the total number $\Sigma_{y=1}^{I} K_y$ of acts are finite, then some stationary strategy is optimal.

■ *Proof.* Omitted; see Blackwell [**12**]. ■

■ Note that Theorem 10.2.1 pertains even when $\beta = 1$ and is therefore relevant for the next section as well. ■

Given that some stationary strategy is optimal, the characterization of and search for an optimal strategy in the infinite-horizon problem can be considerably expedited. Theorem 10.2.2 presents some preliminary results; both parts should be unsurprising in view of Theorem 10.1.1.

Theorem 10.2.2. Under the assumptions of Theorem 10.2.1, and if $\beta < 1$:

(A) For every state x, $\lim_{n\to\infty} v_n(x)$ exists (as a finite number).
(B) Denoting $\lim_{n\to\infty} v_n(x)$ by $v(x)$, the repeated policy $g(\cdot)$ of an optimal stationary strategy must attain for every x the maximum in

$$v(x) = \max_{g\in G(x)} \left[q_x^g + \beta \sum_{y=1}^{I} v(y) p_{xy}^g \right]. \qquad (10.2.1)$$

*■ *Proof.* Part (B) follows from part (A) by taking the limit of each side of (10.1.1) as $n \to +\infty$. To prove part (A), let $M = \max_{(x,g)}[q_x^g]$ and $m = \min_{(x,g)}[q_x^g]$. The present value of receiving M at every stage is the geometric series $\Sigma_{n=1}^{\infty} \beta^{n-1} M = M/(1-\beta)$, which constitutes an upper bound for $v_n(x)$ for every n and x. If $m \geq 0$, it is clear that $v_n(x) \geq v_{n-1}(x)$ for every n and x. Hence, for every x, $\{v_1(x), v_2(x), \ldots \textit{ad inf}\}$ is an increasing sequence bounded above, and therefore $\lim_{n\to\infty} v_n(x)$ exists. If $m < 0$, define $(q_x^g)' = q_x^g - m$ for every x and g, and let $v_n'(x)$ denote the maximum expected value of the n-stage Markovian dpuu with stage returns $(q_x^g)'$, which are nonnegative. By the preceding argument, $\lim_{n\to\infty} v_n'(x)$ exists, and equals $v'(x)$, say, for every x. But

$$v_n(x) = v_n'(x) + m(1-\beta^n)/(1-\beta), \qquad (*)$$

as you may verify in Exercise 10.2.1. Taking limits of both sides of (*) as $n \to +\infty$ yields $v(x) = v'(x) + m/(1-\beta)$, implying existence of $v(x)$. ■

■ In Example 10.1.3, $v(x)$ was denoted by $v(x)$, to stress its role as the limit of $v_n(x)$ as $n \to +\infty$. ■

Despite their apparent similarity, there is a fundamental difference between Equations (10.1.1) and (10.2.1). The former indicated how the optimal strategy could be determined, recursively, for the finite-horizon Markovian dpuu's. But (10.2.1) does *not* immediately indicate how to find an optimal stationary strategy for the infinite-stage Markovian dpuu's, because there is nowhere from which to begin the recursion! One could, of course, try to solve Equation (10.2.1) by performing some value iterations with Equation (10.1.1) and then guess at the $v(x)$'s, but such an approach begs the question as to how rapidly convergence obtains.

Fortunately, a simple procedure exists for solving Equation (10.2.1) and thus obtaining an optimal stationary strategy. It is *Howard's algorithm for* $\beta < 1$, and proceeds by successively determining (1) an estimate of $v(x)$ for every x and (2) an estimate of the optimal policy to use at every stage (given a stationary strategy) by maximizing the right-hand side of (10.2.1) for every x. This estimated policy is then used to reestimate the $v(x)$'s, and so on, until the estimate of the optimal policy cannot be improved upon, in which case the current estimate of the optimal policy is in fact optimal, and the current estimates of the $v(x)$'s are in fact their actual values. Howard's algorithm for $\beta < 1$ is often summarized by depicting a single cycle, as in Figure 10.2.

We can *start* in either box of Figure 10.2. To start in the Policy-Evaluation Operation box, we must guess at the optimal policy and specify the I acts $g^*(x)$, one for each x. To start in the Policy-Improvement Routine box, we must guess at the I expected values $v(x)$; often we take $v(x) = 0$ for every x and thus obtain $g^*(\cdot) = g_1(\cdot) =$ the optimal policy for the one-stage Markovian dpuu, as the initial guess. [Why? Exercise 10.2.1(B).] The algorithm *terminates* once you have either (1) not increased any $v(x)$ from one policy-evaluation operation to the next or (2) obtained the same policy $g^*(\cdot)$ on two successive applications of the policy-improvement routine.

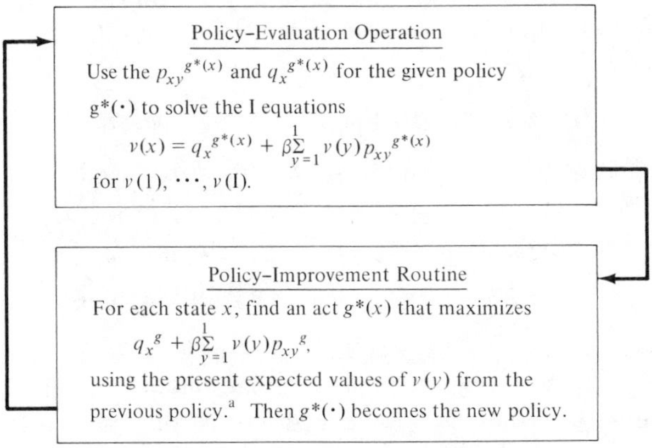

Figure 10.2 One cycle of the Howard algorithm for $\beta < 1$.

[a] If the previous policy's act is "tied" for maximizer, use it in preference to the other maximizer(s), to prevent unnecessary iterations.

Theorem 10.2.3. The Howard algorithm for $\beta < 1$ determines an optimal policy $g^*(\cdot)$, and hence an optimal stationary strategy $\{g^\circ(\cdot), g^\circ(\cdot), \ldots$ *ad inf*$\}$, in a finite number of iterations.

■ *Proof.* Omitted; see Howard [**63**] or Blackwell [**13**], for example. ■

Example 10.2.1

In Example 10.1.3 we obtained $g_n^\circ(0) = 2$ and $g_n^\circ(x) = 0$ for $x > 0$ and $n = 2, 3, 4, 5$. Hence this policy seems a reasonable guess for optimality, and we shall use it as $g^*(\cdot)$ for the first policy-evaluation operation. If $x = 0$, then $p_{xy}^{g^*(x)} = .2, .7$, and $.1$ for $y = 0, 1$, and 2 respectively, because $0 + g^*(0) = 2$. If $x = 1$, then $p_{xy}^{g^*(x)} = .9$ and $.1$ for $y = 0$ and 1 respectively, while for $x \geq 2$, $p_{xy}^{g^*(x)} = .2, .7$, and $.1$ for $y = x - 2$, $x - 1$, and x respectively. Moreover, the expected immediate returns are $q_x^{g^*(x)} = -0.9$, 8.9, 10.1, 9.1, 8.1, and 7.1 for $x = 0, 1, 2, 3, 4$, and 5 respectively. Hence we obtain the system of equations

$$\begin{aligned}
v(0) &= -0.9 + .9[.2v(0) + .7v(1) + .1v(2)] \\
v(1) &= 8.9 + .9[.9v(0) + .1v(1)] \\
v(2) &= 10.1 + .9[.2v(0) + .7v(1) + .1v(2)] \\
v(3) &= 9.1 + .9[.2v(1) + .7v(2) + .1v(3)] \\
v(4) &= 8.1 + .9[.2v(2) + .7v(3) + .1v(4)] \\
v(5) &= 7.1 + .9[.2v(3) + .7v(4) + .1v(5)],
\end{aligned}$$

to be solved for $v(0), \ldots, v(5)$. After a bit of algebra, we find $v(0) = 36.94$, $v(1) = 42.66$, $v(2) = 47.94$, $v(3) = 51.63$, $v(4) = 54.13$, and $v(5) = 55.49$, thus completing the first policy-evaluation operation. In the policy-improvement routine, we use these $v(x)$'s to try to find a better policy. Given $x = 0$, for example, we compute:

$$\begin{aligned}
\text{For } g = 0&: \quad 0 + .9[(1.0)(36.94)] &&= 33.25 \\
g = 1&: \quad 2.4 + .9[(.9)(36.94) + (.1)(42.66)] &&= 36.16 \\
g = 2&: \quad -0.9 + .9[(.2)(36.94) + (.7)(42.66) + (.1)(47.94)] &&= 36.94 \\
g = 3&: \quad -6.4 + .9[(.2)(42.66) + (.7)(47.94) + (.1)(51.63)] &&= 36.13 \\
g = 4&: \quad -11.9 + .9[(.2)(47.94) + (.7)(51.63) + (.1)(54.13)] &&= 34.13 \\
g = 5&: \quad -17.4 + .9[(.2)(51.63) + (.7)(54.13) + (.1)(55.49)] &&= 30.99,
\end{aligned}$$

so that we cannot improve on our former guess of $g^*(0) = 2$. Similar computations for $x = 1, 2, 3, 4$, and 5 yield comparable results; i.e., $g^*(x) = 0$ and the previous policy is unchanged. Thus this policy is optimal and should be used at *every* stage, thereby constituting an optimal stationary strategy. In Example 10.1.3, we denoted the optimal stationary policy by $g_\infty^\circ(\cdot)$ and its expected present values by $v_\infty(x)$, to stress their roles as limits when $n \to +\infty$.

Example 10.2.2

If we had performed no value iterations in Example 10.1.3, it would be reasonable to start by assuming that $v(x) = 0$ for every x and entering the policy-improvement routine. The effect of starting in the policy-improvement routine with all $v(x) = 0$ is to select the policy which maximizes immediate expected return q_x^g in every state. From the table of q_x^g's in Example 10.1.2, the resulting policy is $g(0) = 1$ and $g(x) = 0$ for every $x \ge 1$. Using this policy and entering the policy-evaluation operation results in $v(0) = 29.85$, $v(1) = 36.35$, $v(2) = 42.17$, $v(3) = 46.38$, $v(4) = 49.35$, and $v(5) = 51.15$. Using these values, we return to the policy-improvement routine and thereby obtain the new policy $g^*(0) = 2$, $g^*(x) = 0$ for every $x \ge 1$. Returning to the policy-evaluation operation with this policy yields the system of equations considered in Example 10.2.1, because, there, we had started with this policy as our initial guess. We would obtain the same $v(x)$'s as in Example 10.2.1, and, returning again to the policy-improvement routine, we find that we cannot improve on this policy, thus signaling its optimality.

*■ There is an alternative way of finding an optimal policy for the infinite-horizon problem by using linear programming. There are $\Sigma_{x=1}^{I} K_x$ decision variables z_{gx} = expected, discounted number of stages with state x in which act g is chosen. Naturally, the z_{gx} depend upon the as-yet undetermined stationary strategy $\mathcal{g}$. The total present expected value of $\mathcal{g}$ can then be written as

$$\sum_{x=1}^{I} \sum_{g=1}^{K_x} q_x^g z_{gx}. \tag{10.2.2}$$

Now let $\pi_1, \ldots, \pi_I$ be *positive* numbers which sum to one, with π_x denoting the probability that the initial state is x. These numbers may be chosen arbitrarily; e.g., $\pi_x = 1/I$ is convenient. You can verify in *Exercise 10.2.5 that the z_{gx} must be nonnegative and satisfy

$$\sum_{g=1}^{K_y} z_{gy} - \beta \sum_{x=1}^{I} \sum_{g=1}^{K_x} p_{xy}^g z_{gx} = \pi_y \tag{10.2.3}$$

for $y = 1, \ldots, I$. The linear-programming algorithm consists in maximizing (10.2.2) subject to the constraints (10.2.3) and $z_{gx} \ge 0$ for all (x, g). It can be shown that, in an optimal solution, $z_{g^*x} > 0$ if and only if g^* is an optimal choice for $g^\circ(x)$ in an optimal stationary strategy. This and alternative linear-programming formulations are due to D'Epenoux [24]. ■

Exercise 10.2.1

(A) Derive (*) in the proof of Theorem 10.2.2.

(B) Show that if one starts the Howard algorithm for $\beta < 1$ with the guess $v(x) = 0$ for every x in the policy-improvement routine, then the initial estimate $g^*(\cdot)$ of the optimal policy is an optimal strategy $g_1^\circ(\cdot)$ for the one-stage Markovian dpuu.

Exercise 10.2.2

Rework Example 10.2.1 with the sole changed assumption that (A) $\beta = 0$ and (B) $\beta = .50$.

Exercise 10.2.3

Rework Example 10.2.1 with the sole changed assumptions (1) that for every n, $\tilde{y}_n = 0, 1, 2,$ or 3 with respective probabilities .2, .4, .3, and .1; and (2) that (A) $\beta = .90$, (B) $\beta = 0$, and (C) $\beta = .50$.

Exercise 10.2.4

Find the optimal stationary strategy for the infinite-horizon problem of the toymaker in Exercise 10.1.4 if (A) $\beta = .90$, (B) $\beta = .60$, and (C) $\beta = .20$.

***Exercise 10.2.5**

Verify the assertions about (10.2.2), and explain why the z_{gx} must satisfy (10.2.3).
[*Hint*: What is the intuitive meaning of the constraints (10.2.3)?]

10.3 INFINITE-HORIZON MARKOVIAN dpuu WHEN $\beta = 1$

When $\beta = 1$, $\lim_{n\to\infty} v_n(x)$ generally does not exist as a finite number, and hence the methodology of the previous section for determining an optimal strategy does not apply without modification. Indeed, we must be careful about the meaning of optimality itself when the expected values of many strategies may be infinite.

When $\beta = 1$, it is natural to say that a strategy g is optimal (for the infinite-horizon problem) if it maximizes the *average return* r^g *per stage.* As Theorem 10.2.1 indicates, there exists an optimal stationary strategy under this definition.

But to characterize an optimal stationary strategy, it is first necessary to introduce an ostensibly irrelevant assumption about the p_{xy}^g.

Assumption 10.3.1: Complete Ergodicity.[b] For every policy $g(\cdot)$ under consideration, there is a unique solution $z(1), \ldots, z(I)$ to the system of

---

[b] Complete ergodicity is a property of Markov chains. See Howard [**63**], Feller [**33**], or Karlin [**68**], for example.

equations

$$\sum_{x=1}^{I} p_{xy}^{g(x)} z(x) = z(y), \qquad y = 1, \ldots, I$$

$$\sum_{x=1}^{I} z(x) = 1.$$

■ The asserted solution $z(1), \ldots, z(I)$ does depend upon $g(\cdot)$. Note that the first I equations determine the $z(x)$'s only up to a multiplicative constant, whereas the last equation fixes that constant. ■

Assumption 10.3.1 is always satisfied when $p_{xy}^{g} > 0$ for every x, y, and g. It is also satisfied under the milder condition that

$$P(\text{state } n \text{ stages hence} = y | \text{state now} = x, \text{ policy } g(\cdot) \text{ used every time}) > 0$$

for *some* $n \geq 1$. Note well, however, that the positive-probabilities condition is sufficient, but *not necessary*, for Assumption 10.3.1 to obtain, as the following example shows.

Example 10.3.1

Let $I = 2$ and $p_{xy}^{g(x)}$ be given by $p_{11}^{g(1)} = 1$, $p_{12}^{g(1)} = 0$, $p_{21}^{g(2)} = \frac{1}{2}$, and $p_{22}^{g(2)} = \frac{1}{2}$. Arrange these numbers in matrix format.

$$\mathbf{P} = \begin{bmatrix} 1 & 0 \\ \frac{1}{2} & \frac{1}{2} \end{bmatrix}.$$

Then $P(y = n$ stages hence$|x$ now, $g(\cdot)$ always used) is the (x, y)th element of the n-fold matrix product $(\mathbf{P})^n = \mathbf{P} \cdot \mathbf{P} \cdot \cdots \cdot \mathbf{P}$; and, since

$$(\mathbf{P})^n = \begin{bmatrix} 1 & 0 \\ 1 - (\frac{1}{2})^n & (\frac{1}{2})^n \end{bmatrix}$$

for every n, it follows that the probability of ever getting to state 2 from state 1 is zero. Yet the equations

$$z(1) + (\tfrac{1}{2})z(2) = z(1)$$

$$(\tfrac{1}{2})z(2) = z(2)$$

$$z(1) + \quad z(2) = 1$$

have the unique solution $z(1) = 1$, $z(2) = 0$, and hence Assumption 10.3.1 is satisfied, at least for this policy $g(\cdot)$.

The role of Assumption 10.3.1 is to insure that the average return r^g per stage, using a stationary strategy, is independent of the initial state x. When this assumption fails, more complex procedures are necessary; these will be described in the final comment of this section. But Assumption 10.3.1 is usually satisfied in problems arising from realistic applications.

■ Some authors advise guaranteeing the satisfaction of Assumption 10.3.1 by altering the p_{xy}^g's slightly so that all are positive. See, for example, Bellman and Dreyfus [8]. ■

Theorem 10.3.1 is analogous to Theorem 10.2.2 in that it lays the groundwork for an algorithm to determine an optimal stationary strategy. Compare, especially, Theorem 10.3.1(A) with Theorem 10.2.2(A), to see the role of the average return r per stage given an optimal stationary strategy. It says that, for any n, the *total* expected return $v_n(x)$ from an optimal strategy for the infinite-horizon problem starting in state x is expressible as n times the average return r per stage, plus a constant $w(x)$ reflecting the differential effect of starting in state x vis-à-vis other states, plus a remainder term $e(n, x)$ which becomes insignificant as n becomes very large.

Theorem 10.3.1. If $\beta = 1$ and if I and $\Sigma_{y=1}^{I} K_y$ are finite, then:

(A) There exists a number $r \in (-\infty, +\infty)$, numbers $w(1), \ldots, w(I)$, and numbers $e(n, x)$ such that, for every n and x,

$$v_n(x) = n \cdot r + w(x) + e(n, x),$$

where, for every x, $\lim_{n\to\infty} e(n, x) = 0$.

(B) The repeated policy $g(\cdot)$ of an optimal stationary strategy must attain for every x the maximum in

$$r + w(x) = \max_{g\in G(x)} \left[q_x^g + \sum_{y=1}^{I} w(y) p_{xy}^g \right]. \tag{10.3.1}$$

*■ *Proof.* Proof of part (A) is omitted; see Howard [63]. For part (B), if n is very large, the remainder terms $e(n, x)$ may be ignored and hence

$$v_{n+1}(x) = \max_{g\in G(x)} \left[q_x^g + \sum_{y=1}^{I} v_n(y) p_{xy}^g \right]$$

can be rewritten, using $nr + w(x)$ in place of $v_n(x)$, as

$$\begin{aligned}(n+1)r + w(x) &= \max_{g\in G(x)} \left[q_x^g + \sum_{y=1}^{I} [nr + w(y)] p_{xy}^g \right] \\ &=^1 \max_{g\in G(x)} \left[q_x^g + nr + \sum_{y=1}^{I} w(y) p_{xy}^g \right] \\ &=^2 nr + \max_{g\in G(x)} \left[q_x^g + \sum_{y=1}^{I} w(y) p_{xy}^g \right],\end{aligned}$$

from which (10.3.1) follows by subtracting nr from each side; equality 1 obtains by algebra and the fact that $\Sigma_{y=1}^{I} p_{xy}^g = 1$ for every x and g whereas equality 2 follows by moving nr outside the maximand, since it does not depend upon $g(\cdot)$ overtly, by assumption that r and $w(y)$ are determined from the optimal policy. ■

As for the case of $\beta < 1$, there is a simple procedure for solving Equation (10.3.1) and obtaining an optimal stationary strategy. It is *Howard's algorithm for* $\beta = 1$, summarized in Figure 10.3, and, reminiscent of the algorithm for $\beta < 1$, proceeds by successively determining (1) estimates of r and the $w(x)$'s and (2) an estimate of the optimal policy to use at every stage, by maximizing the right-hand side of (10.3.1) for every x, this estimated policy being used to reestimate r and the $w(x)$'s, and so forth, until the current estimate of the optimal policy cannot be improved upon, in which case it is optimal in fact and should be used at every stage of the infinite-horizon problem. Again, the iterations can be started in either box, with policy evaluation using an initial policy guess, or with policy improvement using initial guesses at r and the $w(x)$'s.

■ As in the algorithm for $\beta < 1$, in the absence of any intuition about an optimal policy, we can start in the policy-improvement routine with $r = w(1) = \cdots = w(I) = 0$ as an initial guess. ■

In the policy-evaluation operation, note that there are I equations in $I + 1$ variables—r, $w(1), \ldots, w(I)$—and hence the solution is not unique. A unique solution is obtained by arbitrarily specifying one of the differential values $w(x)$; one concrete choice is to set $w(1) = 0$.

■ In the policy-improvement routine, note from the policy-evaluation operation that r, $w(1), \ldots, w(I)$ for the unimproved, current policy $g^*(\cdot)$ is given by

$$r = q_x^{g^*(x)} + \sum_{y=1}^{g^*(x)} w(y) p_{xy}^{g^*(x)} - w(x);$$

our objective is to maximize r; and maximizing $q_x^g + \Sigma_{y=1}^I w(y) p_{xy}^g - w(x)$

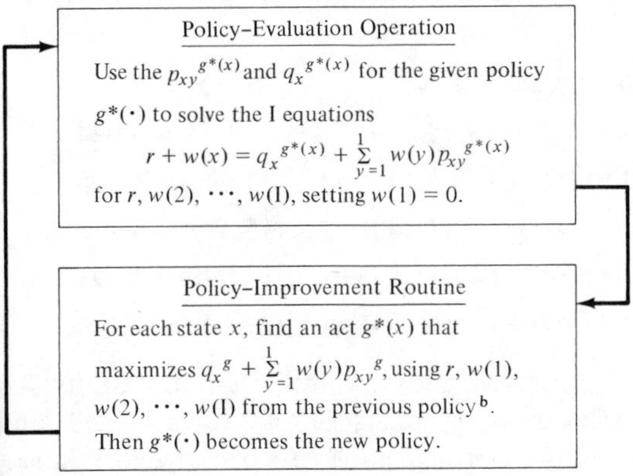

Figure 10.3 One cycle of the Howard algorithm for $\beta = 1$.

[b] If the previous policy's act given x is "tied" for maximizer, use it in preference to the other maximizer(s), to prevent unnecessary iterations.

is equivalent to maximizing $q_x^g + \Sigma_{y=1}^I w(y)p_{xy}^g$. Hence the thrust of the policy-improvement routine is to find a policy with an average return per stage greater than that under the current policy. ■

Theorem 10.3.2. If Assumption 10.3.1 is satisfied for every policy considered, then the Howard algorithm for $\beta = 1$ determines an optimal policy $g°(\cdot)$ and hence an optimal stationary strategy $\{g°(\cdot), g°(\cdot), \ldots, ad\ inf\}$ in a finite number of iterations.

■ *Proof.* Omitted; see Howard **[63]**, Blackwell **[12]**, or Miller and Veinott **[96]**, for example. ■

Example 10.3.2

In the inventory problem, our value iterations for $\beta = .9$ suggest that $g^*(0) = 2$, $g^*(x) = 0$ for $x \geq 1$ is a good policy. We start with that policy in the policy-evaluation-operation box to solve for r, $w(1), \ldots, w(5)$, with $w(0) = 0$:

$$\begin{aligned}
r \qquad &= -0.9 + .7w(1) + .1w(2)\\
r + w(1) &= 8.9 + .1w(1)\\
r + w(2) &= 10.1 + .7w(1) + .1w(2)\\
r + w(3) &= 9.1 + .2w(1) + .7w(2) + .1w(3)\\
r + w(4) &= 8.1 + .2w(2) + .7w(3) + .1w(4)\\
r + w(5) &= 7.1 + .2w(3) + .7w(4) + .1w(5),
\end{aligned}$$

the solution to which is $r = 4.01$, $w(0) = 0$, $w(1) = 5.44$, $w(2) = 11.00$, $w(3) = 15.42$, $w(4) = 18.99$, and $w(5) = 21.63$, thus completing the first policy-evaluation operation. Entering the policy-improvement routine, we use these $w(x)$'s to try to find a better policy. Given $x = 0$, for example, we compute

$$\begin{aligned}
\text{for } g = 0:&\quad 0 + [(1.0)(0)] &&= 0\\
g = 1:&\quad 2.4 + [(.9)(0) + (.1)(5.44)] &&= 3.044\\
g = 2:&\quad -0.9 + [(.2)(0) + (.7)(5.44) + (.1)(11.00)] &&= 4.008\\
g = 3:&\quad -6.4 + [(.2)(5.44) + (.7)(11.00) + (.1)(15.42)] &&= 3.930\\
g = 4:&\quad -11.9 + [(.2)(11.00) + (.7)(15.42) + (.1)(18.99)] &&= 2.993\\
g = 5:&\quad -17.4 + [(.2)(15.42) + (.7)(18.99) + (.1)(21.63)] &&= 1.140,
\end{aligned}$$

so that $g^*(0) = 2$, as in our original guess. Similar computations for $x = 1, 2, 3, 4, 5$ yield comparable results; i.e., $g^*(x) = 0$ for all $x \geq 1$, and therefore our original guess is the optimal policy to use at every stage. Using it yields an average return of \$4.01 per stage in the infinite-horizon problem, and no other policy—indeed, not even a nonstationary strategy—can yield a higher average return per stage.

*■ $r = 4.01$ can be arrived at by another method. The solution $z(1), \ldots, z(I)$ of the equations cited in Assumption 10.3.1 gives the limiting probabilities of the various states, as $n \to +\infty$. That is, $\lim_{n\to\infty} P(y = \text{state } n \text{ stages hence } | x = \text{state now, } g(\cdot) \text{ always used}) = z(y)$, independent of x, but dependent upon $g(\cdot)$. Assumption 10.3.1 ensures that $z(1), \ldots, z(I)$ are independent of the initial state x, thus reflecting the intuitive idea that the initial state's effect on the probabilities of the states n stages hence should wane as n increases. Since the long run happens forever when the horizon is infinite, it is plausible (and can be proved) that $r = \Sigma_{x=1}^{I} q_x^{g(x)} z(x)$, where $z(\cdot)$ and r are both relative to the stationary strategy $\{g(\cdot), g(\cdot), \ldots, \textit{ad inf}\}$. In Example 10.3.1, you may verify that for the optimal policy, $z(0) = 81/160$, $z(1) = 70/160$, $z(2) = 9/160$, and $z(3) = z(4) = z(5) = 0$; and, since $q_0{}^2 = -0.9$, $q_1{}^0 = 8.9$, and $q_2{}^0 = 10.1$, an easy calculation yields $r = (-0.9)(81/160) + (8.9)(70/160) + (10.1)(9/160) = \$4.01$. ■

*■ As for the case of $\beta < 1$, there is a linear-programming algorithm, due to Manne [**94**]. Let z_{gx} denote the probability of (g, x) under a given, as-yet undetermined, stationary strategy. Then the average return per stage is maximized if one maximizes

$$\sum_{x=1}^{I} \sum_{g=1}^{K_x} q_x^g z_{gx}. \tag{10.3.2}$$

Now, the probabilities z_{gx} must satisfy

$$\sum_{g=1}^{K_y} z_{gy} - \sum_{x=1}^{I} \sum_{g=1}^{K_x} p_{xy}^g z_{gx} = 0 \tag{10.3.3}$$

for every y, as well as

$$\sum_{x=1}^{I} \sum_{g=1}^{K_x} z_{gx} = 1 \tag{10.3.4}$$

and $z_{gx} \geq 0$ for every g and x. The linear program consists in maximizing (10.3.2) subject to the constraints (10.3.3), (10.3.4), and nonnegativity $[z_{gx} \geq 0]$. Under an optimal solution, $z_{g^*x} \geq 0$ if and only if g^* is an optimal choice for $g°(x)$ in an optimal stationary strategy. ■

*■ To discuss what occurs if Assumption 10.3.1 may fail, it is necessary to study the behavior of the probabilistic process by which the transitions are made from stage to stage; i.e., the Markov chain. Arrange the I^2 probabilities $p_{xy}^{g(x)}$ for a given stationary strategy $\mathscr{g} = \{g(\cdot), g(\cdot), \ldots \textit{ ad inf}\}$ as an I-by-I matrix $\mathbf{P}^{\mathscr{g}}$, the (x, y)th element of which is $p_{xy}^{g(x)}$. As in Example 10.3.1, $P(y = \text{state } n \text{ stages hence } | x = \text{state now, } g(\cdot) \text{ always used})$ is the (x, y)th element of $(\mathbf{P}^{\mathscr{g}})^n = \mathbf{P}^{\mathscr{g}} \cdot \mathbf{P}^{\mathscr{g}} \cdot \cdots \cdot \mathbf{P}^{\mathscr{g}}$, the matrix product of $\mathbf{P}^{\mathscr{g}}$ with itself for a total of n factors. Now, states x and y are said to *communicate* if there are positive integers m and n such that the (x, y)th element of $(\mathbf{P}^{\mathscr{g}})^m$ and the (y, x)th element of $(\mathbf{P}^{\mathscr{g}})^n$ are both positive; that

is, if transitions from x to y and from y to x can both be made, each possibly in more than one step. A *recurrent chain* $\mathscr{C}$ is a set of states which all communicate with each other, and such (1) that no superset $\mathscr{C}'$ of $\mathscr{C}$ has this property and (2) that $\Sigma\{p_{xy}^{g(x)}: y \in \mathscr{C}\} = 1$ for every $x \in \mathscr{C}$. Every Markov chain with a finite number of states has at least one recurrent chain, and Assumption 10.3.1 is equivalent to assuming that there is *exactly* one recurrent chain for every policy under consideration. It can be shown that if there is more than one recurrent chain, then any two recurrent chains are disjoint (because "communicates with" is a transitive relation). But there may be some states which do not belong to any recurrent chain; they are called *transient* states. In Example 10.3.1, state 1 constitutes a recurrent chain all by itself, and state 2 is transient, a natural description in view of the fact that there is miniscule probability $(\frac{1}{2})^n$ of finding the system in state 2 after a large number n of stages given that it started in state 2. Under the optimal policy for our inventory example,

$$\mathbf{P}^g = \begin{bmatrix} .2 & .7 & .1 & 0 & 0 & 0 \\ .9 & .1 & 0 & 0 & 0 & 0 \\ .2 & .7 & .1 & 0 & 0 & 0 \\ 0 & .2 & .7 & .1 & 0 & 0 \\ 0 & 0 & .2 & .7 & .1 & 0 \\ 0 & 0 & 0 & .2 & .7 & .1 \end{bmatrix},$$

which implies that $\mathscr{C} = \{0, 1, 2\}$ is a recurrent chain whereas 3, 4, and 5 are transient states. It is conceivable, however, to have a $\mathbf{P}^g$ with more than one recurrent chain; e.g.,

$$\mathbf{P}^g = \begin{bmatrix} 1 & 0 & 0 & 0 & 0 \\ 0 & .5 & 0 & 0 & .5 \\ 0 & .3 & .5 & .2 & 0 \\ .1 & 0 & .3 & .5 & .1 \\ 0 & .2 & 0 & 0 & .8 \end{bmatrix},$$

in which $\{1\}$ and $\{2, 5\}$ are recurrent chains and states 3 and 4 are transient. Such $\mathbf{P}^g$'s violate Assumption 10.3.1, in that the system of $I + 1$ equations cited in that assumption has more than one solution. One for the case at hand is $z(1) = 1$, $z(x) = 0$ for $x = 2, 3, 4, 5$, and another is $z(1) = 0$, $z(2) = 2/7$, $z(3) = z(4) = 0$, and $z(5) = 5/7$. In Markovian dpuu's where some $\mathbf{P}^g$'s are suspected of giving rise to more than one recurrent chain, the Howard algorithm becomes much more complex. In the first place, the average return per stage is dependent upon the initial state x and hence must be written as $r(x)$. Then these average returns have the property that $r(x) = r(y)$ whenever x and y belong to the same recurrent chain; and *all* are related by

$$r(x) = \sum_{y=1}^{I} p_{xy}^{g(x)}\, r(y). \tag{10.3.5}$$

When there is only one recurrent chain, (10.3.5) implies that $r(x) = r$ for every state x, even the transient states. The Howard algorithm for $\beta = 1$ with Assumption 10.3.1 fails (or may fail) consists of a policy-evaluation operation and a policy-improvement routine. It is summarized in Figure 10.4; note that the policy-improvement routine makes maximizing the expected average return per stage starting in state x overridingly more important as a criterion than maximizing immediate return plus the expected differential advantage of starting in the various states. In view of the remark preceding Theorem 10.3.2, however, the difference from Figure 10.3 is not great, since maximizing $q_x^g + \Sigma_{y=1}^{I} w(y)p_{xy}^g$ is tantamount to maximizing the unique average return r. Denardo and Fox **[23]** present a linear-programming algorithm which furnishes an alternative to the Howard algorithm of Figure 10.4. ■

Policy-Evaluation Operation

Use the $p_{xy}^{g^*(x)}$ and $q_x^{g^*(x)}$ for the given policy to solve the 2I equations

$$r(x) = \sum_{y=1}^{I} r(y) p_{xy}^{g^*(x)}, \quad x = 1, \cdots, I$$

$$w(x) + r(x) = q_x^{g^*(x)} + \sum_{y=1}^{I} w(y) p_{xy}^{g^*(x)}, \quad x = 1, \cdots, I$$

for all $r(x)$ and $w(x)$, by setting $w(x) = 0$ for one x in each recurrent chain under the policy $g^*(\cdot)$.

Policy-Improvement Routine

Using the $r(x)$ and $w(x)$ for the current policy, find an act $g^*(x)$ which maximizes $\sum_{y=1}^{I} r(y) p_{xy}^g$, breaking any "ties" for maximizer by choosing that which maximizes $q_x^g + \sum_{y=1}^{I} w(y) p_{xy}^g$. If there is still a tie and the current $g^*(x)$ is a candidate, use it. Do this for all x, thus determining a new policy $g^*(\cdot)$. Stop if $g^*(\cdot)$ is unchanged from the previous iteration.

Figure 10.4 One cycle of the Howard algorithm for $\beta = 1$ when Assumption 10.3.1 is not assured.

Exercise 10.3.1

Find an optimal stationary strategy for the infinite-horizon problem of the toymaker in Exercise 10.1.4 and 10.2.4 if $\beta = 1$.

Exercise 10.3.2

Rework Example 10.3.2 with the sole changed assumption that, for every n, $\tilde{y}_n = 0$, 1, 2, or 3 with respective probabilities .2, .4, .3, and .1.

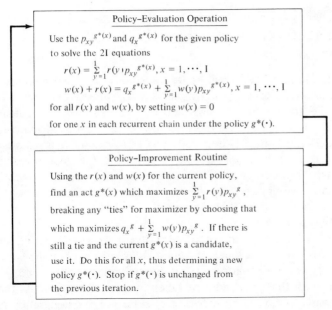

***Exercise 10.3.3**

Find an optimal stationary strategy for the following Markovian dpuu, in which Assumption 10.3.1 may fail to obtain.

State x	Act g	p_{x1}^g	p_{x2}^g	p_{x3}^g	q_x^g
1	1	1	0	0	5
1	2	.5	.5	0	4
1	3	0	.7	.3	1
2	1	.2	.6	.2	6
2	2	0	1	0	2
3	1	.4	.5	.1	7
3	2	1	0	0	1
3	3	0	0	1	8

10.4 SOME GENERALIZATIONS, EXTENSIONS, AND VARIATIONS

Throughout this chapter, the dpuu's under consideration have possessed a number of characteristics which, in certain circumstances, may be restrictive from the standpoint of analytical convenience and/or realism. Among these characteristics are (1) the assumption that the total numbers of states and of acts are finite; (2) decisions and transitions from one stage to the next occur at discrete, equally spaced, and a priori known points in time; (3) the assumption that $\mathscr{D}$'s utility is linear in net present monetary return; and (4) the assumption that the transition probabilities p_{xy}^g are a priori known by $\mathscr{D}$, in the sense that the partial history of the dpuu cannot give $\mathscr{D}$ an incentive to alter them.

Numerous studies have been made of Markovian dpuu's in which the number of states and/or the number of acts available in each state is not necessarily finite, but may be, variously, countably or uncountably infinite. These studies establish that care must be taken in generalizing results that obtain when I is finite and K_y is finite for every state y. For example, in the case of I and K_y finite, Blackwell [**12**] proved that a stationary strategy $\mathscr{g} = \{g(\cdot), g(\cdot), \ldots, \textit{ad inf}\}$ which is optimal in the context of Section 10.3, for $\beta = 1$, must also be optimal in the context of Section 10.2, for β sufficiently close to 1. This is an intuitively appealing result. But in [**44**], Flynn furnishes an example in which this result does *not* hold; it is an example with countably infinitely many states but a finite total number of acts. Worse yet, in a similar problem with finitely many acts and a countable infinity of states, Ross [**122**] shows that no stationary strategy can be very close to optimal! In [**13**], Blackwell defines weakened forms of optimality and commences the investigation of how stationary strategies as a whole fare under various infinite-set assumptions about the states and acts.

It is possible, and often useful, to generalize from the problem studied in Sections 10.2 and 10.3 by assuming that time flows continuously, in that a transition from state x to state y during a small interval of time of length dt has approximate probability $p_{xy}^g dt$ if act g is "in effect" during that interval. Monetary return is earned at the *rate* of q_x^g per unit of time in state x with act g, and a return of r_{xy}^g is earned when the transition to y occurs. Such a model is particularly useful when otherwise the stages would have to be defined as succeeding each other very rapidly in real time and the probability p_{xx}^g of staying in x is very high for most (x, g). Maintenance policy for a randomly failing machine is a frequently cited example. Howard [**63**] furnishes a full analysis and algorithm for such problems when the number of states and acts is finite.

Both discrete-time and continuous-time Markovian dpuu's are special cases of *Markov-renewal decisions*, introduced and studied by Jewell [**67**]. Here, the duration of a stage is uncertain and has a probability distribution which depends in principle not only upon the state x and act g in that stage, but also upon the state y with which the next stage commences. $\mathcal{D}$'s return depends, perhaps probabilistically, upon x, y, g, and the duration t of stage n. Jewell's work, extended and generalized by Lippman [**86**], Denardo [**22**], and others, distinguishes carefully among three "value-iteration"-type cases with finite horizon: the case in which the system operates for a total of n stages (and hence for a random length of time); the case in which the system operates for a fixed length of time (and hence goes through a random number of stages and stops probably during a stage); and the case in which the system is to operate for a given length T of time but stops only at the end of the stage in progress at time T. You should note that such a distinction is not very useful, or even meaningful, in the context of the Markovian model studied in the preceding sections; but the distinction is vital when stages are of random duration. Jewell [**67**] develops Howard-type algorithms for the infinite-horizon versions of the problem, for both the case $\beta < 1$ and the case $\beta = 1$. Fox [**45**], Denardo [**21**], and Denardo and Fox [**23**] extend and furnish linear-programming alternatives to Jewell's algorithms.

A third avenue of generalization consists in relaxing the assumption that $\mathcal{D}$'s utility is linear in net present monetary return. Howard and Matheson [**64**] furnish algorithms very similar to those in Figures 10.2 and 10.3 for the case in which $\mathcal{D}$'s utility function is the constant-risk-aversion exponential. Although the systems of equations for their algorithm are harder to handle than those in this chapter, it is encouraging to learn that aversion to risk can be introduced without the complete destruction of tractability in the infinite-horizon cases. It is very doubtful, however, that the assumption of constant risk aversion could be relaxed at all, and hence the one-parameter exponential utility function is most likely the *only* nonlinear utility function which permits tractable analysis of Markovian dpuu's.

A fourth avenue of generalization is to suppose that the transition probabilities p_{xy}^g are not known to $\mathcal{D}$ a priori, but rather that $\mathcal{D}$ has some prior probability distribution of these quantities. He will then use the observed transitions to revise his judgments about the p_{xy}^g, and by so doing he will affect his choices of g on this and subsequent stages. Martin **[95]** has studied this situation extensively. His analysis clearly shows that the definition of "state" must be expanded from x to include whatever other features of the preceding partial history are relevant to summarize its effect upon $\mathscr{D}$'s current judgments about the p_{xy}^g's. This just goes to illustrate a point that you should always keep in mind; namely, the definition of "state" *follows from*, rather than precedes and implies, a specification of the dependencies in the dpuu. It is all too easy in the complexities of analysis to put the cart before the horse and define "state" before arriving at a careful judgment about what determines the necessary probabilities and returns at any given stage.

■ The Markovian decision problem studied in this chapter is a good example of the fact that restricting attention to a special case of a general model usually permits a person to ask and answer deeper and more penetrating questions than are even meaningful in the context of the general model. To those who declare that a specialist is one who learns more and more about less and less until he knows everything about nothing, we can retort that the generalist can state less and less about more and more until he ends up able to state nothing about everything. ■

IV

SEVERAL-PERSON DECISIONS

"A small knowledge of human nature will convince us that with far the greatest part of mankind, interest is the governing principle, and that almost every man is more or less under its influence."

Alexander Hamilton, 1778

11

MONETARY GROUP DECISION PROBLEMS

For whoso reaps renown above the rest,
with heaps of hate shall surely be opprest.
Sir Walter Raleigh—dedicatory poem
for Gascoigne's *The Steele Glass*

11.1 INTRODUCTION

This is the first of two diffident chapters concerning dpuu's possessed by several individuals rather than the unique decision maker $\mathscr{D}$ of the preceding chapters. Diffidence is appropriate because both chapters assume that all the individuals know the preferences and judgments of the others, and also because both chapters undertake the difficult task of advising all the individuals simultaneously despite their potentially conflicting preferences and judgments. Knowledge cannot be that perfect when people have incentives to misrepresent their preferences and judgments; and universally acceptable advice cannot be that easy to come by, or else interpersonal conflicts would not be as prevalent as they, regrettably, seem to be.

In both chapters, a primary element of concern is conflicting preferences for the ultimate consequence of the individuals' collective choice. Chapter 12 concerns the potentially more general situation in which the individuals have their own sets of pure strategies and can make their strategic choices independently of each other. Here, we assume that the status quo prevails unless the individual group members can agree *unanimously* on a pure strategy, and we suppose that the consequences of the problem are monetary returns rather than the arbitrary consequences in Chapter 12. This sacrifice of generality enables us to focus on the problem of how a financial risk can be shared to the mutual profit of all individuals concerned when no single individual can assume profitably the entire risk by himself.

Somewhat more generally, we assume here that a *group* consists of n individual *members* i and possesses, as a collectivity, a dpuu in normal form with monetary-return consequences $c[s, \sigma]$ of the group's choosing s and $\mathscr{E}$'s choosing σ.

■ The assumption that the group already exists and possesses the dpuu is not unduly restrictive, since within the ensuing analytical framework we can study whether a group *can* form for the purpose of risk sharing, etc. See Example 11.1.1 and Exercises 11.1.1–11.1.4. ■

It will be convenient at times to abbreviate the $\#(S)$-by-$\#(\Sigma)$ table of monetary returns $c[s, \sigma]$ by $\mathbf{c}$, as in Chapter 8.

Now, the group's problem is really twofold. First and obviously, the group must agree on a choice of s, its collective strategy vis-à-vis the rest of the world. We call this the group's *external* subproblem. Second, however, the group must also determine how its ultimate monetary return should be divided among the members, or, more generally, among a number of accounts $h = 1, \ldots, k$ in which the members are interested. What to do with net monetary return accruing to the group is the group's *internal* subproblem.

■ The accounts in which the members are interested would normally include a personal account for each member, in which is "deposited" his personal share of the group's proceeds; but there may be some accounts h in the nature of "retained earnings" and contributions to various charities as well. ■

Unfortunately, it can be shown that the external and internal subproblems can only rarely be analyzed separately without risking the group's adopting a less than maximally beneficial overall strategy. In other words, only rarely is it safe to think of the group as an individual decision maker vis-à-vis the external world, with its own group utility function and its own group probabilities. Example 11.2.5 shows conclusively that a group utility function need not exist, even when all members i agree on all probabilities.

In Section 11.2 we develop the formal structure of the group decision problem and show how no member should object to the group's eliminating from consideration many of the available joint solutions to the external and internal subproblems, so as to focus interpersonal conflict on the remaining joint choices in much the same fashion as we developed the concepts of dominance and admissibility in Chapter 7.

In Section 11.3 we present *one* method of selecting an ultimate joint choice, or equivalently, or resolving the remaining interpersonal conflict within the group. Although the method we present is not easy to apply in practice, it is more naturally applicable than some alternatives that have been suggested.

References on the first three sections of this chapter include Wilson [**144**], Rosing [**121**], Wallace [**138**], Chapter 8 of Raiffa [**113**], LaValle [**77**] and [**79**], and Eck [**29**]. Section 11.4 concludes the chapter with comments on the much more general problem of collective choice, basic references on which include Sen [**129**] and Arrow [**1**].

The following concrete example and related exercises set the stage for the analytical framework in Section 11.2. Example 11.1.1 illustrates both the potential advantages of risk sharing and how these advantages can be analyzed by supposing provisionally that the group in question has formed and possesses the dpuu.

Example 11.1.1

Suppose that Joe possesses the dpuu

	σ^1	σ^2	σ^3
s^1	\$400	\$0	\$−200
s^2	\$0	\$0	\$0

Equivalently, Joe has the option of accepting the lottery s^1 which yields \$400, \$0, or \$−200 if σ^1, σ^2, or σ^3, respectively, occurs. Suppose that Joe's probabilities are .25, .50, and .25 of σ^1, σ^2, and σ^3, respectively, and that his utility function $u(c)$ for monetary returns c is given by

$$u(c) = \begin{cases} c + 80, & c \leq \$-100 \\ c/5, & c \geq \$-100; \end{cases} \tag{11.1.1}$$

you may sketch the graph of $u(\cdot)$ to see that it is continuous, concave, and consists of two linear segments joined at $c = \$-100$ with $u(\$-100) = -20$. It is easy to verify that Joe's expected return $E\{c[s^1, \tilde{\sigma}]\}$ from lottery s^1 is $(.25)(\$400) + (.50)(\$0) + (.25)(\$-200) = \$50$, and that s^2 (= "don't gamble") yields \$0 for sure. Hence Joe would prefer s^1 to s^2 *if* his utility were linear in return. But it is not, and hence we compare $E\{u(c[s^1, \tilde{\sigma}])\} = (.25)(400/5) + (.50)(0/5) + (.25)(-200 + 80) = 20 + 0 - 30 = -10$ with $E\{u(c[s^2, \tilde{\sigma}])\} = u(\$0) = 0$ to conclude that Joe prefers playing safe via s^2 to gambling via s^1. Now enter Sam, whose utility function we shall assume coincides with Joe's, but whose probabilities of σ^1, σ^2, and σ^3 are .40, .20, and .40 respectively. Can Joe and Sam find happiness by forming a group to share the risk inherent in s^1 in such a way that each prefers his share of s^1 to getting nothing? To see that they can, we suppose that they do form a group and, at least tentatively, agree to choose s^1 and split the proceeds equally between them, so that each will receive \$200, \$0, or \$−100 if σ^1, σ^2, or σ^3 occurs, respectively. Joe's utility of this prospect is $E\{u(\tfrac{1}{2}c[s^1, \tilde{\sigma}])\} = (.25)(200/5) + (.50)(0/5) + (.25)(-100/5) = 5$; Sam's utility of his share is $E\{u(\tfrac{1}{2}c[s^1, \tilde{\sigma}])\} = (.40) \times (200/5) + (.20)(0/5) + (.40)(-100/5) = 8$. Each person's utility of their collectively accepting s^1 and splitting the proceeds exceeds his utility, zero, of forgoing the opportunity presented by s^1! Hence Joe should want Sam to join him in a group, and Sam should want to join Joe.

This example will be continued in later sections, because much more general and clever ways of sharing the riskiness of s^1 can be developed. Moreover, the general analysis to follow does not depend in the slightest upon two special features of this example; namely, Joe's and Sam's having identical utility functions, and the assumption that Joe's utility is independent of Sam's return, and vice versa.

Exercise 11.1.1

For Example 11.1.1, find:

(A) Joe's reservation price for his 50 percent of the group dpuu.
(B) Sam's reservation price for his 50 percent of the group dpuu.
(C) Sam's bid ceiling for 50 percent of Joe's dpuu before it becomes a group dpuu.

[*Hint*: Recall that reservation price and bid ceiling were defined in Section 8.3; take $\mathcal{Q}'_S = \{s^1, s^2\}$ and $\mathbf{c}' = \frac{1}{2}c[\cdot, \cdot]$ in each case.]

Exercise 11.1.2

In Example 11.1.1, Sam's probabilities differed from Joe's. Is this what caused the mutual advantageousness of risk sharing? To find out, calculate Sam's $E\{u(\frac{1}{2}c[s^1, \tilde{\sigma}])\}$ under the assumption that his probabilities coincided with Joe's.

Exercise 11.1.3: Diversification via Merger (I)

Suppose that Tom has the one-strategy dpuu consisting of s^1 alone in Example 11.1.1; unlike Joe, he is forced to undertake the risk. Suppose that Jerry has the one-strategy dpuu s^3, defined by

	σ^1	σ^2	σ^3
s^3	\$−200	\$0	\$400

Also suppose that Tom and Jerry each have the utility function (11.1.1), and that each has probabilities .25, .50, and .25 of σ^1, σ^2, and σ^3 respectively. Assume that the σ's mean the same thing in both dpuu's, so that σ^1 implies that s^1 yields \$400 *and* s^3 yields \$−200.

(A) What is Tom's reservation price for $\{s^1\}$?
(B) What is Jerry's reservation price for $\{s^3\}$?
(C) Suppose that Tom and Jerry "merge" their dpuu's into the single dpuu $\{(s^1, s^3)\}$ with total net return the sum of the net returns of s^1 and s^3; that is, $\{(s^1, s^3)\}$ yields \$200, \$0, and \$200 given σ^1, σ^2, and σ^3 respectively. If each takes 50 percent of the merged dpuu, what are their reservation prices for their respective shares of $\{(s^1, s^3)\}$?

■ In merger analysis it is often asserted that a good merger exhibits "synergy" in the sense that in Exercise 11.1.3 one might anticipate that $c[(s^1, s^3), \sigma] \geq c[s^1, \sigma] + c[s^3, \sigma]$ for every σ, with ">" obtaining for some σ's. Exercise 11.1.3 shows that even if there is no synergy, and $c[(s^1, s^3), \sigma] = c[s^1, \sigma] + c[s^3, \sigma]$ for every σ, a merger might still be mutually advantageous if the "owners" of the premerger dpuu's are risk averse and the dpuu's exhibit complementary diversification, in the sense that one dpuu typically yields a good return when the other yields a bad return. The same phenomenon arises when the owners are risk neutral but each evaluates the other's dpuu as better than his own because of differences in probabilities; Exercise 11.1.4 is a case in point. ■

Exercise 11.1.4: Diversification via Merger (II)

Suppose that Tom and Jerry "own" the following risks.

	σ^1	σ^2
s^T	-100	100

	σ^1	σ^2
s^J	100	-100

that Tom and Jerry have probabilities .8 and .3 of σ^1 respectively, and that $u(c) = c$ for all c for each of Tom and Jerry.

(A) What is Tom's reservation price of his dpuu?
(B) What is Jerry's reservation price of his dpuu?
(C) If the two merge their dpuu's on an equal-shares basis, what are Tom's and Jerry's reservation prices for their individual shares of the merged dpuu?

■ Complementary diversification as an incentive to merge is sufficiently powerful to overcome a modicum of "negative synergy": the post-merger return being less than the sum of the pre-merger returns. In Exercise 11.1.4, suppose that $c[(s^T, s^J), \sigma^j] = -20 < -100 + 100 = c[s^T, \sigma^j] + c[s^J, \sigma^j]$ for $j = 1, 2$. Each should still vote for the merger. Why? ■

11.2 FORMAL STRUCTURE OF THE GROUP DECISION PROBLEM

The beginning of this section is quite similar in spirit to that of Section 7.3, which led to the definitions of the attainable utility set and cardinally admissible strategies for an individual dpuu in normal form. The analogous concepts here are called the *achievable* utility set and *Pareto-optimal* group strategies.[a] The remainder of this section concerns ways of characterizing Pareto-optimal strategies, the role of randomization as a

[a] There is an important difference between the problems here and those in Section 7.3, as multidimensionality is caused here by multiple members of the group, whereas in Section 7.3 it was caused by multiple σ's.

technique for compromise in group decisions, and the question of whether "group utility" and "group probability" functions exist.

Nothing need be added to the discussion in Section 11.1 concerning the external decision faced jointly by the n members i of the group, the choice of s. The choice of s is generally interdependent, however, with the group's choice of a sharing rule $\mathbf{r}$, the role of which is to specify how the group's net return c is to be divided among the k accounts h of the group.

Definition 11.2.1. Given the set Σ of pure strategies σ for $\mathscr{E}$ in a group dpuu with monetary-return consequences, and given the group's accounts $h = 1, \ldots, k$, a *sharing rule* for the group is a function $\mathbf{r}$: $R^1 \times \Sigma \to R^k$ with the property that

$$\sum_{h=1}^{k} r_h(c, \sigma) = c$$

for every $(c, \sigma) \in R^1 \times \Sigma$, where, for $h = 1, \ldots, k$, the hth component $r_h(c, \sigma)$ of the k-tuple $\mathbf{r}(c, \sigma)$ denotes the portion of $\$c$ credited to account h if $\mathscr{E}$ chooses σ.

At this point it may seem unnecessarily complex to allow the apportionments $r_h(c, \sigma)$ of c to accounts h to depend upon $\mathscr{E}$'s as-yet unobserved choice of σ. But to allow such dependence in principle is tantamount to permitting the members to capitalize via a priori mutually advantageous "side betting" on their differences of opinion about $\tilde{\sigma}$.

Example 11.2.1

Suppose that $n = 2$ people constitute a group which possesses the completely trivial dpuu in which $\mathbf{c}$ consists of one row, $(0, 0)$; that is, there is only one pure strategy s, and the group is sure to receive \$0 regardless of whether σ^1 or σ^2 occurs. No matter how you divide zero proportionally, you get zero. But suppose member 1 thinks that σ^1 is certain and member 2 thinks σ^2 is certain. A "side bet" of \$1,000, to which each should readily agree, consists in 1's paying \$1,000 to 2 if σ^2 occurs, and 2's paying \$1,000 to 1 if σ^1 occurs. If $k = 2$ and account h is what member $i = h$ receives, then this side bet amounts to a sharing rule such that $r_1(\$0, \sigma^1) = r_2(\$0, \sigma^2) = \$1{,}000$ and $r_1(\$0, \sigma^2) = r_2(\$0, \sigma^1) = \$-1{,}000$. If all that each member cares about is his own return, then this sharing rule makes each feel that he is a sure \$1,000 better off!

■ Note in Example 11.2.1 that $\Sigma_{h=1}^{2} r_h(\$0, \sigma^j) = \$1{,}000 + \$-1{,}000 = \$0$ for $j = 1, 2$, as required by the definition of sharing rules. The requirement that $\Sigma_{h=1}^{k} r_h(c, \sigma) = c$ for every σ simply says that all the group's net return c must be accounted for in the apportionment process. ■

■ Examples 11.2.3 and 11.2.4 also illustrate the advantages to be gained by enabling the group members to side-bet on $\tilde{\sigma}$. ■

■ If $k \geq 2$, there are infinitely many sharing rules and hence arises the specter of the group's having to choose from an infinite set. In some cases this is not technically difficult. In others, it can be simplified by reneging on our usually convenient assumption that money is infinitely divisible, and/or by restricting attention to a small subset of the set of *all* sharing rules **r**. ■

■ We have defined a sharing rule to be a function of *all conceivable* group net returns c rather than a function of only those net returns $c[s, \sigma]$ attainable in the specific external problem. There are two reasons for this degree of generality. First, to adopt the more restrictive definition would make the group's internal subproblem subordinate to its external subproblem. Second, as Theorems 11.2.3 and 11.2.4 indicate, we are often interested in the form of "good" sharing rules under particular assumptions about the group, abstracted from any consideration of specific external subproblems. ■

The group's external problem is to choose s, and its internal problem is to choose **r**. Thus the group's *total* problem is to choose a pair $(s, \mathbf{r})$, which we shall often denote by t (for "total") in what follows.

How should the group evaluate various pure total strategies $t = (s, \mathbf{r})$? Presumably, the group's evaluation of t should reflect the individual members' evaluations of t, and hence we now examine t from the typical member i's point of view.

We shall suppose that member i, as a dedicated alumnus of the preceding chapters, (1) has quantified his judgments about $\tilde{\sigma}$ in the form of probabilities p_{ij} (= i's probability of σ^j) for $j = 1, 2, \ldots, \#(\Sigma)$ and (2) has quantified his preferences for the possible *ultimate* outcomes of the group dpuu [namely, for various vectors $(x_1, \ldots, x_k)$ of increments to the group's accounts], in the form of a utility function u_i: $R^k \to R^1$.

Now, when member i examines $t = (s, \mathbf{r})$, he observes that if σ^j obtains, then: (1) because of s, the group will receive $c[s, \sigma^j]$ as its net monetary return; (2) because of **r**, the vector of increments to the group's accounts will be $\mathbf{r}(c[s, \sigma^j], \sigma^j)$, where σ^j influences the sharing rule $\mathbf{r}(\cdot, \sigma^j)$ as well as the return $c[\cdot, \sigma^j]$; and so (3) his utility will be $u_i(\mathbf{r}(c[s, \sigma^j], \sigma^j))$. Thus i's utility $U_i(t, \mathbf{c})$ of t, *prior to* observing σ^j, is simply *his* expectation of his utility $u_i(\mathbf{r}(c[s, \tilde{\sigma}], \tilde{\sigma}))$. That is, with $t = (s, \mathbf{r})$,

$$U_i(t, \mathbf{c}) = \sum_{j=1}^{\#(\Sigma)} u_i(\mathbf{r}(c[s, \sigma^j], \sigma^j))p_{ij}. \tag{11.2.1}$$

Equation (11.2.1) expresses member i's evaluation of t. The entire membership's evaluation of t may be expressed as the vector $\mathbf{U}(t, \mathbf{c})$ of $U_i(t, \mathbf{c})$'s; that is

$$\mathbf{U}(t, \mathbf{c}) = (U_1(t, \mathbf{c}), \ldots, U_n(t, \mathbf{c})). \tag{11.2.2}$$

Somewhat more generally, the group might decide on a *randomized* total strategy P_T, in which (say) $t^{m_1}, \ldots, t^{m_v}$ are given probabilities

$P_T(t^{m_1}), \ldots, P_T(t^{m_v})$—summing to one—of being chosen by the group. Clearly, member i's utility $U_i(P_T, \mathbf{c})$ of P_T is given by

$$U_i(P_T, \mathbf{c}) = \sum_{\ell=1}^{v} U_i(t^{m_\ell}, \mathbf{c}) P_T(t^{m_\ell}). \tag{11.2.3}$$

Moreover, the group's overall evaluation $\mathbf{U}(P_T, \mathbf{c}) = (U_i(P_T, \mathbf{c}), \ldots, U_n(P_T, \mathbf{c}))$ of P_T clearly satisfies

$$\mathbf{U}(P_T, \mathbf{c}) = \sum_{\ell=1}^{v} \mathbf{U}(t^{m_\ell}, \mathbf{c}) P_T(t^{m_\ell}). \tag{11.2.4}$$

Note from (11.2.4) that if $v = 2$, then $\mathbf{U}(P_T, \mathbf{c})$ lies on the line segment joining $\mathbf{U}(t^{m_1}, \mathbf{c})$ and $\mathbf{U}(t^{m_2}, \mathbf{c})$. In general, (11.2.4) expresses $\mathbf{U}(P_T, \mathbf{c})$ as a convex combination of the vectors $\mathbf{U}(t^{m_1}, \mathbf{c}), \ldots, \mathbf{U}(t^{m_v}, \mathbf{c})$.

Next, we shall suppose that the group is constrained to choose a randomized total strategy from a nonempty set $\mathcal{Q}'_T$ of randomized total strategies; and we define the group's *achievable utility set* $X(\mathcal{Q}'_T, \mathbf{c})$ by

$$X(\mathcal{Q}'_T, \mathbf{c}) = \{\mathbf{U}(P_T, \mathbf{c}) \colon P_T \in \mathcal{Q}'_T\}. \tag{11.2.5}$$

■ The achievable utility set $X(\mathcal{Q}'_T, \mathbf{c})$ *should* remind you of the attainable utility set $X(\mathcal{Q}'_S)$ in Chapter 7. Their roles in the following analyses are very similar, as we shall soon see. ■

If $X(\mathcal{Q}'_T, \mathbf{c})$ were as in Figure 11.1(A), the group's choice would be easy, because no member could do better than under the group's choice of P°_T. But much more frequently, the shape of $X(\mathcal{Q}'_T, \mathbf{c})$ is as in Figure 11.1(B), where members' preferences conflict at least partially.

But Figure 11.1(B) does suggest a useful criterion by which the group can narrow down its choice of a P_T. If we look at $\mathbf{U}(P'_T, \mathbf{c})$ in Figure 11.1(B), it becomes clear that neither member should seriously advocate continued consideration of P'_T, because there exist points in $X(\mathcal{Q}'_T, \mathbf{c})$—and hence corresponding strategies P_T—yielding *each* member more utility

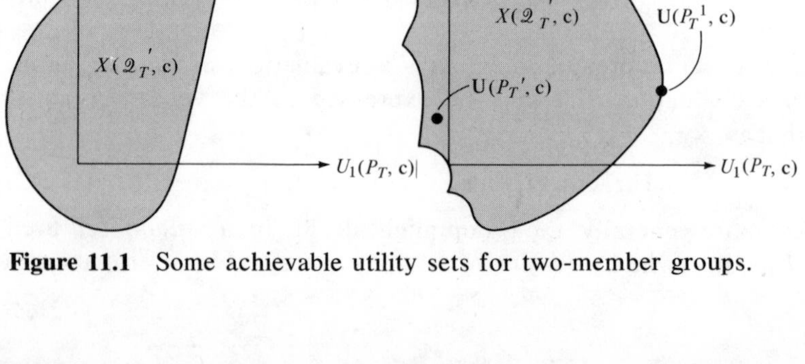

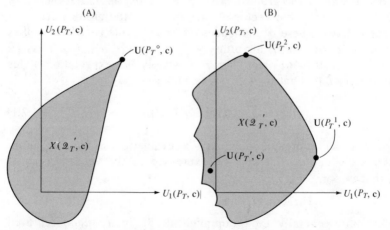

Figure 11.1 Some achievable utility sets for two-member groups.

than does P'_T. Hence, by analogy with Chapter 7, P'_T is *dominated* by other strategies P_T, in the sense that P'_T is consensually inferior. But neither P_T^1 nor P_T^2 is dominated in this sense; and clearly the members have conflicting preferences between P_T^1 and P_T^2. We formalize these familiar notions by saying that P_T^1 and P_T^2 are both *Pareto-optimal* and that P'_T is *Pareto-inferior.*

Definition 11.2.2: Pareto-Optimality. Let $X(\mathcal{Q}'_T, \mathbf{c})$ be given. Then

(A) P'_T is *Pareto-inferior* [with respect to $(\mathcal{Q}'_T, \mathbf{c})$] if there exists some P''_T in $\mathcal{Q}'_T$ such that $U_i(P''_T, \mathbf{c}) \geq U_i(P'_T, \mathbf{c})$ for every i and $U_i(P''_T, \mathbf{c}) > U_i(P'_T, \mathbf{c})$ for at least one i.

(B) P_T is *Pareto-optimal* if it is not Pareto-inferior.

An eminently reasonable thing for the group to do is to eliminate from further consideration all the Pareto-inferior strategies, and hence confine further attention (and controversy!) to the Pareto-optimal strategies.

Example 11.2.2

In Figure 11.1(A), only P°_T is Pareto-optimal; but in Figure 11.1(B), all strategies P_T such that $\mathbf{U}(P_T, \mathbf{c})$ lies on the "northeast border" of $X(\mathcal{Q}'_T, \mathbf{c})$, between $\mathbf{U}(P_T^1, \mathbf{c})$ and $\mathbf{U}(P_T^2, \mathbf{c})$, are Pareto-optimal.

Counseling the group to confine its further attention to the Pareto-optimal strategies is as far as we can go without introducing controversial arbitration mechanisms, since if P'_T and P''_T are both Pareto-optimal and $\mathbf{U}(P''_T, \mathbf{c}) \neq \mathbf{U}(P'_T, \mathbf{c})$, then some group member or members prefer P'_T to P''_T while others prefer P''_T to P'_T. [In Figure 11.1(B), for example, member i prefers P_T^i, for $i = 1, 2$.] One such arbitration mechanism is introduced and examined in the following section.

We turn now to the important topic of how to identify the Pareto-optimal strategies. As in Chapter 7, there is a frequently useful technique based on the Efficiency Theorem in Appendix 3. In general, we shall show that under certain conditions a (pure or randomized) strategy is Pareto-optimal only if it maximizes (over $\mathcal{Q}'_T$) some *weighted average* $\Sigma_{i=1}^n w_i U_i(\cdot, \mathbf{c})$ of the group members' utilities, where the members' relative weights w_i are nonnegative and sum to one. Thus, finding the Pareto-optimal strategies is technically equivalent to finding all cardinally admissible strategies in an individual dpuu, as discussed in Section 7.3, where here the w_i's correspond to the decision maker's probabilities $P_\Sigma(\sigma)$ of the σ's.

Theorem 11.2.1. If $X(\mathcal{Q}'_T, \mathbf{c})$ is convex above, then:

(A) If P'_T is Pareto-optimal, then P'_T is a maximizer in $\mathcal{Q}'_T$ of $\Sigma_{i=1}^n w_i U_i(P_T, \mathbf{c})$ for some $w_1, \ldots, w_n$ such that $\Sigma_{i=1}^n w_i = 1$ and $w_i \geq 0$ for every i.

(B) Every maximizer in $\mathcal{Q}'_T$ of $\Sigma_{i=1}^n w_i U_i(P_T, \mathbf{c})$ is Pareto-optimal if $\Sigma_{i=1}^n w_i = 1$ and $w_i > 0$ for every i.

■ *Proof*. Immediate from the Efficiency Theorem in Appendix 3, by virtue of Definition 11.2.2, which in the language of the Efficiency Theorem defines P_T to be Pareto-optimal in $\mathcal{Q}'_T$ if and only if $\mathbf{U}(P_T, \mathbf{c})$ is efficient in $X(\mathcal{Q}'_T, \mathbf{c})$. ■

■ In practice, one applies Theorem 11.2.1 by maximizing $\Sigma_{i=1}^{n} w_i U_i(P_T, \mathbf{c})$ over $\mathcal{Q}'_T$ for positive relative weights w_i, and then exploring what happens as one or more of the w_i tend to zero. The justification for this procedure is an extension of the Efficiency Theorem to the effect that every efficient point $\mathbf{x}^*$ of a set X is the limit of a sequence of maximizers of $\Sigma_i w_i x_i$ for all w_i positive. ■

■ Two additional restrictions on $X(\mathcal{Q}'_T, \mathbf{c})$ suffice for Pareto-optimal strategies in $\mathcal{Q}'_T$ to exist in profusion. If we require that $X(\mathcal{Q}'_T, \mathbf{c})$ be closed above and locally bounded above, then every P_T in $\mathcal{Q}'_T$ is either Pareto-optimal itself or Pareto-inferior to some *Pareto-optimal* strategy in $\mathcal{Q}'_T$. See Appendix 3. ■

■ There is a signal that $X(\mathcal{Q}'_T, \mathbf{c})$ is not convex above. If, for some set of relative weights $w_1, \ldots, w_n$, there are two maximizers P'_T and P''_T of $\Sigma_{i=1}^{n} w_i U_i(P_T, \mathbf{c})$ in $\mathcal{Q}'_T$, and $\mathbf{U}(P'_T, \mathbf{c}) \neq \mathbf{U}(P''_T, \mathbf{c})$, and if $p\mathbf{U}(P'_T, \mathbf{c}) + (1-p)\mathbf{U}(P''_T, \mathbf{c})$ lies outside $X(\mathcal{Q}'_T, \mathbf{c})$ for some $p \in (0, 1)$, then $X(\mathcal{Q}'_T, \mathbf{c})$ is not convex above. ■

We shall gain some insight into the role of sharing rules in affecting the members' utilities in a group decision by considering cases in which the external strategy s is fixed and hence attention is focussed on an available set $\mathcal{R}$ of sharing rules. As Theorem 11.2.2 shows, convexity of $\mathcal{R}$ and concavity of every member's utility function suffice for application of Theorem 11.2.1.

Theorem 11.2.2. Let $\mathcal{R}$ be a convex set of sharing rules and let $\mathcal{Q}'_T = \{(s, \mathbf{r}): \mathbf{r} \in \mathcal{R}\}$. If $u_i: R^k \to R^1$ is concave for every i, then $X(\mathcal{Q}'_T, \mathbf{c})$ is convex above.

■ *Proof*. It suffices to show that

$$(*) \qquad \mathbf{U}((s, p\mathbf{r}' + [1-p]\mathbf{r}''), \mathbf{c}) \geq p\mathbf{U}((s, \mathbf{r}'), \mathbf{c}) + [1-p]\mathbf{U}((s, \mathbf{r}''), \mathbf{c})$$

whenever $\mathbf{r}', \mathbf{r}'' \in \mathcal{R}$ and $p \in [0, 1]$. First, convexity of $\mathcal{R}$ implies that $p\mathbf{r}' + [1-p]\mathbf{r}'' \in \mathcal{R}$. Second, (*) obtains by virtue of:

$$U_i((s, p\mathbf{r}' + [1-p]\mathbf{r}''), \mathbf{c})$$

$$(1) \qquad = \sum_{j=1}^{\#(\Sigma)} u_i(p\mathbf{r}'(c[s, \sigma^j], \sigma^j) + [1-p]\mathbf{r}''(c[s, \sigma^j], \sigma^j))p_{ij}$$

$$(2) \qquad \geq \sum_{j=1}^{\#(\Sigma)} \{p u_i(\mathbf{r}'(c[s, \sigma^j], \sigma^j)) + [1-p]u_i(\mathbf{r}''(c[s, \sigma^j], \sigma^j))\}p_{ij}$$

$$(3) \qquad = p\sum_{j=1}^{\#(\Sigma)} u_i(\mathbf{r}'(c[s, \sigma^j], \sigma^j))p_{ij} + [1-p]\sum_{j=1}^{\#(\Sigma)} u_i(\mathbf{r}''(c[s, \sigma^j], \sigma^j))p_{ij}$$

$$(4) \qquad = pU_i((s, \mathbf{r}'), \mathbf{c}) + [1-p]U_i(s, \mathbf{r}''), \mathbf{c}),$$

where (1) and (4) are by (11.2.1), (2) is by the assumed concavity of $u_i(\cdot)$, and (3) is rearrangement of summation and factoring. ■

*■ It is productive to think of a sharing rule as a $\#(\Sigma)$-by-k matrix whose (j, h)th component r_{jh} is the increment $r_h(c, \sigma^j)$ to account h if σ^j obtains. Then the set of *all* conceivable sharing rules is the set of all $\#(\Sigma)$-by-k matrices satisfying $\#(\Sigma)$ linear constraints of the form $\Sigma_{h=1}^{k} r_{jh} = c[s, \sigma^j]$. Thus the set $\mathcal{R}$ of *all* sharing rules is easily seen to be convex. So is any set $\mathcal{R}$ of all sharing rules satisfying additional linear equality or inequality constraints, say, constraints of the form that $0 \le r_{jh} \le c[s, \sigma^j]$ for all h whenever $c[s, \sigma^j] \ge 0$ and $0 \ge r_{jh} \ge c[s, \sigma^j]$ for all h whenever $c[s, \sigma^j] \le 0$. ■

■ Concavity of u_i: $R^k \to R^1$ expresses a general aversion to risk on the part of member i, since it implies that i would never prefer an uncertain vector $\mathbf{r}(c[s, \tilde{\sigma}], \tilde{\sigma})$ of account increments to his expectation $\Sigma_{j=1}^{\#(\Sigma)} \mathbf{r}(c[s, \sigma^j], \sigma^j)p_{ij}$ of that vector. When $k = n$ and the accounts are the members' personal shares, and when i's utility function u_i depends solely upon his own share, the argument in Section 5.4 implies that u_i should be concave. ■

Suppose now that $\mathcal{Q}'_T = \{(s, \mathbf{r}): \mathbf{r} \in \mathcal{R}\}$, as in Theorem 11.2.2, for a given external strategy s and a given, convex set $\mathcal{R}$ of sharing rules. Theorems 11.2.1 and 11.2.2 suggest the technique of finding Pareto-optimal sharing rules as functions of positive relative weights $w_1, \ldots, w_n$ by maximizing $\Sigma_{i=1}^{n} w_i U_i((s, \mathbf{r}), \mathbf{c})$. This is often more easily said than done. Then, if $n = 2$, one may want to graph the group evaluations $\mathbf{U}((s, \mathbf{r}), \mathbf{c})$ of the Pareto-optimal sharing rules. If more than one external pure strategy is available, the whole process could be repeated as many times as there are external pure strategies. Figure 11.2 is a sketch of the lines L^1 and L^2, where $L^m = \{\mathbf{U}((s^m, \mathbf{r}), \mathbf{c}): \mathbf{r}$ Pareto-optimal in $\mathcal{R}^m\}$, for a group decision problem in which there are two external pure strategies s^1 and s^2, and each s^m has an associated set $\mathcal{R}^m$ of feasible sharing rules. We assume that both members' utility functions are concave and that each $\mathcal{R}^m$ is convex, so that

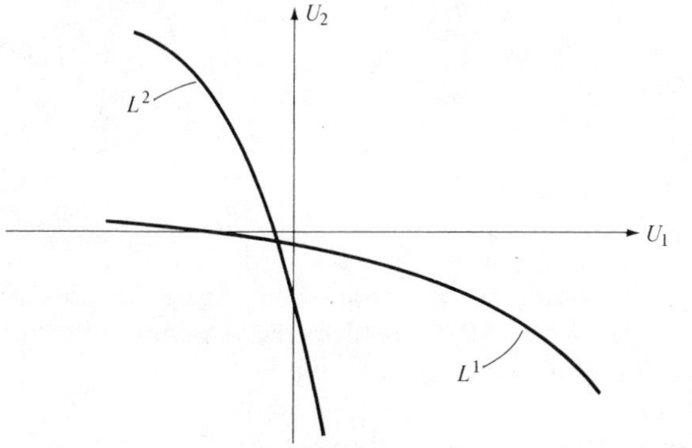

Figure 11.2 Group evaluations of Pareto-optimal pure total strategies.

each of the lines L^m is the northeast boundary of some convex set, as implied by Theorem 11.2.2.

Because the group must choose some total strategy if it forms, and because neither member should object to Pareto-optimality, it would seem to follow that the group should choose some Pareto-optimal t in $T = T^1 \cup T^2$, where $T^m = \{(s^m, \mathbf{r}): \mathbf{r} \in \mathcal{R}^m\}$; more specifically, the group should choose either a point on L^1 to the right of the intersection of L^1 and L^2, or a point on L^2 to the left of the intersection.

But this line of argument neglects the potentially mutually advantageous role of randomization. Let us suppose, further, that if the group does not form, then the status quo has utility zero for each member. Hence the origin in Figure 11.2 is the joint evaluation of the status quo, which will obtain unless the members can agree on some $(s, \mathbf{r})$. But Figure 11.2 indicates that there is no $(s, \mathbf{r})$ which both prefer to the status quo, since whenever member 1 prefers some possible t to the status quo, member 2 prefers the status quo to t, and vice versa. This seems to indicate that the group cannot form. Now examine Figure 11.3, which shows that the randomized strategy $P_T^{1/2}$ placing equal probabilities on $(s^1, \mathbf{r}^1)$ and $(s^2, \mathbf{r}^2)$ is preferred to the status quo by each member—before the randomization is conducted, of course. Hence the group *can* form and contract to use the randomized strategy $P_T^{1/2}$.

Figure 11.3 and the related discussion[b] show that *randomization can play an important role in group decisions as an arbitration, or compromise, mechanism.* Member 1 might convince 2 to join him in the decision depicted in Figure 11.3 by arguing:

"If we form a group, there is a prospect, $(s^1, \mathbf{r}^1)$, in which I will do very well and you will be a bit worse off than under the status quo. But there is another prospect, $(s^1, \mathbf{r}^2)$, in which you will do very well and I

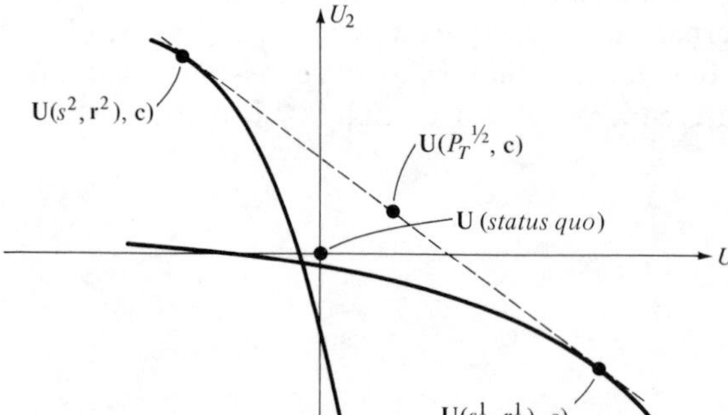

Figure 11.3 Advantages of randomization in the context of Figure 11.2. (*Key*: dashed line $= \{U(P_T^p, \mathbf{c}): 0 \le p \le 1\}$, where P_T^p = randomized strategy selecting $(s^1, \mathbf{r}^1)$ and $(s^2, \mathbf{r}^2)$ with respective probabilities p and $1-p$.)

[b] Which closely approximates that in Chapter 8 of Raiffa [**113**].

will be a bit worse off than under the status quo. We can, however, agree to flip a fair coin to choose between these prospects, and this "gambling prospect" looks attractive to *each* of us because it gives equal chances at a moderate loss and a big gain."

Example 11.2.3 concerns the Pareto-optimal sharing rules for the simple risk-sharing problem of Joe and Sam, introduced in Example 11.1.1. Because of its lengthiness, the example is divided into a number of parts, of which the first is a summary of the results.

Example 11.2.3(A): Joe and Sam (the Results)

Recall from Example 11.1.1 that Joe possesses a lottery s^1 which yields \$400, \$0, or \$ -200 if σ^1, σ^2, or σ^3 respectively occurs. Also recall that Joe is trying to get Sam to share this risk, and that both have the piecewise linear utility function (11.1.1), and hence the preferences of each are exclusively determined by his own return. For brevity, let $y_j = r_1(c[s^1, \sigma^j], \sigma^j) =$ *Joe's* return if σ^j occurs, for $j = 1, 2, 3$, so that Joe's and Sam's shares of \$400 are y_1 and $400 - y_1$ respectively; their shares of \$0 are y_2 and $-y_2$ respectively; and their shares of \$ -200 are y_3 and $-200 - y_3$ respectively.[c] Table A lists this information about Joe's and Sam's "shares" of s^1 if the sharing rule determined by Joe's share vector (y_1, y_2, y_3) is adopted, together with Joe's and Sam's probabilities of the σ's. By using the table in conjunction with the utility function (11.1.1), we can calculate Joe's utility $U_1((s^1, (y_1, y_2, y_3)), \mathbf{c})$ of any total pure strategy $(s^1, (y_1, y_2, y_3))$, and Sam's utility $U_2((s^1, (y_1, y_2, y_3)), \mathbf{c})$ of same. Figure 11.4 plots these utility pairs for the Pareto-optimal sharing rules, which are of four general forms.

Form	*Joe's Share*
I	$(y_1, -100, y_3)$ for $y_1 \leq -100$ and $y_3 \leq -100$
II	$(-100, y_2, -100)$ for $-100 \leq y_2 \leq 100$
III	$(y_1, 100, -100)$ for $-100 \leq y_1 \leq 500$
IV	$(500, y_2, -100)$ for $y_2 \geq 100$

Table A Joe and Sam—Basic Data

State σ^j	σ^1	σ^2	σ^3
J's probability	.25	.50	.25
S's probability	.40	.20	.40
$c[s^1, \sigma^j]$	400	0	-200
J's share	y_1	y_2	y_3
S's share	$400 - y_1$	$-y_2$	$-200 - y_3$

[c] In general, Sam's share is $r_2(c[s^1, \sigma^j], \sigma^j) = c[s^1, \sigma^j] - r_1(c[s^1, \sigma^j], \sigma^j) = c[s^1, \sigma^j] - y_1$, by the requirement that a sharing rule must fully apportion group income.

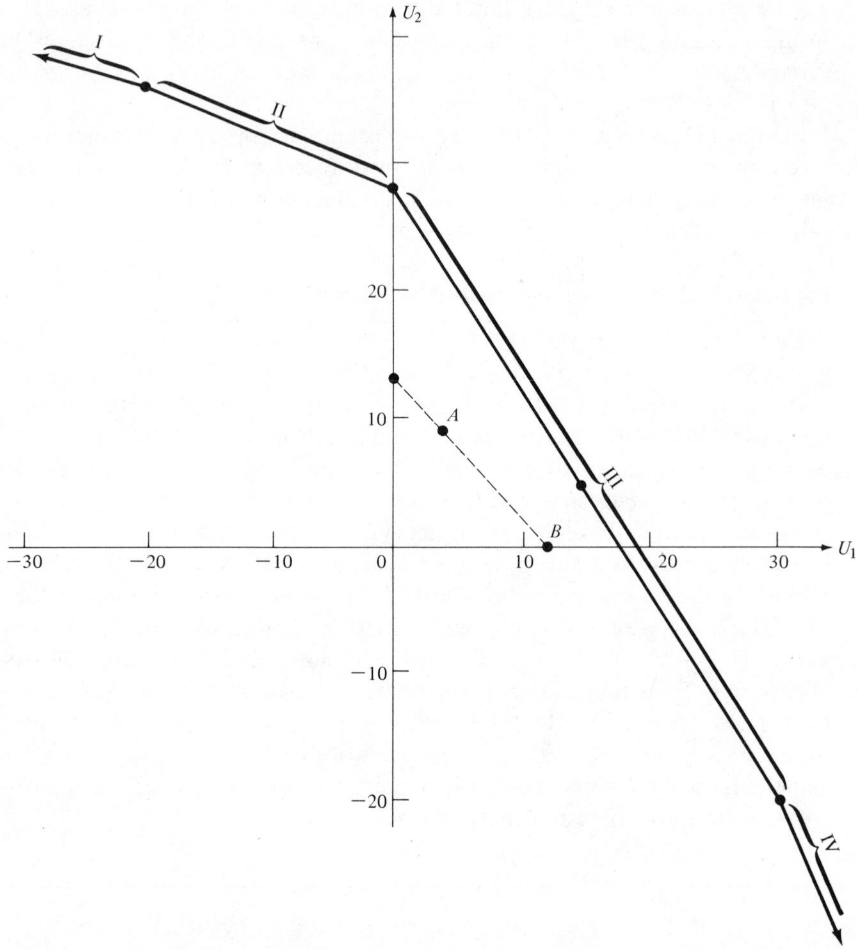

Figure 11.4 Joe and Sam. (*Key*: (1) Dashed line explained in Example 11.2.4. (2) On Segments I, II, III, and IV the sharing rules are of the forms I, II, III, and IV, respectively, described in Example 11.2.3(A).)

Now, if we recall that the status quo for Joe and Sam is \$0 each for sure and that $u(0) = 0$, it is clear from Figure 11.4 that Joe and Sam should select a sharing rule of the form III with the property that $U_1 \geq 0$ and $U_2 \geq 0$. This implies that y_1 cannot exceed 250. Hence Joe and Sam can confine their bargaining to sharing rules of the form $(y_1, 100, -100)$ with $-100 \leq y_1 \leq 250$. Note that the equal-shares split, $(y_1, y_2, y_3) = (200, 0, -100)$ produces point A on Figure 11.4, and hence is *not* Pareto-optimal!

■ Technically, we have not determined the Pareto-optimal sharing rules in their entirety, but rather only how these sharing rules divide the specific outcomes \$400, \$0, and \$ − 200 of s^1. Esthetics dictate, however, that we accept this solecism rather than always refer to, say, the "relevant part" of the Pareto-optimal sharing rules. ■

The remaining parts of this example show how we arrived at the results summarized in Example 11.2.3(A).

Example 11.2.3(B): Joe and Sam (Utilities)

From Table A and Equation (11.1.1), we can calculate Joe's utility $U_1((s^1, (y_1, y_2, y_3)), \mathbf{c})$ of the total pure strategy $(s^1, (y_1, y_2, y_3))$, in which the sharing rule is summarized by Joe's share. Eight cases must be distinguished, according as y_1, y_2, and/or y_3 is less than -100 or not. These eight formulae are listed in Table B. [For example, if $y_3 < -100$ but y_1 and y_2 are each at least -100, then $U_1((s^1, (y_1, y_2, y_3)), \mathbf{c}) = .25[y_1/5] + .50[y_2/5] + .25[y_3 + 80] = .05y_1 + .10y_2 + .25y_3 + 20.$] Similarly, Table C records the formulae for Sam's utility $U_2((s^1, (y_1, y_2, y_3)), \mathbf{c})$ given his probabilities and his utility function (11.1.1). Again, eight cases must be distinguished, according as $400 - y_1$, $-y_2$, and/or $-200 - y_3$ is less than -100 or not; or equivalently, according as $y_1 > 500$, $y_2 > 100$, and/or $y_3 > -100$ or not. [For example, if $y_3 > -100$ but $y_1 \le 500$ and $y_2 \le 100$, then $U_2((s^1, (y_1, y_2, y_3)), \mathbf{c}) = .40[(400 - y_1)/5] + .20[-y_2/5] + .40[(-200 - y_3) + 80] = -.08y_1 - .04y_2 - .40y_3 - 16.$] Although voluminous, these tables provide the data necessary for finding the Pareto-optimal sharing rules for this problem.

Table B Joe's Utility of Total Strategy, by Region

Region #	y_1	y_2	y_3	$U_1((s^1, (y_1, y_2, y_3)), \mathbf{c})$
J1	≥ -100	≥ -100	≥ -100	$.05y_1 + .10y_2 + .05y_3$
J2	≥ -100	≥ -100	< -100	$.05y_1 + .10y_2 + .25y_3 + 20$
J3	≥ -100	< -100	≥ -100	$.05y_1 + .50y_2 + .05y_3 + 40$
J4	< -100	≥ -100	≥ -100	$.25y_1 + .10y_2 + .05y_3 + 20$
J5	≥ -100	< -100	< -100	$.05y_1 + .50y_2 + .25y_3 + 60$
J6	< -100	≥ -100	< -100	$.25y_1 + .10y_2 + .25y_3 + 40$
J7	< -100	< -100	≥ -100	$.25y_1 + .50y_2 + .05y_3 + 60$
J8	< -100	< -100	< -100	$.25y_1 + .50y_2 + .25y_3 + 80$

Table C Sam's Utility of Total Strategy, by Region

Region #	y_1	y_2	y_3	$U_2((s^1, (y_1, y_2, y_3)), \mathbf{c})$
S1	≤ 500	≤ 100	≤ -100	$-.08y_1 - .04y_2 - .08y_3 + 16$
S2	≤ 500	≤ 100	> -100	$-.08y_1 - .04y_2 - .40y_3 - 16$
S3	≤ 500	> 100	≤ -100	$-.08y_1 - .20y_2 - .08y_3 + 32$
S4	> 500	≤ 100	≤ -100	$-.40y_1 - .04y_2 - .08y_3 + 176$
S5	≤ 500	> 100	> -100	$-.08y_1 - .20y_2 - .40y_3$
S6	> 500	≤ 100	> -100	$-.40y_1 - .04y_2 - .40y_3 + 144$
S7	> 500	> 100	≤ -100	$-.40y_1 - .20y_2 - .08y_3 + 192$
S8	> 500	> 100	> -100	$-.40y_1 - .20y_2 - .40y_3 + 160$

Example 11.2.3(C): Joe and Sam (Methodology)

Because of the many different formulae for U_1 and U_2 in Tables B and C, we cannot make effective use of Theorem 11.2.2 in isolating the Pareto-optimal sharing rules.[d] Instead, we shall proceed as follows. As a *first phase*, we shall pick a convenient sharing rule (y_1, y_2, y_3), say, the even split in each state; and we shall look at which *pair* of regions (one for Joe and one for Sam) that rule falls in. We shall look at the pair of utility formulae for that rule in order to find out which y_j we should increase so as to increase Joe's utility U_1 at minimal cost in terms of reducing Sam's utility U_2. We shall then increase that y_j as far as possible until we hit the boundary of a utility-formula region. If we increased that same y_j further, we would be in a different region, where a different pair of formulae for U_1 and U_2 pertained, and hence where the trade-off ratio of Sam's for Joe's utility would be different. We shall compare this new trade-off ratio with the previously determined ratios for increases in the other two y_j's so as to find out, once again, how to increase Joe's utility with minimal reduction in Sam's. We keep on doing this until we are certain that we have a class of Pareto-optimal sharing rules; i.e., that we have reached and are traveling along the northeastern frontier of $X(T, \mathbf{c})$. Then, as *phase two*, we reverse the procedure and, at each step, effect the greatest possible increase in Sam's utility at the cost of minimally reducing Joe's. In this fashion we traverse the entire northeastern frontier of $X(T, \mathbf{c})$. The Pareto-optimal sharing rules fall out as by-products. The next part of the example gives the technical specifics about the trade-off ratio.

Example 11.2.3(D): Joe and Sam (Technicalities of Varying the Sharing Rule)

At any given sharing rule (y_1, y_2, y_3) and given that we contemplate a (small) increase in a certain y_j, we have two formulae from Table B specifying U_1 and U_2 over the range of contemplated changes; say, $U_i = x_{1i}y_1 + x_{2i}y_2 + x_{3i}y_3 + K_i$ for $i = 1, 2$ [= Joe, Sam], where the x_{j1}'s are positive and the x_{j2}'s are negative. To increase Joe's utility by one unit via increasing y_j takes an increase of $1/x_{j1}$ in y_j. That increase causes a change of $x_{j2}(1/x_{j1})$, or x_{j2}/x_{j1}, in Sam's utility. This ratio is, of course, negative. We calculate it for each of the three y_j's which could be increased from the current position (y_1, y_2, y_3), using the appropriate formulae for U_i given each of the three contemplated increases. (Be careful! Moving from one region into another causes at least one of the U_i's to change form.) When increasing Joe's utility at minimal cost to Sam, we increase whichever y_j possesses the largest (i.e., least negative) ratio x_{j2}/x_{j1}. Going backward, in phase two, entails a similar analysis. To reduce Joe's utility one unit by reducing y_j requires a change of $-1/x_{j1}$ in

[d] I have gone through the Theorem 11.2.2 approach, and it is *much* longer than the one we use here!

y_j. This produces an increase of $x_{j2}(-1/x_{j1}) = -x_{j2}/x_{j1}$ in Sam's utility. We want to maximize Sam's gain per unit loss to Joe, and hence we maximize $-x_{j2}/x_{j1}$; or equivalently, we minimize x_{j2}/x_{j1}, the same negative ratio as before.

Example 11.2.3(E) goes through the gory details of the computations. Nonmasochistic readers may simply want to refer again to Figure 11.4 and Example 11.2.3(A) for a final summation.

Example 11.2.3(E): Joe and Sam (Computations)

Phase one. We start with the equal-split sharing rule of Example 11.1.1; namely, $(y_1, y_2, y_3) = (200, 0, -100)$, producing point A on Figure 11.4 as the joint evaluation. Increasing y_1 (by any amount up to $y_1 = 500$) would keep Joe and Sam in regions J1 and S1; $x_{11} = .05$ and $x_{12} = -.08$; and hence the ratio is $-.08/.05 = -1.6$. Increasing y_2 (by any amount up to $y_2 = 100$) would again have Joe and Sam in regions J1 and S1; $x_{21} = .10$ and $x_{22} = -.04$; and hence the ratio is $-.04/.10 = -.4$. Increasing y_3 (by any amount from its current value of -100) would put Joe and Sam in regions J1 and S2; $x_{31} = .05$ and $x_{32} = -.40$; and hence the ratio is $-.40/.05 = -8.0$. Thus the least painful way to increase Joe's utility is to increase y_2 up to $y_2 = 100$. We are now at point B in Figure 11.4, with $(y_1, y_2, y_3) = (200, 100, -100)$—but we do not yet know that this point is on the northeastern frontier of $X(T, \mathbf{c})$. Small increases in y_1 or y_3 from this point would put Joe and Sam in regions J1 and S1, or in J1 and S2, so that for each of these increases the previously calculated ratio -1.6 or -8.0 pertains. But increasing y_2 above 100 puts Joe and Sam in regions J1 and S3; $x_{21} = .10$ and $x_{22} = -.20$; and hence the ratio for increasing y_2 is $-.20/.10 = -2.0$. Hence y_1 should be increased. If we increase y_1 as far as possible without changing regions, we are at $(500, 100, -100)$. From here, increasing y_2 above 100 produces the just calculated ratio of -2.0. Increasing y_3 above -100 produces the previously calculated ratio of -8.0. Increasing y_1 further causes Joe and Sam to be in regions J1 and S4; here, $x_{11} = .05$ and $x_{12} = -.40$, so that the ratio for increasing y_1 is $x_{12}/x_{11} = -.40/.05 = -8.0$. Thus the best thing to do is to increase y_2 above 100 from this point on; Joe and Sam will remain in regions J1 and S3 forever. Now we know that all sharing rules (in which Joe's shares are) of the form $(500, y_2, -100)$ for $y_2 > 100$ are Pareto-optimal, because no further regional boundaries can be encountered, and increasing y_2 is the least painful way to Sam of increasing Joe's utility.

Phase two. We work back, from $(500, y_2, -100)$ with $y_2 > 100$. Decreasing y_1 puts Joe and Sam in regions J1 and S3 and produces the ratio $x_{12}/x_{11} = -.08/.05 = -1.6$. Decreasing y_3 below -100 puts Joe and Sam in regions J2 and S3 and produces the ratio $x_{32}/x_{31} = -.08/.25 = -.32$. Decreasing y_2 (to any point above 100) puts Joe and Sam in regions J1 and S3 and produces the ratio $x_{22}/x_{21} = -.20/.10 = -2.0$. This is the most negative of the ratios, and hence decreasing y_2 is the least painful way (to Joe) of increasing Sam's utility. Decreasing y_2 to the boundary of the region produces the sharing rule (500, 100, −100). From here, decreasing y_1 somewhat puts Joe and Sam in regions J1 and S1 and produces the ratio $x_{12}/x_{11} = -.08/.05 = -1.6$. Decreasing y_2 below 100 somewhat puts Joe and Sam in regions J1 and S1 and produces the ratio $x_{22}/x_{21} = -.04/.10 = -.4$. Decreasing y_3 below −100 puts Joe and Sam in regions J2 and S1 and produces the ratio $x_{32}/x_{31} = -.08/.25 = -.32$. Thus y_1 should be decreased; it can be decreased to −100 without producing a region change. Now we are at (−100, 100, −100) and have determined that all sharing rules (in which Joe's shares are) of the form $(y_1, 100, -100)$ for $-100 \leq y_1 \leq 500$ are Pareto-optimal. From (−100, 100, −100), decreasing y_1 puts Joe and Sam in regions J4 and S1 and produces the ratio $x_{12}/x_{11} = -.08/.25 = -.32$. Decreasing y_2 below 100 puts Joe and Sam in regions J1 and S1 and produces the ratio $x_{22}/x_{21} = -.04/.10 = -.4$. Decreasing y_3 below −100 puts Joe and Sam in regions J2 and S1 and produces the ratio $x_{32}/x_{31} = -.08/.25 = -.32$. Hence y_2 should be decreased. It can be decreased down to −100 without producing a region change, and we thus see that all sharing rules (in which Joe's shares are) of the form $(-100, y_2, -100)$ for $-100 \leq y_2 \leq 100$ are Pareto-optimal. From the new extreme, (−100, −100, −100), decreasing y_1 below −100 puts Joe and Sam in regions J4 and S1 and produces the ratio $x_{12}/x_{11} = -.08/.25 = -.32$. Decreasing y_2 below −100 puts Joe and Sam in regions J3 and S1 and produces the ratio $x_{22}/x_{21} = -.04/.50 = -.08$. Decreasing y_3 below −100 puts Joe and Sam in regions J2 and S1 and produces the ratio $x_{32}/x_{31} = -.08/.25 = -.32$. Hence y_1 *and/or* y_3 may be decreased below −100 indefinitely. Thus all sharing rules (in which Joe's shares are) of the form $(y_1, -100, y_3)$ for $y_1 \leq -100$ and $y_3 \leq -100$ are Pareto-optimal. Figure 11.4 depicts the group evaluations $\mathbf{U}((s^1, (y_1, y_2, y_3)), \mathbf{c})$ of the Pareto-optimal sharing rules, i.e., the northeastern frontier of $X(T, \mathbf{c})$.

We mentioned earlier that the definition of sharing rules, allowing for side-betting on the state, affords the group members an opportunity to do better than they could under proportional sharing rules, even with a system of "antes," when the members have different probabilities of the states. Example 11.2.4 shows that this is the case for Joe and Sam.

Example 11.2.4: Joe and Sam Again

Suppose that the only sharing rules permitted are those in which Joe keeps proportion p of s^1 and sells the remaining proportion, $1-p$, to Sam for \$$x$. Thus $r_1(c[s^1, \sigma^j], \sigma^j) = pc[s^1, \sigma^j] + x$, and $r_2(c[s^1, \sigma^j], \sigma^j) = (1-p)c[s^1, \sigma^j] - x$. As for Example 11.2.3, a case-by-case derivation of U_1 and U_2 is required, and then one must determine which sharing rules of the requisite form are both Pareto-optimal (in relation to the requisite form) and mutually acceptable. The result is that sharing rules in which Sam pays Joe $x = 200p - 100$ for his proportion $1 - p$ are Pareto-optimal and mutually acceptable if $.40 \le p \le 9/14 \doteq .65$. For such sharing rules, $U_1 = 50p - 20$ and $U_2 = -56p + 36$. These (U_1, U_2) pairs are graphed as the dashed line in Figure 11.4, from (0, 13.6) for $p = .40$ to (170/14, 0) for $p = 9/14$.

■ The analysis is less difficult than that for Example 11.2.3. We have four regions in the (p, x) plane for each of U_1 and U_2.

Region	Definition	U_1
J I	$400p + x \le -100$	$50p + x + 80$
J II	$x \le -100 \le 400p + x$	$-30p + .80x + 60$
J III	$-200p + x \le -100 \le x$	$-30p + .40x + 20$
J IV	$-100 \le -200p + x$	$10p + .20x$

Region	Definition	U_2
S I	$400(1-p) - x \le -100$	$-80p - x + 160$
S II	$-x \le -100 \le 400(1-p) - x$	$48p - .68x$
S III	$-200(1-p) - x \le -100 \le -x$	$48p - .52x - 16$
S IV	$-100 \le -200(1-p) - x$	$-16p - .20x + 16$

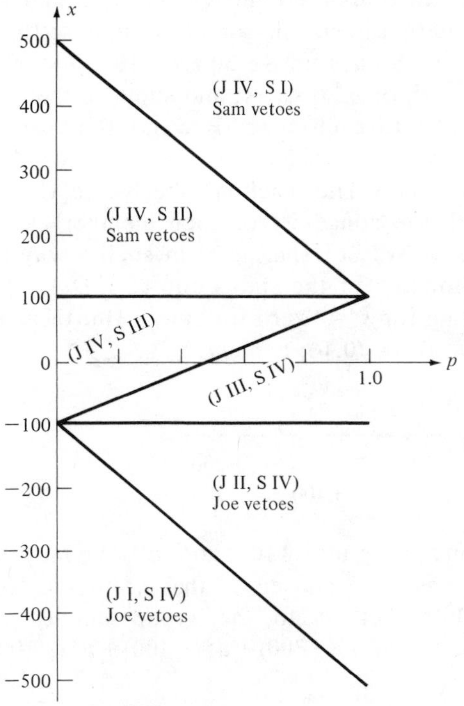

Figure 11.5 Paired regions for Example 11.2.4.

Next, we observe that acceptability to Joe, i.e., $U_1 \geq 0$, prevents consideration of any (p, x) in J I or J II, since all such give rise to $U_1 < 0$. Similarly, Sam would veto all (p, x) in regions S I and S II. This leaves only J III, J IV, S III, and S IV, and only the paired regions (J III, S IV) and (J IV, S III). (See Figure 11.5.) Now, in (J III, S IV), each member wants to reduce p as far as possible for any given x, while in (J IV, S III) each member wants to increase p as far as possible for any given x. Hence we arrive at the boundary line $x = 200p - 100$, where $U_1 = 50p - 20$ and $U_2 = -56p + 36$. Again requiring that $U_1 \geq 0$ and $U_2 \geq 0$ produces the constraint that $.40 \leq p \leq 9/14$. ■

We conclude the section with, first, an example showing that a group cannot in general act as an individual decision maker would, because the members' utility functions cannot always be combined into a "group utility function"; and, second, examples of special cases in which members' preferences and judgments *can* be combined into group utility and probability functions.

Example 11.2.5: Nonexistence of a Group Utility Function

(Attributed to J. W. Pratt in Raiffa [**113**].) Suppose that Al and Bob constitute a two-person group and that their utility functions are defined as follows: Al has $u_A(c) = c$ for $-100 \leq c \leq 200$; Bob has $u_B(c) = c$ for $-200 \leq c \leq 100$, and both functions are concave. Thus Al and Bob are each averse to the risks inherent in sizable gambles. Suppose that s^1 is a potential group decision which pays off \$−300, \$−100, \$+100, or \$+300 given that state σ^1, σ^2, σ^3, or σ^4 obtains; and suppose that each member assesses probability .25 to each state. Because there are no judgmental disagreements, a group probability function would also have to assign probability .25 to each state. Thus each member's $E\{c[s^1, \tilde{\sigma}]\} = 0$, and this fact, together with the concavity of each member's utility function, implies that there is no way of sharing s^1 in such a way that (U_A, U_B) would lie north and/or east of the status quo $(U_A, U_B) = (0, 0)$ obtainable by the group's opting for $s^2 =$ "zero for sure." But there is a way of sharing s^1 so that $(U_A, U_B) = (0, 0)$; namely,

	σ^1	σ^2	σ^3	σ^4
Al	−100	−100	0	+200
Bob	−200	0	+100	+100

Now suppose that there is some group utility function, u^*. We may, and shall, suppose that its origin has been chosen so that $u^*(0) = 0$. Now, the bare, marginal acceptability of s^1 to the group implies that $E\{u^*(c[s^1, \tilde{\sigma}])\} = 0$; that is, $(.25)[u^*(-300) + u^*(-100) + u^*(+100) + u^*(+300)] = 0$, or equivalently,

(a) $$u^*(-300) + u^*(-100) + u^*(+100) + u^*(+300) = 0.$$

Now consider s^3, which yields a return of $\$-100$ given σ^1 or σ^2 and a return of $\$+100$ given σ^3 or σ^4. Concavity of the members' utility functions and the fact that $E\{c[s^3, \tilde{\sigma}]\} = 0$ imply that s^3 is also just marginally acceptable to the group, so that $E\{u^*(c[s^3, \tilde{\sigma}])\} = 0$, which implies that $(.50)[u^*(-100) + u^*(+100)] = 0$, or equivalently,

(b) $$u^*(-100) + u^*(+100) = 0.$$

Now, (a) and (b) together imply

$$u^*(-300) + u^*(+300) = 0,$$

which means that the group should also find just acceptable an s^4 which yields a return of $\$-300$ if σ^1 or σ^2 obtains and a return of $\$+300$ if σ^3 or σ^4 obtains. Since the members' judgments coincide, we are free to combine states yielding the same return and thus to regard s^4 as yielding $\$-300$ if state τ^1 obtains and $\$+300$ if state τ^2 obtains, where τ^1 and τ^2 each have probability .50. But now we can show that there is *no* sharing rule for s^4 which gives each member $U_i \geq 0$, and thus we obtain the contradiction to our assumption that a group utility function exists. If such a sharing rule for s^4 did exist, it would be expressible as

	τ^1	τ^2
s^4	-300	300
Al's share	y_1	y_2
Bob's share	$-300 - y_1$	$+300 - y_2$

Because Al's utility function is concave, acceptability to Al requires that

(c) $$(.50)(y_1 + y_2) \geq 0.$$

Similarly, for Bob,

(d) $$(.50)[(-300 - y_1) + (+300 - y_2)] \geq 0.$$

Reexpressing (d) and comparing it with (c) yield

(e) $$y_2 = -y_1.$$

Now, concavity of Al's utility function requires that $-100 \leq y_1 \leq 200$ and $-100 \leq y_2 \leq 200$, which, together with (e) imply

(f) $$-100 \leq y_1 \leq 100.$$

Similarly, concavity of Bob's utility function requires that $-200 \leq -300 - y_1 \leq 100$ and $-200 \leq +300 - y_2 \leq 100$, which, together with (e), imply

(g) $$-400 \leq y_1 \leq -200.$$

Now no y_1 can satisfy both (f) and (g). Hence no mutually acceptable sharing rule exists, and hence no group utility function u^* exists.

Nonexistence in general of group utility functions (and group probability functions) is a disappointing fact, because only in cases where the group can formulate such functions can it choose an external strategy s without reference to a choice of sharing rule $\mathbf{r}$.

But group utility and probability functions do exist in a few special cases defined in terms of the forms of the members' utility and probability functions. General treatments of necessary and sufficient conditions for the existence of such group functions can be found in Wilson [**144**], Rosing [**121**], and Wallace [**138**]. We shall present two examples, the first being a special case of the second and requiring that each member be concerned only with his own monetary return and have a utility function of the concave-exponential, constant-risk-aversion form.

Theorem 11.2.3. If any conceivable sharing rule may be chosen by the group, if $p_{ij} > 0$ for every i and j, if $k = n$, and if, for $i = 1, \ldots, n$, member i's utility function is of the form

$$u_i(\mathbf{r}) = -e^{-g_i r_i}$$

for some $g_i > 0$ and every $\mathbf{r} \in R^n$, then:

(A) $\mathbf{r}$ is a Pareto-optimal sharing rule if and only if there exist real numbers $\alpha_1, \ldots, \alpha_n$ such that $\Sigma_{h=1}^{n} \alpha_h = 0$ and

$$r_i(c, \sigma^j) = \xi_i \cdot c + \beta_{ij} + \alpha_i \tag{11.2.6}$$

for every i, j, and c, where

$$\xi_i = g_i^{-1} \Big/ \sum_{h=1}^{n} g_h^{-1}$$

and

$$\beta_{ij} = \xi_i \cdot \sum_{h=1}^{n} g_h^{-1} \log_e (p_{ij}/p_{hj}).$$

(B) The choice of sharing rule can be made independently of the choice of external strategy s, in that each member should find the group's choosing s'' at least as desirable as its choosing s' if and only if

$$\sum_{j=1}^{\#(\Sigma)} u^*(c[s'', \sigma^j])p_j^* \geq \sum_{j=1}^{\#(\Sigma)} u^*(c[s', \sigma^j])p_j^*,$$

where

$$p_j^* = \prod_{h=1}^{n} p_{hj}^{\xi_h} \Big/ \sum_{\ell=1}^{\#(\Sigma)} \left[\prod_{h=1}^{n} p_{h\ell}^{\xi_h}\right]$$

and

$$u^*(c) = -e^{-g^* c}$$

for every c, where

$$g^* = \left[\sum_{h=1}^{n} g_h^{-1}\right]^{-1}.$$

■ *Proof.* A special case of Theorem 11.2.4. ■

■ The second hypothesis of Theorem 11.2.3, that every p_{ij} be positive, is necessary for the existence (and closure) of the northeast (or efficient) border of $X(T, \mathbf{c})$, at least when consideration is not restricted to some bounded set of sharing rules. To see that this is so, suppose $p_{11} = 0$ and $p_{21} > 0$. Then 2's utility can be increased at no cost to 1 by forming a side bet in which 2 wins \$$x$ from 1 if σ^1 occurs. Now, 1 is certain that he will not lose x, whereas 2's utility increases. By increasing x, we move vertically in the (U_1, U_2) plane—either arbitrarily far if u_2 is not bounded above, or approaching the asymptote $\lim_{c\to\infty} u_2(c)$ if u_2 is bounded above. ■

As Exercise 11.2.1, you may verify that $\xi_i > 0$ for every i, that $\Sigma_{i=1}^n \xi_i = 1$, and that $\Sigma_{i=1}^n \beta_{ij} = 0$ for every j. These facts and Equation (11.2.6) imply that in groups within the scope of Theorem 11.2.3, all Pareto-optimal sharing rules give i the same *proportional share* ξ_i of group income c and possess the same structure of *side bets* β_{ij} on the states σ^j. Two Pareto-optimal sharing rules differ only in the system of initial "antes" α_i.

Note, further, how (any) member i's share $r_i(c, \sigma^j)$ depends upon the "risk-aversion parameters" $g_1, \ldots, g_n$ and upon the probabilities $p_{1j}, \ldots, p_{nj}$ of σ^j. *Ceteris paribus*, g_i large means that i is more averse to risk than does g_i small; and increasing g_i has two effects. (1) It *reduces* i's proportional share ξ_i of group income c and the absolute magnitude of each of i's net side-bet inflows or outflows β_{ij}. (2) Also *ceteris paribus*, increasing p_{ij} increases (possibly from negative to positive) the net inflow β_{ij} from side bets to i if σ^j occurs.

Moreover, note how the group utility and probability functions u^* and p^* depend upon the parameters of the members. The group utility function u^* is also of the concave-exponential, constant-risk-aversion form; and, if we call g_i^{-1} member i's "risk tolerance" and g^{*-1} the *group*'s risk tolerance, then the group's risk tolerance is just the *sum* of the members' risk tolerances. On the other hand, the group probability p_j^* of σ^j is a (scaled) geometric mean of the members' probabilities $p_{1j}, \ldots, p_{nj}$ of σ^j. *Ceteris paribus*, those members with the highest risk tolerance (and lowest risk aversion) will have the highest weights ξ_h in the geometric mean, and hence their probabilities p_{hj} will have the greatest influence upon p_j^*, since members with low ξ_h will have $p_{hj}^{\xi_h}$ close to 1 in each of the products defining p_j^*.

Finally, note that the hypothesis that any sharing rule may be chosen by the group cannot be dispensed with lightly. If constraints bar sharing rules of the form (11.2.6), then group utility and probability functions need not exist.

■ Wallace [**138**], building on the work of Wilson [**144**] and of Rosing [**121**], derives Theorem 11.2.3 in the context of finding very general conditions for the existence of group utility and probability functions.

Under the assumptions that all conceivable sharing rules are available for choice by the group and that every member is risk averse and is concerned exclusively with his own return, Wallace shows (1) that "uniform decisiveness," in the sense that criss-crossing group evaluations of Pareto-optimal pure total strategies (Figure 11.2) is impossible, is equivalent to the existence of separate group utility and probability functions; (2) that if the members disagree on probabilities, then the *only* situation in which uniform decisiveness is bound to obtain is that of Theorem 11.2.3, in which every member has a concave exponential utility function; and (3) that if consensus on probabilities does obtain, then the agreed-upon probabilities will of course constitute the group probability function, but there will also be a group utility function *only* if *all* members' utility functions belong to the *same* one of the following classes.

$$u_i(\mathbf{r}) = \log_e (r_i - g_i) \text{ for } r_i > g_i;$$
$$u_i(\mathbf{r}) = -(g_i - r_i)^{1-z} \text{ for } r_i < g_i \text{ and } 1 - z > 1;$$
$$u_i(\mathbf{r}) = -(r_i - g_i)^{1-z} \text{ for } r_i > g_i \text{ and } 1 - z < 0;$$
$$u_i(\mathbf{r}) = +(r_i - g_i)^{1-z} \text{ for } r_i > g_i \text{ and } 0 < 1 - z \le 1;$$

and, of course,

$$u_i(\mathbf{r}) = -e^{-g_i r_i}.$$

Thus existence of group utility and probability functions occurs only under rather special circumstances. ■

Theorem 11.2.4 generalizes Theorem 11.2.3 to the case in which the members are mutually concerned, in the sense that the utility of each is (at least potentially) affected by the net return of the other members, and is of the special form

$$u_i(r_1, \ldots, r_n) = -e^{-\sum_{h=1}^{n} g_{ih} r_h} \tag{11.2.7}$$

discussed in Section 5.7. Readers unfamiliar with linear algebra and/or not too interested in this generalization of Theorem 11.2.3 may skip to the exercises.

*■ *Preliminary notation for Theorem* 11.2.4. Let $\mathbf{x}'$ denote the transpose of (column vector) $\mathbf{x}$; let

$u_i(\mathbf{r}) = -e^{-\mathbf{g}_i'\mathbf{r}}$ for $i = 1, \ldots, n$;

$\mathbf{G} = n$th order matrix of which column i is $\mathbf{g}_i$, for $i = 1, \ldots, n$;

$\mathbf{I} =$ identity matrix; $\mathbf{1} =$ vector of ones;

$d^* = \mathbf{1}'\mathbf{G}^{-1}\mathbf{1} = \mathbf{1}'(\mathbf{G}')^{-1}\mathbf{1}$; $g^* = 1/d^*$;

$\mathbf{M} = d^*\mathbf{G}'$;

$$\boldsymbol{\rho} = \mathbf{M}^{-1}\mathbf{1} = [\mathbf{1}'(\mathbf{G}')^{-1}\mathbf{1}]^{-1}(\mathbf{G}')^{-1}\mathbf{1}; \tag{11.2.8}$$
$$\boldsymbol{\xi} = (\mathbf{M}^{-1})'\mathbf{1} = [\mathbf{1}'\mathbf{G}^{-1}\mathbf{1}]^{-1}\mathbf{G}^{-1}\mathbf{1}; \tag{11.2.9}$$

$\boldsymbol{\Psi} = n$th order matrix each row of which is $\boldsymbol{\xi}'$;

$\mathbf{H} = d^*\mathbf{M}^{-1}[\mathbf{I} - \boldsymbol{\Psi}]$;

$b_{ij} = \log_e(p_{ij})$ and $\mathbf{b}_j = (b_{1j}, \ldots, b_{nj})'$;

$$\boldsymbol{\beta}_j = \mathbf{H}\mathbf{b}_j; \tag{11.2.10}$$

$$u^*(c) = -e^{-g^*c} \text{ for every } c; \tag{11.2.11}$$

and

$$p_j^* = \prod_{h=1}^{n} p_{jh}^{\xi_h} \Big/ \sum_{\ell=1}^{\#(\Sigma)} \left[\prod_{h=1}^{n} p_{h\ell}^{\xi_h}\right]. \tag{11.2.12}$$

For Theorem 11.2.4, we make the following *assumptions.*

(1) Any conceivable sharing rule may be chosen by the group.
(2) Every member i's utility function is of the form (11.2.7).
(3) $p_{ij} > 0$ for every i and j.
(4) $\mathbf{G}$ is nonsingular.
(5) $\mathbf{G}^{-1}\mathbf{1} > \mathbf{0}$ *or* $\mathbf{G}^{-1}\mathbf{1} < \mathbf{0}$. ■

***Theorem 11.2.4.** If the group satisfies Assumptions (1)–(5), then:

(A) $\mathbf{r}$ is a Pareto-optimal sharing rule if and only if there exists $\boldsymbol{\alpha} \in R^n$ such that $\mathbf{1}'\boldsymbol{\alpha} = 0$ and

$$\mathbf{r}(c, \sigma^j) = \boldsymbol{\rho} \cdot c + \boldsymbol{\beta}_j + \boldsymbol{\alpha}$$

for every j and \$$c$, where $\boldsymbol{\rho}$ and $\boldsymbol{\beta}_j$ are as defined by (11.2.8) and (11.2.10) respectively.

(B) External strategy s may be chosen independently of sharing rule, inasmuch as (11.2.11) and (11.2.12) are a group utility function and a group probability function respectively.

■ *Proof.* A lengthy application of Theorem 11.2.2; see LaValle [**79**]. ■

■ The first alternative in Assumption (5), that $\mathbf{G}^{-1}\mathbf{1} > \mathbf{0}$, implies that d^* and hence g^* are positive, and thus from (11.2.11) that the group always prefers more net return to less. On the other hand, $\mathbf{G}^{-1}\mathbf{1} < \mathbf{0}$ implies that $g^* < 0$, and hence that the group prefers *less* income to more! Such a group should self-destruct in short order. ■

■ As in the Theorem 11.2.3 definition, it is easy to verify that $\Sigma_{h=1}^n \xi_h = 1$ and $\xi_h > 0$ for every h, due to Assumption (5). But the proportional shares of group income are now given by $\rho_1, \ldots, \rho_n$; and, while $\Sigma_{h=1}^n \rho_h = 1$, it is not true that all the ρ_h must be positive. ■

■ Even if $\mathbf{G}^{-1}\mathbf{1} > \mathbf{0}$ and the group prefers more income to less, it is possible for some of the g_{ih} to be negative. A negative g_{ih} means that i prefers that h receive *less* income; i.e., i dislikes h. Since we need no assumption that $\mathbf{G}$ is symmetric, we need not suppose that i's antipathy toward h is reciprocated. Exercise 11.2.2 affords you the opportunity to work out some such details for a two-person group. ■

■ Because group utility and probability functions exist for groups satisfying the hypotheses of Theorem 11.2.4, it follows that the group need never resort to randomization. ■

■ It remains to be seen whether multivariate generalizations due to Rothblum [**123**] of the nonexponential classes of utility functions cited above in connection with Wallace's work [**138**] give rise to group utility functions when there is a consensus on probabilities. ■

Exercise 11.2.1

Prove that $\Sigma_{i=1}^{n} \xi_i = 1$, $\xi_i > 0$ for every i, and $\Sigma_{i=1}^{n} \beta_{ij} = 0$ for every j, where ξ_i and β_{ij} are as defined for (A) Theorem 11.2.3 and (B) Theorem 11.2.4, and for this part also prove that $\Sigma_{i=1}^{n} \rho_i = 1$.

Exercise 11.2.2

Let $k = n = 2$, $u_1(\mathbf{r}) = -e^{-ar_1 - br_2}$, and $u_2(\mathbf{r}) = -e^{-cr_1 - dr_2}$.

(A) Find g^*, $\boldsymbol{\rho}$, $\boldsymbol{\xi}$, and $\boldsymbol{\beta}_j$ in terms of a, b, c, and d.

(B) Specialize (A) to the following cases:

 (i) $a = .10$, $b = 0$, $c = 0$, $d = .20$

 (ii) $a = .10$, $b = .05$, $c = .10$, $d = .20$

 (iii) $a = .10$, $b = -.05$, $c = -.10$, $d = .20$

Exercise 11.2.3

Given the Pareto-optimal sharing rules for Joe and Sam in Figure 11.4 and Example 11.2.3(A), plot Sam's reservation price of his share as a function of Joe's reservation price of his (Joe's) share.

Exercise 11.2.4

Recall the Tom and Jerry merger problem of Exercise 11.1.3.

(A) Assuming, as in Exercise 11.1.3, that each has the utility function (11.1.1), find the Pareto-optimal sharing rules for the merged dpuu $\{(s^1, s^3)\}$, and plot Tom's reservation price of his share as a function of Jerry's reservation price of his (Jerry's) share. What is the status quo if no merger takes place?

(B) Redo (A) under the assumption that $u_i(r_T, r_J) = -e^{-g_i r_i}$ for $i \in \{T, J\} = \{\text{Tom}, \text{Jerry}\}$.

Exercise 11.2.5

Redo the pertinent parts of Example 11.2.3 under the altered assumption that Joe's probabilities of σ^1, σ^2, and σ^3 are .40, .50, and .10 respectively. Also plot Sam's reservation price for his share as a function of Joe's reservation price for his (Joe's) share; how does this compare with your answer to Exercise 11.2.3?

Exercise 11.2.6

Assume that $k = n$ and $u_i(\mathbf{r}) = r_i$ for $i = 1, \ldots, n$. Also assume that, for every j and every possible c, every share must lie on the same side of zero as the amount c to be shared. Let $\mathcal{R}$ be this set of sharing rules, let $\#(S) < \infty$ and $\#(\Sigma) < \infty$, and let $\mathcal{Q}'_T$ be as in Theorem 11.2.2, the hypothesis of which is amply satisfied because $\mathcal{R}$ is bounded and the u_i are concave. Let $w_1, \ldots, w_n$ be as in Theorem 11.2.2. Show that $\mathbf{r}$ is Pareto-optimal only if

$$r_i(c, \sigma^j) = \begin{cases} c, & c > 0 \text{ and } w_i p_{ij} = \max\,[w_h p_{hj}\colon h = 1, \ldots, n] \\ c, & c < 0 \text{ and } w_i p_{ij} = \min\,[w_h p_{hj}\colon h = 1, \ldots, n] \\ 0, & \text{otherwise.} \end{cases}$$

[*Hint*: Let $r_{ij} = r_i(c, \sigma^j)$, and try to maximize $\Sigma_{i=1}^{n} \Sigma_{j=1}^{\#(\Sigma)} w_i r_{ij} p_{ij}$ subject to all constraints.]

■ In Exercise 11.2.6, note that each group member i who knows the form of the Pareto-optimal sharing rules has, in any specific context $c[s, \cdot]$, the incentive to set $p_{ij} = 1$ for the j producing the greatest return $c[s, \sigma^j]$! Such misrepresentation of judgments, and similar misrepresentation of preferences, are to be expected and account for the discouraging degree of difficulty in applying the material in this and the following chapter to real problems. ■

11.3 ONE APPROACH TO SELECTING A PARETO-OPTIMAL TOTAL STRATEGY

In Section 11.2 we noted that the group can profitably confine its attention to *Pareto-optimal* total strategies, but we did not address the vexing problem of how to choose one of the generally infinitely many such strategies.

One way of choosing a Pareto-optimal strategy is to have the group apply Theorem 11.2.1 with a vengeance, by undergoing a haggling process which results in the group's choosing relative weights $w_1, \ldots, w_n$ for the members' utilities. Christenson [**19**] furnishes axioms implying this approach, which appears implicit in some of Wilson's analysis [**144**].

This is not the approach we shall examine here. Before proceeding, however, we pause to stress that whenever there are many Pareto-optimal strategies preferable to the status quo, the choice of one over another cannot be unanimous. By the very nature of Pareto-optimality, increasing one member's utility must be at the cost of decreasing at least one other member's; see Figure 11.1(B), for example. Hence any approach to strategy selection which goes beyond simply advising that the chosen strategy be Pareto-optimal must necessarily incorporate some resolution of an internal bargaining process among the members.

With that point clearly understood, we proceed to suppose that the group is willing to confine its internal bargaining to the *simpler* problem of how *sure* group returns $\$x$ should be allocated to the accounts $h = 1, \ldots, k$ rather than flounder around with the actual group decision problem with all its side-betting, sharing-rule, external-strategy, and randomization complexities. Then we shall apply the solution of the simpler problem in constructing a solution to the actual problem.

■ This focussing of decision makers' attentions on ancillary, hypothetical problems is similar in spirit to the methods suggested in Chapters 3, 5, and 6 for quantifying preferences and judgments. ■

Our approach assumes that *the group shall determine how to allocate any sure group return* $x *to the accounts* h = 1, . . . , k *if no side betting is permitted.* That determination can be represented by a function $\mathbf{a}: R^1 \to R^k$, where $\mathbf{a}(x) = (a_1(x), a_2(x), \ldots, a_k(x))$ and $a_h(x)$ is the portion of x allocated to account h, for $h = 1, \ldots, k$. Such a function $\mathbf{a}(\cdot)$ will be called an *allocation* function if it possesses some properties that seem reasonable from the group-decision viewpoint.

Definition 11.3.1: Allocation Functions. A function $\mathbf{a}: R^1 \to R^k$ is an allocation function for a given n-member group with utility functions $u_i: R^k \to R^1$ if

(A) $\Sigma_{h=1}^{k} a_h(x) = x$ for every $x \in R^1$.
(B) $a_h(0) = 0$ for $h = 1, \ldots, k$.
(C) If $x'' > x'$, then $u_i(\mathbf{a}(x'')) \geq u_i(\mathbf{a}(x'))$ for every i, and $u_i(\mathbf{a}(x'')) > u_i(\mathbf{a}(x'))$ for at least one i.
(D) $u_i(\mathbf{a}(\cdot))$ is continuous, for every i.

■ (A) requires that any allocation function must fully allocate any group income to the accounts. (B) says that, since side betting is not permitted by an allocation function, such a function leaves the status quo intact if there is no group income. (C) requires that the allocation function be compatible with the members' preferences in the sense that every member should be just as happy with the group's having more income, x'', than less, x', to allocate; and someone should in fact be happier with more rather than less. (D) is a reasonable continuity condition, easily satisfied if every $u_i(\cdot)$ and every $a_h(\cdot)$ is continuous. ■

Example 11.3.1: Proportional Shares (Stockownership)

Suppose, in the simple risk-sharing context, that $k = n$ and every member i's utility function depends solely upon his own share of group income. If $u_i: R^1 \to R^1$ is strictly increasing and continuous, and if $a_i(x) = a_i \cdot x$ for $i = 1, \ldots, n$, where $a_1, \ldots, a_n$ are nonnegative real numbers summing to one, then $\mathbf{a}(\cdot) = (a_1(\cdot), \ldots, a_n(\cdot))$ is an allocation function, satisfying (A)–(D) of Definition 11.3.1—as you may verify in Exercise 11.3.1. Such an allocation function is naturally called a *propor-*

tional-shares allocation function. It corresponds to small partnerships, corporations, and syndicates in which the members have potentially differing proportional ownerships and hence differing proportional claims to the fruits of the group endeavors.

The allocation function determined by the group gives rise naturally to a *trajectory* $\mathbf{u}(\mathbf{a}(\cdot)) = \{(u_1(\mathbf{a}(x)), \ldots, u_n(\mathbf{a}(x))): x \in R^1\}$ in the n-space of the members' conceivable utility positions. If, as we shall assume, this trajectory cuts through the efficient (i.e., northeast) boundary of $X(\mathcal{Q}'_T, \mathbf{c})$, we declare the (Pareto-optimal) strategy corresponding to the intersection point to be the *allocation-function-determined solution*, or $\mathbf{a}(\cdot)$-*solution*, to the group decision problem. The point of intersection of the trajectory $\mathbf{u}(\mathbf{a}(\cdot))$ with the efficient boundary of $X(\mathcal{Q}'_T, \mathbf{c})$ will be denoted by $\mathbf{U}^{\mathbf{a}}(\mathcal{Q}'_T, \mathbf{c})$. See Figure 11.6.

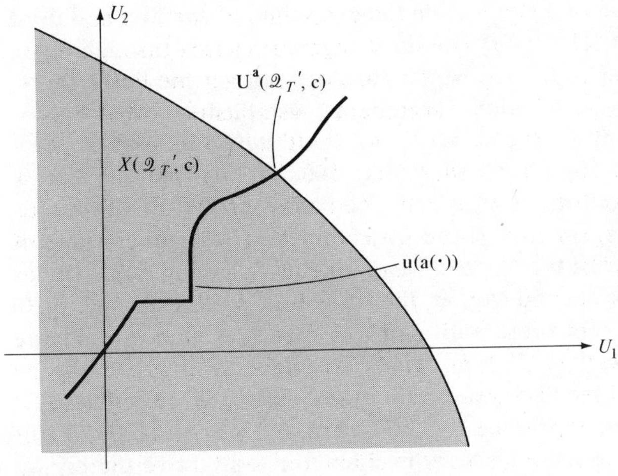

Figure 11.6 Sketch of the $\mathbf{a}(\cdot)$—solution concept.

In less formal terminology, we are counseling the group to find a Pareto-optimal strategy that will give *every* member of the group the same satisfaction (i.e., utility) as he would reap from the group's having some sure return \$$x$ to allocate via the already-bargained-for allocation function. In this way the group uses the simpler context of choosing an allocation function for sure, non-side-bettable returns \$$x$ to solve its actual decision problem.

Example 11.3.2: Joe and Sam

Suppose that Joe and Sam haggle for awhile and decide on a proportional-shares allocation function in which $a_1(x) = a \cdot x$ and $a_2(x) =$

$(1-a)\cdot x$, where $a_1(x)$ and $a_2(x)$ are Joe's and Sam's shares of x respectively, and $0<a<1$. We derive the trajectory $\mathbf{u}(\mathbf{a}(\cdot))$ by examining $\mathbf{a}(\cdot)$ in conjunction with (11.1.1). First, suppose that Joe's proportion a does not exceed ½. There are three cases to consider. *Case 1*: $x \le -100/a$. Then $ax \le -100$ and $(1-a)x \le -100$, so that $u_1(a\cdot x) = ax+80$, or $x=(u_1-80)/a$; and $u_2((1-a)\cdot x) = (1-a)x+80 = [(1-a)/a](u_1-80)+80$, for all $u_1 \le -20$. *Case 2*: $-100/a \le x \le -100/(1-a)$. Then $ax \ge -100$ but $(1-a)x \le -100$, so that $u_1(ax) = ax/5$, or $x = 5u_1/a$, and $u_2((1-a)a) = (1-a)x+80 = 5[(1-a)/a]u_1+80$, for all u_1 such that $-20 \le u_1 \le -20a/(1-a)$—corresponding to $-100/a \le x \le -100/(1-a)$. *Case 3*: $x \ge -100/(1-a)$. Then $ax \ge -100$ and $(1-a)x \ge -100$, so that $u_1(ax) = ax/5$, or $x=5u_1/a$, and $u_2((1-a)x) = (1-a)x/5 = [(1-a)/a]u_1$, for all $u_1 \ge -20a/(1-a)$. Thus $\mathbf{u}(\mathbf{a}(\cdot))$ is composed of three connected line segments, the first going down to the southwest from $(u_1, u_2) = (-20, 80-100(1-a)/a)$ with slope $(1-a)/a$, the third going up to the northeast from $(-20a/(1-a), -20)$ with the same slope $(1-a)/a$ as the first, and the second connecting the endpoints of the first and third segments, with slope $5(1-a)/a$. The third segment passes through $(0,0)$. All this is for the case $a \le$ ½, in which Joe does not get the lion's share. Now, if $a \ge$ ½, a similar argument establishes that $u_2 = [(1-a)/a](u_1-80)+80$ for all $u_1 \le -100a/(1-a)+80$, $u_2 = [(1-a)/(5a)](u_1-80)$ for all u_1 such that $100a/(1-a)+80 \le u_1 \le -20$, and $u_2 = [(1-a)/a]u_1$ for all $u_1 \ge -20$. You may prove all of this as Exercise 11.3.2. Thus, for $a \ge$ ½, the trajectory $\mathbf{u}(\mathbf{a}(\cdot))$ again consists of three line segments with the "lowest" and "highest" having equal slopes and connected by the second. Again, the trajectory passes through $(0,0)$ and goes northeast from there with slope $(1-a)/a$. A glance at Figure 11.4 suffices to show that $\mathbf{U}^{\mathbf{a}}(T, \mathbf{c})$ exists, is unique, and can easily be found once a is specified. The reasoning is as follows. Since segment III of Figure 11.4 can be expressed by the equation $U_2 = -1.6U_1+28$ (for $0 \le U_1 \le 30$), and since the trajectory is characterized (above the origin) by $U_2 = [(1-a)/a]U_1$, we solve for U_1 to find $U_1 = 28a/(1+0.6a)$. But on segment III Joe's utility of his share $(y_1, 100, -100)$ is given by the formulae in Table B for region J1; namely, $U_1 = .05y_1 + .10(100) + .05(-100) = .05y_1+5$. Equating $.05y_1+5$ with $28a/(1+0.6a)$ and solving for y_1 yields $y_1 = (500a-100)/(1+0.6a)$, for $0 \le a \le 1$. Thus we have found Joe's (and hence also Sam's) Pareto-optimal shares of the group income $c[s^1, \tilde{\sigma}]$ as functions of Joe's proportional allocation a under the agreed-upon allocation function.

We shall now show that, even though group utility and probability functions need not exist, a group using the $\mathbf{a}(\cdot)$-solution can still behave with consistency vis-à-vis the external world, in the sense that it has a real-valued measure of the relative desirabilities to the group of various opportunities presented by the external world. The content of the following analysis is very similar to that in Sections 8.2 and 8.3.

Definition 11.3.2: Opportunities et al

(A) An *opportunity* is a pair $(\mathcal{Q}'_T, \mathbf{c}')$ consisting of (1) a nonempty and closed set $\mathcal{Q}'_T$ of total strategies for the group; and (2) a monetary-return function $\mathbf{c}' = c': S \times \Sigma \to R^1$ specifying the group's net monetary return given s and σ.

(B) $\mathbf{U^a}(\mathcal{Q}'_T, \mathbf{c}')$ denotes the utility position in $X(\mathcal{Q}'_T, \mathbf{c}')$ of the $\mathbf{a}(\cdot)$-solution for the opportunity $(\mathcal{Q}'_T, \mathbf{c}')$.

(C) We denote by $\mathbf{c}' + x$ the return function $\mathbf{c}''$ defined, for every s and σ, by $c''[s, \sigma] = c'[s, \sigma] + x$.

We shall impose some seemingly stringent but actually not unduly restrictive assumptions about the behavior of $\mathbf{U^a}(\mathcal{Q}'_T, \mathbf{c}' + x)$ as a function of x. For every opportunity $(\mathcal{Q}'_T, \mathbf{c}')$ under consideration, we assume that:

(1) For every $x \in R^1$, $\mathbf{U^a}(\mathcal{Q}'_T, \mathbf{c}' + x)$ exists and is unique.

(2) The function $\mathbf{U^a}(\mathcal{Q}'_T, \mathbf{c}' + x)$ on $(x \in)$ R^1 to R^n is
   (a) continuous and
   (b) strictly increasing, in the sense that if $x'' > x'$, then $\mathbf{U^a}(\mathcal{Q}'_T, \mathbf{c}' + x'') \geq \mathbf{U^a}(\mathcal{Q}'_T, \mathbf{c}' + x')$ but $\mathbf{U^a}(\mathcal{Q}'_T, \mathbf{c}' + x'') \neq \mathbf{U^a}(\mathcal{Q}'_T, \mathbf{c}' + x')$.

(3) The function $\mathbf{U^a}(\mathcal{Q}'_T, \mathbf{c}' + x)$ on $(x \in)$ R^1 to R^n coincides with the trajectory $\mathbf{u}(\mathbf{a}(\cdot))$.

■ The uniqueness part of (1) is an immediate consequence of the fact that $\mathbf{u}(\mathbf{a}(\cdot))$ is increasing [Definition 11.3.1(C)], whereas the efficient boundary of $X(\mathcal{Q}'_T, \mathbf{c}' + x)$ is decreasing in the sense of being composed of tangents to hyperplanes with nonnegative outward normals $\mathbf{w}$. The existence part of (1) and the other assumptions are primarily consequences of having a sufficient diversity of preferences within the group and a richness of available sharing rules. It can be shown that, in the simple risk-sharing context with $k = n$ and every member's utility function u_i depending solely upon his own share, if every u_i is strictly increasing, continuous, and concave, if every $p_{ij} > 0$, and if any conceivable sharing rule can be coupled with any external strategy s, then all assumptions are satisfied. ■

Before proceeding further, we pause to make a fundamental observation: A group committed to the $\mathbf{a}(\cdot)$-solution concept will evince unanimity of preference between any two opportunities $(\mathcal{Q}'_T, \mathbf{c}')$ and $(\mathcal{Q}''_T, \mathbf{c}'')$. To see that this is so, just compare the utility positions $\mathbf{U^a}(\mathcal{Q}'_T, \mathbf{c}')$ and $\mathbf{U^a}(\mathcal{Q}''_T, \mathbf{c}'')$. Because every $\mathbf{U^a}$ lies on the trajectory $\mathbf{u}(\mathbf{a}(\cdot))$, it follows immediately from Definition 11.3.1(C) ("the trajectory goes north and/or east") that one of three conditions must obtain. Either

$$\mathbf{U^a}(\mathcal{Q}'_T, \mathbf{c}') \geq \& \neq \mathbf{U^a}(\mathcal{Q}''_T, \mathbf{c}''), \tag{11.3.1a}$$

in which case the group prefers $(\mathcal{Q}'_T, \mathbf{c}')$ to $(\mathcal{Q}''_T, \mathbf{c}'')$; or

$$\mathbf{U^a}(\mathcal{Q}'_T, \mathbf{c}') \leq \& \neq \mathbf{U^a}(\mathcal{Q}''_T, \mathbf{c}''), \tag{11.3.1b}$$

in which case the group prefers $(\mathcal{Q}''_T, \mathbf{c}'')$ to $(\mathcal{Q}'_T, \mathbf{c}')$; or

$$\mathbf{U}^{\mathbf{a}}(\mathcal{Q}'_T, \mathbf{c}') = \mathbf{U}^{\mathbf{a}}(\mathcal{Q}''_T, \mathbf{c}''), \tag{11.3.1c}$$

in which case the group is completely indifferent between $(\mathcal{Q}'_T, \mathbf{c}')$ and $(\mathcal{Q}''_T, \mathbf{c}'')$.

■ The symbols "$\geq \& \neq$" and "$\leq \& \neq$" mean exactly what they seem to mean. For real numbers, $a \geq \& \neq b$ if and only if $a > b$, but for vectors, $\mathbf{a} \geq \& \neq \mathbf{b}$ does not imply that $\mathbf{a} > \mathbf{b}$. For example, $(2, 1) \geq \& \neq (1, 1)$ but $(2, 1) \not> (1, 1)$, since $\mathbf{a} > \mathbf{b}$ if and only if $a_i > b_i$ for all i. So $\mathbf{a} \geq \& \neq \mathbf{b}$ if $a_i \geq b_i$ for all i and $a_i > b_i$ for some i, whereas $\mathbf{a} \leq \& \neq \mathbf{b}$ if $a_i \leq b_i$ for all i and $a_i < b_i$ for some i. ■

Now suppose that the group currently possesses an opportunity $(\mathcal{Q}'_T, \mathbf{c}')$ but could exchange it for another, $(\mathcal{Q}''_T, \mathbf{c}'')$, at a cost of \$$x$. Should it make the exchange? If it does not, then it will implement the $\mathbf{a}(\cdot)$-solution to $(\mathcal{Q}'_T, \mathbf{c}')$ and the members' utility position will be $\mathbf{U}^{\mathbf{a}}(\mathcal{Q}'_T, \mathbf{c}')$. But if it does make the exchange, then the effective opportunity that it will possess will be $(\mathcal{Q}''_T, \mathbf{c}'' - x)$, and by implementing the $\mathbf{a}(\cdot)$-solution the members' utility position will be $\mathbf{U}^{\mathbf{a}}(\mathcal{Q}''_T, \mathbf{c}'' - x)$. Clearly, the group *should exchange at cost* \$$x$ if

$$\mathbf{U}^{\mathbf{a}}(\mathcal{Q}''_T, \mathbf{c}'' - x) \leq \& \neq \mathbf{U}^{\mathbf{a}}(\mathcal{Q}'_T, \mathbf{c}'), \tag{11.3.2a}$$

whereas the group *should not exchange at cost* \$$x$ if

$$\mathbf{U}^{\mathbf{a}}(\mathcal{Q}''_T, \mathbf{c}'' - x) \leq \& \neq \mathbf{U}^{\mathbf{a}}(\mathcal{Q}'_T, \mathbf{c}'). \tag{11.3.2b}$$

If the two sides of (11.3.2) are equal, then the group should be completely indifferent between exchanging and not exchanging at cost \$$x$. Moreover, by Assumptions (2b) and (3), there exists a *maximum justifiable* cost, x^*, of exchanging $(\mathcal{Q}'_T, \mathbf{c}')$ for $(\mathcal{Q}''_T, \mathbf{c}'')$. Like its counterpart in Chapter 8, we call x^* the *exchange value* and denote it by $V(\mathcal{Q}'_T, \mathbf{c}' \to \mathcal{Q}''_T, \mathbf{c}'')$ for explicitness.

Definition 11.3.3. The *value* $V(\mathcal{Q}'_T, \mathbf{c}' \to \mathcal{Q}''_T, \mathbf{c}'')$ *of exchanging* $(\mathcal{Q}'_T, \mathbf{c}')$ *for* $(\mathcal{Q}''_T, \mathbf{c}'')$ is defined implicitly by

$$\mathbf{U}^{\mathbf{a}}(\mathcal{Q}''_T, \mathbf{c}'' - V(\mathcal{Q}'_T, \mathbf{c}' \to \mathcal{Q}''_T, \mathbf{c}'')) = \mathbf{U}^{\mathbf{a}}(\mathcal{Q}'_T, \mathbf{c}').$$

Compare Definition 11.3.3 with Definition 8.3.1. The only difference is that here the utilities are vectorial, one component for each member. Exchange value in the group-decision context plays exactly the same role, and possesses the same qualitative properties, as it does in the individual-decision context, and hence the discussion in Chapter 8 need not be duplicated here. The relatives and variants of exchange value in Chapter 8 can also be defined in the group context, provided that care is taken to define the status quo, which need not be zero if the group has the opportunity to make side bets. (See Example 11.2.1.) The interested reader may make the translation from $\mathcal{U}$ to $\mathbf{U}^{\mathbf{a}}$ of all material in Section 8.3 and ascertain the extent to which the definitions and assertions there carry over for a group under the $\mathbf{a}(\cdot)$-solution concept. Although interesting, such results should not obscure the main point; viz., that under the $\mathbf{a}(\cdot)$-solution concept the group can rank all opportunities according to how far up the trajectory $\mathbf{u}(\mathbf{a}(\cdot))$ their utility positions $\mathbf{U}^{\mathbf{a}}$ lie.

We conclude the section with a derivation of some explicit results for groups satisfying the hypotheses of Theorem 11.2.4. Recall from the final remark in Section 11.2 that if any sharing rule (satisfying Definition 11.2.1, of course) can be coupled with any external strategy s, then existence of group utility and probability functions implies that the group need never resort to randomizing its choice of total strategy $t = (s, \mathbf{r})$. This means that, unless outside constraints require randomization, attention can be confined to opportunities of the form $T' = S' \times \mathcal{R}$, where $\mathcal{R}$ is the set of all conceivable sharing rules. But even greater simplification is possible: Because of the group utility and probability functions, the efficient (i.e., northeast) boundary of $X(T', \mathbf{c}')$ *coincides with* the efficient boundary of $X[\{s^\circ\} \times \mathcal{R}, \mathbf{c}')$, where s° is any external strategy optimal in S' under the group utility and probability functions. This fact justifies our restricting attention in the Theorem 11.2.4 context to opportunities in which $T = \{s\} \times \mathcal{R}$, and which therefore differ only in the group's monetary net-return function.

Theorem 11.3.1

(A) If $T = \{s\} \times \mathcal{R}$, where $\mathcal{R}$ is the set of all conceivable sharing rules, then $\mathbf{r}$ is the $\mathbf{a}(\cdot)$-solution Pareto-optimal sharing rule for a group satisfying the hypotheses of Theorem 11.2.4 if and only if

$$\mathbf{r}(c, \sigma^j) = \boldsymbol{\rho} \cdot c + \boldsymbol{\beta}_j + \mathbf{a}(\mathbf{1}'(\mathbf{G}')^{-1}\mathbf{z}) - (\mathbf{G}')^{-1}\mathbf{z},$$

where $\mathbf{z} = (z_1, \ldots, z_n)'$ and[e]

$$z_i = -\log_e\left[\sum_{j=1}^{\#(\Sigma)} \exp\left(-\mathbf{g}_i'[\boldsymbol{\rho} \cdot c[s, \sigma^j] + \boldsymbol{\beta}_j]\right)p_{ij}\right]$$

for $i = 1, \ldots, n$.

(B) Let $\mathbf{z}^1$ and $\mathbf{z}^2$ be the $\mathbf{z}$-vectors as defined in (A) for the opportunities $T^1 = \{s^1\} \times \mathcal{R}$ and $T^2 = \{s^2\} \times \mathcal{R}$ respectively. Under the hypotheses of (A), the group's value of exchanging T^1 for T^2 is $\mathbf{1}'(\mathbf{G}')^{-1}(\mathbf{z}^2 - \mathbf{z}^1)$.

■ *Proof*: *Of* (A). By Theorem 11.2.4, every Pareto-optimal sharing rule is of the form

$$r_h(c, \sigma^j) = \rho_h \cdot c + \beta_{hj} + \alpha_h$$

for some $\alpha_1, \ldots, \alpha_n$ which sum to zero. Member i's expected utility of $c[s, \tilde{\sigma}]$ under such a sharing rule is easily seen to be given by

$$\sum_{j=1}^{\#(\Sigma)}\left[-\exp\left(-\sum_{h=1}^{n} g_{ih}[\rho_h \cdot c[s, \sigma^j] + \beta_{hj} + \alpha_h]\right)\right]p_{ij}$$

$$= \left[-\exp\left(-\sum_{h=1}^{n} g_{ih}\alpha_h\right)\right] \cdot \sum_{j=1}^{\#(\Sigma)} \exp\left(-\sum_{h=1}^{n} g_{ih}[\rho_h \cdot c[s, \sigma^j] + \beta_{hj}]\right)p_{ij}$$

$$= [-\exp(-\mathbf{g}_i'\boldsymbol{\alpha})] \cdot \exp(-z_i), \tag{*}$$

---

[e] $\exp(x) = e^x$ for every real number x.

given the definition of z_i in (A). But member i's utility of the group's having \$$x$ to allocate according to $\mathbf{a}(\cdot)$ is

$$u_i(\mathbf{a}(x)) = -\exp(-\mathbf{g}_i'\mathbf{a}(x)). \tag{**}$$

At the $\mathbf{a}(\cdot)$-solution, (*) and (**) must be equal for every i and some x, and hence so must be the natural logs of the negatives of these expressions; that is,

$$-z_i - \mathbf{g}_i'\boldsymbol{\alpha} = -\mathbf{g}_i'\mathbf{a}(x),$$

implying readily that $\mathbf{G}'\boldsymbol{\alpha} = \mathbf{G}'\mathbf{a}(x) - \mathbf{z}$, or equivalently, that

$$\boldsymbol{\alpha} = \mathbf{a}(x) - (\mathbf{G}')^{-1}\mathbf{z}, \tag{†}$$

for some appropriate x. But the components of $\boldsymbol{\alpha}$ sum to zero; that is,

$$0 = \mathbf{1}'\boldsymbol{\alpha} = \mathbf{1}'\mathbf{a}(x) - \mathbf{1}'(\mathbf{G}')^{-1}\mathbf{z},$$

whereas $\mathbf{1}'\mathbf{a}(x) = x$ by Definition 11.3.1(A). Hence it follows that

$$x = \mathbf{1}'(\mathbf{G}')^{-1}\mathbf{z}, \tag{††}$$

which, when substituted back into (†), yields part (A).

Of (B). In addition to $\mathbf{z}^1$ and $\mathbf{z}^2$, define $\mathbf{z}^y$ as in part (A) for the opportunity whose net return function is $c[s^2, \cdot] - y$. You may verify as Exercise 11.3.3 that

$$z_i^y = z_i^2 - \mathbf{g}_i'\boldsymbol{\rho} \cdot y,$$

and hence that

$$\mathbf{z}^y = \mathbf{z}^2 - \mathbf{G}'\boldsymbol{\rho}y. \tag{†*}$$

Now, (††) gives the reservation price of an opportunity vis-à-vis the status quo; and, by virtue of Definition 11.3.3, the group's exchange value is that value of y such that $c[s^2, \cdot] - y$ and $c[s^1, \cdot]$ are equally desirable and have the same reservation prices. Hence (†*) implies

$$\begin{aligned}\mathbf{1}'(\mathbf{G}')^{-1}\mathbf{z}^1 &= \mathbf{1}'(\mathbf{G}')^{-1}\mathbf{z}^y = \mathbf{1}'(\mathbf{G}')^{-1}(\mathbf{z}^2 - \mathbf{G}'\boldsymbol{\rho}y)\\ &= \mathbf{1}'(\mathbf{G}')^{-1}\mathbf{z}^2 - \mathbf{1}'\boldsymbol{\rho}y = \mathbf{1}'(\mathbf{G}')^{-1}\mathbf{z}^2 - y,\end{aligned}$$

the final equality being immediate from (11.2.8) and the penultimate equality following from elementary linear algebra. Hence

$$y = \mathbf{1}'(\mathbf{G}')^{-1}\mathbf{z}^2 - \mathbf{1}'(\mathbf{G}')^{-1}\mathbf{z}^1,$$

from which (B) is immediate. ■

Exercise 11.3.1: Proportional-Shares Allocation Functions

Prove that if for every i, member i's utility function depends solely on his own share; $u_i(\cdot)$ is strictly increasing and continuous; $a_i(x) = a_i \cdot x$ for all x; $a_1, \ldots, a_n$ are nonnegative and sum to one; and $\mathbf{a}(\cdot) = (a_1(\cdot), \ldots, a_n(\cdot))$, then $\mathbf{a}(\cdot)$ satisfies (A)–(D) of Definition 11.3.1 and is therefore an allocation function.

Exercise 11.3.2: Joe and Sam

(A) Derive the asserted form of the trajectory $\mathbf{u}(\mathbf{a}(\cdot))$ in Example 11.3.2 for the case $a \geq 1/2$.

(B) Graph the trajectory $\mathbf{u}(\mathbf{a}(\cdot))$ corresponding to (i) $a = 1/4$, (ii) $a = 1/2$, (iii) $a = 3/4$.

(C) Show that the group's reservation price, defined as the value of exchanging the status quo, zero point with side bets forbidden, as a function of a is $140/(1 + 0.6a)$. Why is this small when Joe's proportional share under the allocation function is large?

(D) Suppose as in Example 11.2.4 that Joe and Sam were restricted to proportional sharing rules plus antes, in which context the dashed line in Figure 11.4 is the efficient boundary representing mutually acceptable sharing rules of the form in which Sam pays Joe $200p - 100$ for the proportion $1 - p$ of the proceeds of s^1, where $.40 \leq p \leq 9/14$. In the context of Example 11.3.2, find p and $200p - 100$ as functions of Joe's proportional share a under the allocation function $\mathbf{a}(x) = (ax, [1 - a]x)$.

(E) In the context of part (D), find the group reservation price as a function of a.

Exercise 11.3.3

In the proof of Theorem 11.3.1(B), show that

$$z_i^y = z_i^2 - \mathbf{g}_i'\mathbf{p} \cdot y.$$

11.4 COLLECTIVE CHOICE IN GENERAL

Our focus in this chapter has been on a very special type of problem concerning how individuals with varying preferences and judgments can form a group, arbitrate their differences in some reasonable fashion, and thereby improve their individual situations over what would obtain if the opportunity to "collectivize" had not arisen.

■ Our problem is very special in several respects, among which are the assumption that consequences are all describable entirely in terms of monetary return, and the absence of explicit consideration of options open to the individuals if one or more decide to reject the ultimate group decision. ■

The *general* problem of collective choice has as long a history as man has been a political animal, but only in recent years has it been formulated in such a way as to underscore its inherent complexity.

Take the problem of collective choice to be that of specifying how *any* set of individuals' preference orderings (of some set of options)—one for each member of the group—can be reasonably synthesized into a *group* preference ordering (of this same set of options). A mechanism for synthesizing the individuals' preference orderings is called a *social welfare function.* Clearly, there can be bad social welfare functions as well as good ones: suppose, for example, that all individuals prefer option x to option y, whereas the social welfare function declares the group to prefer y to x!

Hence the primary task of the collective choice theorist would seem to be the definition of "good" social welfare functions—as those which possess certain a priori desirable properties. A recent excellent exposition of this endeavor is that by Sen [**129**], who summarizes much of the work initiated and stimulated by Arrow [**1**], who in 1951 exploded a bombshell.

Arrow stated some seemingly very reasonable properties for any social welfare function, and then proved that no social welfare function satisfying all of them could exist!

■ The key properties are as follows. (See Sen [**129**].)

(1) The social welfare function must yield a group preference ordering for *every conceivable* set of individual members' preference orderings—so that disparate interests of the individual members could not prevent existence of a group preference ordering. ["Unrestricted Domain"]

(2) If everyone prefers x to y, then the social welfare function must have the group preferring x to y as well. ["Weak Pareto Principle"]

(3) The group preference ordering of a set of alternatives, as dictated by the social welfare function, must depend only upon the members' preference orderings over *that* set of alternatives and not upon their preferences over that set together with others. For example, group preference between x and y must depend only upon the members' preferences between x and y and not upon where they individually place some third option z in relation to x and y. ["Independence of Irrelevant Alternatives"]

(4) The social welfare function should be nondictatorial, in the sense that there should not be any member i with the property that the group prefers x to y whenever i prefers x to y, irrespective of the preferences of the other members. ["Nondictatorship"] ■

It may be hard to believe that even the "majority-rule" social welfare function fails to possess reasonable properties, but its limitations have been known long before Arrow's work. In fact, group preferences determined by majority vote may be intransitive!

■ To see this, suppose that there are three individuals—1, 2, and 3—and three options—x, y, and z. Suppose that 1 prefers x to y and y to z; that 2 prefers y to z and z to x; that 3 prefers z to x and x to y; and that each individual is transitive, with 1 preferring x to z, 2 preferring y to x, and 3 preferring z to y. Group preferences determined by majority vote are: x preferred to y by two to one, y preferred to z by two to one—and z preferred to x by two to one! ■

And this frustrating note appears to be a good one on which to close this first of two diffident chapters. Interested readers may consult Sen [**129**] for further information on the collective choice problem.

12

A GLIMPSE AT GAME THEORY

> You philosophers work only on paper, which suffers all, while I, a poor empress, I work on human skins, which are considerably more irritable and ticklish.
>
> Catherine the Great, to Diderot

A: INTRODUCTION

12.1 WHAT IS GAME THEORY?

Game theory is most productively (but modestly) regarded as the study of the implications of various definitions of rational behavior in several-person choice situations. One of the desired outcomes of reading this chapter should be a skepticism about pat characterizations of certain forms of behavior as irrational.

We shall approach game theory in much the same way as a businessman or a statesman approaches theoretical economics. The businessman typically views theoretical economics as a body of general statements about tendencies, trends, cause-effect relationships, etc., that should be kept in mind when grappling with his decisions. The statesman typically views theoretical economics as furnishing indications about how the system itself can or should be changed.

Both these viewpoints are appropriate with respect to game theory. For purposes of individual decision making by a player in a game, the theory does not prescribe action in the same unambiguous sense that we have studied in, say, Parts I–III of this book. Rather, game theory furnishes a *language* and some *concepts* within which the interacting strategic choices of different persons may be examined. This language, and these concepts, should be used by each of the given individuals concerned as judgmental inputs for his individual decision problem of deciding on a strategy.

From an overall policy viewpoint, game theory furnishes insights into collusive or competitive behavior in interpersonal choice situations, insights which can be useful as guidelines to redesigning the incentives in these situations so as to produce better social behavior (according to given criteria).

If these prefatory comments seem less positive and less specific than usual, it is because game theory itself bears the heavy burden of speaking to, or for, all the *players*, or *persons*, under consideration in the situation. Almost immediately, three problems arise from this breadth of focus.

First, and least serious, is the problem of how to structure the decision situation logically; two players may each have to choose an act in ignorance of the other's choice, and hence the Chapter 2 guideline about drawing decision trees from $\mathscr{D}$'s viewpoint breaks down when there is more than one $\mathscr{D}$. But this is easily remedied, in Section 12.2, by supplementing our previous work on tree construction.

Second, and very serious, is a pervasive need to make sweeping assumptions about how the players react to each other and to the ambiguities of the strategic situation. In Section 12.3 we examine two approaches, the first being one in which each player might take a rather conservative viewpoint that the others are out to get him. But we see that if everyone takes this viewpoint, then one or more of the players may have an incentive to do something other than be conservative, and hence "everybody being conservative" should not be anticipated—except perhaps in the case of complete and total competition discussed in Section 12.4. The second approach is to seek a strategy for each player with the property that the resulting collection of strategies gives none of the players an incentive to choose something other than his member of that collection. The trouble with this approach is that there may be *many* such collections, and a given player may hence be left awash on the sea of ambiguity after all.

Third, and also very serious from the standpoint of ready applicability, is the *complete-information* assumption: that every player knows the structure of the situation and also every other player's utility function and judgmental probabilities on $\mathscr{E}$-moves. Although Harsanyi [**52**] has made a major contribution toward weakening this assumption, his approach is difficult to apply in complex situations. The vast preponderance of game theory at present relies on the complete-information assumption, which is an undeniable drawback of the first magnitude to prescriptive applicability by nonomniscient players. Nevertheless, we shall assume in this chapter that all situations under discussion do satisfy the complete-information assumption, which is rendered more specific in Section 12.2.

Before previewing the sections of this chapter it is necessary to develop some terminology. A *game* is a situation in which there are at least two *players*, or *persons*, each as worthy of attention and counsel as was our heretofore ubiquitous $\mathscr{D}$, and in which $\mathscr{E}$ may be involved as well. We typically denote the players by i, for $i = 1, \ldots, n$, and we use $\mathscr{N}$ to denote the *set* $\{1, \ldots, n\}$ of *all* players. The term *participant* refers to either $\mathscr{E}$ or some player i. A game is said to be *noncooperative* if interplayer collusion is forbidden. More specifically, we take a noncooperative game to be one in which players cannot even communicate with each other, much less decide on joint courses of action. The other extreme is to suppose that communication is completely unrestricted and that the players can enter into any mutually binding agreements they choose to; games "played" under such conditions are called *cooperative.*

There is a big difference between $n = 2$ players and games with $n > 2$ players, since in the latter case a proper subset $\mathscr{C}$ of the set $\mathscr{N}$ of all players can, under the cooperative assumption, get together and form a *coalition* to further their own interests. We denote the typical coalition by $\mathscr{C}$.

Finally, some approaches to cooperative games make an assumption to the effect that *utility is unrestrictedly transferable* from any player to any other player, *with conservation*: the utility one gives up equals the utility the other gains. This assumption is most plausible in the case where consequences are entirely monetary and every player's utility function depends only upon his own net monetary return and is expressible by that monetary return itself. A shorthand description of this assumption is to say that "side payments are permitted."

■ The sharing rules and allocation functions discussed in Chapter 11 constitute a much less crude sort of "side payments" assumption. There, we did not need to fix unessential parameters of the group members' utility functions; but here it is clear that changing the scale of a player's utility function completely destroys the "conservation" of utility in any side payments to which he is party. ■

The effect of assuming that side payments can or cannot be made is great or small, depending upon the particular game context. The question is irrelevant in noncooperative games, because the players have to communicate in order to arrange side payments. In cooperative games with $n = 2$ players, there is some effect on the set of possible utility-vector outcomes of the game. But when $n > 2$ there is a great effect on the way in which we represent what is attainable by the various possible coalitions.

Now for the preview. In Section 12.2 we discuss games in extensive form, games in normal form, and how to reexpress in normal form a game given in extensive form. Part B of this chapter, Sections 12.3 and 12.4, concerns noncooperative games. Section 12.3 introduces the two distinct and less than completely satisfactory approaches to players' behavior in such games that we alluded to earlier in this introduction. Section 12.4 concerns the classic case, the $n = 2$ person zerosum games, in which the two approaches to players' behavior coincide and in which game theory has a great deal to say about how each player *should* behave, given that each player knows the other is intelligent.

Part C concerns cooperative games, in which communication and collusion are completely unrestricted. Section 12.5 concerns two-person cooperative games, about which many interesting things can be said. It is, however, only a sketchy introduction. Section 12.6 provides a glimpse at cooperative games with more than two players by sketching the theory of games in characteristic-function form with side payments; here, the emphasis is on the role of coalitions.

Since this chapter can make no pretensions about exhaustiveness even as an introduction, we hereby advise the interested reader that there are several very stimulating books on this subject, among which are Rapoport's accessible and zestful two-volume work [**117**], the classic treatise [**106**] by von Neumann and Morgenstern, Owen [**108**], and Luce and Raiffa [**92**]. These books contain many references to work we do not cite here. See also the Princeton volumes [**26**], [**27**], [**75**], and [**135**], the Morgenstern *Festschrift* [**131**] edited by Shubik, and current and past volumes of the *International Journal of Game Theory*, the *Journal of Conflict Resolution*, and *Behavioral Science.*

12.2 GAMES IN EXTENSIVE AND NORMAL FORM

In Chapter 2 we introduced a basic guideline for constructing decision trees: sequence things in the order in which $\mathscr{D}$ will experience them. This guideline went unchallenged in Chapter 11, because although we had several interested individuals as group members, it was the group as a whole that had to choose an act in each $\mathscr{D}$-move; in other words, each $\mathscr{D}$-move was "owned by" the group as a whole.

But in this chapter the non-$\mathscr{E}$-moves are owned by different individuals each warranting separate consideration, and this means that, unfortunately, the Chapter 2 basic guideline for arborculture must be supplemented. The basic difficulty arises from having to express a participant's partial or total *ignorance* about the "preceding" choices of other participants. For instance, if each of players 1 and 2 must independently choose an act in ignorance of the other's choice, then the Chapter 2 rule says that 1's move should precede copies of 2's move,[a] and also that 2's move should precede copies of 1's move![b]

Example 12.2.1: "Throwing Fingers"

Suppose that each of two players must decide whether to "throw" one or two fingers, that the players will throw fingers *simultaneously*, that player 1 will win \$10 from player 2 if an even total number of fingers is thrown and 2 wins \$10 from 1 if the total is odd, and that $u_i(c_i, c_2) = c_i$ for all (c_1, c_2) in R^2 and $i = 1, 2$, where c_i denotes i's winning in the game. This game is structurally familiar to almost everyone and bears poignant witness to the need to supplement established arborculture so as to accommodate effective simultaneity of choices in games.

Other than on the problem of appropriate sequencing and the representation of ignorance, little need be said concerning game trees. A *game tree*

[a] So as to represent 1's ignorance of 2's choice.

[b] So as to represent 2's ignorance of 1's choice.

is a tree in which

(1) Each move is owned by one of the participants, namely, $\mathscr{E}$ and the players $1, \ldots, n$.
(2) Each endpoint has an attached consequence which describes everything relevant to each and every one of the players $1, \ldots, n$ about that possible history of the game.

Ignorance on the part of participant $i \in \{\mathscr{E}, 1, \ldots, n\}$ at some point in the game really means that i has to make a choice from one of his moves but does not know from which of several moves this choice will be made. Clearly, all such moves in question must be *qualitatively identical*, in that they are all *copies of the same move but attached to different partial paths* through the tree. The set of *all* such moves in question, for i at a given point in the game, is called an *information set* of i.

■ If participant i will know what has transpired before a given move of his, then that move constitutes an information set all by itself. Thus it is clear that a participant in a game may have *many* information sets. ■

Information sets are easily denoted in game trees. We enclose the base points (□ for $i = 1, \ldots, n$ and ○ for $i = \mathscr{E}$) of all moves constituting an information set of i with a dotted loop, and write "i" prominently some-

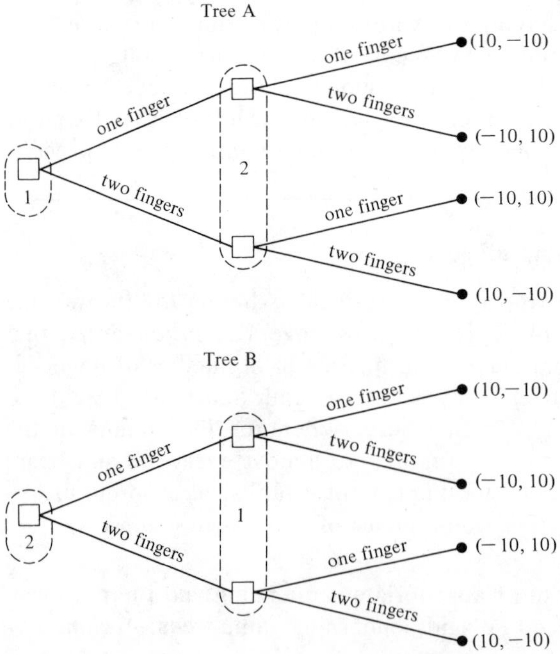

Figure 12.1 Game trees for "throwing fingers." (*Key*: In (x_1, x_2) attached to an endpoint, $x_i = i$'s utility of the consequence attached to that endpoint. [In this case, $x_i = i$'s net return.])

where inside the loop. See Figure 12.1, which gives two game trees, each of which is an appropriate tree representation of "throwing fingers."

A more elaborate version of "throwing fingers" will serve as an adequate vehicle for many subsequent illustrative purposes.

Example 12.2.2: Elaborate Throwing Fingers

Suppose that the basic rules of the game are as given in Example 12.2.1, with the following additions. First, 1 has \$40 in his pocket and 2 has \$15 in his. Second, unless each player throws two fingers, $\mathscr{E}$ will determine whether a robber (= player 3) will have the opportunity of holding up either 1 or 2 (but not both); and $\mathscr{E}$'s determination does not depend otherwise upon 1's and 2's choices. Third, players 1, 2, and 3 assess (and reveal to the other players) probabilities .9, .7, and .2 respectively to the event that $\mathscr{E}$ does tip 3 off unless 1 and 2 each throw two fingers. Fourth, if 3 is tipped off, he will have to choose his victim in ignorance of their throws, except that insofar as his being tipped off ensures that at least one of 1 and 2 threw one finger. Fifth, if c_i again denotes i's net return from the game, then we assume that $u_i(c_1, c_2, c_3) = c_i$ for $i = 1, 2, 3$. This more elaborate game is depicted in Figure 12.2.

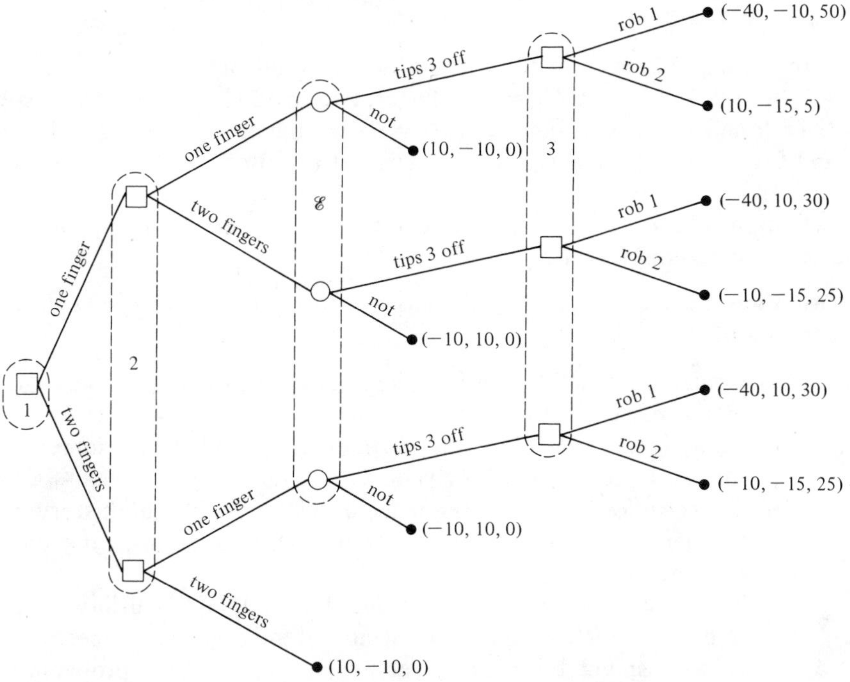

Figure 12.2 Elaborate "throwing fingers." (*Key*: In (x_1, x_2, x_3) attached to an endpoint, x_i = i's utility of the consequence attached to that endpoint. [In this case, x_i = i's net return.])

In Example 12.2.2, the decision by player 2 to throw one finger amounts in Figure 12.2 to choosing "one finger" in *each* of the two moves comprising 2's information set. His ignorance as to 1's choice forbids his adapting his choice to 1's choice.

This interpretation of information sets in game trees obtains in general. *Qualitatively identical branches must be chosen in all moves within any given information set*, since the placement of copies of the same move at different positions within the tree is a manifestation of the participant's ignorance about his position when called upon to choose.

But this suggests the subsequently convenient notation

$$G_{it} = \{g_{it}^1, g_{it}^2, \ldots, g_{it}^{\#(G_{it})}\} \tag{12.2.1}$$

for i's typical, tth information set: By choosing g_{it}^j, participant i is really choosing branch j in *each* of the moves comprising his tth information set.

Thus we have two ways of looking at information sets: (1) pictorially, as a set of moves which to their owner are indistinguishable in the course of the game; and (2) operationally, as a set of choices which the participant can make at some point in the game. The latter viewpoint and its concomitant notation will come in handy when we discuss reexpressing in normal form a game in extensive form.

Example 12.2.3

In Figure 12.2, each of the four participants has only one information set. For $i = 1$, we have $G_{11} = \{$one finger, two fingers$\}$; for $i = 2$, we have $G_{21} = \{$one finger, two fingers$\}$; for $i = 3$, we have $G_{31} = \{$rob 1, rob 2$\}$; and for $i = \mathscr{E}$, we have $G_{\mathscr{E}1} = \{$tip 3 off, not (tip 3 off)$\}$.

Definition 12.2.1 summarizes all features of a game of complete information in extensive form.

Definition 12.2.1. A *game of complete information in extensive form* consists of:

(A) a game tree, each move in which is assigned to one and only one of the participants $i \in \{\mathscr{E}, 1, \ldots, n\}$.
(B) an assignment of each of participant i's moves to some information set G_{it} of i, for each $i \in \{\mathscr{E}, 1, \ldots, n\}$.
(C) a description of the consequence of each possible full history of the game and an assignment of that consequence to the corresponding endpoint.
(D) an assessment by each *player* $i \in \{1, \ldots, n\}$ of his utility function $u_i\colon C \to R^1$ on the set C of all potential consequences.
(E) an assessment by each player $i \in \{1, \ldots, n\}$ of his probability function on each of $\mathscr{E}$'s information sets $G_{\mathscr{E}t}$.
(F) full knowledge of (A)–(E) on the part of each player $i \in \{1, \ldots, n\}$.

■ Part (F) of Definition 12.2.1 is essential for the game to be of complete information. By contrast, a game of incomplete information is one in which some of the players may be unclear about the structure of the game tree and/or other players' utility and/or probability functions. Harsanyi describes [**52**] how all such uncertainties may be represented as uncertainties about other players' utility functions. ■

As one might suspect, games in extensive form can be somewhat cumbersome for purposes of strategic analysis. The intuitive clarity conveyed by the extensive form of a game is bought at the cost of lost notational compactness. Fortunately, just as with individual decision problems, there is a normal form. Its advantages in general game theory are closely akin to the advantages of the normal form in individual decision analysis.

A given game in extensive form can be reexpressed in normal form by a very slight adaptation of the techniques introduced in Chapter 7 for individual dpuu's. The only adaptations necessary are (1) noting that here we have $n+1$ participants—$\mathscr{E}, 1, \ldots, n$—instead of the two—$\mathscr{D}$ and $\mathscr{E}$—in Chapter 7; and (2) treating the information sets $G_{i1}, \ldots, G_{iT(i)}$ of a participant i here just as we treated the moves of a player in Chapter 7. In particular, a complete pure strategy of participant i is a set $\{g_{i1}, g_{i2}, \ldots, g_{iT(i)}\}$ consisting of one choice g_{it} from each of i's information sets G_{it}.

Example 12.2.4

In Figure 12.2 and in Example 12.2.3, each participant has only one information set, and hence, we can define the sets $\Sigma = G_{\mathscr{E}1}$ and $S_i = G_{i1}$ for $i = 1, 2, 3$ of pure strategies for each of the participants, where the G_{i1}'s are as defined in Example 12.2.3.

Now, once the set Σ of all pure strategies σ for $\mathscr{E}$ and the n sets S_i of all pure strategies s_i for player $i \in \{1, \ldots, n\}$ have been defined, it follows, just as in Chapter 7, that each $(s_1, \ldots, s_n, \sigma)$ in $S_1 \times \cdots \times S_n \times \Sigma$ specifies a unique path through the game tree, and thus a unique consequence $c[s_1, \ldots, s_n, \sigma]$.

■ The scorekeeper argument remains valid. The scorekeeper locates the owner of the initial move, traces his choice in that move, locates the owner of the move attached to the branch chosen in the initial move, traces the stated choice in that move, and so on. That he ultimately arrives at a consequence unambiguously follows from the facts that every move in the game tree belongs to some information set, that every information set is owned by some participant, and that a choice from an information set specifies a choice from each move contained in that information set. ■

In the sequel we will not vary the players' judgments about $\mathscr{E}$'s choices in his information sets $G_{\mathscr{E}t}$, and hence we can simplify matters by eliminating $\mathscr{E}$ from the game altogether, through computing appropriate expectations of the players' utilities. To see how this can be done, we must note first that the complete-information assumption, Definition 12.2.1(F), implies that *every player knows*:

(1) all the sets $\Sigma, S_1, \ldots, S_n$ of pure strategies.
(2) each player i's utility $u_i(c[s_1, \ldots, s_n, \sigma])$, for every $(s_1, \ldots, s_n, \sigma)$ in $S_1 \times \cdots \times S_n \times \Sigma$.
(3) each player i's probability function $P_{\Sigma i}(\cdot)$ on Σ.

Hence each player i's utility evaluation

$$U_i(s_1, \ldots, s_n) = \sum_{j=1}^{\#(\Sigma)} u_i(c[s_1, \ldots, s_n, \sigma^j])P_{\Sigma i}(\sigma^j) \qquad (12.2.2)$$

of every $(s_1, \ldots, s_n)$ in $S_1 \times \cdots \times S_n$ is common knowledge in the set $\mathscr{N} = \{1, \ldots, n\}$ of players.

Example 12.2.5

In "elaborate throwing fingers" let s_i^1 = "i throws one finger" and s_i^2 = "i throws two fingers" for $i = 1, 2$; let s_3^k = "3 robs k" for $k = 1, 2$; and let σ^1 = "$\mathscr{E}$ tips 3 off," with σ^2 = "$\mathscr{E}$ does not tip 3 off." The preceding data for this game imply the following tabulation of utilities (u_1, u_2, u_3), where $u_i = u_i(c[s_1, \ldots, s_n, \sigma])$.

Table A

		s_3^1		s_3^2	
		s_2^1	s_2^2	s_2^1	s_2^2
σ^1	s_1^1	(−40, −10, 50)	(−40, 10, 30)	(10, −15, 5)	(−10, −15, 25)
	s_1^2	(−40, 10, 30)	(10, −10, 0)	(−10, −15, 25)	(10, −10, 0)
σ^2	s_1^1	(10, −10, 0)	(−10, 10, 0)	(10, −10, 0)	(−10, 10, 0)
	s_1^2	(−10, 10, 0)	(10, −10, 0)	(−10, 10, 0)	(10, −10, 0)

Now the role of $\mathscr{E}$ can be eliminated by applying (12.2.2) to this table, thus obtaining the following table:

Table B

	s_3^1		s_3^2	
	s_2^1	s_2^2	s_2^1	s_2^2
s_1^1	(−35, −10, 10)	(−37, 10, 6)	(10, −13.5, 1)	(−10, −7.5, 5)
s_1^2	(−37, 10, 6)	(10, −10, 0)	(−10, −7.5, 5)	(10, −10, 0)

We stress the fact that this table is essentially known to players 1–3, since each can perform such calculations as

$$U_1(s_1^{\ 1}, s_2^{\ 1}, s_3^{\ 1}) = (.9)u_1(c[s_1^{\ 1}, s_2^{\ 1}, s_3^{\ 1}, \sigma^1]) + (.1)u_1(c[s_1^{\ 1}, s_2^{\ 1}, s_3^{\ 1}, \sigma^2])$$
$$= (.9)(-40) + (.1)(10) = -35,$$
$$U_2(s_1^{\ 2}, s_2^{\ 1}, s_3^{\ 2}) = (.7)u_2(c[s_1^{\ 2}, s_2^{\ 1}, s_3^{\ 2}, \sigma^1]) + (.3)u_2(c[s_1^{\ 2}, s_2^{\ 1}, s_3^{\ 2}, \sigma^2])$$
$$= (.7)(-15) + (.3)(10) = -7.5,$$
$$U_3(s_1^{\ 1}, s_2^{\ 1}, s_3^{\ 1}) = (.2)u_3(c[s_1^{\ 1}, s_2^{\ 1}, s_3^{\ 1}, \sigma^1]) + (.8)u_3(c[s_1^{\ 1}, s_2^{\ 1}, s_3^{\ 1}, \sigma^2])$$
$$= (.2)(50) + (.8)(0) = 10.$$

Since our intent is to study the strategic potentials of the players $i \in \mathcal{N}$, elimination of $\mathscr{E}$ from explicit consideration via (12.2.2) is a harmless simplification of the description of the game in normal form. Thus no essential generality is sacrificed by the following definition.

Definition 12.2.2. A *game of complete information in normal form* consists of:

(A) a nonempty set S_i consisting of all pure strategies s_i for player i, for each player $i \in \mathcal{N}$.
(B) a utility function U_i: $S_1 \times \cdots \times S_n \to R^1$ for each player $i \in \mathcal{N}$.
(C) full knowledge of (A) and (B) on the part of each player $i \in \mathcal{N}$.

■ Part (C) of Definition 12.2.2 is necessary for the game in question to be of complete information. ■

Because of the clumsiness and frequently necessary arbitrariness associated with formulating games in extensive form, it is often desirable to *begin* by defining the pure-strategy sets $S_1, \ldots, S_n$ and Σ, and determining the consequences $c[s_1, \ldots, s_n, \sigma]$, and to proceed via (12.2.2) in simplifying matters.

■ When $\mathscr{E}$ is not a participant, then it is clear that (12.2.2) is irrelevant and that

$$U_i(s_1, \ldots, s_n) = u_i(c[s_1, \ldots, s_n]) \tag{12.2.3}$$

for every $(s_1, \ldots, s_n)$ in $S_1 \times \cdots \times S_n$. This same fact follows by prefixing the real initial move in the game by a dummy $\mathscr{E}$-move $\{\sigma^*\}$ and applying (12.2.2), since necessarily $P_{\Sigma i}(\sigma^*) = 1$ for every player $i \in \mathcal{N}$. ■

When $n = 2$, the strategy sets and utilities of the players are conveniently representable in a tabular format, called a "bimatrix," in which row ℓ denotes player 1's pure strategy $s_1^{\ \ell}$, column m denotes player 2's pure strategy $s_2^{\ m}$, and the (ℓ, m)th position of the table contains the pair $(U_1(s_1^{\ \ell}, s_2^{\ m}), U_2(s_1^{\ \ell}, s_2^{\ m}))$ of players' utilities.

Example 12.2.6

It is clear in Example 12.2.1 that each of players 1 and 2 in the simple "throwing fingers" game has exactly two pure strategies, s_i^1 = "one finger" and s_i^2 = "two fingers." The bimatrix representation of "throwing fingers" in normal form is thus

	s_2^1	s_2^2
s_1^1	$(10, -10)$	$(-10, 10)$
s_1^2	$(-10, 10)$	$(10, -10)$

Note that $\mathscr{E}$ is not a participant, and hence, no expectations according to (12.2.2) were required. Also note that this same table is gotten from either of the two equally representative game trees in Figure 12.1!

We close this section with some definitions and notation useful in dealing with randomized strategies and cooperative behavior in games of complete information in normal form.

Let $\mathscr{N}$ be the set $\{1, \ldots, n\}$ of all players (excluding $\mathscr{E}$!) in the game, and let $\mathscr{C}$ be a nonempty subset of $\mathscr{N}$. If the players i belonging to $\mathscr{C}$ have the ability to specify their *joint* choices of pure strategies, then attention is focused on the set

$$S_{\mathscr{C}} = \times_{i \in \mathscr{C}} S_i \tag{12.2.4}$$

of all such "joint pure strategies" $s_{\mathscr{C}}$. For obvious reasons, a nonempty subset $\mathscr{C}$ of the set $\mathscr{N}$ of all players is called a *coalition*, as noted in Section 12.1.

Now suppose that $\mathscr{C}_1, \ldots, \mathscr{C}_M$ are coalitions and constitute a partition of $\mathscr{N}$. Then every player i belongs to one and only one of the coalitions $\mathscr{C}_m$: There are no conflicting loyalties, and at worst a given player i is all alone in a coalition $\{i\}$. Since a player i's pure strategy s_i is dictated by the joint pure strategy $s_{\mathscr{C}_m}$ of the coalition $\mathscr{C}_m$ to which i belongs, it follows that every player's pure strategy is determined by the M-tuple $(s_{\mathscr{C}_1}, \ldots, s_{\mathscr{C}_M})$ of joint pure strategies of the M coalitions.

■ Mathematically, it is easy to verify that if $\mathscr{C}_1, \ldots, \mathscr{C}_M$ constitute a partition of $\mathscr{N}$, then $S_{\mathscr{C}_1} \times \cdots \times S_{\mathscr{C}_M}$ is obtainable from $S_1 \times \cdots \times S_n$ by (1) rearrangement of factors (so as to list all players in $\mathscr{C}_1$ first, then all in $\mathscr{C}_2$, and so on); and (2) associating the factors by parenthesizing. For instance, if $\mathscr{N} = \{1, 2, 3\}$, $\mathscr{C}_1 = \{1, 3\}$, and $\mathscr{C}_2 = \{2\}$, then $S_{\mathscr{C}_1} \times S_{\mathscr{C}_2} = \{((s_1, s_3), s_2): s_1 \in S_1, s_2 \in S_2, s_3 \in S_3\}$. Clearly, knowing (s_1, s_2, s_3) in $S_1 \times S_2 \times S_3$ enables one to determine the corresponding $((s_1, s_3), s_2)$ in $s_{\mathscr{C}_1} \times S_{\mathscr{C}_2}$ easily, and vice versa. ■

An individual player i is no less free to choose his pure strategy s_i by randomization than was an individual decision maker, $\mathscr{D}$, in Chapter 7. We denote by P_i the typical randomized strategy for player i; that is,

$$P_i = (P_i(s_i^1), \ldots, P_i(s_i^{\#(S_i)})), \tag{12.2.5a}$$

with $P_i(s_i^{j_i})$ denoting the probability with which i's chosen randomization will select strategy $s_i^{j_i}$. Moreover, we define $\mathcal{Q}_i$ by

$$\mathcal{Q}_i = \{P_i\colon P_i \text{ is a randomized strategy for } i\}. \tag{12.2.6a}$$

These definitions are so akin to their analogues in Chapter 7 that further comment is unnecessary.

But a coalition $\mathscr{C}$ is equally free to randomize its choice of a joint pure strategy, and we accordingly define

$$P_{\mathscr{C}} = (P_{\mathscr{C}}(s_{\mathscr{C}}^{1}), \ldots, P_{\mathscr{C}}(s_{\mathscr{C}}^{\#(S_{\mathscr{C}})})) \tag{12.2.5b}$$

and

$$\mathcal{Q}_{\mathscr{C}} = \{P_{\mathscr{C}}\colon P_{\mathscr{C}} \text{ is a randomized strategy for } \mathscr{C}\}. \tag{12.2.6b}$$

■ If coalition $\mathscr{C}$ is allowed to form and binding contracts are possible, then the members of $\mathscr{C}$ can do at least as well in concert as independently. To see this, note that independent choices of strategies P_i by the members of $\mathscr{C}$ determine a $P_{\mathscr{C}}$ via the obvious implication of unrelated randomizations that if $s_i^{j_i}$ is i's pure strategy in $s_{\mathscr{C}}$, then

$$P_{\mathscr{C}}(s_{\mathscr{C}}) = \prod_{i\in\mathscr{C}} P_i(s_i^{j_i}). \tag{12.2.7}$$

But most randomized strategies $P_{\mathscr{C}}$ for coalition $\mathscr{C}$ do not satisfy (12.2.7) for every $s_{\mathscr{C}}$ in $S_{\mathscr{C}}$ and some given randomized strategies P_i for the $\#(\mathscr{C})$ members i of $\mathscr{C}$. ■

■ The truth of this last sentence is clarified by noting that if the members of $\mathscr{C}$ are denoted by i_t for $t = 1, 2, \ldots, \#(\mathscr{C})$, then an arbitrary randomized strategy $P_{\mathscr{C}}$ for $\mathscr{C}$ requires the specification of

$$\#(S_{\mathscr{C}}) - 1 = \prod_{t=1}^{\#(\mathscr{C})} \#(S_{i_t}) - 1$$

probabilities, whereas strategies $P_{\mathscr{C}}$ obtainable via (12.2.7) from some $P_{i_1}, \ldots, P_{i_{\#(\mathscr{C})}}$ require specification of $\#(S_{i_t}) - 1$ probabilities for each t, and therefore a total of

$$\sum_{t=1}^{\#(\mathscr{C})} \#(S_{i_t}) - \#(\mathscr{C})$$

probabilities. Now,

$$\sum_{t=1}^{\#(\mathscr{C})} \#(S_{i_t}) - \#(\mathscr{C}) < \prod_{t=1}^{\#(\mathscr{C})} \#(S_{i_t}) - 1$$

whenever $\#(\mathscr{C}) \geq 2$ and at least two members of $\mathscr{C}$ have at least two pure strategies apiece. Hence $\mathcal{Q}_{i_1} \times \cdots \times \mathcal{Q}_{i\#(\mathscr{C})}$ may be identified with a usually drastically proper subset of $\mathcal{Q}_{\mathscr{C}}$. ■

Finally, note that if coalitions $\mathscr{C}_1, \ldots, \mathscr{C}_M$ constitute a partition of $\mathcal{N}$, then the M strategies $P_{\mathscr{C}_1}, \ldots, P_{\mathscr{C}_M}$ of the coalitions determine a probability function $P_{\mathcal{N}}(\cdot|P_{\mathscr{C}_1}, \ldots, P_{\mathscr{C}_M})$ on $S_1 \ldots S_n$ in a natural manner.

■ Define $T_m: S_1 \times \cdots \times S_n \to S_{\mathscr{C}_m}$ by

$$T_m(s_1, \ldots, s_n) = (s_{i1}, \ldots, s_{i\#(\mathscr{C})}) \tag{12.2.8}$$

for $m = 1, \ldots, M$; that is, $T_m(\cdot)$ simply picks out of $(s_1, \ldots, s_n)$ the pure strategies of the members i_t of coalition $\mathscr{C}_m$. Then $P_{\mathscr{N}}(\cdot|P_{\mathscr{C}_1}, \ldots, P_{\mathscr{C}_M})$ is given by

$$P_{\mathscr{N}}(s_{\mathscr{N}}|P_{\mathscr{C}_1}, \ldots, P_{\mathscr{C}_M}) = \prod_{m=1}^{M} P_{\mathscr{C}_m}(T_m(s_{\mathscr{N}})) \tag{12.2.9}$$

for every $s_{\mathscr{N}} = (s_1, \ldots, s_n)$ in $S_{\mathscr{N}} = S_1 \times \cdots \times S_n$. It really is harder to write down than to grasp. ■

This suggests a conveniently compact notation for each player's utility of a complete set of coalitions' randomized strategies.

Definition 12.2.3. Suppose that $\mathscr{C}_1, \ldots, \mathscr{C}_M$ are coalitions which constitute a partition of $\mathscr{N} = \{1, \ldots, n\}$. Then i's utility $U_i(P_{\mathscr{C}_1}, \ldots, P_{\mathscr{C}_M})$ of $(P_{\mathscr{C}_1}, \ldots, P_{\mathscr{C}_M})$ is given by

$$U_i(P_{\mathscr{C}_1}, \ldots, P_{\mathscr{C}_M}) = \sum_{j=1}^{\#(S_{\mathscr{N}})} U_i(s_{\mathscr{N}}^j) P_{\mathscr{N}}(s_{\mathscr{N}}^j | P_{\mathscr{C}_1}, \ldots, P_{\mathscr{C}_M})$$

for every $(P_{\mathscr{C}_1}, \ldots, P_{\mathscr{C}_M})$ in $\mathscr{Q}_{\mathscr{C}_1} \times \cdots \times \mathscr{Q}_{\mathscr{C}_M}$.

Example 12.2.7

Consider "elaborate throwing fingers" as represented by Table B in Example 12.2.5. Suppose that $\mathscr{C}_1 = \{1, 2\}$ and $\mathscr{C}_2 = \{3\}$. Then $S_{\mathscr{C}_1} = \{(s_1^1, s_2^1), (s_1^1, s_2^2), (s_1^2, s_2^1), (s_1^2, s_2^2)\}$, whereas $S_{\mathscr{C}_2} = S_3 = \{s_3^1, s_3^2\}$. Let $P_{\mathscr{C}_2}$ be specified by $P_{\mathscr{C}_2}(s_3^1) = .80$, and $P_{\mathscr{C}_1}$ by $P_{\mathscr{C}_1}(s_1^1, s_2^1) = .10$, $P_{\mathscr{C}_1}(s_1^1, s_2^2) = .20$, and $P_{\mathscr{C}_1}(s_1^2, s_2^1) = .30$. Thus, also $P_{\mathscr{C}_1}(s_1^2, s_2^2) = .40$ and $P_{\mathscr{C}_2}(s_3^2) = .20$. Then

$$\begin{aligned} U_1(P_{\mathscr{C}_1}, P_{\mathscr{C}_2}) &= (-35)(.10)(.80) + (-37)(.20)(.80) + (-37)(.30)(.80) \\ &\quad + (\ 10)(.40)(.80) + (\ 10)(.10)(.20) + (-10)(.20)(.20) \\ &\quad + (-10)(.30)(.20) + (\ 10)(.40)(.20) \\ &= -14.40. \end{aligned}$$

Exercise 12.2.1: Partitioning One of *i*'s information Sets Tends to Increase the Number of His Pure Strategies

(A) Suppose that in "elaborate throwing fingers" 2's information set is subdivided into two information sets each consisting of one move, thus indicating that 2 can choose his number of fingers after learning the number chosen by 1. Show that 2 now has four pure strategies, and find the normal-form table analogous to Table B in Example 12.2.5.

(B) Suppose that "elaborate throwing fingers" is as depicted in Figure 12.2, except that 3's information set is partitioned into three information sets each consisting of one move. Show that 3 now has eight pure strategies, and find the normal-form table analogous to Table A in Example 12.2.5.

Exercise 12.2.2.

Reexpress in normal form the game in extensive form depicted in (A) Figure 12.3, (B) Figure 12.4, and (C) Figure 12.5.

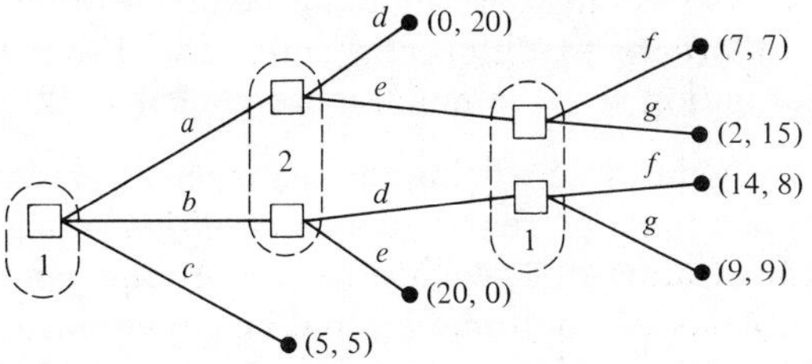

Figure 12.3 Game tree for Exercise 12.2.2(A).

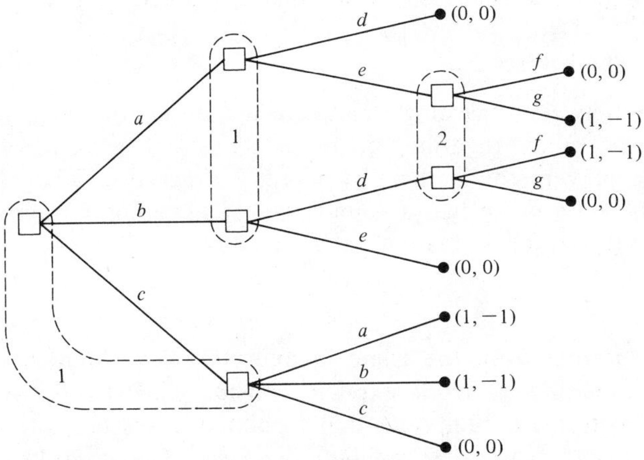

Figure 12.4 Game tree for Exercise 12.2.2(B).

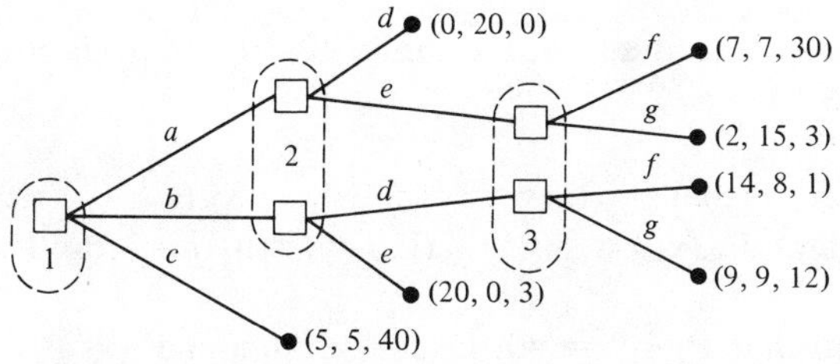

Figure 12.5 Game tree for exercise 12.2.2(C).

■ It appears odd to have an information set (such as 1's first in Figure 12.4) in which some move is a possible successor to another move in that same information set. Such information sets are the exception rather than the rule in practice to date, but one possible justification for them is described by Isbell [**66**]. Suppose (1) that actual selection and implementation of an act in a move of participant i is the task of an *agent* employed by i; (2) that orders to the agents of i are transmitted by a central *strategist*; and (3) that an information set of i is a set of moves whose agents must all receive the *same* order from the strategist, because of time pressures, impossibility of obtaining better information or of transmitting different orders to different agents, etc. From this point of view the odd information sets are much more plausible. ■

■ In [**74**], Kuhn regards an agent of i as being the manager of an entire information set, in which case i can be regarded as a *coalition of agents* who either employ a central strategist or get together before the game to coordinate strategy. If problems of coordination and/or communication dictate that two or more agents must make the same choices, then they may be regarded as one and their individual information sets may be coalesced. Again, there might result some information set containing a move which is a possible successor to another contained in it. ■

Exercise 12.2.3: Think Piece

The difference between regarding a player as a coalition of agents (see the preceding remark) and regarding coalitions as sets of players is that in the latter the players retain their own utility functions. What difference would there be if we had a sound way of assigning an agreed-upon coalition utility to each coalition strategy $P_{\mathscr{C}}$?

Exercise 12.2.4

Reexpress in normal form the game in extensive form depicted in Figure 12.6, eliminating $\mathscr{E}$ from explicit consideration by applying (12.2.2) together with the assumptions that 1 publicly assesses $P_{\mathscr{E}1}(\text{go}) = .75$ and $P_{\mathscr{E}1}(\text{here}) = .50$, while 2 publicly assesses $P_{\mathscr{E}2}(\text{go}) = .40$ and $P_{\mathscr{E}2}(\text{here}) = .10$.

Exercise 12.2.5

Let $\mathscr{C}_1$ and $\mathscr{C}_2$ be as described in Example 12.2.7 for "elaborate throwing fingers." Find:

(A) $U_i(P_{\mathscr{C}_1}, P_{\mathscr{C}_2})$ given that $P_{\mathscr{C}_1}(s_1{}^1, s_2{}^1) = .60$, $P_{\mathscr{C}_1}(s_1{}^1, s_2{}^2) = .10$, $P_{\mathscr{C}_1}(s_1{}^2, s_2{}^1) = .10$, and $P_{\mathscr{C}_2}(s_3{}^1) = .50$, for (i) $i = 1$, (ii) $i = 2$, and (iii) $i = 3$.
(B) $U_i(P_1, P_2, P_3)$ given that $P_1(s_1{}^1) = .40$, $P_2(s_2{}^1) = .70$, and $P_3(s_3{}^1) = .50$, for (i) $i = 1$, (ii) $i = 2$, and (iii) $i = 3$.

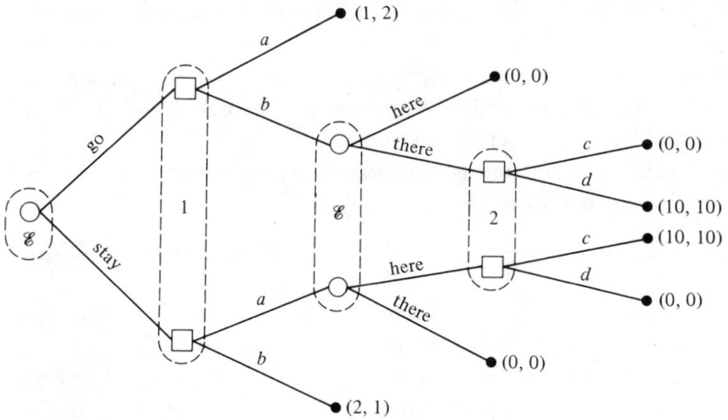

Figure 12.6 Game tree for Exercise 12.2.4.

■ That three numbers are required in (A) to specify $P_{\mathscr{C}_1}$ whereas only two are required in (B) to specify both P_1 and P_2 illustrates our earlier remark that $\mathscr{Q}_{i_1} \times \cdots \times \mathscr{Q}_{i_{\#(\mathscr{C})}}$ may be regarded as a usually drastically proper subset of $\mathscr{Q}_{\mathscr{C}}$ for $\mathscr{C} = \{i_1, \ldots, i_{\#(\mathscr{C})}\}$. ■

Exercise 12.2.6

Show that $\mathscr{Q}_i$ can be identified with $\mathscr{Q}_{\{i\}}$ for every $i \in \mathscr{N}$.

B: NONCOOPERATIVE GAMES

12.3 MAXIMIN AND EQUILIBRIUM APPROACHES

Suppose that a given n (≥ 2) person game has been expressed in normal form, and that each player knows (1) all n sets $S_1, \ldots, S_n$ of pure strategies; and (2) all n utility functions U_i: $S_1 \times \cdots \times S_n \to R^1$. Further suppose that *no communication or cooperation* between or among the players is allowed. Then it makes no sense to consider side payments and the issue of whether utility can be "transferred with conservation" from one player to another. It also makes no sense to consider whether some of the players, say 1 and 2, can get together and make a joint choice of strategies, say (s_1, s_2), so as to fare better than they could anticipate doing independently, say by independently choosing s_1 and s_2.

■ Planning side payments and joint strategy choices can be very hazardous to say the least without the ability to communicate and to sign binding agreements. These are precisely the societal characteristics absent in a noncooperative game. ■

Instead, we shall have to consider solution concepts not involving any joint actions.

In this section we introduce two solution concepts, *maxmin* and *equilibrium*. The former was introduced in exercises in Chapter 7, but the latter has not been introduced in previous chapters.

The maximin solution concept is *semipessimistically conservative*, because it amounts to each player i's assuming:

(1) that his choice of *randomized* strategy P_i will be revealed to all the other players.
(2) that the *pure* strategy s_i which will be selected by the randomization will *not* be revealed to anyone but himself.
(3) that all the other players will band together in a coalition $\overline{\{i\}} = \bar{i}$ for purposes of choosing a jointly randomized strategy $P_{\bar{i}}$ so as to do i the greatest possible harm.

■ It should be clear that the inclusion of (2) makes (1)–(3) only *semi*pessimistic, because it is clearly preferable to i to leave his malevolent colleagues at least partially in the dark about what he will do. ■

■ Our writing $\bar{i}$ instead of $\overline{\{i\}}$ is purely for purposes of typographical convenience. ■

If i is *really* convinced of the validity of these assumptions, then he should choose that strategy whose revelation to his malevolent colleagues in $\bar{i}$ will leave him ultimately best off. More precisely, he should *first* evaluate each of his randomized strategies P_i by

$$U_i(P_i, \mathcal{Q}_{\bar{i}}) = \min_{P_{\bar{i}} \in \mathcal{Q}_{\bar{i}}} [U_i(P_i, P_{\bar{i}})], \tag{12.3.1}$$

for every P_i in $\mathcal{Q}_i$, and *then* choose any strategy $P_i^\dagger$ such that

$$U_i(P_i^\dagger, \mathcal{Q}_{\bar{i}}) = \max_{P_i \in \mathcal{Q}_i} [U_i(P_i, \mathcal{Q}_{\bar{i}})]. \tag{12.3.2}$$

Definition 12.3.1. We say that $P_i^\dagger$ is a *maxmin* strategy for player i if $P_i^\dagger$ satisfies (12.3.2).

■ Note that the definition of a maximin strategy for player i disregards the utilities of the other players i', because the assumption that they will act to do i the greatest possible damage means that *they* will disregard *their own* utilities and concentrate on minimizing i's utility. ■

Before we discuss examples of maximin strategies, we shall examine how to find them. By inspecting (12.3.1) it is not hard to arrive at the very useful conclusion that

$$U_i(P_i, \mathcal{Q}_{\bar{i}}) = \min_{s_{\bar{i}} \in S_{\bar{i}}} [U_i(P_i, s_{\bar{i}})]. \tag{12.3.3}$$

■ Since $U_i(P_i, P_{\bar{i}})$ is a probability-weighted average of $U_i(P_i, s_{\bar{i}}^1), \ldots, U_i(P_i, s_{\bar{i}}^{\#(S_{\bar{i}})})$, in which $U_i(P_i, s_{\bar{i}}^j)$ has relative weight $P_{\bar{i}}(s_{\bar{i}}^j)$, it follows that $U_i(P_i, P_{\bar{i}})$ is minimized by choosing $P_{\bar{i}}$ so as to place probability one on any $s_{\bar{i}}^\dagger$ such that $U_i(P_i, s_{\bar{i}}^\dagger) = \min_{s_{\bar{i}} \in S_{\bar{i}}} [U_i(P_i, s_{\bar{i}})]$. ■

From (12.3.3) it is easy to develop a graphical procedure for finding $P_i^\dagger$ when $\#(S_i) = 2$, since in this case every P_i in $\mathscr{Q}_i$ may be written in the form

$$P_i = [p, 1-p] \qquad (12.3.4)$$

for some $p = P_i(s_i^1)$ in $[0, 1]$. Set up a graph with abscissa p and ordinate U_i; then graph the line segment $\{U_i([p, 1-p], s_{\bar{i}}): 0 \le p \le 1\}$ for each $s_{\bar{i}}$ in $S_{\bar{i}}$. For each p now locate the lowest point on any of these line segments; that point is $U_i([p, 1-p], \mathscr{Q}_{\bar{i}})$. Heavily draw over the locus $\{U_i([p, 1-p], \mathscr{Q}_{\bar{i}}): 0 \le p \le 1\}$ of these points. Then $P_i^\dagger(s_i^1) = p^\dagger$ is the abscissa corresponding to any maximum value on the heavily drawn line, and $P_i = [p^\dagger, 1-p^\dagger]$.

■ That $\{U_i([p, 1-p], s_{\bar{i}}): 0 \le p \le 1\}$ is a line segment is clear from the fact that, given any $s_{\bar{i}}$ in $S_{\bar{i}}$,

$$U_i([p, 1-p], s_{\bar{i}}) = pU_i(s_i^1, s_{\bar{i}}) + (1-p)U_i(s_i^2, s_{\bar{i}}) \qquad (12.3.5)$$

for every p in $[0, 1]$. ■

Example 12.3.1

In Figure 12.7 we determine that player 1's maximin strategy for "simple throwing fingers" as given in Example 12.2.6 in normal form is $\mathbf{P}_1^\dagger = [1/2, 1/2]$. Here, $\bar{1} = 2$, and hence all we had to do was graph $U_1[(p, 1-p], s_2^1) = 10p + (-10)(1-p)$ and $U_1([p, 1-p], s_2^2) = (-10)p + 10(1-p)$ as functions of p. Precisely the same approach results in the discovery that $P_2^\dagger = [1/2, 1/2]$ too.

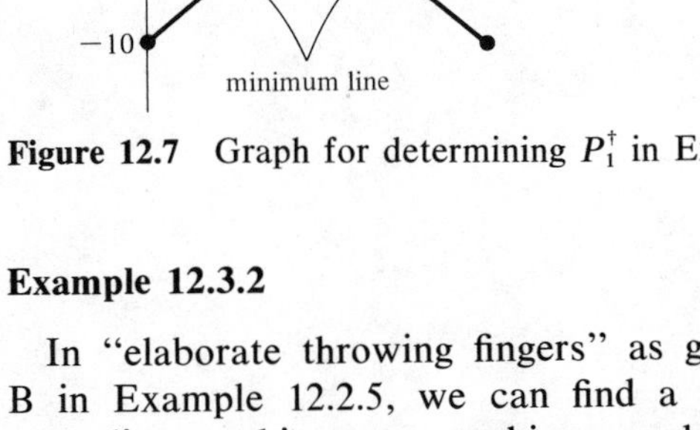

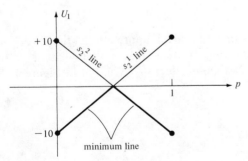

Figure 12.7 Graph for determining $P_1^\dagger$ in Example 12.2.6.

Example 12.3.2

In "elaborate throwing fingers" as given in normal form by Table B in Example 12.2.5, we can find a maximin strategy for player 1 according to this same graphic procedure, since 1 has only two pure

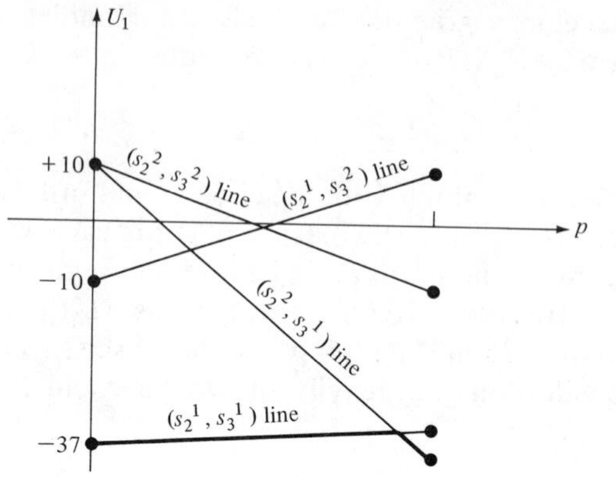

Figure 12.8 Graph for determining $P_{\bar{1}}^{\dagger}$ in Example 12.2.5.

strategies. But now $S_{\bar{1}} = \{(s_2^1, s_3^1), (s_2^1, s_3^2), (s_2^2, s_3^1), (s_2^2, s_3^2)\}$. In Figure 12.8 we graph

$$U_1([p, 1-p], (s_2^1, s_3^1)) = (-35)p + (-37)(1-p),$$
$$U_1([p, 1-p], (s_2^1, s_3^2)) = (10)p + (-10)(1-p),$$
$$U_1([p, 1-p], (s_2^2, s_3^1)) = (-37)p + (10)(1-p),$$

and

$$U_1([p, 1-p], (s_2^2, s_3^2)) = (-10)p + (10)(1-p)$$

as functions of p, thus demonstrating that $P_{\bar{1}}^{\dagger} = [47/49, 2/49]$ and $U_1(P_1^{\dagger}, \mathcal{Q}_{\bar{1}}) = -1719/49 \doteq -35.1$. A similar analysis for player 2 yields $P_{\bar{2}}^{\dagger} = [5/17, 12/17]$ and $U_2(P_2^{\dagger}, \mathcal{Q}_{\bar{2}}) = -315/34$. See Figure 12.9. For player 3, we find that *every* strategy in $\mathcal{Q}_3$ is maximin, and $U_3(P_3^{\dagger}, \mathcal{Q}_{\bar{3}}) = 0$. This is so because $\{1, 2\}$'s choice of (s_1^2, s_2^2) shuts 3 out of the game altogether. See Figure 12.10.

An important by-product of this graphical approach is that we can readily determine the most damaging strategy that $\bar{i}$ can visit on i. In Figure 12.10 it is bvious that $\{1, 2\}$ should select (s_1^2, s_2^2), but in Figure 12.9

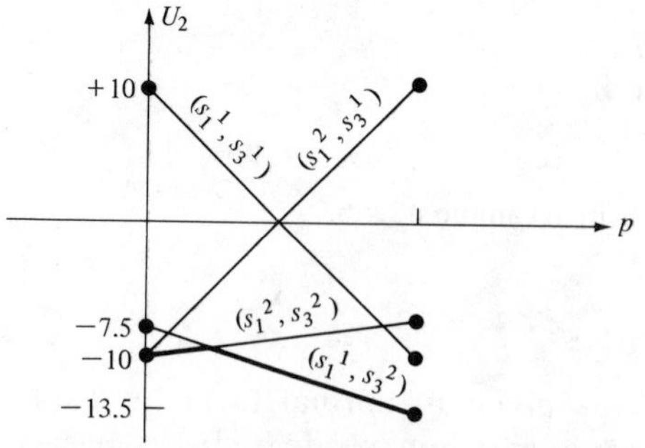

Figure 12.9 Graph for determining $P_{\bar{2}}^{\dagger}$ in Example 12.2.5.

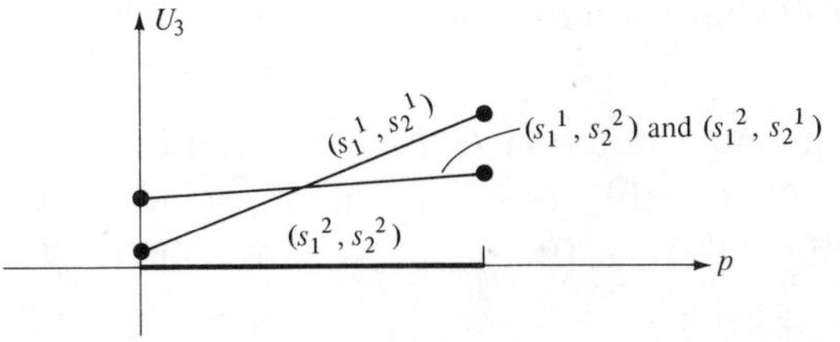

Figure 12.10 Graph for determining $P_3^\dagger$ in Example 12.2.5.

either (s_1^2, s_3^2) or (s_1^1, s_3^2) may be chosen. Similarly, in Figure 12.8 it follows that $\{2, 3\}$ may choose either (s_2^1, s_3^1) or (s_2^2, s_3^1) to do 1 the greatest possible harm.

But the ambiguity in Figures 12.8 and 12.9 regarding $\bar{i}$'s most punishing choices is detrimental to $\bar{i}$'s purpose. If 1 should get wind of $\{2, 3\}$'s choice of (s_2^1, s_3^1) in Figure 12.8, he should abandon $P_1^\dagger$ and choose s_1^1 for certain to attain $U_1([1, 0], (s_2^1, s_3^1)) = -35$; if 1 learns that $\{2, 3\}$ chose (s_2^2, s_3^1), then he should choose s_1^2 for certain to attain $U_1([0, 1], (s_2^2, s_3^1)) = +10$. In either case, $\{2, 3\}$ has failed to hold 1 down to $U_1(P_1^\dagger, \mathscr{Q}_{\bar{1}}) \doteq -35.1$.

But if we now allow $\bar{i}$ to be semipessimistically conservative and assume that i can only ascertain $\bar{i}$'s *randomized* strategy, then $\bar{i}$ *can* hold i down to $U_i(P_i^\dagger, \mathscr{Q}_{\bar{i}})$, in the sense that $\bar{i}$ has a *minimax* strategy $P_{\bar{i}}^{\dagger\dagger}$ with the properties that

$$\max_{P_i \in \mathscr{Q}_i} [U_i(P_i, P_{\bar{i}}^{\dagger\dagger})] = \min_{P_{\bar{i}} \in \mathscr{Q}_{\bar{i}}} [\max_{P_i \in \mathscr{Q}_i} [U_i(P_i, P_{\bar{i}})]], \qquad (12.3.6)$$

and

$$\max_{P_i \in \mathscr{Q}_i} [U_i(P_i, P_{\bar{i}}^{\dagger\dagger})] = U_i(P_i^\dagger, \mathscr{Q}_{\bar{i}}). \qquad (12.3.7)$$

■ A minimax strategy for $\bar{i}$ is what we called a least favorable strategy for $\mathscr{E}$ in Chapter 7. Equations (12.3.6) and (12.3.7) are in fact the celebrated "minimax theorem" of von Neumann [**105**]. Equivalently, the minimax theorem says that

$$\max_{P_i} [\min_{P_{\bar{i}}} [U_i(P_i, P_{\bar{i}})]] = \min_{P_{\bar{i}}} [\max_{P_i} [U_i(P_i, P_{\bar{i}})]]; \qquad (12.3.8)$$

that is, the order in which the maximization with respect to P_i and the minimization with respect to $P_{\bar{i}}$ are performed is irrelevant. We shall prove this theorem as a corollary of the equilibrium theorem, Theorem 12.3.1, and discuss it further in Section 12.4. For now, it suffices to note that the left-hand side of (12.3.8) represents the *floor* which i can establish for his utility, whereas the right-hand side of (12.3.8) represents the *ceiling* to i's utility which $\bar{i}$ can construct. Thus (12.3.8) says that the floor and the ceiling coincide if i chooses a maximin strategy and $\bar{i}$ chooses a minimax strategy. ■

Finding a minimax strategy for $\bar{i}$ is not difficult in the graphical approach developed above. There are four cases.

CASE 1. The (heavy line) minimum function $U_i([p, 1-p], \mathscr{Q}_{\bar{i}})$ is a *non-decreasing* function of p. Then $P_i^\dagger = [1, 0]$, and $P_{\bar{i}}^{\dagger\dagger}$ is any randomized strategy which places probability zero on every $s_{\bar{i}}$ such that $U_i([1,0], s_{\bar{i}}) > U_i([1,0], \mathcal{Q}_{\bar{i}})$.

CASE 2. The (heavy line) minimum function $U_i([p, 1-p], \mathcal{Q}_{\bar{i}})$ is a *non-increasing* function of p. Then $P_i^{\dagger} = [0, 1]$, and $P_{\bar{i}}^{\dagger\dagger}$ is any randomized strategy which places probability zero on every $s_{\bar{i}}$ such that $U_i([0,1], s_{\bar{i}}) > U_i([0,1], \mathcal{Q}_{\bar{i}})$.

CASE 3. The (heavy line) minimum function $U_i([p, 1-p], \mathcal{Q}_{\bar{i}})$ is maximized at every p in an interval $[p^-, p^+] \subset [0, 1]$ of positive width (e.g., Figure 12.10). This means that some of the $s_{\bar{i}}$ lines are horizontal. Then $P_{\bar{i}}^{\dagger\dagger}$ is any randomized strategy which places probability zero on every $s_{\bar{i}}$ such that $U_i([p^-, 1-p^-], s_{\bar{i}}) > U_i([p^-, 1-p^-], \mathcal{Q}_{\bar{i}})$.

CASE 4. The (heavy line) minimum function $U_i([p, 1-p], \mathcal{Q}_{\bar{i}})$ is maximized at the unique value $p^{\dagger}$ in $(0, 1)$—the roof-peak case (e.g., Figures 12.8 and 12.9). Let $s_{\bar{i}}^{\uparrow}$ be a strategy with line of *increasing* slope passing through the peak, and let $s_{\bar{i}}^{\downarrow}$ be a strategy with line of *decreasing* slope passing through that peak. Then a minimax strategy $P_{\bar{i}}^{\dagger\dagger}$ is obtained by setting $P_{\bar{i}}^{\dagger\dagger}(s_{\bar{i}}) = 0$ for every $s_{\bar{i}} \notin \{s_{\bar{i}}^{\uparrow}, s_{\bar{i}}^{\downarrow}\}$, and by choosing $q = P_{\bar{i}}^{\dagger\dagger}(s_{\bar{i}}^{\uparrow})$ and $1 - q = P_{\bar{i}}^{\dagger\dagger}(s_{\bar{i}}^{\downarrow})$ to satisfy

$$q = \frac{U_i([0,1], s_{\bar{i}}^{\downarrow}) - U_i([1,0], s_{\bar{i}}^{\downarrow})}{[U_i([1,0], s_{\bar{i}}^{\uparrow}) - U_i([1,0], s_{\bar{i}}^{\downarrow})] + [U_i([0,1], s_{\bar{i}}^{\downarrow}) - U_i([0,1], s_{\bar{i}}^{\uparrow})]}. \tag{12.3.9}$$

These choices result in $U_i([p, 1-p], P_{\bar{i}}^{\dagger\dagger})$ being a horizontal-line function of p, the constant value being $U_i(P_i^{\dagger}, \mathcal{Q}_{\bar{i}})$. Here, $P_{\bar{i}}^{\dagger\dagger}$ must be genuinely randomized.

Example 12.3.3

We have noted that $\{1, 2\}$'s minimax strategy in Example 12.2.5 is to choose (s_1^2, s_2^2) for certain. This results from ascertaining that Figure 12.10 falls within the purview of case 3, and that (s_1^2, s_2^2) is the only $s_{\bar{3}}$ such that $U_3([p^-, 1-p^-], s_{\bar{3}}) = 0$. Figures 12.8 and 12.9 fall within the purview of case 4. In Figure 12.8, we see that $P_{\bar{1}}^{\dagger\dagger}(s_2^1, s_3^2) = P_{\bar{1}}^{\dagger\dagger}(s_2^2, s_3^2) = 0$, $s_{\bar{1}}^{\uparrow} = (s_2^1, s_3^1)$, $s_{\bar{1}}^{\downarrow} = (s_2^2, s_3^1)$, and hence from (12.3.9) that $q = P_{\bar{1}}^{\dagger\dagger}(s_2^1, s_3^1) = [10-(-37)]/([-35-(-37)] + [10-(-37)] = 47/49$, so that $P_{\bar{1}}^{\dagger\dagger}(s_2^2, s_3^1) = 1 - q = 2/49$. In Figure 12.9, we see that $P_{\bar{2}}^{\dagger\dagger}(s_1^1, s_3^1) = 0 = P_{\bar{2}}^{\dagger\dagger}(s_1^2, s_3^1)$, $s_{\bar{2}}^{\uparrow} = (s_1^2, s_3^2)$, $s_{\bar{2}}^{\downarrow} = (s_1^1, s_3^2)$, and hence from (12.3.9) that $q = [-7.5-(-13.5)]/([-7.5-(-13.5)] + [-7.5-(-10.0))] = 12/17 = P_{\bar{2}}^{\dagger\dagger}(s_1^2, s_3^2)$, so that $P_{\bar{2}}^{\dagger\dagger}(s_1^1, s_3^2) = 5/17$.

There is more to be said about finding maximin strategies for i and minimax strategies for $\bar{i}$, but we shall defer further discussion to the following section and proceed now to introduce some famous two-person games as examples.

Example 12.3.4: "Battle of the Sexes"

Suppose that a husband and a wife must *independently* decide how to spend the evening. The same two options are open to each: s_i^1 = "go to the football game" and s_i^2 = "go to the Women's Lib meeting." Since their choices must be made independently, and in ignorance of the choice made by the spouse, each has two pure strategies. Suppose that their utilities are as given by the following table, which reflects preferences for the spouse's company to the context.

	s_2^1	s_2^2
s_1^1	(10, 5)	(0, 0)
s_1^2	(0, 0)	(5, 10)

It is clear that 1's maximin strategy is $P_1^\dagger = [1/3, 2/3]$, 2's maximin strategy is $P_2^\dagger = [2/3, 1/3]$, 1's minimax strategy is $P_1^{\dagger\dagger} = P_2^{\dagger\dagger} = [2/3, 1/3]$, and 2's minimax strategy is $P_2^{\dagger\dagger} = P_1^{\dagger\dagger} = [1/3, 2/3]$. But there is something very unstable about this situation. Suppose that 1 and 2 have provisionally settled on using their maximin strategies and that each correctly suspects this to be true of the other. Then 1 sees that $U_1(s_1^1, P_2^\dagger) = 20/3 \doteq 6.67$ whereas $U_1(P_1^\dagger, P_2^\dagger) = 10/3 \doteq 3.33$. Hence 1 is motivated to choose s_1^1 for sure. But then 2 is motivated to choose s_2^1 for sure. Now no one is motivated to switch his (or her) choice unilaterally. But if we start from $(P_1^\dagger, P_2^\dagger)$ by noting that 2 sees that $U_2(P_1^\dagger, s_2^2) = 20/3 \doteq 6.67$ whereas $U_2(P_1^\dagger, P_2^\dagger) = 10/3 \doteq 3.33$, then it follows that 2 is motivated to choose s_2^2 for sure, and hence 1 is motivated to choose s_1^2 for sure, resulting in (s_1^2, s_2^2) as the strategy pair from which neither is motivated to switch unilaterally. Hence, whether we end up at (s_1^1, s_2^1) or (s_1^2, s_2^2) depends upon whom we start the argument for. For the present we leave the "battle of the sexes" with three observations.

(1) Each player's choosing his maximin strategy and expecting the others to do so too may not be such a sound solution concept after all, because nothing guarantees that one (or more) of the players in a general game might not wish to choose something else given that his colleagues are expected to choose their maximin strategies.

(2) In the "battle of the sexes," if the players can *communicate* with each other, player 1 would have a strong incentive to *commit* himself *credibly* to s_1^1 before the spouse, 2, could commit herself credibly to s_2^2, and vice versa—but such commitments have to be credible to be worthwhile.

(3) It would be nice if the spouses were permitted to agree bindingly to randomize between (s_1^1, s_2^1) and (s_1^2, s_2^2), since so doing with equal probabilities would result in each having a (prerandomization) utility of 7.50 and would reflect the obviously equal strategic potentials of the two players in this game.

Example 12.3.5

"*Chicken*," as formerly played by juvenile delinquents, involves two players' driving cars toward each other on a collision course; the driver who swerves first (always to the right) is "chicken" and is subjected to the opprobrium of the gang. Suppose that the drivers i must each make the swerve $(= s_i^1)$ or not-swerve $(= s_i^2)$ decision only once and in ignorance of the choice of the other driver. Supposing that the whole game is laughed off if both swerve, that a sole nonswerver is gang hero, and that being "chicken" is preferable to being in a collision, the normal form of "chicken" appears as in the following table.

	s_2^1	s_2^2
s_1^1	(5, 5)	(2, 10)
s_1^2	(10, 2)	(0, 0)

It is clear that i's maximin strategy is $s_1^1 =$ "swerve" for each of $i = 1, 2$. But each is motivated to choose $s_i^2 =$ "not swerve" if he thinks the other will play safe. Player 1 would switch to s_1^2, and there it would stop; but if 2 starts the process, he would switch to s_2^2 and then 1 would not want to switch. But if both switch, each ends up terribly. Here again we see that if interplayer *communication* is permitted, then each has an incentive to *commit* himself *credibly* to not swerving $(= s_i^2)$ before his opponent can make such a commitment; and that if *joint* commitments were permitted, then it appears reasonable to suppose that 1 and 2 might agree to randomize in such a way that (s_1^1, s_2^2) and (s_1^2, s_2^1) are each selected with probability 1/2.

Example 12.3.6: "Prisoners' Dilemma"

A district attorney offers each of two prisoners i the opportunity to confess $(= s_i^2)$ and tells him that his alleged accomplice is just about to confess. If neither confesses, (s_1^1, s_2^1), then both will be convicted on a petty charge and i will get x_i months in jail (for $i = 1, 2$). If 1 alone confesses, (s_1^2, s_2^1), he goes scot free while the book is thrown at 2, who gets $y_2 > x_2$ months in jail, and vice versa: if 2 alone confesses, he goes free while 1 gets $y_1 > x_1$ months in jail. If both confess, then each serves a moderate term $z_i \in (x_i, y_i)$. Provided that the utility of each suspect is strictly decreasing in the length of his own term and does not depend at all upon his accomplice's term, the normal-form table would be similar to the following.

	s_2^1	s_2^2
s_1^1	(5. 5)	(0, 10)
s_1^2	(10, 0)	(2, 2)

It is easy to see that each player i's maximin and minimax strategies both consist of confessing, s_i^2. Communication per se is of no help, since a unilateral commitment cannot affect the other player's motivation to confess. Confessing is preferable to not confessing *regardless* of what the other prisoner does. But a *binding* compact to choose (s_1^1, s_2^1) makes a great deal of sense, if it were permissible.

■ In all three preceding examples, what really matters qualitatively is the *ranking* of the utilities for the given players, and not the specific utility numbers themselves. Each of these games can be written in the form

	s_2^1	s_2^2
s_1^1	(A_1, A_2)	(B_1, C_2)
s_1^2	(C_1, B_2)	(D_1, D_2)

A "battle of the sexes" game is characterized by $A_1 > D_1 > \max[B_1, C_1]$ and $D_2 > A_2 > \max[B_2, C_2]$; a "chicken" game, by $C_i > A_i > B_i > D_i$ for each i; and a "prisoners' dilemma," by $C_i > A_i > D_i > B_i$ for each i. The superiority of s_i^2 over s_i^1 in the "prisoners' dilemma" is independent of the particular values of A_i, B_i, C_i, and D_i, as long as their ranking is $C_i > A_i > D_i > B_i$ for $i = 1, 2$. In any "chicken," s_i^1 is i's maximin; s_i^2 is i's minimax; and each is motivated to choose s_i^2 if he suspects the other will choose his maximin strategy. In any "battle of the sexes," you may verify as Exercise 12.3.1 that

$$P_i^{\dagger}(s_i^1) = (D_i - C_i)/([A_i - C_i] + [D_i - B_i]) \qquad (12.3.10a)$$

for $i = 1, 2$, and

$$P_i^{\dagger\dagger}(s_i^1) = (D_j - B_j)/([A_j - C_j] + [D_j - B_j]) \qquad (12.3.10b)$$

for $(i, j) \in \{(1, 2), (2, 1)\}$. ■

■ The "prisoners' dilemma" and "chicken" games are important simplified models of conflict with possibility of beneficial cooperation. Important references on them are Schelling **[125]** and **[126]**, Luce and Raiffa **[92]**, and Rapoport and Chammah **[118]**. Much work has been devoted to what occurs in *repetitions* of one of these games (and utility being additive over the repetitions): Do the players develop some sort of tacit cooperative behavior via a mutual learning (and teaching) process? How do different sorts of individuals behave in these games? *Should* some sort of tacit collusion develop according to the dictates of some particular definition of rational behavior? (If a "prisoners' dilemma" is played an a priori known finite number of times, neither player has any incentive to depart from the maximin strategy of "confessing" ($= s_i^2$) every time—provided that he expects the other player to confess every time regardless of what he, i, does. But empirical findings, in **[118]** for instance, indicate that people *do not* tend to choose s_i^2 every time, and hence the proviso in the preceding sentence is shaky. Each of many pairs of players seems to adopt a "tit for tat" strategy which results, after some jockeying, in choices of s_1^1 and s_2^1 on every repetition, unless $2A_i < B_i + C_i$ for each i, in which case the most mutually advantageous tacit collusion results in alternation between (s_1^1, s_2^2) and (s_1^2, s_2^1), since each player's average per game is thereby maximized.) *Behavioral Science* and the *Journal of Conflict Resolution* have published many papers on these and related games; the interested reader is invited to browse through current and past volumes of these publications. ■

■ The "prisoners' dilemma" is a simple model of the "fallacy of composition" in economics: If everyone in society does what is best for himself (e.g., save a great deal), then each will end up worse off (e.g., in a bad depression) than he would if everyone had done what he should *not* have (e.g., spent as if money were going out of style). Voluntary production quotas, voluntary import quotas, voluntary refraining from price cutting, and other economic situations are more complicated cousins of the "prisoners' dilemma," which is itself sufficiently complex to humble anyone who has a glib definition of rational behavior in interacting decision contexts. ■

The preceding examples and remarks have brought out a number of interesting features of games. The remainder of this section follows up on one; namely, *the* n-*tuple* $(P_1^\dagger, \ldots, P_n^\dagger)$ *of players' maximin strategies is not satisfactorily stable*, because one or more players may have an incentive to choose some nonmaximin strategy if he suspects that his colleagues in the game will choose their maximin strategies.

We therefore seek an n-tuple $(P_1^\circ, \ldots, P_n^\circ)$ of strategies which *is* stable, in the sense that if each player i had tentatively announced a commitment to his component P_i°, then no player would be induced to reconsider his commitment. Such an n-tuple is called an *equilibrium* n-tuple, an *equilibrium point*, or a *Nash equilibrium*.

Definition 12.3.2. We say that $(P_1^\circ, \ldots, P_n^\circ)$ in $\mathcal{Q}_1 \times \cdots \times \mathcal{Q}_n$ is an *equilibrium point* if

$$U_i(P_1^\circ, \ldots, P_n^\circ) = \max_{P_i \in \mathcal{Q}_i} [U_i(P_1^\circ, \ldots, P_{i-1}^\circ, P_i, P_{i+1}^\circ, \ldots, P_n^\circ)]$$

for every $i \in \mathcal{N}$.

Example 12.3.7

In the "prisoners' dilemma" game, there is only one equilibrium point, namely, (s_1^2, s_2^2), or, more formally, ([0, 1], [0, 1]). Since s_i^2 (= "confess," "violate quota," "cut price," etc.) is preferable to s_i^1 regardless of what the other player does, neither is induced to deviate from s_i^2 once he is informed or guesses that the other is doing likewise.

Example 12.3.8

In two-person games where each player has two pure strategies, suppose that *each of* $P_1^{\dagger\dagger}$ *and* $P_2^{\dagger\dagger}$ *is genuinely randomized.* Then $U_2(s_2^{\,1}, P_1^{\dagger\dagger}) = U_2(s_2^{\,2}, P_1^{\dagger\dagger})$, and hence *every* P_2 is a "best response" to $P_1^{\dagger\dagger}$; in particular, $P_2^{\dagger\dagger}$ maximizes $U_2(P_2, P_1^{\dagger\dagger})$. Likewise, $U_1(s_1^{\,1}, P_2^{\dagger\dagger}) = U_1(s_1^{\,2}, P_2^{\dagger\dagger})$, and hence $P_1^{\dagger\dagger}$ (and every other P_1) maximizes $U_1(P_1, P_2^{\dagger\dagger})$. Therefore each of $P_1^{\dagger\dagger}$ and $P_2^{\dagger\dagger}$ is a "best response" to the other, and hence $(P_1^{\dagger\dagger}, P_2^{\dagger\dagger})$ is an equilibrium point. This implies that in the "battle of the sexes" there are three equilibria; namely, $([1,0],[1,0])$, $([0,1],[0,1])$—check these—*and* $([2/3, 1/3], [1/3, 2/3])$. In each of these strategy pairs, prior announcement of the pair would induce no player who believed the other would stick to his member of that pair to modify his. The utility pairs corresponding to these three equilibria are $(10, 5)$, $(5, 10)$, and $(10/3, 10/3)$ respectively. Here we see too that one equilibrium may leave *each* player *worse* off than some other equilibrium. This is because the definition of an equilibrium contemplates only *unilateral* deviations by the players.

Example 12.3.9

In "chicken" neither $P_1^{\dagger\dagger}$ nor $P_2^{\dagger\dagger}$ is genuinely randomized, and $(P_1^{\dagger\dagger}, P_2^{\dagger\dagger})$ is decidedly not an equilibrium point. There are two pure-strategy equilibria; namely, $([1,0], [0,1])$ and $([0,1], [1,0])$. But there is an equilibrium in genuinely randomized strategies: $([2/7, 5/7], [2/7, 5/7])$. Check that neither player can profitably deviate from $[2/7, 5/7]$ if the other sticks to it. The utility pairs corresponding to the three equilibria are $(2, 10)$, $(10, 2)$, and $(140/49, 140/49)$ respectively. Here at least, i's utility of the equilibrium in genuinely randomized strategies is higher than his utility of the equilibrium in which he knuckles under by swerving.

■ In Exercise 12.3.2 you may generalize the results in Examples 12.3.8 and 12.3.9 to the general (A_i, B_i, C_i, D_i) battle-of-the-sexes and chicken games. ■

Actually solving for equilibria in n-person games is very difficult, but we know that at least one equilibrium point $(P_1^\circ, \ldots, P_n^\circ)$ exists in $\mathcal{Q}_1 \times \cdots \times \mathcal{Q}_n$. This is the gist of the *Nash Equilibrium Theorem.*

Theorem 12.3.1. If every player's set S_i of pure strategies is finite, there is at least one equilibrium point $(P_1^\circ, \ldots, P_n^\circ)$.

*■ *Proof.* Due to Nash [**101**]. The basic idea is to define a function $T\colon \mathcal{Q}_1 \times \cdots \times \mathcal{Q}_n \to \mathcal{Q}_1 \times \cdots \times \mathcal{Q}_n$ in such a way that $(P_1^\circ, \ldots, P_n^\circ)$ is an equilibrium point if and only if it is a fixed point of $T(\cdot)$—that is, if and only if $T(P_1^\circ, \ldots, P_n^\circ) = (P_1^\circ, \ldots, P_n^\circ)$—and also to show that $T(\cdot)$ has a fixed point by verifying that it satisfies the hypotheses of the Brouwer

Fixed Point Theorem (in Appendix 3). We will write $T(\cdot)$ in the form

$$T(P_1, \ldots, P_n) = (T_1(P_1, \ldots, P_n), \ldots, T_n(P_1, \ldots, P_n)),$$

with $T_i(P_1, \ldots, P_n) \in \mathcal{Q}_i$ for every i. Hence, it suffices to define each $T_i(\cdot)$. This is done by first defining

$$B_i(j; P_1, \ldots, P_n) = \max\,[0, U_i(P_1, \ldots, P_{i-1}, s_i^j, P_{i+1}, \ldots, P_n) \\ - U_i(P_1, \ldots, P_n)],$$

which is the amount (if any) by which i can better his utility by switching from his strategy P_i in $(P_1, \ldots, P_n)$ to s_i^j for sure, given that the others stick to their strategies in $(P_1, \ldots, P_n)$. Now we define the jth component $T_i^j(P_1, \ldots, P_n)$—i's new probability of selecting s_i^j—by

$$T_i^j(P_1, \ldots, P_n) = \frac{P_i(S_i^j) + B_i(j; P_1, \ldots, P_n)}{1 + \sum_{k=1}^{\#(S_i)} B_i(k; P_1, \ldots, P_n)},$$

for every $j \in \{1, \ldots, \#(S_i)\}$, every $i \in \mathcal{N}$, and every $(P_1, \ldots, P_n) \in \mathcal{Q}_1 \times \cdots \times \mathcal{Q}_n$. It is clear that these numbers are nonnegative and yield one when summed with respect to j. Hence $T_i(P_1, \ldots, P_n)$ does belong to $\mathcal{Q}_i$. Moreover, $T_i\colon \mathcal{Q}_1 \times \cdots \times \mathcal{Q}_n \to \mathcal{Q}_i$ is continuous, because (1) $U_i(\cdot)$ and thus also $B_i(s_i^j; \cdot)$ are continuous in $(P_1, \ldots, P_n)$; (2), therefore the numerator and denominator of each $T_i^j(\cdot)$ are continuous; (3), the denominator does not vanish; (4), therefore every component $T_i^j(\cdot)$ of $T_i(\cdot)$ is continuous; and (5) therefore $T_i(\cdot)$ is itself continuous on $\mathcal{Q}_1 \times \cdots \times \mathcal{Q}_n$. Now, $\mathcal{Q}_i$ is a compact, convex, and nonempty subset of $R^{\#(S_i)}$ for each i, and hence $\mathcal{Q}_1 \times \cdots \times \mathcal{Q}_n$ is itself compact, convex, and nonempty. Thus $T(\cdot)$ and $\mathcal{Q}_1 \times \cdots \times \mathcal{Q}_n$ meet the requirements of $r(\cdot)$ and X in the Brouwer Fixed Point Theorem (in Appendix 3), which therefore yields the conclusion that $T(\cdot)$ has at least one fixed point. It now remains for us to show that an equilibrium point in the game is a fixed point of $T(\cdot)$ and vice versa. Suppose that $(P_1^\circ, \ldots, P_n^\circ)$ is an equilibrium point. Then $B_i(j; P_1^\circ, \ldots, P_n^\circ)$ must be zero for every j and every i—no player i could improve upon P_i° by selecting any of his pure strategies s_i^j, and hence improvement by selecting any other *randomized* strategy P_i' is also impossible. But if every $B_i(j; P_1^\circ, \ldots, P_n^\circ) = 0$, then every $T_i^j(P_1^\circ, \ldots, P_n^\circ) = P_i^\circ(s_i^j)$ and hence $T(P_1^\circ, \ldots, P_n^\circ) = (P_1^\circ, \ldots, P_n^\circ)$. Thus an equilibrium point in the game is a fixed point of $T(\cdot)$. It remains to see that if $(P_1^\circ, \ldots, P_n^\circ)$ is a fixed point of $T(\cdot)$, then it is also an equilibrium point in the game. We first note that for every i there must exist at least one j such that both $P_i^\circ(s_i^j) > 0$ and $B_i(j; P_1^\circ, \ldots, P_n^\circ) = 0$, because the fact that, by definition,

$$U_i(P_1^\circ, \ldots, P_n^\circ) = \sum_{j=1}^{\#(S_i)} U_i(P_1^\circ, \ldots, P_{i-1}^\circ, s_i^j, P_{i+1}^\circ, \ldots, P_n^\circ) P_i^\circ(s_i^j)$$

implies that

(*) $$U_i(P_1^\circ, \ldots, P_n^\circ) < U_i(P_1^\circ, \ldots, P_{i-1}^\circ, s_i^j, P_{i+1}^\circ, \ldots, P_n^\circ)$$

cannot obtain for every j with $P_i^\circ(s_i^j) > 0$. Hence, there exists at least one j with $P_i^\circ(s_i^j) > 0$ and "$<$" replaced by "$\geq$" in (*), and for this j we have $B_i(j; P_1^\circ, \ldots, P_n^\circ) = 0$. But since $(P_1^\circ, \ldots, P_n^\circ)$ is assumed to be a fixed point of $T(\cdot)$, we have for this j that

(**) $$P_i^\circ(s_i^j) = \frac{P_i^\circ(s_i^j) + 0}{1 + \sum_{k=1}^{\#(S_i)} B_i(k; P_1^\circ, \ldots, P_n^\circ)},$$

which implies that $\Sigma_{k=1}^{\#(S_i)} B_i(k; P_1^\circ, \ldots, P_n^\circ) = 0$. Hence $B_i(j; P_1^\circ, \ldots, P_n^\circ) = 0$ for *every* $j \in \{1, \ldots, \#(S_i)\}$, because they are all nonnegative; and this means that i cannot improve upon P_i° given the others' strategies in $(P_1^\circ, \ldots, P_n^\circ)$. Since this argument obtains for every $i \in \mathcal{N}$, we conclude that no player i can unilaterally improve on his component of $(P_1^\circ, \ldots, P_n^\circ)$. Hence $(P_1^\circ, \ldots, P_n^\circ)$ is an equilibrium point of the game. ■

■ The preceding proof does not point the way to a simple method of *calculating* equilibria in general n-person games. We shall comment on this generally difficult subject in the following section and give you the opportunity of handling an easy, special but important case as Exercise 12.3.8. ■

Theorem 12.3.1 and its proof tell us nothing immediately useful about *how many* equilibrium points $(P_1^\circ, \ldots, P_n^\circ)$ there are in a given game in normal form. Furthermore, we have seen that "battle of the sexes" and "chicken" each have three equilibria whereas "prisoners' dilemma" has only one. There is no easy result concerning the number of equilibria in a game, but some discussion of this subject is productive.

If the equilibrium concept is to be of much value to an individual player i in a noncooperative game of complete information, in normal form, with communication forbidden, then he should be able to choose his components P_i° of an equilibrium point $(P_1^\circ, \ldots, P_n^\circ)$ and rest content. But even if this player thinks that the others will do likewise, he may be in trouble because of the existence of multiple equilibria and the resulting potential ambiguity regarding which strategies the others will *actually* choose. Hence i might be better advised to quantify his judgments regarding the actual strategy choices of the others and then behave as an individual decision maker.

■ *If* these judgments are represented by the $n-1$ other components $P_1^\circ, \ldots, P_{i-1}^\circ, P_{i+1}^\circ, \ldots, P_n^\circ$ of some equilibrium point, then Definition 12.3.2 implies, of course, that i cannot do better *as an individual decision maker* than choose *his* component P_i° of that equilibrium point. Therein lies the appeal of the equilibrium concept. But the "if" is a big one. ■

We shall illustrate with two examples the potentially frustrating ambiguity of the equilibrium concept from the individual player's point of view.

Example 12.3.10

Consider the following *game of coordination,* in which the players have nonconflicting preferences but in which explicit communication is forbidden.

	s_2^1	s_2^2	s_2^3	s_2^4
s_1^1	(8, 8)	(0, 0)	(0, 0)	(0, 0)
s_1^2	(0, 0)	(8, 8)	(0, 0)	(0, 0)
s_1^3	(0, 0)	(0, 0)	(3, 3)	(0, 0)
s_1^4	(0, 0)	(0, 0)	(0, 0)	(8, 8)

There are four equilibria in pure strategies, (s_1^j, s_2^j) for $j = 1, 2, 3, 4$; and every pure strategy for a given player is his component of *some* equilibrium. Hence the equilibrium concept is of no help in choosing his strategy unilaterally. That is because the equilibria are not "interchangeable." The equilibria in a set $\{(P_1^t, \ldots, P_n^t): t \in T\}$ of equilibria in a game are said to be *interchangeable* if any n-tuple $(P_1^*, \ldots, P_n^*)$ which results from each player i's selecting P_i^* to be in his corresponding set $\{P_i^t: t \in T\}$ must also be an equilibrium point. In this game, (s_1^j, s_2^k) is not an equilibrium point unless $j = k$, and hence the equilibria in $\{(s_1^j, s_2^j): j = 1, 2, 3, 4\}$ are not interchangeable. But in this game, there is something special about (s_1^3, s_2^3), something *prominent* which distinguishes it from the other equilibria. Schelling has shown in **[126]** that actual players usually succeed in making independent choices of s_i^3 and thus in achieving an equilibrium point. He reports similar successes in much more complicated experiments involving, say, two players meeting somewhere on a map with five houses, three thickets, four cornfields, and one bridge. You have probably guessed correctly that the one bridge was a *prominent* meeting place. In the specific game considered here, each player's maximin strategy is $P_i^\dagger = [3/17, 3/17, 8/17, 3/17]$, with $U_i(P_i^\dagger, \mathcal{Q}_{\bar{i}}) = 24/14$, which you may deduce from Exercise 12.3.3, and hence player 1 should choose s_1^3 as an individual decision maker if his probability p^3 of 2's choosing s_2^3 exceeds 8/14 and his probability p^j of 2's choosing s_2^j does not exceed $3p^3/8$ for any $j \neq 3$.

Example 12.3.11

Even if equilibria in a set are interchangeable, they need not be "equivalent." The equilibria in a set $\{(P_1^t, \ldots, P_n^t): t \in T\}$ of equilibria are said to be *equivalent* if no player i's utility $U_i(P_1^t, \ldots, P_n^t)$ depends upon t, that is, if all the equilibria in the set yield each player the same utility. In Example 12.3.10 (game of coordination) the equilibria in $\{(s_1^j, s_2^j): J = 1, 2, 4\}$ are equivalent, but (s_1^3, s_2^3) is not equivalent to any of the others. Consider the following game in which each player "determines the other's utility."

	s_2^1	s_2^2
s_1^1	(2, 2)	(0, 2)
s_1^2	(2, 0)	(0, 0)

Here, *every* (P_1, P_2) is an equilibrium pair, but the equilibria in $\mathcal{Q}_1 \times \mathcal{Q}_2$ are obviously not equivalent.

In Exercise 12.3.8 and the remark following it, we show that certain sorts of games have equilibria of certain convenient forms (e.g., equilibria which consist solely of pure strategies).

To summarize this section, we note that each player's choosing a maximin strategy and expecting the others also to do so may induce some players to choose nonmaximin strategies. On the other hand, the equilibrium points $(P_1^\circ, \ldots, P_n^\circ)$, which do not present this unsatisfactory inducement to renege, may have flaws of their own. In the absence of communication, it requires a great leap of faith on the part of each player to assume that the others will choose their components of an equilibrium point, even if the game has only one equilibrium point; but if this leap of faith is not made by a given player, then he has no incentive to use the equilibrium concept in choosing his strategy. And when equilibria in a game are not interchangeable, the equilibrium concept is downright ambiguous from an individual player's point of view.

■ The equilibrium concept assumes greater importance in repetitions of a given game over time, under the assumption that each player's overall utility is the sum of his utilities in the individual games. Such a model also strengthens the instability objection to the maximin solution concept $(P_1^\dagger, \ldots, P_n^\dagger)$. But our previous remark about repetitions of "prisoners' dilemma" indicates that equilibrium is not a good *descriptor* of what is likely to happen in such cases, and the low utility of each player under the equilibrium solution suggests that better *prescriptive* advice can be developed too. Furthermore, in "battle of the sexes," for instance, common sense suffices to show that few wives would continue to choose s_2^1 in each of 100 repetitions so as to maintain the (s_1^1, s_2^1) equilibrium. ■

Exercise 12.3.1

Derive (A) Equation (12.3.10a) and (B) Equation (12.3.10b).

Exercise 12.3.2

Consider the two-by-two, two-person game bimatrix with entries (A_1, A_2), (B_1, C_2), (C_1, B_2), and (D_1, D_2) introduced in the remark following Example 12.3.6.

(A) Find all three equilibria in "battle of the sexes," in which

$$A_1 > D_1 > \max[B_1, C_1] \text{ and } D_2 > A_2 > \max[B_2, C_2].$$

(B) Find all three equilibria in "chicken," in which

$$C_i > A_i > B_i > D_i \text{ for each } i.$$

***Exercise 12.3.3**

Consider a two-person "game of coordination" in which $U_i(s_1{}^j, s_2{}^k) = a_{ik}$ if $j = k$ and equals zero otherwise, for $i = 1, 2$ and $k = 1, 2, \ldots, K$, where every a_{ik} is positive.

(A) Show that each player's maximin strategy $P_i^\dagger$ is given by $P_i^\dagger(s_i{}^k) = 1/[a_{ik} \Sigma_{j=1}^K a_{ij}^{-1}]$ for $k = 1, \ldots, K$ and $i = 1, 2$.
(B) Show that each player's minimax strategy $P_i^{\dagger\dagger}$ is the *other* player's maximin strategy.
(C) Show that $(P_1^{\dagger\dagger}, P_2^{\dagger\dagger})$ is an equilibrium point.

[*Hint*: $P_1^\dagger$ should have the property that $U_1(P_1^\dagger, s_2{}^k)$ does not depend upon k, and $P_2^{\dagger\dagger}$ should have the property that $U_1(s_1{}^k, P_2^{\dagger\dagger})$ does not depend upon k. Thus, $a_{ik}P_i^\dagger(s_i^k) = a_{i1}P_i^\dagger(s_i^1)$ for $k = 1, \ldots, K$. Placing higher probability than $P_i(s_i{}^k)$ on s_i^k would tempt the "opponent" to reduce his probability of choosing his kth strategy.]

■ There are many, many equilibrium points besides the $(s_1{}^k, s_2{}^k)$'s and $(P_1^{\dagger\dagger}, P_2^{\dagger\dagger})$ when K is large. The pair of minimax strategies for the "principal minor" bimatrix, formed by deleting some rows and the *corresponding* columns of the original K-by-K bimatrix produces an equilibrium point, once probability zero is assigned to the deleted rows and columns. In Example 12.3.10, for instance, ([3/11, 0, 8/11, 0], [3/11, 0, 8/11, 0]) is an equilibrium point. ■

Exercise 12.3.4

Find an equilibrium point $(P_1^\circ, P_2^\circ, P_3^\circ)$ for "elaborate throwing fingers," as given by Table B in Example 12.2.5.
[*Hint*: $P_3^\circ = [1, 0]$, since $s_3{}^1$ is a better response by 3 than $s_3{}^2$ to everything except $(s_1{}^2, s_2{}^2)$, and $s_3{}^1$ is as good as $s_3{}^2$ in response to $(s_1{}^2, s_2{}^2)$.]

Exercise 12.3.5

Consider the following two-person games in normal form.

Game 1

	s_2^1	s_2^2
s_1^1	(2.00, 3.80)	(.75, .80)
s_1^2	(.50, .60)	(4.25, 2.60)

Game 2

	s_2^1	s_2^2
s_1^1	(0, 20)	(7, 7)
s_1^2	(0, 20)	(2, 15)
s_1^3	(14, 8)	(20, 0)
s_1^4	(9, 9)	(20, 0)
s_1^5	(5, 5)	(5, 5)

(A) Find $P_1^\dagger$, $P_2^\dagger$, and three equilibria in game 1.
(B) Find $P_1^\dagger$, $P_2^\dagger$, and one equilibrium in game 2.

***Exercise 12.3.6**

Prove that if S_i is a finite set for every $i \in \mathcal{N}$, then:

(A) A maximin strategy P_i for player i exists, for every $i \in \mathcal{N}$.
(B) The set $\mathcal{Q}_i^\dagger = \{P_i^\dagger : P_i^\dagger \text{ is a maximin strategy for } i\}$ is convex.

[*Hints*: (1) $U_i(P_i, s_{\bar{i}})$ is a linear function of P_i.
(2) $U_i(\cdot, \mathcal{Q}_{\bar{i}})$ is the pointwise minimum of a finite number of linear functions and is therefore continuous and concave on $\mathcal{Q}_i$.
(3) $\mathcal{Q}_i$ is nonempty and compact.]

Exercise 12.3.7: "Cutthroat Auction"[c]

Suppose that players 1 and 2 are concerned only with their own net monetary returns and have utility functions linear in return. An object of value v_i to player $i \in \{1, 2\}$ is to be auctioned to the higher bidder in the usual manner, in which the bidders successively cry out successively higher bids until the current underdog no longer wishes to escalate. Then the high bidder (say, i) pays the amount b_i of his latest bid and gets the object, for a net return of $v_i - b_i$; the other bidder gets nothing. But, *unlike* usual auctions, the low bidder loses his last bid—say, to reflect his unsuccessful commitment in the game. No collusion is allowed; each v_i is a multiple of a "basic unit" h; player 2 will be awarded the object for nothing unless 1 bids; each v_i is positive; and "preemption" is not allowed: the first bid must be h (by player 1) and successive bids must differ by h. A pure strategy for i consists of a *maximum* bid s_i beyond which he will not bid. According to the rules of the game, s_1 is either zero or an *odd* positive multiple of h, and s_2 is either zero or an *even* positive multiple of h.

(A) Show that $U_2(0, 0) = v_2$, but otherwise

$$U_i(s_1, s_2) = \begin{cases} v_i - (s_j + h), & s_i > s_j \\ -s_i, & s_i \leq s_j \end{cases}$$

for $i = 1, 2$ and $j \neq i$.

[c] This game was apparently formulated by M. Shubik.

(B) Show that $s_i = 0$ is maximin for each player i.
(C) Show that the following sorts of pure-strategy pairs are equilibria:
  (i) $(0, s_2)$ for any $s_2 \geq v_1 - h$.
  (ii) $(s_1, 0)$ for any $s_1 \geq v_2 - h$.

■ These results indicate a strong incentive on the part of each player to commit himself credibly to a limit so high that the other bidder could not win except by bidding more than the value of the object to him. Whoever succeeds in being first to "preempt" with such a commitment has succeeded indeed. ■

■ Suppose that the cutthroat auction is modified by having the lower bidder i pay a fixed sum $L_i \geq 0$ instead of his last bid s_i. One might interpret L_i as the value of commitments elsewhere lost because of i's failure to win the bidding in this game, given that i actually entered the bidding with $s_i > 0$. This suggests the "domino theory." Then it follows that every limit s_i exceeding $v_i + L_i$ is *dominated by* $s_i - 2h$, in the sense that $U_i(s_i, s_j) \leq U_i(s_i - 2h, s_j)$ for every s_j, with strict inequality "$<$" obtaining for some s_j. ■

■ Try selling a dollar bill at a cocktail party, with $h = 10$ cents, under the rules of Exercise 12.3.7 with the one modification for $n > 2$ players being that only the second highest bidder has to pay the amount of his final bid. The trapped look of the 90-cent bidder when he realizes that he should bid \$1.10, and the startled looks of the spectators when he does so, are sights to behold. ■

Exercise 12.3.8

A game in extensive form is said to be a *game of perfect information*[d] if every information set consists of only one move.

(A) Verify Kuhn's Corollary 1 in **[74]** that every game of perfect information has at least one equilibrium point $(P_1^\circ, \ldots, P_n^\circ)$ in which *every* P_i° is a *pure* strategy.
[*Hint*: Work backward from the *players*' terminal moves, pruning all but one act branch in each move and recording the utility n-tuple of the remaining (optimal-for-the-move-owner) branch as the "evaluation" of that move. As usual, take component-by-component expectations over the branches of $\mathscr{E}$-moves, using player i's judgments for the component-i expectations.]
(B) A chess game ends in a tie if neither player has achieved "checkmate" in a specified, finite number of moves (in the parlor-game sense of "move"). Verify that chess is a game of perfect information. Can you conceive of how huge the description of each player's component of an equilibrium strategy would be, and how

[d] [*Caution*: This terminology conflicts with "perfect information" as defined in statistical decision theory.]

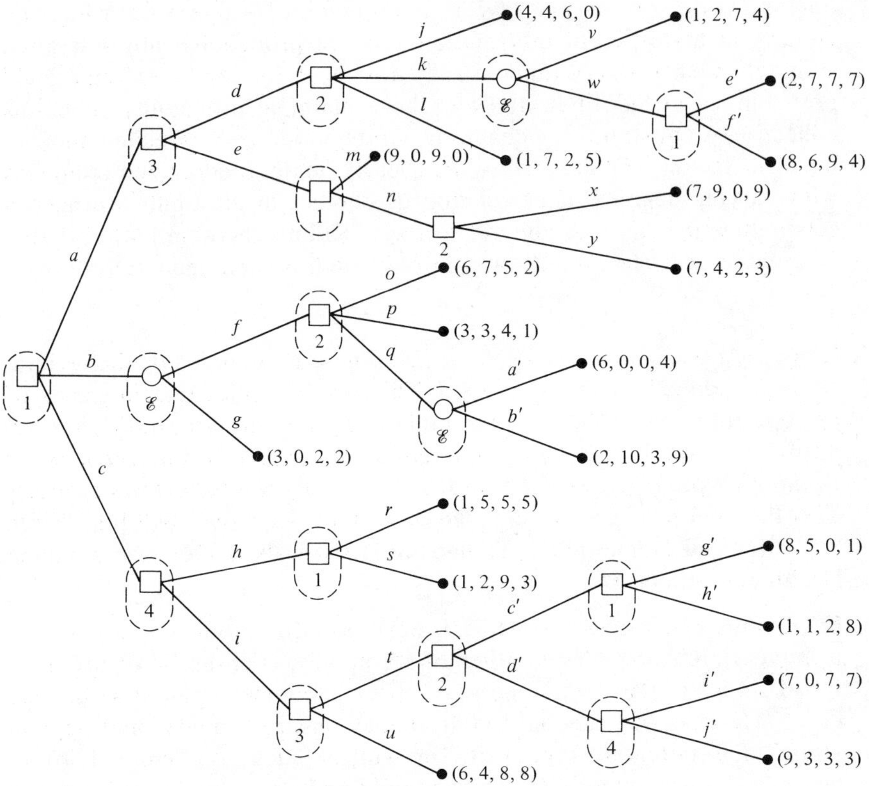

1's probabilities in $\mathscr{E}$-moves: $P(f) = .2, P(v) = .6$, and $P(a') = .5$.
2's probabilities in $\mathscr{E}$-moves: $P(f) = .3, P(v) = .4$, and $P(a') = .3$.
3's probabilities in $\mathscr{E}$-moves: $P(f) = .1, P(v) = .7$, and $P(a') = .1$.
4's probabilities in $\mathscr{E}$-moves: $P(f) = .8, P(v) = .1$, and $P(a') = .9$.

Figure 12.11 Game for Exercise 12.3.8(C).

hard it would be to determine an equilibrium point for chess?

(C) Find *all* pure-strategy equilibria in the game depicted in Figure 12.11.

■ A game in extensive form is said to be a *game of perfect recall* if every player remembers what he did in previous information sets, in the sense that never do we have two information sets $G_{it'}$ and $G_{it''}$ and a choice $g^*_{it'}$—construed as a set of branches, one from each of the moves in $G_{it'}$—such that some move in $G_{it''}$ is reachable given i's previous choice of $g^*_{it'}$ and another move in $G_{it''}$ is *not* reachable given i's previous choice of $g^*_{it'}$. (Such information sets would indicate that at $G_{it''}$ player i must have forgotten whether he chose $g^*_{it'}$ in $G_{it'}$, since otherwise $G_{it''}$ would have to be partitioned into two information sets, one consisting of all moves reachable from $g^*_{it'}$ and the other consisting of the remaining moves. Remember, an information set represents ignorance!) A *be-*

havioral strategy for player i is a randomized strategy in which i's choices in his different information sets are probabilistically unrelated (= independent). Every randomized strategy P_i for i induces "marginal" probabilities in his different G_{it}'s; these marginal probability functions collectively constitute a behavioral strategy for i; it is called the *behavioral strategy* P_i^b *induced by* P_i. Clearly, many randomized strategies with varying degrees of correlation of choices in different information sets induce the same behavioral strategy. Kuhn's theorem [**74**] says that *in a game of perfect recall* (and in only such a game, generally),

$$U_i(P_1^{\,b}, \ldots, P_n^{\,b}) = U_i(P_1, \ldots, P_n)$$

for every $i \in \mathcal{N}$ and every $(P_1, \ldots, P_n)$, where P_i^b denotes the behavioral strategy induced by P_i for every i. This result implies that in games of perfect recall the players can randomize on an information-set-by-information-set basis. Thus *in a game of perfect recall, there is an equilibrium point* $(P_1^\circ, \ldots, P_n^\circ)$ *in which every* P_i° *is a behavioral strategy.* This theorem has been generalized in various ways by Isbell [**66**], Birch [**9**], Otter and Dunne [**107**], Thompson [**132**], Dalkey [**20**], and Aumann [**4**], among others. ■

■ Obviously, in a game of perfect recall no information set can contain a move which precedes another of its moves, such as in Figure 12.4. Isbell shows in [**66**] that some very queer phenomena can arise in such cases. He calls our randomized strategies *linear* strategies and demonstrates the potential superiority in games (such as Figure 12.4) of strategies which involve (1) i's randomizing in each of his information sets and (2) i's choosing a joint randomization over such "local" randomizations. In Figure 12.4, for instance, he shows that 1 can guarantee a utility of only 1/2 by randomizing in the usual "linear" manner, placing probability 1/2 on each of the pure strategies (a, e) and (b, d); but 1 can guarantee a utility of 9/16 by placing probability 1/2 on each of the following pairs of locally randomized (behavioral) strategies.

$$\textit{pair 1} \quad \begin{bmatrix} P(a) = 3/4, P(b) = 0, P(c) = 1/4 \\ P(d) = 0, P(e) = 1 \end{bmatrix}$$

$$\textit{pair 2} \quad \begin{bmatrix} P(a) = 0, P(b) = 3/4, P(c) = 1/4 \\ P(d) = 1, P(e) = 0 \end{bmatrix}.$$

This line of analysis is very interesting, but so complex in application as to make one think twice before accepting information sets containing two moves with one a possible successor to the other! ■

Exercise 12.3.9: Dominance and Iterated Dominance

Just as with an individual decision maker in Chapter 7, no player i has an incentive to choose a strategy P_i which is *dominated*, in the sense that there exists another strategy P_i^* in $\mathcal{Q}_i$ such that $U_i(P_i^*, s_{\bar{i}}) \geq U_i(P_i, s_{\bar{i}})$ for every $s_{\bar{i}}$ in $S_{\bar{i}}$ and ">" obtains for some $s_{\bar{i}}$. A strategy P_i of player i is said to be *strongly dominated* if i has another strategy P_i^* such that $U_i(P_i^*, s_{\bar{i}}) > U_i(P_i, s_{\bar{i}})$ for every $s_{\bar{i}}$ in $S_{\bar{i}}$.

(A) Show that if P_i is strongly dominated, then P_i is neither maximin for i nor i's component of any equilibrium point.

(B) Construct an example in which $n = 2$ to illustrate the fact that a dominated, but not strongly dominated, strategy P_1 may be maximin for 1 and also 1's component of an equilibrium point.

(C) *Iterated Dominance.* Consider the iterative procedure in which:
  (1) Each player i eliminates all his dominated pure strategies, leaving a subset $S_i^{(1)}$ of S_i.
  (2) In the reduced game with pure-strategy sets $S_1^{(1)}, \ldots, S_n^{(1)}$, each player eliminates all his dominated pure strategies, leaving a subset $S_i^{(2)}$ of $S_i^{(1)}$ and hence also of S_i.

  And so on, until no player can eliminate any more pure strategies. Show that there exists an equilibrium point $(P_1^\circ, \ldots, P_n^\circ)$ in the original game such that, for every player i, P_i° assigns probability zero to every strategy eliminated at some point in the iterative procedure.

(D) Find an equilibrium point in the following game.

	s_2^1	s_2^2	s_2^3	s_2^4
s_1^1	(10, 2)	(9, 5)	(9, 6)	(5, 2)
s_1^2	(15, 10)	(8, 18)	(5, 10)	(0, 25)
s_1^3	(10, 10)	(4, 20)	(3, 5)	(5, 0)
s_1^4	(12, 8)	(20, 9)	(0, 9)	(7, 15)
s_1^5	(17, 4)	(10, 3)	(3, 5)	(8, 7)

■ Applying iterated dominance to the fullest extent possible in this game yields a "battle of the sexes." ■

12.4 TWO-PERSON ZEROSUM GAMES

The most extreme situation in which the players' interests conflict is a two-person game in which the players' preferences are so divergent that they cannot agree on anything essential to the situation. More precisely, let "$\succsim_i$" denote "at least as desirable from i's standpoint as." By *totally opposing preferences* we mean that

$$\ell'' \begin{Bmatrix} \succ_1 \\ \sim_1 \\ \prec_1 \end{Bmatrix} \ell' \quad \text{if and only if} \quad \ell' \begin{Bmatrix} \succ_2 \\ \sim_2 \\ \prec_2 \end{Bmatrix} \ell'' \tag{12.4.1}$$

for any two lotteries ℓ' and ℓ'' with outcomes in the set C of potential consequences $c[s_1, s_2]$ of the players' strategy choices in the game.[e]

Definition 12.4.1. A game with two players whose preferences are totally opposing is called a *two-person zerosum game.*

The name "zerosum" derives from the fact that by choosing inessential parameters[f] of player 2's utility function $u_2(\cdot)$ appropriately, we have

$$u_2(c) = -u_1(c) \tag{12.4.2}$$

for every c in $C = \{c[s_1, s_2]: s_1 \in S_1, s_2 \in S_2\}$.

■ Pick two reference consequences c^- and c^+ such that $c^+ >_1 c^-$. By (12.4.1) with $\ell'' =$ "c^+ for sure" and $\ell' =$ "c^- for sure," we have $c^- >_2 c^+$. Therefore we may choose $u_2(c^-) = -u_1(c^-)$ and $u_2(c^+) = -u_1(c^+)$ and thus obtain $u_2(c^-) > u_2(c^+)$, as necessary. Now suppose that $c^+ >_1 c >_1 c^-$ and $c \sim_1 \ell^*(p; c^+, c^-)$. By (12.4.1), $c \sim_2 \ell^*(p; c^+, c^-)$, so that $u_2(c) = u_2(c^-) + p[u_2(c^+) - u_2(c^-)] = -(u_1(c^-) + p[u_1(c^+) - u_1(c^-)]) = -u_1(c)$. Similar arguments show that (12.4.2) obtains when $c >_1 c^+$ and $c^- >_1 c$. ■

Example 12.4.1

Simple "throwing fingers" is a two-person zerosum game, as is the game depicted in Figure 12.4.

Since we can recapture 2's utility function via (12.4.2) from 1's, we shall omit mention of $u_2(\cdot)$ in this section, because (12.4.2) implies that 2's *objective of maximizing* U_2 (i.e., helping himself) *is equivalent to the objective of minimizing* U_1 (i.e., hurting 1).

Theorem 12.4.1. In a two-person zerosum game, a maximin strategy for i is also a minimax strategy for i, and vice versa.

■ *Proof.* Since $U_2(P_1, P_2) = -U_1(P_1, P_2)$ for every (P_1, P_2) in $\mathcal{Q}_1 \times \mathcal{Q}_2$, and since $\max_t [-x(t)] = -\min_t [x(t)]$ and $\min_t [-x(t)] = -\max_t [x(t)]$, it follows immediately that

$$\begin{aligned} &-U_2(P_1^\dagger, \mathcal{Q}_2) \\ &= U_1(P_1^\dagger, \mathcal{Q}_2) \\ &= \max_{P_1 \in \mathcal{Q}_1} [\min_{P_2 \in \mathcal{Q}_2} [U_1(P_1, P_2)]] \\ &= \max_{P_1 \in \mathcal{Q}_1} [\min_{P_2 \in \mathcal{Q}_2} [-U_2(P_1, P_2)]] \\ &= -\min_{P_1 \in \mathcal{Q}_1} [\max_{P_2 \in \mathcal{Q}_2} [U_2(P_1, P_2)]] \\ &= -U_2(P_1^{\dagger\dagger}, \mathcal{Q}_2), \end{aligned}$$

implying that $U_2(P_1^\dagger, \mathcal{Q}_2) = U_2(P_1^{\dagger\dagger}, \mathcal{Q}_2)$ and hence that $P_1^\dagger$ is a minimax

[e] These consequences will be lotteries on Σ if $\mathscr{C}$ is a participant.

[f] Parameters which add a constant to all utilities or multiply all utilities by a positive constant are inessential.

strategy for 1 against 2 as well as maximin for himself. Interchange the roles of 1 and 2 above to obtain the same result for player 2. ■

The fact that maximin strategies are also minimax is very useful. In the remainder of this section we shall use $P_1^\dagger$ to mean a maximin strategy for 1 and $P_2^\dagger$ to mean a minimax strategy for 2, realizing that $P_2^\dagger$ is also maximin for 2 by virtue of Theorem 12.4.1. With attention confined to 1's utility $U_1(\cdot, \cdot)$, we note that semipessimistically cautious behavior means maximin for 1 and minimax[g] for 2.

The "a priori floor" $U_1^{\,1}$ that 1 can ensure via choosing a maximin strategy cannot exceed the "a priori ceiling" $U_1^{\,2}$ which 2 can impose on 1 by choosing a minimax strategy.

Lemma 12.4.1. Let

$$U_1^{\,1} = \max_{P_1 \in \mathcal{Q}_1} [\min_{P_2 \in \mathcal{Q}_2} [U_1(P_1, P_2)]]$$

and

$$U_1^{\,2} = \min_{P_2 \in \mathcal{Q}_2} [\max_{P_1 \in \mathcal{Q}_1} [U_1(P_1, P_2)]].$$

Then

$$U_1^{\,1} \le U_1^{\,2}.$$

■ *Proof.* Since

$$\min_{P_2 \in \mathcal{Q}_2} [U_1(P_1, P_2)] \le U_1(P_1, P_2) \le \max_{P_1 \in \mathcal{Q}_1} [U_1(P_1, P_2)]$$

for every (P_1, P_2) in $\mathcal{Q}_1 \times \mathcal{Q}_2$, it follows that

$$(*) \qquad \min_{P_2 \in \mathcal{Q}_2} [U_1(P_1, P_2)] \le \max_{P_1 \in \mathcal{Q}_1} [U_1(P_1, P_2)]$$

for every (P_1, P_2) in $\mathcal{Q}_1 \times \mathcal{Q}_2$. Hence the maximum $U_1^{\,1}$ of the left-hand side of (*) cannot exceed the minimum $U_1^{\,2}$ of the right-hand side of (*). ■

■ If the randomized strategies P_1 and P_2 are restricted to be *pure*, then we may have $U_1^{\,1} < U_1^{\,2}$, which also can occur in games where one or both players have infinitely many pure strategies. In fact, it was to prevent $U_1^{\,1} < U_1^{\,2}$ that genuinely randomized strategies were originally introduced. ■

Corollary 12.4.1. In every two-person zerosum game in which S_1 and S_2 are both finite, there exists at least one equilibrium pair (P_1°, P_2°).

■ *Proof.* Immediate from Theorem 12.3.1. ■

Theorem 12.4.2 follows readily from Lemma 12.4.1 and Corollary 12.4.1. Its essence is that there is no conflict in a two-person zerosum game between the equilibrium and maximin solution concepts; or equivalently, neither player has any incentive not to be semipessimistically conservative if he suspects that the other player will be the same.

---

[g] "Maximin" is an abbreviation for *maximum minimorum*; "minimax," for *minimum maximorum*. A generalization which covers both and embodies the real objectives of each player is "optipes," for *optimum pessimorum*—the best of the worst.

Theorem 12.4.2. In a two-person zerosum game with S_1 and S_2 each finite,

(A) $U_1^1 = U_1^2$.

(B) (P_1°, P_2°) is an equilibrium point if and only if P_1° is maximin for 1 and P_2° is minimax for 2.

(C) There exists a number y, a strategy P_1° in $\mathcal{Q}_1$, and a strategy P_2° in $\mathcal{Q}_2$ such that

$$\sum_{j=1}^{\#(S_1)} U_1(s_1^j, s_2^k)P_1^\circ(s_1^j) \geq y \qquad \text{for every } k \in \{1, \ldots, \#(S_2)\},$$

and

$$\sum_{k=1}^{\#(S_2)} U_1(s_1^j, s_2^k)P_2^\circ(s_2^k) \leq y \qquad \text{for every } j \in \{1, \ldots, \#(S_1)\}.$$

■ *Proof.* (A) and (B) both follow from the relation chain

$$\begin{aligned}
&U_1^2\\
&= \min_{P_2\in\mathcal{Q}_2}[\max_{P_1\in\mathcal{Q}_1}[U_1(P_1, P_2)]] && \text{(Definition of } U_1^2)\\
&\leq \max_{P_1\in\mathcal{Q}_1}[U_1(P_1, P_2^\circ)] && \text{(Definition of min)}\\
&= U_1(P_1^\circ, P_2^\circ) \text{ whenever } P_1^\circ \text{ is a best response to } P_2^\circ\\
&= \min_{P_2\in\mathcal{Q}_2}[U_1(P_1^\circ, P_2)] \text{ whenever } P_2^\circ \text{ is a best response to } P_1^\circ\\
&\leq \max_{P_1\in\mathcal{Q}_1}[\min_{P_2\in\mathcal{Q}_2}[U_1(P_1, P_2)]] && \text{(Definition of max)}\\
&= U_1^1. && \text{(Definition of } U_1^1)
\end{aligned}$$

Hence $U_1^2 \leq U_1^1$, which together with $U_1^2 \geq U_1^1$ from Lemma 12.4.1 implies (A). It is clear from this chain that (P_1°, P_2°) is an equilibrium point if and only if the two middle equalities obtain, in which case P_1° is maximin for 1 and P_2° is minimax for 2. To prove (C), let $y = U_1^1 = U_1^2$, and suppose that P_1° is maximin for 1 and P_2° is minimax for 2. [Hence, (P_1°, P_2°) is an equilibrium point.] If the first equality of (C) failed for some k, say k^*, then 2's best response would be to place probability one on $s_2^{k^*}$ and thus hold 1's utility (the left-hand side) down to less than $U_1^1 = y$. A similar argument obtains for the second, "$\leq$" inequality of (C). ■

Finding an equilibrium point in a two-person zerosum game amounts, therefore, to finding a maximin strategy for 1 and a minimax strategy for 2. Each player's adopting the maximin philosophy does *not* lead to the instability in the form of incentives to deviate from maximin that we observed in other sorts of games in Section 12.3. Furthermore, in two-person zerosum games there is no incentive whatsoever for the players to communicate or to collaborate; whatever a player could accomplish by doing so would require the other player's acting against his own interests, which is not to be expected.

Another very pleasant and important consequence of Theorem 12.4.2(B) is that ambiguity cannot arise from multiple equilibria. Since *any* maximin strategy for 1 and *any* minimax strategy for 2 jointly constitute an equilibrium point, neither player need second-guess the other's strategy choice.

■ In the terminology of Example 12.3.10, *any* two equilibria are interchangeable. Moreover, uniqueness of $y = U_1^{\ 1} = U_1^{\ 2}$ implies that all equilibria are equivalent. Furthermore, by virtue of *Exercise 12.3.6, the set $\{P_1^\dagger\colon P_1^\dagger \text{ is maximin for } 1\} \times \{P_2^\dagger\colon P_2^\dagger \text{ is minimax for } 2\}$ of all equilibria is a convex set. ■

The discussion in Section 12.3 on finding maximin and minimax strategies is pertinent here to finding equilibria, by virtue of Theorem 12.4.2(B), with $i = 1$ and $\bar{i} = 2$. The geometric approach there is of immediate pertinence, and so is the geometric approach in Chapter 7 for characterizing maximin strategies for 1 [and for conveniently locating them when $\#(S_2) = 2$]. In the Chapter 7 approach, P_2° is just a least favorable strategy for $\mathscr{E}$ from $\mathscr{D}$'s viewpoint, since $\mathscr{E}$ and $\mathscr{D}$ there play the roles of 2 and 1, respectively.

■ The Section 12.3 approach to finding minimax strategies when $\#(S_1) = 2$ can be adapted to finding maximin strategies when $\#(S_2) = 2$. For each s_1 in S_1 draw the line $\{U_1(s_1, [p, 1-p])\colon 0 \leq p \leq 1\}$; then heavily draw over the (uppermost-segments) line $\{\max_{s_1 \in S_1} [U_1(s_1, [p, 1-p])]\colon 0 \leq p \leq 1\}$; find a p° which *minimizes* this piecewise linear function; declare $P_2^\circ = [p^\circ, 1-p^\circ]$; and use the now obvious analogues of the four cases described in Section 12.3 to find P_1°. Whenever we said "$\bar{i}$" and "minimize" in Section 12.3, we say "1" and "maximize" in the context of this remark. ■

■ By the same token, the geometric approach in Chapter 7 for finding maximin strategies when $\#(S_2) = 2$ can be adapted to finding minimax strategies when $\#(S_1) = 2$. ■

Thus, it is not too hard to find P_1° and P_2° when $\#(S_1) = 2$ and/or $\#(S_2) = 2$. But when each player has more than two pure strategies, this task is very much more complicated except in special cases such as *Exercise 12.3.3. But the *numerical* task is *easy* for anyone who knows how easily "linear programming" problems can be solved by the simplex method.

We close this section with a description of how (P_1°, P_2°) can be found in a two-person zerosum game by linear programming, and how an equilibrium point (P_1, P_2)—not necessarily a maximin-minimax pair—can be found in a two-person nonzerosum game by quadratic programming. Readers who are not famimilar with one or both of these subjects should skip the corresponding remarks.

*■ Theorem 12.4.2(C) furnishes a strong hint as to how to solve for P_1° and P_2°. As a preliminary, we want to ensure that 1's floor, y, is *positive*,

as will be true if he has some pure strategy s_1^* such that $U_1(s_1^*, s_2^k)$ is positive for all k. If there is no such s_1^* with U_1 as originally defined, add a constant to all values $U_1(s_1^j, s_2^k)$ to produce such a s_1^*. This addition amounts to changing $u_1(\cdot)$ by an inessential parameter and it therefore has no strategic relevance. Now let $x_j = P_1(s_1^j)/y$ for $j = 1, \ldots, \#(S_1)$ and $z_k = P_2(s_2^k)/y$ for $k = 1, \ldots, \#(S_2)$. We consider *2's problem* first. Since 2 wants to minimize 1's ceiling y, and [since $P_2(s_2^1), \ldots, P_2(s_2^{\#(S_2)})$ sum to one]

$$\sum_{k=1}^{\#(S_2)} z_k = \sum_{k=1}^{\#(S_2)} P_2(s_2^k)/y = 1/y,$$

it follows that 2 wants to *maximize* $\sum_{k=1}^{\#(S_2)} z_k$. But by the facts that every $P_2(s_2^k)$ must be nonnegative and y must be positive, we see that $z_k \geq 0$ for every k. Furthermore, the "$\leq$" inequalities in Theorem 12.4.2(C) can be divided on each side by y to yield

$$\sum_{k=1}^{\#(S_2)} U_1(s_1^j, s_2^k) z_k \leq 1 \qquad \text{for every } j \in \{1, \ldots, \#(S_1)\}.$$

Hence *2's linear programming problem* is

maximize

$$\sum_{k=1}^{\#(S_2)} z_k$$

subject to

$$\sum_{k=1}^{\#(S_2)} U_1(s_1^j, s_2^k) z_k \leq 1 \qquad \text{for every } j \in \{1, \ldots, \#(S_1)\}$$

and

$$z_k \geq 0 \qquad \text{for every } k \in \{1, \ldots, \#(S_2)\}.$$

Knowledge of y is not required for solving this problem, but obviously $y = 1/\sum_{k=1}^{\#(S_2)} z_k^\circ$, where $\mathbf{z}^\circ = (z_1^\circ, \ldots, z_{\#(S_2)}^\circ)$ is the optimal solution. Moreover, it is clear that $P_2^\circ = y \cdot z^\circ$. Now, player *1's problem* is very similar. Since he wants to maximize his floor, y, it follows that he wants to minimize

$$\sum_{j=1}^{\#(S_1)} x_j = \sum_{j=1}^{\#(S_1)} P_1(s_1^j)/y = 1/y;$$

that x_j must be nonnegative for every j; and that the "$\geq$" inequalities of Theorem 12.4.2(C) are tantamount to

$$\sum_{j=1}^{\#(S_1)} U_1(s_1^j, s_2^k) x_j \geq 1 \qquad \text{for every } k \in \{1, \ldots, \#(S_2)\}.$$

Hence *1's linear programming problem* is

minimize

$$\sum_{j=1}^{\#(S_1)} x_j$$

subject to

$$\sum_{j=1}^{\#(S_1)} U_1(s_1^j, s_2^k)x_j \geq 1 \qquad \text{for every } k \in \{1, \ldots, \#(S_2)\}$$

and

$$x_j \geq 0 \qquad \text{for every } j \in \{1, \ldots, \#(S_1)\}.$$

Let $\mathbf{x}^\circ$ be an optimal solution to this problem. Then $y = 1/\Sigma_{j=1}^{\#(S_1)} x_j^\circ$ and $P_1^\circ = y \cdot \mathbf{x}^\circ$. Now a cursory examination of 1's and 2's problems indicates that they are dual to each other, and hence only one need be solved by the simplex method, in which the dual solution is readily obtainable from the final tableau of the primal problem. [If a constant K was added to all of 1's utilities to produce the $U_1(s_1^j, s_2^k)$'s in these dual linear programming problems, then the y obtained in them is larger than the y with the original utilities by exactly K.] ■

*■ In [97], Mills shows how finding equilibrium points in n-person games may be formulated in terms of mathematical programming problems. We shall sketch this approach for $n = 2$, in which case the result is a certain sort of quadratic programming problem. We need to assume a bit of linear algebra. Let $\mathbf{U}_i$ denote the matrix whose (j, k)th element is $U_i(s_1^j, s_2^k)$; add constants K_1 and K_2 to all values of $\mathbf{U}_1$ and $\mathbf{U}_2$ respectively so that all will be nonnegative. Regard i's strategy P_i as a *column* vector unless transposed, denoted by P_i'. Then $U_i(P_1, P_2) = P_1'\mathbf{U}_iP_2$, for example. Let $\mathbf{1}$ denote a column vector every component of which is one. The basic result is that (P_1°, P_2°) is an equilibrium point of the game if and only if $\mathbf{z}^\circ = (P_1^{\circ\prime}, P_2^{\circ\prime}, y_1^\circ, y_2^\circ)'$ is an optimal solution to the problem of choosing (P_1', P_2', y_1, y_2) so as to

maximize

$$P_1'(\mathbf{U}_1 + \mathbf{U}_2)P_2 - y_1 - y_2$$

subject to

$$\begin{aligned} \mathbf{U}_1P_2 &\leq y_1\mathbf{1} \\ P_1'\mathbf{U}_2 &\leq y_2\mathbf{1}' \\ P_i &\geq \mathbf{0} \qquad \text{for } i = 1, 2 \\ \mathbf{1}'P_i &= 1 \qquad \text{for } i = 1, 2 \\ y_i &\geq 0 \qquad \text{for } i = 1, 2. \end{aligned}$$

Moreover, the optimal value of the maximand is zero. This optimization problem can be reformulated by a simple elaboration of notation as a quadratic programming problem of the form

maximize

$$\mathbf{z}'\mathbf{D}\mathbf{z} - \mathbf{c}'\mathbf{z}$$

subject to

$$\mathbf{A}\mathbf{z} \begin{Bmatrix} \leq \\ = \end{Bmatrix} \mathbf{b}$$

$$\mathbf{z} \geq \mathbf{0};$$

but be careful; it is not a quadratic programming of the easily solvable sort that arose in Chapter 4 in connection with the portfolio problem. Subsequent references on the computation of equilibrium points in two-person nonzerosum games are Lemke [**82**] and Lemke and Howson [**83**]. ■

Exercise 12.4.1

A two-person *game of opposing interests* is one in which

$$c'' \begin{Bmatrix} >_1 \\ \sim_1 \\ <_1 \end{Bmatrix} c' \quad \text{if and only if} \quad c' \begin{Bmatrix} >_2 \\ \sim_2 \\ <_2 \end{Bmatrix} c''$$

for every c' and c'' in the set C of potential consequences $c[s_1, s_2]$ of the players' strategy choices in the game.[h]

(A) Show that a two-person zerosum game is a game of opposing interests.

(B) Consider the two-person game with *consequence* table

	s_2^1	s_2^2
s_1^1	c^1	c^3
s_1^2	c^3	c^2

in which the players' utilities are $u_1(c^1) = 10$, $u_1(c^2) = 25$, $u_1(c^3) = 16$, $u_2(c^1) = 100$, $u_2(c^2) = 50$, and $u_2(c^3) = q$ for some q greater than 50 and less than 100.

(i) Show that for every such q the game is one of opposing interests.

(ii) Show that only for $q = 80$ are the players' interests *totally* opposed, and hence only for $q = 80$ may the game be regarded as zerosum.

(C) A *saddlepoint* in a game of opposing interests with typical consequence $c[s_1^j, s_2^k]$ is a consequence $c[s_1^*, s_2^*]$ such that

$$c[s_1^*, s_2^*] \gtrsim_1 c[s_1^j, s_2^*]$$

for every $j \in \{1, \ldots, \#(S_1)\}$ and

$$c[s_1^*, s_2^k] \gtrsim_1 c[s_1^*, s_2^*]$$

for every $k \in \{1, \ldots, \#(S_2)\}$. Show that whatever 2's specific, numerical utility function is, (s_1^*, s_2^*) is an equilibrium point if $c[s_1^*, s_2^*]$ is a saddlepoint.

■ Luce and Raiffa [**92**, pp. 64 and 65] describe the Battle of the Bismarck Sea in World War II as a game with opposing interests which

[h] Again, these consequences will be lotteries on Σ if $\mathscr{E}$ is a participant.

has a saddlepoint. Allied and enemy commanders chose s_1^* and s_2^* respectively, as the theory would anticipate. This game-theoretic interpretation was due to Haywood **[57]**, who also discussed a crucial decision by Generals Bradley and von Kluge at Normandy. ■

Exercise 12.4.2

Find P_1° and P_2° in the two-person zerosum game with the following table of 1's utilities.

	s_2^1	s_2^2
s_1^1	1	t
s_1^2	0	2

as functions of t, for $t \in (-\infty, +\infty)$.

Exercise 12.4.3

Find P_1° and P_2° in the two-person zerosum game with the following table of 1's utilities.

	s_2^1	s_2^2	s_2^3	s_2^4
s_1^1	1	4	3	2
s_1^2	2	1	4	3
s_1^3	3	2	1	4
s_1^4	4	3	2	1

***Exercise 12.4.4**

Consider a "hide-and-seek," two-person zerosum game in which each player has K pure strategies and

$$U_1(s_1^j, s_2^k) = \begin{cases} a_k, & j = k \\ 0, & j \neq k. \end{cases}$$

In Exercise 12.3.3 the maximin and minimax strategies for 1 and 2 respectively were characterized under the assumption that $a_k > 0$ for every k. Suppose that $a_1 \geq a_2 \geq \cdots \geq a_K$, with the first K' of the a_k's positive, the next $K'' - K'$ zero, and the final $K - K''$ negative. Assume that $0 < K' \leq K'' < K$.

(A) Show that P_1 is a maximin strategy for 1 if and only if $P_1(s_1^j) = 0$ for every $j > K''$.

(B) Show that P_2 is a minimax strategy for 1 if and only if $P_2(s_2^k) = 0$ for every $k \leq K'$.

C: COOPERATIVE GAMES

12.5 TWO-PERSON COOPERATIVE GAMES

In Sections 12.3 and 12.4 we have examined the difficulties which arise in noncooperative games, where communication and collusion are forbidden. It is therefore to be hoped that cooperative games do not present such problems. We shall see, however, that other problems arise when the players are allowed to communicate freely and to enter into mutually binding agreements.

This section is largely restricted to *two*-person cooperative games, for two basic reasons. (1) When $n > 2$ the role of *coalitions* becomes very important, as some players may have incentives to gang up on other players. (2) When $n > 2$ it makes much more difference whether or not side payments (with conservation) are permitted. Some of the analysis in this section generalizes readily to more than two-person games, as we shall indicate. Most of the analysis is equally pertinent to the "side payments" and the "no side payments" cases; we shall generally assume the latter but remark on what modifications are appropriate when the former pertains.

Our first task will be to construct a geometric model for two-person cooperative games. Then we discuss some reasonable requirements which any agreed-upon outcome should have; these requirements usually do *not* single out a unique solution to the game. Hence we introduce more requirements which *do* suffice in this regard. Finally, we point out certain inadequacies of the earlier requirements from the standpoint of the individual players, and we construct the *extended Nash solution*, in which the concept of "optimal threat strategies" is central.

We hasten to point out that there are *many* other approaches to finding unique solutions to two-person cooperative games; Chapter 6 of Luce and Raiffa [**92**] is a good place to begin looking for more.

Suppose that a given two-person game has been expressed in normal form, with pure-strategy sets S_i and players' utility functions $U_i\colon S_1 \times S_2 \to R^1$ for $i = 1, 2$—just as in previous sections. Once each player i has chosen his pure strategy s_i, player 1 will experience utility $U_1(s_1, s_2)$ and player 2 will experience utility $U_2(s_1, s_2)$. Hence *both* players' evaluations of the *strategy* pair (s_1, s_2) are summarized by the *numerical* pair $(U_1(s_1, s_2), U_2(s_1, s_2))$. Geometrically, this means that to each $s_{\mathcal{N}} = (s_1, s_2)$ in $S_{\mathcal{N}} = S_1 \times S_2$, there corresponds a point

$$\mathbf{U}_{\mathcal{N}}(s_{\mathcal{N}}) = (U_1(s_{\mathcal{N}}), U_2(s_{\mathcal{N}}))$$

in the $U_1 - U_2$ plane, with U_1 as abscissa and U_2 as ordinate.

More generally, suppose that 1 and 2 agree to use a jointly randomized strategy $P_{\mathcal{N}}$. Then the players' evaluations of $P_{\mathcal{N}}$ are given by the pair

$$\mathbf{U}_{\mathcal{N}}(P_{\mathcal{N}}) = (U_1(P_{\mathcal{N}}), U_2(P_{\mathcal{N}})), \tag{12.5.1}$$

in which $P_{\mathcal{N}} = P_{\mathcal{N}}(s_{\mathcal{N}}^1), \ldots, P_{\mathcal{N}}(s_{\mathcal{N}}^{\#(S_{\mathcal{N}})})$,

$$U_i(P_{\mathcal{N}}) = \sum_{j=1}^{\#(S_{\mathcal{N}})} U_i(s_{\mathcal{N}}^j)P_{\mathcal{N}}(s_{\mathcal{N}}^j), \tag{12.5.2}$$

and we note that every $s_{\mathcal{N}}^j$ is of the form (s_1, s_2) for some (s_1, s_2) in $S_1 \times S_2 = S_{\mathcal{N}}$.

But (12.5.1) and (12.5.2) readily imply that the set $H = \{\mathbf{U}_{\mathcal{N}}(P_{\mathcal{N}}): P_{\mathcal{N}} \in \mathcal{Q}_{\mathcal{N}}\}$ of *all* numerical pairs of players' evaluations of jointly randomized strategies $P_{\mathcal{N}}$ is simply the convex hull of the set of all points $\mathbf{U}_{\mathcal{N}}(s_{\mathcal{N}})$. That is,

$$\mathbf{H} = \text{CONV}[\mathbf{U}_{\mathcal{N}}(s_{\mathcal{N}}): s_{\mathcal{N}} \in S_{\mathcal{N}}]. \tag{12.5.3}$$

■ Deriving this fact is an easy task, relegated to you as Exercise 12.5.1. ■

Definition 12.5.1

(A) The set H defined by (12.5.3) is called the *evaluation set* [or *feasible set*, or *payoff space*].
(B) Each point $\mathbf{U}$ in H is called a (*joint*) *evaluation* [or *outcome*].

■ This definition generalizes readily to more than two players, since $\mathcal{N}$ is then $\{1, \ldots, n\}$ for $n > 2$ instead of $\{1, 2\}$. In place of (12.5.1) we have $\mathbf{U}_{\mathcal{N}}(\mathbf{P}_{\mathcal{N}}) = (U_1(\mathbf{P}_{\mathcal{N}}), \ldots, U_n(\mathbf{P}_{\mathcal{N}}))$; Equations (12.5.2) and (12.5.3) are unchanged. In every case (including $n = 2$), the joint evaluations are n-tuples and H is closed, convex, and nonempty subset of R^n. ■

■ This terminology is quite consistent with that in Chapter 11, except that the role of joint pure strategies s here is played by strategy-sharing rule pairs there. ■

■ In the side-payments (with conservation) case, it makes good sense to build the side payments (if any) into the definition of an evaluation and hence of H. Thus we would consider each player's evaluation of randomizations over ($s_{\mathcal{N}}$, side-payment) pairs; but the moneylike nature of the players' utilities in the side-payments (with conservation) case implies that there is a much easier approach. Since utility is freely transferable with conservation, it is clear that the players should agree to choose an $s_{\mathcal{N}}$ which maximizes the sum of their utilities $U_i(s_{\mathcal{N}})$, so as to produce the biggest possible "pot" to be split. Let $y_{\max}$ denote this maximum sum, as $s_{\mathcal{N}}$ ranges over $S_{\mathcal{N}}$. Also let $y_{\min}$ denote the *smallest* possible "pot" to be split. Then it is easy to see that

$$H = \left\{(U_1, \ldots, U_n): y_{\min} \leq \sum_{i=1}^{n} U_i \leq y_{\max}\right\}, \tag{12.5.4}$$

once the possibility of randomization over ($s_{\mathcal{N}}$, side-payment) pairs is taken into account. Any H given by (12.5.4) is a doubly infinite, diagonal strip such as that in Figure 12.13. ■

Example 12.5.1

Consider the following game in normal form.

	s_2^1	s_2^2
s_1^1	(4, 3)	(10, 2)
s_1^2	(5, 10)	(3, 0)

The evaluation set H for this game is depicted in Figure 12.12. The dotted lines in this figure will be explained momentarily. Figure 12.13 shows H for this game under the assumption that side payments can be made with conservation, in which case we clearly see that $3 \leq U_1 + U_2 \leq 15$.

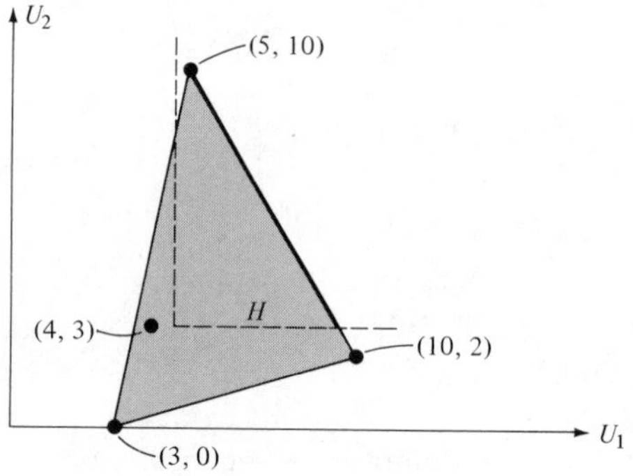

Figure 12.12 H for Example 12.5.1 without side payments.

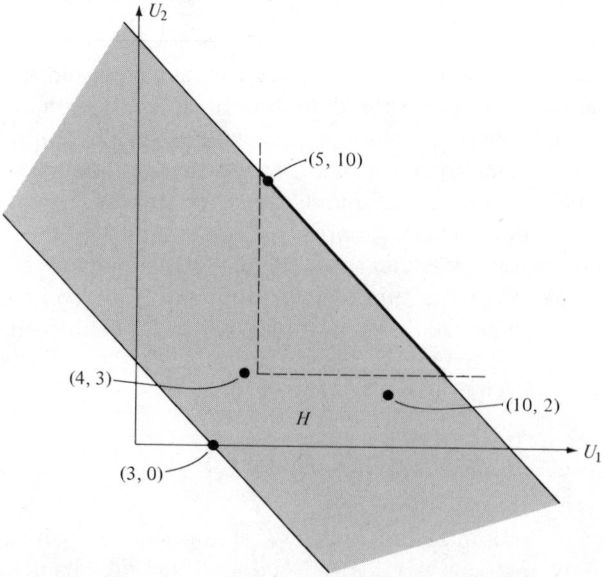

Figure 12.13 H for Example 12.5.1 with side payments.

The dotted lines in Figures 12.12 and 12.13 are easily explained. They enclose all joint evaluations such that each player i receives at least the utility $U_i(P_i^\dagger, \mathcal{Q}_{\bar{i}})$ which he can guarantee for himself by independently choosing a maximin strategy $P_i^\dagger$. Why should a player agree to receive less from any contract than $U_i(P_i^\dagger, \mathcal{Q}_{\bar{i}})$?

Definition 12.5.2. A point $\mathbf{U}$ in H is called *individually rational* if $U_i \geq U_i(P_i^\dagger, \mathcal{Q}_{\bar{i}})$ for each player i.

Example 12.5.2

Regardless of whether or not side payments can be made with conservation, player 1's maximin evaluation of the game in Example 12.5.1 is $U_1(P_1^\dagger, \mathcal{Q}_2) = 4.75$, for $P_1^\dagger = [1/4, 3/4]$, whereas $U_2(P_2^\dagger, \mathcal{Q}_1) = 3.00$, for $P_2^\dagger = [1, 0]$. Hence $\mathbf{U}$ is individually rational if and only if $U_i \geq 4.75$ and $U_2 \geq 3.00$, or equivalently, if and only if $\mathbf{U}$ is neither below the horizontal dotted line ($U_2 = 3.00$) nor left of the vertical dotted line ($U_1 = 4.75$).

It certainly appears reasonable to suppose that neither player would agree to receive less than he can guarantee for himself, acting without collusion.

Another reasonable assumption is that the players should not jointly bind themselves to some decision with joint evaluation $\mathbf{U}^\downarrow$ (say) if there is another decision with joint evaluation $\mathbf{U}^\uparrow$ (say) such that $\mathbf{U}^\uparrow$ is "north and/or east" of $\mathbf{U}^\downarrow$, meaning that each player does at least as well in $\mathbf{U}^\uparrow$ as in $\mathbf{U}^\downarrow$, and at least one of the players does *better* in $\mathbf{U}^\uparrow$ than in $\mathbf{U}^\downarrow$. This line of thinking is quite familiar from Chapter 7 and, even more, from Chapter 11. It motivates the following definition.

Definition 12.5.3

(A) A joint evaluation $\mathbf{U}^\downarrow$ in H is said to be *inefficient* if there is another joint evaluation $\mathbf{U}^\uparrow$ in H such that $U_i^\uparrow \geq U_i^\downarrow$ for each i and $U_i^\uparrow > U_i^\downarrow$ for at least one i.

(B) A joint evaluation $\mathbf{U}^\uparrow$ in H is said to be *efficient*, or *Pareto-optimal*, if it is not inefficient.

(C) A strategy, or strategy *cum* side payment, is said to be *jointly undominated* if its joint evaluation is Pareto-optimal.

Example 12.5.3

The Pareto-optimal joint evaluations in Figure 12.12 are precisely those which lie on the line segment joining the joint evaluations (5, 10) and (10, 2). The corresponding jointly undominated strategies are those randomized strategies which assign probability zero to $(s_1{}^1, s_2{}^1)$ and to $(s_1{}^2, s_2{}^2)$. In Figure 12.13, the same game but with side payments allowed, every $\mathbf{U}$ such that $U_1 + U_2 = 15$ is Pareto-optimal. Hence the Pareto-optimal subset of H in Figure 12.13 is the entire line $\{(U_1, U_2)\colon U_1 + U_2 =$

15}; each point on that line is the joint evaluation of the decision to choose (s_1^2, s_2^1) and then make a side payment of x from 2 to 1 for an ultimate evaluation of $(5+x, 10-x)$. Note that $\{(5+x, 10-x): -\infty < x < \infty\} = \{(U_1, U_2): U_1 + U_2 = 15\}$.

■ The Pareto-optimal subset $H°$ of H is an old friend from Chapter 11. Its role and justification here are the same as before. ■

These two mild restrictions on the anticipated joint evaluation to be agreed on in a cooperative game can be applied in conjunction. The result of doing so is still another subset of H, called the negotiation set.

Definition 12.5.4. The *negotiation set* of a game with evaluation set H is the set of all $\mathbf{U}$ in H which are both individually rational and Pareto-optimal.

Example 12.5.4

In Figure 12.12 the negotiation set is the heavily drawn part of the line segment joining (5, 10) and (10, 2). Although (10, 2) is Pareto-optimal, player 2 should never agree to the corresponding contract for (s_1^1, s_2^2) because he can guarantee himself a utility of 3 by choosing s_2^1, with no help from player 1. Similarly, in Figure 12.13 the negotiation set is the heavily drawn part of the line $\{(U_1, U_2): U_1 + U_2 = 15\}$.

■ Negotiation sets are readily definable for n-person cooperative games, since (1) Definition 12.5.2 applies *verbatim* to n-person games and (2) so does Definition 12.5.3. In fact, we phrased these definitions so that they *would* remain germane in games with $n > 2$ players! ■

It is with (essentially) the negotiation set that von Neumann and Morgenstern [**106**] stopped, alleging that although the theory at this point fails to single out a *unique* joint evaluation, the negotiation set is about as far as one can push matters without introducing rather controversial features such as relative bargaining skills and social customs. This is indeed the case!

The remainder of this section concerns the ramifications of introducing some additional axioms purporting to characterize desirable properties of procedures which determine unique solutions to cooperative games. The axioms are those of Nash [**99**], [**100**]. Chapter 6 of Luce and Raiffa [**92**] and Chapter VII of Owen [**108**] provide much fuller discussion and derivations than we do here.

First, we shall temporarily abstract away from the players' strategies and their roles in determining H, as well as the criteria of individual rationality and Pareto-optimality. The situation for which the basic Nash solution is defined is the "abstract bargaining game," consisting of a set H of evaluations $\mathbf{U}$ of possible "bargains" or "trades," together with a distinguished evaluation $\mathbf{U}^*$ of the status quo ante which results if the players cannot agree on some jointly preferred evaluation.

Definition 12.5.5. Let $n \geq 2$. An *abstract* n-*person bargaining game* is a pair $(H, \mathbf{U}^*)$ in which H is a nonempty, compact, convex subset of R^n, and $\mathbf{U}^*$ is a point in H.

From the preceding discussion it is clear how one should define the efficient, or Pareto-optimal, evaluations in H. In the context of the abstract bargaining game, it is also clear that an evaluation $\mathbf{U}$ in H is to be regarded as individually rational if $U_i \geq U_i^*$ for each player i. The negotiation set in H is again the set of all individually rational and Pareto-optimal evaluations.

Suppose that we consider a function $\mathcal{T}(\cdot, \cdot)$ on the set of all abstract bargaining games $(H, \mathbf{U}^*)$ such that $\mathcal{T}(H, \mathbf{U}^*)$ is the "*t*heoretically sanctioned," unique evaluation[i] in $(H, \mathbf{U}^*)$. For short, we call $\mathcal{T}(H, \mathbf{U}^*) = (\mathcal{T}_1(H, \mathbf{U}^*), \ldots, \mathcal{T}_n(H, \mathbf{U}^*))$ the *solution* to $(H, \mathbf{U}^*)$. What properties should $\mathcal{T}(\cdot, \cdot)$ have—in some idealistic sense? Nash's four axiomatized properties are as follows.

Axiom 1: (Pareto-Optimality and Individual Rationality). $\mathcal{T}(H, \mathbf{U}^*)$ should belong to the negotiation set in H.

Axiom 2: (Independence of Irrelevant Alternatives). If $\mathbf{U}^* \in H'$, $H' \subset H$, and $\mathcal{T}(H, \mathbf{U}^*) \in H'$, then $\mathcal{T}(H', \mathbf{U}^*) = \mathcal{T}(H, \mathbf{U}^*)$.

■ Axiom 2 really says that the solution to a bargaining game remains the same as *unchosen* (evaluations of) potential bargains are eliminated from H. A more descriptive name, apparently due to Harsanyi, is "irrelevance of unchosen alternatives." This axiom is a reasonable description of what actually goes on in many real bargaining situations, where potential bargains cannot be reintroduced once they have been rejected, and hence where the ultimate outcome is determined by a process of successive elimination. ■

Axiom 3: (Invariance with Respect to Utilities' Scales and Origins). Suppose that $(H', \mathbf{U}^*)$ is obtained from $(H, \mathbf{U}^*)$ by changing the scales and origins of the players' utilities; that is, there exist real numbers $b_1, \ldots, b_n$ and *positive* real numbers $a_1, \ldots, a_n$ such that

(i) $$U_i^{*\prime} = a_i U_i^* + b_i \text{ for every } i \in \mathcal{N},$$

and

(ii) $(U_1, \ldots, U_n) \in H$ if and only if $(a_1 U_1 + b_1, \ldots, a_n U_n + b_n) \in H'$.

Then $\mathcal{T}_i(H', \mathbf{U}^{*\prime}) = a_i \mathcal{T}_i(H, \mathbf{U}^*) + b_i$ for every $i \in \mathcal{N}$.

■ Axiom 3 says that the arbitrary choices of inessential utility parameters by the individual players should have no effect on the qualitatively defined solution to the abstract bargaining game. It is eminently reasonable so long as one avoids demanding that the a_i's and

[i] Luce and Raiffa term: arbitrated solution.

b_i's have been so chosen as to facilitate *interpersonal comparisons of utility* in the form of questions such as "will this hurt 1 more than it helps 2," etc. Such questions and their object are different from simply asserting that for some choices of the a_i's and b_i's side payments may be made with conservation. ■

Axiom 4: (Symmetry). Suppose that

(i) $U_1^* = \cdots = U_n^*$ in $\mathbf{U}^* = (U_1^*, \ldots, U_n^*)$, and
(ii) If $\mathbf{U} = (U_1, \ldots, U_n)$ belongs to H, then so does every n-tuple formed by rearranging the components of $\mathbf{U}$.

Then $\mathcal{T}_1(H, \mathbf{U}^*) = \cdots = \mathcal{T}_n(H, \mathbf{U}^*)$.

■ Axiom 4 states that *if* the abstract bargaining game is completely symmetric in that (i) players all have the same utility of the status quo ante and (ii) players all have the same relatively advantageous trading positions, *then* the solution should not distinguish among the players. This is sort of an equal-treatment-of-players axiom, and, as such, is rather questionable. It requires that nothing external to the bargaining situation such as the players relative statūs in the world have any influence on the theoretically sanctioned outcome. ■

The remarkable consequence of these four axioms is that there is a *unique* solution function $\mathcal{T}(\cdot, \cdot)$ which satisfies them, and that $\mathcal{T}(H, \mathbf{U}^*)$ always exists. If there is an evaluation $(U_1, \ldots, U_n)$ in H such that $U_i > U_i^*$ for every i, then $\mathcal{T}(H, \mathbf{U}^*)$ satisfies the equation

$$\prod_{i=1}^{n} (\mathcal{T}_i(H, \mathbf{U}^*) - U_i^*) = \max \left[\prod_{i=1}^{n} (U_i - U_i^*) : \mathbf{U} \in H,\ U_i > U_i^* \text{ for every } i \right]. \tag{12.5.5}$$

Definition 12.5.6. The solution $\mathcal{T}(\cdot, \cdot)$ which uniquely satisfies Axioms 1–4 is called the *Nash solution* to the abstract bargaining game.

■ Proving that there is one and only one solution $\mathcal{T}(\cdot, \cdot)$ which satisfies Axioms 1–4, and proving that this solution satisfies (12.5.5) when there is some $\mathbf{U}$ in H with $U_i > U_i^*$ for every i, is involved. For $n = 2$, a proof is sketched in Luce and Raiffa [**92**, pp. 127–128], and a detailed proof is given in Owen [**108**, pp. 142–145]. The general proof for any $n \geq 2$ can be found in Harsanyi [**53**]. ■

A by-product of the derivation of the Nash solution when $n = 2$ is a graphical procedure for locating $\mathcal{T}(H, \mathbf{U}^*)$. The derivation shows that if $\mathcal{T}(H, \mathbf{U}^*)$ lies inside a line segment of the Pareto-optimal set, then the slope of this segment is *the negative of* the slope of the line segment joining $\mathbf{U}^*$ to $\mathcal{T}(H, \mathbf{U}^*)$. Thus one can draw two lines back into H at each vertex where two Pareto-optimal line segments meet, one each with slope negative to that of one of the adjoining Pareto-optimal segments.[j] The two lines corresponding to each such vertex form a "cone" in H; if $\mathbf{U}^*$ lies in this cone, then $\mathcal{T}(H, \mathbf{U}^*)$ is the vertex of the cone. If $\mathbf{U}^*$ does not lie in any such cone, then $\mathcal{T}(H, \mathbf{U}^*)$ lies inside a Pareto-optimal segment, *which* segment being clear from where $\mathbf{U}^*$ lies in relation to adjacent cones, and hence the line of appropriate slope may be extended from $\mathbf{U}^*$ to the Pareto-optimal set to find $\mathcal{T}(H, \mathbf{U}^*)$.

■ Suppose the drawing of cones establishes that $\mathcal{T}(H, \mathbf{U}^*)$ lies inside a segment of slope $-m$, with equation $mU_1 + U_2 = K$ for some real number K. Then the line of slope $+m$ passing through $\mathbf{U}^*$ has equation $mU_1 - U_2 = mU_1^* - U_2^*$. Therefore the intersection of these two lines is given by

$$\mathcal{T}_1(H, \mathbf{U}^*) = (1/2)U_1^* + (1/[2m])(K - U_2^*) \qquad (12.5.6a)$$

and

$$\mathcal{T}_2(H, \mathbf{U}^*) = (1/2)U_2^* + (1/2)(K - mU_1^*). \qquad (12.5.6b)$$

Thus exact rather than graphically only approximate Nash solutions are obtainable. ■

Example 12.5.5

In Figure 12.14 we graph H for some abstract bargaining game and indicate the Nash solution $\mathcal{T}(H, \mathbf{U}^*)$ for various status quo ante points $\mathbf{U}^*$. This figure shows, for example, that $\mathcal{T}(H, (3, 0))$ is in segment 2, the equation for which is $2U_1 + U_2 = 17$, so that $K = 17$ and $m = 2$. Hence (12.5.6) implies that $\mathcal{T}(H, (3, 0)) = (23/4, 11/2)$.

We now return from abstract bargaining games to two-person cooperative games, which indicate players' strategies but not necessarily a status quo ante point. The Nash solution to bargaining games is translated into a solution to two-person cooperative games by showing how to reformulate a two-person cooperative game as an abstract bargaining game. Naturally, the evaluation set H of the cooperative game is to be the set H of the abstract bargaining game. It remains "only" to define a status quo ante $\mathbf{U}^*$ in an intelligent fashion.

There are two main approaches to choosing $\mathbf{U}^*$. The first is what we shall call the *Nash-maximin* approach:[k] we define $\mathbf{U}^* = (U_1^*, U_2^*)$ by

$$U_i^* = U_i(P_i^\dagger, \mathcal{Q}_{\bar{i}}) \qquad (12.5.7)$$

for each player i.

Now, one always calculates each player's maximin evaluation $U_i(P_i^\dagger, \mathcal{Q}_{\bar{i}})$

[j] At the uppermost and rightmost vertices of the Pareto-optimal set, there is only one adjoining Pareto-optimal segment. Hence at these vertices draw only one line of appropriate slope back into H. If $\mathbf{U}^*$ is "southeast of" the line at the rightmost vertex, that vertex is $\mathcal{T}(H, \mathbf{U}^*)$; if $\mathbf{U}^*$ is "northwest of" the line at the uppermost vertex, that vertex is $\mathcal{T}(H, \mathbf{U}^*)$.

[k] Luce and Raiffa term: Shapley solution.

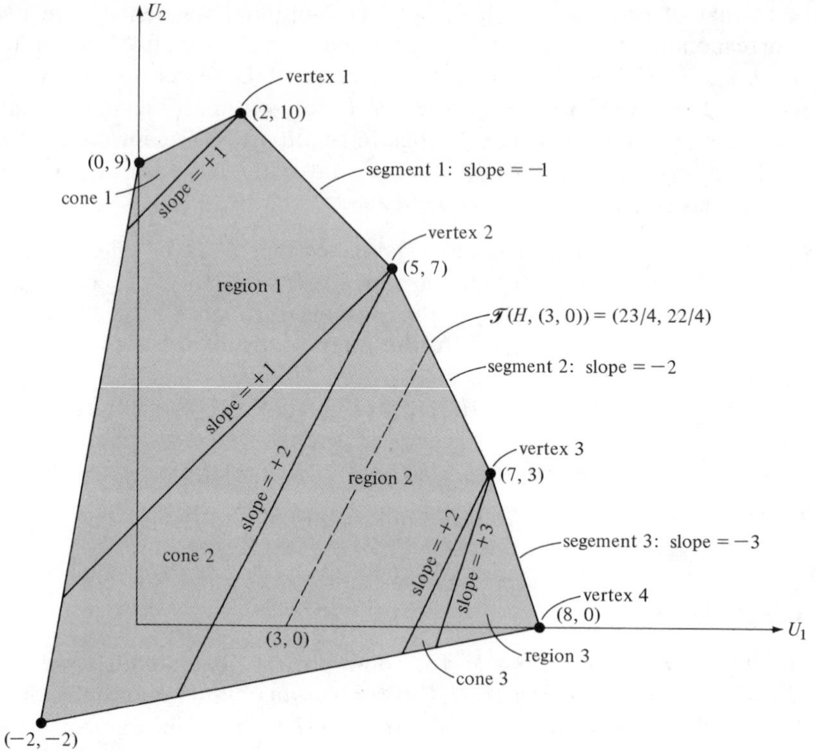

Figure 12.14 Abstract bargaining game in Example 12.5.5. (*Key*: $\mathcal{T}(H, \mathbf{U}^*) =$ vertex i if $\mathbf{U}^*$ is in cone i. If $\mathbf{U}^*$ is in region i, then $\mathcal{T}(H, \mathbf{U}^*)$ is on segment i.)

anyway in order to determine the negotiation set; choosing the pair $(U_1(P_1^\dagger, \mathcal{Q}_2), U_2(P_2^\dagger, \mathcal{Q}_1))$ is therefore a very easy way of determining $\mathbf{U}^*$.

Example 12.5.6

For the game in Example 12.5.1 without side payments, depicted in Figure 12.12, $\mathbf{U}^* = (4.75, 3.00)$ from Example 12.5.2. The slope of the only Pareto-optimal segment in Figure 12.12 is $-8/5 = -1.6 = -m$, and the equation of that line is $1.6U_1 + U_2 = 18 = K$. Hence $\mathcal{T}(H, (4.75, 3.00)) = (7.0625, 6.7000)$, by (12.5.6). But if side payments with conservation are permitted, then the equation of the Pareto-optimal line is $U_1 + U_2 = 15 = K$, with $m = 1$; in this case, (12.5.6) yields $\mathcal{T}(H, (4.75, 3.00)) = (8.375, 6.625)$. Note how in this case 1 and 2 disagree about the desirability of permitting side payments!

The Nash-maximin approach is relatively easy to determine, but there are indications that it does not reflect adequately the players' relative strategic strengths. Consider the following classical example, reported in Owen [**108**] and elsewhere.

Example 12.5.7

Player 1 is an industrialist with only one pure strategy and player 2 is a worker who can either follow orders (s_2^1) or sabotage his machine (s_2^2) by thrusting his arm into it. Assume the following utility bimatrix.

	s_2^1	s_2^2
s_1^1	(10, 0)	(0, − 100,000)

Clearly, $U_1(P_1^\dagger, \mathscr{Q}_2) = U_1(s_1^1, s_2^2) = 0$ and $U_2(P_2^\dagger, \mathscr{Q}_1) = U_2(s_1^1, s_2^1) = 0$. Hence $\mathbf{U}^* = (0, 0)$ according to the Nash-maximin approach. But if side payments with conservation are permitted, then the Pareto-optimal set is the line with equation $U_1 + U_2 = 10$, and (12.5.6) readily yields $\mathscr{T}(H, \mathbf{U}^*) = (5, 5)$. Now, this is quite poor performance on the part of the Nash-maximin approach, because in order to *implement* his "threat" of s_2^2 unless he is awarded half the profit, the worker must accept $-100,000$ for himself. Thus s_2^2 is not a *credible* threat.

This insenstivity of the Nash-maximin approach to credibility of the threat to choose a maximin strategy is not so typical of the *extended Nash solution*, introduced in **[100]**. Nash assumes a two-stage "demand" model, with the following stages.

(1) Players 1 and 2 *simultaneously commit* themselves to "threat strategies" P_1 and P_2 respectively, these strategies to be implemented in case their (subsequently made) "demands" t_1 and t_2 are not mutually attainable, as will happen if $(t_1, t_2) \notin H$.

(2) After players 1 and 2 have informed each other as to their chosen threat strategies, they *simultaneously* make respective demands t_1 and t_2. If $(t_1, t_2) \in H$, then each receives his demand; but if $(t_1, t_2) \notin H$, then 1 and 2 must implement their respective threat strategies, a situation which, before they randomize, produces the joint evaluation $(U_1(P_1, P_2),\ U_2(P_1, P_2))$.

Two additional axioms suffice to determine that the demands should be chosen in such a way that (t_1, t_2) is the Nash solution to the bargaining game $(H, (U_1(P_1, P_2), U_2(P_1, P_2)))$.

■ Briefly and informally, Axiom 5 requires that if we restrict a player i by reducing the number of his "threat" strategies, then his utility of the ultimate result of the game is not increased; Axiom 6 requires that if a player is restricted to a single (randomized) threat strategy, then his opponent can also be restricted to a single randomized threat strategy, without increasing i's utility of the ultimate result above what it was before restricting the opponent. ■

For each pair (P_1, P_2) of "threat strategies," define $t_i(P_1, P_2)$ by

$$t_i(P_1, P_2) = \mathscr{T}_i(H, (U_1(P_1, P_2), U_2(P_1, P_2))), \qquad (12.5.8)$$

where $\mathcal{T}(\cdot, \cdot)$ denotes the *Nash* solution to abstract two-person bargaining games. By reasoning recursively, it follows, from the fact that the demands (t_1, t_2) should be the Nash solution to $(H, (U_1(P_1, P_2), U_2(P_1, P_2)))$, that the functions $t_i: \mathcal{Q}_1 \times \mathcal{Q}_2 \rightarrow R^1$ capture the essence of the *first* stage, in which the players simultaneously choose "threats" P_1 and P_2.

Now, $(t_1(P_1, P_2), t_2(P_1, P_2))$ belongs to the Pareto-optimal subset of H for every (P_1, P_2) in $\mathcal{Q}_1 \times \mathcal{Q}_2$. As we move to the right along a Pareto-optimal set in R^2, we move *down*; and hence 1's attempt to increase t_1 is tantamount to attempting to decrease t_2. Therefore the first-stage game, in which the players simultaneously choose threat strategies P_1 and P_2 and evaluate their efficacy by $t_1(P_1, P_2)$ and $t_2(P_1, P_2)$ respectively, is a game of opposing interests, rather like a two-person zerosum game.[1] In fact, we can ignore $t_2(\cdot, \cdot)$ and subsequently concentrate on $t_1(\cdot, \cdot)$, which 1 wants to maximize and 2 wants to minimize.

This game differs from a two-person zerosum game in two computationally important respects. First, in this game i's *pure*-strategy set is $\mathcal{Q}_i$, the set of all his *randomized* strategies in the original game; it is therefore an *infinite* set of pure strategies, and so Theorem 12.3.1 and Corollary 12.4.1 do not seem to apply as stated. Second, it is fallacious to suppose that

$$t_1(P_1, P_2) = \sum_{i=1}^{\#(S_1)} \sum_{j=1}^{\#(S_2)} t_1(s_1{}^i, s_2{}^j) P_1(s_1{}^i) P_2(s_2{}^j)$$

in general.

Nevertheless, it can be shown that the nice properties of solutions to two-person zerosum games do also obtain for the threat game. Nash shows [**100**] that there exist "optimal" threat strategies P_1° and P_2° such that:

(1) Each of P_1° and P_2° is optimal against the other; that is,

$$t_1(P_1^\circ, P_2^\circ) = \max_{P_1 \in \mathcal{Q}_1} [t_1(P_1, P_2^\circ)],$$

and similarly for P_2°.

(2) Each P_i° is both maximin and minimax for i; that is,

$$\min_{P_2 \in \mathcal{Q}_2} [t_1(P_1^\circ, P_2)] = \max_{P_1 \in \mathcal{Q}_1} [\min_{P_2 \in \mathcal{Q}_2} [t_1(P_1, P_2)]]$$

and

$$\max_{P_2 \in \mathcal{Q}_2} [t_2(P_1^\circ, P_2)] = \min_{P_1 \in \mathcal{Q}_1} [\max_{P_2 \in \mathcal{Q}_2} [t_2(P_1, P_2)]],$$

and similarly for P_2°.

(3) *All* such equilibrium pairs (P_1°, P_2°) of threat strategies are both equivalent and interchangeable.

The import of these nice properties is that 1 need "only" find some strategy P_1° such that

$$\min_{P_2 \in \mathcal{Q}_2} [t_1(P_1^\circ, P_2)] = \max_{P_1 \in \mathcal{Q}_1} [\min_{P_2 \in \mathcal{Q}_2} [t_1(P_1, P_2)]],$$

and similarly for 2. The resulting pair (P_1°, P_2°) is necessarily in equilibrium,

---

[1] See Exercise 12.4.1.

and all such equilibria yield the same solution point on the Pareto-optimal set.

We put "only" in quotes because this task is somewhat complicated unless we assume that utility is freely transferable with conservation, in which case it can be shown[m] that (P_1°, P_2°) is any maximin-minimax pair in a corresponding two-person zerosum *game of relative advantage*, in which player 1's utility of (s_1, s_2) is assumed to be his "relative advantage" $U_1(s_1, s_2) - U_2(s_1, s_2)$, for every (s_1, s_2) in $S_1 \times S_2$. Then, once (P_1°, P_2°) is known, it is easy to calculate $U_1(P_1^\circ, P_2^\circ)$ and $U_2(P_1^\circ, P_2^\circ)$, and then find the Nash solution to the abstract bargaining game $(H, (U_1(P_1^\circ, P_2^\circ), U_2(P_1^\circ, P_2^\circ)))$.

Example 12.5.8

In the game of Figure 12.13 and Example 12.5.1, with side payments (with conservation) permitted, the associated game of relative advantage has the following matrix of utilities (= "advantages to 1").

	s_2^1	s_2^2
s_1^1	1	8
s_1^2	−5	3

Clearly, $(P_1^\circ, P_2^\circ) = ([1, 0], [1, 0])$, and hence $(U_1(P_1^\circ, P_2^\circ), U_2(P_1^\circ, P_2^\circ)) = (4, 3)$. It follows readily that $(t_1(P_1^\circ, P_2^\circ), t_2(P_1^\circ, P_2^\circ)) = \mathcal{T}(H, (4, 3)) = (8, 7)$, in which 2 fares a little better than he did in Example 12.5.6 under the Nash-maximin approach, which yielded the evaluation (8.375, 6.625).

Example 12.5.9

For the industrialist-worker game of Example 12.5.7, the associated game of relative advantage has the one-row matrix [10, +100,000], from which $(P_1^\circ, P_2^\circ) = ([1], [1, 0])$ is obvious. Hence $(U_1(P_1^\circ, P_2^\circ), U_2(P_1^\circ, P_2^\circ)) = (10, 0)$; and since this is Pareto-optimal it cannot be improved upon. Therefore the extended Nash solution does not ignore the incredibility of 2's threat to use s_2^2.

There is one other situation in which finding the extended Nash solution is not too difficult, even when no side payments are permitted. That is when the original two-person game is *symmetric*, in the sense that:

(i) $\#(S_1) = \#(S_2)$.
(ii) $U_1(s_1^i, s_2^j) = U_2(s_1^j, s_2^i)$ for every i and every j.

In this case, the symmetry axiom implies that $t_1(P_1^\circ, P_2^\circ) = t_2(P_1^\circ, P_2^\circ)$. Since only one point on the Pareto-optimal set has equal coordinates, that point is the extended Nash solution.

---

[m] Owen [**108**, p. 151], for example.

The remainder of this section is an optional remark on how one might obtain the extended Nash solution when side payments are not allowed and when the game is not symmetric. But as a closing comment for *all* readers, we reiterate that there are *many* reasonable ways of obtaining unique solutions to two-person cooperative games. We have barely scratched the surface here.

*■ Every two-person game in which each player has a finite number of pure strategies has an evaluation set H of which the Pareto-optimal subset $H°$ is a union of k line segments[n] L_i^* having respective, negative slopes $-m_i$, for some $k \geq 1$. Suppose that these segments have been numbered as in Figure 12.14, with $m_1 < m_2 < \cdots < m_k$. Each segment L_i^* is a subset of a doubly infinite line L_i with equation $m_i U_1 + U_2 = K_i$, for some real number K_i. Along L_i^* the players' utilities are *transferable via randomization*, not with conservation unless $m_i = 1$, but rather at the constant rate of m_i utility units gained by 2 per utility unit lost by 1. The essence of the following procedure is to imagine that utility is transferable at that exchange rate in the entire game, and then to find that (necessarily unique) segment (if any) L_i^* which contains the extended Nash solution to the game with $H° = L_i$ (the whole line).

This procedure is described in a more general context by Raiffa [**114**]. Let $\mathbf{t}^i$ denote the Nash solution to the abstract bargaining game $(H^i, (U_1(P_1^i, P_2^i), U_2(P_1^i, P_2^i)))$, where $H^i = \{(U_1, U_2): m_i U_1 + U_2 \leq K_i\}$ and (P_1^i, P_2^i) is a maximin-minimax pair in the two-person zerosum game whose matrix has the typical entry $m_i U_1(s_1, s_2) - U_2(s_1, s_2)$—the obvious generalization of the relative-advantage game matrix. Now, $\mathbf{t}^i$ necessarily lies on L_i, but it need not lie on the segment $L_i^* \subset L_i$. If it does, then $\mathbf{t}^i$ is the desired extended Nash solution. If $\mathbf{t}^i$ is to the northwest of L_i^* on L_i, then try $\mathbf{t}^{i-1}$ to see if it lies on $L_{i-1}^* \subset L_{i-1}$. If so, then $\mathbf{t}^{i-1}$ is the extended Nash solution; but if $\mathbf{t}^{i-1}$ lies to the southeast of L_{i-1}^* on L_{i-1}, then the extended Nash solution that we seek is the vertex at which L_{i-1}^* and L_i^* meet. On the other hand, if $\mathbf{t}^i$ is to the southeast of L_i^* on L_i, then try $\mathbf{t}^{i+1}$ to see if $\mathbf{t}^{i+1}$ lies on L_{i+1}^* in L_{i+1}. If so, then $\mathbf{t}^{i+1}$ is the desired extended Nash solution; but if $\mathbf{t}^{i+1}$ lies northwest of L_{i+1}^* on L_{i+1}, then the extended Nash solution is the vertex at which L_i^* and L_{i+1}^* meet. Naturally, $\mathbf{t}^1$ northwest of L_1^* along L_1 implies that the northwest endpoint of L_1^* is the extended Nash solution, and $\mathbf{t}^k$ southeast of L_k^* along L_k implies that the southeast endpoint of L_k^* is the desired solution. If one starts with a reasonable guess as to the correct segment, then the number of $\mathbf{t}^i$'s that he needs to compute will be small. Otherwise, one might end up computing $\mathbf{t}^1$, going southeast and computing $\mathbf{t}^2$, going southeast again and computing $\mathbf{t}^3$, and so on, *or* computing $\mathbf{t}^k$, going northwest and computing $\mathbf{t}^{k-1}$, going northwest again and computing $\mathbf{t}^{k-2}$, etc. ■

[n] Unless $H°$ consists of a single point, in which case that point obviously *is* the extended Nash solution.

***Example 12.5.10**

In the game of Example 12.5.1 and Figure 12.12, the Pareto-optimal set consists of only one segment, L_1^*, on the line with equation $1.6U_1 + U_2 = 18$, so that $m_1 = 1.6$ and $K_1 = 18$. We calculate a maximin-minimax pair (P_1^1, P_2^1) for the zerosum game with matrix $1.6U_1(s_1, s_2) - U_2(s_1, s_2)$; namely,

	s_2^1	s_2^2
s_1^1	3.4	14.0
s_1^2	−2.0	4.8

Clearly, $(P_1^1, P_2^1) = ([1, 0], [1, 0])$ and $(U_1(P_1^1, P_2^1), U_2(P_1^1, P_2^2)) = (4, 3)$. The line of slope 1.6 passing through (4, 3) intersects L_1^* at $(107/16, 73/10) = (6.6875, 7.3000)$, which is therefore the extended Nash solution.

***Example 12.5.11**

Suppose that the evaluation set H in Figure 12.14 arose from the following two-person game.

	s_2^1	s_2^2	s_2^3
s_1^1	(5, 7)	(7, 3)	(2, 10)
s_1^2	(−2, −2)	(0, 9)	(8, 0)

On segment L_1^*, $m_1 = +1$ and hence the associated relative-advantage game has the following matrix.

	s_2^1	s_2^2	s_2^3
s_1^1	−2	4	−8
s_1^2	0	−9	8

for which a maximin-minimax pair is $(P_1^1, P_2^1) = ([3/5, 2/5], [13/15, 2/15, 0])$, which implies that $(U_1(P_1^1, P_2^1), U_2(P_1^1, P_2^1)) = (185/75, 275/75)$; it follows that $\mathbf{t}^1 = \mathcal{T}(H^1, (185/75, 275/75)) = (5.40, 6.60)$. This is on L_1 to the southeast of the vertex endpoint (5, 7) of L_1^*. Hence we go to segment L_2^*, where $m_2 = +2$, and hence where the associated relative-advantage game has the following matrix.

	s_2^1	s_2^2	s_2^3
s_1^1	3	11	−6
s_1^2	−2	−9	16

for which a maximin-minimax pair is $(P_1^2, P_2^2) = ([2/3, 1/3], [22/27, 0, 5/27])$, which implies that $(U_1(P_1^2, P_2^2), U_2(P_1^2, P_2^2)) = (236/81, 364/81)$; it follows that $\mathbf{t}^2 = \mathcal{T}(H^2, (236/81, 364/81)) = (55/12, 94/12) \doteq (4.68, 7.83)$. This is on L_2 to the northwest of the vertex endpoint $(5, 7)$ of L_2^*. Hence the extended Nash solution is the vertex $(5, 7)$.

Exercise 12.5.1

Derive (12.5.3); that is, prove that

$$\{\mathbf{U}_{\mathcal{N}}(P_{\mathcal{N}}): P_{\mathcal{N}} \in \mathcal{Q}_{\mathcal{N}}\} = \mathrm{CONV}[\mathbf{U}_{\mathcal{N}}(s_{\mathcal{N}}): s_{\mathcal{N}} \in S_{\mathcal{N}}].$$

Exercise 12.5.2. Determine:

(A) all individually rational joint evaluations,
(B) all Pareto-optimal joint evaluations, in the "chicken" game

	s_2^1	s_2^2
s_1^1	$(5, 5)$	$(2, x)$
s_1^2	$(x, 2)$	$(0, 0)$

for every $x > 5$, assuming:
(i) that utility is freely transferable with conservation—the side-payments case;
(ii) that no side payments are permitted.

(C) Use the Nash-maximin approach to find a unique solution to this game, as a function of x for $x > 5$, for each of side-payment cases (i) and (ii).

Exercise 12.5.3

Find the extended Nash solution to the two-person game with the following bimatrix:

	s_2^1	s_2^2	s_2^3
s_1^1	$(0, 9)$	$(7, 3)$	$(2, 10)$
s_1^2	$(-2, -2)$	$(5, 7)$	$(8, 0)$

(A) assuming that utility is freely transferable with conservation.
*(B) assuming that no side payments are permitted.

Exercise 12.5.4

Find the extended Nash solution to the two-person game with the following bimatrix:

	s_2^1	s_2^2	s_2^3
s_1^1	$(5, 7)$	$(7, 3)$	$(8, 0)$
s_1^2	$(-2, -2)$	$(0, 9)$	$(2, 10)$

(A) assuming that utility is freely transferable with conservation.
*(B) assuming that no side payments are permitted.

***Exercise 12.5.5**

Find the extended Nash solution to the two-person game with bimatrix as given in Exercise 12.3.9(D):
(A) assuming that utility is freely transferable with conservation.
(B) assuming that no side payments are permitted.

12.6 COALITIONS AND n-PERSON COOPERATIVE GAMES

In games with more than two players, some may gang up on others, whereas in two-person games "ganging up" is a solitary activity because the only nonempty coalitions are $\{1\}$, $\{2\}$, and $\{1, 2\}$. The phenomenon of nontrivial coalitions is an important one and causes a sharp distinction between $n = 2$ and $n > 2$ in cooperative game theory.

It is fair to say that there is no single, unified theory of n (> 2) person cooperative games; rather, there are several different theories, each corresponding to one reasonable way of accounting for the potentialities of the coalitions. Most of these theories are (at least, and well) sketched in Owen [**108**, Chapters VIII–X] and surveyed by Lucas [**88**], [**89**]; we shall sketch only one, the "characteristic-function" approach introduced by von Neumann and Morgenstern [**106**] for games with side payments and by Aumann and Peleg [**7**] for games without side payments.[o]

As in Section 12.5, our attention will be focused more on joint evaluations $\mathbf{U}$ in H rather than on the strategies which have the joint evaluations. Such a focus is perfectly reasonable, since one can always "read back" from $\mathbf{U}$ in H to the joint strategy, or joint strategy with side-payment accounts, which has that joint evaluation $\mathbf{U}$.

We therefore seek "good" joint evaluations $\mathbf{U}$. Now, in Section 12.5 we interpreted "good" to mean (1) that no single player i would veto $\mathbf{U}$ on the grounds that he could do better than his component U_i of $\mathbf{U}$ all by himself (individual rationality) and (2) that the set $\mathcal{N}$ of all players would not veto $\mathbf{U}$ on the grounds that some other evaluation $\mathbf{U}^*$ was at least as good as $\mathbf{U}$ for everybody and better than $\mathbf{U}$ for somebody (Pareto-optimality). In other words, $\mathbf{U}$ is individually rational if no player i could veto it, and $\mathbf{U}$ is Pareto-optimal if all players together have no incentive to veto it. The negotiation set is, therefore, the set of all joint evaluations which should not be vetoed (1) by any individual player and (2) by all players together.

Now suppose that the veto power is extended to all coalitions $\mathscr{C}$ in some set COLIT of coalitions, which we shall assume includes $\mathcal{N}$ and all one-player coalitions $\{i\}$. Then we should be able to narrow down the set of

[o] See Aumann [**5**] for a more recent survey of the theory of cooperative games without side payments.

"good," meaning nonvetoable, joint evaluations. This is the route that we shall take here.

First, we define $H = \{\mathbf{U}_{\mathcal{N}}(P_{\mathcal{N}}): P_{\mathcal{N}} \in \mathcal{Q}_{\mathcal{N}}\}$ exactly as before, and we define *individual rationality* of an n-tuple $\mathbf{U} = (U_1, \ldots, U_n)$ also as before: $\mathbf{U}$ is individually rational if $U_i \geq U_i(P_i, \mathcal{Q}_{\bar{i}})$ for every $i \in \mathcal{N}$. But in n-person theory it is convenient to use a slightly weaker condition than Pareto-optimality. Instead, we shall say that an n-tuple $\mathbf{U}$ in H is *group-rational* if there is no n-tuple $\mathbf{U}^*$ in H such that $U_i^* > U_i$ for every $i \in \mathcal{N}$.

■ Recall that $\mathbf{U} \in H$ is Pareto-optimal if there is no $\mathbf{U}^* \in H$ such that $U_i^* \geq U_i$ for every $i \in \mathcal{N}$ and $U_i^* > U_i$ for at least one $i \in \mathcal{N}$. Clearly, $\mathbf{U}$ is group-rational if it is Pareto-optimal, but $\mathbf{U}$ can be group-rational without being Pareto-optimal; see Figure 12.15, in which the horizontal and vertical line segments are comprised of group-rational but not Pareto-optimal joint evaluations. ■

■ In games with side payments, in which utility is freely transferable with conservation, a joint evaluation is Pareto-optimal if it is group-rational, because in all such games the upper boundary of H is a (line, plane, or) hyperplane with equation $\Sigma_{i=1}^{n} U_i = K$ and hence contains no "eastern" and/or "northern" faces. ■

Because our set COLIT of permissible coalitions will always be assumed to include $\mathcal{N}$ and all the single-player coalitions $\{i\}$, and because individual and group rationality seem very reasonable requirements of "good" joint evaluations, we have a special name for joint evaluations which are both individually rational and group rational. Such joint evaluations are called *imputations*.

Now we shall examine the role of an arbitrary coalition $\mathscr{C}$ in COLIT. Intuitively, $\mathscr{C}$ has a "meaningful veto" of a utility n-tuple $\mathbf{U}$ in R^n if there is another utility n-tuple $\mathbf{U}^*$ in R^n with two properties: (1), *every* member of $\mathscr{C}$ *prefers* $\mathbf{U}^*$ to $\mathbf{U}$, in that $U_i^* > U_i$ for every i $\mathscr{C}$; (2), coalition $\mathscr{C}$ *can attain* its members' utility components U_i^* of $\mathbf{U}^*$ and hence "enforce" their preference for $\mathbf{U}^*$ over $\mathbf{U}$.

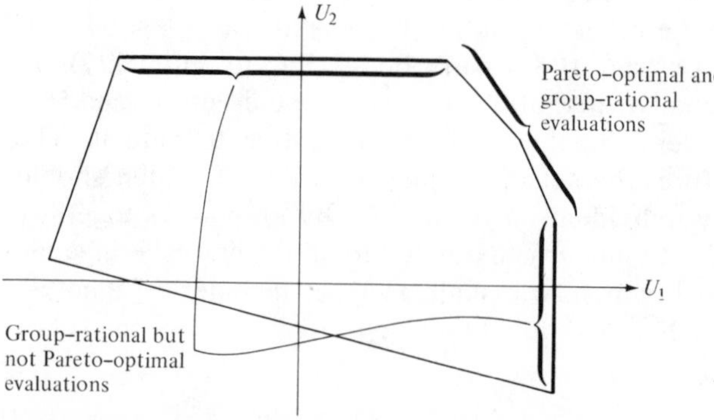

Figure 12.15 Group-rational and Pareto-optimal sets for a game with $n = 2$.

Definition 12.6.1

(A) We denote by $v(\mathscr{C})$ the set of all $\mathbf{U}$ in R^n such that $\mathscr{C}$ can attain its members' components of $\mathbf{U}$.
(B) We say that $\mathscr{C}$ *is effective for* $\mathbf{U}$ if $\mathbf{U} \in v(\mathscr{C})$.
(C) The function $v(\cdot)$ which maps coalitions $\mathscr{C}$ to subsets of R^n is called a *characteristic function.*

The expression "can attain" in Definition 12.6.1(A) must be made more precise. There are two rather obvious ways of doing so in games where COLIT includes all complements $\bar{\mathscr{C}}$ of its member coalitions other than $\mathscr{N}$. *Alpha-effectiveness* of $\mathscr{C}$ for $\mathbf{U}$ means that $\mathscr{C}$ has some coalition strategy that ensures each member's attaining at least his component U_i of $\mathbf{U}$ regardless of what the complementary coalition $\bar{\mathscr{C}}$ does. *Beta-effectiveness* of $\mathscr{C}$ for $\mathbf{U}$ means that $\mathscr{C}$ has a response to each strategy $P_{\bar{\mathscr{C}}}$ of $\bar{\mathscr{C}}$ which attains for each member of $\mathscr{C}$ at least his component U_i of $\mathbf{U}$. Roughly and intuitively, alpha-effectiveness is reminiscent of maximin and beta-effectiveness of equilibrium.

Definition 12.6.2. Let $\mathbf{U} = (U_1, \ldots, U_n)$.

(A) $\mathscr{C}$ is *alpha-effective* for $\mathbf{U}$ if there exists a strategy $P^{\dagger}_{\mathscr{C}}[\mathbf{U}]$ in $\mathscr{Q}_{\mathscr{C}}$ such that

$$U_i \leq U_i(P^{\dagger}_{\mathscr{C}}[\mathbf{U}], P_{\bar{\mathscr{C}}})$$

for every $i \in \mathscr{C}$ and for every $P_{\bar{\mathscr{C}}} \in \mathscr{Q}_{\bar{\mathscr{C}}}$.
(B) $\mathscr{C}$ is *beta-effective* for $\mathbf{U}$ if, for every $P_{\bar{\mathscr{C}}} \in \mathscr{Q}_{\bar{\mathscr{C}}}$, there exists a strategy $P^{\circ}_{\mathscr{C}}[\mathbf{U}, P_{\bar{\mathscr{C}}}]$ in $\mathscr{Q}_{\mathscr{C}}$ such that

$$U_i \leq U_i(P^{\circ}_{\mathscr{C}}[\mathbf{U}, P_{\bar{\mathscr{C}}}], P_{\bar{\mathscr{C}}})$$

for every $i \in \mathscr{C}$.

In games with side payments forbidden, it may be difficult to determine the sets $v_\alpha(\mathscr{C})$ and $v_\beta(\mathscr{C})$ of all $\mathbf{U}$ for which $\mathscr{C}$ is respectively alpha-effective and beta-effective. But when utility is freely transferable with conservation, this task is straightforward: since the members of $\mathscr{C}$ can make side payments, they should take as their objective the maximization of their joint income $\sum_{i \in \mathscr{C}} U_i$, in which case $\mathscr{C}$ is alpha-effective for $\mathbf{U}$ if and only if

$$\sum_{i \in \mathscr{C}} U_i \leq v^{\#}(\mathscr{C}), \tag{12.6.1}$$

where

$$v^{\#}(\mathscr{C}) = \max_{P_{\mathscr{C}} \in \mathscr{Q}_{\mathscr{C}}} \left[\min_{P_{\bar{\mathscr{C}}} \in \mathscr{Q}_{\bar{\mathscr{C}}}} \left[\sum_{i \in \mathscr{C}} U_i(P_{\mathscr{C}}, P_{\bar{\mathscr{C}}}) \right] \right]. \tag{12.6.2}$$

Similarly, $\mathscr{C}$ is beta-effective for $\mathbf{U}$ if and only if

$$\sum_{i \in \mathscr{C}} U_i \leq \min_{P_{\bar{\mathscr{C}}} \in \mathscr{Q}_{\bar{\mathscr{C}}}} \left[\max_{P_{\mathscr{C}} \in \mathscr{Q}_{\mathscr{C}}} \left[\sum_{i \in \mathscr{C}} U_i(P_{\mathscr{C}}, P_{\bar{\mathscr{C}}}) \right] \right]. \tag{12.6.3}$$

But Corollary 12.4.1 implies that the right-hand sides of (12.6.2) and (12.6.3) are equal, since they are the maximin and minimax values of the "two-person" zerosum game in which $\mathscr{C}$ plays the role of player 1 and $\Sigma_{i\in\mathscr{C}} U_i(P_{\mathscr{C}}, P_{\bar{\mathscr{C}}})$ plays the role of $U_1(P_1, P_2)$. Hence alpha- and beta-effectivenesses amount to the same thing for games with side payments, and furthermore,

$$v(\mathscr{C}) = \left\{\mathbf{U}: \mathbf{U} \in R^n, \sum_{i\in\mathscr{C}} U_i \le v^{\#}(\mathscr{C})\right\} \tag{12.6.4}$$

for such games, which implies that everything can be done in terms of the numbers $v^{\#}(\mathscr{C})$ instead of the sets $v(\mathscr{C})$.

Example 12.6.1

In "elaborate throwing fingers," as expressed by Table B in Example 12.2.5, suppose that COLIT consists of all nonempty coalitions and that utility is freely transferable with conservation. Then $v^{\#}(\{i\})$ is just i's maximin value, and so we recall from Example 12.3.2 that $v^{\#}(\{1\}) = -1719/49$, $v^{\#}(\{2\}) = -315/34$, and $v^{\#}(\{3\}) = 0$. Also, $v^{\#}(\mathscr{N})$ is just the maximum utility sum, and hence $v^{\#}(\{1, 2, 3\}) = 0$. To obtain $v^{\#}(\{1, 2\})$, we rearrange Table B in Example 12.2.5 and record only 1's and 2's utilities and their sum, obtaining

	s_3^1	s_3^2
(s_1^1, s_2^1)	$(-35, -10) \to -45$	$(10, -13.5) \to -3.5$
(s_1^1, s_2^2)	$(-37, 10) \to -27$	$(-10, -7.5) \to -17.5$
(s_1^2, s_2^1)	$(-37, 10) \to -27$	$(-10, -7.5) \to -17.5$
(s_1^2, s_2^2)	$(10, -10) \to 0$	$(10, -10) \to 0,$

from which it follows immediately that $v^{\#}(\{1, 2\}) = 0$. To find $v^{\#}(\{1, 3\})$, we obtain the similar table of 1's and 3's utilities and their sums.

	s_2^1	s_2^2
(s_1^1, s_3^1)	$(-35, 10) \to -25$	$(-37, 6) \to -31$
(s_1^1, s_3^2)	$(10, 1) \to 11$	$(-10, 5) \to -5$
(s_1^2, s_3^1)	$(-37, 6) \to -31$	$(10, 0) \to 10$
(s_1^2, s_3^2)	$(-10, 5) \to -5$	$(10, 0) \to 10$

When this is regarded as a zerosum-game matrix, we readily determine that a maximin strategy for $\{1, 3\}$ is $[0, 15/31, 0, 16/31]$ and that $v^{\#}(\{1, 3\}) = 85/31$. To find $v^{\#}(\{2, 3\})$, we form the table

	s_1^1	s_1^2
(s_2^1, s_3^1)	$(-10, 10) \to 0$	$(10, 6) \to 16$
(s_2^1, s_3^2)	$(-13.5, 1) \to -12.5$	$(-7.5, 5) \to -2.5$
(s_2^2, s_3^1)	$(10, 6) \to 16$	$(-10, 0) \to -10$
(s_2^2, s_3^2)	$(-7.5, 5) \to -2.5$	$(-10, 0) \to -10,$

implying that a maximin strategy for $\{2, 3\}$ is $[26/42, 0, 16/42, 0]$ and that $v^{\#}(\{2, 3\}) = 256/42$.

As noted previously, matters are not nearly so simple in games without side payments. But some general properties of characteristic functions can be stated.

Theorem 12.6.1. Let $v(\cdot)$ be either the alpha-effectiveness or the beta-effectiveness characteristic function of an n-person game. Then:

(A) For every nonempty coalition $\mathscr{C}$, $v(\mathscr{C})$ is a nonempty, closed, and convex set.
(B) If $\mathbf{U}'' \in v(\mathscr{C})$ and $U'_i \leq U''_i$ for every $i \in \mathscr{C}$, then $\mathbf{U}' = (U'_1, \ldots, U'_n)$ also belongs to $v(\mathscr{C})$.
(C) $\mathbf{U} \in v(\mathscr{N})$ if and only if there is some $\mathbf{U}^* \in H$ such that $U_i \leq U^*_i$ for every $i \in \mathscr{N}$.
(D) If $\mathscr{C}_1 \cap \mathscr{C}_2 = \emptyset$, then $v(\mathscr{C}_1) \cap v(\mathscr{C}_2) \subset v(\mathscr{C}_1 \cup \mathscr{C}_2)$.

■ *Proof.* Omitted. See Aumann [5]. ■

■ Except possibly for the convexity part of (A), these properties are of immediate intuitive desirability. (B) simply says that if $\mathscr{C}$ can attain its members' components of $\mathbf{U}''$, then it can also attain its members' components of any $\mathbf{U}'$ which is no better than $\mathbf{U}''$ for any member of $\mathscr{C}$. Similarly, (C) says that $v(\mathscr{C})$ is everything in H and everything "south and/or west" of H. (D) expresses the idea that if two nonoverlapping coalitions are both effective for $\mathbf{U}$, then their "merger," $\mathscr{C}_1 \cup \mathscr{C}_2$, continues to be effective for $\mathbf{U}$; that is, in unity there is, if not strength, at least not debility. ■

For games with side payments, Theorem 12.6.1(D) is equivalent to the *superadditivity property*

$$v^{\#}(\mathscr{C}_1) + v^{\#}(\mathscr{C}_2) \leq v^{\#}(\mathscr{C}_1 \cup \mathscr{C}_2) \text{ if } \mathscr{C}_1 \cap \mathscr{C}_2 = \emptyset \tag{12.6.5}$$

of $v^{\#}$, which is what von Neumann and Morgenstern [**106**] define as the characteristic function, since their work assumes that utility is transferable with conservation.

We are now prepared to define precisely what is meant by saying that $\mathbf{U}$ is meaningfully vetoable, or dominated. Essentially, $\mathbf{U}$ is dominated if there is a permissible coalition $\mathscr{C}$ and a $\mathbf{U}^*$ such that $\mathscr{C}$ is effective for $\mathbf{U}^*$ and the members of $\mathscr{C}$ all prefer $\mathbf{U}^*$ to $\mathbf{U}$.

Definition 12.6.3: Domination

(A) $\mathbf{U}^*$ dominates $\mathbf{U}$ via $\mathscr{C}$, written $\mathbf{U}^*\ \mathrm{DOM}_{\mathscr{C}}\mathbf{U}$, if and only if
  (i) $\mathbf{U}^* \in v(\mathscr{C})$, and
  (ii) $U^*_i > U_i$ for every $i \in \mathscr{C}$.
(B) $\mathbf{U}^*$ *dominates* $\mathbf{U}$, written $\mathbf{U}^*$ DOM $\mathbf{U}$, if and only if there is some coalition $\mathscr{C}$ in COLIT such that $\mathbf{U}^*\ \mathrm{DOM}_{\mathscr{C}}\ \mathbf{U}$.

■ As a relation between pairs of n-tuples, DOM is very ill behaved. It is not even transitive! To see this, suppose that $n = 3$, side payments are permitted, COLIT includes all nonempty coalitions, $V^{\#}(\{i\}) = 0$ for every

i, and $v^{\#}(\{1,2\}) = v^{\#}(\{1,3\}) = v^{\#}(\{2,3\}) = v^{\#}(\{1,2,3\}) = 1$. This game is called "divide the dollar," because it may be regarded as a situation in which any two of the players who can decide between themselves how to divide a dollar can commandeer the entire dollar. Now, $(0, \frac{1}{4}, \frac{3}{4})$ DOM $(\frac{3}{4}, 0, \frac{1}{4})$—via $\{2,3\}$; $(\frac{3}{4}, 0, \frac{1}{4})$ DOM $(\frac{1}{2}, \frac{1}{2}, 0)$—via $\{1,3\}$; but $(0, \frac{3}{4}, \frac{1}{4})$ does *not* DOM $(\frac{1}{2}, \frac{1}{2}, 0)$, as transitivity would require. In fact, only player 3 prefers $(0, \frac{1}{4}, \frac{3}{4})$ to $(\frac{1}{2}, \frac{1}{2}, 0)$, and 3 is not effective for $(0, \frac{1}{4}, \frac{3}{4})$ because he cannot guarantee his share $\frac{3}{4}$, which exceeds $v^{\#}(\{3\}) = 0$. Indeed, $(\frac{1}{2}, \frac{1}{2}, 0)$ dominates $(0, \frac{1}{4}, \frac{3}{4})$, via $\{1,2\}$! ■

Let us summarize briefly. We have introduced characteristic functions to represent effectiveness of coalitions, and we have defined domination in terms of characteristic functions. To say that **U*** DOM **U** means that **U** is meaningfully vetoable, and hence, presumably, not good. Conversely, it makes sense intuitively to say that **U** *is* good if it is not meaningfully vetoable, i.e., if it is undominated.

Definition 12.6.4. The *core* of a game is the set of all undominated joint evaluations.

Note that in at least two respects the core of a game depends tacitly upon more than its normal form. First, whether side payments are permitted makes a great deal of difference, not only in determinining the shape of H and $V(\mathcal{N})$, but also in raising the issue in the no-side-payments case of whether the alpha-effectiveness, the beta-effectiveness, or some other characteristic function should be used. Second, the specification of the set COLIT of permissible coalitions is also crucial.

■ It is easy to see that $v(\{i\}) = \{\mathbf{U}\colon U_i \le v^{\#}(\{i\})\}$, where $v^{\#}(\{i\})$ is some number which should not be less than $U_i(P_i^{\dagger}, \mathcal{Q}_{\bar{i}})$, which $\{i\}$ can attain all by himself given the active opposition of $\bar{i}$. Therefore, since COLIT is always assumed to include all single-player coalitions, the core is a subset of all individually rational joint evaluations. Equivalently, if **U** is not individually rational, then it is dominated via $\{i\}$ for some $i \in \mathcal{N}$, and hence **U** is not in the core. Similarly, if **U** is not group rational, then **U** is dominated via $\mathcal{N}$, because $\mathcal{N}$ is always assumed to be in COLIT, and hence **U** is not in the core. Therefore the core is a subset of the set of all imputations. ■

If sets COLIT^1 and COLIT^2 of coalitions are such that every coalition in COLIT^1 is a coalition in COLIT^2, or equivalently, $\text{COLIT}^1 \subset \text{COLIT}^2$, then their corresponding cores CORE^1 and CORE^2 satisfy the reverse inclusion $\text{CORE}^2 \subset \text{CORE}^1$, because COLIT^2 has at least as many (and more if $\text{COLIT}^1 \ne \text{COLIT}^2$) coalitions as COLIT^1 via which any given **U** can be dominated.

But now for an unpleasant surprise. Quite often it happens that the core of a game is empty when COLIT is taken (as it usually is) to be the set of *all* nonempty coalitions. But if CORE $= \emptyset$, then *all* joint evaluations are meaningfully vetoable!

Example 12.6.2

In Example 12.6.1 we easily conclude that if $\mathbf{U} = (U_1, U_2, U_3) \in$ CORE, then the following inequalities must hold, given the COLIT consists of all nonempty coalitions.

	Inequality	*To prevent being dominated via*
(1)	$U_1 \geq -1719/49$	$\{1\}$
(2)	$U_2 \geq -315/34$	$\{2\}$
(3)	$U_3 \geq 0$	$\{3\}$
(4)	$U_1 + U_2 \geq 0$	$\{1, 2\}$
(5)	$U_1 + U_3 \geq 85/31$	$\{1, 3\}$
(6)	$U_2 + U_3 \geq 256/42$	$\{2, 3\}$
(7)	$U_1 + U_2 + U_3 \geq 0$	$\{1, 2, 3\}$.

But to be in H, it is necessary that

(8) $$U_1 + U_2 + U_3 \leq 0.$$

Now, (7) and (8) imply

(9) $$U_1 + U_2 + U_3 = 0.$$

Clearly, (9), (3), and (4) imply

(10a) $$U_3 = 0,$$

(10b) $$U_1 + U_2 = 0.$$

But (10a), (5), and (6) imply

(11a) $$U_1 \geq 85/31$$

(11b) $$U_2 \geq 256/42.$$

Now, (11a) and (11b) can be added to produce

(12) $$U_1 + U_2 \geq 85/31 + 256/42.$$

But (12) and (10b) are inconsistent. Hence there is *no* $\mathbf{U}$ which satisfies (1)–(7). Therefore the core is empty!

Example 12.6.3

Suppose in the context of Examples 12.6.1 and 12.6.2 that player 1 is not allowed to enter into a coalition with 2 against 3 or with 3 against 2. Then COLIT = {{1}, {2}, {3}, {2, 3}, {1, 2, 3}}, and CORE is defined by inequalities (1)–(3) and (6)–(8) in Example 12.6.2. From Equation (9) there we write

(13) $$U_3 = -(U_1 + U_2);$$

substituting (13) into (3) yields $-(U_1 + U_2) \geq 0$, or

(14) $$U_2 \leq -U_1.$$

Substituting (13) into (6) yields $U_2 - (U_1 + U_2) \geq 256/42$, or

$$U_1 \leq -256/42. \tag{15}$$

Combining (15) with (1) and (14) with (2) yields

$$-1719/49 \leq U_1 \leq -256/42 \tag{16a}$$

$$-315/34 \leq U_2 \leq -U_1, \tag{16b}$$

which, together with (13), do constitute a consistent set of inequalities, so that CORE $\neq \emptyset$. For example, $(-10, 0, 10)$ CORE.

■ We might interpret COLIT in Example 12.6.3 as being arrived at in the following manner. Player 2 is a rather shady person, sufficiently scruffy, in fact, not to be above colluding with a robber, 3. Player 1, however, is a rather high-principled, albeit realistic person, who certainly would not gang up with the robber against 2 and who would not collude with his "opponent" in the throwing-fingers game, because to do so would be a sticky wicket. But he would join 2 and 3 if they were in cahoots against him. Note that CORE for this COLIT allows 2 and 3 to exploit 1, perhaps as extremely as with an imputation in which $U_1 = -1719/49$. ■

Example 12.6.4

Now suppose that each of 1 and 2 may collude with 3 against the other finger-thrower, but that they cannot collude with each other against the robber. Hence to COLIT of Example 12.6.3 we add $\{1, 3\}$ but *not* $\{1, 2\}$. This means that CORE here is constrained by the additional inequality (5) in Example 12.6.2, which in view of (13) readily reduces to

$$U_2 \leq -85/31, \tag{17}$$

a constraint on U_2 but not on U_1 or U_3. (This should not be too surprising.) Since $-85/31 < -U_1$ for every U_1 satisfying (16b), it follows that CORE here is the set of all (U_1, U_2, U_3) satisfying (13), (16a), and

$$-315/34 \leq U_2 \leq -85/31. \tag{16c}$$

Note how every imputation in CORE gives the robber a positive return amounting at to least $85/31 + 256/42 \doteq 8.84$. The moral seems to be that if honest people are not allowed to collude with each other, but only with criminals, then it is the criminals who stand to benefit. But we should not generalize excessively.

■ In the context of Examples 12.6.2–12.6.4, it would seem reasonable for 1 and 2 to "pass a law" forbidding unilateral collusion with 3, so that COLIT $= \{\{1\}, \{2\}, \{1, 2\}, \{1, 2, 3\}\}$. Here, it is not difficult to verify that CORE $= \{(U_1, -U_1, 0)\colon -1719/49 \leq U_1 \leq 315/34\}$, which you may verify in Exercise 12.6.1. ■

Because cores are often empty when all nonempty coalitions are permitted, it is necessary to seek a less restrictive characterization of "good" joint evaluations. One such gives rise to *stable sets* of imputations.

Definition 12.6.5: Stable sets. A set H' of imputations is said to be:

(A) *externally stable*, if for every $\mathbf{U} \notin H'$ there is a $\mathbf{U}^* \in H'$ such that $\mathbf{U}^*$ DOM $\mathbf{U}$;
(B) *internally stable*, if no $\mathbf{U}'$ in H' dominates any other $\mathbf{U}''$ in H';
(C) a *stable set*, or a *von Neumann-Morgenstern solution*, if H' is both externally and internally stable.

Von Neumann and Morgenstern introduce the concept of stable sets in **[106]**, for the side-payments case, as embodiments of social norms. A given game may have *many* stable sets, as Examples 12.6.6 and 12.6.7 indicate, or it may have *none*, as Lucas **[91]** (see also **[90]**) has recently demonstrated for the side-payments case. If it has many, then one can argue that the social norm determines the stable set, and then the relative bargaining talents of the players determine the ultimate imputation in that stable set.

Example 12.6.5

In "elaborate throwing fingers," suppose that COLIT consists of all nonempty coalitions, and hence CORE $= \emptyset$. We shall show, however, that $H' = \{(U_1, -U_1, 0)\colon\ -1719/49 \leq U_1 \leq 315/34\}$ is a stable set. H' is internally stable, since all imputations in H' give 3 zero, making 3 indifferent among them, whereas between 1 and 2 the situation posed by a choice from H' is strictly competitive. Hence domination via $\{1, 3\}$, $\{2, 3\}$, or $\{1, 2, 3\}$ is out of the question between imputations in H', as is domination via $\{1, 2\}$. To show that H' is externally stable, note that if $\mathbf{U} \notin H'$, then $\mathbf{U} = (U_1, U_2, U_3)$ with $U_3 > 0$. Therefore, $\mathbf{U}$ is dominated via $\{1, 2\}$ by some imputation in H' obtained from $\mathbf{U}$ by giving each of 1 and 2 a positive share of U_3. Intuitively, this stable set makes good sense in this game, because it implies, working back to the original tree, that 1 and 2 agree to throw two fingers each and thereby effectively freeze the robber out.

Example 12.6.6

In the "divide-the-dollar" game as specified in the remark following Definition 12.6.3, assume that utility is transferable with conservation and that COLIT consists of all nonempty coalitions. Then CORE $= \emptyset$, as you may verify in Exercise 12.6.1. But let $H' = \{(.50, .50, 0), (0, .50, .50), (.50, 0, .50)\}$. You may easily verify that none of these three imputations dominates any of the others, and hence that H' is internally stable. To show that H' is externally stable, let $\mathbf{U} = (U_1, U_2, U_3)$ and suppose, without loss of generality, that $U_1 \geq U_2 \geq U_3$. If $U_1 > .50$, then $U_2 + U_3 < .50$, and $\mathbf{U}$ is dominated via $\{2, 3\}$ by $(0, .50, .50)$. If $U_1 = .50$, then either

$U_2 = .50$ and $\mathbf{U} \in H'$, or U_2 and U_3 are both less than .50, in which case $\mathbf{U}$ is again dominated via $\{2, 3\}$ by $(0, .50, .50)$. Finally, if $U_1 < .50$, then $U_2 < .50$ also, and $\mathbf{U}$ is dominated via $\{1, 2\}$ by $(.50, .50, 0)$. Similar arguments obviously pertain for all other orderings of U_1, U_2, and U_3. Hence H' is also externally stable. Therefore H' is a stable set.

As we have noted, some games have a veritable plethora of stable sets. "Divide the dollar" is among them.

Example 12.6.7

In "divide the dollar," with side payments and COLIT consisting of all nonempty coalitions, let $0 \le x < .50$ and define

$$H'_{1x} = \{(x, U_2, 1 - U_2 - x): 0 \le U_2 \le 1 - x\};$$
$$H'_{2x} = \{(U_1, x, 1 - U_1 - x): 0 \le U_1 \le 1 - x\};$$

and

$$H'_{3x} = \{(U_1, 1 - U_1 - x, x): 0 \le U_1 \le 1 - x\}.$$

Now it is not difficult to show[p] that H'_{ix} is a stable set for *every* $x \in [0, .50)$ and every $i \in \{1, 2, 3\}$. But every imputation in this game belongs to some such H'_{ix}!

What is more serious than nonuniqueness of stable sets is the fact that, as noted above, there are games which have no stable set. From 1944, the publication date of the first edition of von Neumann and Morgenstern [**106**], until 1967, when Lucas' example [**91**] appeared, it was not known whether stable sets existed in all cases. Von Neumann and Morgenstern evidently considered the existence question of great importance: "There can be, of course, no concessions as regards existence [of stable sets]. If it should turn out that our requirements concerning a solution [= stable set] are, in any special case, unfulfillable,—this would certainly necessitate a fundamental change in the theory."[q]

Several other solution concepts for cooperative n-person games have been formulated. Most are based on the characteristic-function form of the game or on some generalization thereof, such as the Thrall-Lucas model of games in partition-function form, Luce's ψ-stability model, the Aumann-Maschler bargaining sets, Shapley's (and others') valuation models which ultimately determine a unique imputation, etc. The second volume of Rapoport [**117**], Owen [**108**], Lucas' surveys [**88**] and [**89**], and the *International Journal of Game Theory*, as well as the more recent Princeton volumes [**26**], [**27**], and [**135**] present much of this work and contain many references.

In general, the valuation models always produce a unique imputation, so

[p] Owen [**108**, pp. 166–167].

[q] [**106**, p. 42].

that existence and uniqueness are not issues per se; but uniqueness is typically obtained at the cost of imposing assumptions about the bargaining process or some idealized model of it that do not necessarily correspond to what one might expect a group of players to accept, even in principle. The other models typically pose existence problems; in fact, Lucas' example cited above is a modification of an example of a game in "partition-function" form with no stable set.

Depending upon one's point of view, such facts are either discouraging or challenging. At the very least, however, it is now clear that one should be diffident about characterizing the behavior of several interacting decision makers as rational or irrational!

Exercise 12.6.1

(A) Verify that CORE for "elaborate throwing fingers" with side payments when COLIT $= \{\{1\}, \{2\}, \{1,2\}, \{1,2,3\}\}$ coincides with the stable set H' in Example 12.6.5 (in which COLIT consists of all nonempty coalitions).

(B) Show that CORE of "divide the dollar" when COLIT consists of all nonempty coalitions is empty.

Exercise 12.6.2

Verify in Example 12.6.1 the superadditivity of $v^{\#}$ [i.e., show that (12.6.5) is always satisfied].

Exercise 12.6.3

An n-person cooperative game with side payments is called *inessential* if equality always obtains in (12.6.5). If COLIT consists of all nonempty coalitions in such a game, show that its core consists of only one joint evaluation. What is it?

Exercise 12.6.4

In an n-person game, possibly with side payments forbidden, of what joint evaluations does the core consist if:

(A) COLIT $= \{\{i\}: i = 1, \ldots, n\}$.

(B) COLIT $= \{\{1\}, \ldots, \{n\}, \mathcal{N}\}$.

(C) COLIT $= \{\mathcal{N}\}$.

[*Note*: (A) and (C) violate the assumption we have always made that every $\{i\}$ and $\mathcal{N}$ always belong to COLIT. The correct answer to this question shows why we make such an assumption.]

APPENDICES

Appendix 1

GREEK ALPHABET

Alpha	A, α	Nu	N, ν
Beta	B, β	Xi	Ξ, ξ
Gamma	Γ, γ	Omicron	O, o
Delta	Δ, δ	Pi	Π, π
Epsilon	E, ϵ	Rho	P, ρ
Zeta	Z, ζ	Sigma	Σ, σ
Eta	H, η	Tau	T, τ
Theta	Θ, θ	Upsilon	Υ, υ
Iota	I, ι	Phi	Φ, ϕ
Kappa	K, κ	Chi	X, χ
Lambda	Λ, λ	Psi	Ψ, ψ
Mu	M, μ	Omega	Ω, ω

Appendix 2

REVIEW OF BASIC CONCEPTS AND TERMINOLOGY OF SET THEORY

Readers with a prior exposure to set theory need only to skim everything up to our discussion and examples of functions and of relations and orders. Others will wish to read everything more carefully; for them, the best advice is to regard the following material (1) as primarily establishing a notational shorthand for familiar concepts and (2) as applying this shorthand in making some rather obvious statements when translated into everyday language (for example, De Morgan's Laws).

A *set* is a collection of mutually distinguishable objects called *elements*. Sets are usually denoted by capital letters, such as X: elements are denoted by lower-case letters, such as x. We write

$$x \in X$$

to mean that x *is an element of* X, or that x *belongs to* X. If x is *not* an element of X, we write

$$x \notin X.$$

Example A2.1. If x is a person and X is the set of all U.S. Citizens, then $x \in X$ if and only if x is a U.S. citizen.

Notation for Sets. If X is a set which consists of only a few elements, then X is often written by listing its elements and enclosing the list in braces. For example, if X consists of the elements x^1, x^2, x^3, and x^4, then we could write

$$X = \{x^1, x^2, x^3, x^4\}.$$

In such a listing, duplicating the symbol for an element does not produce a different set. Thus

$$\{x^1, x^2, x^3, x^4\} = \{x^1, x^1, x^2, x^3, x^3, x^3, x^4, x^4\}.$$

But when a set X consists of many, even infinitely many, elements, it is not practical to try to denote it by listing all its elements. Rather, we denote X as the set of all elements which satisfy a given requirement, or possess a given property. Let $S(x)$ be some statement which is either unambiguously true or unambiguously false about any element x. Then the set X of all x's for which $S(x)$ is *true* is written as

$$X = \{x: S(x)\};$$

here, the left brace is read as "the set of all"; the colon, as "such that"; and the right brace, as "is true." The colon is sometimes replaced with a vertical slash.

Example A2.2. Throughout this book we use R^k to denote the set of all k-*tuples* $x = (x_1, x_2, \ldots, x_k)$ of real numbers. Two such k-tuples are regarded as different even when they consist of the same numbers but in different orders of listing. Since we have defined R^k to be the set of *all* such k-tuples $(x_1, \ldots, x_k)$ with $-\infty < x_i < +\infty$ for every i, we may write

$$R^k = \{(x_1, \ldots, x_k): -\infty < x_i < +\infty \text{ for every } i\}.$$

Sometimes two or more statements of the form $S(x)$ are required to specify X. If so, we write

$$X = \{x: S_1(x), S_2(x), \ldots, S_m(x)\}$$

to mean that X is the set of all x's such that $S_1(x)$ is true, $S_2(x)$ is true, . . . , and $S_m(x)$ is true.

Example A2.3. Let $S_1(x) =$ "x is over 30 years old" and $S_2(x) =$ "you can trust x." Then the set of all trustworthy x's over 30 is precisely

$$\{x: S_1(x), S_2(x)\}.$$

Example A2.4. Suppose that I is a set of elements i and that to each i is associated an element x^i. Then the set X of all elements x associated with some i belonging to I is denoted by

$$X = \{x^i: i \in I\}.$$

Subsets. When one is considering a given set X, it often happens that he is interested in sets which consist of only some of the elements of X. Such sets are called *subsets* of X. We write

$$Y \subset X$$

to denote that Y is a subset of X. With a little manipulation of words it is obvious that Y is a subset of X if and only if every element of Y is an element of X. (But there may be elements of X which are not elements of Y.) If some element of Y is *not* an element of X, then Y is not a subset of X, and we write

$$Y \not\subset X.$$

We say that $Y = X$ if and only if X and Y consist of exactly the same elements, that is, if and only if both $Y \subset X$ and $X \subset Y$.

Example A2.5. Let $X = \{x\text{: } x \text{ is a positive real number}\}$ and $Y = \{x: x \in R^1, x > 0\}$. Then $X = Y$. This set is obviously a subset of the set R^1 of all real numbers (positive, negative, and zero).

Occasionally we say that X is a *superset* of Y to mean that Y is a subset of X.

Example A2.6. Every element of X is an element of X, trivially; and thus every set X is a subset of itself. It is called the *improper* subset of X to signify the straining of the associated connotations.

Empty Set. In order to make certain relationships among sets valid for all sets, it is necessary to introduce the notion of the *empty* set, or *void* set, denoted by $\emptyset$ (a Scandinavian letter, not the Greek phi).

Example A2.7. Suppose that $S_1(x)$ is true if and only if $S_2(x)$ is false. Then

$$\emptyset = \{x: S_1(x), S_2(x)\}.$$

For instance, $\{x$: x is a cow, x is a man$\} = \emptyset$ if we rule out mythological cases of dubious authenticity.

The empty set $\emptyset$ is a subset of every set X (because there is no element of $\emptyset$ which fails to belong to X)! Often a mathematical deduction breaks down only when applied to the empty set, and hence we frequently use the terminology *nonempty* set X, or *nonvoid* set X.

Union of Sets. If X_1 and X_2 are sets, we define the *union* $X_1 \cup X_2$ of X_1 and X_2 to be the set consisting of all elements which belong to X_1 or to X_2 or to both. Hence

$$X_1 \cup X_2 = \{x: x \in X_1 \text{ and/or } x \in X_2\}.$$

More generally, if X_i is a set for every element i of a nonempty *index* set I, then we define the union $\cup_{i \in I} X_i$ of all the X_i's by

$$\cup_{i \in I} X_i = \{x: X \in X_i \text{ for at least one } i \text{ in } I\}.$$

Intuitively, the union of a collection of sets is the set formed by aggregating all the elements of the given sets into one big set.

■ The order of aggregation is immaterial, because with a little thought it is clear that "union" is "commutative" and "associative," in the sense that

$$X \cup Y = Y \cup X \qquad \text{(commutativity)}$$

for all sets X and Y, and

$$(X \cup Y) \cup Z = X \cup (Y \cup Z) \qquad \text{(associativity)}$$

for all sets X, Y, and Z. Moreover, the empty set contributes nothing to a union, in that

$$X \cup \emptyset = X$$

for every set X. ■

In the special case where $I = \{1, 2, \ldots, n\}$, we usually write "$\cup_{i=1}^{n} X_i$" instead of "$\cup_{i \in I} X_i$ for $I = \{1, \ldots, n\}$." The space saving is obvious, as is the reduction in pedantry.

■ Suppose that $X_1, X_2, \ldots, X_n$ are n given sets, and that $Y_i = \cup_{j=i}^{n} X_j$ for every i in $\{1, \ldots, n\}$. Then $Y_n \subset Y_{n-1} \subset \cdots \subset Y_1$. Moreover, $Y_i \cup Y_{i+1} = Y_i$. (Why?) More generally, if $X \subset Y$, then $X \cup Y = Y$, since anything X could contribute to the union is already contributed by Y. ■

Intersection of Sets. If X_1 and X_2 are sets, we define the *intersection* $X_1 \cap X_2$ of X_1 and X_2 to be the set of all elements which belong to *both* X_1 and X_2. Hence

$$X_1 \cap X_2 = \{x: x \in X_1, x \in X_2\}.$$

More generally, if X_i is a set for every element i of a nonempty index set I, then we define the *intersection* $\cap_{i \in I} X_i$ of all the X_i's by

$$\cap_{i \in I} X_i = \{x: x \in X_i \text{ for every } i \text{ in } I\}.$$

The intersection of a collection of sets is the set of elements common to all the given sets. If *no* element is common to all the X_i's then $\cap_{i \in I} X_i = \emptyset$.

■ As with unions of sets, the order of determining commonality of elements is immaterial. On a little reflection it is clear that "intersection" is "commutative" and "associative," in the sense that

$$X \cap Y = Y \cap X \qquad \text{(commutativity)}$$

for all sets X and Y, and

$$(X \cap Y) \cap Z = X \cap (Y \cap Z) \qquad \text{(associativity)}$$

for all sets X, Y, and Z. Taking the intersection of any set X with the empty set $\emptyset$ is pure death, in that

$$X \cap \emptyset = \emptyset$$

for every set X. ■

Again as with unions, we simplify notation by writing "$\cap_{i=1}^{n} X_i$" in place of "$\cap_{i \in I} X_i$ for $I = \{1, \ldots, n\}$."

■ Let $X_1, \ldots, X_n$ be given and $Y_i = \cup_{j=i}^{n} X_j$ for every i. Then $Y_i \cap Y_{i+1} = Y_{i+1}$. (Why?) More generally, if $X \subset Y$, then $X \cap Y = X$, since the elements common to both X and Y are precisely the elements of X whenever X is a subset of Y. ■

■ **"Distributivity" of Unions and Intersections with Respect to Each Other.** Everyone knows that if x, y, and z are real numbers, then $x \cdot (y + z) = (x \cdot y) + (x \cdot z)$, but that $x + (y \cdot z)$ does not equal $(x + y) \cdot (x + z)$ except by happenstance (when $x = 0$ or $x + y + z = 1$). Because $x \cdot (y + z) = (x \cdot y) + (x \cdot z)$, we say that "$\cdot$" is distributive over "+." For sets X, Y, and Z and operations "$\cup$" and "$\cap$," *each* operation is distributive over the other.

$$X \cap (Y \cup Z) = (X \cap Y) \cup (X \cap Z)$$

and

$$X \cup (Y \cap Z) = (X \cup Y) \cap (X \cup Z)$$

for any sets X, Y, and Z. More generally,

$$X \cap (\cup_{i \in I} Y_i) = \cup_{i \in I} (X \cap Y_i)$$

and

$$X \cup (\cap_{i \in I} Y_i) = \cap_{i \in I} (X \cup Y_i). \blacksquare$$

A bit more terminology: Given a set X_i for every i in a nonempty index set I, we say that the X_i's are *mutually exclusive*, or *disjoint*, if $X_i \cap X_j = \emptyset$ whenever $i \neq j$. In other words, the X_i's are mutually exclusive if no two of them have any element in common.

Venn Diagrams. A commonly used device of pretending that sets are regions of the plane is a considerable aid to intuition in thinking about sets. In Figure A2.1 we depict three sets, X, Y, and Z, together with $X \cap Y \cap Z$. Such a picture is called a *Venn diagram.* In this one, $X \cup Y \cup Z$ is the entire unshaded portion of the square. (Assume for definiteness that X, Y, and Z do not include their boundary points.)

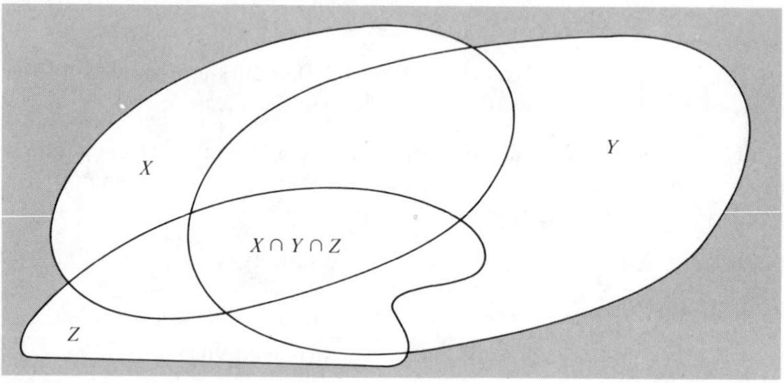

Figure A2.1 A Venn diagram.

Universal Set. Within a given framework of discussion, it frequently happens that all sets under consideration are subsets of some given set, which is then called the *universal* set, the *universe* of discourse, or some such global term. For instance, the interval $(-\infty, +\infty)$ is the universal set when dealing with sets of real numbers; and sure events are universal sets when discussing events in (= subsets of) them.

Complement. Suppose that X is a universal set and that Y is a subset of X. The *complement* of Y, or complement (in X) of Y, is denoted by $\bar{Y}$ and defined as the set of all elements of X which do *not* belong to Y. Thus

$$\bar{Y} = \{x: x \in X, x \notin Y\}$$

for every subset Y of X. Complements are also denoted by superscript c's, but we shall avoid so doing.

■ **Relative Complement, or Set Difference.** Suppose that Y and Z are subsets of a universal set X. The complement of Y relative to Z, or complement of Y in Z, or difference $Z - Y$, is written variously as $Z - Y$, $C_Z Y$, $Z \sim Y$, and $Z \backslash Y$, and defined by

$$Z - Y = \{x: x \in Z, x \notin Y\}.$$

Note that this definition makes no use of the universal set X; one which does is

$$Z - Y = Z \cap \bar{Y}.$$

Universal sets, and complements, are always understood as being relative to a given framework of discourse. ■

■ **De Morgan's Laws.** With a little thought it is clear that the following two-set equalities obtain.

$$Z - (\cup_{i \in I} Y_i) = \cap_{i \in I}(Z - Y_i)$$

and

$$Z - (\cap_{i \in I} Y_i) = \cup_{i \in I}(Z - Y_i).$$

When Z is the universal set, these formulae become

$$\overline{\bigcup_{i\in I} Y_i} = \bigcap_{i\in I} \overline{Y_i}$$

and

$$\overline{\bigcap_{i\in I} Y_i} = \bigcup_{i\in I} \overline{Y_i}$$

respectively. ■

Partitions. Suppose that X_i is a set for every i in a nonempty index set I, and that X is a set. We say that the X_i's *constitute a partition of* X if they are mutually exclusive and also *collectively exhaustive of* X, in the sense that

$$\bigcup_{i\in I} X_i = X.$$

Intuitively, the X_i's constitute a partition of X if they are the pieces which result from some process of carving X up.

Example A2.8. Suppose that X is the set of all tax returns for a given year. The IRS might wish to partition X for audit purposes into

$X_1 = \{x$: net taxable income on $x \leq \$10{,}000\}$,
$X_2 = \{x$: net taxable income on $x \in (\$10{,}000, \$25{,}000]\}$, and
$X_3 = \{x$: net taxable income on $x > \$25{,}000\}$. Clearly, X_1, X_2, and X_3 constitute a partition of X.

Cartesian Products. Let X and Y be two nonempty sets. The *Cartesian product* $X \times Y$ of X and Y (in that order!) is the set of all *ordered* pairs (x, y) of elements, in which the first member belongs to X and the second member belongs to Y. Formally,

$$X \times Y = \{(x, y): x \in X, y \in Y\}.$$

More generally, the Cartesian product $X_1 \times \cdots \times X_n$, or $\times_{i=1}^{n} X_i$ of $X_1, \ldots, X_n$ (in that order) is defined by

$$\times_{i=1}^{n} X_i = \{(x_1, \ldots, x_n): x_i \in X_i \text{ for every } i\}.$$

The individual sets X_i in a Cartesian product are often (and naturally) called *factors* of the Cartesian product.

Example A2.9. The set R^k defined in Example A2.2 may be regarded as the Cartesian product $\times_{i=1}^{k} X_i$ in which every factor X_i is the set R^1, or $(-\infty, +\infty)$, of all real numbers, since every element of R^k is a k-tuple of real numbers. Alternatively, suppose that $j + m = k$; then $R^k = R^j \times R^m$, provided that we drop parentheses so as to write

$$((x_1, \ldots, x_j), (x_{j+1}, \ldots, x_k)) = (x_1, \ldots, x_j, x_{j+1}, \ldots, x_k).$$

Functions. Suppose that X and Y are nonempty sets. Rigorously speaking, a *function* f with *domain* X and *codomain* Y (or "range in" Y, or "values in" Y), written $f: X \to Y$, is by definition *a subset of* $X \times Y$ with the property that for each X there is precisely one y such that (x, y) belongs to f. In everyday terminology, we denote that y such that (x, y) f by $f(x)$. Hence, rigorously speaking,

$$f = \{(x, f(x)): x \in X\},$$

where $f(x) \in Y$ for every x in X. To avoid confusion, we usually denote functions with the symbolism $f(\cdot)$ when the domain and codomain are clear from the context.

Example A2.10. Let X be the set of all people and Y be the set of all dates. There is a function $f: X \to Y$ associating with each x his or her birthday $f(x)$. Every x has one and only one birthday $f(x)$, whether he knows it or not. But the birthday *function* is the set of *all* (person x, x's birthday) pairs.

■ **Composition of Functions.** Suppose that X, Y, and Z are sets, that $f_1: X \to Y$ is a function, and that $f_2: Y \to Z$ is a function. The *composition* $f_2 \circ f_1: X \to Z$ of f_2 with f_1 is defined by

$$f_2 \circ f_1 = \{(x, f_2)[f_1(x)]: x \in X\}.$$

We also write $f_2 \circ f_1$ in the more explicit form $f_2[f_1(\cdot)]$. ■

Example A2.11. Any function $x: \{1, \ldots, k\} \to (-\infty, +\infty)$ can be written pedantically as

$$x = \{(1, x(1)), (2, x(2)), \ldots, (k, x(k))\}.$$

It is clear that there is a natural correspondence between x and the k-tuple $(x_1, \ldots, x_k)$ of real numbers such that $x_i = x(i)$ for each i. This means that the set R^k of *all* k-tuples $(x_1, \ldots, x_k)$ of real numbers may be regarded as the set of *all functions* x with domain $\{1, \ldots, k\}$ and codomain $(-\infty, +\infty)$.

Example A2.12. The set of all functions with domain Z and codomain X is denoted by X^Z. When $X = (-\infty, +\infty) = R = R^1$, and $Z = \{1, \ldots, k\}$, we obtain $R^{\{1,\ldots,k\}}$ and abbreviate it to R^k, as before. More generally, the set R^Z of all real-valued functions[a] on a set Z is called a *real vector space*; its elements enjoy relationships not generally definable when the codomain is not $(-\infty, +\infty)$.

(a) The *sum* $\mathbf{x}' + \mathbf{x}''$ of two elements $\mathbf{x}'$ and $\mathbf{x}''$ of R^Z is *defined* to be the function such that $(\mathbf{x}' + \mathbf{x}'')(z) = \mathbf{x}'(z) + \mathbf{x}''(z)$ for every z in Z.

(b) The *scalar multiple* $t\mathbf{x}$ of an element $\mathbf{x}$ of R^Z by a real number t is *defined* to be the function such that $(t\mathbf{x})(z) = t \cdot \mathbf{x}(z)$ for every z in Z.

When $Z = \{1, \ldots, k\}$ and every $\mathbf{x}$ is written as $(x_1, \ldots, x_k)$, these definitions reduce to the usual vectorial definitions

(a_k) $\mathbf{x}' + \mathbf{x}'' = (x_1' + x_1'', x_2' + x_2'', \ldots, x_k' + x_k'')$

and

(b_k) $t\mathbf{x} = (tx_1, tx_2, \ldots, tx_k)$.

These operations are illustrated in Figure A2.2 for $k = 2$. One further definition in connection with real vector spaces. The *line segment* $L(\mathbf{x}', \mathbf{x}'')$ *joining* two elements $\mathbf{x}'$ and $\mathbf{x}''$ of R^z is defined to be the set of all elements of the form $t\mathbf{x}' + (1 - t)\mathbf{x}''$ for some t in the interval $[0, 1]$. That is,

$$L(\mathbf{x}', \mathbf{x}'') = \{t\mathbf{x}' + (1 - t)\mathbf{x}'': \mathbf{x}' \in R^Z, \mathbf{x}'' \in R^Z, 0 \le t \le 1\}.$$

Example A2.13. Suppose that X_i is a nonempty set for every element i of a nonempty index set I. The set $\mathbf{X}$ of all functions $\mathbf{x}: I \to \bigcup_{i \in I} X_i$ such that $\mathbf{x}(i) \in X_i$ for every i in I corresponds in a natural way to the notion of a Cartesian product, but it is more general in that we have not supposed that the set I has

[a] "Real-valued" means "codomain R^1."

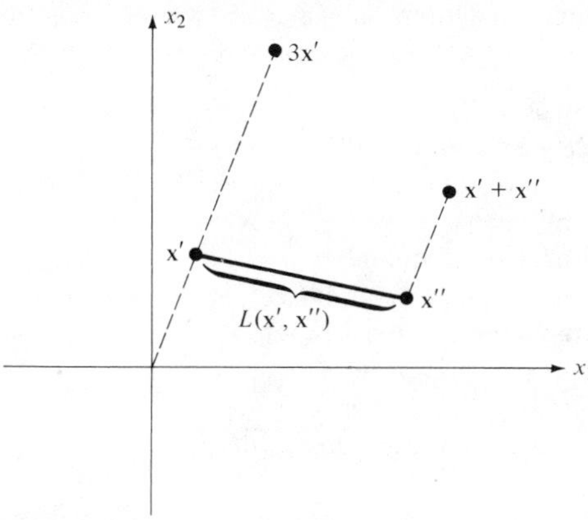

Figure A2.2 Vector operations in R^2.

any particular order. For instance, if $I = \{1, 2, 3\}$, $X_1 = \{a, b\}$, $X_2 = \{a, c\}$, and $X_3 = \{b, c\}$, then $\cup_{i=1}^{3} X_i = \{a, b, c\}$ and there are 27 functions $\mathbf{x}^j: \{1, 2, 3\} \to \{a, b, c\}$, of which only eight satisfy the requirement that $\mathbf{x}^j(i) \in X_i$ for every i. These eight are as follows.

j	$x^j(1)$	$x^j(2)$	$x^j(3)$	Corresponding element of $X_1 \times X_2 \times X_3$
1	a	a	b	(a, a, b)
2	a	a	c	(a, a, c)
3	a	c	b	(a, c, b)
4	a	c	c	(a, c, c)
5	b	a	b	(b, a, b)
6	b	a	c	(b, a, c)
7	b	c	b	(b, c, b)
8	b	c	c	(b, c, c).

Relations and Orders. Let X be a set. A *relation R in X*, or between elements of X, is a subset of $X \times X$. Any subset of $X \times X$ defines a relation in X. We write xRy and say "x is related to y via R," or "x is R-related to y" if and only if $(x, y) \in R$.

Example A2.14. The subset "$\geq$" of $(-\infty, +\infty) \times (-\infty, +\infty) = R^2$ corresponding to the relation "greater than or equal" between real numbers [= elements of $(-\infty, +\infty)$] is simply $\{(x, y): x \geq y\}$. See Figure A2.3.

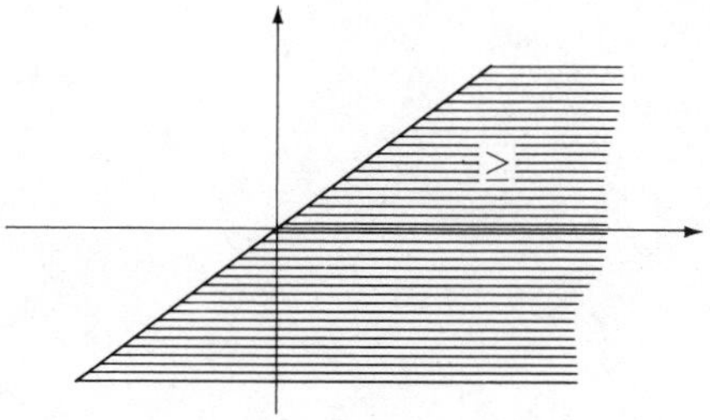

Figure A2.3 The relation $\geq$ in $(-\infty, +\infty)$.

General relations are not of much interest. *Orders* are special types of relations, ones which possess certain useful properties. We say that a relation R in X is:

(1) *reflexive* if xRx for every x in X;
(2) *transitive* if xRz whenever xRy and yRz;
(3) a *quasi order* if R is reflexive and transitive;
(4) *complete* if xRy and/or yRx for every (x, y) in $X \times X$;
(5) a *weak order* if R is a quasi order and complete;
(6) *antisymmetric* if $x = y$ whenever both xRy and yRx;
(7) a *partial order* if R is an antisymmetric quasi order;
(8) a *linear order* if R is a complete partial order;
(9) *symmetric* if yRx whenever xRy;
(10) an *equivalence relation* if R is a symmetric quasi order.

Example A2.15. The relation "$\succsim$", "at least as desirable as," in the set C of consequences of a dpuu is assumed to be a weak order in this book. (Some scholars of preference theory assume that "$\succsim$" is only a quasi order.) The relation "$\sim$", "is equally as desirable as," is assumed to be an equivalence relation here.

Example A2.16. The relation "$\geq$" between real numbers is reflexive, antisymmetric, transitive, and complete, and is therefore a linear order.

Example A2.17. The relation "$\geq$" in R^k, or between real k-tuples, is defined by $(x'_1, \ldots, x'_k) \geq (x''_1, \ldots, x''_k)$ if and only if $x'_i \geq x''_i$ for every i. It is easy to verify that "$\geq$" is reflexive, antisymmetric, and transitive, but *not* complete: neither $(1, 2) \geq (2, 1)$ nor $(2, 1) \geq (1, 2)$, for instance. Hence "$\geq$" in R^k is only a partial order. It is called the *usual* partial order of R^k.

Relations between Orders. Let R_1 and R_2 be relations in a set X. We say that R_1 *implies* R_2 if xR_2y whenever xR_1y. It is clear that R_1 implies R_2 if and only if $R_1 \subset R_2$, each R being regarded as a subset of $X \times X$. When R_1 and R_2 are orders or equivalence relations, we also say that R_1 *is at least as fine as* R_2 if R_1 implies R_2.

Example A2.18. The *identity* relation R in X, defined by xRy if and only if $x = y$, is clearly an equivalence relation. Because any equivalence relation Q in X is reflexive, it follows immediately that R implies Q, or that R is at least as fine as Q. For instance, $c' \sim c''$ if $c' = c''$. Hence "$=$" implies "$\sim$."

Appendix 3

CONVEXITY AND RELATED TOPICS IN ANALYSIS

This appendix is a sketchy introduction to convex sets, convex and concave functions, and related topics of importance in decision analysis. Its only prerequisite is Appendix 2, on elementary set theory. Readers wishing a more complete introduction to convex analysis are referred to Rockafellar [**120**]. Choquet [**18**] is a good reference on continuity and related subjects. Finally, readers whose appetites are whetted by this appendix are encouraged to progress to a good text on mathematical programming, such as Mangasarian [**93**] and Zangwill [**150**], where extensive applications and extensions of the material introduced here are brought to bear on optimization problems.

A *real vector space* R^Z was defined in Example A2.12 as the set of all functions $\mathbf{x}: Z \rightarrow (-\infty, +\infty)$. Such functions are usually called *vectors*. Sums $\mathbf{x}' + \mathbf{x}''$ and scalar products $t\mathbf{x}$ of vectors in R^Z were also defined in that example, as was the line segment $L(\mathbf{x}', \mathbf{x}'')$ joining $\mathbf{x}'$ and $\mathbf{x}''$.

Convex Sets. A subset X of R^Z is *convex* if $L(\mathbf{x}', \mathbf{x}'') \subset X$ whenever both $\mathbf{x}'$ and $\mathbf{x}''$ belong to X. Intuitively, X is convex if it includes all points on the line segment joining any two of its points, and thus, has no indentations. Figure A3.1 depicts a convex and a nonconvex subset of R^2.

■ The set $\mathcal{P}$ of all probability functions on events in an $\mathscr{E}$-move Ω can be regarded as a convex set in a real vector space R^Z. [If Ω contains k elements, then each probability function corresponds to the vector $\mathbf{P} = (P(\omega^1), \ldots, P(\omega^k))$ of probabilities of the elementary events.] That $\mathcal{P}$ is a convex set is clear from the fact that $t\mathbf{P}^1 + (1-t)\mathbf{P}^2$ is the probability function assigning probability $tP^1(\Omega') + (1-t)P^2(\Omega')$ to every event Ω' in Ω whenever $\mathbf{P}^1$ and $\mathbf{P}^2$ are in $\mathcal{P}$ and $t \in [0, 1]$. ■

Convex Functions. Let X be a convex set. A real-valued function $r(\cdot)$ on X is said to be *convex* if

$$r(t\mathbf{x}' + (1-t)\mathbf{x}'') \leq tr(\mathbf{x}') + (1-t)r(\mathbf{x}'') \qquad \text{(A3.1)}$$

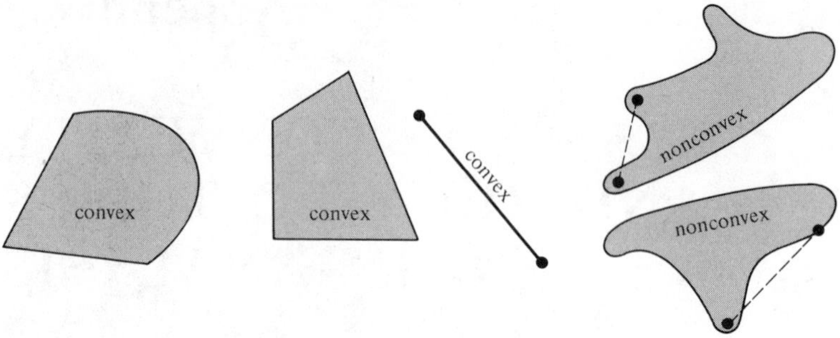

Figure A3.1 Convex and nonconvex sets in R^2.

whenever $\mathbf{x}'$ and $\mathbf{x}''$ belong to X and $t \in [0, 1]$. In other words, the chord joining two points on the graph $(\cdot, r(\cdot))$ of $r(\cdot)$ always lies on or above the corresponding function values.

■ The chord interpretation of the definition of a convex function is tantamount to an alternative definition, equivalent to the given one: $r(\cdot)$ is convex if the subset $\{(\mathbf{x}, z): \mathbf{x} \in X, z \geq r(\mathbf{x})\}$ of $X \times (-\infty, +\infty)$ is convex. ■

Concave Functions. Let X be a convex set. A real-valued function $r(\cdot)$ on X is said to be *concave* if

$$r(t\mathbf{x}' + (1-t)\mathbf{x}'') \geq tr(\mathbf{x}') + (1-t)r(\mathbf{x}'') \tag{A3.2}$$

whenever $\mathbf{x}'$ and $\mathbf{x}''$ belong to X and $t \in [0, 1]$. In other words, the chord joining two points on the graph of $r(\cdot)$ lies on or below the corresponding function values.

■ Alternatively, $r(\cdot)$ is concave if $\{(\mathbf{x}, z): \mathbf{x} \in X, z \leq r(\mathbf{x})\}$ is convex. ■

■ It is clear that $r(\cdot)$ is concave if and only if $-r(\cdot)$ is convex. ■

Figure A3.2 depicts a convex function, a concave function, and a function neither convex nor concave. All functions are defined on $X = [1, 2]$. Note that convex and concave functions need not be continuous.

The importance of convex functions in decision analysis devolves from their extremely good behavior in minimization problems: "Local" minimizers, such as $x = 3/2$ for $r_3(\cdot)$ in Figure A3.2, cannot fail to be "global" minimizers as well when $r(\cdot)$ is convex. Similarly, "local" maximizers of *concave* functions are also "global" maximizers.

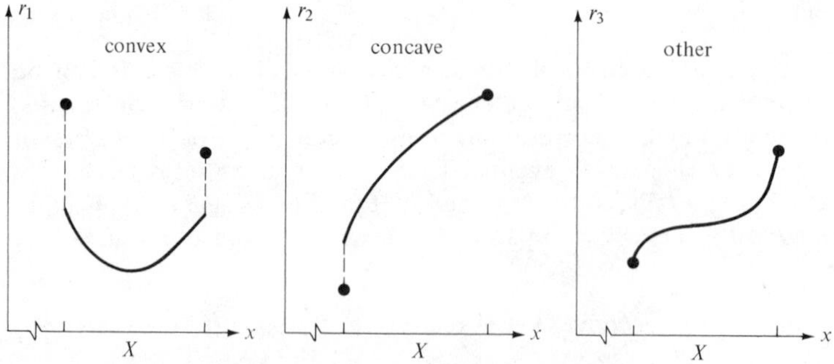

Figure A3.2 Convex, concave, and other functions.

Quasi-Convex Functions. Let X be a convex set. A real-valued function $r(\cdot)$ on X is said to be *quasi convex* if

$$r(t\mathbf{x}' + (1-t)\mathbf{x}'') \leq \max\,[r(\mathbf{x}'), r(\mathbf{x}'')] \tag{A3.3}$$

whenever $\mathbf{x}'$ and $\mathbf{x}''$ belong to X and $t \in [0, 1]$.

■ An equivalent definition of quasi convexity is: $r(\cdot)$ is quasi convex on X if $\{\mathbf{x}: \mathbf{x} \in X, r(\mathbf{x}) \leq z\}$ is convex for every $z \in (-\infty, +\infty)$. (Check that the empty set $\emptyset$ and R^Z are convex sets!) A proof that this definition is equivalent to the original one may be found in Zangwill **[150]**. ■

Quasi-Concave Functions. Let X be a convex set. A real-valued function $r(\cdot)$ on X is said to be *quasi concave* if

$$r(t\mathbf{x}' + (1-t)\mathbf{x}'') \geq \min\,[r(\mathbf{x}'), r(\mathbf{x}'')] \tag{A3.4}$$

whenever $\mathbf{x}'$ and $\mathbf{x}''$ belong to X and $t \in [0, 1]$.

■ An equivalent definition of quasi concavity is: $r(\cdot)$ is quasi concave on X if $\{\mathbf{x}: \mathbf{x} \in X, r(\mathbf{x}) \geq z\}$ is convex for every $z \in (-\infty, +\infty)$. ■

■ Clearly, $r(\cdot)$ is quasi concave if and only if $-r(\cdot)$ is quasi convex. ■

It is easy to see that if $r(\cdot)$ is convex (or, respectively, concave), then $r(\cdot)$ is also quasi convex (or, respectively, quasi concave); but the converse does *not* obtain. Figure A3.3 exhibits a quasi-concave function which is obviously not concave.

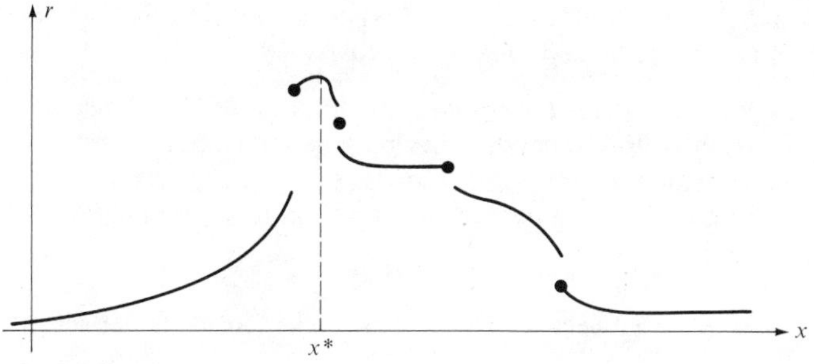

Figure A3.3 A quasi-concave function.

Quasi-concave functions are important in decision analysis because they possess the property that if $\mathbf{x}^*$ is a maximizer of $r(\cdot)$ on X and if $\mathbf{x}' \in L(\mathbf{x}^*, \mathbf{x}'')$, then $r(\mathbf{x}') \geq r(\mathbf{x}'')$. [This is immediate from (A3.4) and the fact that $\mathbf{x}^*$ is a maximizer of $r(\cdot)$ on X if and only if $r(\mathbf{x}^*) \geq r(\mathbf{x})$ for every $\mathbf{x} \in X$; in particular, $r(\mathbf{x}^*) \geq r(\mathbf{x}'')$, and hence $\min\,[r(\mathbf{x}^*), r(\mathbf{x}'')] = r(\mathbf{x}'')$.] In other words, the value of $r(\cdot)$ is nonincreasing anywhere as one travels in a straight line in any direction away from $\mathbf{x}^*$. Check this in Figure A3.3. Naturally, quasi-convex functions possess analogous properties; replace "maximizer," "≥," "min," and "nonincreasing" in the preceding sentences with "minimizer," "≤," "max," and "nondecreasing."

Convex Hulls. Let X be a subset of R^Z. The *convex hull* CONV$[X]$ of X is defined to be the set of all vectors of the form $\Sigma_{i=1}^{j} t_i\mathbf{x}^i$, where $\mathbf{x}^i \in X$ for every i, $t_i > 0$ for every i, $\Sigma_{i=1}^{j} t_i = 1$, and j is a (finite!) positive integer.

■ Alternatively, CONV[X] is the smallest convex set which includes X. In fact, it is the intersection of all convex sets which include X. ■

■ It is easy to see that $X = \text{CONV}[X]$ if and only if X is a convex set. ■

In Figure A3.4, the shaded region depicts CONV[X], where X is the set of vectors in R^2 represented by heavy dots.

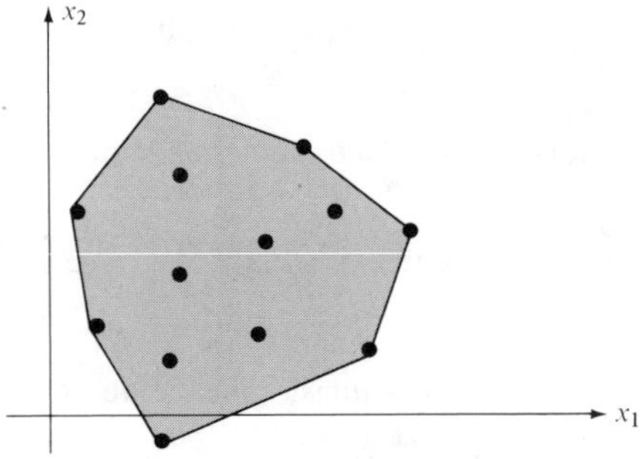

Figure A3.4 The convex hull of a finite set in R^2.

■ An interesting and occasionally useful fact about convex hulls in R^k is *Carathèodory's Theorem*: If $X \subset R^k$ and $\mathbf{x} \in \text{CONV}[X]$, then there exists a set $\{\mathbf{x}^1, \ldots, \mathbf{x}^j\}$ consisting of $j \leq k+1$ vectors in X, and there exist j positive numbers t_i such that $\Sigma_{i=1}^{j} t_i = 1$ and $\mathbf{x} = \Sigma_{i=1}^{j} t_i \mathbf{x}^i$. What is new here is that any point $\mathbf{x}$ in CONV[X] is a weighted average of *at most* $k+1$ $\mathbf{x}^i$'s. ■

Other useful facts require introducing properties such as closedness, distance, and continuity. The remainder of this appendix develops the requisite concepts and is slightly more advanced than the preceding rudiments. To keep the requisite additional sophistication to a bare minimum, we *assume from here on that* $\mathrm{R^Z = R^k}$ *for a (finite!) positive integer* k.

Distance. The distance $|\mathbf{x}' - \mathbf{x}''|$ between two vectors $\mathbf{x}'$ and $\mathbf{x}''$ in R^k is defined by

$$|\mathbf{x}' - \mathbf{x}''| = \max_i \, [|x_i' - x_i''|], \tag{A3.5}$$

where $|x_i' - x_i''|$ denotes the absolute value of $x_i' - x_i''$.

e-**Neighborhood of** *x*. Let $\mathbf{x} \in R^k$ and $e > 0$. The *e-neighborhood* N(**x**, e) *of* **x** is defined by

$$N(\mathbf{x}, e) = \{\mathbf{x}' \colon \mathbf{x}' \in R^k, |\mathbf{x} - \mathbf{x}'| < e\}. \tag{A3.6}$$

Three e-neighborhoods of (1, 1) R^2 are depicted in Figure A3.5; e-neighborhoods in R^2 are squares of side length $2e$ centered at $\mathbf{x}$.

Boundary Points of a Set. Suppose that $X \subset R^k$. A vector $\mathbf{x}$ in R^k is said to be a boundary point of X, or to be an element of the set X^b of *all* boundary points of X, if neither $N(\mathbf{x}, e) \cap X$ nor $N(\mathbf{x}, e) \cap \bar{X}$ is empty for any positive real number e. Thus, it is clear from Figure A3.5 that (1, 1) is a boundary point of $X = \{\mathbf{x} \colon \mathbf{x} \in R^2, (x_1)^2 + (x_2)^2 < 2\}$ even though $(1, 1) \notin X$; this set X includes *none* of its boundary points. On the other hand, two sets, namely, $\emptyset$ and R^k, in R^k have *no*

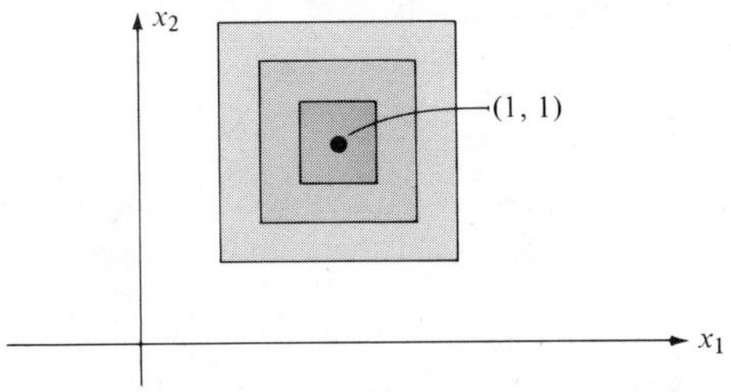

Figure A3.5 e-Neighborhoods of (1, 1).

boundary points at all. Other sets are *all* boundary; for example, any line segment in R^2 is all boundary. (Why?)

More Types of Sets. Let X be a subset of R^k. We say that X is:

(1) *open*, if $X \cap X^b = \emptyset$ [i.e., X contains none of its boundary points];
(2) *closed*, if $X^b \subset X$ [i.e., X contains all of its boundary points];
(3) *bounded*, if $X \subset N(\mathbf{0}, e)$ for some $e \in (0, +\infty)$ [i.e., if *every* $\mathbf{x}$ in X is within some fixed, finite distance e of the zero vector $\mathbf{0}$ in R^k];
(4) *compact*, if X is closed and bounded.

■ Recall that $\bar{X}$ denotes the complement of X (in R^k). It is easy to see that $X^b = (\bar{X})^b$, and hence, that X is closed if and only if $\bar{X}$ is open. ■

■ It is an easy exercise to show that *if* X *is a finite subset of* R^k, *then CONV*[X] *is compact.* ■

■ Let $\mathcal{P}$ denote the set of all probability functions on a finite $\mathscr{E}$-move Ω, each such function represented by the vector $\mathbf{P} = (P(\omega^1), \ldots, P(\omega^k))$ of probabilities assigned to the elementary events in Ω. Then $\mathcal{P}$ is compact. In fact, $\mathcal{P}$ is the convex hull of the set of k "sure probabilities" $\mathbf{P}^i$ in which $P^i(\omega^i) = 1$ and $P^i(\omega^j) = 0$ for $j \neq i$. ■

Figure A3.6 depicts an open set, a closed set, and a set neither closed nor open, all in R^2. In this figure, light portions of boundaries signify the *exclusion* of those points and heavy portions signify inclusion of those points, in X.

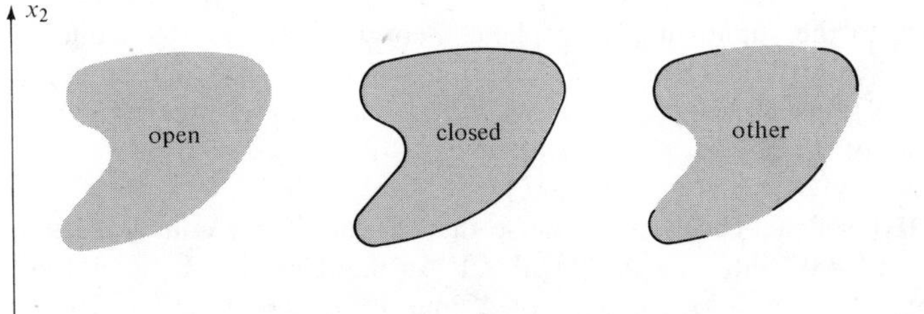

Figure A3.6 Open, closed, and other sets in R^2.

Hyperplanes. A subset H of R^k is called a *hyperplane* if

$$H = \left\{ \mathbf{x}: \mathbf{x} \in R^k, \sum_{i=1}^{k} a_i x_i = z \right\} \tag{A3.7}$$

for some *nonzero* vector $\mathbf{a} = (a_1, \ldots, a_k)$ in R^k and some real number z. Sometimes one writes $H(\mathbf{a}, z)$ for explicitness. Hyperplanes in R^2 are straight lines (unbounded in each direction).

Supporting Hyperplanes. Let X be a convex subset of R^k and let $\mathbf{x}^*$ belong to the boundary X^b of X. The *supporting hyperplane theorem* states that there exists a hyperplane $H(\mathbf{a}, z)$ in R^k such that

(1) $$\mathbf{x}^* \in H(\mathbf{a}, z);$$

and

(2) $$\sum_{i=1}^{k} a_i x_i \le z \text{ for every } \mathbf{x} = (x_1, \ldots, x_k) \text{ in } X.$$

That is, all of the convex set X lies on one side or the other of $H(\mathbf{a}, z)$, which is called a supporting hyperplane of X at $\mathbf{x}^*$. If X is closed, so that $\mathbf{x}^*$ is sure to belong to X, then "tangent hyperplane" is a more graphic term.

■ Since $H(\mathbf{a}, z) = H(-\mathbf{a}, -z)$ (why?), it follows that (2) above could be replaced by

(2′) $$\sum_{i=1}^{k} a_i x_i \ge z \text{ for every } \mathbf{x} = (x_1, \ldots, x_k) \text{ in } X. \blacksquare$$

Figure A3.7 depicts some supporting hyperplanes (lines) to a convex subset of R^2.

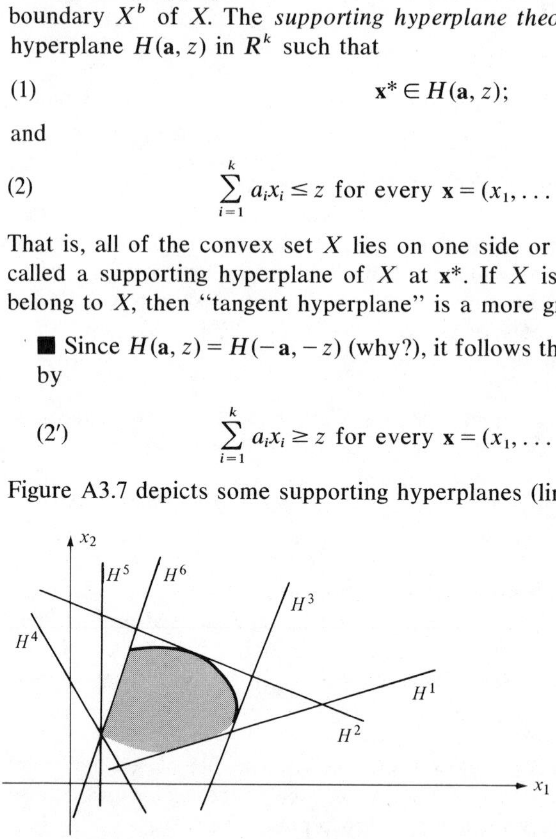

Figure A3.7 Supporting hyperplanes in R^2.

■ As a corollary to the supporting hyperplane theorem, it is easy to deduce a corollary to Carathèodory's Theorem; namely, if CONV[X] is closed, then every point in $(\text{CONV}[X])^b$ is a "weighted average" $\Sigma_{i=1}^{j} a_i \mathbf{x}^i$ of j = at most k (rather than $k+1$) points of X. ■

The Separating Hyperplane Theorem. Suppose that X and Y are convex sets in R^k such that $Y - Y^b \neq \emptyset$ and $(X - X^b) \cap (Y - Y^b) = \emptyset$. (That is, Y is not all boundary; and there is no "interior-overlap" of X and Y.) Under these conditions, the *separating hyperplane theorem* states that there exists a hyperplane $H(\mathbf{a}, z)$ in R^k such that X and Y "lie on opposite sides of $H(\mathbf{a}, z)$," in the sense that

$\Sigma_{i=1}^k a_i x_i \leq z$ for every $\mathbf{x} = (x_1, \ldots, x_k)$ in X and $\Sigma_{i=1}^k a_i y_i \geq z$ for every $\mathbf{y} = (y_1, \ldots, y_k)$ in Y.

■ We can replace the stated inequalities with $\Sigma_{i=1}^k a_i x_i \geq z$ for every $\mathbf{x} = (x_1, \ldots, x_k)$ in X and $\Sigma_{i=1}^k a_i y_i \leq z$ for every $\mathbf{y} = (y_1, \ldots, y_k)$ in Y; the reasoning is just the same as in the comment regarding the supporting hyperplane theorem. Furthermore, we can reverse the roles of X and Y: either one must have interior points (be not all boundary). ■

The *efficiency theorem* is a useful corollary of the separating hyperplane theorem. A point $\mathbf{x}^*$ of a set X in R^k is called *efficient* (in X) if there is no point $\mathbf{x}'$ in X such that $\mathbf{x}' \geq \mathbf{x}^*$ but $\mathbf{x}' \neq \mathbf{x}^*$; or equivalently, if there is no $\mathbf{x}'$ in X such that $x_i' \geq x_i^*$ for every i and $x_i' > x_i^*$ for at least one i in $\{1, \ldots, k\}$. The relation "$\geq$" is the usual partial order of R^k, defined in Example A2.17.

■ Let $Y(\mathbf{x}) = \{\mathbf{x}' : \mathbf{x}' \in R^k, \mathbf{x}' \geq \mathbf{x}\}$. Then $\mathbf{x}^*$ is efficient in X if and only if $X \cap Y(\mathbf{x}^*) = \{\mathbf{x}^*\}$. (Why?) ■

A subset X of R^k is said to be *locally bounded above* if for every $\mathbf{x}$ in X there exists a point $\mathbf{y} = \mathbf{y}(\mathbf{x})$ in R^k with the property that $\mathbf{y}(\mathbf{x}) \geq \mathbf{x}'$ for every $\mathbf{x}'$ in X such that $\mathbf{x}' \geq \mathbf{x}$.

■ Define $Y(\mathbf{x})$ as in the preceding remark. Then X is locally bounded above if and only if $Y(\mathbf{x}) \cap X$ is bounded for every $\mathbf{x}$ in X. (Why?) ■

Let $X^- = \{\mathbf{x}' : \mathbf{x}' \in R^k, \mathbf{x}' \leq \mathbf{x}$ for some $\mathbf{x} \in X\}$. Then X^- can be regarded as the "filling in of X below."

■ $\mathbf{x}^*$ is efficient in X if and only if $\mathbf{x}^*$ is efficient in X^-. (Why?) ■

A subset X of R^k is said to be *closed above* if X^- is closed, and *convex above* if X^- is convex.

■ In R^2, if X lacks only part of its "southwest" boundary, then X is closed above. If X "balloons" to the "northeast" but "scallops" to the "southwest," then X is convex above. ■

■ Figure A3.8 illustrates the restrictiveness of local boundedness above, closedness above, and convexity above. ■

Let $W = \{\mathbf{w} : \mathbf{w} = (w_1, \ldots, w_k) \in R^k,\ w_i \geq 0$ for every i, $\Sigma_{i=1}^k w_i = 1\}$, and let $W^+ = \{\mathbf{w} : \mathbf{w} \in W,\ w_i > 0$ for every $i\}$. A fairly general version of the *efficiency theorem* states that:

(A) If X is closed above and locally bounded above, then to every $\mathbf{x} \in X$ there corresponds at least one *efficient* point $\mathbf{x}^*(\mathbf{x}) \in X$ such that $\mathbf{x}^*(\mathbf{x}) \geq \mathbf{x}$.

(B) If X is convex above, then

  (i) if $\mathbf{x}^*$ is efficient in X, there exists a point $\mathbf{w} \in W$ such that

$$\sum_{i=1}^k w_i x_i^* = \max\left[\sum_{i=1}^k w_i x_i : \mathbf{x} = (x_1, \ldots, x_k) \in X\right]; \qquad (\dagger)$$

  (ii) if $\mathbf{w} \in W^+$ is given, every $\mathbf{x}^* = (x_1^*, \ldots, x_k^*)$ which satisfies (†) is efficient.

(C) If X is convex above, closed above, and locally bounded above, and if $\mathbf{w} \in W$ is given, then at least one $\mathbf{x}^* = (x_1^*, \ldots, x_k^*)$ which satisfies (†) is efficient.

(D) If X is the convex hull of a finite number of points in R^k, then every $\mathbf{x}^*$ efficient in X satisfies (†) for some $w \in W^+$.

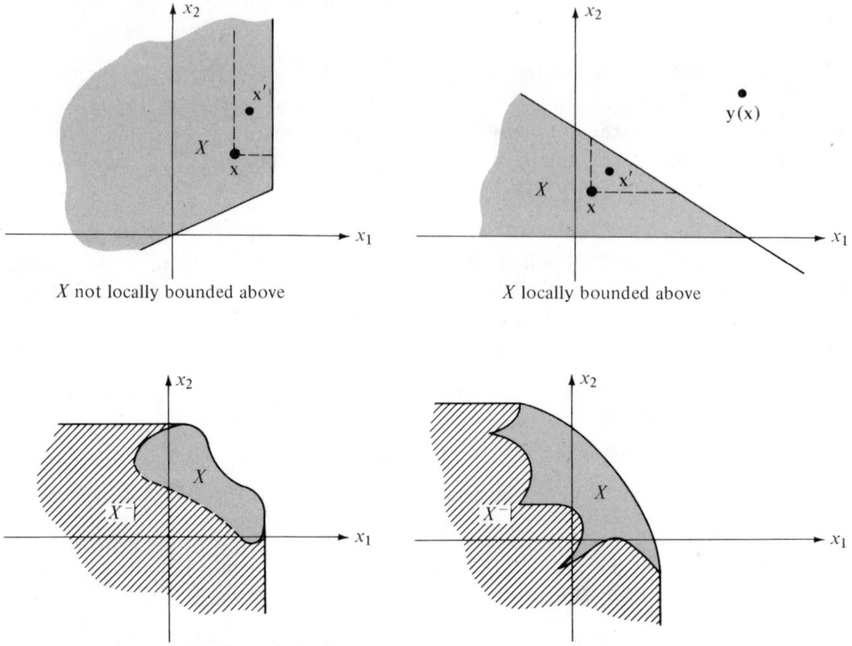

Figure A3.8 Local boundedness, closedness, and convexity above.

■ *Proof*: *Of* (*A*). Given $\mathbf{x}$, the following iterative procedure produces an efficient $\mathbf{x}^*(\mathbf{x})$ in (at most) k steps. *Step 0*: Define $\mathbf{x}^0 = \mathbf{x}$. *Step h* ($h = 1, \ldots, k$): choose $\mathbf{x}^h \in X$ so as to maximize $x_h{}^h$ subject to the constraints that $x_i^h \geq x_i^{h-1}$ for $i \neq h$. Closedness above and local boundedness above imply existence and finiteness respectively of each such maximizer $\mathbf{x}^h$. Performance of steps $1, 2, \ldots, k$ in that order produces a sequence $\mathbf{x}^0 \leq \mathbf{x}^1 \leq \mathbf{x}^2 \leq \cdots \leq \mathbf{x}^k$, with $\mathbf{x}^k$ clearly efficient. Define $\mathbf{x}^*(\mathbf{x}) = \mathbf{x}^k$.

Of (*Bi*). $\mathbf{x}^*$ is efficient in X if and only if $Y(\mathbf{x}^*) \cap X = \{\mathbf{x}^*\}$, and hence $[Y(\mathbf{x}^*) - (Y(\mathbf{x}^*))^b] \cap [X - X^b] = \emptyset$. Moreover, $Y(\mathbf{x}^*)$ is not all boundary. Each of X^- and $Y(\mathbf{x}^*)$ is convex. Hence a separating hyperplane $H(\mathbf{w}, z)$ exists. Clearly, $\mathbf{w} \geq 0$ or else $H(\mathbf{w}, z)$ would cut into the interior of $Y(\mathbf{x}^*)$; and $\mathbf{w} \neq \mathbf{0}$. Hence both $\mathbf{w}$ and z can be multiplied by a common scale factor to produce $\Sigma_{i=1}^k w_i = 1$. The asserted maximum (†) follows from the facts that $\Sigma_{i=1}^k w_i x_i \leq z$ for every $\mathbf{x} \in X$ and $\Sigma_{i=1}^k w_i x_i^* = z$.

Of (*Bii*). Let $\mathbf{w} \in W^+$ be given, and define $X^\circ(\mathbf{w}) = \{\mathbf{x}' : \mathbf{x}' \in X,\ \mathbf{x}'$ satisfies (†) for $\mathbf{w}\}$. Suppose $\mathbf{x}' \in X^\circ(\mathbf{w})$ is inefficient. Then there exists $\mathbf{x}^* \in X$ such that $\mathbf{x}^* \geq \mathbf{x}'$ and $\mathbf{x}^* \neq \mathbf{x}'$. But $\mathbf{w} > \mathbf{0}$ implies that $\Sigma_{i=1}^k w_i x_i^* > \Sigma_{i=1}^k w_i x_i$, contradicting the assumption that $\mathbf{x}' \in X^\circ(\mathbf{w})$. Hence $\mathbf{x}'$ must be efficient.

Of (*C*). With $X^\circ(\mathbf{w})$ as defined in the proof of (Bii), let $\mathbf{x}' \in X^\circ(\mathbf{w})$. If $\mathbf{x}^* X$, $\mathbf{x}^* \geq \mathbf{x}'$, and $\mathbf{x}^* \neq \mathbf{x}'$, then $\mathbf{x}^* \in X^\circ(\mathbf{w})$ by the argument for (Bii). The procedure in the proof of (A) will locate such an $\mathbf{x}^*$ if $\mathbf{x}'$ is not itself efficient. Proof of (D) is omitted, as it entails development of considerable information about convex hulls. ■

Semicontinuity and Continuity. Suppose X is a nonempty subset of R^k and $r: X \to (-\infty, +\infty)$. Then $r(\cdot)$ is said to be:

(1) *upper semicontinuous*, if for every z in $(-\infty, +\infty)$ the set $\{\mathbf{x}: r(\mathbf{x}) < z\}$ is open in R^k.
(2) *lower semicontinuous*, if for every z in $(-\infty, +\infty)$ the set $\{\mathbf{x}: r(\mathbf{x}) > z\}$ is open in R^k.
(3) *continuous*, if $r(\cdot)$ is both lower and upper semicontinuous.

Figure A3.9 depicts upper and lower semicontinuous functions on R^1. Note the locations of the function values, indicated by dots, at the discontinuity points in the figure for each of the otherwise similar functions.

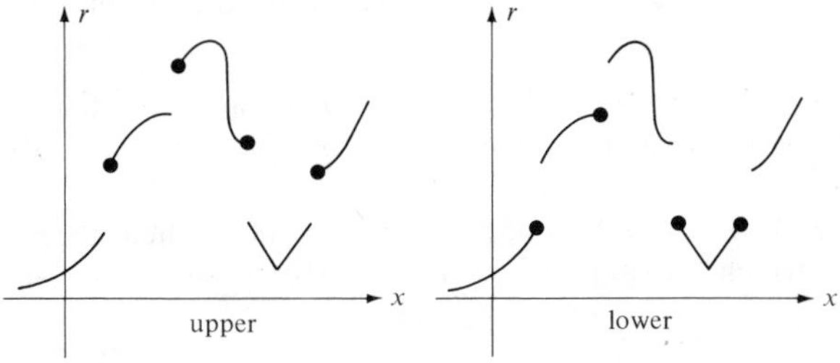

Figure A3.9 Upper and lower semicontinuous functions.

■ It can be shown that every convex function $r(\cdot)$ on a closed convex set X in R^k is upper semicontinuous, the discontinuities (if any) being on the boundary X^b of X. Similarly, every concave function on a closed convex set X is lower semicontinuous with discontinuities (if any) being on the boundary. (See Figure A3.2.) A convex or concave function on an *open* convex set is necessarily continuous. But, from Figure A3.3, it should be obvious that quasi-concave and quasi-convex functions need not be semicontinuous. ■

■ Semicontinuous functions often arise from limit operations on continuous functions. Let $r_i: X \to (-\infty, +\infty)$ be a continuous function for every i in a nonempty index set I, and suppose that there are functions $r_*: X \to (-\infty, +\infty)$ and $r^*: X \to (-\infty, +\infty)$ such that $r_*(\mathbf{x}) \le r_i(\mathbf{x}) \le r^*(\mathbf{x})$ for every $\mathbf{x} \in X$ and every $i \in I$. Then there exist functions $r_U: X \to (-\infty, +\infty)$ and $r_L: X \to (-\infty, +\infty)$ such that

(1) $r_L(\mathbf{x}) \le r_i(\mathbf{x}) \le r_U(\mathbf{x})$ for every $\mathbf{x} \in X$ and every $i \in I$.
(2) If $s_L: X \to (-\infty, +\infty)$ and $s_U: X \to (-\infty, +\infty)$ also satisfy (1) in place of r_L and r_U respectively, then $s_L(\mathbf{x}) \le r_L(\mathbf{x})$ for every $\mathbf{x}X$, and $s_U(\mathbf{x}) \ge r_U(\mathbf{x})$ for every $\mathbf{x} \in X$.

Intuitively, r_L is the *greatest* function lying on or below all the r_i's, and r_U is the *least* function lying on or above all the r_i's. If all the r_i's are continuous, then r_L is *upper* semicontinuous and r_U is *lower* semicontinuous. (More generally, if all the r_i's are upper semicontinuous, then r_L is also upper semicontinuous; and if all of the r_i's are lower semicontinuous, then r_U is also lower semicontinuous.) For instance, suppose that $X = (-\infty, +\infty)$ and that for $i = 1, 2, \ldots,$ *ad inf*,

$$r_i(x) = \begin{cases} 0, & x \le 0 \\ 1 - e^{-ix}, & x > 0. \end{cases}$$

Every r_i is continuous, but

$$r_U(x) = \begin{cases} 0, & x \leq 0 \\ 1, & x > 0, \end{cases}$$

and is thus lower semicontinuous. But if I is a *finite* set and every r_i is continuous, then so are r_L and r_U. ■

■ It should be obvious that $r(\cdot)$ is upper semicontinuous if and only if $-r(\cdot)$ is lower semicontinuous. ■

Weierstrass[a] Extremum Theorem. It is clearly desirable that minimizers and maximizers of functions defined on sets X (of, say, acts) exist. Sufficient conditions for the existence of minimizers and maximizers are furnished by the Weierstrass Extremum Theorem, which states that:

(1) If X is compact and $r: X \rightarrow (-\infty, +\infty)$ is lower semicontinuous, then there is an element $\mathbf{x}^*$ of X which minimizes $r(\cdot)$, in that $r(\mathbf{x}^*) \leq r(\mathbf{x})$ for every $\mathbf{x} \in X$.

(2) If X is compact and $r: X \rightarrow (-\infty, +\infty)$ is upper semicontinuous, then there is an element $\mathbf{x}^*$ of X which maximizes $r(\cdot)$, in that $r(\mathbf{x}^*) \geq r(\mathbf{x})$ for every $\mathbf{x} \in X$.

Recall that a subset X of R^k is compact if it is closed and bounded.

Brouwer[b] Fixed Point Theorem. Suppose that X is a compact, convex, and nonempty subset of R^k and that $r: X \rightarrow X$ is continuous, in that if $\{\mathbf{x}^i: i = 1, 2, \ldots$ *ad inf*$\}$ have $\lim_{i\rightarrow+\infty} \mathbf{x}^i = \mathbf{x}^*$ in X, then $\{r(\mathbf{x}^i): i = 1, 2, \ldots$ *ad inf*$\}$ have $\lim_{i\rightarrow+\infty} r(\mathbf{x}^i) = r(\mathbf{x}^*)$ in X. The Brouwer Fixed Point Theorem states that under these assumptions there is a point $\mathbf{x}^f$ in X such that $r(\mathbf{x}^f) = \mathbf{x}^f$. In other words, r does not move $\mathbf{x}^f$, which is therefore called a *fixed point* of r. In Figure A3.10, we take $k = 1$ and $X = [0, 1]$. Then *any* continuous function $r: [0, 1] \rightarrow [0, 1]$ must touch the diagonal line $r = x$ at least once—as you can readily convince yourself—and a fixed point of $r(\cdot)$ is just such a point of contact.

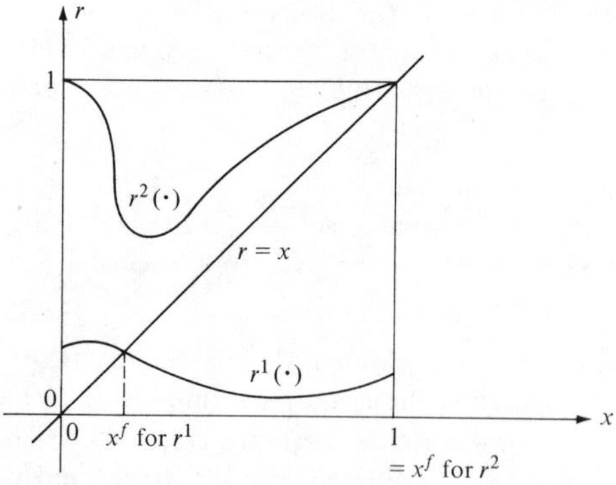

Figure A3.10 Brouwer Fixed Point Theorem for $k = 1$.

---

[a] Pronounced "vie'-ehr-shtrahs."
[b] Rhymes with "flower."

Appendix 4

BISECTION TECHNIQUE FOR FINDING ROOTS

In Chapter 5 and elsewhere arises the problem of trying to solve an equation of the form

$$r(x) = 0 \tag{A4.1}$$

for x, where $r(\cdot)$ is a real-valued function of a real argument x. A solution x^* of this equation is called a *root* of $r(\cdot)$.

This appendix describes a very simple procedure, called the *bisection technique*, for finding a tolerable approximation to x^*. The bisection technique is a good procedure because it requires the calculation of function values $r(x)$ for only a relatively few values of the argument x. Although there are even better techniques according to this criterion, they are much less simple; see Wilde [**143**] for an exposition of alternatives to the bisection technique.

■ Our interest in reducing the number of function-value calculations stems from the extreme complexity of the functions in whose roots we are interested. In information evaluation, for example, we wish to find the root x_K^* of the function

$$r(x) = E_{\tilde{\zeta}|e}\{\max\,[E_{\theta|e,\zeta,a}\{u(v(a,\tilde{\theta}) - x)|e, \tilde{\zeta}, a\}: a \in A(e, \tilde{\zeta})]|e\} - K,$$

where K is an expected utility—of optimal action without information if x_K^* is the prior value of e-information; and of optimal action with e-information if x_K^* is the certainty equivalent of the nonconstant cost of e-information. ■

In order to apply the bisection technique, it is necessary to make two assumptions about the nature of $r(\cdot)$.

Assumption 1. It is known that a root x^* of $r(\cdot)$ belongs to the (finite-length) interval (A, B).
Assumption 2. If x' and x'' in $[A, B]$ satisfy $x' < x^* < x''$, then $r(x') < 0$ and $r(x'') > 0$. [Here again, x^* denotes the root of $r(\cdot)$ in (A, B).]

Assumption 1 is necessary because the fundamental idea behind the bisection technique is the judicious and successive discarding of ends of intervals, so as to narrow in on x^*. Assumption 2 is also necessary in our developing a criterion for discarding ends of intervals.

■ Some functions $t(\cdot)$ satisfy $t(x')>0$ and $t(x'')<0$ whenever $x'<x^*<x''$. To apply the bisection technique to them, simply define $r(x)=-t(x)$ for every x. A root x^* of $r(\cdot)$ is obviously a root of $t(\cdot)$. ■

■ It is clear from Assumption 2 and the preceding remark that any strictly increasing or strictly decreasing function on $[A, B]$ is amenable to the bisection technique. Moreover, if $r(\cdot)$ is strictly increasing with $r(A)<0$ and $r(B)>0$, and if $r(\cdot)$ is continuous on $[A, B]$, then $r(\cdot)$ has a unique root x^* in (A, B), hence Assumption 1 obtains as well. ■

We record the bisection technique for a function $r(\cdot)$ satisfying Assumptions 1 and 2 in the flow chart of Figure A4.1. The box with the blackened corners indicates the onerous function-value calculation.

■ Equalities in the flow chart are to be understood in the FORTRAN sense. Thus "$A=x$" means "discard anything that happens to be in the A location and replace it with x." ■

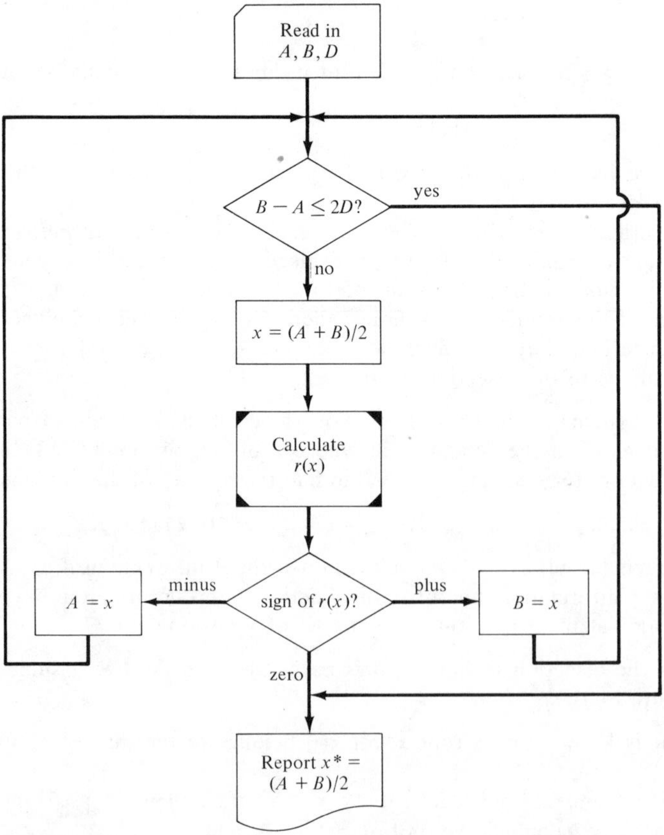

Figure A4.1 Flow chart for bisection technique.

Although the flow chart is largely self-explanatory, we shall describe the procedure briefly.

(1) Choose a tolerable error D within which you wish to locate x^*; and record D together with A and B such that $r(A) < 0$ and $r(B) > 0$.

(2a) If $B - A \leq 2D$, then $(A + B)/2$ is within D of x^* because x^* is in (A, B). Hence, stop, and declare that $x^* = (A + B)/2$.

(2b) But if $B - A > 2D$, choose $x = (A + B)/2$ as a test value and calculate $r(x)$.

(3a) If $r(x) = 0$, then $x^* = x = (A + B)/2$, without further ado.

(3b) But if $r(x) < 0$, then $x^* > x$ and you can discard the interval $[A, x)$. This is done by replacing the old A with x and returning to step (2a).

(3c) But if $r(x) > 0$, then $x^* < x$ and you can discard the interval $(x, B]$. This is done by replacing the old B with x and returning to step (2a).

■ After the nth iteration, the length of the remaining interval is only $1/2^n$ times the length of the original interval. Hence you narrow in on x^* rapidly. In fact, you can *predetermine* the (maximum) number of function-value calculations required. Since the remaining interval after the nth iteration is of length (original) $(B - A)/2^n$, it fails to exceed $2D$ if and only if

$$2^n \geq (B - A)/(2D). \tag{A4.2}$$

The (maximum) number of function-value calculations is the smallest non-negative integer n satisfying inequality (A4.2). ■

Example A4.1. Suppose that we wish to approximate the root x^* of $r(x) = x - 90$ and know that $x^* \in (0, 100)$; and that we wish our approximation to be within ¼ of the actual root. (This problem is trivial to solve by analytical methods, but that is all to the good in furnishing a clear example!) By (A4.2) with $A = 0$, $B = 100$, and $D = .25$, we see that eight function-value calculations will be required. In this case, rounding all calculations to four decimal places does not increase the number; and they are:

(1) Set $x = (0 + 100)/2 = 50$; calculate $r(50) = -40$. Hence new $A = 50$ and B remains 100.

(2) Set $x = (50 + 100)/2 = 75$; calculate $r(75) = -25$. Hence new $A = 75$ and B remains 100.

(3) Set $x = (75 + 100)/2 = 87.5$; calculate $r(87.5) = -2.5$. Hence new $A = 87.5$ and B remains 100.

(4) Set $x = (87.5 + 100)/2 = 93.75$; calculate $r(93.75) = 3.75$. Hence new $B = 93.75$ and A remains 87.5.

(5) Set $x = (87.5 + 93.75)/2 = 90.625$; calculate $r(90.625) = .625$. Hence new $B = 90.625$ and A remains 87.5.

(6) Set $x = (87.5 + 90.625)/2 = 89.0625$; calculate $r(89.0625) = -.9375$. Hence new $A = 89.0625$ and B remains 90.625.

(7) Set $x = (89.0625 + 90.625)/2 = 89.8438$; calculate $r(89.8438) = -.1562$. Hence new $A = 89.8438$ and B remains 90.625.

(8) Set $x = (89.8438 + 90.625)/2 = 90.2344$; calculate $r(90.2344) = .2344$. Hence new $B = 90.2344$ and A remains 89.8438. STOP: $B - A = .3906$, which fails to exceed $2D = .5000$. Hence report $x^* = (89.8438 + 90.2344)/2 = 90.0391$.

■ Fox and Landi have shown [**46**] that the bisection technique is *optimal*, in the following sense. Let $\#(D|t, r(\cdot))$ denote the number of values $r(x)$ which have to be calculated in order to determine the root x^* of $r(\cdot)$ to within D, given that technique t is used to choose the test values x. Also let $n(D|t)$ denote the *maximum* of $\#(D|t, r(\cdot))$ over all functions $r(\cdot)$ which satisfy Assumptions 1 and 2. Then $n(D|t)$ is a conservative measure of the *in*efficiency of technique t. Fox and Landi (essentially) show that $t =$ "bisection" *minimizes* $n(D|t)$ for every $D > 0$. ■

■ If we define $n(D|t)$ as the maximum of $\#(D|t, r(\cdot))$ only over those $r(\cdot)$ which are *convex* as well as which satisfy Assumptions 1 and 2, then another, somewhat more complicated technique minimizes $n(D|t)$ for every $D > 0$. This technique is due to Gross and Johnson [**48**] and is also described in Bellman and Dreyfus [**8**, pp. 156–167]. The Gross-Johnson technique is usually *much* better than the bisection technique when $r(\cdot)$ is *known* to be *convex*. [But if $t(\cdot)$ is *concave* and $t(x') > 0 > t(x'')$ whenever $x' < x^* < x''$, then define $r(\cdot) = -t(\cdot)$; it follows that $r(\cdot)$ is convex and that the Gross-Johnson procedure is applicable.] ■

Appendix 5

REFERENCES

1. Arrow, K. J., *Social Choice and Individual Values.* 2d ed. New York: John Wiley & Sons, 1963. (First edition, 1951.)
2. ———, *Aspects of the Theory of Risk Bearing.* Helsinki: The Academic Bookstore, 1965.
3. Aumann, R. J., The Core of a Cooperative Game without Side Payments. *Transactions of the American Mathematical Society,* **98** (1961), 539–552.
4. ———, Mixed and Behavior Strategies in Infinite Extensive Games. In Dresher, M., *et al.* [26].
5. ———, A Survey of Cooperative Games without Side Payments. In Shubik, M., [131].
6. ———, and M. Maschler, The Bargaining Set for Cooperative Games. In Dresher, M., *et al.* [26].
7. ———, and B. Peleg, Von Neumann-Morgenstern Solutions to Cooperative Games without Side Payments. *Bulletin of the American Mathematical Society,* **66** (1960), 173–179.
8. Bellman, R. E., and S. E. Dreyfus, *Applied Dynamic Programming.* Princeton, N.J.: Princeton University Press, 1962.
9. Birch, B. J., On Games with Almost Complete Information. *Proceedings of the Cambridge Philosophical Society,* **51** (1955), 275–287.
10. Birnbaum, A., On the Foundations of Statistical Inference (with ensuing discussion). *Journal of the American Statistical Association,* **65** (1962), 269–326.
11. ———, On Durbin's Modified Principle of Conditionality. *Journal of the American Statistical Association,* **65** (1970), 402–403.
12. Blackwell, D., Discrete Dynamic Programming. *Annals of Mathematical Statistics,* **33** (1962), 719–726.
13. ———, Discounted Dynamic Programming. *Annals of Mathematical Statistics,* **36** (1965), 226–235.

14. Brown, R. V., A. S. Kahr, and C. Peterson, *Decision Analysis for the Manager.* New York: Holt, Rinehart and Winston, 1974.
15. Brunk, H. D., *An Introduction to Mathematical Statistics.* 2d ed. Waltham, Mass.: Blaisdell Publishing Co., 1965. (First edition, 1960.)
16. Cabot, V. L., On the Structure of the Convex Hull of a Finite Number of Points in E^n. Part I: Theory. (Unpublished manuscript)
17. Chandler, D., *The Campaigns of Napoleon.* New York: The Macmillan Company, 1966.
18. Choquet, G., *Topology.* New York: The Academic Press, 1966.
19. Christenson, C. J., *Strategic Aspects of Competitive Bidding for Corporate Debt Securities.* Boston: Division of Research, The Harvard Business School, 1965.
20. Dalkey, N., Equivalence of Information Patterns and Essentially Determinate Games. In H. W. Kuhn and A. W. Tucker [75].
21. Denardo, E. V., On Linear Programming in a Markov Decision Problem. *Management Science,* **16** (1970), 281–288.
22. ———, Markov Renewal Programming with Small Interest Rates. *Annals of Mathematical Statistics,* **42** (1971), 477–496.
23. ———, and B. L. Fox, Multichain Markov Renewal Programs. *SIAM Journal of Applied Mathematics,* **16** (1968), 468–487.
24. D'Epenoux, F., Sur un Probleme de Production et de Stockage dans l' a leatiore. English translation in *Management Science,* **10** (1963), 98–108.
25. Derman, C., *Finite State Markovian Decision Processes.* New York: The Academic Press, 1970.
26. Dresher, M., L. S. Shapley, and A. W. Tucker (eds), *Advances in Game Theory* (Annals of Mathematics Studies No. 52). Princeton, N.J.: Princeton University Press, 1964.
27. ———, A. W. Tucker, and P. Wolfe (eds), *Contributions to the Theory of Games,* Vol III (Annals of Mathematics Studies No. 39). Princeton, N.J.: Princeton University Press, 1957.
28. Durbin, J., On Birnbaum's Theorem and the Relation Between Sufficiency, Conditionality and Likelihood. *Journal of the American Statistical Association,* **65** (1970), 395–398.
29. Eck, R. D., *Some Group Decision-Theoretic Results with Particular Reference to Acquisitions and Mergers.* (Ph.D. Dissertation) New Orleans, La.: The Graduate School, Tulane University, 1973.
30. Edwards, W., H. Lindman, and L. J. Savage, Bayesian Statistical Inference for Psychological Research. *Psychological Review,* **70** (1953), 193–242.
31. Farquhar, P., Fractional Hypercube Decompositions of Multiattribute Utility Functions. *Department of Operations Research Technical Reports,* **222** (1974) (Cornell University, Ithaca, N.Y.).
32. ———, A Fractional Hypercube Decomposition Theorem for Multiattribute Utility Functions. To appear in *Operations Research.*
33. Feller, W., *An Introduction to Probability Theory and its Applications,* Vol I. New York: John Wiley & Sons, 1950.
34. Ferguson, T. S., *Mathematical Statistics: A Decision-Theoretic Approach.* New York: The Academic Press, 1967.
35. Fishburn, P. C., Convex Stochastic Dominance with Continuous Distribution Functions. To appear in *Journal of Economic Theory.*
36. ———, Convex Stochastic Dominance with Finite Consequence Sets. To appear in *Theory and Decision.*

37. ———, Lexicographic Orders, Utilities, and Decision Rules: A Survey. *Management Science,* **20** (1974), 1442–1471.
38. ———, von Neumann-Morgenstern Utility Functions on Two Attributes. *Operations Research,* **22** (1974), 35–45.
39. ———, A Study of Independence in Multivariate Utility Theory. *Econometrica,* **37** (1969), 107–121.
40. ———, Utility Theory. *Management Science,* **14** (1968), 335–378.
41. ———, Methods of Estimating Additive Utilities. *Management Science,* **13** (1967), 435–453.
42. ———, Independence in Utility Theory with Whole Product Sets. *Operations Research,* **13** (1965), 28–45.
43. ———, and I. H. LaValle, The Effectiveness of Ordinal Dominance in Decision Analysis. *Operations Research,* **22** (1974), 177–180.
44. Flynn, J., Averaging *vs.* Discounting in Dynamic Programming: A Counterexample. *Annals of Statistics,* **2** (1974), 411–413.
45. Fox, B. L., Markov-Renewal Programming by Linear Fractional Programming. *SIAM Journal of Applied Mathematics,* **14** (1966), 1418–1430.
46. ———, and D. M. Landi, Searching for the Multiplier in One-Constraint Optimization Problems. *Operations Research,* **18** (1970), 253–262.
47. Geoffrion, A. M., Jr., Proper Efficiency and the Theory of Vector Maximization. *Journal of Mathematical Analysis and Applications,* **22** (1968), 618–630.
48. Gross, O. A., and S. M. Johnson, Sequential Minimax Search for a Zero of a Convex Function. *Mathematical Tables and Other Aids to Computation,* **13** (1959), 44–51.
49. Hadar, J., and W. R. Russell, Rules for Ordering Uncertain Prospects. *American Economic Review,* **59** (1969), 25–34.
50. Halton, J. H., A Retrospective and Prospective Survey of the Monte Carlo Method. *SIAM Review,* **12** (1970), 1–63.
51. Hammond, J. S., Simplifying the Choice Between Uncertain Prospects. *Management Science,* **20** (1974), 1047–1072.
52. Harsanyi, J. C., Games with Incomplete Information Played by 'Bayesian' Players, Part I: The Basic Model; Part II: Bayesian Equilibrium Points; Part III: The Basic Probability Distribution of the Game. *Management Science,* **14** (1967), 159–182, 320–334, 486–502.
53. ———, A Bargaining Model for the Cooperative *n*-Person Game. In A. W. Tucker and R. D. Luce [135].
54. Hartigan, J. A., The Likelihood and Invariance Principles. *Journal of the Royal Statistical Society, Series B,* **29** (1967), 533–539.
55. Hausner, M., Multidimensional Utilities. In R. M. Thrall *et al.* [134].
56. Hays, W. L., and R. L. Winkler, *Statistics: Probability, Inference, and Decision.* New York: Holt, Rinehart and Winston, 1970.
57. Haywood, O. G., Jr., Military Decision and Game Theory. *Operations Research,* **2** (1954), 365–385.
58. Hildebrand, F. B., *Introduction to Numerical Analysis.* New York: McGraw-Hill Book Company, 1956.
59. Hildreth, C., Bayesian Statisticians and Remote Clients. *Econometrica,* **31** (1963), 422–438.
60. Hirschleifer, J., Investment Decision Under Uncertainty. *Quarterly Journal of Economics,* **LXXIX** (1965), 509–536.
61. Hogarth, R. M., Cognitive Processes and the Assessment of Subjective

Probability Distributions (with ensuing discussion). *Journal of the American Statistical Association*, **70** (1975), 271–294.
62. Howard, R. A., *Dynamic Probabilistic Systems*. New York: John Wiley & Sons, 1971.
63. ———, *Dynamic Programming and Markov Processes*. Cambridge, Mass.: The M.I.T. Press, 1960.
64. ——— and J. E. Matheson, Risk-Sensitive Markovian Decision Processes. *Management Science*, **18** (1972), 356–369.
65. ———, ———, and K. L. Miller (eds.), *Readings in Decision Analysis*. Menlo Park, Calif.: Decision Analysis Group, Stanford Research Institute, 1974.
66. Isbell, J. R., Finitary Games. In M. Dresher *et al.* [27].
67. Jewell, W. S., Markov Renewal Programming I, II. *Operations Research*, **11** (1963), 938–971.
68. Karlin, S., *A First Course in Stochastic Processes*. New York: The Academic Press, 1966.
69. Keeney, R. L., Multiplicative Utility Functions. *Operations Research*, **22** (1974), 22–34.
70. ———, Risk Independence and Multiattributed Utility Functions. *Econometrica*, **41** (1973), 27–34.
71. ———, Utility Functions for Multiattributed Consequences. *Management Science*, **18** (1972) 276–287.
72. ———, Utility Independence and Preferences for Multiattributed Consequences. *Operations Research*, **19** (1971), 875–893.
73. ———, Note on Multiattribute Constant Risk Aversion. (Unpublished manuscript)
74. Kuhn, H. W., Extensive Games and the Problem of Information. In H. W. Kuhn and A. W. Tucker [75].
75. ———, and A. W. Tucker (eds.), *Contributions to the Theory of Games*, Vol. II (Annals of Mathematics Studies No. 28). Princeton, N.J.: Princeton University Press, 1953.
76. LaValle, I. H., On Admissibilities and Bayesness When Risk Attitude But Not the Preference Ranking Is Permitted to Vary. In J. L. Cochrane and M. Zeleny (eds.), *Multiple Criteria Decision Making*. Columbia: The University of South Carolina Press, 1973.
77. ———, A Solution Concept for Group Decision Problems. *Graduate School of Business Reports*, **64** (1970). (Tulane University, New Orleans, La.)
78. ———, *An Introduction to Probability, Decision, and Inference*. New York: Holt, Rinehart and Winston, 1970.
79. ———, Group Decisions in Which Members Are Mutually Concerned, Risk Averse, and Have Multivariate Constant Risk Aversion. *Graduate School of Business Reports*, **20** (1969). (Tulane University, New Orleans, La.)
80. ———, On Multivariate Constant Risk Aversion. *Graduate School of Business Reports*, **10** (1968). (Tulane University, New Orleans, La.)
81. ———, On Cash Equivalents and Information Evaluation in Decisions Under Uncertainty. *Journal of the American Statistical Association*, **63** (1968), 252–290.
82. Lemke, C. E., Bimatrix Equilibrium Points and Mathematical Programming. *Management Science*, **11** (1965), 681–689.
83. ———, and J. T. Howson, Jr., Equilibrium Points of Bimatrix Games. *SIAM Journal of Applied Mathematics*, **12** (1964), 413–423.

84. Lindgren, B. W., *Statistical Theory.* 2d ed., New York: The Macmillan Company, 1968. (First edition, 1960.)
85. Lindley, D. V., *Bayesian Statistics: A Review.* Regional Conference Series No. 2. Philadelphia: Society for Industrial and Applied Mathematics, 1971.
86. Lippman, S. A., Maximal Average-Reward Policies for Semi-Markov Decision Processes with Arbitrary State and Action Space. *Annals of Mathematical Statistics,* **42** (1971), 1717–1726.
87. Loève, M., *Probability Theory.* 3d ed. Princeton, N.J.: D. Van Nostrand Company, 1963.
88. Lucas, W. F., An Overview of the Mathematical Theory of Games. *Management Science,* **18** (1972), P3–P19.
89. ———, Some Recent Developments in *n*-Person Game Theory. *SIAM Review,* **13** (1971).
90. ———, The Proof That a Game May Not Have a Solution. *Transactions of The American Mathematical Society,* **137** (1969), 219–229.
91. ———, A Game with No Solution. Santa Monica, Calif.: RAND Corp. Memorandum RM-5518-PR, 1967.
92. ———, and H. Raiffa, *Games and Decisions.* New York: John Wiley & Sons, 1957.
93. Mangasarian, O. L., *Nonlinear Programming.* New York: McGraw-Hill Book Company, 1969.
94. Manne, A., Linear Programming and Sequential Decisions. *Management Science,* **6** (1960), 259–267.
95. Martin, J. J., *Bayesian Decision Problems and Markov Chains.* New York: John Wiley & Sons, 1967.
96. Miller, B. L., and A. F. Veinott, Jr., Discrete Dynamic Programming with a Small Interest Rate. *Annals of Mathematical Statistics,* **40** (1969), 366–370.
97. Mills, H., Equilibrium Points in Finite Games. *SIAM Journal of Applied Mathematics,* **8** (1960), 397–402.
98. Mood, A. M., and F. K. Graybill, *Introduction to the Theory of Statistics.* New York: McGraw-Hill Book Company, 1963.
99. Nash, J. F., The Bargaining Problem. *Econometrica,* **18** (1950), 155–162.
100. ———, Two-Person Cooperative Games. *Econometrica,* **21** (1953), 128–140.
101. ———, Equilibrium Points in *n*-Person Games. *Proceedings of the National Academy of Sciences, U.S.A.,* **36** (1950), 48–49.
102. Naylor, T. H., *Computer Simulation Experiments with Models of Economic Systems.* New York: John Wiley & Sons, 1971.
103. ———, J. L. Balintfy, T. S. Burdick, and K. Chu, *Computer Simulation Techniques.* New York: John Wiley & Sons, 1966.
104. Nemhauser, G. L., *Introduction to Dynamic Programming.* New York: John Wiley & Sons, 1966.
105. von Neumann, J., Zur Theorie der Gesellschaftsspiele. *Mathematische Annalen,* **100** (1928), 295–320.
106. ———, and Morgenstern, O., *Theory of Games and Economic Behavior.* 3d ed. Princeton, N.J.: Princeton University Press, 1953. (First edition, 1944.)
107. Otter, R., and J. J. Dunne, Games with Equilibrium Points. *Proceedings of the National Academy of Sciences, U.S.A.,* **39** (1953), 310–314.
108. Owen, G., *Game Theory.* Philadelphia: W. B. Saunders Co., 1968.
109. Pfanzagl, J., A General Theory of Measurement Applications to Utility. *Naval Research Logistics Quarterly,* **6** (1959), 283–294.
110. Pollak, R. A., The Risk Independence Axiom. *Econometrica,* **41** (1973), 35–40.

111. Pratt, J. W., Risk Aversion in the Small and in the Large. *Econometrica*, **32** (1964).
112. ———, H. Raiffa, and R. O. Schlaifer, *Introduction to Statistical Decision Theory*. (preliminary edition) New York: McGraw-Hill Book Company, 1965.
113. Raiffa, H., *Decision Analysis: Introductory Lectures on Choices Under Uncertainty*. Reading, Mass.: Addison-Wesley Publishing Company, 1968.
114. ———, Arbitration Schemes for Generalized Two-Person Games. In H. W. Kuhn and A. W. Tucker [75].
115. ———, and R. O. Schlaifer, *Applied Statistical Decision Theory*. Boston: Division of Research, The Harvard Business School, 1961.
116. RAND Corp., *A Million Random Digits with 100,000 Normal Deviates*. Glencoe, Ill.: The Free Press, 1955.
117. Rapoport, A., *Two-Person Game Theory: The Essential Ideas* (Vol I); *N-Person Game Theory: Concepts and Applications* (Vol II). Ann Arbor: The University of Michigan Press, 1966 (Vol I), 1970 (Vol II).
118. ———, and A. M. Chammah, *Prisoners' Dilemma: A Study of Conflict and Cooperation*. Ann Arbor: The University of Michigan Press, 1965.
119. Roberts, H., Informative Stopping Rules About Population Size. *Journal of the American Statistical Association*, **62** (1967), 763–775.
120. Rockafellar, R. T., *Convex Analysis*. Princeton, N.J.: Princeton University Press, 1970.
121. Rosing, J., The Formation of Groups for Cooperative Decision Making under Uncertainty. *Econometrica*, **38** (1970), 430–448.
122. Ross, S. M., On the Nonexistence of ϵ-Optimal Randomized Stationary Policies in Average-Cost Markov Decision Models. *Annals of Mathematical Statistics*, **42** (1971), 1767–1768.
123. Rothblum, U. G., Multivariate Constant Risk Posture. *Technical Report 43* (1974), Department of Operations Research. Stanford University, California.
124. Savage, L. J., Comments on a Weakened Principle of Conditionality. *Journal of the American Statistical Association*, **65** (1970), 399–401.
125. Schelling, T. C., *Arms and Influence*. New Haven, Conn.: Yale University Press, 1966.
126. ———, *The Strategy of Conflict*. Cambridge, Mass.: Harvard University Press, 1960.
127. Schlaifer, R. O., *Computer Programs for Elementary Decision Analysis*. Cambridge, Mass.: Harvard University Press, 1971.
128. ———, *Analysis of Decisions under Uncertainty*. New York: McGraw-Hill Book Company, 1969.
129. Sen, A. K., *Collective Choice and Social Welfare*. San Francisco: Holden-Day, 1970.
130. Shachtman, R. H., Generation of the Admissible Boundary of a Convex Polytope. *Operations Research*, **22** (1974), 151–159.
131. Shubik, M. (ed), *Essays in Mathematical Economics in Honor of Oskar Morgenstern*. Princeton, N.J.: Princeton University Press, 1967.
132. Thompson, G. L., Signaling Strategies in n-Person Games. In H. W. Kuhn and A. W. Tucker [75].
133. Thrall, R. M., Applications of Multi-dimensional Utility Theory. In R. M. Thrall, *et al.* [134].
134. ———, C. H. Coombs, and R. L. Davis (eds.), *Decision Processes*. New York: John Wiley & Sons, 1954.

135. Tucker, A. W., and R. D. Luce, *Contributions to the Theory of Games*, Vol IV. (Annals of Mathematics Studies No. 40). Princeton, N.J.: Princeton University Press, 1959.
136. Tucker, H. G., *A Graduate Course in Probability*. New York: The Academic Press, 1967.
137. Walker, M. R., Determination of the Convex Hull of a Finite Set of Points. Master of Science in Operations Research Thesis, Chapel Hill: The University of North Carolina, 1973.
138. Wallace, D. L., Uniform Decisiveness in Group Decision Making. (Unpublished manuscript) Boston: The Harvard Business School, 1974.
139. von Wartenburg, Gen. Count Y., *Napoleon as a General*. West Point, N.Y.: U.S. Military Academy. (First edition, 1902.)
140. Weiss, L., *Statistical Decision Theory*. New York: McGraw-Hill Book Company, 1961.
141. White, D. J., *Dynamic Programming*. San Francisco: Holden-Day, 1969.
142. Whitmore, G. A., Third-Degree Stochastic Dominance. *American Economic Review*, **60** (1970), 457–459.
143. Wilde, D. J., *Optimum Seeking Methods*. Englewood Cliffs, N.J.: Prentice-Hall, 1964.
144. Wilson, R., The Theory of Syndicates. *Econometrica*, **36** (1968), 119–132.
145. Winkler, R. L., The Assessment of Prior Distributions in Bayesian Analysis. *Journal of the American Statistical Association*, **62** (1967), 776–800.
146. ———, The Quantification of Judgment: Some Methodological Suggestions. *Journal of the American Statistical Association*, **62** (1967), 1105–1120.
147. ———, Scoring Rules and the Evaluation of Probability Assessors, *Journal of the American Statistical Association*, **64** (1969), 1073–1078.
148. ———, *Introduction to Bayesian Inference and Decision*. New York: Holt, Rinehart and Winston, 1972.
149. Yu, P. L., Cone Convexity, Cone Extreme Points, and Nondominated Solutions to Decision Problems with Multiobjectives. *Journal of Optimization Theory and Applications*, **14** (1974), 319–377.
150. Zangwill, W., *Nonlinear Programming*. Englewood Cliffs, N.J.: Prentice-Hall, 1969.
151. Zellner, A., *An Introduction to Bayesian Inference in Econometrics*. New York: John Wiley & Sons, 1971.

INDEX